D1429555

Encyclopedia of
Plant Physiology

New Series Volume 12A

Editors

A. Pirson, Göttingen
M. H. Zimmermann, Harvard

Physiological Plant Ecology I

Responses to the Physical Environment

Edited by

O.L. Lange P.S. Nobel C.B. Osmond H. Ziegler

Contributors

M. Aragno H. Bauer P. Benecke J.A. Berry O. Björkman
M.M. Caldwell G.S. Campbell S. Ichikawa S.W. Jeffrey
L. Kappen W. Larcher K.J. McCree D.C. Morgan M. Neushul
P.S. Nobel R.R. van der Ploeg J.K. Raison P.W. Rundel
F.B. Salisbury H. Smith P.L. Steponkus W.N. Wheeler

With 110 Figures

Springer-Verlag Berlin Heidelberg New York 1981

Professor Dr. O.L. LANGE
Lehrstuhl für Botanik II der Universität
Mittlerer Dallenbergweg 64, 8700 Würzburg/FRG

Professor P.S. NOBEL
Department of Biology
Division of Environmental Biology
of the Laboratory of Biomedical and Environmental Sciences
University of California
Los Angeles, California 90024/USA

Professor C.B. OSMOND
Department of Environmental Biology
Research School of Biological Sciences
Australian National University
Box 475, Canberra City 2601/Australia

Professor DR. H. ZIEGLER
Institut für Botanik und Mikrobiologie der Technischen Universität
Arcisstraße 21, 8000 München 2/FRG

ISBN 3-540-10763-0 Springer-Verlag Berlin Heidelberg New York
ISBN 0-387-10763-0 Springer-Verlag New York Heidelberg Berlin

Library of Congress Cataloging in Publication Data. Main entry under title: Physiological plant ecology. (Encyclopedia of plant physiology; new ser., v. 12 A) Bibliography: p. Includes indexes. 1. Plant physiology. 2. Botany–Ecology. I. Lange, O.L. (Otto Ludwig) II. Aragno, M. III. Series. QK711.2.E5 new ser., vol. 12A 581.1s 81-9348 [QK754] [581.5] AACR2.

Typesetting, printing and bookbinding: Universitätsdruckerei H. Stürtz AG, Würzburg.
2131/3130-543210

List of Contributors

M. Aragno
Institut de Botanique de l'Université
Laboratoire de Microbiologie
Case Postale 2
CH-2000 Neuchâtel, 7/Switzerland

H. Bauer
Institut für Botanik
Universität Innsbruck
Sternwartestraße 15
A-6020 Innsbruck/Austria

P. Benecke
Institut für Bodenkunde und Waldernährung
Universität Göttingen
Büsgenweg 2
D-3400 Göttingen/FRG

J.A. Berry
Department of Plant Biology
Carnegie Institution of Washington
Stanford, California 94305/USA

O. Björkman
Department of Plant Biology
Carnegie Institution of Washington
Stanford, California 94305/USA

M.M. Caldwell
Department of Range Science and the Ecology Center
Utah State University
Logan, Utah 84322/USA

G.S. Campbell
Department of Agronomy and Soils
Washington State University
Pullman, Washington 99164/USA

S. Ichikawa
Department of Regulation Biology
Laboratory of Genetics
Faculty of Science
Saitama University
Urwa 338/Japan

S.W. Jeffrey
CSIRO Division of Fisheries and Oceanography
Cronulla, N.S.W. 2230/Australia

L. Kappen
Botanisches Institut
Lehrstuhl Ökophysiologie
Olshausenstraße 40–60
D-2300 Kiel/FRG

W. Larcher
Institut für Botanik
Universität Innsbruck
Sternwartestraße 15
A-6020 Innsbruck/Austria

K.J. McCree
Department of Soil and Crop Sciences
Texas A & M University
College Station, Texas 77843/USA

D.C. Morgan
Department of Botany
Adrian Building
University of Leicester
University Road
Leicester LE1 7 RH/United Kingdom

M. Neushul
Marine Science Institute
University of California
Santa Barbara, California 93106/USA

P.S. Nobel
Department of Biology
University of California
Los Angeles, California 90024/USA

R.R. van der Ploeg
Institut für Bodenkunde und Waldernährung
Universität Göttingen
Büsgenweg 2
D-3400 Göttingen/FRG

J.K. Raison
CSIRO Division of Food Research
North Ryde, N.S.W./Australia

P.W. Rundel
Department of Ecology and
Evolutionary Biology
University of California
Irvine, California 92717/USA

F.B. Salisbury
Department of Plant Science, UMC 48
Utah State University
Logan, Utah 84322/USA

H. Smith
Department of Botany
Adrian Building
University of Leicester
University Road
Leicester LE1 7 RH/United Kingdom

P.L. Steponkus
Department of Agronomy
Cornell University
609 Bradfield Hall
Ithaca, New York 14853/USA

W.N. Wheeler
Department of Biological Sciences
Simon Fraser University
Burnaby, B.C. V5A 1 S6/Canada

Contents Part A

Contents Part B–D

Part B: Physiological Plant Ecology II
Water Relations and Carbon Assimilation

Part C: Physiological Plant Ecology III
Responses to the Chemical and Biological Environment

Part D: Physiological Plant Ecology IV
Ecosystem Processes: Mineral Cycling, Productivity and Man's Influence

Symbols, Abbreviations and Units

Some specialized symbols are defined and used only in particular Chapters. If no definition is given, then the symbols or abbreviations have the meanings indicated below. The International System of Units (SI) has been generally adapted throughout. However, concentrations are expressed in mM as well as the SI unit, mol m^{-3} (1 mM=1 mol m^{-3}); water potentials are expressed in bars as well as the SI unit, pascals (1 bar=0.1 MPa).

For convenience and to provide uniformity, superscripts are used to indicate locations and subscripts designate species, quantities, or conditions. For example, $r^{a}_{CO_2}$ is the resistance of the air boundary layer to CO_2 diffusion, T^{leaf}_{min} is the minimum leaf temperature, and Ψ^{xylem}_{P} is the pressure potential component of the total water potential ($\Psi = \Psi_P + \Psi_\pi + \Psi_m$) in the xylem.

Symbol or Abbreviation	Description	Unit
A	area	m^2
a	absorptance	
a	activity	mM or mol m^{-3}
a	superscript for air boundary layer	
ADP	adenosine diphosphate	
AMP	adenosine monophosphate	
ATP	adenosine triphosphate	
C_3	photosynthetic carbon reduction cycle (Calvin cycle)	
C_4	dicarboxylic acid pathway of photosynthesis	
CAM	crassulacean acid metabolism	
chl	chlorophyll	
c	concentration	mM or mol m^{-3}
D	diffusion coefficient	$m^2\ s^{-1}$
DCMU	3-(3′,4′-dichlorophenyl)-1,1-dimethylurea	
DNA	deoxyribonucleic acid	
d	diameter	m
d	day	
E	Einstein	
EDTA	ethylenediaminetetraacetic acid	
e	emittance	
F	Faraday	96,487 A s mol^{-1}
Fd	ferredoxin	
g	conductance	m s^{-1} or mol $m^{-2}\ s^{-1}$ [a]

Symbol or Abbreviation	Description	Unit
h	hour	
ha	hectare	
i	superscript for inside (generally intracellular)	
IR	infrared	
J_j	net flux of species j	mol m^{-2} s^{-1}
J_v	volume flow	m s^{-1}
K	thermal conductivity coefficient	W m^{-1} k^{-1}
K	degrees on Kelvin scale	
K_i	inhibitor constant	mM or mol m^{-3}
K_m	Michaelis-Menten constant	mM or mol m^{-3}
k	first order rate constant	s^{-1}
L_P	hydraulic conductivity coefficient	m s^{-1} bar^{-1} or m s^{-1} Pa^{-1}
l	length	m
ln	logarithm to base e	
log	logarithm to base 10	
M	molecular weight	g mol^{-1}
m	subscript for matric potential	
max	subscript for maximum value	
mes	superscript for mesophyll	
min	subscript for minimum value	
N	Avogadro's number	6.023×10^{23} molecules mol^{-1}
N_j	mole fraction of species j	
NAD^+	nicotinamide-adenine dinucleotide, oxidized	
NADH	nicotinamide-adenine dinucleotide, reduced	
$NADP^+$	nicotinamide-adenine dinucleotide phosphate, oxidized	
NADPH	nicotinamide-adenine dinucleotide phosphate, reduced	
n	number of moles	
o	superscript for outside (generally ambient)	
P	subscript for hydrostatic pressure potential	
p	partial pressure in gas phase	bar or Pa
PAR	photosynthetically active radiation (generally 400 to 700 nm)	E m^{-2} s^{-1}
PEP	phosphoenolpyruvate	
Pi	inorganic phosphate	
PS I, PS II	Photosystem I, Photosystem II	
Q_{10}	temperature coefficient	
R	gas constant	8.314 J mol^{-1} K^{-1}
RNA	ribonucleic acid	
RuBP	ribulose-1,5-bisphosphate	

Symbol or Abbreviation	Description	Unit
r	resistance	$s\ m^{-1}$ or $m^2\ s\ mol^{-1}$ [a]
r	radius	m
r	reflectance	
SD	standard deviation	
SEM	standard error of the mean	
T	temperature	°C (or K, especially indicating differences in temperature)
t	time	s
UV	ultraviolet	
V	volume	m^3
V_{max}	maximum reaction velocity	$mol\ s^{-1}$ or $mM\ s^{-1}$
v	velocity	$m\ s^{-1}$
v	reaction velocity	$mol\ s^{-1}$ or $mM\ s^{-1}$
wt	weight (e.g. dry wt)	
wv	subscript for water vapor	
γ	activity coefficient	
λ	wavelength	nm
μ	chemical potential	$J\ mol^{-1}$
π	subscript for osmotic potential	
ρ	density	$kg\ m^{-3}$
σ	reflection coefficient	
σ	Stefan-Boltzmann constant	$5.67 \times 10^{-8}\ W\ m^{-2}\ K^{-4}$
ψ	water potential	bar or Pa

[a] Diffusional gas fluxes are most commonly represented by $J_j = g_j\, \Delta c_j = \Delta c_j / r_j$, where g_j can be in $m\ s^{-1}$. Alternatively, $J_j = g_j\, \Delta N_j = \Delta N_j / r_j$, where g_j can be in $mol\ m^{-2}\ s^{-1}$

WOOD (1925) who studied salinity and water relations of Australian halophytes; and TURESSON (1922) who introduced genecology as a tool in ecological plant physiology. They were followed by the true founders of the discipline who must include HUBER (e.g., 1935), LUNDEGÅRDH (e.g., 1924), PISEK (e.g., PISEK and CARTELLIERI 1939), STOCKER (e.g., 1928), and WALTER (e.g., 1931) in Europe; MAXIMOV (e.g., 1923) in Russia; EVENARI (e.g., EVENARI and RICHTER 1937) and OPPENHEIMER (e.g., 1932) in Israel; DAUBENMIRE (e.g., 1947) in the United States; and BEADLE (e.g., 1952) in Australia. The majority of the contributors to the present volumes are, figuratively speaking, the offspring of these pioneers, at most only three or four generations removed from SCHIMPER.

The measurement of climate on a scale relevant to plant responses was a limiting factor in the development of ecological plant physiology. In spite of the pioneering studies in microclimatology by KRAUS (1911) and GEIGER (1927) progress is still influenced by the availability of adequate methods for quantitative field investigations. Precise measurements of the physical plant environment, as well as of plant metabolic performance in nature, are prerequisites for ecological conclusions. However, the investigator is confronted with many technical and methodological problems. Continuous and irregular changes in weather render exact determinations of natural plant responses very difficult. Frequently, harsh field environments tax the stability of the most robust and sophisticated equipment. Even when designed equipment functions appropriately, we frequently struggle with logistic problems and are limited in our ability to grasp stochastic elements hidden in the data thus obtained.

Only recently have techniques been developed for the accurate measurement of temperature of plant leaves (Chap. 1, Vol. 12A), or of the amount of photosynthetically active radiation which they receive (Chap. 2, Vol. 12A). Although transpirational water loss was one of the first plant responses to be studied under field conditions, suitable methods have been found only in the last two decades. Earlier methods, reviewed by STOCKER (1956) and widely used by ecological physiologists, were widely known to be inadequate and were extremely suspect among laboratory physiologists. Such shortcomings in available methods severely limited the credibility and development of ecological physiology. In the last 20 years, these shortcomings have been largely eliminated and modern techniques for CO_2 and water vapor exchange permit continuous monitoring of stomatal processes and the rates of transpiration and photosynthesis under natural conditions (see Vol. 12B). Performance of aboveground plant parts in nature can now be adequately monitored to suit many purposes. Monitored data have been used as a basis for constructing and testing models of energy budgets (Chap. 1, Vol. 12A), stomatal behavior (Chaps. 6 and 7, Vol. 12B), photosynthetic productivity (Chap. 16, Vol. 12B; Chaps. 5 to 10, Vol. 12D), and growth (Chap. 4, Vol. 12D). Some limitations, such as our inability to simulate turbulence effects or windspeed (Chap. 15, Vol. 12A) in measurement chambers, remain a challenge. Presently available technology is inadequate in other areas as well. Measurement of light, turbulence, and carbon exchange in aquatic systems has only been accomplished with primitive means. Shortcomings in measurements of plant processes and plant environment belowground are acute (NYE and TINKER 1977), and the possibilities for imaginative develop-

Introduction: Perspectives in Ecological Plant Physiology

O.L. LANGE, P.S. NOBEL, C.B. OSMOND, and H. ZIEGLER

The objective of ecological plant physiology is to explain processes in plant ecology, such as plant performance, survival, and distribution, in physiological, biophysical, biochemical, and molecular terms. It is a relatively young discipline and we are aware of only a few basic textbooks that have been written (LARCHER 1973, 1980a, b; KREEB 1974; BANNISTER 1976; ETHERINGTON 1975, 1978). Physiological plant ecology was first reviewed as a separate discipline by BILLINGS (1957), but the knowledge since gained in ecological-physiological studies has been rapidly integrated into and has strongly influenced the parent disciplines of plant physiology and ecology. Nearly a century ago the necessity of ecophysiological work was recognized by SCHIMPER in the introduction to his classical text (1898) *Pflanzengeographie auf physiologischer Grundlage* and was expressed succinctly: „Nur wenn sie in engster Fühlung mit der experimentellen Physiologie verbleibt, wird die Oekologie der Pflanzengeographie neue Bahnen eröffnen können, denn sie setzt eine genaue Kenntnis der Lebensbedingungen der Pflanzen voraus, welche nur das Experiment verschaffen kann.“ (In the original English translation: "the oecology of plant-distribution will succeed in opening new paths on condition only that it leans closely on experimental physiology, for it presupposes an accurate knowledge of the conditions of the life of plants which experiment alone can bestow.") When introducing our perspectives as editors of four volumes dealing with physiological plant ecology in the New Series of the Encyclopedia of Plant Physiology, one of our tasks should be to consider the extent to which SCHIMPER's prediction has been fulfilled. We will do this by briefly viewing the discipline in retrospect before analyzing the present status of the discipline and concluding with some of our expectations for the future.

Although the anatomical, phytogeographical, and ecological views of HABERLANDT (1884), SCHIMPER (1898), WARMING (1896), and CLEMENTS (1907) challenged plant biologists of the early 20th century with an array of speculation and hypotheses about plant response and adaptation to habitat factors, further development along these lines was slow for some time. For the first quarter century, ecological research remained mainly descriptive, with little serious correlation of plant distribution and habitat factors, and physiological research remained confined to the laboratory. For a time it seemed that some were determined to drive a philosophical wedge between physiology and ecology, and much mutual misunderstanding and distrust were generated. There were notable exceptions, of course, and the pioneers who took physiological experimentation to the field included the plant chemists at the Carnegie Institution's Desert Research Laboratory (RICHARDS 1915; SPOEHR 1919); FITTING (1911) and SHREVE (1914) who investigated plant water relations under desert conditions;

ment of experimental approaches are immense. Although there has been great progress in our understanding of stomatal control of water loss from plants, the measurement and evaluation of crucial water relations parameters in the plant cell, such as turgor pressure or osmotic potential, remain relatively difficult (Chaps. 1 and 2, Vol. 12B). To a large extent the same remains true for plant ionic relations. Progress in both fields is limited by the technology available to measure important parameters at an appropriate level.

However, these limitations have not inhibited development of important new theories of regulation and function. One such development concerns the concept of optimal use of limiting resources and the physiological basis of the compromises required. Models have been used to examine how plants may manage their water resources so as to maximize carbon intake and maintain growth without incurring damage due to drought. Theoretical analysis of the regulation of plant carbon gain in relation to water use (Chap. 17, Vol. 12B) will stimulate new research, and this theory may soon be broadened to embrace aspects of mineral nutrition limitation. Understanding of the fundamental compromise, described by STOCKER (1952), as the path plants must steer on their Homeric journey between the Scylla of death by hunger and the Carybdis of death by thirst, now seems within reach. Similar examinations of plant/insect interactions may help to identify the costs in primary productivity of defense mechanisms based on secondary chemical compounds (Chap. 17, Vol. 12B), or of nectar in order to ensure pollination by attracting insects or birds (Chap. 15, Vol. 12C).

The end result of these advances has been that ecological physiology is no longer a somewhat suspect and scarcely accepted offshoot of physiology or ecology. As a separate discipline, it can rightly claim to have stimulated new physiological and ecological research in many fields. These include the role of membranes in response to extreme temperatures (Chaps. 10 and 12, Vol. 12A) and the physiological and biochemical processes in different pathways of photosynthetic carbon metabolism (Chap. 15, Vol. 12B). Use of modeling techniques in plant physiology holds much promise, as does the incorporation of physiological response into ecological models. In all of these respects SCHIMPER's predictions may have been fulfilled.

This brief summary of objectives and of progress is no different from that which might be found in the introduction to any such volume or series of volumes. Nevertheless, it is worth devoting a little space to discussing what we perceive as the special problems, and the special challenge, of physiological plant ecology. These both result from the breadth of the discipline which by definition seeks to link the physiological science of general plant life functions with ecology which is concerned with the habitats, modes of life, and functional relationships of the different types of organisms to their surroundings. In physiological ecology, it is possible to make useful observations on almost any scale; it is much more difficult to evaluate the significance of these observations in a relevant context. For example, HABERLANDT's speculation on the division of labor between mesophyll and bundle-sheath cells in the leaves of some plants might be assessed for ecological purposes with the aid of a hand lens in a good herbarium. From a biochemical and physiological viewpoint, a wealth of further detail is available as to the mechanisms and function of this particular leaf anatomy and the C_4 pathway of photosynthesis. There may be many functional relationships and possible adaptive physiological traits linking ecological success and the presence or absence of this important physiological attribute but few of these have been proven.

We have long accepted, of course, that physiological responses of all kinds at each stage of growth have been selected to secure the long-term objectives of plant survival, growth, and reproduction. Generally, however, we have not succeeded in formally identifying the adaptive function of physiological responses, particularly because the future environmental circumstances of an individual plant can only be determined in a statistical sense. Very few would disagree with DOBZHANSKY (1970) that the more spectacular findings in biology are likely to be made at the molecular level and, indeed, it is evident that man's imagination has been more stirred by the success of the biochemist than the problems of the ecologist (PASSIOURA 1976). If understanding of a unique kind can be expected from physiological ecology, it must be to account in physiological terms for the ways plants are adapted to maintain themselves in a community for long periods of time (ECKARDT 1975). This means that we cannot only be preoccupied with events at the molecular level but must also broach broader questions.

Two views of plant adaptation emerge from the broad scale of enquiry upon which physiological ecology is based (OSMOND et al. 1980). These two views reflect parallel differences found in the ecological sub-disciplines of autecology and synecology. At one extreme we find the physiologist preoccupied with states and processes within relatively small scales of space and time which lead to understanding of adaptation in terms of the way component processes are fitted together for optimal performance of the individual organism in a particular habitat. At the other we find the ecologist preoccupied with much larger-scale states and processes in which the survival of the individual organism is but one component of the performance of the larger system, the ecosystem. For most ecologists, adaptation may be a product of any property of an organism which confers a reproductive advantage relative to its competitors.

The two views of adaptation lead to a preoccupation with the physiology of performance within the limits to survival on the one hand, and the biology of survival on the other. Few physiologists concern themselves much with the relationships between performance and survival via different reproductive processes. Few ecologists are in a position to specify key problems of this type, and only a rapid development of plant population ecology will bridge this gap (HARPER 1977; GRIME 1979). Such studies should be extended from the more convenient, short-lived populations examined so far, in which inter-organism competition dominates the population response, to longer-lived populations in which environmental determination of performance and survival can be more clearly established (NOBLE 1977). Quantitative analysis in a population sense of the physiological basis of ecotypic and genotypic differentiation is a powerful tool (ANTONOVICS et al. 1971; CLOUGH et al. 1979) which has not yet had much impact on ecological physiology.

Population ecologists will almost certainly confirm that the period of seedling establishment (Chap. 12, Vol. 12B) is the most vulnerable in the life cycle of terrestrial plants. They will also help identify the strategies and goals as they are perceived by organisms of diverse life forms (Chap. 18, Vol. 12B). Such studies will be the foundation for the integration of performance and survival. They will serve to focus attention on physiological processes which are the key to ecological relationships. The integration of performance and survival which is a central challenge in physiological ecology has been most succinctly expressed by SCOTT (1974): "The existence of a living plant (in a particular habitat) demonstrates two things about its environmental relationships. Firstly, that none of the environmental variables, either alone or in combination, has exceeded the tolerance range of the plant at any stage during its lifetime; and secondly, that while growth has presumably varied with environmental fluctuations, the plant has made net growth. Thus, there are two types of analytical problems; the first concerned with defining the tolerances, and the second, the performance (e.g., growth) under the fluctuating environmental conditions within these tolerances. The potential tolerances and responses are determined by the genome of the plant, but its response at a particular instant will be conditioned by its past environmental interactions. Because of the known difference in response between plants, the analysis of a vegetation/environment complex should probably be attempted at the level of the species or appropriate physiological-genetic entity."

A pragmatic extension of this view from autecology into the realm of holism by WUENSCHER (1974) suggests that the number of significant environmental variables can usually be reduced to manageable proportions and responses determined experimentally: "To fully define a species' niche it is necessary, then, to include not only the limits of survival of the species along environmental gradients, but the level of response within these limits ... Complete specification of a species' niche would include every possible response to every environmental variable. Realistically this is impossible, and the best that can be achieved is an open-ended partial niche description. This practical limitation is not serious, however, since under natural conditions an organism's survival is determined by a few critical factors which actually impinge upon it and significantly influence its biological processes."

This, of course, has been the success story of ecological physiology and perhaps the notion that it may facilitate meaningful integration of autecology and synecology is one of its special fascinations. For the most part, interactions between single plant processes and isolated environmental variables have been explored in organisms from relatively extreme habitats. In spite of this, our evidence of adaptation, even of adaptation to extreme conditions, is in most cases far from complete. Modern technologies facilitate the quantification of practically all the speculations and hypotheses of SCHIMPER, HABERLANDT, and MAXIMOV, and provide the means to generate a great many more. Quantitation is clearly the first step in validation of any hypothesis as to adaptive significance, but all too frequently the exercise rests prematurely. To demonstrate adaptive significance of a physical or other trait requires evidence that it confers superior fitness in one genotype compared to another. Rarely in physiological ecology are comparisons made with the unsuccessful genotype and rarely is it shown that traits actually confer a competitive or reproductive advantage. It is also necessary to specify, by means of experiment, the appropriate perspectives of space and time for any trait of presumed adaptive significance. ECKARDT (1975) suggests that traits, such as large leaf area, which may increase the prospects for survival through rapid assimilation early in the life cycle may later contribute to its demise, perhaps by prematurely exhausting limited water resources. As has already been emphasized, the rules of optimization and compromise within either concept of adaptation are largely unknown at the present time.

It would have been inappropriate, and impossible, to have overtly impressed the above prejudices upon the contributors to these volumes. The reader will detect some evidence of a flow from molecular to organismic levels in the organization of the four volumes, and at times within chapters. It has been an interesting experience attempting to coordinate the essentially encyclopedic style of many European contributors, especially those with German as a mother tongue, with the more selective "review" style of contributors in the New World, most of whom share English as a mother tongue. There is, we hope, some balance between the discussion of performance and survival attributes and their physiological basis. The New Series of the Encyclopedia has already covered many fundamental aspects of plant physiology which are relevant to ecological physiology. Photosynthesis, for example, is well covered in two published volumes of the Encyclopedia (TREBST and AVRON 1977; GIBBS and LATZKO 1979), as are fundamental aspects of ion absorption and transport (ZIMMERMANN and MILBURN 1975; LÜTTGE and PITMAN 1976; STOCKING and HEBER 1976), and a volume on inorganic nutrition is in preparation. Detailed biochemical and physiological treatment of these topics is outside the immediate scope of the present volumes which it is hoped will complement the above. On the other hand, aspects of plant water relations, which were discussed by DAINTY (1976) and RASCHKE (1979) in earlier volumes of this series are developed more fully here (see Vol. 12B). The present volumes are also intended to update some, but by no means all, of those aspects of ecological plant physiology covered by the series Ecological Studies (e.g., DI CASTRI and MOONEY 1973; POLJAKOFF-MAYBER and GALE 1975; LANGE et al. 1976; GUDERIAN 1977).

Almost any aspect of plant physiology could be held to have ecological implications and hence to merit a place in these volumes. We have done our best to balance current fashion with what is perhaps obscure but almost certainly relevant in modern physiology and ecology. Only the next decades will determine how appropriate or misplaced our judgement was. We are confident, however, that these volumes will help to remove the crippling legacy of agronomic practice in plant physiology. Given our best efforts, the future of mankind will not be one in which unlimited resources can be applied to the husbandry of plants for food or fodder. The notion that, given sufficient human ingenuity, the concept of limiting factors could be rendered irrelevant, has been proven false. Ecological perspectives in plant physiology and its applications are increasingly important to our future.

Ecological plant physiology is presently teaching us the molecular basis of biological limitations and promises insights into the accounting and optimization of biological activity in complex systems with finite resources. If anthropocentric, teleological terminology is necessary to articulate these hypotheses for experimental analysis, it may also facilitate their acceptance in other fields of human endeavor. Appropriately, these volumes will conclude with chapters addressing the physiological bases of ecosystem disorders, some of them a direct consequence of the enthusiastic application of earlier generations of imaginative plant physiology (Chaps. 11 and 12, Vol. 12D). It is a particular pleasure for us to note that the final chapter (Chap. 16, Vol. 12D) has been contributed by one of the foremost holists in plant ecology, who also contributed the only review published by the Annual Review of Plant Physiology under the title *Physiological Ecology* (BILLINGS 1957).

References

Antonovics J, Bradshaw AD, Turner RG (1971) Heavy metal tolerance in plants. Adv Ecol Res 7:1–85

Bannister P (1976) Introduction to physiological plant ecology. Blackwell, Oxford London

Beadle NCW (1952) Studies on halophytes. I. The germination of the seeds and establishment of seedlings of five species of *Atriplex* in Australia. Ecology 33:49–62

Billings WD (1957) Physiological ecology. Annu Rev Plant Physiol 8:375–392

Castri di F, Mooney HA (1973) Mediterranean type ecosystems – origin and structure. Ecological Studies 7. Springer, Berlin Heidelberg New York

Clements FE (1907) Plant physiology and ecology. Holt, New York

Clough JM, Teeri JA, Alberti RS (1979) Photosynthetic adaptation of *Solanum dulcamara* L. to sun and shade environments. I. A comparison of sun and shade populations. Oecologia 38:13–21

Dainty J (1976) Water relations of plants cells. In: Lüttge U, Pitman MG (eds) Encylopedia of plant physiology new series, Vol 2, Part A. Springer, Berlin Heidelberg New York, pp 12–35

Daubenmire RF (1947) Plants and environment. A textbook of plant autecology. Wiley, New York

Dobzhansky TW (1970) Genetics of the evolutionary process. Columbia Univ Press, New York

Eckardt FE (1975) Functioning of the biosphere at the primary production level – objectives and achievements. In: Cooper JP (ed) Photosynthesis and productivity in different environments, IBP 3, Cambridge Univ Press, Cambridge, pp 173–185

Etherington JR (1975) Environment and plant ecology. Wiley, New York London

Etherington JR (1978) Plant physiological ecology. Edward Arnold, London

Evenari M, Richter R (1937) Physiological-ecological investigations in the wilderness of Judaea. J Linn Soc London 51:333–351

Fitting H (1911) Die Wasserversorgung und die osmotischen Druckverhältnisse der Wüstenpflanzen. Ein Beitrag zur ökologischen Pflanzengeographie. Z Bot 3:209–275

Geiger R (1927) Das Klima der bodennahen Luftschicht. Vieweg, Braunschweig

Gibbs M, Latzko E (1979) Photosynthesis II. Photosynthetic carbon metabolism and related processes. Encyclopedia of plant physiology new series, Vol 6. Springer, Berlin Heidelberg New York

Grime JP (1979) Plant strategies and vegetation processes. Wiley, New York London

Guderian R (1977) Air pollution. Phytotoxicity of acidic gases and its significance in air pollution control. Ecological Studies 22. Springer, Berlin Heidelberg New York

Haberlandt G (1884) Physiologische Pflanzenanatomie. Engelmann, Leipzig

Harper JL (1977) Population biology of plants. Academic Press, London New York

Huber B (1935) Der Wärmehaushalt der Pflanzen. Abhandlungen und Vorträge über Grundlagen und Grundfragen der Naturwissenschaft und Landwirtschaft. 17:1–148

Kraus G (1911) Boden und Klima auf kleinstem Raum – Versuch einer exakten Behandlung des Standortes auf dem Wellenkalk. Fischer, Jena

Kreeb K (1974) Ökophysiologie der Pflanzen. Eine Einführung. In: Bausteine der modernen Physiologie 1. Fischer, Stuttgart

Lange OL, Kappen L, Schulze E-D (1976) Water and plant life, problems and modern approaches. Ecological Studies 19. Springer, Berlin Heidelberg New York

Larcher W (1973) Ökologie der Pflanzen. 1. Aufl, Ulmer, Stuttgart

Larcher W (1980a) Physiological plant ecology. Springer, Berlin Heidelberg New York

Larcher W (1980b) Ökologie der Pflanzen auf physiologischer Grundlage. 3. Aufl, Ulmer, Stuttgart

Lundegårdh H (1924) Der Kreislauf der Kohlensäure in der Natur. Ein Beitrag zur Pflanzenökologie und zur landwirtschaftlichen Düngungslehre. Fischer, Jena

Lüttge U, Pitman MG (1976) Transport in Plants II. Part A, Cells, Part B, Tissues and organs. Encyclopedia of plant physiology new series, Vol 2. Springer, Berlin Heidelberg New York

Maximov NA (1923) Physiologisch-ökologische Untersuchungen über die Dürreresistenz der Xerophyten. Jahrb Wiss Bot 62:128–144

Noble IR (1977) Long term biomass dynamics in an arid chenopod shrub community at Koonamore, South Australia. Aust J Bot 25:639–653

Nye PH, Tinker PB (1977) Solute movement in the soil-root system. Blackwell, Oxford

Oppenheimer HR (1932) Zur Kenntnis der hochsommerlichen Wasserbilanz mediterraner Gehölze. Ber Dtsch Bot Ges 50a:185–245

Osmond CB, Björkman O, Anderson DJ (1980) Physiological processes in plant ecology: towards a synthesis with *Atriplex*. Ecological Studies 36. Springer, Berlin Heidelberg New York

Passioura JB (1976) The control of water movement through plants. In: Wardlaw IF, Passioura JB (eds) Transport and transfer processes in plants. Academic Press, London New York, pp 373–380

Pisek A, Cartellieri E (1939) Zur Kenntnis des Wasserhaushaltes der Pflanzen: IV. Bäume und Sträucher. Jahrb Wiss Bot 88:22–68

Poljakoff-Mayber A, Gale J (1975) Plants in Saline Environments. Ecological Studies 15. Springer, Berlin Heidelberg New York

Raschke K (1979) Movements of Stomata. In: Haupt W, Feinleib ME (eds) Encyclopedia of plant physiology new series, Vol 7. Springer, Berlin Heidelberg New York, pp 383–441

Richards HM (1915) Acidity and gas exchanges in cacti. Carnegie Inst Washington Publ 208

Schimper AFW (1898) Pflanzen-Geographie auf physiologischer Grundlage. Fischer, Jena

Scott D (1974) Description of relationships between plants and environment. In: Billings WD, Strain BR (eds) Vegetation and environment. Handbook of vegetation science, Part 6. Junk, The Hague, pp 49–69

Shreve EB (1914) The daily march of transpiration in a desert perennial. Carnegie Inst Washington Publ 194

Spoehr HA (1919) The carbohydrate economy of cacti. Carnegie Inst Washington Publ 287

Stocker O (1928) Der Wasserhaushalt ägyptischer Wüsten- und Salzpflanzen vom Standpunkt einer experimentellen und vergleichenden Pflanzengeographie aus. Bot Abh 13:1–200. Fischer, Jena

Stocker O (1952) Grundriß der Botanik. Springer, Berlin Göttingen Heidelberg

Stocker O (1956) Meßmethoden der Transpiration. In: Handbuch der Pflanzenphysiologie. Encyclopedia of plant physiology, Vol. 3. Springer, Berlin Göttingen Heidelberg, pp 293–311

Stocking CR, Heber U (1976) Transport in plants III. Intracellular interactions and transport processes. Encyclopedia of plant physiology new series, Vol 3. Springer, Berlin Heidelberg New York

Trebst A, Avron M (1977) Photosynthesis I. Photosynthetic electron transport and photophosphorylation. Encyclopedia of plant physiology new series, Vol. 5. Springer, Berlin Heidelberg New York

Turesson G (1922) The genotypic response of the plant species to the habitat. Hereditas 3:211–350

Walter H (1931) Die Hydratur der Pflanze und ihre physiologisch-ökologische Bedeutung (Untersuchungen über den osmotischen Wert). Fischer, Jena

Warming E (1896) Lehrbuch der ökologischen Pflanzengeographie. Eine Einführung in die Kenntnis der Pflanzenvereine. Borntraeger, Berlin

Wood JG (1925) The selective absorption of chloride ions; and the absorption of water by leaves in the genus *Atriplex*. Aust J Exp Biol Med Sci 2:45–56

Wuenscher JE (1974) The ecological niche and vegetation dynamics. In: Billings WD, Strain BR (eds) Vegetation and environment. Handbook of vegetation science, Part 6. Junk, The Hague, pp 39–45

Zimmermann MH, Milburn JA (1975) Transport in plants I. Phloem transport. Encyclopedia of plant physiology new series, Vol. 1. Springer, Berlin Heidelberg New York

1 Fundamentals of Radiation and Temperature Relations

G.S. CAMPBELL

CONTENTS

1.1 Introduction

Most of the chapters in this volume describe a plant's response to some imposed sets of environmental conditions. Implicit in these descriptions is the assumption that the environmental conditions at the plant are known. In laboratory and carefully monitored field studies, environmental conditions at the plant may be known, but in most cases these conditions must be inferred from measurements made at some distance from the plant. In this chapter methods will be developed for predicting environmental conditions at the plant from climatic data that are more generally available. We will focus primarily on those environmental variables that determine radiant flux at the plant and plant temperature. Once the radiant flux at the surface of a plant part is known, photosynthetic or photomorphogenic responses can be predicted from physiological models.

Likewise, when plant temperature is known, rates of photosynthesis and respiration can be found from temperature-dependent models of these processes.

Since plant temperature is determined in part by the radiant environment of the plant, first consideration will be given to fundamentals of radiative energy transport. Other components of the plant energy balance will then be included, and the energy balance will be used to infer plant temperature and transpiration rate. In this treatment, attention will first be given to energy exchange for a single leaf. The discussion will then be expanded to describe plant canopy exchange processes.

1.2 Radiation Fundamentals

Radiative transfer of energy is important to plants for at least three reasons: (1) the photosynthetic rate of a plant is related to the quantity of photosynthetically active radiation absorbed by the plant, (2) the temperature of a plant is, in part, determined by rates of absorption and emission of radiation, and (3) photomorphogenic response of the plant is determined by absorption of radiant energy in specific wavebands (see, for instance, Chaps. 3, 4, 5 and 6, this Vol.). Adequate specification of the properties of environmental radiation influencing each of these processes requires a knowledge of both the radiant flux density and the wavelength distribution of the radiation.

1.2.1 Spectral Distribution of Environmental Radiation

Radiation in natural plant environments originates either at the sun or at terrestrial objects. Both solar and terrestrial sources have spectral distributions of energy which are closely approximated by the Planck law:

$$i_B(\lambda)=\frac{2\pi\,hc^2}{\lambda^5[\exp(hc/\lambda k\Theta)-1]} \qquad (1.1)$$

where $i_B(\lambda)$ ($W\,m^{-3}$) is the spectral emittance, λ is wavelength (m), c is the speed of light ($3\cdot 10^8$ m s^{-1}), h is Planck's constant ($6.63\cdot 10^{-34}$ J s^{-1}), k is the Boltzmann constant ($1.38\cdot 10^{-23}$ J k^{-1}), and Θ is the temperature in Kelvin. Equation (1.1) may be used to determine that maximum spectral emittance occurs at a wavelength of 480 nm for a 6,000 K source (approximately the radiant temperature of the sun's surface) and 10 μm for terrestrial objects at 288 K. Equation (1.1) may also be integrated to determine that less than 2% of solar radiant energy occurs at wavelengths longer than 4 μm and less than 0.2% of terrestrial radiation occurs at wavelengths shorter than 4 μm (LIST 1971, p. 412). Four micrometers is therefore a convenient (though arbitrary) dividing point between short-wave radiation, which in natural environments comes only from the sun, and long-wave radiation, which is emitted by terrestrial sources.

Short-wave radiation with wavelengths shorter than 700 nm is photosynthetically useful. This radiation is termed photosynthetically active radiation PAR (see Chap. 2, this Vol.). It is often expressed as a photon flux density, but may also be expressed as an energy flux density. The ratio of energy in the PAR waveband to total solar radiation, η, is around 0.5, and is nearly independent of atmospheric conditions. SZEICZ (1974) obtained $\eta = 0.50 \pm 0.03$ for southern England, and STANHILL and FUCHS (1977) obtained $\eta = 0.521 \pm 0.127$ for Israel from 1307 half-hourly observations. Exclusion of data with solar elevation angles less than 10° gave $\eta = 0.492 \pm 0.051$, in excellent agreement with SZEICZ. Further subdivision of solar radiation into even narrower wavebands is sometimes necessary for specific studies, as is shown in Chapter 4 of this Volume, but to determine photosynthesis a division into PAR and near infrared is usually adequate, and to calculate an energy budget, a single waveband is sufficient.

1.2.2 Short-Wave Radiation in the Plant Environment

The quantity of short-wave radiation absorbed by a plant depends on both source characteristics and absorber orientation. Short-wave radiation is either direct or diffuse. For most purposes, the sun can be considered a point source, in which case the direct irradiance ($W m^{-2}$) of a flat surface is given by LAMBERT's cosine law:

$$S_b = S_p \cos\theta \tag{1.2}$$

where S_p is the flux density of radiation measured on a plane perpendicular to the beam, and θ is the angle between a normal to the surface and the beam. Diffuse short-wave is solar radiation scattered or reflected to the plant by sky, clouds, and surrounding objects. COULSON (1975) gives information on radiance distributions for clear and overcast skies. In principle the diffuse sky irradiance of a surface could be computed from a known radiance distribution for the sky. This is often not practical, however, and diffuse irradiance is generally either measured or computed from beam flux estimates.

Short-wave radiation is usually measured using a pyranometer with a flat, horizontal collecting surface (COULSON 1975). The measurement is therefore the total hemispherical short-wave radiation for a horizontal surface:

$$S_t = S_b + S_d = S_p \sin\phi + S_d \tag{1.3}$$

where ϕ is the solar elevation angle, and S_d is the diffuse irradiance of a horizontal surface. The total irradiance must be separated into direct and diffuse components before it can be used to compute irradiances for nonhorizontal surfaces.

A scheme similar to that outlined by LIST (1971, p. 420) is used to separate total radiation into direct and diffuse components when $S_b \neq 0$. We first assume that diffuse radiation is some fraction, α, of the difference between potential (i.e., without scattering) direct radiation on a horizontal surface, and S_b:

$$S_d = \alpha(\beta S_p^0 \sin\phi - S_b). \tag{1.4}$$

where S_p^0 is the solar constant (1,360 $W m^{-2}$). The factor, β, accounts for absorption of the solar beam by atmospheric moisture, ozone, and other constituents (KONDRATYEV 1969). Combining Eq. (1.4) with (1.3), we can write

$$S_b = (S_t - \alpha\beta S_p^0 \sin\phi)/(1-\alpha) \tag{1.5}$$

and

$$S_d = \alpha(\beta S_p^0 \sin\phi - S_t)/(1-\alpha). \tag{1.6}$$

Data obtained by WESLEY and LIPSCHUTZ (1976) are consistent with these equations, and give values of $\beta = 0.79$ and $\alpha = 0.64$ for the Chicago area. The values suggested by LIST (1971) are $\beta = 0.91$ and $\alpha = 0.5$. We would expect these constants to vary somewhat with atmospheric conditions, but the range of variation is not known at present.

From Eq. (1.5) it is readily seen that, when $S_b = 0$, $S_t = \alpha\beta S_p^0 \sin\phi$. From the WESLEY and LIPSCHUTZ (1976) data, the product $\alpha\beta = 0.5$. The values suggested by LIST (1971) give $\alpha\beta = 0.46$. It therefore seems reasonable to assume that $S_d = S_t$ for $S_t < 0.5\, S_p^0 \sin\phi$.

When no measurements of S_t are available, estimates of S_b and S_d can still be obtained. Detailed procedures for this are given by MCCULLOUGH and PORTER (1971), but estimates accurate enough for many ecological studies can be made using GATES (1962):

$$S_p = S_p^0 \exp\left[-0.089\,(mp/p_0)^{0.75} - 0.174\,(mw/20)^{0.6} - 0.083\,(dm)^{0.9}\right] \tag{1.7}$$

where m is the airmass number, or ratio of slant-path to vertical path length through the atmosphere, p is atmospheric pressure, p_0 is sea level pressure (1.013 bar), w is the total precipitable atmospheric water vapor (mm) in the zenith direction, and d is the atmospheric dust concentration. (GATES gives dust concentrations as particles cm^{-3}, but comparison of his typical values, given below, with typical values given by other authors, show his to be about three orders of magnitude too low. More probable units for GATES' numbers are therefore particles mm^{-3}).

If atmospheric refraction is neglected, $m = 1/\sin\phi$. Precipitable water vapor ranges from 1 to 100 mm (ROBINSON 1966). Typical mid-latitude values are $w \simeq 30$ mm for warm, moist summer conditions and $w \simeq 10$ mm for cool winter conditions. GATES gives values for d which range from 0.2 to 3.0. He gives, as typical values for clear days, 0.6 to 1.0. Values for polluted atmospheres near cities may be in the range 1.4 to 2.0. Equation (1.7) predicts values of S_p that are within a few $W m^{-2}$ of the mean curves for clear sky radiation given by PALTRIDGE and PLATT (1976) and WESLEY and LIPSCHUTZ (1976). Once S_p is known, S_d and S_t can be found from Eqs. (1.4) and (1.3).

Table 1.1. Values of the constants a and b for computing diffuse radiation from clouds [Eq. (1.8)] according to cloud type

Cloud type	a	b
Cirrus	0.871	−0.020
Cirrostratus	0.923	−0.089
Altocumulus	0.556	−0.053
Altostratus	0.413	−0.004
Stratocumulus	0.368	−0.045
Stratus	0.252	−0.100
Nimbostratus	0.119	0.226
Fog	0.163	0.031

When the sun is obscured by clouds, all the radiation is diffuse. The diffuse flux under these conditions can be calculated using a formula from HAURWITZ (1948):

$$S_{dc} = S_{t0}\, a\, e^{bm} \tag{1.8}$$

where S_{t0} is the total radiation on a horizontal surface in the absence of clouds, and a and b are empirical constants. Values for the constants, a and b, are given in Table 1.1.

The plant may also receive reflected short-wave radiation from the ground or surrounding objects. The irradiance at a downward-facing horizontal flat surface is

$$S_r = \bar{\rho}\, S_t \tag{1.9}$$

where $\bar{\rho}$ is the albedo or surface reflection coefficient for short-wave. Typical values for $\bar{\rho}$ are 0.15–0.25 for vegetation, 0.1–0.15 for dark, wet soil, and 0.2–0.3 for dry, light-colored soil (LIST 1971). An extensive table of albedos is also given by MONTEITH (1973).

1.2.3 Computing Short-Wave Irradiance of Plant Surfaces

Once S_p, S_d, and S_r are known, the irradiance is computed from the solar elevation and azimuth, and from the leaf inclination and azimuth angles. (Leaf inclination is the angle between the leaf axis and a horizontal; leaf azimuth angle is the angle between the horizontal projection of a normal to the leaf and a specified reference direction.) The solar elevation angle is given by

$$\sin\phi = \sin\delta \sin\lambda + \cos\delta \cos\lambda \cos 15(t_d - t_{sn}) \tag{1.10}$$

where δ is the solar declination (degrees), λ is the geographic latitude (degrees), t_d is time of day (hours), and t_{sn} is the time of solar noon. The solar declination is a function of day of the year and can be found in LIST (1971), Table 169, or computed from (SWIFT 1976):

$$\sin \delta = 0.39785 \sin [278.9709 + 0.9856 \text{ J} + 1.9163 \sin (356.6153 + 0.9856 \text{ J})] \tag{1.11}$$

where J is the Julian day (day of the year starting with January 1 = 1). Equation (1.11) gives declinations accurate to better than ±0.1 degree. The time of solar noon can be determined from local observations, or computed as outlined by MCCULLOUGH and PORTER (1971) or PALTRIDGE and PLATT (1976).

The solar azimuth angle is computed from (LIST 1971):

$$\sin \alpha = -\cos \delta \sin 15 (t_d - t_{sn}) / \cos \phi \tag{1.12}$$

or

$$\cos \alpha = -(\sin \delta - \sin \lambda \sin \phi) / \cos \lambda \cos \phi, \tag{1.13}$$

where α is measured from due south. The angle, θ in Eq. (1.1), between the solar beam and a normal to a plane that is inclined at an angle, γ, relative to a horizontal plane, and with the projection of the normal to the plane having an azimuth, β, referenced to south, is

$$\cos \theta = \cos \gamma \sin \phi + \sin \gamma \cos \phi \cos (\alpha - \beta). \tag{1.14}$$

Equation (1.14) can be used to compute direct irradiance of leaf surfaces, or to compute average irradiance of canopies on sloping terrain. Simple algorithms for these computations can be found in SWIFT (1976).

The diffuse and reflected irradiance of a plane also have an angular dependence, but it is much less pronounced than for direct irradiance, and can often be neglected. In the presence of direct radiation, diffuse short-wave from the sky is 10% to 20% of total radiation. Typical surface albedos are also in the range 10% to 20%, so S_d is about the same size as S_r. Under these conditions leaf angle does not affect diffuse irradiance. In situations requiring the computation of angular dependence of diffuse and reflected irradiance, the following equations from MONTEITH (1973) can be used:

$$\Phi_{\text{upper surface}} = S_d \cos^2(\gamma/2) + S_r \sin^2(\gamma/2) \tag{1.15}$$

$$\Phi_{\text{lower surface}} = S_d \sin^2(\gamma/2) + S_r \cos^2(\gamma/2), \tag{1.16}$$

where γ is the inclination angle of the plane with respect to horizontal.

1.2.4 Long-Wave Radiation in the Plant Environment

The long-wave radiation ($\lambda > 4$ μm) received by the plant is emitted by terrestrial sources around the plant, and the plant emits radiant energy to its surroundings.

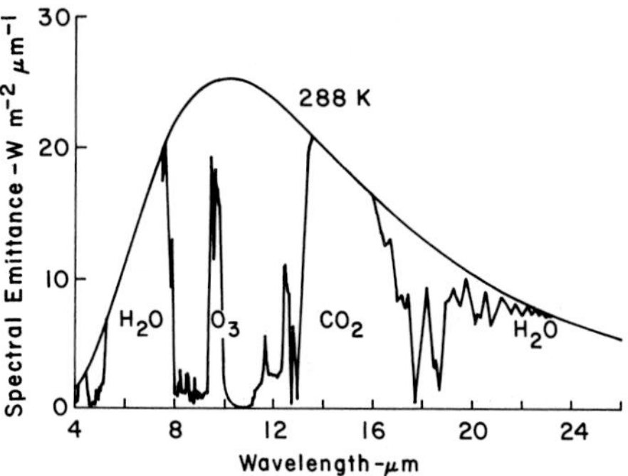

Fig. 1.1. Emittance spectrum for a 288 K black body (*upper line*) and for the Earth's atmosphere. Atmospheric gases principally responsible for each of the emission bands are indicated. (CAMPBELL 1977)

The radiant emittance of a source is proportional to the fourth power of its Kelvin temperature (Stefan-Boltzmann law):

$$\Phi_B = \sigma \Theta^4 \tag{1.17}$$

where Φ is the radiant emittance of a full radiator, or blackbody source, Θ is T+273.16, T is temperature in degrees Celsius, and σ is the Stefan-Boltzmann constant ($5.67 \cdot 10^{-8}$ W m^{-2} K^{-4}). If an emitting surface or body is not a full radiator (i.e., does not emit the maximum possible radiation at all wavelengths) it is called a gray body emitter, and its radiant emittance is computed from

$$\Phi = \varepsilon \Phi_B = \varepsilon \sigma \Theta^4 \tag{1.18}$$

where ε is the emissivity of the surface or body. Typical values of ε for long-wave radiation from leaves range from 0.94–0.99 (MONTEITH 1973). Long-wave emissivities of most other materials in the natural surroundings of plants are also in this range, whereas emissivities of metals are much lower (SIMONSON 1975).

Long-wave radiation emitted by the sky can also be computed using Eq. (1.18), but the emission spectrum of the sky, unlike that of most terrestrial surfaces, is not continuous in the long-wave region. Figure 1.1 compares a blackbody emission spectrum at a typical terrestrial temperature to the spectral emittance of the atmosphere at that temperature. The principal long-wave emitting (and absorbing) gases in the atmosphere are CO_2, H_2O, and O_3. Because the atmospheric H_2O concentration varies, its emissivity (ε) also varies. BRUTSAERT (1975) derived a simple relationship between vapor concentration of the air at shelter-height (c_{va}, g m^{-3}) and ε_a:

$$\varepsilon_a = 0.58\ c_{va}^{1/7}. \tag{1.19}$$

Since vapor density is correlated with air temperature, ε_a can also be expressed as a function of air temperature with reasonable accuracy (IDSO and JACKSON 1969).

For typical air temperatures, the expression

$$\varepsilon_a = 0.72 + 0.005\ T_a \tag{1.20}$$

gives adequate estimates of clear sky emissivity (CAMPBELL 1977). When the sky is cloudy, an emissivity can be calculated from UNSWORTH and MONTEITH (1975):

$$\varepsilon_{ac} = (1 - 0.84\ C)\varepsilon_a + 0.84\ C \tag{1.21}$$

where C is the mean fraction of cloud cover.

The computation of long-wave radiation exchange between a plant or plant part and its surroundings can be quite complicated. Even for the simplest case of a single plane leaf suspended over a soil surface of uniform temperature, and under a sky of known temperature, the angular distribution of long-wave radiation from the sky (UNSWORTH and MONTEITH 1975) and the angle dependence of the leaf emissivity can lead to uncertainties. More complicated radiative exchange situations are dealt with using view factor analysis (BIRD et al. 1960), but require extensive knowledge of the geometry of the system and the radiance of the surroundings. Since temperature of the plant is usually similar to the temperature of the emitting surroundings, it is generally sufficient to take the long-wave irradiance of the plant as the average of long-wave flux from the sky and ground using Eq. (1.18). Long-wave emittance of the plant is also computed using Eq. (1.18). For inclined plane surfaces, Eqs. (1.15) and (1.16) can be used by substituting the appropriate long-wave sky and ground irradiances for the short-wave values given.

1.2.5 Absorption, Transmission, and Reflection of Radiation

Radiation incident on the plant will be absorbed, transmitted, or reflected. Since all of the radiation incident on a plant follows one of these paths, the following relation holds:

$$a(\lambda) + r(\lambda) + t(\lambda) = 1. \tag{1.22}$$

Here, $a(\lambda)$, $r(\lambda)$, and $t(\lambda)$ are the absorptivity, reflectivity, and transmissivity at a particular wavelength. It is convenient to define mean values for these quantities over extended wavebands, or for particular radiation sources, so that the total amount of absorbed solar radiation can be computed. The reflection coefficient, r, is defined as (MONTEITH 1973)

$$r = \frac{\int i(\lambda)\, r(\lambda)\, d\lambda}{\int i(\lambda)\, d\lambda} \tag{1.23}$$

where $i(\lambda)$ is the spectral emittance of the source [Eq. (1.1)]; the integration is performed over all wavelengths in the waveband of interest. Similar definitions can be written for the transmission coefficient, t, and the absorption coefficient, a. While these are useful quantities, it must be borne in mind that they depend on the wavelength distribution of the source. The absorption coefficient for a white surface may therefore be 0.1 for short-wave radiation, but could be 0.95 for long-wave. In subsequent discussion, the subscripts s and L will be used to indicate the source for which the coefficient is defined. Ross (1975) gives optical properties of a "mean" green leaf as $r_s=0.30$, $t_s=0.20$, $a_s=0.50$. The quantities differ markedly between visible and near infrared radiation: $r_{PAR}=0.09$, $t_{PAR}=0.06$, $a_{PAR}=0.85$ and $r_{NIR}=0.51$, $t_{NIR}=0.34$, $a_{NIR}=0.15$. In the long-wave, a_L of a typical leaf is around 0.95, with $t_L=0$ and $r_L=0.05$.

1.2.6 Measurement of Radiative Fluxes in the Plant Environment

The relationships presented in the previous sections are useful for modeling purposes and for making "engineering estimates" of radiant fluxes at plants. They will be used in the energy budget analyses to be presented in the next section. They should not in most cases be considered adequate substitutes for measured radiant fluxes, as is perhaps obvious from the number of assumptions needed to simplify some of the equations to manageable proportions. This being the case, instruments and techniques for measuring radiant fluxes in the plant environment should be mentioned briefly.

Short-wave fluxes are easily measured with a pyranometer (COULSON 1975). Pyranometers with thermopile sensors and glass domes respond uniformly to almost all wavelengths present in short-wave radiation. Total irradiance of a plane can be measured by holding the pyranometer parallel to the plane. The pyranometer can also be used to determine S_p by orienting it perpendicular to the solar beam and subtracting the diffuse irradiance (measured with the sensor shaded by a small disk a meter or more from the radiometer) from the total. Silicon sensors (KERR et al. 1967) are also suitable for measuring incident solar radiation, but should not be used for measurements of reflected short-wave because of their limited spectral response.

The long-wave flux from the leaf surface and to the leaf from surrounding surfaces is conveniently measured using an infrared thermometer. An IR thermometer measures radiant flux, but uses internal circuitry to infer temperature using Eq. (1.18). Infrared thermometers are available from a number of commercial sources. Some have characteristics which make them undesirable for work with plants. The spectral bandpass of filters used in most low temperature IR thermometers is between 6 and 20 μm, but in some the short wavelength cutoff is below 2 μm. These latter are therefore subject to substantial potential error from reflected short-wave radiation. Some units also provide an adjustment for surface emissivity. Such an adjustment should be used with caution, since the radiant flux measured by the thermometer is the sum of reflected and emitted radiation: $\Phi_T=\Phi_e+\Phi_r$. Each of these can be written in terms of a temperature using Eq. (1.18) to give

$$\Theta_T^4=\varepsilon\Theta_s^4+(1-\varepsilon)\,\Theta_i^4 \tag{1.24}$$

where Θ_T is the Kelvin temperature indicated by the thermometer, Θ_s is the true surface temperature, and Θ_i is the mean radiant temperature of the surroundings. The surface is assumed to have $t_L=0$ so that $r_L=1-a_L=1-\varepsilon$. The second term in Eq. (1.24) can be written as $\Theta_i=\Theta_s+\Delta\Theta$, where $\Delta\Theta=\Theta_i-\Theta_s$, and the expression $(\Theta_s+\Delta\Theta)^4$ can be approximated by the first two terms of its expansion (since $\Delta\Theta \ll \Theta$) so that

$$\Theta_T^4=\varepsilon\Theta_s^4+(1-\varepsilon)\,\Theta_s^4+4(1-\varepsilon)\,\Theta_s^3\,\Delta\Theta=\Theta_s^4+4(1-\varepsilon)\,\Theta_s^3\,\Delta\Theta. \qquad (1.25)$$

The last term on the right hand side of Eq. (1.24) is an error term which is zero when $\varepsilon=1$ or when $\Delta\Theta=0$. Since the error term is not uniquely related to ε, it is impossible to compensate for it using only an adjustment for ε. The temperature difference between a leaf surface and its surroundings is extremely variable, but in typical situations, the error resulting from ignoring the error term and assuming $\varepsilon=1$ is less than 0.5 K.

An IR thermometer is usually not suitable for measuring incoming long-wave from the sky because of its limited spectral response. The filters in most IR thermometers are purposely chosen to be transparent in the same waveband as the atmosphere (Fig. 1.1). This reduces atmospheric interference for temperature measurements of terrestrial objects, but it also prevents the thermometer from being used to measure sky radiation unless the sky is fully overcast so that spectral distribution of the radiation becomes unimportant to the measurement. Sky long-wave radiation can be measured either with a carefully constructed pyrgeometer or by measuring total incoming radiation and subtracting incoming short-wave. A detailed procedure for making these measurements is given by CAMPBELL et al. (1978).

Short-wave optical properties of some leaves are given by GATES et al. (1965), EHLERINGER and BJÖRKMAN (1978), and SINCLAIR and THOMAS (1970). Techniques for making these measurements are also given by these authors. Long-wave emissivities of some leaves are given by IDSO et al. (1969). These can be determined using an IR thermometer and Eq. (1.24). The procedure is given by IDSO et al. (1969).

1.3 The Energy Budget of a Leaf

Leaf temperature determines the rate of physiological processes within the leaf (see, for instance, Chaps. 10 through 14, this Vol.). It also affects the rate of water loss from a leaf. The leaf temperature, however, is not directly controlled by the plant, but is determined by the rate of receipt, loss, and storage of energy by the leaf. An analysis of energy inputs and losses is called an energy budget. RASCHKE (1956) was the first to give a detailed analysis of a leaf energy budget and apply this analysis to an ecological problem. This type of analysis has been extended by GATES and others (see, for example, GATES and SCHMERL 1975) and applied to a number of important questions of ecophysiology.

1.3.1 Conservation of Energy

The first law of thermodynamics, or law of conservation of energy states that the sum of energy fluxes at a surface must be zero. For a leaf we can write

$$R_{abs} - L_{oe} - H - \lambda E - q + M = 0 \tag{1.26}$$

where R_{abs} is the flux density ($W\ m^{-2}$) of absorbed long- and short-wave radiation at the leaf, L_{oe} is the flux density of emitted radiation, H is the sensible heat loss, E is the evaporation rate ($g\ m^{-2}s^{-1}$), λ is the latent heat of vaporization for water (2,450 $J\ g^{-1}$ at 20° C), q is the heat flux density representing storage, and M is the energy release or storage by chemical reactions (photosynthesis or other metabolism).

All of the quantities in Eq. (1.26) have dimensions of energy per unit time per unit area, but one must be careful in defining the area to be used for the computations. Processes such as photosynthesis, respiration, or heat storage are typically thought of on a "per unit leaf area" basis rather than on a total leaf surface area basis (2× leaf area for thin leaves). Energy exchange, however, is usually easiest to compute on the basis of a representative unit area of surface. It is therefore important to specify the basis used for computations. For our computations here, all quantities in Eq. (1.26) will be per unit projected area.

1.3.2 Components of the Energy Budget

The average flux density of absorbed radiation for a leaf is

$$R_{abs} = a_s (S_p \cos\theta + S_d + S_r) + 2 a_L\, \varepsilon_e\, \sigma \Theta_a^4 . \tag{1.27}$$

where a_L and a_S are the long- and short-wave absorption coefficients for the leaf, S_p, S_d, and S_r are direct, diffuse, and reflected short-wave irradiances of the leaf surfaces, θ is the angle between a normal to the leaf and the direct beam, ε_e is the effective emissivity of the surroundings and Θ_a is air temperature (Kelvin). Using the subscript, g, to indicate ground temperature and emissivity,

$$\varepsilon_e = (\varepsilon_g\, \Theta_g^4 + \varepsilon_a\, \Theta_a^4) / 2\, \Theta_a^4 . \tag{1.28}$$

The average flux density of emitted radiation is

$$L_{oe} = 2 \varepsilon_L\, \sigma \Theta_L^4 \tag{1.29}$$

where ε_L is the leaf emissivity. The sensible heat flux density is

$$H = \rho c_p (T_L - T_a) / r_H . \tag{1.30}$$

Here ρ is the density of air (1,204 g m^{-3} at 20° C and 1.013 bar), c_p is specific heat of air (1.01 J g^{-1} K^{-1}), and r_H is the mean resistance to heat transfer of the air boundary layer surrounding the leaf.

The evaporation rate for the leaf is directly proportional to the vapor concentration difference (g m^{-3}) between the leaf (c_{vL}) and the air (c_{va}) and inversely proportional to the resistance to vapor transfer:

$$E=(c_{vL}-c_{va})/r_v. \tag{1.31}$$

Equation (1.31) is strictly correct only for isothermal systems, since vapor pressure, and not vapor concentration, drives evaporation in systems at constant pressure. However errors resulting from use of Eq. (1.31) for nonisothermal systems are generally smaller than experimental errors in typical leaf environments. The leaf vapor concentration (c_{vL}) is the concentration of vapor in equilibrium with liquid water at the evaporating surface of leaves. If leaf water potential and leaf temperature are known, c_{vL} can be computed from

$$c_{vL}=c_{vL}^0 \exp(M_w \Psi/R\Theta) \tag{1.32}$$

where c_{vL}^0 is the saturation vapor density at leaf temperature, M_w is the molecular weight of water (0.018 kg/mol), R is the gas constant ($8.31 \cdot 10^{-2}$ bar kg mol^{-1} K^{-1}), Θ is Kelvin temperature, and Ψ is water potential of the leaf (bar) (CAMPBELL 1977). For water potentials above −15 bar (typical of mesophytic leaves) c_{vL} is within 2% of c_{vL}^0, so c_{vL} is usually assumed to equal c_{vL}^0. Water potentials of some species, particularly halophytes, possibly can be low enough to reduce transpiration, compared to that of a leaf at zero potential.

Tables of c_v^0 are available in LIST (1971), or values accurate to 0.1% over the range $-253<\Theta<333$ K (−20° to +60° C) can be computed from

$$c_v^0=[\exp(31.3716-6014.79/\Theta-0.00792495\,\Theta)]/\Theta. \tag{1.33}$$

The total resistance for vapor diffusion, r_v, is a combination of a boundary layer resistance, r_{va}, and a surface (stomatal and cuticle) resistance, r_{vs} (cf. Vol. 12b, Chaps. 6, 7). Care must be used in combining these resistances to determine r_v, since r_{vs} varies from place to place on the leaf. If the conductance for vapor transfer from one area of the leaf is

$$g_i=1/(r_{vs\,i}+r_{va\,i}) \tag{1.34}$$

then

$$r_v=n/\sum g_i.$$

Typically, two resistances are measured, an abaxial r_{vs}^{ab} and an adaxial, r_{vs}^{ad}. The boundary layer resistance is usually assumed equal on both sides of the leaf. The total resistance for vapor is therefore

$$r_v = \frac{(r_{vs}^{ab} + r_{va})(r_{vs}^{ad} + r_{va})}{r_{vs}^{ab} + r_{vs}^{ad} + 2r_{va}}. \tag{1.35}$$

When leaves are hypostomatous, and cuticular transpiration can be neglected, then Eq. (1.35) simplifies to

$$r_v = (r_{vs}^{ab} + r_{va}). \tag{1.36}$$

Where leaves are amphistomatous, with resistances equal on both sides

$$r_v = (r_{vs} + r_{va})/2. \tag{1.37}$$

The heat storage term in Eq. (1.26) is usually considered negligible for thin leaves. MONTEITH (1973) calculates that q is about two orders of magnitude smaller than the short-wave flux for a leaf warming on a clear morning. When the sun is appearing or disappearing behind clouds, he computes q values as high as 100 W m^{-2}, which would not be negligible. In succulents, storage may also be an important component of the energy budget. LEWIS and NOBEL (1977) were able to solve the energy budget equation for a barrel cactus, using a numerical scheme to compute storage and flow of heat in the plant. This approach makes solution of the energy budget equation considerably more complicated, but may be required for some studies.

MONTEITH (1973) gives order-of-magnitude computations for the size of M (energy storage or release by metabolism) in relation to other terms in the energy budget. He concludes that M is generally comparable in magnitude with the measurement error in other terms of the energy budget, but on calm overcast nights, could be comparable with net radiation for a crop. In species which produce metabolic heat to control the temperature of an organ at a value higher than ambient, M is obviously not negligible, and may approach values typical of homeotherms. Skunk cabbage (*Symplocarpus foetidus*) is an example of such a species (KNUTSON 1974).

1.3.3 Boundary Layer Resistances of Leaves

Equations (1.30) and (1.31) contain terms for boundary layer resistance. The boundary layer resistance for heat or mass transfer is computed using formulas from the engineering literature (e.g., BIRD et al. 1960). Transport can be by either free or forced convection. In nature, forced convection is more common, and is the only transport mechanism that will be discussed here.

The transport equations used by engineers are derived from theory assuming laminar flow of fluid over a smooth, rectangular flat plate. The following dimensionless numbers are correlated to determine the transport coefficients:

the Reynolds number:

$$Re = ud/\nu \tag{1.38}$$

the Nusselt number:

$$Nu = Hd/[\rho c_p D_H (T_s - T_a)] \tag{1.39}$$

the Sherwood number (for water vapor):

$$Sh = Ed/[D_v (c_{vs} - c_{va})] \tag{1.40}$$

the Prandtl number:

$$P_r = \nu / D_H \tag{1.41}$$

and the Schmidt number:

$$Sc = \nu / D_v . \tag{1.42}$$

Here, u is fluid velocity, d is the length of the plate in the direction of fluid flow, ν is fluid viscosity, D_H is thermal diffusivity (m^2/s), D_v is water vapor diffusivity, H is heat flux density, and E is vapor flux density. The resistances, defined in terms of these quantities are

$$r_{Ha} = d/D_H \, Nu \tag{1.43}$$

and

$$r_{va} = d/D_v \, Sh. \tag{1.44}$$

SIMONSON (1975) shows how to derive the following relationship for a flat plate from theory:

$$Nu = 0.664 \, Re^{1/2} \, Pr^{1/3} \tag{1.45}$$

$$Sh = 0.664 \, Re^{1/2} \, Sc^{1/3}. \tag{1.46}$$

If these are combined with Eqs. (1.43) and (1.44), and the diffusion coefficients evaluated at 20° C, the resistances for one side of a flat plate become (in $s \, m^{-1}$)

$$r_{Ha} = 307 \, (d/u)^{1/2} \tag{1.47}$$

$$r_{va} = 284 \, (d/u)^{1/2}. \tag{1.48}$$

The constants in Eqs. (1.47) and (1.48) decrease with increasing temperature at a rate of 0.28% per degree and increase with increasing pressure at a rate of 5.8% per 100 mbar, so the constants should be adjusted if these equations are used outside the range $0 < T < 40°$ C and $900 < p < 1{,}013$ mbar.

Equations (1.47) and (1.48) are for laminar flow over smooth, rectangular flat plates. They may be used for leaves in the atmosphere by adjusting the constants for free-stream turbulence and the characteristic dimension for leaf shape (cf. Chap. 15, this Vol.).

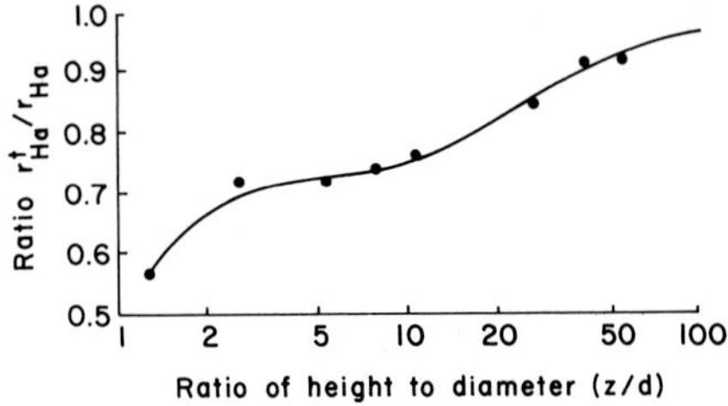

Fig. 1.2. Effect of free-stream turbulence on resistance to heat transport from heated spheres in natural wind. (Data from MITCHELL 1976)

Free-stream turbulence can dramatically affect the boundary layer resistance of a leaf. The factors which influence the resistance most strongly are leaf orientation with respect to the air flow, turbulence intensity of the flow, and scale of turbulence relative to the size of the leaf. Free stream turbulence apparently only affects boundary layer resistance of flat plates when the flow is not parallel to the plate surface (KESTIN 1966). Since leaves are not flat, and cannot remain parallel to air flow when directions fluctuate widely, this condition is probably seldom met in nature. It may, however, explain some of the discrepancies that appear in literature reports of boundary layer resistances of model leaves. For plates which are not flat, angles of attack which are not zero, and for spherical or cylindrical surfaces, free stream turbulence decreases boundary layer resistance. The relationships in Eqs. (1.47) and (1.48) apparently remain valid except that the constants are lowered (KESTIN 1966; SCHLICHTING 1971). The extent to which they are lowered depends on the turbulence intensity of the free stream (NOBEL 1974) and the scale of turbulence relative to the characteristic dimension of the surface (PEARMAN et al. 1972; MITCHELL 1976).

For metal disks exposed to natural turbulence in the field, PEARMAN et al. (1972) found that resistances to heat transfer averaged 70% of those given by Eq. (1.47). The range of variation in their measured heat transfer rates indicated that resistances ranged from 50% to 100% of those predicted by Eq. (1.47). Figure 1.2, adapted from MITCHELL (1976), shows how the boundary layer resistance of spheres in natural wind changes with height above the surface. The dominant eddy size in natural turbulence is a function of height (z). Figure 1.2 shows that as the eddy size approaches the characteristic dimension of the sphere (d), boundary layer resistance is reduced to nearly half the value given by Eq. (1.47). Over the range, $2 < z/d < 10$ (perhaps fairly typical of plant environments) $r^t_{Ha}/r_{Ha} \simeq 0.7$. At greater heights, where eddies are still larger, r^t_{Ha}/r_{Ha} approaches 1. From these measurements, it appears that free stream turbulence, on the average, reduces resistance by around 30%, so resistances given by Eqs. (1.47) and (1.48) should be multiplied by 0.7 for field conditions.

For heat transfer from cylinders and spheres, MITCHELL suggests the correlation

$$\mathrm{Nu} = 0.34\ \mathrm{Re}^{0.6} \tag{1.49}$$

with the characteristic dimension taken as the cube root of the volume. However, for Reynolds numbers typical of plants, Eq. (1.45) with characteristic dimension equal to the diameter, fits the data as well. NOBEL (1974) also found that an exponent of 0.5 on Re works well for cylindrical plant parts.

Equations (1.47) and (1.48) are for rectangular plates, where d is the length of the plate in the direction of wind flow. The characteristic dimension for a leaf is the length of a rectangular plate that would have the same boundary layer resistance as the leaf. We therefore define d as

$$d=\left[\frac{\int_0^w y(x)\,dx}{\int_0^w [y(x)]^{1/2}\,dx}\right]^2 \tag{1.50}$$

where w is the width of the leaf in the x direction, perpendicular to wind flow, and y is the length in the direction of wind flow (CAMPBELL 1977). For typical leaf shapes, d is around 70% of the maximum leaf dimension in the direction of wind flow (TAYLOR and GATES 1970).

1.3.4 Linearization of the Energy Budget Equation

Several of the terms in the energy budget equation are functions of leaf temperature. It should be possible, therefore, to determine leaf temperature from leaf characteristics and environmental conditions. The problem is that L_{oe} and E are nonlinear functions of leaf temperature. While in principle the solution of nonlinear equations is not difficult, in practice it is worthwhile to try to find an explicit equation for T_L to visualize more easily the effects of environment on leaf temperature.

An equation for leaf temperature can be derived by approximating L_{oe} and E by the values they would have if the leaf were at air temperature, and then adding a correction. Equation (1.29) therefore becomes

$$L_{oe}=2\varepsilon_L\,\sigma\Theta_L^4\simeq 2\varepsilon_L\,\sigma\Theta_a^4+2\rho c_p(T_L-T_a)/r_r \tag{1.51}$$

where r_r is a *radiative* resistance given by

$$r_r=\rho c_p/4\varepsilon\,\sigma\bar{\Theta}^3. \tag{1.52}$$

The temperature, $\bar{\Theta}$, in Eq. (1.52) is the average of leaf and air temperature, but since leaf temperature is unknown, air temperature must be used as a first approximation. For improved estimates of leaf temperature, $\bar{\Theta}$ can be found by iteration.

Applying the same reasoning to Eq. (1.31) and assuming $c_{vL}=c_{vL}^0$, we can write (CAMPBELL 1977)

$$E=(c_{vL}^0-c_{va})/r_v\simeq(c_{va}^0-c_{va})/r_v+s(T_L-T_a)/r_v \tag{1.53}$$

where s $(=dc_v/dT)$ is the slope of the saturation vapor concentration function at $(T_L+T_a)/2$. As with Θ, s is set equal to its value at air temperature as a first approximation. Improved estimates may be obtained by iteration.

The transformation used in Eq. (1.53) was first used by PENMAN (1948) to develop his well-known evapotranspiration equation. Values for s can be conveniently computed from FUCHS et al. (1978):

$$s=(c_v^0/\Theta)\,[(\lambda M_w/R\Theta)-1] \tag{1.54}$$

where M_w is the molecular weight of water, λ is the latent heat of vaporization, and R is the gas constant.

Combining Eqs. (1.26), (1.51), and (1.53), and solving for leaf temperature, we obtain

$$T_L=T_a+\frac{r_e\gamma^*}{\rho c_p(s+\gamma^*)}\left[R_{ni}-\frac{\lambda(c_{va}^0-c_{va})}{r_v}\right] \tag{1.55}$$

where R_{ni} is isothermal net radiation (MONTEITH 1973):

$$R_{ni}=R_{abs}-2\varepsilon_L\,\sigma\Theta_a^4. \tag{1.56}$$

R_{abs} is the short-wave absorption coefficient multiplied by all short-wave radiation incident on the top and bottom of the leaf plus the long-wave absorption coefficient multiplied by the incident long-wave radiation. The equivalent resistance, r_e is defined by

$$r_e=\frac{r_{Ha}\,r_r}{2(r_{Ha}+r_r)}, \tag{1.57}$$

where the factor, 2, results from exchange at both surfaces of the leaf. The apparent psychrometer constant, γ^*, is the thermodynamic psychrometer constant ($0.495\ \mathrm{g\,m^{-3}\,K^{-1}}$ at 20° C) multiplied by the ratio of vapor to heat resistance:

$$\gamma^*=\gamma r_v/r_e. \tag{1.58}$$

These same equations can be combined in a different way to determine the transpiration rate for a leaf:

$$\lambda E=\frac{sR_{ni}}{s+\gamma^*}+\frac{\rho c_p(c_{va}^0-c_{va})}{r_e\,(s+\gamma^*)} \tag{1.59}$$

which is a modified version of the Penman-Monteith formula for computing transpiration (MONTEITH 1964).

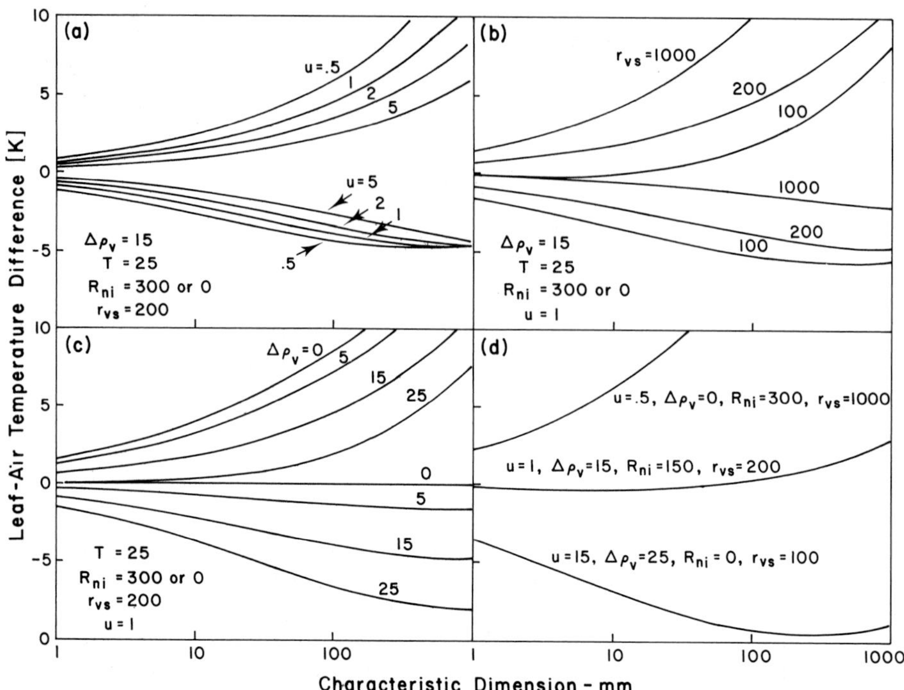

Fig. 1.3a–d. Difference between leaf and air temperature as a function of leaf characteristic dimension and **a** wind speed, **b** stomatal resistance and **c** vapor density deficit. The *upper* and *lower curves* in **d** are intended to represent extremes in environmental variables which will produce extremes in departure leaf from air temperature. The *middle curve* represents a typical combination of these variables. (For explanation of symbols and units see text)

1.3.5 Predictions from the Energy Budget Equation

To use the energy budget equations for the solution of problems in ecophysiology, it is usually necessary to combine energy budget equations with photosynthesis and respiration models (LOMMEN et al. 1971; TAYLOR and SEXTON 1972; CAMPBELL 1977). It is useful, however, to use the energy budget equation to answer some simple questions.

The assumption is often made that leaf temperature and air temperature are equal. Figure 1.3 shows the difference between leaf and air temperature as a function of various environmental and physiological variables (ranges of leaf temperature in relation to heat resistance are discussed in Chap. 14, this Vol.). The range of the variables used in Fig. 1.3 is intended to represent the range of these variables available to a "typical" leaf. Figure 1.3 indicates that small leaves are generally within a degree or so of air temperature. Larger leaves can be at temperatures substantially different from air temperature if they are subjected to high radiation loads and have high stomatal resistance, or are exposed to low radiation loads with low resistance. The more typical case of moderate radiation load with a typical (KÖRNER et al. 1979) diffusion

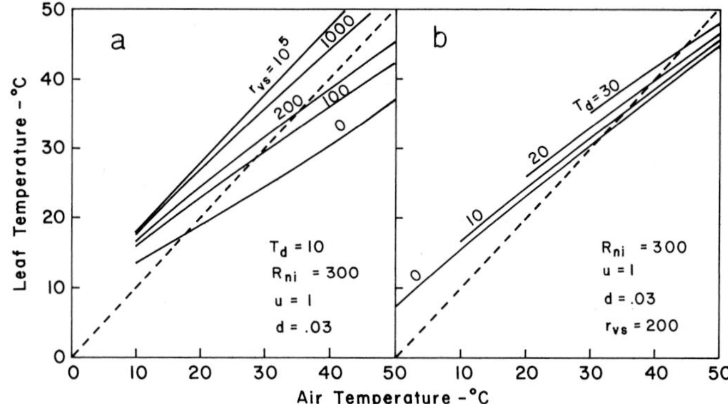

Fig. 1.4a, b. Leaf temperature as a function of air temperature and **a** stomatal resistance or **b** dew point temperature when vapor density (dew point temperature) does not change with air temperature

resistance (middle line in Fig. 1.3d) shows little difference between leaf and air temperature over a wide range of leaf sizes. It therefore appears that leaf temperature may often be assumed to equal air temperature. One must use such a generalization with caution, however, since important exceptions are often reported. For example, SALISBURY and SPOMER (1964) found that leaf temperatures of alpine plants were as much as 20 °C above air temperature under some conditions. The sheltered locations in which alpine plants tend to grow and the cushion growth habit of many alpine species tend to maximize heat transfer resistance, thus maximizing leaf temperature.

At the other extreme, where minimizing leaf temperature would be advantageous, EHLERINGER and MOONEY (1978) found that hairs on the leaf of the desert shrub, *Encelia farinosa* reduced the leaf short-wave absorption coefficient (and therefore R_{ni}) sufficiently so that leaves stayed several degrees below air temperature at mid-day.

The picture resulting from Fig. 1.3 is somewhat artificial since environmental variables in nature tend to be correlated. Typically the air vapor density stays relatively constant over a day, while T_a (and therefore c^0_{va}) increases to T_{max} and then decreases. This causes the vapor concentration deficit to increase with temperature. Figure 1.4 shows the effect of air temperature on leaf temperature when c_{va} (or dew point temperature) remains constant. The lowest curve in Fig. 1.4a represents the behavior of a naturally ventilated wet-bulb. Note that for low to intermediate stomatal resistance, leaf temperature increases less rapidly than air temperature, and that there is some air temperature at which $T_L = T_a$. LINACRE (1964) observed this behavior from data obtained by a number of researchers. He suggested that the crossover point is around 33 °C. From Fig. 1.4 we see that the crossover point depends on several variables and is particularly sensitive to changes in c_{va}, r_{vs}, and R_{ni}. The consistency in the crossover point observed by LINACRE must therefore be somewhat fortuitous.

The energy budget equation can also be used to predict the effect of wind on transpiration rate. GATES (1968) showed that increasing wind speed can

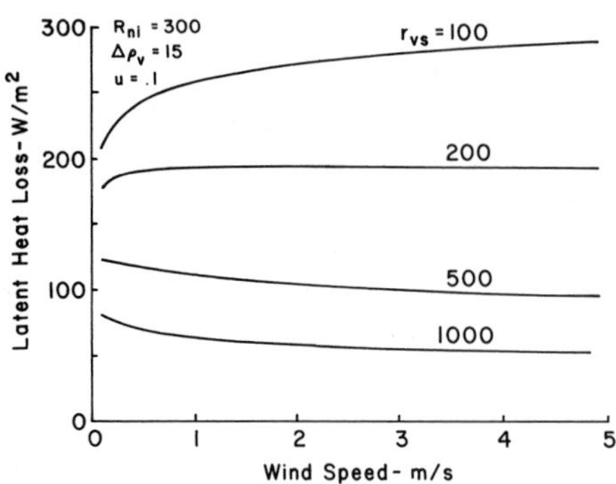

Fig. 1.5. Latent heat loss from a leaf as a function of wind speed and stomatal resistance showing that increasing wind can either increase or decrease transpiration

either increase or decrease transpiration, depending on environment and leaf properties. Figure 1.5 shows this for a "typical" leaf. When r_{vs} is low, increasing wind decreases boundary layer resistance and increases transpiration [Eq. (1.59)]. When r_{vs} is high, increasing wind cools the leaf, decreasing its vapor density and rate of transpiration. Perhaps the most interesting feature of Fig. 1.5 is that there exists a value of r_{vs} such that wind has almost no effect on transpiration rate. For the environment simulated in Fig. 1.5, that resistance is around 300 s m^{-1}, a common value for plants in natural environments (KÖRNER et al. 1979).

1.4 The Canopy Environment

In the discussion so far we have assumed a knowledge of wind, radiation, air temperature, vapor density, etc. in the environment of a single leaf. Most leaves, however, are within canopies made up of other leaves and plant parts. The presence of this canopy influences the environment of the leaf, so that the air temperature, wind speed, vapor density, and radiation around the leaf are not equal to the values normally measured above the canopy. It is, of course, possible to measure all of these variables within the canopy, and such measurements are important for some studies, but it is frequently necessary to estimate environmental parameters from more readily available climatic data that do not include measurements within the canopy. The material which follows will suggest methods for accomplishing this extrapolation.

1.4.1 Wind in Plant Canopies

Wind blowing over a crop is slowed by the drag of the crop (wind as an ecological factor is discussed in Chap. 15, this Vol.). If the momentum flux above the crop is constant with height, then mean wind speed (averaged over 15–30 min) will vary logarithmically with height according to the relationship (BUSINGER 1975; THOM 1975):

$$u(z) = (u^*/k)\ln[(z + z_m - d)/z_m]. \quad (1.60)$$

Here u^* is the friction velocity (m/s), z is height, z_m is a roughness parameter for momentum, k is von Karman's constant (0.4), and d is the zero plane displacement. Equation (1.60) actually describes the wind profile above a flat surface at height, d, with a roughness, z_m. Mean wind profiles above a canopy are described by this equation if values for z_m and d, appropriate to the canopy, can be found. Values of z_m and d for typical closed canopies are

$$d = 0.77\ h$$

and

$$z_m = 0.13\ h, \quad (1.61)$$

where h is crop height. (Most authors combine $d - z_m$ to make a single constant, $d' = 0.64\ h$). Equation (1.60) describes only wind speed above the canopy. It predicts that wind speed will be zero at $z = d$ within the crop. Measurements would, of course, show that this is not the case. Therefore, another equation must be used for wind within the canopy. BUSINGER (1975) used

$$u(z) = u(h)\exp[a(z/h - 1)] \quad (1.62)$$

where a is a crop attenuation coefficient. Typical values for a are 2 to 4 for closed canopies.

Equation (1.62) is useful primarily within the upper $^2/_3$ of the canopy. Below this, the drag of the soil surface slows the wind, and a second logarithmic equation, similar to Eq. (1.60), but with $d = 0$ and z_m appropriate to the soil surface, is used.

Equation (1.60) is technically correct only for a neutral atmosphere ($dT/dz = 0$), but can usually be used with negligible error over crop canopies if the height of the wind measurement is within 1–2 m of the top of the canopy. The method for making stability corrections is outlined elsewhere (BUSINGER 1975; THOM 1975; CAMPBELL 1977).

To predict wind speed in the canopy using these equations, one would need to know the height of the canopy and the wind speed above the canopy. Equations (1.60) and (1.61) would then be used to find u^*. Once u^* is known, u(h) can be computed using these same equations. Wind speed in the upper $^2/_3$ of the canopy is then computed using Eq. (1.62). If estimates of u in the lower $^1/_3$ of the canopy are needed, these can be obtained using u(h/3) from Eq. (1.62), $d = 0$, and a new z_m, to find u^* within the canopy [Eq. (1.60)]. Equation (1.60) with the new canopy u^* and other parameters is used to predict u ($0 < z < h/3$).

1.4.2 Temperature and Vapor Concentration in Plant Canopies

Mean temperature and vapor concentration profiles above canopies follow the same functional form as do wind profiles. The equations are (CAMPBELL 1977):

$$T(z) = T(d) - (H/\rho c_p\, k u^*) \ln[(z + z_H - d)/z_H] \quad (1.63)$$

$$c_{va}(z) = c_{va}(d) - (E/k u^*) \ln[(z + z_v - d)/z_v] \quad (1.64)$$

where $T(d)$ and $c_{va}(d)$ are average temperature and vapor concentration at the exchange surface, and z_H and z_v are the roughness parameters for heat and vapor.

The temperature and vapor concentration within the canopy depends on the distribution of sinks or sources within the canopy. These, in turn, depend on how radiation is absorbed within the canopy and how wind speed varies with depth. Some analytical solutions to this very difficult problem are given by COWAN (1968a), but numerical models are perhaps better suited to determining temperature and vapor concentration profiles within actual canopies. WAGGONER (1975) and NORMAN (1978) review such models, and compare the predictions of the models with actual measurements. Model predictions of temperature or vapor concentration profiles generally agree with measurements to within a few tenths of a degree or g m^{-3}, respectively. The models show temperatures within the canopy to be within a degree of the temperature at the top of the canopy, but vapor concentrations increase by several g m^{-3} with depth. The temperature predicted from Eq. (1.63) (or even the air temperature) may therefore be quite close to the temperature of air in the canopy. However, the vapor concentration in the canopy will usually be significantly higher than $c_{va}(h)$ (which is essentially equal to c_{va} measured at any height above the canopy).

1.4.3 Radiation in Plant Canopies

The environmental variable which is perhaps most difficult to predict within a canopy is radiation. This is partly because the characteristics of the radiation are altered so markedly by the canopy. Both the quantity and quality of short-wave radiation change with depth, and the canopy itself is a source as well as a sink for long-wave radiation. In addition, short-wave radiation has both direct and diffuse components, and these components follow different extinction laws with depth. These complexities make it very difficult to give a detailed description of radiation within a canopy, but simplifying assumptions can be used which will allow the problem to be solved and still will give reasonably accurate results. The subject of radiative transfer in plant canopies has been reviewed in several places recently (NORMAN 1975; ROSS 1975), so only the simplest aspects will be treated here.

If we assume that leaves in a canopy are opaque and randomly distributed in space, and that the leaf area index (projected leaf area per unit ground area) is L, then the average flux density from beam radiation below a layer of leaves of thickness dL is

$$S_b(L + dL) = -S_b(L)\, dL \quad (1.65)$$

where $S_b(L)$ is the flux density above the layer. This average is made up of sunflecks at full sun irradiance, and shadows at near zero irradiance. If Eq. (1.65) is integrated over the canopy, we obtain the Bouguer law:

$$S_b(z) = S_b(h) \exp[-K\,L(z)]. \tag{1.66}$$

The downward cumulative leaf area index, L(z) in Eq. (1.66), is defined as

$$L(z) = \int_z^h a(z)\,dz \tag{1.67}$$

where a(z) is the area density function (area per unit volume) for the canopy. Thus L(z) is zero at $z=h$, and increases to a maximum, L(0) at $z=0$. The extinction coefficient, K, depends on solar elevation angle, ϕ, and leaf inclination angle. If leaves have random inclination, then

$$K_{random} = 1/(2\sin\phi). \tag{1.68}$$

If leaves have uniform inclination of α, then

$$\text{for } \phi \geqq \alpha \quad K_{inclined} = \cos\alpha$$

and

$$\text{for } \phi \leqq \alpha \quad K_{inclined} = [1 + 2(\tan\beta - \beta)/\pi]\cos\alpha \tag{1.69}$$

$$\text{where } \cos\beta = \tan\phi/\tan\alpha.$$

For horizontal leaves, $K=1$, and for vertical leaves, $K=2/(\pi\tan\phi)$.

Equation (1.66) indicates that a canopy transmission coefficient for beam radiation can be defined as

$$\tau_b = \exp[-K\,L(z)], \tag{1.70}$$

which is also the probability for penetration of a beam at angle ϕ through a canopy of leaf area index L. The average diffuse flux at any level in the canopy can be found if a diffuse transmission coefficient for the canopy can be computed. For a random canopy, and uniform sky radiance, the diffuse transmission coefficient is (Ross 1975):

$$\tau_d = 2\int_0^{\pi/2} \tau_b \cos\phi \sin\phi \, d\phi. \tag{1.71}$$

Figure 1.6 shows τ_d as a function of leaf area index for various canopy structures. Tabulated values of τ_d are given by Ross (1975).

Using Eqs. (1.70) and (1.71), and values for $S_b(h)$ and $S_d(h)$, short-wave flux densities within canopies can be estimated at any particular time. However, for many ecological purposes, one is primarily interested in average flux densities

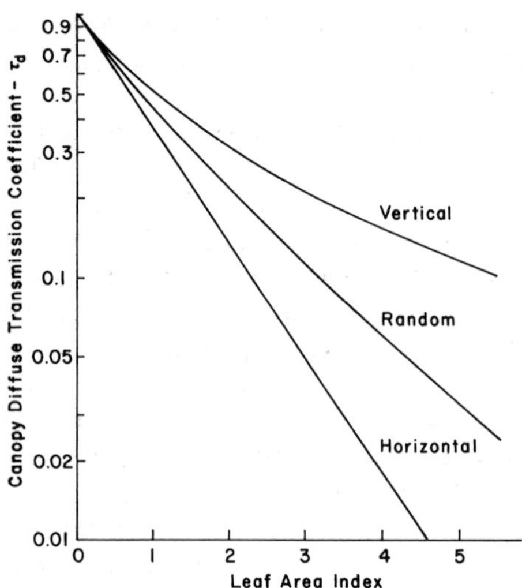

Fig. 1.6. Diffuse transmission coefficient as a function of leaf area index and canopy structure for random canopies

over periods of a day or more. FUCHS et al. (1976) showed that good estimates of daily average short-wave irradiance within a sorghum canopy could be computed from

$$\bar{S}_t(z) = \bar{S}_t(h)\ \tau_d\,, \tag{1.72}$$

where the overbars signify daily averages. Since a large fraction of $\bar{S}_t$ is diffuse in many climates, and the direct radiation integrated over a day has an elevation angle distribution similar to diffuse radiation, this appears to be a useful approximation for many studies.

The assumptions used in deriving Eqs. (1.70) and (1.71) impose limitations on their use, which, in some circumstances, may be critical. Perhaps the most serious are the assumptions of random leaf distribution and opaque leaves. Two approaches have been used to deal with the problem of nonrandom leaf distribution. ACOCK et al. (1970) and NILSON (1971) proposed the use of a "clumpiness factor", λ_0 such that

$$\tau_b = \exp[-\lambda_0 K\, L(z)]. \tag{1.73}$$

For random canopies, $\lambda_0 = 1$. When $\lambda_0 > 1$, the canopy is over-dispersed (new leaves tend to fill gaps in the canopy), and when $\lambda_0 < 1$ the leaves are clumped. ROSS (1975) shows data indicating that λ_0 is a function of solar elevation angle as well as canopy structure. Since there apparently is no simple way of determining λ_0 a priori, its practical value is limited.

Another approach, used by MANN et al. (1980) and NORMAN and WELLES (1981) to predict light penetration in clumped vegetation, is to treat the canopy as a collection of ellipsoids, each of which has a transmission coefficient given by Eq. (1.70). The problem is to determine, for each elevation and azimuth angle, the distance a ray travels within the ellipsoids. This treatment is useful for isolated plants, rows, or vegetation clumped around individual axes of a plant. In the limit of high ellipsoid density, this model becomes identical to the random canopy model.

The assumption that leaves are opaque is correct for long-wave radiation, and close for PAR. Our earlier estimates for a typical green leaf gave a PAR absorption coefficient of around 0.85. However, only about 15% of the near infrared and half of the total short-wave are absorbed. The radiation that is not absorbed is scattered toward the sky and ground, and results in canopy radiation profiles that depart from the exponential absorption law. COWAN (1968b) derived the differential equations describing canopy diffuse radiation, and was able to solve them to determine irradiance profiles for uniform leaf distributions knowing only the optical properties of the leaves, the canopy transmission coefficient, and the reflection coefficient for the underlying surface. The equations are readily solved numerically for more realistic canopy structures. Approaches similar to that of COWAN (1968b), but which divide the canopy into thicker layers, and treat both direct and diffuse radiation have been used by FUCHS (1972), NORMAN and JARVIS (1975), and NORMAN (1978). The model proposed by NORMAN (1978) is probably the simplest and best model available at the present time for computing direct and diffuse radiation in uniform canopies.

1.4.4 Long-Wave Radiation in Plant Canopies

Transfer of long-wave radiation in canopies follows laws similar to those for short-wave. Canopy transmission coefficients are computed using Eq. (1.71), where the assumption that leaves are opaque is valid. An added complication comes from the fact that the leaves are sources of long-wave radiation. Because of this, it is not possible to write explicit equations for long-wave fluxes within the canopy, but linearizing the long-wave exchange equation, it is possible to write a set of implicit equations which can be solved numerically for leaf temperatures and long-wave fluxes within the canopy (NORMAN 1978). These complications may, however, be unnecessary for most eco-physiological studies. WAGGONER (1975) shows good agreement between simulated and measured canopy temperatures using the simple assumption that long- and short-wave radiation are both attenuated exponentially, so that the net radiation profile in a canopy is given by

$$R_n(L) = \tau_b \, R_n(h). \tag{1.74}$$

1.4.5 Measurement of Canopy Transmission Coefficients

For random canopies, the canopy transmission coefficient can often be obtained with adequate accuracy by observing the dominant inclination angle of leaves, estimating or measuring the leaf area index, and computing τ_b and τ_d from Eqs. (1.70) and (1.71). These values are less easily computed for nonrandom canopies, and are often best obtained by direct measurement. The simplest method for measuring τ_b is to place a meter stick on the ground under a canopy and determine the fraction of its length that is sunlit (MILLER 1969; ADAMS and ARKIN 1977). MILLER (1969) made measurements in a deciduous forest holding the meter stick in 16 directions and averaging the results. ADAMS and ARKIN (1977) found that, for row crops, a single measurement with the meter stick across a row gave results which were as precise as an average of 21 measurements with the meter stick along the row. They also found that the meter stick method was as accurate as overhead photographs or spatially averaged PAR measurements.

The disadvantage of this very simple measurement is that each measurement gives only one value for τ. To know how τ_b varies with time of day and time of year would take many measurements under clear-day conditions. Even with these measurements, the computation of τ_d for a nonrandom canopy would not be possible since data are needed for azimuths and elevations that are not traversed by the sun. A more elegant method for measuring τ_b is to make hemispherical photographs of the canopy and sky from ground level using a fisheye lens (ANDERSON 1966, 1971; BONHOMME and CHARTIER 1972; FUCHS et al. 1976). The fisheye lens projects the hemisphere above the camera onto the film plane in such a way that τ_b for each elevation and azimuth angle can be determined. The canopy photograph is divided into small areas, and an average τ_b for each area (fraction of the sky seen in that area) is determined. The information from the photograph can then be used to determine τ_b for any time of the day or year, and can be integrated using Eq. (1.71) to find τ_d. Details regarding methods for making and analyzing the photographs are given by ANDERSON (1971), BONHOMME and CHARTIER (1972), and FUCHS et al. (1976).

1.5 Simulation of Radiation and Temperature in Plant Canopies

An objective of this chapter is to provide the relationships needed to predict the temperature and irradiance of leaves and other plant parts from a knowledge of environmental variables. The basic relationships between plant temperature, plant irradiance, and environmental variables have been presented. Plant temperatures and irradiances are actually predicted using environmental variables as input to a model composed of these basic relationships, and then using the model to simulate plant behavior. This model can be linked with other models to predict, for example, dry matter production or transpiration. The equations in earlier sections of this chapter have been presented in a manner to make their use in models as simple as possible. A detailed model, perhaps of the type discussed by WAGGONER (1975), or NORMAN (1978), would predict temperature and irradiance with sufficient accuracy for most ecological studies. It should be noted, however, that this level of complexity is not always necessary. The simplest model may allow us to assume $T_L = T_a$ and leaf irradiance equals S_t. Such a model may be adequate for many purposes. At the next level of complexity, one may treat the crop as a "big leaf" and use the equations derived earlier in this chapter for single leaves, for the entire crop. SINCLAIR et al. (1976) compared predictions from a simple model, which treats the canopy as a single photosynthesizing and transpiring surface, with a detailed model. Transpiration rates from the two models agreed to within 10% and photosynthesis rates to within 5%. The advantage of this simplification is that all leaves are assumed to be at one temperature, which is computed from Eq. (1.63), and to carry on photosynthesis at a rate determined by the quantity of radiation absorbed by the entire canopy.

The most complete canopy models, which give detailed temperature and radiation distributions within the canopy, are considerably more complex than either of those just mentioned. Because radiation, vapor concentration and temperature interact, the model must somehow solve for all simultaneously. Since the heat exchange of any layer in the canopy will be determined by the temperature of its neighbors, the solution must be implicit. Details of such a model are presented by WAGGONER (1975) and NORMAN (1978).

1.6 Future Research Needs

A cursory look at the various contributions to this volume should convince any reader that the relatively simple equations which describe radiation and temperature at a plant are much better understood than the very complex processes which occur within the plant. At the present time, it does not appear that our progress in ecophysiology is limited by our ability to measure or predict radiation or temperature. ELSTON and MONTEITH (1975), in a recent review of micrometeorological contributions to plant ecology, suggested seven or eight areas for future research. Of these, only one was primarily a problem in micrometeorology. The rest involved measuring plant responses to imposed environmental conditions. The micrometeorological area which they feel deserves additional study is the extension of the essentially one-dimensional analysis presented here to two or three dimensions. The availability of large computers and efficient numerical schemes for solving flow equations in two and three dimensions will almost certainly lead to attempts at solving two- and three-dimensional canopy problems.

Perhaps the most pressing need in the area of micrometeorology and plant ecology is for ecologists to understand and apply state-of-the-art measurement and modeling techniques to their research so that research results can be interpreted in a general context. Certainly the use of radiation and temperature simulation models will increase in the future.

While our knowledge of the physical processes which occur in canopies probably does not limit progress in ecology, some research in this area should continue. Wind profiles within canopies are predicted reasonably accurately, almost independent of assumptions about the transport coefficient within the canopy (THOM 1975). The size of the transport coefficient is important, however, in determining rates of heat and mass transport within the canopy. A more thorough understanding of transport processes within canopies is clearly needed.

Our ability to predict radiant fluxes within canopies has progressed rapidly within the past few years. New models which deal with clumped vegetation will need further study and modification, but appear to deal adequately with most of the complexities of real canopies. The interfacing of these canopy radiation models with photosynthesis models will still take some work (NORMAN 1979, cf. also Chapters in Vols. 12B and 12D).

Finally, effects of scale of turbulence on transport from leaves and other plant parts in a turbulent freestream need further study. It is quite possible that transport resistances will always be uncertain in nature, but the levels of uncertainty should be determined.

Acknowledgments. I would like to thank Dr. Susan Riha and my wife Judy who gave valuable editorial assistance. Dr. John Norman reviewed parts of the canopy section and made useful suggestions.

References

Acock B, Thornley JHM, Wilson JW (1970) Spatial variation of light in the canopy. In: Setlik I (ed) Prediction and measurement of photosynthetic productivity. PUDOC, The Netherlands pp 91–102

Adams JE, Arkin GF (1977) A light interception method for measuring row crop ground cover. Soil Sci Soc Am J 41:789–792

Anderson MC (1966) Some problems of simple characterization of the light climate in plant communities. In: Brainbridge R, Evans GC, Rackham O (eds) Light as an ecological factor. Blackwell, Oxford, pp 77–90

Anderson MC (1971) Radiation and crop structure. In: Šesták Z, Čatský J, Jarvis PG (eds) Plant photosynthetic production. W. Junk, The Hague pp 412–466

Bird RB, Stewart WE, Lightfoot EN (1960) Transport phenomena. Wiley and Sons, New York

Bonhomme R, Chartier P (1972) The interpretation and automatic measurement of hemispherical photographs to obtain sunlit foliage area and gap frequency. Isr J Agric Res 22:53–61

Brutsaert W (1975) On a derivable formula for long-wave radiation from clear skies. Water Resour Res 11:742–744

Businger JA (1975) Aerodynamics of vegetated surfaces. In: deVries DA, Afgan NH (eds) Heat and mass transfer in the biosphere. Wiley and Sons, New York pp 139–165

Campbell GS (1977) An introduction to environmental biophysics. Springer, Berlin, Heidelberg, New York

Campbell GS, Mugaas JN, King JR (1978) Measurement of long-wave radiant flux in organismal energy budgets: a comparison of three methods. Ecology 59:1277–1281

Coulson KL (1975) Solar and terrestrial radiation. Academic Press, New York

Cowan IR (1968a) Mass, heat and momentum exchange between stands of plants and their atmospheric environment. QJR Meteorol Soc 94:523–544

Cowan IR (1968b) The interception and absorption of radiation in plant stands. J Appl Ecol 5:367–379

Ehleringer JR, Björkman O (1978) Pubescence and leaf spectral characteristics in a desert shrub, *Encelia farinosa*. Oecologia 36:151–162

Ehleringer JR, Mooney HA (1978) Leaf hairs: effects on physiological activity and adaptive value to a desert shrub. Oecologia 37:183–200

Elston J, Monteith JL (1975) Micrometeorology and ecology. In: Monteith JL (ed) Vegetation and the atmosphere Vol. 1. Academic Press, New York pp 1–12

Fuchs M (1972) The control of the radiation climate of plant communities. In: Hillel D (ed) Optimizing the soil physical environment toward greater crop yields. Academic Press, New York pp 173–191

Fuchs M, Stanhill G, Moreshet S (1976) Effect of increasing foliage and soil reflectivity on the solar radiation balance of wide-row grain sorghum. Agron J 68:865–871

Fuchs M, Campbell GS, Papendick RI (1978) An analysis of sensible and latent heat flow in a partially frozen unsaturated soil. Soil Sci Soc Am J 42:379–385

Gates DM (1962) Energy Exchange in the Biosphere. Harper and Rowe, New York

Gates DM (1968) Transpiration and leaf temperature. Ann Rev Plant Physiol 19:211–238

Gates DM, Schmerl RB (eds) (1975) Ecological Studies 12. Perspectives in biophysical ecology. Springer, Berlin, Heidelberg, New York

Gates DM, Keegan HN, Schleter JC, Weidner VR (1965) Spectral properties of plants. Appl Opt 4:11–20

Haurwitz B (1948) Insolation in relation to cloud type. J Meteorol 5:110–113

Idso SB, Jackson RD (1969) Thermal radiation from the atmosphere. J Geophys Res 74:5397–5403

Idso SB, Jackson RD, Ehrler WL, Mitchell ST (1969) A method for determination of infrared emittance of leaves. Ecology 50:899–902

Kerr JP, Thurtell GW, Tanner CB (1967) An integrating pyranometer for climatological observer stations and mesoscale networks. J Appl Meteorol 6:688–694

Kestin J (1966) The effect of free stream turbulence in heat transfer. Adv Heat Transfer 3:1–32

Knutson RM (1974) Heat production and temperature regulation in eastern skunk cabbage. Science 186:746–747

Kondratyev KY (1969) Radiation in the atmosphere. Academic Press, New York

Körner C, Scheel JA, Bauer H (1979) Maximum leaf diffusive conductance in vascular plants. Photosynthetica 13:45–82

Lewis DA, Nobel PS (1977) Thermal energy exchange model and water loss of a barrel cactus, *Ferocactus acanthodes*. Plant Physiol 60:609–616

Linacre ET (1964) A note on a feature of leaf and air temperature. Agric Meteorol 1:66–72

List RJ (1971) Smithsonian Meteorological Tables. Smithsonian Institution Press, Washington, DC

Lommen PW, Schwintzer CR, Yocum CS, Gates DM (1971) A model describing photosynthesis in terms of gas diffusion and enzyme kinetics. Planta 98:195–220

Mann JE, Curry GL, DeMichele DW, Baker DN (1980) Light penetration in a row crop with random plant spacing. Agron J 72:131–142

McCullough EC, Porter WP (1971) Computing clear day solar radiation spectra for the terrestrial ecological environment. Ecology 52:1008–1015

Miller PC (1969) Tests of solar radiation models in three forest canopies. Ecology 50:878–885

Mitchell JW (1976) Heat transfer from spheres and other animal forms. Biophys J 16:561–569

Monteith JL (1964) Evaporation and environment. Symp Soc Exp Biol 19:205–234

Monteith JL (1973) Principles of Environmental Physics. Edward Arnold, London

Nilson T (1971) A theoretical analysis of the frequency of gaps in plant stands. Agric Meteorol 8:25–38

Nobel PS (1974) Boundary layers of air adjacent to cylinders. Plant Physiol 54:177–181

Norman JM (1975) Radiative transfer in vegetation. In: deVries DA, Afgan NH (eds) Heat and mass transfer in the biosphere. John Wiley, New York pp 187–205

Norman JM (1978) Modeling the complete crop canopy. In Barfield BJ, Gerber JF (eds) Modification of the aerial environment of plants. American Society of Agricultural Engineers, St. Joseph, Mich.

Norman JM (1980) Interfacing leaf and canopy light interception models. In: Hesketh

JD (ed) Predicting photosynthate production and use for ecosystem models Vol. 2. CRC Press, Boca Raton, Florida

Norman JM, Jarvis PG (1975) Photosynthesis in sitka spruce [*Picea sitchensis* (Bong.) Carr.] V. Radiation penetration theory and a test case. J Appl Ecol 12:839–878

Norman JM, Welles JM (1981) Radiative transfer in an array of canopies. Agron J

Paltridge GW, Platt CMR (1976) Radiative processes in meteorology and climatology. Elsevier, New York

Pearman GI, Weaver HL, Tanner CB (1972) Boundary layer heat transfer coefficients under field conditions. Agric Meteor 10:83–92

Penman HL (1948) Natural evaporation from open water, bare soil and grass. Proc R Soc London Ser A 194:220–245

Raschke K (1956) Über die physikalischen Beziehungen zwischen Wärmeübergangszahl, Strahlungsaustausch, Temperatur und Transpiration eines Blattes. Planta 48:200–237

Robinson N (1966) Solar radiation. Elsevier, New York

Ross J (1975) Radiative transfer in plant communities. In: Monteith JL (ed) Vegetation and the Atmosphere Vol. 1. Academic Press, New York pp 13–55

Salisbury FB, Spomer GG (1964) Leaf temperatures of alpine plants in the field. Planta 60:497–505

Schlichting (1971) A survey of some recent research investigations on boundary layers and heat transfer. J Appl Mech 38:289–300

Simonson JR (1975) Engineering Heat Transfer. MacMillan, London

Sinclair R, Thomas DA (1970) Optical properties of leaves of some species in arid South Australia. Aust J Bot 18:261–273

Sinclair TR, Murphy CE, Knoerr KR (1976) Development and evaluation of simplified models for simulating canopy photosynthesis and transpiration. J Appl Ecol 13:813–829

Stanhill G, Fuchs M (1977) The relative flux density of photosynthetically active radiation. J Appl Ecol 14:317–322

Swift LW (1976) Algorithm for solar radiation on mountain slopes. Water Resour Res 12:108–112

Szeicz G (1974) Solar radiation for plant growth. J Appl Ecol 11:617–636

Taylor SE, Gates DM (1970) Some field methods for obtaining meaningful leaf diffusion resistances and transpiration rates. Oecol Plant 5:105–113

Taylor SE, Sexton OJ (1972) Some implications of leaf tearing in Musaceae. Ecology 53:143–149

Thom AS (1975) Momentum, mass and heat exchange of plant communities. In: Monteith JL (ed) Vegetation and the atmosphere Vol. 1. Academic Press, New York pp. 57–110

Unsworth MH, Monteith JL (1975) Long-wave radiation at the ground I. Angular distribution of incoming radiation. QJR Meteorol Soc. 101:13–24

Waggoner PE (1975) Micrometeorological models. In: Monteith JL (ed) Vegetation and the atmosphere Vol. 1. Academic Press, New York pp 205–228

Wesley ML, Lipschutz RC (1976) An experimental study of the effects of aerosols on diffuse and direct solar radiation received during the summer near Chicago. Atmos Environ 10:981–987

2 Photosynthetically Active Radiation

K.J. McCree

CONTENTS

2.1 Introduction

It is well known that plants vary in the sensitivity of the photosynthetic apparatus to radiation of different wavelengths (Rabinowitch 1951). This would seem to eliminate any possibility of a unique definition of photosynthetically active radiation (PAR) for use in ecophysiology. Gabrielsen (1960), in his review of photosynthetic action spectra for the original edition of this encyclopedia, gave two basic reasons why this is not so: the variations in spectral response become important only when (1) the spectrum of the radiation is changing, and (2) the photosynthetic rate is limited by the amount of radiation available.

The spectral quality of daylight is relatively constant within the photosynthetically active waveband (400–700 nm) (Ross 1975), and for much of the time the radiation level is well above the linear part of the photosynthesis response curve. Thus, for many ecophysiological studies, it is possible to define photosynthetically active radiation without taking into account the spectral sensitivity of the photosynthetic apparatus. A purely physical definition suffices. This can be demonstrated through a simple set of calculations, based on known variations in the action spectrum for photosynthesis and in the spectral quality of the radiation that is likely to be encountered.

In the field, variations in spectral quality are encountered within plant canopies. Much greater variations occur when artificial light sources are used. Thus, the most critical test of the adequacy of a physical definition of PAR comes when the laboratory data are extrapolated to the field. Since the time of Gabrielsen's review, the adequacy of such a definition has been demonstrated for a wide range of natural and artificial spectra, using new information on the action spectrum for photosynthesis in higher plants.

There has also been increasing awareness of the need for rigorous treatment of the physical aspects of ecophysiology (CAMPBELL 1977; MONTEITH 1973). The use of "psychophysical" light units such as the lux in plant growth studies, deplored by a succession of physical scientists, is now becoming a rarity. PAR meters are commercially available and have been widely accepted. At least one professional society has adopted a standard set of terms relating to PAR (SHIBLES 1976) and banned the use of the lux in its journals (DYBING 1977). Perhaps, with the publication of this new series, photosynthetically active radiation will "come of age" as the primary measure of available light in ecophysiological studies (McCREE 1973).

This review consists of a brief introduction to the principles of radiometry and their application to PAR measurements, a review of recent measurements of the spectral quantum yield of photosynthesis in higher plants, as they relate to the definition of PAR, and a discussion of sources of error of PAR measurements.

2.2 Principles of Radiometry

Radiometry (CIE 1970) is the measurement of the properties of *radiant energy* (SI unit: joule, J), which is one of the many interchangable forms of energy. The rate of flow of radiant energy, in the form of an electromagnetic wave, is called the *radiant flux* (unit: watt, W; $1\ W = 1\ J\ s^{-1}$). Radiant flux can be measured as it flows from the *source* (the sun, in natural conditions), through one or more reflecting, absorbing, scattering and transmitting *media* (the Earth's atmosphere, a plant canopy) to the *receiving surface* of interest (a photosynthesizing leaf).

The radiometric term for the radiant flux incident on the receiving surface from all directions, per unit area of surface is *irradiance* (unit: $W\ m^{-2}$). When a parallel beam of radiation of given cross-sectional area spreads over a flat surface, the area that it covers is inversely proportional to the cosine of the angle between the beam and a plane normal to the surface. Therefore, the irradiance due to the beam is proportional to the cosine of the angle. A radiometer whose response to beams coming from different directions follows the same relationship is said to be "cosine-corrected."

In radiometry, the intensity is a property of the source, not the receiving surface. The *radiant intensity* is the flux leaving a point on the source, per unit solid angle of space surrounding the point. *Irradiance* is the correct radiometric term for the property that is commonly referred to as the "light intensity." The use of the word "light" (radiation that is visible to humans) is inappropriate in plant research. The terms "ultraviolet light" and "infrared light" clearly are contradictory.

The *absorptance* is the fraction of the incident flux that is absorbed by a medium. *Reflectance* and *transmittance* are equivalent terms for the fractions that are reflected or transmitted.

All the properties of the radiant flux depend on the *wavelength* of the radiation. The prefix *spectral* is added when the wavelength dependency is being described. Thus, the *spectral irradiance* is the irradiance at a given wavelength, per unit wavelength interval. The irradiance within a given waveband is the integral of the spectral irradiance with respect to wavelength.

Many radiation phenomena are best treated by considering the radiant energy to be carried in discrete bundles called *photons*. The quantity (*quantum*) of energy carried by each photon is equal to $h\nu$, where h is Planck's constant ($6.63 \cdot 10^{-34}$ J s) and ν is the frequency of the electromagnetic wave ($=c/\lambda$, where c is the speed of the wave, $3.00 \cdot 10^{8}$ m s^{-1}, and λ is its wavelength; when λ is in nm, the quantum energy is $1.99 \cdot 10^{-16}/\lambda$ joules). The *quantum yield* of a photochemical reaction is the number of molecules activated, divided by the number of quanta (photons) absorbed (=number of moles activated per mole of photons absorbed).

In photobiology, a mole (Avagadro's number, $6.022 \cdot 10^{23}$) of photons is commonly referred to as an einstein (E) (not an SI unit – INCOLL et al. 1977). One einstein of monochromatic radiation carries an energy of $1.20 \cdot 10^{8}/\lambda$ joules. Thus, a radiant energy flux of one W is equivalent to a photon flux (quantum flux of $\lambda/(1.20 \cdot 10^{8}) = 8.35 \cdot 10^{-9}\,\lambda$ E s^{-1}.

In photometry, which is the measurement of the properties of radiation evaluated according to its *visual effect* (CIE 1970), *luminous* flux is substituted for radiant flux, and the term *illuminance* replaces irradiance.

The absolute accuracy of radiometric measurements is much less than that of most other physical measurements. An accuracy of $\pm 5\%$ or better can be achieved only in a standardizing laboratory. A routine calibration is unlikely to be better than $\pm 10\%$ in absolute terms. Systematic differences between instruments that differ substantially in their radiometric properties (particularly in spectral sensitivity) can be much greater than this.

2.3 The Radiometric Measurement of PAR

The need to define more clearly the radiation that is active in photosynthesis was recognized by RABINOWITCH (1951) when he compiled his review of photosynthesis studies. Rabinowitch suggested that photosynthetically active radiation be defined as radiation with a wavelength less than 700 nm, and that its irradiance be measured with a thermopile fitted with an infrared-absorbing filter (MCCREE 1966). Since the primary action of radiation in photosynthesis is photochemical, it is reasonable to expect *equal numbers of absorbed photons* of different wavelengths to be equally effective, not *equal quantities of energy*. Rabinowitch indicated how the absorbed flux of photons could be calculated from the irradiance in the photosynthetic waveband (photosynthetic irradiance), or the illuminance, when the spectral characteristics of the radiation were known.

Gaastra (1959) provided a detailed set of such calculations for a variety of different artificial light sources, in combination with three different spectral absorptances of the leaves (representing leaves with low, average, and high absorptances of PAR). Although the primary purpose was to provide conversion factors that could be used to convert radiometric or photometric measurements to absorbed photon fluxes, the same calculations provided an estimate of adequacy of the irradiance in the waveband 400 to 700 nm as a measure of true photosynthetically active radiation, assuming that the photosynthetic response of a leaf would be proportional to the absorbed photon flux density in the same waveband. Formally stated:

$$PI = \int_{400}^{700} I_\lambda \, d\lambda \tag{2.1}$$

$$PPFD_a = 8.35 \cdot 10^{-9} \int_{400}^{700} I_\lambda \, a_\lambda \, \lambda \, d\lambda \tag{2.2}$$

$$P = 8.35 \cdot 10^{-9} \int_{400}^{700} I_\lambda \, a_\lambda \, \lambda \, \phi_\lambda \, d\lambda \tag{2.3}$$

$$= \phi \quad PPFD_a \quad \text{if } \phi \text{ is independent of } \lambda \text{, as assumed by Gaastra.} \tag{2.4}$$

Where PI = photosynthetic irradiance (Wm^{-2}), $PPFD_a$ = absorbed photosynthetic photon flux density ($Em^{-2}\,s^{-1}$), P = photosynthetic rate (mol $m^{-2}\,s^{-1}$), I_λ = spectral irradiance ($Wm^{-2}\,nm^{-1}$), a_λ = spectral absorptance of leaf, ϕ_λ = spectral quantum yield of photosynthesis (mol E^{-1} absorbed), and λ = wavelength (nm).

From these equations, it can be seen that P/PI depends on I_λ and a_λ, while $P/PPFD_a$ does not (if ϕ is assumed to be independent of λ). Therefore, when I_λ is varied by changing the light source, the amount by which P/PI varies is a measure of the error involved in assuming that P is proportional to PI rather than to $PPFD_a$. The variation in P/PI is a function of a_λ.

On this basis, Gaastra (1959) concluded that the photosynthetic irradiance should be within about 20% of the true PAR value for most artificial light sources and most leaf absorptances. He tested this conclusion by measuring the actual photosynthetic rates per unit of irradiance (P/PI) in radiation from incandescent, fluorescent, and mercury lamps. For sugar beet leaves, P/PI in fluorescent radiation was 80%, and in mercury radiation 70%, of the value in incandescent radiation, while the calculations had indicated that $PPFD_a/PI$ should be 89% and 80%, respectively, of the incandescent value. Gaastra attributed the differences between measured and calculated percentages to low efficiency of energy transfer from blue-absorbing pigments, and to a slight increase in stomatal diffusion resistance. Calculations with spectra of natural radiation, both above and within plant canopies (Gaastra 1968), indicated that P/PI should vary by about 10% in the field, for any one leaf. Absolute values of P/PI were calculated to be about twice as great for a strongly absorbing leaf as for a weakly absorbing leaf.

2.4 Physiological Basis for Defining PAR

GAASTRA (1959) assumed that the quantum yield of photosynthesis (ϕ_λ) was independent of wavelength over a range of 400–700 nm. At the time, very few such data were available for higher plant leaves (GABRIELSEN 1960). In the intervening years, several sets of data have been published, and these will now be critically reviewed.

It is important to note that the relevant quantity is the *yield per absorbed quantum*. To calculate this quantity, it is necessary to measure both the yield per incident quantum (commonly referred to as the quantized action spectrum) and the spectral absorptance of the leaf. Action spectra cannot be interpreted without knowledge of the spectral absorptance. The absorptance must be measured with an integrating sphere to eliminate scattering effects (RABINOWITCH 1951).

A second requirement is that the wavelength interval (bandwidth) be sufficiently small to reveal the structure in the quantum yield curve. A bandwidth of 25 nm or less is appropriate, and can be achieved with a monochromator or narrow-band interference filters.

BJÖRKMAN (1968) measured the spectral quantum yield of photosynthesis in leaves of three species of higher plant (*Solidago virgaurea*, *Mimulus cardinalis*, and *Plantago lanceolata*) grown under controlled-environment conditions. The photosynthetic rate was measured at an ambient CO_2 concentration of 300–310 $\mu l\, l^{-1}$ and a leaf temperature of 22 °C. The quantum yield was found to be 0.074 mol CO_2 E^{-1} absorbed between 670 and 540 nm, decreasing to 0.062 between 520 and 450 nm, and rapidly decreasing at wavelengths greater than 670 nm. The data did not cover the whole of the photosynthetically active waveband.

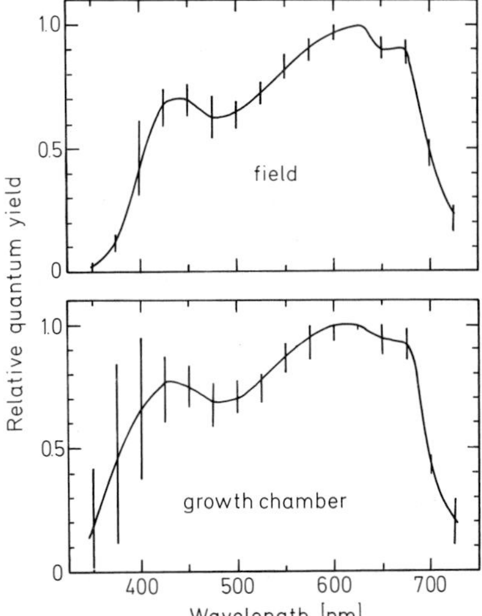

Fig. 2.1. Spectral quantum yield of photosynthesis (ϕ_λ) according to MCCREE (1972a). *Curves* are averages for 22 species of crop plant, grown either in the field or in a growth chamber. *Bars* show range of values obtained

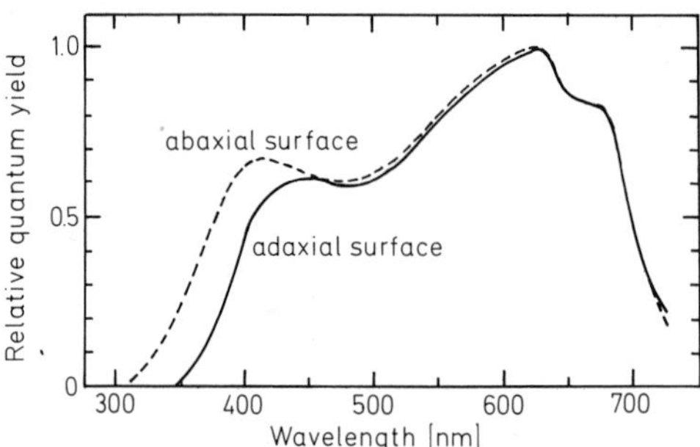

Fig. 2.2. Spectral quantum yield of photosynthesis (ϕ_λ) for a squash leaf (*Cucurbita pepo* L.), with either the abaxial surface or the adaxial surface facing the source of radiation. (McCree 1972a)

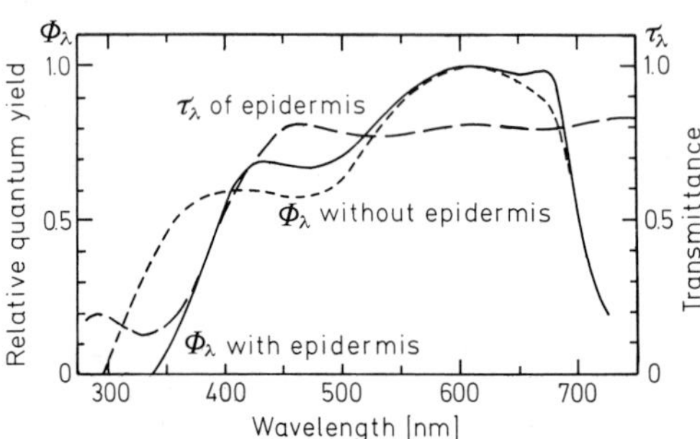

Fig. 2.3. Spectral quantum yield of photosynthesis (ϕ_λ) for a cabbage leaf (*Brassica oleracea* L.) with and without the (adaxial) epidermis, and spectral transmittance (τ_λ) of the epidermis. (McCree and Keener 1974)

The data of McCree (1972a) are more comprehensive. The quantum yield was measured for all wavelengths that are photosynthetically active (350 to 750 nm, extended to 300 nm by McCree and Keener 1974). Data were obtained for leaves of 22 species of crop plant. The standard measuring conditions were: 350 μl l^{-1} CO_2, leaf temperature 28 °C, 75% relative humidity. To minimize nonlinearity of the photosynthetic response, the measurements were made at a constant photosynthetic rate, rather than a constant photon flux density. Tests on selected samples showed that the relative yield at different wavelengths was unaffected by CO_2 concentration or leaf temperature.

In all cases, the spectral quantum yield curve (Fig. 2.1) showed two broad maxima, centered at 620 and 440 nm, with a shoulder at 670 nm. The relative height of the blue maximum and the shortwave cutoff wavelength varied with the specimen. Samples from field-grown plants tended to have relatively lower blue and ultraviolet responses than samples from growth-chamber-grown plants. Also, samples of asymmetric dicot leaves showed higher ultraviolet responses when irradiated from the abaxial side (Fig. 2.2). The response in the ultraviolet was shown to be largely determined by the spectral transmittance of the epidermis (Fig. 2.3).

McCree provided data that resolved the problem of the apparent anomaly (Gabrielsen 1960) of Hoover's (1937) well-known curve for the action spectrum of photosynthesis in wheat leaves. Wheat plants grown under the conditions specified by Hoover were as pale green as a lettuce leaf. The abnormally low absorptance in the green, coupled with an abnormally high quantum yield

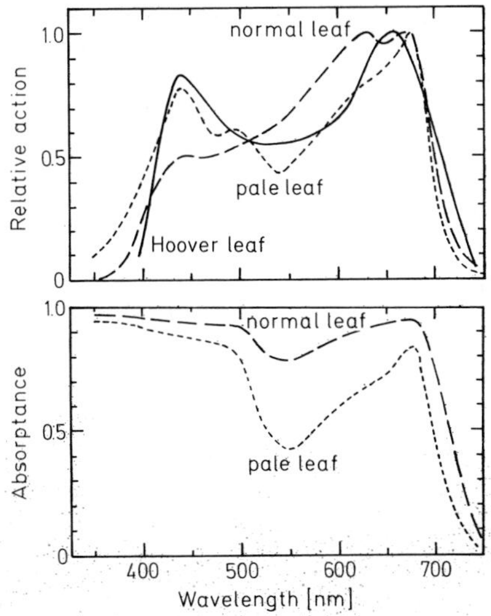

Fig. 2.4. Action spectrum for photosynthesis (relative photosynthetic rate per unit of radiant energy flux incident) and spectral absorptance, for a normal field-grown wheat leaf (*Triticum aestivum* L. em. Thell.), and for a pale green wheat leaf grown under conditions similar to those used by HOOVER (1937). Hoover's action spectrum is shown for comparison. (MCCREE 1972a)

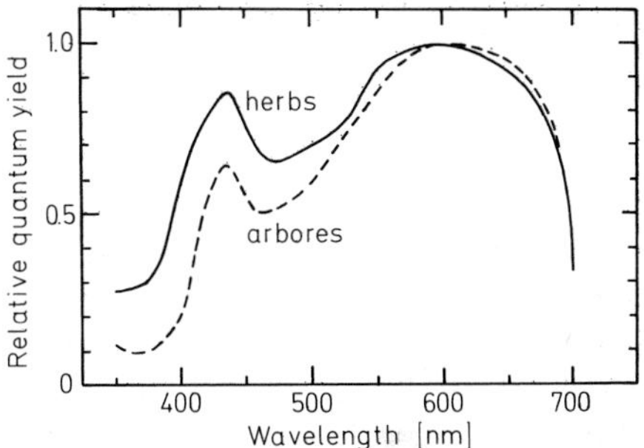

Fig. 2.5. Spectral quantum yield of photosynthesis according to INADA (1976). *Curves* are averages for 26 species of herbaceous crop plants and 7 arboraceous species

in the blue, caused the action spectrum to show the characteristically strong double peak of the Hoover curve (Fig. 2.4).

A second set of comprehensive quantum yield data was obtained by INADA (1976). Conditions and techniques were similar to those used by MCCREE (1972a). Data were obtained for leaves of 33 species, including 26 herbaceous crop species and 7 trees. Three of the crop samples and all of the tree samples were taken from field-grown plants, while the rest were taken from plants grown in a greenhouse with nutrient solution.

The average spectral quantum yield curve (Fig. 2.5) was very similar in shape to that found by MCCREE (1972a). Systematic differences between crop plants and trees were similar to those previously found between field-grown and growth-chamber-grown crop plants (Fig. 2.1). In some of the tree samples the height of the blue peak was only 50% of that of the red peak. This was attributed to absorption of quanta by inactive pigments.

The effect of variation in chlorophyll content on the action spectrum was confirmed, a typical "Hoover-type" double-peaked curve being obtained for

Table 2.1. Ratio of the photosynthetic rate of a leaf, measured in light from various sources, to the photosynthetic rate of the same leaf calculated from the quantum yield and the absorbed photon flux, using Eq. (2.3). (McCree 1971)

Leaf	Light source				
	Incand.	Mercury	Metal halide	Fluor.	Average for leaf
Oat	0.93	0.92	1.01	0.90	0.94
Castor bean	0.92	0.96	0.95	0.90	0.93
Cucumber	0.95	0.99	0.93	0.97	0.96
Maize	1.00	1.05	1.01	1.10	1.04
Sorghum	1.09	1.09	1.07	1.04	1.07
Amaranth	1.00	1.07	0.89	0.97	0.98
Average for source	0.98	1.01	0.98	0.98	0.99

lettuce leaves, and a typical "Gabrielsen-type" single-peaked curve for the tree leaves. The curves for leaves of crop plants were intermediate. Both action and quantum yield of middle wavelengths (500 to 600 nm) were greatly reduced in purple leaves, due to inactive absorption by anthocyanin. The spectral quantum yield was not affected by the color of the radiation under which the plants were grown (Inada 1977).

When the spectral quantum yield data are used to develop definitions of PAR, it is assumed that the yields of different wavelengths are independent and additive, so that the photosynthetic rate in "white light" is equal to the integral of the products of the yield and absorbed photon flux density over the active waveband [Eq. (2.3)]. It is well known that the quantum yield at some wavelengths can be "enhanced" by simultaneous irradiation at other wavelengths (Björkman 1968; Inada 1978a; Myers 1971). Nevertheless, McCree (1971) found that the actual photosynthetic rate agreed with the value calculated according to Eq. (2.3), within the limits of accuracy of his test ($\pm 7\%$) (Table 2.1). In addition, enhancement by "white" light of the yield in monochromatic radiation was found to be negligible (Inada 1978b; McCree 1972a). It is evident that additivity can be assumed to hold with sufficient accuracy for the purpose in hand (Gabrielsen 1960).

2.5 The Photon Flux as a Measure of PAR

Since the spectral quantum yield of photosynthesis varies with wavelength, the photosynthetic response per photon absorbed will not be independent of the spectrum of the light source, as was assumed by Gaastra [Eq. (2.4)]. Thus, the absorbed photosynthetic photon flux density ($PPFD_a$) will not be a perfect measure of PAR. Moreover, the photosynthetic irradiance (PI) may be less adequate than it appeared from Gaastra's calculations (Gaastra 1959).

Table 2.2. Range of variations in the photosynthetic rate per unit of photosynthetic irradiance (P/PI), absorbed photosynthetic photon flux density ($P/PPFD_a$) and incident photosynthetic photon flux density ($P/PPFD_i$), calculated for a range of natural and artificial light sources. (McCREE 1972b)

	Range of variation		
	P/PI	$P/PPFD_a$	$P/PPFD_i$
Average field-grown plant	$\pm 16\%$	$\pm 8\%$	$\pm 7\%$
Average growth-chamber-grown plant	$\pm 11\%$	$\pm 4\%$	$\pm 4\%$

McCREE (1972b) re-evaluated the problem, substituting his measured ϕ_λ values in Eq. (2.3), and extending the wavelength limits to 360–760 nm. He also calculated $P/PPFD_i$, where $PPFD_i$ is the incident photon flux density:

$$PPFD_i = 8.35 \cdot 10^{-9} \int_{400}^{700} I_\lambda \, \lambda \, d\lambda \; (E m^{-2} s^{-1}) \tag{2.5}$$

The results of the calculations are summarized in Table 2.2. The variations in $P/PPFD_a$ and $P/PPFD_i$ with light source are much less than the variations in P/I. McCREE concluded that $PPFD_i$ was clearly superior to PI as a measure of PAR. For example, if the irradiance measure were used, the photosynthetic rate in daylight would be seriously overestimated (+19%) by data taken in incandescent radiation, while the error would be reduced to +8% if the incident photon flux density were used.

Basing the definition on the absorbed photon flux density or the average quantum yield would not reduce the error further, and would have the undesirable effect of introducing biological variables into the definition. Both the irradiance and the incident photon flux density are purely physical quantities.

The error would be slightly greater for field-grown than for growth-chamber-grown plants, because the average spectral quantum yield curve (Fig. 2.1) deviates more from the "ideal" shape (constant between 400 and 700 nm and zero outside those limits) in the case of the field grown plants. According to the data obtained by INADA (Fig. 2.5), the deviation would be even greater for trees. Variations in the spectral irradiance within plant canopies would produce errors comparable with those indicated in Table 2.2. For example, the photosynthetic rate of a leaf at the bottom of a dense canopy, where the spectral changes are the greatest, would be underestimated by 7%, compared with that of the same leaf at the top of the canopy, when the $PPFD_i$ measure is used.

Extending the integration limits for PI to 380–710 nm, as was suggested by NICHIPOROVICH (1960), would also increase the error, because the quantum yield decreases rapidly at wavelengths below 400 nm and above 700 nm. Therefore, a definition of PAR based on the limits 380–710 nm is inferior to one based on 400–700 nm.

Like Gaastra, McCree also tested his conclusion by measurements. For an oat leaf, P/PI in mercury radiation was found to be 74% of that in incandescent radiation, while $P/PPFD_i$ was 85% of that value. The 15% drop in $P/PPFD_i$ was fully explained by the low quantum yield in the blue compared with that in the red.

Thus Gaastra's (1968) conclusion that the irradiance in the waveband 400–700 nm (photosynthetic irradiance, PI) is an adequate measure of PAR needs to be modified in the light of recent data. It has been demonstrated, both through calculations based on the spectral quantum yield of photosynthesis, and through measurements of leaf photosynthetic rates, that there can be serious systematic errors when the irradiance measure is used. It now appears that the photon flux density in the 400–700 nm waveband (photosynthetic photon flux density, PPFD) is a more acceptable though not perfect measure.

2.6 Spectral Errors of PAR Measurements

In theory, it is possible to tailor the spectral response of any radiometer to match the ideal response (a constant response to equal radiant energy fluxes between 400 and 700 nm for PI measurements, a constant response to equal photon fluxes in the same waveband for PPFD measurements). In practice, the response will always deviate from the ideal. The error that this introduces into PAR measurements can be determined by calculating the ratio of the broadband response to the true value of the PI or PPFD, as follows

$$R_{PI} = \int I_\lambda S_\lambda \, d\lambda, \qquad (2.6)$$

where R_{PI} = response of the PI meter (arbitrary meter units), I_λ = spectral irradiance ($W\,m^{-2}\,nm^{-1}$), and S = spectral sensitivity of PI meter (arbitrary meter units per unit of I_λ) or

$$R_{PPFD} = \int PFD_\lambda \, S'_\lambda \, d\lambda, \qquad (2.7)$$

where R_{PPFD} = response of the PPFD meter (arbitrary meter units), PFD_λ = spectral photon flux density ($E\,s^{-1}\,m^{-2}\,nm^{-1}$), and S'_λ = spectral sensitivity of PPFD meter (arbitrary meter units per unit of PFD_λ).

By analogy with the previous calculations [Eqs. (2.1)–(2.4)], it can be seen that the variation in the ratio R_{PI}/PI or $R_{PPFD}/PPFD$ when I_λ is varied is a measure of the error due to the spectral inadequacy of the instrument. Examples of such calculations for specific instruments can be seen in papers by Biggs et al. (1971), Federer and Tanner (1966), Gaastra (1968), McCree (1966), McPherson (1969), Norman et al. (1969), and Rvachev et al. (1963).

The importance of the spectral error depends entirely on the range of spectral irradiances I_λ that is likely to be encountered. Within the photosynthetically active waveband, the spectral quality of daylight is virtually independent of weather conditions (McCree and Keener 1974; Ross 1975). Hence, the spectral sensitivity of the meter is relatively unimportant when daylight is being measured. In fact, for many purposes it is sufficient to take the PI as 50% of the shortwave (global) irradiance as measured with a pyranometer (SZEICZ 1974), although this could result in a systematic overestimate of the PI for very clear conditions and an underestimate for very cloudy conditions (GAASTRA 1968; MCCREE 1966). The error probably does not exceed 10% either way.

The spectral irradiance changes with the depth in the plant canopy (ROSS 1975), but the changes are largely confined to the far red and infrared parts of the spectrum (cf. Chap. 1, this Vol.). Most leaves are highly opaque to PAR of all wavelengths, while transmitting about 50% of the infrared radiation between 750 and 1,350 nm (GAUSMAN et al. 1973). Therefore, any radiometer with an appreciable infrared response will overestimate the PI or PPFD within the canopy, compared with that above it (GAASTRA 1968). The error will be greatest when the sensor is immediately beneath a leaf. If the leaf is unusually transparent, the response of the sensor within the PAR waveband can also become important (FEDERER and TANNER 1966).

The greatest potential for error arises with data taken under artificial light sources, especially when these are compared with those taken in the field. Compared with daylight, most fluorescent lamps are green-yellow biased, and incandescent lamps are red-infrared biased. Therefore, sensors that deviate greatly from the ideal in these spectral regions should be particularly suspect. From the papers cited above, it appears that a spectral error of $\pm 20\%$ is not uncommon, with a lower limit for the most carefully designed instruments of $\pm 5\%$. It is good practice to specify both the instrument used and the type of lamp in some detail.

Assuming now that the radiometer has the ideal spectral response, the expected ratio of the response of a PI meter to that of a PPFD meter can be calculated using equations analogous to those already presented. This ratio can be used to convert from one system to the other. The conversion factor is specific to the light source. Examples of such conversion factors are given in Table 2.3.

Table 2.3. Approximate conversion factors for various light sources. (MCCREE 1972b) (PAR waveband 400–700 nm)

	Light source					
	Daylight	Metal halide	Sodium (HP)	Mercury	White fluor.	Incand.
To convert	Multiply by					
$W\ m^{-2}$(PAR) to $\mu E\ s^{-1}\ m^{-2}$(PAR)	4.6	4.6	5.0	4.7	4.6	5.0
klux to $\mu E\ s^{-1}\ m^{-2}$(PAR)	18	14	14	14	12	20
klux to $W\ m^{-2}$(PAR)	4.0	3.1	2.8	3.0	2.7	4.0

It will be seen that the conversion factor 1 W m^{-2}(PAR)=4.6 μE s^{-1} m^{-2} (PAR) can be used for most "daylight-type" sources (including xenon). The factor 5.0 should be used for "yellow" sources such as sodium or incandescent lamps. Included in the table are factors for converting illuminances in kilolux to PI or PPFD. These factors are much more variable and should be considered very approximate, since spectral errors are much more likely in this case. As a rule, it is better to measure the PPFD with a meter that is designed for this purpose than to convert the readings of other types of meters.

Spectral errors can of course be eliminated by measuring with a spectroradiometer and calculating the PI or PPFD (Norris 1968). However, the cost of the instrument and the complexity of the measurement are sufficient to deter most ecophysiologists. The spectral response of photosynthesis is by now so well defined, and the art of tailoring the spectral response of radiometers so far advanced, that the PAR measurement can now be made with sufficient accuracy without resorting to I_λ measurements. This is not necessarily true of other photobiological reponses (cf. Chap. 4, this Vol.).

2.7 Cosine Response Errors of PAR Measurements

The second most common error of radiometers is failure to follow the "cosine law," according to which the true flux density produced by a beam of radiation on a surface is proportional to the cosine of the angles of incidence. All radiometers deviate from the true cosine response at high angle of incidence (low angles of elevation). Standard procedure (Kubín 1971) is to fit the radiometer with a plate of opal glass or plastic, which acts as the primary receiving surface, and which is then viewed by the sensor itself through the spectrally correcting filters. The opal plate is often shaped to improve the response at low angles of elevation. Even better correction can be achieved by using an integrating sphere instead of a plate. Radiometers are routinely calibrated at normal incidence only.

The importance of the cosine error depends on the frequency with which radiation from low angles of elevation is likely to be encountered. This happens at the beginning and at the end of every day in field measurements. However, the effect of cosine error on daily totals of radiation received can usually be neglected. If data at low elevation angles are important, the cosine response should be determined with a collimated beam and corrections applied. It is good practice to level the receiving surface with a bubble-level.

The use of the so-called "spherical" sensor that measures the flux into a point, rather than the flux per unit surface area, has not become common in ecophysiology, despite repeated attempts to introduce it. The problem is that plants are neither points nor plane horizontal surfaces, so that neither measure can be said to "represent" a plant or community of plants. Nevertheless, it is possible to develop theoretical or empirical relationships between the surface irradiance and the flux absorbed by the plants in a given geometrical situation (Monteith 1973). This approach is preferable to introducing a second measure based on the flux into a point, at least until it can be demonstrated that it simplifies the relationships.

The radiation field inside a plant canopy is so complicated, geometrically as well as spectrally (MONSI et al. 1973), that it is not possible to describe it accurately with a single number. For many purposes, it may be sufficient to average the PI or PPFD with the so-called "tube" or "line" radiometer that measures the irradiance over an extended surface area. The cosine errors of such instruments are generally greater than those of small fully corrected radiometers, but the error can be minimized by taking measurements in two directions at right angles (ANDERSON 1971).

Other errors that have traditionally been encountered in radiometers, such as those due to nonlinearity, temperature sensitivity, or instability (short-term or long-term), have decreased in importance with the advent of quality-controlled commercial instruments that use solid-state circuitry. Incorrect spectral response or inadequate cosine correction remain as likely sources of error, no matter what form the radiometer make take.

2.8 Conclusions

The photon flux density in the waveband 400 to 700 nm can be considered an adequate measure of photosynthetically active radiation for most ecophysiological studies. According to the data that have been summarized here, the actual photosynthetic response of a leaf can be expected to be within ±7% of the stated value, independent of the nature of the light source or leaf, when the photon measure is used. Theoretically, systematic differences between different sets of published data should not exceed ±7% after universal adoption of this definition of PAR.

The practical realization of PAR measurements will involve additional errors, which can be expected to be comparable in magnitude. Thus, the overall accuracy of PAR measurements is unlikely to exceed ±10%. Even this level of accuracy cannot be achieved without great care in the design, calibration, and use of the instruments.

Finally, it should be noted that photosynthetically active radiation is nothing more than a measure of the radiation available for *photosynthesis*. It should not be treated as a universal measure of the radiation available for *growth*. Unfortunately, there is no such measure. Radiation affects many different growth processes, each of which should be treated independently, as will be evident from a reading of Chapters 4 to 7 in this Volume.

References

Anderson MC (1971) Radiation and crop structure. In: Šesták Z, Čatský J, Jarvis PG (eds) Plant photosynthetic production: manual of methods. W Junk, The Hague, pp 412–456

Biggs WW, Edison AR, Eastin JD, Brown KW, Maranville JW, Clegg MD (1971) Photosynthesis light sensor and meter. Ecology 52:125–131

Björkman O (1968) Further studies on differentiation of photosynthetic properties in sun and shade ecotypes of *Solidago virgaurea.* Physiol Plant 21:84–99

Campbell GS (1977) An introduction to environmental biophysics. Springer, Berlin Heidelberg New York

CIE (Commission Internationale de l'Eclairage) 1970 International Lighting Vocabulary, 3rd edn. Bureau Central de la CIE, Paris

Dybing CD (1977) Letter from the editor. Crop Sci 17:ii

Federer CA, Tanner CB (1966) Sensors for measuring light available for photosynthesis. Ecology 47:654–657

Gaastra P (1959) Photosynthesis of crop plants as influenced by light, carbon dioxide, temperature and stomatal diffusion resistance. Meded Landbouwhogesch Wageningen 59(13):1–68

Gaastra P (1968) Radiation measurements for investigations of photosynthesis under natural conditions. In: Eckardt FE (ed) Functioning of terrestrial ecosystems at the primary production level. UNESCO, Paris, pp 467–478

Gabrielsen EK (1960) Lichtwellenlänge und Photosynthese. In: Ruhland W (ed) Handbuch der Pflanzenphysiologie. Springer, Berlin, Heidelberg, New York, Vol. V/2, pp 49–78

Gausman HW, Allen WA, Wiegand CL, Escobar DE, Rodriquez RR, Richardson AJ (1973) The leaf mesophylls of twenty crops, their light spectra, and optical and geometrical parameters. Tech Bull No 1465, Agric Res Service, US Department Agric, Washington DC

Hoover WH (1937) The dependence of carbon dioxide assimilation in a higher plant on wavelength of radiation. Smithsonian Misc Coll 95(2):1–13

Inada K (1976) Action spectra for photosynthesis in higher plants. Plant Cell Physiol 17:355–365

Inada K (1977) Effects of leaf color and the light quality applied to leaf-developing period on the photosynthetic response spectra in crop plants. Proc Crop Sci Soc Japan 46:37–44

Inada K (1978a) Photosynthetic enhancement spectra in higher plants. Plant Cell Physiol 19:1007–1017

Inada K (1978b) Spectral dependence of photosynthesis in crop plants. Acta Hortic 87:177–184

Incoll LD, Long SP, Ashmore MR (1977) SI units in publications in plant science. Curr Adv Plant Sci 9:331–343

Kubín Š (1971) Measurement of radiant energy. In: Šesták Z, Čatský J, Jarvis PG (eds) Plant photosynthetic production: manual of methods. W Junk, The Hague, pp 702–763

McCree KJ (1966) A solarimeter for measuring photosynthetically active radiation. Agric Meteorol 3:353–366

McCree KJ (1971) Significance of enhancement for calculations based on the action spectrum for photosynthesis. Plant Physiol 49:704–706

McCree KJ (1972a) The action spectrum, absorptance and quantum yield of photosynthesis in crop plants. Agric Meteorol 9:191–216

McCree KJ (1972b) Test of current definitions of photosynthetically active radiation against leaf photosynthesis data. Agric Meteorol 10:443–453

McCree KJ (1973) A rational approach to light measurement in plant ecology. Curr Adv Plant Sci 3(4):39–43

McCree KJ, Keener ME (1974) Effect of atmospheric turbidity on the photosynthetic rates of leaves. Agric Meteorol 13:349–357

McPherson HG (1969) Photocell-filter combinations for measuring photosynthetically active radiation. Agric Meteorol 6:347–356

Monsi M, Uchijima Z, Oikawa T (1973) Structure of foliage canopies and photosynthesis. Ann Rev Ecology Syst 4:301–327

Monteith JL (1973) Principles of environmental physics. Edward Arnold, London

Myers J (1971) Enhancement studies in photosynthesis. Ann Rev Plant Physiol 22:289–312

Nichiporovich AA (1960) Conference on measurement of visible radiation in plant physiology, agrometeorology and ecology. Fiziol Rast 7:744–747

Norman JM, Tanner CB, Thurtell GW (1969) Photosynthetic light sensor for measurements in plant canopies. Agron J 61:840–843

Norris KH (1968) Evaluation of visible radiation for plant growth. Ann Rev Plant Physiol 19:490–499

Rabinowitch EE (1951) Photosynthesis and related processes: Vol 2 Part 1, Spectroscopy and fluorescence of photosynthetic pigments: Kinetics of photosynthesis. Interscience, New York, pp 837–844

Ross J (1975) Radiative transfer in plant communities. In: Monteith J (ed) Vegetation and the atmosphere Vol 1. Academic Press, New York, pp 13–55

Rvachev VP, Berdnikov VF, Vaschchenko VI (1963) Physical basis of measurements of the energy of photosynthetically active radiation by selective detectors. Fiziol Rast 10:598–602

Shibles R (1976) Committee report: Terminology relating to photosynthesis. Crop Sci 16:437–439

Szeicz G (1974) Solar radiation for plant growth. J Appl Ecol 11:617–636

3 Responses to Different Quantum Flux Densities

O. Björkman

CONTENTS

3.1 Introduction

Growth of autotrophic plants is directly and dramatically influenced by the intensity of light – the driving force of photosynthesis – which provides nearly all of the carbon and chemical energy needed for plant growth. Moreover, light intensity (quantum flux density) is perhaps the most conspicuous environmental variable with which plants must cope. Contrasting terrestrial habitats may differ by at least two orders of magnitude in the daily quantum flux available for photosynthesis. In any habitat the quantum flux density varies seasonally, diurnally, and spatially (such as within a canopy of a given plant stand).

For these reasons the response to quantum flux density (PAR, see previous Chapter of this Volume) of photosynthesis and primary productivity has been the subject of much investigation for over a century. That plants occupying sunny habitats (sun plants) are generally capable of higher photosynthetic rates at high quantum flux densities than plants restricted to shaded locations (shade plants) has been recognized for about half a century. GABRIELSEN in the previous series of Encyclopedia of Plant Physiology (Vol. V/2, pp. 8–78) and RABINOWITCH (1951) have provided comprehensive reviews on the light dependence of photosynthesis. More recent reviews on various aspects of photosynthetic response and adaptation to light have been provided by BJÖRKMAN (1973), BOARDMAN (1977), WILD (1979), and OSMOND et al. (1980). Other aspects of the effect of quantum flux density on plant growth have been treated in reviews by BLACKMAN (1956, 1968).

3.2 Photosynthetic Gas Exchange Characteristics

Light dependence curves for photosynthesis typical for sun and shade plants grown under light regimes found in their respective habitats are shown in Fig. 3.1 A. At low quantum flux densities photosynthesis (P) is linearly dependent on quantum flux density (I) and the efficiency of light utilization, or the quantum yield ($\phi_i = dP/dI$), is constant and maximal. At higher quantum flux densities

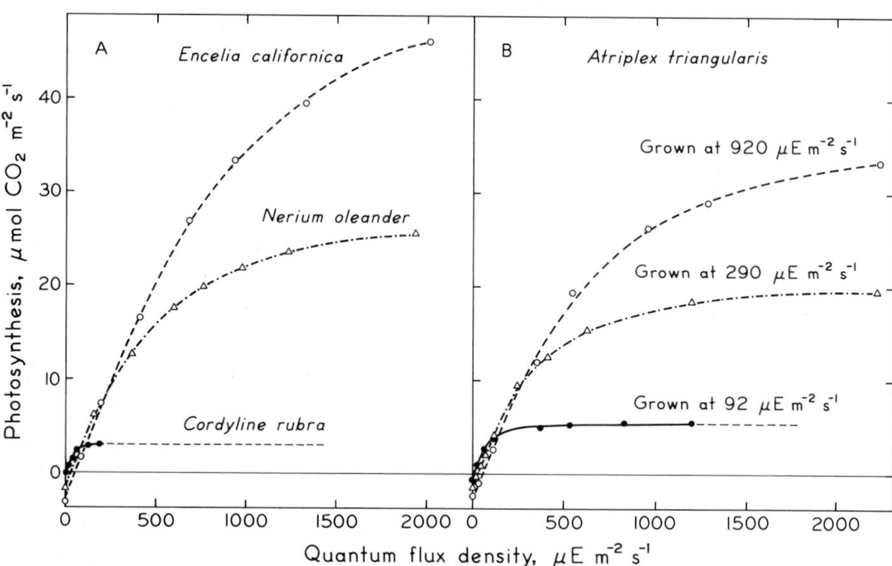

Fig. 3.1 A, B. Rate of net photosynthesis as a function of incident quantum flux density (400–700 nm) for **A** the sun species *Encelia californica* and *Nerium oleander*, grown under natural daylight (~40 E m^{-2} day^{-1}) and the shade species *Cordyline rubra* grown in its native rain forest habitat (~0.3 E m^{-2} day^{-1}) and **B** for the sun species *Atriplex triangularis*, grown under three different light intensity regimes. All measurements were made in air of normal CO_2 and O_2 partial pressures and a leaf temperature of 25 or 30 °C. (Data for *E. californica*, EHLERINGER and BJÖRKMAN 1978; *N. oleander*, BJÖRKMAN unpublished; *C. rubra*, BJÖRKMAN et al. 1972a; *A. triangularis*, BJÖRKMAN et al. 1972b)

the increase in P is less than proportional to the increase in I (partial light saturation) and ultimately, P fails to increase with increasing I (light saturation).

It is evident from Fig. 3.1 A that great differences exist in the light dependence of photosynthesis between sun plants such as *Nerium oleander* and *Encelia californica* and shade plants such as *Cordyline rubra*. In the latter plant, because of the very low rate of dark respiration, extremely low levels of I are sufficient to balance respiration (light compensation). The light compensation thus occurs at much lower quantum flux densities and the rate of net photosynthesis at low light levels is considerably higher in the shade plant than in the sun plants which have relatively high respiration rates. However, light saturation is reached early in the shade plant, and at values of I greater than about 100 $\mu E\ m^{-2}\ s^{-1}$ (equivalent to about 5% of maximum daylight) the sun plants, still operating in the linear part of the light curve, have a higher rate of net photosynthesis than the shade plant. In *E. californica* photosynthesis has not yet fully reached light saturation at 2,000 $\mu E\ m^{-2}\ s^{-1}$ (approximately equivalent to maximum daylight), whereas for *C. rubra* this quantum flux density is about 20-fold higher than needed for saturation. Sustained exposure to quantum flux densities in excess of that required to saturate photosynthesis (supersaturation) may lead to a time-dependent inactivation of photosynthesis (photoinhibition). The region in which such injurious photoinhibition occurs in *C. rubra* is indicated in Fig. 3.1 A by the broken horizontal line. The kinds of light responses shown in Fig. 3.1 A are clearly *adaptive*, since they permit the shade plant to function efficiently under the low quantum flux densities that prevail in its habitat and enable the sun plants to make effective use of moderate and high quantum flux densities.

3.2.1 Light Acclimation

Sun Plants. In the preceding paragraph we compared the photosynthetic light dependence of plants which in nature are found in open sunny locations but not in deep shade (obligate sun plant) with those of a plant which is capable of growing in extreme shade habitats and is not found in open sunny locations (obligate shade plant). The sun and the shade plants had been grown under natural light regimes of a sunny exposed habitat (40 $E\ m^{-2}\ day^{-1}$) and a rain-forest floor (0.3 $E\ m^{-2}\ day^{-1}$), respectively. The light dependence characteristics of each plant thus represent a combination of both genetic and environmental influences. Studies on identical genotypes grown under different light regimes, and comparisons between sun and shade leaves of the same individual plant, demonstrate that the light response characteristics of a given plant or individual leaf may be strongly modified by the growth light regime.

As a rule, growth of sun plants (obligate and facultative) in the shade results in photosynthetic light dependence characteristics tending toward those of obligate shade plants. An example of this type of response is shown in Fig. 3.1 B for *Atriplex triangularis*, a species which may occupy sunny beaches as well as partially shaded locations in salt marshes dominated by tall grasses and sedges. When grown under a high light regime, leaves of this species have

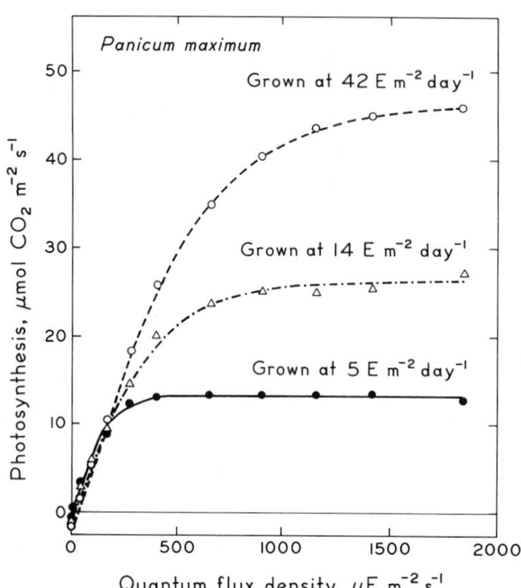

Fig. 3.2. Effect of different growth light regimes on the light dependence of net photosynthesis in the C_4 species *Panicum maximum*. Measurements were made in air of normal CO_2 and O_2 partial pressures at a leaf temperature of 30 °C. Adapted from Ludlow and Wilson (1971). Quantum flux density values are estimates based on the illuminance values given by these authors

high light-saturated photosynthetic rates, and relatively high rates of dark respiration. Both of these rates show a strong decline with decreasing light regime. Leaves of plants grown at 5.3 E m^{-2} day^{-1} have rates of light-saturated photosynthesis and dark respiration that are only about 20% of those grown at 53 E m^{-2} day^{-1} and, accordingly, the quantum flux density required for compensation of dark respiration is only about 20% as high as in the plants grown at the higher light regime. Nevertheless, the minimum daily radiation receipt required for significant growth in *A. triangularis* is at least an order of magnitude higher than that for shade plants such as *Cordyline rubra* and *Alocasia macrorrhiza,* and the dark respiration rates and light compensation points of these shade species, determined in their native rain forest habitat (Björkman et al. 1972a), are much lower than in *A. triangularis* grown near its shade tolerance limit (Björkman et al. 1972b).

Photosynthetic light acclimation (environmentally induced adaptation) appears to occur to a smaller or greater extent in all sun plants for which such studies have been made. A summary of published results on the effect of different light growth regimes on the light-saturated rate of photosynthesis for many sun and shade species of C_3 plants is shown in Table 3.1. Many of these species will be referred to throughout this chapter. Although only a few C_4 species have been included in studies on light acclimation, the results with the C_4 grass *Panicum maximum* by Ludlow and Wilson (1971), shown in Fig. 3.2, *Zea mays* (Louwerse and v.d. Zweerde 1977), *Sorghum bicolor* and the C_4 dicot, *Amaranthus retroflexus* (Singh et al. 1974), show that at least some C_4 plants are capable of photosynthetic light acclimation qualitatively similar to the situation in C_3 sun plants.

Shade Plants. Although it is widely recognized among horticulturalists that obligate shade plants may suffer damage to their leaves, grow poorly, and even die when attempts are made to grow them at high irradiance levels, there is a paucity of controlled studies on the effect of light regime on photosynthetic characteristics of shade plants. The lack of a favorable response in such plants to high irradiance levels could in part be related to correlated factors such as adverse water relations. However, there is strong evidence that obligate shade plants have an intrinsically low potential for photosynthetic light acclimation, and it seems likely that their high susceptibility to high light injury is largely a consequence of their inherent low ability to increase their capacity for effective utilization of high quantum flux densities for photosynthesis.

In a comparative study of the potential for photosynthetic light acclimation in ecotypes of *Solidago virgaurea,* native to open habitats and to densely shaded forest floors (BJÖRKMAN and HOLMGREN 1963), the "shade ecotypes" were unable to acclimate to a high light growth regime, whereas the response of the "sun ecotypes" was similar to that generally found in sun plants. When cloned individuals of sun and shade ecotypes were grown at two different light regimes (6.6 and 33.1 E m^{-2} day^{-1}) with ample water and nutrient supply, the sun clones grew very slowly under the lower light regime but grew rapidly under the high light regime. By contrast, the shade clones exhibited a relatively rapid growth under the lower light regime, whereas the growth of these clones was retarded under the high light regime. As shown in Fig. 3.3, unlike the situation with the sun clones, growth of the shade clones under the high light regime did not result in an increased capacity for light-saturated photosynthesis. Moreover, in the shade clone, high quantum flux density during growth caused a reduction in the quantum yield of photosynthesis at rate-limiting quantum flux densities. The latter effect is undoubtedly attributable to photoinhibition (Sect. 3.5). Later work (BJÖRKMAN 1968b) showed that the failure of shade clones of *Solidago* to adapt to high light levels is associated with a low capacity of such plants to increase the level of ribulose-1,5-bisphosphate carboxylase (and perhaps also other components which potentially determine the capacity for light-saturated photosynthesis).

GAUHL (1969, 1970, 1976) found that certain shade and sun clones of *Solanum dulcamara* differed in their potential photosynthetic light acclimation resembling the situation in *Solidago virgaurea.* GAUHL also showed that fully expanded leaves of a *Solanum* sun clone, developed at 5.3 E m^{-2} day^{-1}, were able to increase their light-saturated photosynthetic capacity upon transfer to 24.3 E m^{-2} day^{-1} and that there was a parallel increase in protein synthesis and RuBP carboxylase activity. Leaves of a *Solanum* shade clone lacked the ability to increase the photosynthetic capacity upon transfer of the plants from the low to the high light regime and the protein content and RuBP carboxylase activity also did not increase.

Other evidence in support of the conclusion that obligate shade plants have a limited capacity for photosynthetic acclimation to high quantum flux densities includes studies on umbrophytic fern sporophytes of *Pteris cretica* (HARIRI and BRANGEON 1977; HARIRI and PRIOUL 1978), gametophytes of the tree fern *Cibotium glaucum* (FRIEND 1975), and leaves of *Selaginella unicata,* a lower vascular umbrophyte (JAGELS 1970). Growth at even moderate quantum flux densities resulted in a decrease rather than an increase in the capacity for

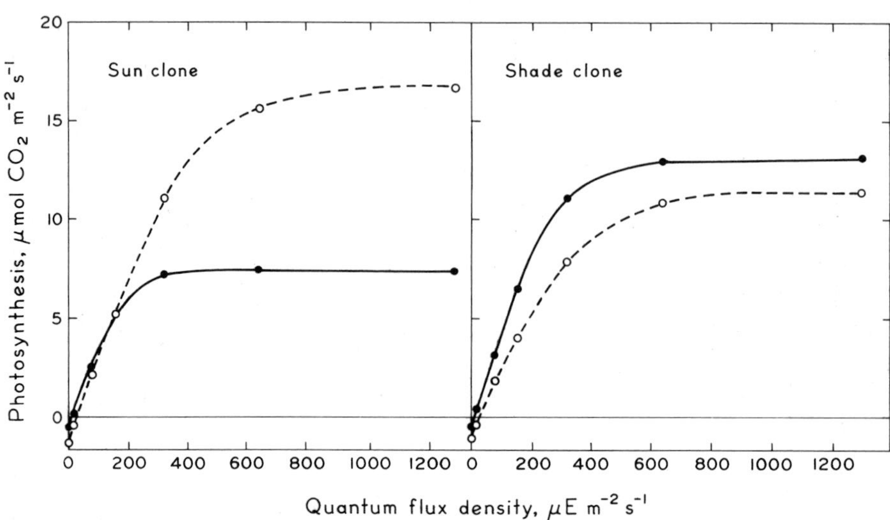

Fig. 3.3. Rate of photosynthesis as a function of incident quantum flux density in clones of sun and shade ecotypes of *Solidago virgaurea*, grown at 6.6 (●) and 33.1 (○) E m^{-2} day^{-1}. Measurements were made under 320 μbar CO_2, 210 mbar O_2, and a leaf temperature of 22 °C. (Adapted from BJÖRKMAN and HOLMGREN 1963)

light-saturated photosynthesis, frequently accompanied by chlorophyll bleaching and other adverse effects. In *Pteris cretica* which is able to grow under extreme shade, both the light-limited and the light-saturated photosynthetic rates were highest in plants grown under the smallest quantum dose used in these experiments (~1–3 E m^{-2} day), and an increase in daily quantum flux by a factor of only 4 to 5 caused a reduction in photosynthetic rate (HARIRI and PRIOUL 1978). A shade plant type of response, although less extreme than in *Pteris*, was also observed with *Fragaria vesca* (CHABOT and CHABOT 1977) and *F. virginiana* (JURIK et al. 1979). In these species an increase in the daily quantum flux from 2–3 to about 10 E m^{-2} day^{-1} had only a relatively small effect on light-saturated photosynthetic rate and a further increase to about 35 E m^{-2} day^{-1} caused a decline in this rate.

3.3 Factors Determining Photosynthetic Performance in Weak Light

It seems evident that a prerequisite for success in a low light environment is that the leaves must be able to trap the available light and to convert it into chemical energy with the highest possible efficiency. At the same time, respiratory losses and the cost of producing and maintaining the photosynthetic system relative to the gain in photosynthate production must be kept as low as possible. Another consideration is that the highest possible fraction of the

Table 3.1. Comparison of light-saturated net photosynthetic rates of different C_3 species, grown under various light regimes. Shade species are indicated by an *asterisk*. Rates were determined in air of normal CO_2 and O_2 pressures and a leaf temperature of 20 to 30 °C. Total daily quantum flux density values are estimates for the wave band 400–700 nm. The letters C, N, and H following the quantum flux density values denote artificial light, natural day light, and native habitat, respectively. The right part of the table refers to the correlation analysis shown in Table 3.4 between photosynthetic rate and specific leaf weight (SLW), chlorophyll content (Chl), P-700, soluble protein (Prot.) and ribulose 1,5-bisphosphate carboxylase (RuBPC)

Species	Growth light regime, $E\ m^{-2}\ day^{-1}$	Photo-synthesis, $\mu mol\ m^{-2}\ s^{-1}$	Parameters used for correlation analysis in Table 3.4					Reference
			SLW	Chl	P-700	Prot.	RuBPC	
Abutilon theophrasti	4.9 C	6.5		×	×			Patterson et al. (1978)
	17.2 C	9.7		×	×			
	40.5 C	11.6		×	×			
**Adenocaulon bicolor*	~0.5 H	1.4		×		×	×	Björkman (1968a)
**Alocasia macrorrhiza*	~0.3 H	3.2	×	×	×	×		Boardman et al. (1972)
**Aralia californica*	~0.5 H	2.0		×		×	×	Björkman (1968a)
Atriplex triangularis	5.3 C	6.9	×	×	×	×	×	Björkman et al. (1972b)
	36.3 C	24.0	×	×	×	×	×	
	53.0 C	35.3	×	×	×	×	×	
Betula verrucosa	3.1 C	5.0	×	×	×			Öquist et al. (1981)
	15.1 C	8.8	×	×	×			
	36.7 C	11.0	×	×	×			
Camissonia brevipes	~45 N	37.0	×	×	×			Armond and Mooney (1978)
**Cordyline rubra*	~0.3 H	3.3	×	×	×	×	×	Boardman et al. (1972)
Datura meteloides	~45 N	19.0	×	×	×			Armond and Mooney (1978)
Echinodorus berteroi	~45 H	13.5		×		×	×	Björkman (1968a)

Table 3.1 (continued)

Species	Growth light regime, E m^{-2} day^{-1}	Photo-synthesis, μmol m^{-2} s^{-1}	Parameters used for correlation analysis in Table 3.4					Reference
			SLW	Chl	P-700	Prot.	RuBPC	
Eichhornia crassipes	4.5 C	10.1	×	×	×	×	×	PATTERSON and DUKE (1979)
	16.1 C	16.8	×	×	×	×	×	
	37.8 C	18.5	×	×	×	×	×	
Encelia californica	~40 N	38.0	×	×		×		EHLERINGER and BJÖRKMAN (1978)
**Fragaria virginiana*	3.5 C	5.5	×					JURIK et al. (1979)
	36.6 C	6.5	×					
Glycine max	10.9 C	13.0	×	×	×			BUNCE et al. (1977)
	36.7 C	15.9	×	×	×			
	~13.5 N	12.6					×	SINGH et al. (1974)
	~22.5 N	17.5					×	
	~31.5 N	17.7					×	
	~45.0 N	22.1					×	
Gossypium hirsutum	4.9 C	4.2		×	×			PATTERSON et al. (1978)
	17.2 C	7.6		×	×			
	40.5 C	13.8		×	×			
	24.8 C	17.7	×	×	×			PATTERSON et al. (1977)
	~42.0 N	27.1	×	×	×			
Larrea divaricata	~48.0 N	23.0	×	×				MOONEY et al. (1978)
Lupinus sparsiflorus	~45.0 N	28.0	×	×	×			ARMOND and MOONEY (1978)
Medicago sativa	~10.0 C	17.7	×					PEARCE and LEE (1979)
	~25.0 C	30.3	×					
Mimulus cardinalis	~45.0 H	23.7		×		×	×	BJÖRKMAN (1968a)
Nerium oleander	40.0 N	22.0	×	×		×		BJÖRKMAN et al. (1978)
Perityle emoryi	~45.0 N	19.0	×	×	×			ARMOND and MOONEY (1978)

Species	Growth light regime, $E\ m^{-2}\ day^{-1}$	Photo-synthesis, $\mu mol\ m^{-2}\ s^{-1}$	Parameters used for correlation analysis in Table 3.4					Reference
			SLW	Chl	P-700	Prot.	RuBPC	
Phaseolus atropurpureus	4.6	8.2	×					LUDLOW and WILSON (1971)
	13.9	12.6	×					
	42.2	22.7	×					
Phaseolus vulgaris	~ 3.6 N	10.0	×	×				POWLES (1979)
	~15.0 N	21.0	×	×				
	~60.0 N	40.0	×	×				
Plantago lanceolata	~45.0 H	14.2		×		×	×	BJÖRKMAN (1968a)
Silene alba	~ 3.0 C	7.6		×				WILMOT and MOORE (1973)
	~ 15.0 C	13.4		×				
Silene dioica	~ 3.0 C	3.7		×				WILMOT and MOORE (1973)
	~ 15.0 C	4.9		×				
Solanum dulcamara	5.3 C	11.8		×		×	×	GAUHL (1976)
	24.3 C	18.2		×		×	×	
	~ 1.6 N	2.5	×	×	×			CLOUGH et al. (1979)
	~40.0 N	8.3	×	×	×			
Solidago virgaurea	6.6 C	8.7	×	×				BJÖRKMAN and HOLMGREN (1963)
	33.1 C	16.1	×	×				
*	6.6 C	11.6	×	×				
*	33.1 C	11.3	×	×				
	5.3 C	12.1				×	×	BJÖRKMAN (1968b)
	26.5 C	17.0				×	×	
*	5.3 C	11.0				×	×	
*	26.5 C	10.3					×	
Trillium ovatum	~0.5 H	2.0		×		×	×	BJÖRKMAN (1968a)
Triticum aestivum	~13.5 N	6.9					×	SINGH et al. (1974)
	~22.5 N	6.8					×	
	~31.5 N	9.5					×	
	~45.0 N	11.4					×	

photosynthate must be reinvested into photosynthetic tissue, i.e., that allocation to nonphotosynthetic tissue be kept as low as possible.

3.3.1 Chlorophyll Content and Light Absorption

The efficiency by which light may be absorbed by a leaf depends on the content of chlorophyll per unit leaf area. The higher the chlorophyll content the greater is the proportion of the incident light which is absorbed by the leaf. However, the fractional absorptance is less than proportional to increase in pigment content. At any given wavelength (λ), the relationship between fractional absorptance (a) and pigment is given by the expression

$$a_\lambda = 1 - \exp(k_\lambda l_\lambda c)$$

where k_λ is a constant, l_λ the effective optical pathlength of the light in the leaf and c the pigment content. This relationship is strictly valid only for a homogeneous system such as extracted chlorophyll in an acetone solution. A leaf is a highly heterogeneous system and the chlorophyll protein complexes in vivo have different light absorption characteristics than extracted chlorophyll. Hence, the values for k_λ differ considerably from the extinction coefficient for a chlorophyll extract, and in part because of multiple light reflection and scattering within a leaf, the effective light path (l_λ) is greater than that given by the thickness of the leaf mesophyll and it varies with both wavelength and leaf anatomy. A leaf is therefore a more efficient absorber than a chlorophyll solution with the same chlorophyll content. Glabrous leaves with normal chlorophyll content (400 to 600 mg chl m^{-2}) absorb about 80 to 85% (a=0.80 to 0.85) of the daylight in the waveband 400 to 700 nm (e.g., Gabrielsen 1960; McCree 1972; Björkman 1968b). A threefold increase in chlorophyll content from about 250 to 750 mg m^{-2} causes only a very small increase (2–3%) in absorptance in the blue and the red where chlorophyll has a very high specific absorption; the main effect of an increased chlorophyll content is in the green and far-red regions where chlorophyll has a low absorption coefficient. For example, a light green leaf with 250 mg chl m^{-2} absorbs about 60% of the light at the absorptance minimum at 550 nm; a dark green leaf with 750 mg chl m^{-2} absorbs about 82% at this wavelength. The average absorptances over the waveband 400–700 nm for the light green and the dark green leaves are approximately 73 and 87% respectively. It should be emphasized that although an increase in chlorophyll content in this range by no means results in a proportional increase in absorptance, it may nevertheless confer a significant advantage under conditions where the quantum flux density is low and thus severely limits the photosynthetic rate.

Reflective coatings of leaves, such as the presence of a dense pubescence on the upper leaf epidermis, may obviously alter the absorptance dramatically. For example, heavily pubescent leaves of the desert species *Encelia farinosa* absorb as little as 30% of the solar radiation in the waveband 400 to 700 nm, while glabrous *Encelia* leaves of equal chlorophyll content absorb about 84%

(EHLERINGER et al. 1976). The quantum yield for photosynthesis at limiting quantum flux densities expressed on an incident light basis (ϕ_i) was directly proportional to the absorptance, while the light-saturated rate of photosynthesis was the same in pubescent and glabrous leaves (EHLERINGER and BJÖRKMAN 1978).

Chlorophyll Content and Chloroplast Morphology in Sun and Shade Plants. The average chlorophyll content for 49 sun and shade plants listed in Table 3.1, grown under a wide range of light regimes, was 485 mg chl m^{-2}. A more detailed analysis of this data base and the data shown in Table 3.5 led to the following conclusions. For a given sun plant, chlorophyll content per leaf area tends to remain relatively constant over a wide range of growth light levels but severe shading may cause significant decrease in chlorophyll content in some sun plants (e.g., *Solanum dulcamara* and *Betula verrucosa,* shown in Table 3.5). It appears that such differences may be more pronounced in young developing leaves than in fully expanded mature ones. However, obligate shade plants grown under deep shade often have at least as high chlorophyll content as do sun plants grown at high light levels; high chlorophyll content appears to be especially frequent among evergreen shade plants native to tropical forests (e.g., *Cordyline rubra,* Table 3.5).

Chloroplasts of shade plants such as *Alocasia macrorrhiza, Cordyline rubra,* and *Lomandra longifolia* have unusually large grana stacks containing as many as 100 thylakoids per granum. The grana are irregularly oriented in a single chloroplast and not in one plane as they usually are in sun plant chloroplasts. The proportion of lamellae-forming grana and the ratio of thylakoid membranes to stroma is greater than in sun plants (BOARDMAN et al. 1975; GOODCHILD et al. 1972). The extensive grana formation in the shade plants is presumably a means by which these plants are able to attain a high chlorophyll content per unit leaf area in spite of their having a relatively low ratio of chloroplast volume to leaf area. There is also evidence that growth of sun plants such as *Atriplex triangularis* in weak light induces an increase in thylakoid stacking density and the thylakoid membrane to stroma ratio, tending toward but not reaching the situation found in the rain forest species (BJÖRKMAN et al. 1972b). As discussed in Section 3.5, exposure of shade plants to high light levels may have detrimental effects on the photosynthetic system, resulting in a reduction in the chlorophyll content.

It is well known that growth under low light levels tends to result in an enrichment in chlorophyll b relative to chlorophyll a and that shade plants grown in deep shade tend to have a markedly lower chla/chlb ratio than do sun plants grown under a high light level (e.g., EGLE 1960). Many, more recent reports have confirmed these earlier findings. Since all of the chlorophyll b is considered to belong to the light-harvesting Chlab–protein (LHChl) complex (THORNBER 1975), it is probable that such differences in chla/chlb ratio reflect a difference in the proportion of the LHchl complex to the total chlorophyll complement of the chloroplast. Since it is further considered that the LHchl complex is primarily associated with photosystem (PS) II (BUTLER 1977), it may be that the shade plants have a higher ratio of PS II to PS I reaction centers than sun plants (Sect. 3.4.3). Such differences in the ratio of the two photosystems do not seem unreasonable in view of the recent results of MELIS and BROWN (1980a), showing that this ratio is not necessarily unity but may vary between different organisms and may also be subject to seasonal or other environmentally induced changes in a given plant. A possible function of an increased PS II/PS I ratio in shade plants is to provide a more balanced energy distribution between the two photosystems in shaded habitats such as forest floors which, because of the filtering effect of the forest canopy, have a very high proportion of far-red light, effective only in excitation of PS I. Such changes in PS II/PS I ratio could also explain the tendency of shade plants to have a slightly higher ratio of total chlorophyll to P-700 (Table 3.5), a phenomenon some authors (e.g., ALBERTE et al. 1976; HARVEY 1980) have interpreted as an increase in photosynthetic unit size (Sect. 3.4.3).

3.3.2 Quantum Yield of Photosynthesis in C_3 and C_4 Plants

The quantum yield expressed on the basis of incident quanta, ϕ_i, may be defined as $\phi_i = a \cdot \phi_a$; where ϕ_a is the efficiency of utilization of absorbed quanta (cf. also Chap. 2, this Vol.). Hence, if ϕ_a is constant, then the efficiency of utilization of low incident quantum flux densities would be proportional to leaf absorptance. One would expect, on theoretical grounds, ϕ_a to be the same among plants that use identical mechanisms for energy conversion in the photoacts and identical biochemical pathways for CO_2 fixation, since both the efficiency of ATP and $NADPH_2$ production, as well as the requirement for these compounds in the fixation and reduction of one CO_2 molecule, would then presumably also be identical. Experimentally determined values for ϕ_a are in substantial agreement with these expectations. No significant differences in ϕ_a were found between sun- and shade-grown sun ecotypes of *Solidago virgaurea* (BJÖRKMAN 1968b) or *Solanum dulcamara* (GAUHL 1976), or between sun and shade leaves of sun species such as *Mimulus cardinalis* (HIESEY et al. 1971), *Atriplex triangularis* (BJÖRKMAN et al. 1972b) and *Betula verrucosa* (ÖQUIST et al. 1980). Also, inspection of numerous light dependence curves for photosynthesis reported in the literature, such as those shown in Figs. 3.1 and 3.2, do not indicate that there exist marked differences in ϕ_i between sun and shade leaves of the same species, nor does there appear to be any marked difference in ϕ_i between extreme shade plants grown in deep shade and sun plants grown under a high light regime (e.g., BJÖRKMAN 1968a; BJÖRKMAN et al. 1972a, b; see Fig. 3.1). Comparison of ϕ_a for a number of C_3 and C_4 species (EHLERINGER and BJÖRKMAN 1977) did not indicate any significant variation in ϕ_a within each photosynthetic type when measurements were made under identical conditions. Finally, as was already demonstrated by GABRIELSEN (1948) (Fig. 3.4), variation in ϕ_i among plants having widely different chlorophyll contents (ranging from 21 mg chl m^{-2} in an *aurea* mutant of *Sambucus canadensis* to 870 mg chl m^{-2} in *Populus canadensis*) can be largely accounted for by the effect of chlorophyll content on leaf absorptance alone.

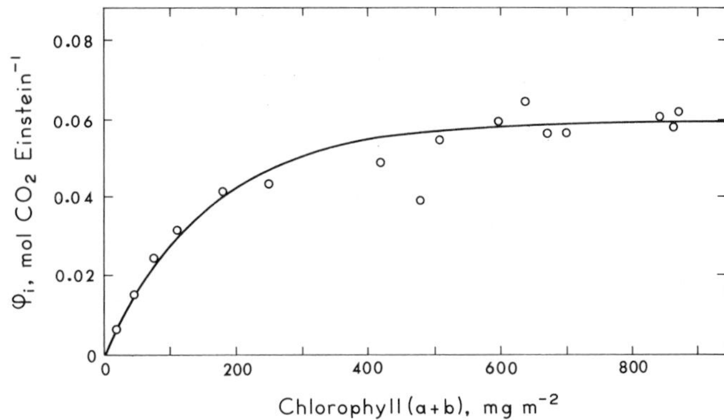

Fig. 3.4. Quantum yield of photosynthetic CO_2 uptake for leaves of differing chlorophyll contents. Each *data point* represents a different C_3 species or variety. (Adapted from GABRIELSEN 1948)

It seems justified to conclude that ϕ_a is essentially the same in normal healthy leaves of higher plants, regardless of species and growth light regime. Exceptions to this general rule are that (1) depending on conditions there may be differences in ϕ_a between C_3 and C_4 plants (see below), (2) ϕ_a may be strongly reduced in shade plants subjected to excessively high light levels (Sect. 3.5), or (3) in plants subjected to other stresses such as low leaf water potential (MOHANTY and BOYER 1976; MOONEY et al. 1977) and excessively high or low temperature (for a review see BERRY and BJÖRKMAN 1980).

Since the reactions involved in the trapping of quanta and in their conversion into early photoproducts are intrinsically independent of variations in factors such as temperature and CO_2 pressure, one might also expect that ϕ_a would be independent of these factors. As shown by EHLERINGER and BJÖRKMAN (1977) and KU and EDWARDS (1978), and as illustrated in Fig. 3.5, ϕ_a determined at low O_2 pressure is indeed essentially independent of temperature and CO_2 pressure in both C_3 and C_4 plants. Under these conditions, C_4 plants have about one-third lower ϕ_a than do C_3 plants. This is in accordance with the higher energy cost for CO_2 fixation via the C_4 pathway. (C_4 photosynthesis theoretically requires 5 ATP and 2 NADP, while C_3 photosynthesis requires 3 ATP and 2 NADP per CO_2 fixed.)

In the presence of normal O_2 pressure ϕ_a in C_3 plants is, however, strongly CO_2-dependent (Fig. 3.5 left). In normal air (210 mbar O_2, 330 μbar CO_2), ϕ_a is lower than it is at low O_2 pressure or high CO_2 pressure. ϕ_a also becomes temperature-dependent, declining by about 30% when the temperature is raised from 15 to 35 °C. EHLERINGER and BJÖRKMAN (1977) attributed these interactive effects of O_2 pressure, CO_2 pressure, and temperature on ϕ_a to their influence on the balance between the carboxylase and oxygenase activities of the bifunctional enzyme RuBP carboxylase-oxygenase. Under conditions where the rate of NADPH and ATP generation is strictly light-limited, oxygenation of RuBP will inevitably result in a lowering of ϕ_a. The absence of a significant effect of temperature, and of CO_2 and O_2 pressure on ϕ_a in C_4 plants is consistent

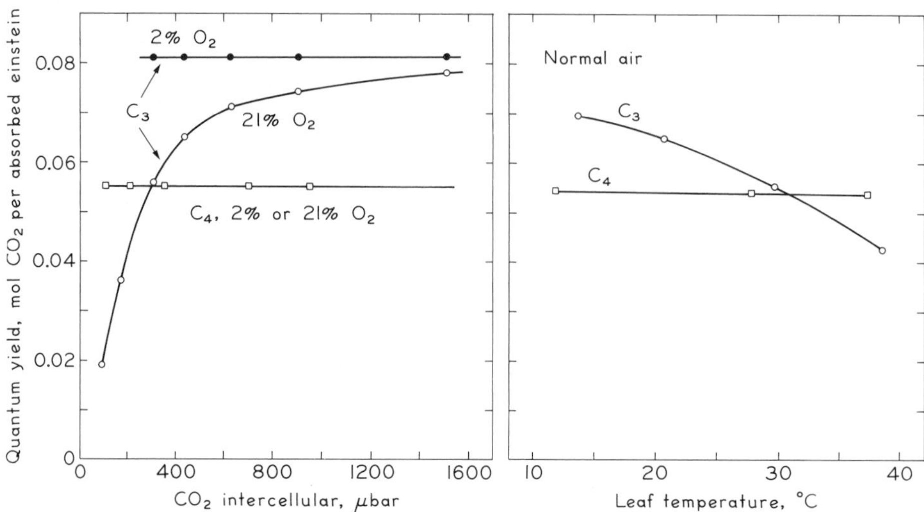

Fig. 3.5. Effect of intercellular CO_2 pressure and leaf temperature on the quantum yield of photosynthetic CO_2 uptake (ϕ_a) in C_3 and C_4 plants. The CO_2 and temperature dependence of ϕ_a was determined at 30 °C and 325 μbar CO_2, 210 mbar O_2, respectively. (Based on data of EHLERINGER and BJÖRKMAN 1977)

with the concept that C_4 photosynthesis functions as a CO_2-concentrating mechanism, permitting the RuBP carboxylase/oxygenase in C_4 plants to operate at a CO_2 pressure sufficiently high to inhibit the oxygenase reaction.

The above results have important implications on the efficiency of light utilization by C_3 and C_4 plants in nature. High temperature evidently leads to a decrease in the quantum yield in C_3 plants but has no such effect in C_4 plants. However, C_3 plants are superior at low temperatures and only above 30 °C does this situation start to be reversed. This may provide an explanation for the finding that C_4 plants are rarely found in densely shaded habitats even at lower latitudes, and could at least in part also account for the scarcity of C_4 plants at high latitudes, among winter annuals, or in many other cool situations where a large fraction of the leaves in the canopy operates under rate-limiting quantum flux densities for much of the day (cf. EHLERINGER, 1978).

3.3.3 Cost of the Photosynthetic and Respiratory Systems

For a plant photosynthesizing under very low quantum flux densities it is of utmost importance that the cost of producing and maintaining its leaves in relation to the rate of photosynthesis it may achieve under these conditions be kept as low as possible. While plants subjected to high light levels in order to make efficient use of the available light must have a high content of certain chloroplast components needed to cope with a rapid flow of electrons, one would expect that plants photosynthesizing in weak light could achieve maximum efficiency of light utilization with a considerably lower content of these components. The relationship between these components and photosynthetic capacity at high light levels will be discussed in Section 3.4. Here they will be considered in the context of cost to the plant.

Leaves of shade plants have considerably less total protein per total chlorophyll (and per leaf area) than do sun leaves of sun plants. A comparison between leaves of 10 shade species, growing in their native rain forest or redwood forest habitats with those of 13 sun species, growing in a range of open habitats (GOODCHILD et al. 1972), showed that the soluble protein content of shade plant leaves was only 15 to 25% of that of sun plant leaves with similar chlorophyll contents. The soluble protein content also falls with decreased light level for a given sun plant. For example, *Atriplex triangularis*, grown at 5.3 E m^{-2} day^{-1}, had only 50% as much soluble protein as those grown at 53 E m^{-2} day^{-1} with little difference in chlorophyll content (BJÖRKMAN et al. 1972b). A major part of these savings in soluble protein is attributable to a much lower content of RuBP carboxylase in shade plants (Table 3.6). Presumably, also the content of other enzymes of photosynthetic carbon metabolism is reduced in shade plants. That this may be the case is indicated by the lower activities of several such enzymes in leaves of C_4 plants, grown at a low in comparison with high light regime (Sect. 3.4.1), although the contribution of these enzymes to the total soluble protein fraction is not known.

It seems probable that shade plants may also have a somewhat lower content of nonsoluble proteins associated with the chloroplast membranes. It is generally considered that roughly one-half of the total chloroplast protein is associated with these membranes, but little is known about the effect of light on the level of these proteins, or how it might differ between sun and shade plants. Significant savings may also be realized by the reduced content in shade leaves of photosynthetic electron carrier proteins (Sect. 3.4.2). The protein component of the chlorophyll-protein complexes and the chloroplast membrane lipids constitute a large fraction of the total investment in chloroplast constituents (e.g., PARK and PON 1963). A reduction in the amount of these proteins and lipids in relation to chlorophyll in shade plants could thus confer substantial savings. However, no comparisons have been made between shade and sun plants as to the content of these constituents, and in the absence of contrary evidence it seems reasonable to assume that the content of these constituents is directly proportional to the chlorophyll content of the leaf. It might also seem that some savings could be achieved if shade plants had fewer photosynthetic reaction centers in relation to the total chlorophyll content of the leaf. However, as discussed in Section 3.4.3, there is no evidence that substantial differences exist between sun and shade plants in this respect, and estimates show that the additional cost of maintaining a constant ratio of reaction centers to chlorophyll, instead of reducing this ratio in proportion to the reduction in electron carriers or RuBP carboxylase, is very small, amounting to 1% or less of the total protein content of the leaf.

As mentioned earlier, shade plants have a very low rate of dark respiration, a characteristic which is of paramount importance in the maintenance of a positive carbon balance in shaded habitats. Although such low respiratory rates could be caused by regulation of respiration (primarily determined by the demand for ATP required for heterotrophic biosynthesis), it seems likely that, at least in part, it is associated with a lower content of respiratory machinery than in sun plants. If the latter situation exists, then the shade plants would presumably incur a significantly lower cost also in the production and maintenance of the respiratory system.

Further studies are obviously needed before an accurate quantitative assessment can be made of the relative energy costs incurred in producing and maintaining the photosynthetic and respiratory systems of sun and shade leaves. However, it is clear that shade leaves are considerably less costly than sun leaves because of their lower content of a number of components which are important to the capacity of photosynthesis at high light levels, but have little or no influence on the rate of photosynthesis at low light levels.

3.3.4 Distribution of Dry Matter in the Plant

The productivity of a plant, especially in a low light environment, depends not only upon the net rate by which the photosynthesizing leaves are able to acquire energy and carbon on the basis of unit leaf area, but it is also strongly dependent on the total leaf area displayed to the light, as well as on the allocation of photosynthate to photosynthetic versus nonphotosynthetic tissue. In a low light environment it is obviously imperative that the photosynthetically active area per total plant mass be as high as possible, while at the same time mutual shading among the leaves be minimized.

Maximization of photosynthetically active area per total plant mass may be achieved in several different ways. One category of such adjustment involves an increased specific leaf area, i.e., the ratio of leaf area (L) to total leaf dry matter (W_L). Another category involves an increased leaf weight ratio, i.e., the fraction of the total plant mass (W_P) which is distributed to leaf material (W_L). The relationship between the different growth parameters is given by the following expression

$$RGR = NAR \times LAR = NAR \times LWR \times SLW$$

where RGR = relative growth rate $\left(=\frac{1}{W_P}\frac{dW_P}{dt}\right)$; NAR = net assimilation rate $\left(=\frac{1}{L}\frac{dW_P}{dt}\right)$; LAR = leaf area ratio $\left(=\frac{L}{W_P}\right)$; LWR = leaf weight ratio $\left(=\frac{W_L}{W_P}\right)$; and SLW = specific leaf area $\left(=\frac{L}{W_L}\right)$.

NAR = mean daytime photosynthetic rate × hours of light − mean respiration rate × total hours.

Examples of the first category of leaf area adjustment are given in Fig. 3.8 which shows the effect of daily quantum flux on specific leaf weight, W_L/L (i.e., the reciprocal of specific leaf area) in *Atriplex triangularis* and *Fragaria virginiana*. Numerous studies show that this type of response is almost universal among both sun and shade species, although the extent of such changes may show marked species differences, and shade plants do not necessarily have particularly high specific leaf areas (see Sect. 3.4.5). It seems probable that the major factor contributing to an increased specific leaf area in response to shading is a reduction in several components of the photosynthetic system which govern the capacity at high quantum flux densities. However, it should be emphasized that changes in specific leaf area are also likely to involve changes in the proportion of photosynthetically inactive to photosynthetically active *leaf* material. Although little quantitative information exists on this subject, rough estimates indicate that a reduction in the pool of photosynthate, mainly sugars and starch, could perhaps account for up to a 20% increase in specific leaf area. A reduction in other photosynthetically inactive components such as epidermal tissue, cell walls, and vascular tissue could perhaps cause a similar increase. Since the energy load, the requirement for mechanical strength, and the rate of water transport are much reduced in shaded situations, such savings are unlikely to impose any significant disadvantageous effects.

Most studies seem to indicate that although shading tends to increase the proportion of the total dry matter which is distributed to the leaves (leaf weight ratio, W_L/W_P), generally this influence is small, especially in comparison with the effect on specific leaf area (e.g., BLACKMAN and BLACK 1959 for *Helianthus annuus* and *Lathyrus maritimus;* COOPER 1967 for *Medicago sativa* and *Lotus corniculatus;* DOLEY 1978 for *Eucalyptus grandis;* EVANS and HUGHES 1961 for *Impatiens parviflora;* HUGHES and COCKSHULL 1971 for *Chrysanthemum morifolium;* LEDIG et al. 1970, and LOACH 1970 for seedlings of a number of tree species). However, this behavior does not extend to all species. For example, LOACH (1970) showed that in shade-tolerant *Acer rubrum* and the moderately shade-tolerant *Quercus rubra* there was a substantial increase in leaf weight ratio with decreased light level, and WHITEHEAD (1973) reported that in shade-tolerant *Filipendula ulmaria* and *Iris pseudacorus* as the light level was decreased a proportionately greater amount of photosynthate was devoted to the formation of leaf material. Similarly, in *Chamaenerion angustifolium,* owing to an increased leaf weight ratio in response to shading, the same amount of carbon was accumulated over a period of time by plants grown at 100%, 70%, and 40% daylight (MYERSCOUGH and WHITEHEAD 1966). By contrast, in shade-tolerant species such as *Datura stramonium* (WHITEHEAD 1969) and *Amaranthus spinosus* (SINGH and GOPAL 1973), there was little or no diversion of assimilates to compensate for the lowered photosynthetic rate and, as shown below, in the shade-intolerant *Helianthus annuus* shading actually decreased the leaf area ratio.

Apparently, increased allocation to leaf growth in response to shading may largely be at the expense of root growth. A reduced root weight ratio may not be harmful in shaded locations with adequate nutrient levels and favorable water relations but is likely to have serious consequences where this is not the case. In the study with the shade-tolerant species, *Filipendula ulmaria,* referred to in the preceding paragraph, the increased allocation to leaf growth in response to shading was accompanied by a corresponding decrease in allocation to root growth and shading of *Chamaenerion angustifolium* reduced the root weight

Table 3.2. Growth parameters for shade-tolerant *Impatiens parviflora* and shade-intolerant *Helianthus annuus* under different levels of shading. (Data for *I. parviflora* and *H. annuus* derived from EVANS and HUGHES 1961 and HIROI and MONSI 1963, respectively)

Growth parameter		Relative quantum flux density, percent[a]				
		100	50	22	10	5
Net assimilation rate, $g\ m^{-2}\ week^{-1} \times 10^{-3}$	*Impatiens*	61	52	31	20	12
	Helianthus	68	55	29	20	5
Specific leaf area, $m^{-2}\ g^{-1} \times 10^{-3}$	*Impatiens*	32	42	53	71	80
	Helianthus	26	32	43	41	36
Leaf weight ratio, $g\ g^{-1}$	*Impatiens*	0.41	0.43	0.44	0.45	0.45
	Helianthus	0.61	0.57	0.54	0.47	0.46
Leaf area ratio, $m^{-2}\ g^{-1} \times 10^{-3}$	*Impatiens*	13.2	18.0	23.5	32.0	36.0
	Helianthus	16.4	19.0	22.0	19.0	17.0
Relative growth rate, $g\ g^{-1}\ week^{-1}$	*Impatiens*	0.80	0.93	0.73	0.64	0.43
	Helianthus	1.10	1.01	0.63	0.37	0.09

[a] Estimated daily quantum flux densities (100%) were comparable in the two experiments (40 to 45 E $m^{-2}\ day^{-1}$)

ratio to less than half that of unshaded plants. In the studies with *Impatiens parviflora,* discussed in following paragraphs, root weight ratio showed a consistent decrease with shading, ranging from 0.44 in full daylight to 0.31 in 5% daylight. However, only a part of this saving was allocated to leaf growth, for shading caused an increase in stem weight ratio from 0.15 in full daylight to 0.24 in 5% daylight. It should also be emphasized that the response of root weight ratio to shading is likely to be strongly influenced by nutrient and water relations.

The comparison between the responses to shading in the woodland, facultative shade species, *Impatiens parviflora,* and the obligate sun species, *Helianthus annuus* (Table 3.2), provides a good illustration of the importance of compensating changes in leaf area ratio. These results were obtained with young seedlings during a 3- to 4-week period following expansion of the first foliage leaves, grown in the field under different degrees of shading imposed by screens.

The net assimilation rates of these two species were rather similar over a wide range of daily irradiance receipts. Only at the heaviest shade was there a pronounced difference in net assimilation rate, presumably attributable to a lower respiration rate in *Impatiens* than in *Helianthus.*

Shading resulted in an increased specific leaf area in both species. In *Helianthus* maximum specific leaf weight (165% of unshaded controls) was reached under the 22% light regime; further shading led to a decline in specific leaf area and under the 5% light regime it was only about one-third higher than under unshaded conditions. By contrast, *Impatiens* continued to increase its specific leaf area with shading so that under the 5% light regime the value was about 2.5 times that of the unshaded controls.

Table 3.3. Comparisons of relative growth rates and the proportion of dry weight growth of the leaves to the whole plant ($\Delta W_L/\Delta W_P$) in 4- to 5-week-old plants of *Helianthus annuus* (shade-intolerant), *Phaseolus aureus* (moderately shade-tolerant) and *Impatiens parviflora* (shade-tolerant). (After HIROI and MONSI 1964)

	Relative quantum flux density, percent			
	100	50	22	10
Relative growth rate				
Helianthus	1.01	0.71	0.28	Negative
Phaseolus	0.92	0.81	0.62	0.34
Impatiens	0.80	0.93	0.73	0.54
($\Delta W_L/\Delta W_P$)				
Helianthus	0.31	0.38	0.16	None
Phaseolus	0.59	0.47	0.46	0.35
Impatiens	0.47	0.45	0.47	0.55

In addition to these large species differences in the response to shading with respect to specific leaf area there were also smaller, but important differences in the response of dry matter distribution to the leaves. In *Impatiens*, shading resulted in a slight but significant increase in the leaf weight ratio (about 10% greater in dense shade) whereas in *Helianthus*, shading caused a progressive decline in photosynthate allocation to the leaves, so that in dense shade the leaf weight ratio was only about 75% of that found in the unshaded controls. The combination of these changes in specific leaf area and leaf weight ratio resulted in large differences in leaf area ratio. *Impatiens* plants, grown under the 5% light regime, had almost three times as high leaf area ratio as the unshaded control plants, whereas there was no significant difference in leaf area ratio between *Helianthus* plants grown under the 5% and the 100% light regimes.

These differences in the responses of specific leaf weight and leaf weight ratio to shading provide an explanation for the large difference in shade tolerances between the species. Because of these compensating changes in *Impatiens*, the relative growth rate under the 10% light regime was as high as 80% of that under the 100% regime, even though the corresponding net assimilation rate was only 33% of the unshaded control. In *Helianthus*, the effect of shading on relative growth rate was nearly the same as the effect on net assimilation.

The data for *H. annuus*, shown in Table 3.2, represent means for the period between the first and the fourth week after germination. The shade intolerance of this species becomes increasingly pronounced with advancing plant age. During the fourth and fifth week the relative growth rate fell sharply under the lower light treatments and, as shown in Table 3.3, under the 22% regime it fell to about 28% of that for the unshaded plants and became negative under the 10% (and 5%) regimes. This time-dependent decline in relative growth rate was associated with a decreased allocation of photosynthate to new leaf growth (expressed as $\Delta W_L/\Delta W_P$), ultimately leading to premature senescence of the leaves. By comparison, the moderately tolerant *Phaseolus aureus* maintained relatively high growth rates and allocation to new leaf growth under the 22% regime and these parameters were still positive under the 10% regime. In the highly shade-tolerant *I. parviflora,* there was little time-dependent decline in relative growth rate even under the 10% regime and allocation of photosynthate to new leaf growth remained at a high level.

3.4 Factors Determining Photosynthetic Performance in Strong Light

The leaf factors which limit the photosynthetic rate of a leaf at high quantum flux densities are different from those which limit the rate at low quantum flux densities. One would expect, on theoretical grounds, light-saturated photosynthesis (P_s) to be independent of those factors which determine the efficiency of light absorption (i.e., chlorophyll and carotenoid pigments) and of the conversion of the absorbed light into early photoproducts. Thus, the main effect of a decreased chlorophyll content (or the presence of a reflective coating on the leaf surface) would be to increase the incident light level required to reach light saturation but it would not affect P_s. As was already recognized by WILLSTÄTTER and STOLL (1918) and discussed in detail by GABRIELSEN (1960), species variation in P_s cannot be explained by differences in chlorophyll content. More

Table 3.4. Correlation and linear regression between light-saturated photosynthesis in normal air (Y variable), μmol CO_2 m^{-2} s^{-1} and chlorophyll content, P-700 content, RuBP carboxylase activity, soluble protein content, and specific leaf dry weight among various C_3 species grown under different regimes. LL=grown under low, HL=grown under high, light regime. Su=sun species, Sh=shade species. R=coefficient of correlation, N=number of pairs, $\bar{X}$=mean of X-variable, $\bar{Y}$=mean of Y-variable (photosynthetic rate), M=slope and B=Y-intercept of regression line

X Variable	Treatment, plant types	R	N	$\bar{X} \pm SD$	$\bar{Y} \pm SD$	M	B
Chlorophyll (a+b), mg m^{-2}	LL, HL, Su, Sh	0.35	49	485±120	14.1±9.7	0.029	0.03
	HL, Su	0.21	19	534± 89	23.0±8.8	0.021	11.9
	LL, Su, Sh	0.31	17	453+148	5.9+3.4	0.0072	2.7
P-700, μmol m^{-2}	LL, HL, Su, Sh	0.32	27	1.36±0.44	14.2±9.3	6.78	5.0
RuBP carboxylase, μmol m^{-2} s^{-1}	LL, HL, Su, Sh	0.96	23	18.87±11.47	13.5±8.0	0.664	1.01
Soluble protein, g m^{-2}	LL, HL, Su, Sh	0.86	20	4.96±3.06	14.6±10.2	2.06	0.45
Spec. leaf wt., g dry wt. m^{-2}	LL, HL, Su, Sh	0.31	38	39.3±21.8	16.5±10.1	0.144	10.87
	HL, Su	−0.11	16	50.9±22.2	25.5±8.4	−0.040	27.5
	LL, Su, Sh	−0.01	11	24.7±12.2	6.8±3.4	−0.002	6.9

Table 3.5. Comparison of the effect of growth light regime on the contents of chlorophyll (a+b) and P-700 and on light-saturated net photosynthetic rate in different C_3 species. Total daily quantum flux densities are estimates for the wave band 400–700 nm. The letters C, N, and H following the quantum flux density values denote controlled artificial light, natural daylight, and native habitat, respectively

Species	Growth light regime, E m^{-2} day^{-1}	Chlorophyll content, mg chl m^{-2}	P-700 content, µmol m^{-2}	chl/P-700 mol mol^{-1}	Photosynthesis, µmol s^{-1} m^{-2}	Photosynthesis, µmol s^{-1} g^{-1} chl	Photosynthesis, µmol s^{-1} $µmol^{-1}$ P-700	Reference
Abutilon theophrasti	4.9 C	396	1.23	379	6.5	16.4	5.3	Patterson et al. (1978)
	17.2 C	495	1.39	366	9.7	19.5	7.0	
	40.5 C	543	1.68	335	11.6	21.4	6.9	
Alocasia macrorrhiza	~ 0.3 H	580	1.38	470	3.2	5.5	2.3	Boardman et al. (1972)
Atriplex triangularis	5.3 C	470	1.48	354	6.9	14.7	4.7	Björkman et al. (1972b)
	36.3 C	570	1.79	354	24.0	42.1	13.4	
	53.0 C	570	1.72	369	35.3	61.9	20.5	
Betula verrucosa	3.1 C	233	0.62	419	5.0	21.5	8.1	Öquist et al. (1981)
	15.1 C	285	0.58	545	8.8	30.9	15.2	
	36.7 C	375	0.64	656	11.0	29.3	17.2	
Camissonia brevipes	~45 N	550	1.18	514	37.0	67.3	31.3	Armond and Mooney (1978)
Cordyline rubra	~ 0.3 H	870	2.11	460	3.3	4.7	1.6	Boardman et al. (1972)
Datura meteloides	~45 N	520	1.01	574	19.0	36.5	18.8	Armond and Mooney (1978)
Eichhornia crassipes	4.5 C	523	1.18	480	10.1	19.3	8.6	Patterson and Duke (1979)
	16.1 C	627	1.42	493	16.8	26.8	11.8	
	37.8 C	479	1.00	602	18.5	38.6	18.5	
Glycine max	10.9 C	440	1.54	320	13.0	29.5	8.4	Bunce et al. (1977)
	36.7 C	440	1.73	285	15.9	36.1	9.2	
Gossypium hirsutum	28.4 C	350	1.31	300	17.7	50.6	13.5	Patterson et al. (1977)
	~42.0 N	560	2.44	260	27.1	48.4	11.1	
	4.9 C	415	1.12	403	4.2	10.1	3.8	Patterson et al. (1978)
	17.2 C	452	1.50	370	7.6	16.8	5.1	
	40.5 C	496	1.79	341	13.8	27.8	7.7	
Lupinus sparsiflorus	~45 N	590	1.77	371	28.0	47.5	15.8	Armond and Mooney (1978)
Perityle emoryi	~45 N	340	0.99	383	19.0	55.9	19.2	Armond and Mooney (1978)
Sinapis alba	0.5 C	215	0.98	209	–	–	–	Wild et al. (1973)
	53.0 C	209	0.94	210	–	–	–	
Solanum dulcamara	~ 1.6 N	250	0.77	361	2.5	10.0	3.2	Clough et al. (1979)
	~40.0 N	410	1.37	343	8.3	20.2	6.1	

recent studies support the conclusions reached in earlier work. As shown in Table 3.4, there is a poor correlation between chlorophyll content and P_s, determined for 49 different plants of sun and shade species grown under various light regimes, and as shown in Table 3.5, differences in P_s between low light-grown shade species and high light-grown sun species are sometimes even more pronounced when expressed on the basis of chlorophyll rather than leaf area (e.g., *Cordyline rubra* versus *Atriplex triangularis*). Although in the same species, growth under different light regimes in some instances tended to cause a parallel shift in chlorophyll content and P_s (e.g., in the experiments with *Gossypium hirsutum* by PATTERSON et al. 1977), generally the differences in P_s between sun and shade leaves of the same species are largely unrelated to chlorophyll content.

Some authors have contended that the largest proportion of the difference in P_s between different species and between plants of the same species, grown under high and low light, is related to changes in the number of photochemical reaction centers per unit leaf area expressed in terms of the content of P-700, the reaction center of PS I (e.g., ALBERTE et al. 1976; CLOUGH et al. 1979; Table 3.5). However, the correlation between P_s and P-700 content based on determinations on 27 different plants is similar to that obtained between P_s and chlorophyll content (Table 3.4). Also, the range of values, shown in Table 3.5, for P_s expressed on the basis of P-700 (1.6 to 31.3) is rather greater than that for P_s expressed on a leaf area basis (4.7 to 56). Although in some instances, such as in the data for *Abutilon theophrasti* and *Solanum dulcamara*, differences in P_s between plants grown under high and low light levels are somewhat smaller when expressed on the basis of P-700 than when expressed on a leaf area basis, no such trends are seen in other cases such as *Atriplex triangularis, Eichhornia crassipes*, and *Betula verrucosa*. These results show that the capacity of component steps of photosynthesis which determine the absorption and trapping of light quanta probably has little or no impact on the light-saturated photosynthetic rate.

Leaf factors which may be expected potentially to determine P_s in normal air may be divided into three categories: (1) the capacity of enzymic steps of photosynthetic carbon metabolism, (2) the capacity for photosynthetic electron transport and photophosphorylation, and (3) the conductance to diffusion of CO_2 from the ambient air to the chloroplasts. The first two categories of factors may influence P_s both at saturating and normal CO_2 pressure, whereas the third category can only affect P_s under conditions of limiting CO_2 pressures. Changes in any one of the three categories of factors are likely to be associated with changes in leaf anatomical characteristics.

3.4.1 Enzymes of Photosynthetic Carbon Metabolism

Enzymes involved in the fixation and reduction of CO_2 make up the bulk of the protein content of the chloroplast stroma. A major fraction of this protein is one single enzyme, RuBP carboxylase. Because of the relatively low affinity of RuBP carboxylase for CO_2, and because of the additional function of this enzyme as an RuBP oxygenase, the CO_2-fixing activity of the enzyme at normal atmospheric CO_2 and O_2 pressures in many cases appears to be just about sufficient to support the light-saturated rate of CO_2 fixation by intact leaves in spite of the large amount of enzyme protein present in the chloroplasts. For these reasons, RuBP carboxylase has been implicated as a potential rate-limiting enzyme, and the variation among species in and the effect of different growing conditions on the level of this enzyme has been widely studied during the past decade.

As was mentioned in a previous section of this article, shade plants generally have a much lower level of RuBP carboxylase than sun plants and the concomitantly lower protein content of shade leaves was considered to confer important savings to the plant without affecting the photosynthetic rate at low light intensities. It was also suggested that the failure of obligate shade species to attain high rates of light-saturated photosynthesis may at least in part be attributable to a genetically low capacity to synthesize high levels of RuBP carboxylase.

Table 3.6 compares the activity of RuBP carboxylase with the light-saturated photosynthetic rate in normal air determined on the same leaves for a number of C_3 sun and shade plants, grown under different light regimes. Only those studies, which have been made with similar enzyme assay methods and conditions have been included in Table 3.6. There is very high degree of correlation ($R=0.96$) between RuBP carboxylase level and P_s for this group of plants and there is also a close relationship between these two parameters for any given species grown under different light regimes. Similarly close relationships between the effect of different growth light regimes on RuBP carboxylase and P_s have been found in studies on a number of other C_3 species, not included in Table 3.6, e.g., *Phaseolus vulgaris* (Powles 1979), *Sinapis alba* (Wild and Höhler 1978), *Hordeum vulgare* (Blenkinsop and Dale 1974), and *Lolium multiflorum* (Reyss and Prioul 1975).

Close correlations between changes in P_s and RuBP carboxylase have also been observed in studies where factors other than growth light regime were used as the experimental variable. Reduction in nitrogen nutrient supply caused a similar decline in P_s and RuBP carboxylase level (Medina 1970, 1971). Partial defoliation of plants resulted in a parallel increase in P_s and RuBP carboxylase activity in the remaining leaves (Neales et al. 1971) and Woolhouse (1967) showed that decline of P_s with increasing leaf age was accompanied by a similar decline in the content of Fraction-I protein which is synonymous with RuBP carboxylase. Some authors (e.g., Blenkinsop and Dale 1974; Wild and Höhler 1978) have argued that although there was a close correlation between P_s and RuBP carboxylase activity, the latter was too high to limit P_s. However, these authors compared the maximum velocity of the enzyme reaction at saturating concentrations of CO_2 (and RuBP) with the net photosynthetic rate of leaves at normal ambient CO_2 pressure. It is probable that if they had taken into account the K_m (CO_2) for the enzyme and compared the calculated rate of enzyme activity at CO_2 pressures corresponding to those occurring in photosynthesizing leaves exposed to normal air with P_s, they would have found that the enzyme activities were rather lower than P_s.

Table 3.6. Comparison of light-saturated photosynthetic rate (in normal air) and RuBP carboxylase activity of sun and shade plants grown at different light regimes. Enzyme activity has been standardized to 25 °C using an activation energy of −14.2 kcal mol^{-1} K^{-1}. H=natural habitat, N=natural daylight, C=controlled artificial light

Species	Growth light regime E m^{-2} day^{-1}	Photosynthesis µmol m^{-2} s^{-1}	RuBP carboxylase activity at 25 °C µmol m^{-2} s^{-1}	Reference
Adenocaulon bicolor	~0.5 H	1.4	3.3	BJÖRKMAN (1968a)
Aralia californica	~0.5 H	2.0	3.3	BJÖRKMAN (1968a)
Atriplex triangularis	5.3 C	6.9	11.7	BJÖRKMAN et al. (1972b)
	36.3 C	24.0	36.0	
	53.0 C	35.3	48.5	
Echinodorus berteroi	~45 H	13.5	29.3	BJÖRKMAN (1968a)
Glycine max	~13.5 N	12.6	11.1	SINGH et al. (1974)
	~22.5 N	17.5	20.0	
	~31.5 N	17.7	22.1	
	~45.0 N	22.1	25.5	
Mimulus cardinalis	~45.0 N	23.7	33.3	BJÖRKMAN (1968a)
Plantago lanceolata	~45.0 H	14.2	19.6	BJÖRKMAN (1968a)
Solanum dulcamara, sun clone	5.3 C	11.7	15.6	GAUHL (1976)
	24.3 C	21.7	34.9	
Solidago virgaurea, sun clones	5.3 C	12.1	19.0	BJÖRKMAN (1968b)
	26.5 C	17.0	25.0	
Solidago virgaurea, shade clones	5.3 C	11.0	13.5	
	26.5 C	10.3	16.3	
Trillium ovatum	~0.5 H	2.0	3.3	BJÖRKMAN (1968a)
Triticum aestivum	~13.5 N	6.9	6.8	SINGH et al. (1974)
	~22.5 N	6.8	9.4	
	~31.5 N	9.5	12.0	
	~45.0 N	11.4	14.5	

It is noteworthy in this connection that recent studies in which the pool size of RuBP during photosynthesis was determined at different light intensities and CO_2 and O_2 pressures (COLLATZ 1978; HITZ and STEWART 1980) suggest that at high CO_2 pressures the rate of RuBP regeneration rather than RuBP carboxylase activity may become rate-limiting to P_s. Hence, as the CO_2 pressure is increased limitation of the CO_2 fixation rate may gradually shift from RuBP carboxylase to other component steps such as electron transport (Sect. 3.4.2). This would inevitably result in a P_s versus CO_2 pressure relationship which deviates from that predicted on the basis of the Michaelis-Menten equation for a single substrate reaction and may well provide an explanation for the shape of the experimentally determined relationship between P_s and intercellular CO_2 pressure. In the past, such deviations of actual CO_2 response curves from those based on simple analog models have often been interpreted to mean that conductance to physical diffusion of CO_2 between the leaf intercellular spaces imposes a major limitation to P_s.

Leaves of C_4 plants tend to have higher rates of light-saturated photosynthesis in normal air than do those of C_3 plants. This difference becomes especially marked when P_s is related to RuBP carboxylase level which generally is markedly lower in C_4 than in C_3 plants. These observations are in agreement with the concept that the C_4 pathway serves as a metabolic CO_2 concentrating mechanism which enables the RuBP carboxylase to operate at or near CO_2 saturation even at normal CO_2 and O_2 pressures. Hence, although the RuBP carboxylase is relatively low, P_s may be relatively high. Another consequence of the CO_2-concentrating mechanism is that in C_4 plants the responses of P_s to light intensity and temperature resemble that exhibited by C_3 plants at elevated CO_2 pressures. (For reviews, see BERRY and BJÖRKMAN 1980 and OSMOND et al. 1980.) Although few experiments have been made to determine the effect of growth light regime on RuBP carboxylase level in C_4 plants, it appears that it is similar to that found in C_3 plants (SINGH et al. 1974).

Little information is available concerning the effect of different growth light regimes on enzymes of photosynthetic carbon metabolism other than RuBP carboxylase. A large part of the difference in soluble protein content between sun and shade plants (Sect. 3.3.3) and the high correlation between soluble protein and P_s (Table 3.4) may well be caused by differences in RuBP carboxylase level. It seems unlikely, however, that this enzyme alone could account for all of the difference in soluble protein content, suggesting that sun leaves may have an increased level of some other carbon metabolism enzymes as well. Some support for this supposition is provided by the finding that in the C_3 species *Lolium multiflorum*, grown under different light regimes, the activities of carbonic anhydrase and PEP carboxylase increased in approximately the same proportion as RuBP carboxylase (REYSS and PRIOUL 1975) and in the two C_4 species *Amaranthus palmeri* and *Zea mays*, the activities of several enzymes associated with the C_4 pathway of CO_2 fixation increased in response to increased light during growth (HATCH et al. 1969).

3.4.2 Electron Transport Capacity

Increased rate of carboxylation requires an increased rate of pyridine nucleotide reduction and ATP formation. Both of these compounds are needed to convert the carboxylation product, 3-phosphoglycerate, to sugars, starch, and other endproducts of photosynthesis and to regenerate the CO_2 acceptor, RuBP. In C_3 plants photosynthesizing in normal air, additional reducing power and ATP are required in the metabolism of phosphoglycerate, formed in oxygenation of RuBP. In C_4 plants, the oxygenation presumably occurs at a very low rate, but, instead, additional ATP is required to regenerate the initial CO_2 acceptor, phosphoenolpyruvate. Increased photosynthetic rate may therefore require that the increased carboxylation capacity of sun as compared with shade plants be accompanied by an increased capacity for photosynthetic electron transport.

It is now well established that increased quantum flux density during growth substantially increases the capacity for light-saturated electron transport (on a chlorophyll basis) in a number of unrelated sun species of higher plants such as *Atriplex triangularis* (BJÖRKMAN et al. 1972b), *Betula verrucosa* (ÖQUIST et al. 1981), *Phaseolus vulgaris* (POWLES 1979), *Picea sitchensis* (LEWANDOWSKA et al. 1976), *Sinapis alba* (WILD et al. 1975) and *Teucrium scorodonia* (MOUSSEAU et al. 1967) as well as in the green alga, *Scenedesmus obliquus* (FLEISCHHACKER and SENGER 1978). It is further evident that obligate shade plants such as *Alocasia macrorrhiza* and *Cordyline rubra* have much lower electron transport capacities than high light-grown sun species such as *Atriplex triangularis* (BJÖRKMAN et al. 1972b; BOARDMAN et al. 1972).

A number of studies show that such differences in electron transport capacity are associated with an increased content of certain carriers of the electron transport chain. BJÖRKMAN et al. (1972b) and BOARDMAN et al. (1972) found that the amount of cytochrome f (on a chlorophyll basis) was about 2.1 times higher in chloroplasts of *Atriplex triangularis* plants grown at 50.3 E m^{-2} day^{-1} than from plants grown at 5.3 E m^{-2} day^{-1} and about 3.4 times higher than in leaves of shade-grown *Alocasia macrorrhiza* (Table 3.7). These results are consistent with the results (BJÖRKMAN et al. 1972b) that large differences exist in the light-saturated capacity but there are no significant differences in the quantum yield for electron transport between *Atriplex* leaves grown under different light regimes.

GRAHL and WILD (1975) reported similar differences in cytochrome f content and correlation between this content and electron transport capacity in *Sinapis alba* leaves grown under low and high light regimes (Table 3.7). Recent studies on plants of 14 different genera, representing three unrelated families, showed that high light intensity during growth in all cases resulted in a substantial increase in the ratio of cytochrome f to chlorophyll (WILD 1979). Increased light level during growth also increases the content of cytochrome b-559 and b-563 which are also associated with photosynthetic electron transport although these changes apparently are not quite as large as the change in cytochrome f content (BJÖRKMAN et al. 1972b; WILD et al. 1973).

Other components of the electron transport chain that increase with the light level during growth include the intermediate pool, P, which is probably

Table 3.7. Comparison of the effect of growth light regime on the molar ratio of chl (a+b) to Cyt f and molar ratios of P-700 and Cyt f to P-700

Species	Growth light regime, E m^{-2} day	chl/cyt f	P-700/cyt f	Chl/P-700	Reference
Alocasia macrorrhiza	~0.3 H	1,120	~2.5	470	BOARDMAN et al. (1972)
Atriplex triangularis	5.3 C 36.3 C 53.0 C	689 509 330	2.2 1.6 1.0	354 354 369	BJÖRKMAN et al. (1972)
Asarum europaeum	"Low"	–	~3.0	–	GRAHL and WILD (1975)
Betula verrucosa	3.1 C 15.1 C 36.7 C	1,346 615 605	3.2 1.1 0.9	419 545 656	ÖQUIST et al. (1981)
Cordyline rubra	~0.3 H	950	~2.5	460	BOARDMAN et al. 1972
Galium odorata	"Low"	–	~3.0	–	GRAHL and WILD (1975)
Scenedesmus obliquus	2.0 C 11.1 C	848 334	~2.5 ~1.0	~330 ~310	FLEISCHHACKER and SENGER (1978)
Sinapis alba	0.5 C 53.0 C	750 280	2.5–3.0 1.0	250 250	GRAHL and WILD (1975)

identical to the reactive plastoquinone pool involved in electron transport (BJÖRKMAN et al. 1972b), the total content of lipoquinones (GRAHL and WILD 1972, 1973; LICHTENTHALER 1971) and ferredoxin (WILD et al. 1972). Very recent results indicate that the content of ATP synthetase also increases with increased light level during growth (BERZBORN and ROOS 1980). As mentioned previously, growth light regime had only a small effect on the content of P-700 and there were no marked differences between sun and shade species in this respect (Table 3.5). As a consequence of the differential effect of light regime on P-700 and cytochrome f the molar ratio between those two components may be as high as 3:1 in shade leaves, while it is approximately 1:1 in sun leaves (Table 3.7).

It is noteworthy that studies on the unicellular green alga, *Scenedesmus obliquus*, yielded results similar to those obtained with higher plants (FLEISCHHACKER and SENGER 1978; SENGER and FLEISCHHACKER 1978). In the cultures grown under a high light level, the light- and CO_2-saturated photosynthetic rate per chlorophyll was three times higher than in the cultures grown under a low light regime, but the quantum yields were equal. Electron transport capacity and RuBP carboxylase level changed in parallel to the photosynthetic capacity, PS II-driven electron transport showing the closest correspondence with photosynthesis. The molar content of cytochrome f was 2.5 times larger and that of reactive plastoquinone more than 3 times larger in the high light intensity cultures. There was no significant difference in the contents of P-700 (Table 3.7) or in PS II reaction centers (estimated from "titrating" the inhibition of PS II activity with DCMU).

3.4.3 A Model of Sun and Shade Leaves

The illustrative model of the possible organization of the photosynthetic system in sun and shade species shown in Fig. 3.6 attempts to take into account the experimental results discussed above. Since cytochrome f appears to be the electron carrier present in the smallest amount, it is considered here that the number of electron transport chains is equal to the number of cytochrome f molecules. Prominent features of this model are that in the shade leaf, each chain is served by a larger number of PS II and PS I units and the ratio of PS II to PS I units is higher than in the sun leaf. As a result, the total amount of chlorophyll serving each chain is much greater and the LHchl complex constitutes a somewhat larger proportion of the total chlorophyll in the shade than in the sun leaves. Since the LHchl complex contains all of the chl b, an increase in the proportion of this complex leads to a somewhat higher overall chl a/chl b ratio. A higher PS II/PS I ratio would also tend to increase the ratio of total chl to P-700. An alternative explanation for the observed shift in the chl a/chl b and chl/P-700 ratios is that the amount of LHchl complex per PS II increases with decreased growth light level with no concomitant change in the amount of PS II.

An increased PS II/PS I (or LHchl/PS I) ratio may be related to an increased proportion of densely stacked grana thylakoids relative to stroma thylakoids.

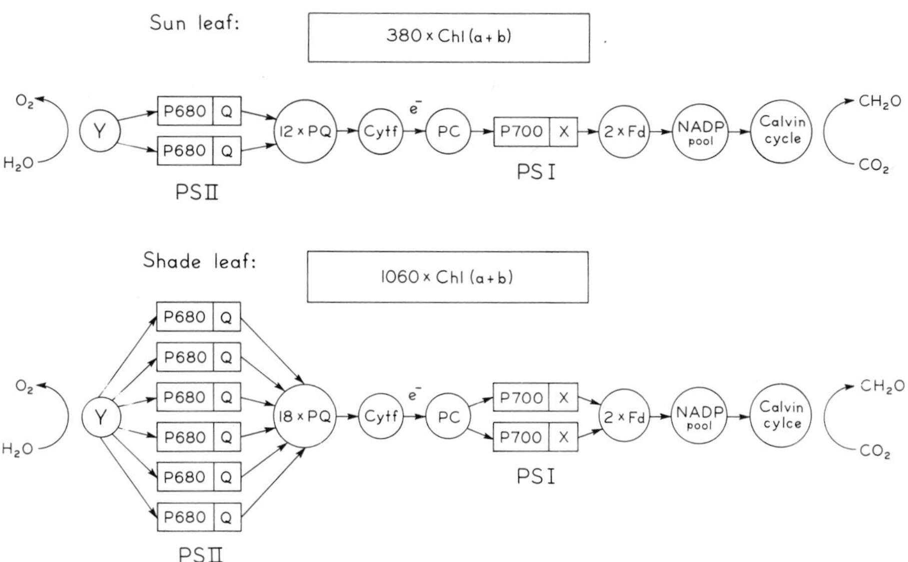

Fig. 3.6. Schematic representation of possible organization of the photosynthetic system in sun and shade leaves. Contents of the various components are per electron transport chain. The PS II/PS I ratios of 2 and 3 for the sun and the shade leaves, respectively, are arbitrary. These numbers were chosen to indicate that the ratios between the two photosystems need not be unity and that shade leaves may have a higher PS II/PS I ratio than sun leaves. It is assumed that the number of chl a molecules associated with the "core" of each of the photosystems, the ratio of LHchl to PS II, and the ratio of chl a to chl b in the LHchl complex are the same in sun and shade leaves. This yields overall chl a/chl b ratios of 2.80 and 2.53 for the sun and shade leaves, respectively. *Y* water splitting complex; *P-680* reaction center of PS II; *Q* primary electron acceptor of PS II; *PQ* plastoquinone pool, Cyt f cytochrome f; *PC* plastocyanin; *P-700* reaction center of PS I; *X* primary electron acceptor of PS I; *Fd* ferredoxin

The grana thylakoids are widely considered to contain all of LHchl, PS II (and much of the PS I) whereas the stroma thylakoids only contain PS I and the chl a associated with it. As mentioned in Section 3.3.1 shade-leaf chloroplasts evidently have a greater development of the grana and a smaller proportion of stroma thylakoids than sun-leaf chloroplasts (Ballantine and Forde 1970; Björkman et al. 1972b; Boardman et al. 1974; Boardman et al. 1975; Charles-Edwards and Ludwig 1975; Crookston et al. 1975; Goodchild et al. 1972; Mache and Loiseaux 1973; Prioul 1973; and Skene 1974).

On a leaf area basis (assuming approximately equal chlorophyll contents), a leaf of a typical sun plant may thus have about 2.5 times as many electron transport chains, slightly more PS I and slightly less PS II than a leaf of a typical shade plant.

Photosynthetic Unit Size. There is considerable contradiction in the literature concerning the effect of growth light level on the size of the photosynthetic unit (PSU). This apparent contradiction largely stems from the varying definitions of PSU unit size. The concept of photosynthetic unit was originally formulated by GAFFRON and WOHL (1936) on the basis of EMERSON and ARNOLD's (1932a, b) determinations of the maximum yield of photosynthesis per flash of light. The PSU was defined as the minimum of chlorophyll molecules required for the evolution of one O_2 or the uptake of one CO_2 molecule, inferring that a large number of chlorophyll molecules act in cooperation to collect enough quanta to drive photosynthesis once. Much later the PSU was considered as an entity consisting of a photosynthetic reaction center together with an antenna of light-harvesting chlorophyll molecules. In the belief that there is a 1:1:1 stoichiometry between the PS II reaction center, cytochrome f and the PS I reaction center (P-700), PSU size has often been expressed as the ratio of total chlorophyll to P-700. More recently, PSU size has been defined as the number of chlorophyll molecules per electron transport chain or per cytochrome f (WILD 1979).

If PSU size is defined as the number of chlorophyll molecules serving each reaction center then no differences in PSU size are readily apparent between sun and shade leaves. In this case it is necessary to specify whether PS II or PS I reaction centers are being considered (especially if the PS II/PS I ratio is variable) and the fraction of the chlorophyll that is connected to each of the two photosystems must also be taken into account. MELIS and BROWN (1980b) propose the term photochemical unit (PCU), defining the number of antenna chlorophyll molecules that transfer their excitation energy to a specific reaction center.

According to the model shown in Fig. 3.6, the sun and the shade leaf do not differ in the ratio between the sum of LHchl and chl a_{II} (i.e., that portion of chl a which is closely associated with PS II) and P-700. The shift in the ratio of total chlorophyll to P-700 is caused by a shift in ratio between the two photosystems rather than by a change in the PSU size (cf. Sect. 3.3.1).

If, on the other hand, PSU size is defined as the number of chlorophyll molecules per electron transport chain (cytochrome f) then, obviously, there exist large differences in PSU size between sun and shade leaves.

3.4.4 Stomatal Conductance

Stomata are effectively variable valves which control the diffusion of water vapor from the intercellular spaces in the leaf to the ambient atmosphere; hence the stomata inevitably also govern the diffusion of CO_2 in the opposite direction. During photosynthesis the CO_2 pressure in the intercellular spaces (p_i) will therefore be lower than that of the ambient air (p_a). The relationship is given by expression

$$p_i = p_a - P\frac{D}{g}$$

where P is the rate of net photosynthesis, g the stomatal conductance to the diffusion of water vapor and D the ratio between the diffusivities of CO_2 and water vapor (~ 1.56). (For the present purpose the conductance of the boundary

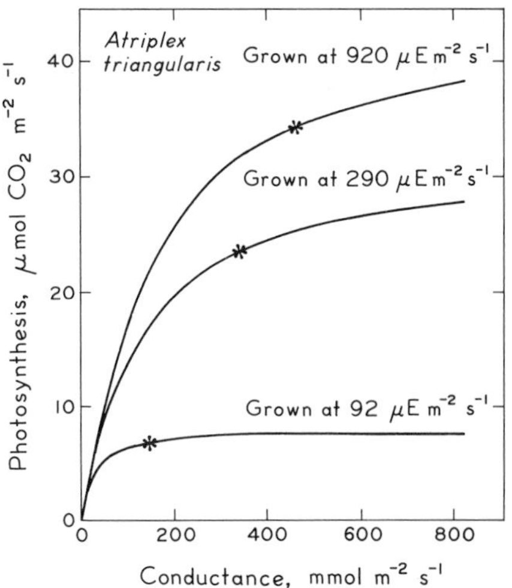

Fig. 3.7. Rate of net photosynthesis as a function of leaf conductance in intact leaves of *Atriplex triangularis*, grown under three different light regimes. *Asterisks* indicate the actual conductance assumed by each leaf in saturating light, 25 °C, 325 μbar CO_2, and 210 mbar O_2. *Solid curves* were calculated from the experimentally determined relationships between P_s and intercellular CO_2 pressure for each leaf. (After BJÖRKMAN et al. 1972b)

layer is included in the stomatal conductance, g). It follows that under conditions where the CO_2 pressure is insufficient to saturate photosynthesis, g will influence the photosynthetic rate through its effect on p_i. As discussed earlier, the rate of net photosynthesis by C_3 plants at normal atmospheric CO_2 and O_2 pressures is partially limited by CO_2 even at low light intensities (Fig. 3.5), but the limitation increases with light intensity and is maximal at light saturation. Consequently, also the effect of stomatal conductance on photosynthesis should be maximal at light saturation.

Figure 3.7 shows the relationship between P_s and g in normal air for *Atriplex triangularis*, grown under three different light regimes. As shown by the solid curves, the differences in P_s between the plants grown under the different light regimes are present over a very wide range of g. This is in accordance with the conclusion that large differences exist in intrinsic photosynthetic capacity between the plants. The actual values of P_s and g, indicated by asterisks, varied in concert with the intrinsic photosynthetic capacity, so that in each case g imposed a relatively small limitation to P_s. In other words, although differences in g clearly cannot account for the differences in P_s between the different plants, an increased value of g in response to increased growth light level is nevertheless an important factor since it permits an essentially full expression of the increased intrinsic photosynthetic capacity. In *Atriplex*, the differences in the value of g between the different growth light regimes were accompanied by corresponding differences in stomatal frequency (BJÖRKMAN et al. 1972b). The leaves developed under 52 E m^{-2} day^{-1} had almost three times as many stomata (per unit leaf area) on the upper epidermis and almost twice as many on the lower epidermis as the leaves developed under 5.3 E m^{-2} day^{-1}.

In their work with rain forest shade plants BJÖRKMAN et al. (1972a) concluded that although stomatal conductance was generally lower than commonly found in C_3 sun plants it imposed only a small limitation on photosynthesis. In the native habitat stomatal conductance in *Alocasia macrorrhiza* responded to variations in light intensity in a manner that enabled photosynthesis to proceed at a rate closer to that which would have occurred at an infinitely high value of g, and even at saturating light g was sufficiently high to permit almost full expression of intrinsic photosynthetic capacity. Such short-term adjustments of stomatal aperture and the long-term adjustments of stomatal frequency tend to maintain a relatively constant intercellular CO_2 pressure (WONG 1979).

3.4.5 Leaf Structure

It has long been recognized that leaf structure may be strongly influenced by the light level during growth (e.g., STAHL 1883; HESSELMAN 1904). As mentioned earlier, increased light level results in an increased overall leaf thickness and specific leaf weight. There is an increased development of the epidermis, the vascular system and the parenchyma. In a sun leaf of a typical C_3 plant the mesophyll is characterized by the presence of a well-developed palisade parenchyma with a high proportion of long columnar cells. In a typical shade leaf the mesophyll cells tend to be round or highly irregular in shape, and the total number of cells across a leaf section is often smaller than in the sun leaf.

There are two major considerations pertaining to the relationship between leaf anatomy and photosynthetic capacity. One concerns the amount of photosynthetic apparatus per unit leaf area, the other concerns the internal area of exposed mesophyll cells (A_{mes}) per unit leaf area (A). To the extent that increases in leaf thickness or specific leaf weight reflect increases in the content of potentially rate-limiting catalysts (such as RuBP carboxylase and electron carriers) per unit leaf area, light-saturated photosynthetic capacity would be expected to increase as well. Both the CO_2-limited and the CO_2-saturated photosynthetic rate may be affected by such changes. The costs and benefits of such changes in sun and shade situations were considered in Section 3.3.3.

An increase A_{mes}/A ratio would increase the surface area available for diffusive transfer of CO_2 from the intercellular spaces to CO_2 acceptor sites in the mesophyll cells (cf. EL SHARKAWY and HESKETH 1965; HOLMGREN 1968; NOBEL et al. 1975; NOBEL 1976, 1977; TURRELL 1936). On the assumption that conductance to such diffusive transfer is potentially limiting to P_s, an increased A_{mes}/A ratio would tend to increase P_s under conditions of limiting CO_2 pressures. It would obviously not affect P_s at saturating CO_2 pressures.

Relationship Between Specific Leaf Weight and P_s. The above considerations have been implicit in many studies attempting to relate P_s with leaf thickness, fresh weight per leaf area, or specific leaf weight (SLW). Using results obtained by WILLSTÄTTER and STOLL (1918) for a number of different species, MCCLENDON (1962) contended that if certain species were excluded P_s showed a linear relationship with the ratio of fresh weight to leaf area. It should be pointed out that these early measurements of P_s were made at high CO_2 pressures. Positive correlations between P_s determined at normal CO_2 pressure and leaf thickness or SLW are often found among leaves of the same species developed under

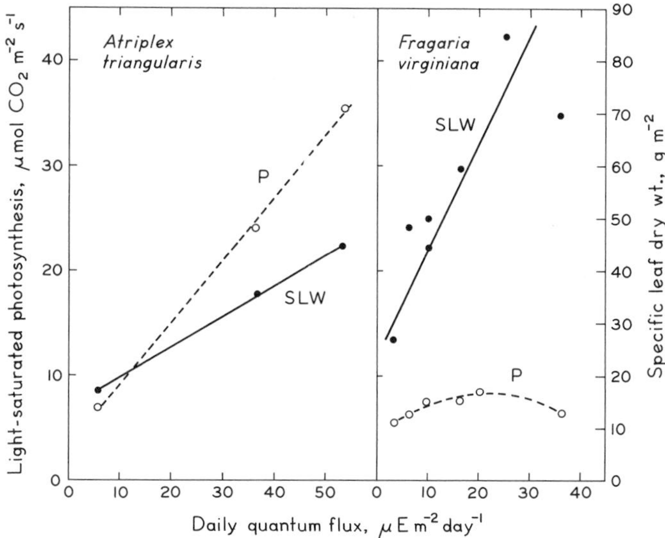

Fig. 3.8. Effect of growth light regime on light-saturated photosynthesis (*P*) and specific leaf dry weight (*SLW*) in *Atriplex triangularis* (data from BJÖRKMAN et al. 1972b) and *Fragaria virginiana* (data from JURIK et al. 1979)

different light regimes (e.g., BJÖRKMAN and HOLMGREN 1963; BJÖRKMAN et al. 1972b; BOWES et al. 1972; CHARLES-EDWARDS and LUDWIG 1975; HOLMGREN 1968; LOUWERSE and V.D. ZWERDE 1977; LUDLOW and WILSON 1971; also see Table 3.8). An example of such a relationship is shown for *Atriplex triangularis* in Fig. 3.8. In this case both P_s and SLW increased linearly with daily quantum flux during growth, although photosynthesis was considerably more affected than SLW. In other species such as *Betula verrucosa*, P_s and SLW are affected to a similar degree (Table 3.8). By contrast, in certain other species such as *Fragaria virginiana* (Fig. 3.8), *F. vesca* and *Glycine max* (Table 3.8), and shade clones of *Solidago virgaurea* (BJÖRKMAN and HOLMGREN 1963; HOLMGREN 1968) P_s failed to increase in parallel with increases in SLW but instead declined at high light levels. For example, in *F. virginiana*, growth regimes extending from 3.5 to 36 E m^{-2} day^{-1} had only a small influence on P_s in spite of its very pronounced effect on SLW (Fig. 3.8). This type of response may well be characteristic of shade plants as a result of their failure to increase the content of potentially rate-limiting catalysts in response to increased light levels and possible damage to the photosynthetic system at high light intensities (Sect. 3.5).

An analysis of the P_s and SLW values for 38 different species and growth light regimes, listed in Table 3.1, indicated that there is only a weak correlation between these two parameters (Table 3.4). Deletion of the data for high light-grown shade plants does not improve the correlation, and no correlation is obtained among different high light-grown sun plants or among low light-grown sun or shade plants. This is not surprising since different species may exhibit large differences in the development of nonphotosynthetic structures such as

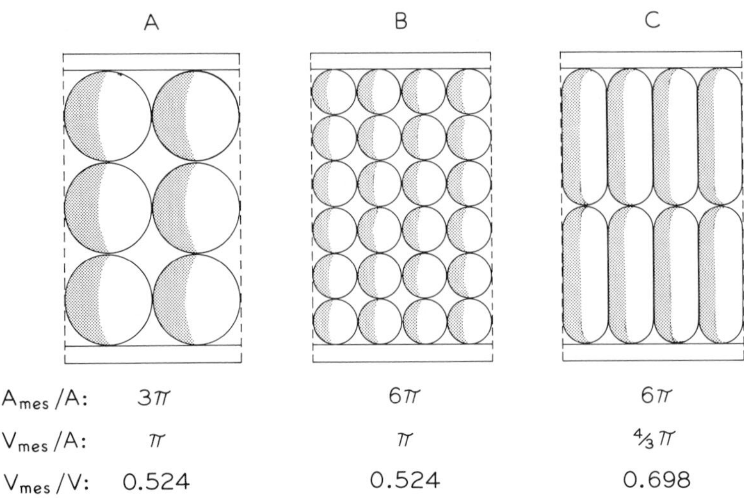

Fig. 3.9 A–C. Geometric representation of the effect of cell size and shape on the ratio A_{mes}/A, V_{mes}/A and V_{mes}/V in a leaf. The cells are arranged in an orthogonal array; those shown in **A** and **B** are spheres, those in **C** are cylinders with hemispherical ends. The radii in **A** are twice those in **B** and **C.** (Adapted from NOBEL 1980)

epidermal cells, cuticle, cell walls, and vascular tissue, and may also differ in the content of starch, sugars, and inorganic solutes.

Relationship Between Mesophyll Volume, Mesophyll Area and P_s. In the absence of changes in cell size and shape an increased thickness of the leaf mesophyll will inevitably be accompanied by a proportional increase in the ratio of mesophyll volume (V_{mes}) and mesopyll area (A_{mes}) to leaf area (A), whereas the ratio of V_{mes} and A_{mes} to leaf volume (V) would remain constant. As illustrated in Fig. 3.9, changes in mesophyll cell size and shape may have considerable effects on some of these parameters even in the absence of a change in mesophyll thickness. For example, a reduction in cell diameter to one-half results in a doubling of A_{mes}/A but it has no effect on V_{mes}/A or V_{mes}/V (Fig. 3.9 A, B). A change in cell shape from spherical (Fig. 3.9 B) to cylindrical with hemispherical ends (Fig. 3.9 C) has no influence on A_{mes}/A but it causes a significant increase in V_{mes}/A and V_{mes}/V.

The increased thickness of the mesophyll effected by increased light level during leaf development is generally attributable to such elongation of the cells in the upper layers of the mesophyll and to an increase in the number of cells across the leaf section. This cell elongation gives rise to the pronounced palisade parenchyma which is characteristic of sun leaves of C_3 plants. The main result is an increased packing density (V_{mes}/V) and thus a reduced proportion of intercellular air spaces.

There is relatively little quantitative information on the effect of light level on cell diameter. Cursory examination of published micrographs or camera lucida drawings of leaf sections from sun and shade leaves of *Atriplex triangularis* (BJÖRKMAN et al. 1972b), *Fragaria vesca* (CHABOT and CHABOT 1977), *F. virginiana*

(CHABOT et al. 1979), *Mimulus cardinalis* (HIESEY et al. 1971), and a number species of deciduous trees (HANSON 1917) does not reveal any obvious differences in cell diameter (minor axis). BALLANTINE and FORDE (1970) observed little influence of growth light level on mesophyll cell size in *Glycine max*. WILSON and COOPER (1969) found that cell size was significantly influenced by light level only in two of the ten genotypes of *Lolium* studied by them, and in these two genotypes leaves from the lower light treatments had smaller cells than those from the high light treatment, and WILD (1979) commented that shade leaves of *Sinapis alba* had markedly *smaller* cells than sun leaves. SENGER and FLEISCHHACKER (1978) found that the average cell size in the unicellular green alga *Scenedesmus obliquus* was about 50% greater when grown at 11.1 E m^{-2} day^{-1} as compared with 2.0 E m^{-2} day^{-1}.

These data suggest that increasing light level may have a greater effect on V_{mes}/A than on A_{mes}/A. Direct measurements of these two parameters in *Fragaria vesca* (CHABOT and CHABOT 1977) and *F. virginiana* (CHABOT et al. 1979) lead to similar conclusions. For example, in the latter species, V_{mes}/A increased about 100% while A_{mes}/A increased by about 50% when the light was increased from 6.5 to 20 E m^{-2} day^{-1}. Similarly, NOBEL et al. (1975) and NOBEL (1980) concluded that the 4.5-fold increase in A_{mes}/A which occurred in *Plectranthus parviflorus* leaves when the light level was increased from 0.8 to 35 E m^{-2} day^{-1} (Table 3.8) was primarily a result of an increase in cell length of the upper two palisade layers with little change in cell diameter. For the spongy mesophyll, which contributed about 35% of the total A_{mes}/A, the number of cells per unit leaf area increased by a factor of 2.4, but the average cell diameter *increased* by 2.6%. Studies on *Hyptis emoryi* (NOBEL 1976) and the fern *Adiantum decorum* (NOBEL 1977) yielded qualitatively similar results (Table 3.8).

The relationship between light-saturated photosynthetic rate and A_{mes}/A has been extensively studied by NOBEL and coworkers (NOBEL et al. 1975; NOBEL 1976, 1977, 1980). In these studies, photosynthesis was measured under conditions of CO_2 limitation, optimum temperature and, to avoid complications caused by oxygenation of RuBP and subsequent photorespiratory metabolism, the O_2 pressure was kept at a low level (0 to 10 mbar O_2). These workers found that in any given species light-saturated photosynthetic rate was closely correlated with A_{mes}/A. As shown in Table 3.8, in *Plectranthus parviflorus* the light-saturated photosynthetic rate per unit leaf area increased by a factor of 3.5 when the growth light level was increased from 0.8 to 35 E m^{-2} day^{-1}; A_{mes}/A increased by a factor of 4.5 over the same range and consequently the photosynthetic rate on the basis of A_{mes} decreased by 23%. Similar results were obtained with *Hyptis emoryi* and the fern *Adiantum decorum*. According to the calculations of NOBEL and coworkers the light-saturated photosynthetic rate per unit mesophyll cell area, determined at a given limiting CO_2 pressure in the intercellular spaces, remained essentially constant over a wide range of growth light regimes, leading these authors to conclude that the increase in photosynthetic rate on leaf area basis within each species is caused by an increase in exposed mesophyll cell area and not from changes in the intrinsic photosynthetic capacity of cells. It seems likely, however, that a similar constancy

Table 3.8. Comparison of the effect of growth light regime on leaf anatomical characteristics and photosynthesis in different species

	Growth light regime, E m^{-2} day^{-1}	Leaf thickness μm	Spec. leaf wt., g m^{-2}	A_{mes}/A m^2 m^{-2}	Photosynthesis mol s^{-1}		Reference
					m^{-2} A	m^{-2} A_{mes}	
Adiantum decorum	"Low"	48	–	3.8	3.6[c]	0.94[c]	NOBEL (1977)
	"Medium"	89	–	7.5	7.0[c]	0.93[c]	
	"High"	140	–	11.7	9.0[c]	0.77[c]	
Betula verrucosa	3.1 C	540	15	26	5.0	0.19	ÖQUIST et al. (1981)
	15.1 C	650	23	31	8.8	0.28	
	36.7 C	830	33	43	11.0	0.26	
Fragaria vesca	1.2 C	74	80[b]	9	3.0	0.33	CHABOT and CHABOT (1977)
	8.1 C	107	144[b]	16	3.9	0.24	
	35.1 C	126	163[b]	26	3.7	0.14	
Fragaria virginiana	3.5 C	121	27	16	5.6	0.35	JURIK et al. (1979)
	36.6 C	153	70	29	6.5	0.22	
Glycine max	10.8 C	146	19	22	13.2	0.60	BUNCE et al. (1977)
	36.7 C	214	43	32	14.3	0.45	
Hyptis emoryi	7 H[a]	130	–	14	6.4[c]	0.45[c]	NOBEL (1976)
	16 H[a]	225	–	24	9.5[c]	0.39[c]	
	27 H[a]	270	–	29	9.9[c]	0.36[c]	
	44 H[a]	300	–	34	10.5[c]	0.31[c]	
	70 H[a]	340	–	39	13.2[c]	0.34[c]	
Plectranthus parviflorus	0.8 C	290	–	11	2.1[c]	0.19[c]	NOBEL et al. (1975)
	3.6 C	420	–	17	3.1[c]	0.18[c]	
	5.8 C	560	–	23	3.8[c]	0.16[c]	
	12.5 C	680	–	29	4.9[c]	0.17[c]	
	19.1 C	820	–	37	5.7[c]	0.15[c]	
	35.0 C	1,020	–	50	7.3[c]	0.15[c]	

[a] Estimated from relationship between leaf length and total daily irradiance
[b] g fresh weight m^{-2} leaf area
[c] Measured in an O_2 free atmosphere

of photosynthetic rate may have been obtained if mesophyll volume rather than mesophyll area had served as the basis in these comparisons. It should also be noted that the correspondence between photosynthetic rate and mesophyll area is considerably poorer for other species listed in Table 3.8. In *Fragaria vesca, F. virginiana,* and *Glycine max,* increases in light level caused large increases in A_{mes}/A but there was only a small effect on photosynthetic rate. In *F. vesca* an increase in light level from 8.1 to 35.1 E m^{-2} day^{-1} resulted in a small decline in photosynthetic rate in spite of a 63% increase in mesophyll area. As mentioned earlier, this decline in photosynthetic rate at high levels may be related to the shade species character of *F. vesca* (CHABOT and CHABOT 1977).

Although there is no evidence that the average diameter of the mesophyll cells decreases with increased light level in any given species, large differences in this respect are found between different species. For example, owing to the much smaller average cell diameter *Adiantum decorum* and *Hyptis emoryi* have greater A_{mes}/A ratios than *Plectranthus parviflorus* for any given mesophyll thickness (NOBEL 1976, 1977). Presumably the former species therefore also have correspondingly greater A_{mes}/V_{mes} ratios. On the assumption that diffusion of CO_2 from the intercellular spaces to the chloroplasts exerts a major limitation to photosynthesis, smaller cell size should thus lead to an increase in photosynthetic rate. The observation by WILSON and COOPER (1967, 1970) that a negative correlation exists between photosynthetic rate and mesophyll size among *Lolium* genotypes is consistent with this assumption.

There is no indication, however, that the A_{mes}/A ratio is a major determinant of photosynthetic rate between the species listed in Table 3.8. For example, at equal photosynthetic rates in the two species *Adiantum decorum* (small cells) has about sixfold greater A_{mes}/A ratios than *Plectranthus parviflorus* (large cells). Also, recent studies by G. HARVEY (personal communication) demonstrate that although shade leaves of the obligate shade species *Asarum caudatum* and sun leaves of *Atriplex triangularis* had similar A_{mes}/A (and V_{mes}/A) ratios, the light-saturated photosynthetic rate in normal air was several times higher in the *A. triangularis* leaves. Similar differences in photosynthetic rate were also found under conditions of saturating CO_2 pressures both with intact leaves and with suspensions of separated mesophyll cells prepared from the leaves.

Conclusions. Increased light level during growth results in an elongation of the cells in the upper layers of the mesophyll and an increase in the number of cells across the leaf section, possibly accompanied by an increase in the average diameter of these cells. These changes result in increased V_{mes}/A and A_{mes}/A ratios and contribute to the increase in leaf thickness and specific leaf weight. The changes occur in both sun species and shade species. In sun species the photosynthetic rate on the basis of unit leaf area tends to increase with increasing V_{mes}/A and A_{mes}/A ratio over a wide range of light levels, but in obligate shade species increases in these ratios often are not accompanied by corresponding increases in photosynthetic rate. This suggests that increases in photosynthetic rate can only occur if increases in V_{mes}/A and A_{mes}/A are accompanied by increases in the amount of RuBP carboxylase, electron transport components, and other constituents determining the photosynthetic rate at the chloroplast level.

It seems likely that in plants which have a high potential for acclimation to different light levels, the amount of these components per mesophyll area or mesophyll volume remains more or less constant over a wide range of light regimes. The alternative mode of response would be an increase in the concentration of these components in the chloroplasts or cells as the light level during growth is increased. However, it is probable that there is an intrinsic upper limit to which such increases in concentration may occur; there is evidence, for example, that RuBP carboxylase concentration in chloroplasts approaches the solubility limit of this protein. It is also possible that the spatial distribution of the CO_2 acceptor sites on the enzyme may be important to efficient capture of CO_2, especially at limiting CO_2 pressures, and that a certain relationship between the number of carboxylation sites, chloroplast area, and cell wall area is necessary for efficient utilization of low intercellular CO_2 concentrations. The concomitant increase in the amount of above-mentioned components of the photosynthetic system and in the area of exposed mesophyll cell walls may serve to maintain a proper balance between these factors.

There is little evidence, however, that differences in light-saturated rate between different species can be attributed to differences in the A_{mes}/A ratio, or that diffusive transfer of CO_2 between the intercellular spaces and the chloroplasts imposes a major limitation to photosynthesis in C_3 plants grown in their native habitats or under light regimes simulating those prevailing in their respective habitats. These conclusions are also supported by carbon isotope ratios determined on a wide range of C_3 plants. Of the carbon in normal atmospheric CO_2, about 1.1% has mass 13 and the remainder has mass 12. It is well known that during photosynthesis there is a discrimination against uptake of the heavier isotope and as a result the plant becomes enriched in the lighter isotope. The basis of this discrimination lies primarily with the RuBP carboxylase reaction which causes a discrimination in the range 27 to $38^0/_{00}$ (O'LEARY 1981). Discrimination in gaseous diffusion of CO_2 is only about $4.4^0/_{00}$ (CRAIG 1953) and that for diffusion in an aqueous solution is less than $0.5^0/_{00}$ (O'LEARY 1981). Using an average value of $32^0/_{00}$ for discrimination by RuBP carboxylase, an external CO_2 pressure of 330 μbar and an intercellular pressure of 220 μbar, it may be calculated that the overall discrimination would be $22.8^0/_{00}$ (G. FARQUHAR, J.A. BERRY and M.H. O'LEARY, in preparation). This assumes that the resistance to diffusive transfer between the intercellular spaces and the CO_2 fixation sites in the chloroplasts is negligible. If this resistance were considerable, say about the same order as the sum of the stomatal and boundary layer resistance, then a CO_2 gradient of 110 μbar would develop between the intercellular spaces and the CO_2 fixation sites and the overall discrimination would fall to about $13^0/_{00}$. If the resistance were sufficiently large to cause the CO_2 pressure inside the cell to fall to the compensation point (as some authors assume), then the overall discrimination would be in the order of $7^0/_{00}$. The actual overall discrimination by C_3 plants is about $22.5^0/_{00}$ and there is surprisingly little variation between species or environmental conditions, strongly indicating that the physical resistance to diffusion of CO_2

between the intercellular spaces and the CO_2 fixation sites is small (see Chap. 15, Vol. 12B).

3.5 Damage by Excess Light

It has been recognized for over a century that exposure of photosynthetic organisms to light intensities in excess of those normally encountered during growth can result in severe damage to photosynthetic tissues (e.g., VON MOHL 1837; BATALIN 1874). This damage was mostly observed as bleaching of the chlorophyll from the exposed leaf tissue but experiments by URSPRUNG (1917) indicated that exposure of leaves to excess light could also cause an inhibition of starch formation and even disappearance of starch from the leaves. This phenomenon, termed solarization, was further explored by HOLMAN (1930) who concluded that the failure of leaves, exposed to excessive light regimes, to produce starch resulted from a direct inhibitory effect of excess light energy on photosynthesis (photoinhibition).

3.5.1 Relationship Between Photoinhibition and Chlorophyll Bleaching

MONTFORT (1941) found that while both chlorophyll bleaching and inhibition of photosynthesis were observed when certain aquatic shade plants were transferred to full sunlight, the inhibition of photosynthesis preceded any detectable change in chlorophyll content. Similar results have been obtained with unicellular green algae exposed to excess light regimes (e.g., KOK 1956; KANDLER and SIRONVAL 1959; BELAY and FOGG 1979).

Photoinhibition and chlorophyll bleaching have also been observed in leaves of land plants exposed to high light levels. As mentioned in Section 3.2, BJÖRKMAN and HOLMGREN (1963) found that leaves of clones of *Solidago virgaurea* ecotypes from shaded habitats displayed substantially reduced ϕ_i when grown under a high light regime in comparison with controls maintained at lower light regimes (Fig. 3.3). In some of the shade clones growth under a high light regime resulted in appreciable bleaching of chlorophyll contained in the uppermost palisade cell layers. Further analysis showed that the reduction in ϕ_i could not be accounted for by the reduced absorptance resulting from chlorophyll bleaching (BJÖRKMAN 1968b). A similar inhibition of ϕ_i was obtained when certain shade ecotype clones of *Solanum dulcamara* were grown under a high light regime (GAUHL 1976). Photoinhibition was also observed in mature leaves of shade ecotypes of *Solidago virgaurea* developed under a low light regime and then transferred for one week to a high light regime (BJÖRKMAN and HOLMGREN 1963). These treatments resulted in an inhibition of both P_s and ϕ_i, as well as a reduction of the chlorophyll content. When plants returned from the higher to the lower light regime, leaves which had suffered severe inhibition of photosynthesis and chlorophyll bleaching showed an essentially complete recovery of photosynthetic activity within one week, but the chlorophyll content did not show any appreciable increase. Powles and Critchley (1980) found that exposure to full sunlight of intact leaves of *Phaseolus vulgaris* plants grown under low light regimes resulted in substantial inhibition of both ϕ_i and P_s. The photoinhibition was also reflected at the chloroplast level as a substantial inhibition of electron flow through PS II, but it was not accompanied by any detectable change in the leaf chlorophyll content. Finally, Jones and Kok (1966a, b) and Satoh (1970a) could not detect any change in chlorophyll content following exposure of isolated spinach chloroplasts to intense light even when such exposure resulted in an almost complete inactivation of electron transport reactions.

The above experiments, in which intact attached leaves of shade-adapted higher plants, algal cells, and isolated chloroplasts were exposed to super-saturating quantum flux densities, show that such treatments result in a substantial inhibition of photosynthetic reactions and that the extent of inhibition is dependent upon both the intensity and the duration of exposure. They further demonstrate that chlorophyll bleaching and photoinhibition are different processes. The primary event is an inhibition of the photosynthetic reactions; chlorophyll bleaching is a secondary reaction occurring only after photosynthesis is severely inhibited. In spite of this overwhelming evidence to the contrary, the assumption that chlorophyll bleaching is a cause of photoinhibition is still often encountered in the current literature.

3.5.2 Mode and Mechanism of Photoinhibition

Photoinhibition is manifested in a reduction of both the quantum yield and light-saturated capacity of photosynthesis of whole leaves and cells and in an inactivation of electron transport and photophosphorylation by isolated chloroplasts. Both PS I and PS II activity are affected, but the kinetics and the dependence of the inactivation on factors such as temperature, O_2 concentration, and electron donors and acceptors are not the same for the two activities (Jones and Kok 1966a, b; Satoh 1970a, b, c). These and other results suggest that there are at least two types of photoinhibition.

The action spectrum for photoinhibition indicates that quanta absorbed by the photosynthetic pigments are responsible for the inactivation (Jones and Kok 1966a; Satoh 1970a). The former authors favored the view that photoinhibition is caused by an excess of excitation energy at the photosynthetic reaction centers. They reasoned that quanta causing the inactivation are transferred to these centers instead of being disposed of as heat (and to some extent as fluorescence) in the light-harvesting pigment complex and that photoinhibition is caused by a secondary reaction at these centers which destroys either the centers themselves or a related moiety.

Oxygen can give rise to excited states such as singlet oxygen and free radical species such as :OH. These states and radicals are potentially extremely destructive to photosynthetic membranes and they could play a primary role in photoinhibition (Foyer and Hall 1980). Chloroplast membranes evidently possess protective mechanisms against these destructive agents, but the capacity of these mechanisms may be exceeded at very high quantum flux densities and their effectiveness might also depend on other factors such as temperature and oxygen pressure.

3.5.3 Interactions with Other Factors

In photosynthesizing cells photosynthesis itself diverts quanta, thus reducing the excess excitation energy at the reaction centers, which is presumably the primary cause of photoinhibition. On these grounds one might expect that restrictions on photosynthetic rate by factors such as reduced CO_2 pressure or suboptimal temperature would increase the susceptibility to photoinhibition. Experimental results are largely in agreement with these expectations.

Oxygen pressure is another factor which might be expected to influence the susceptibility to photoinhibition for at least two very different reasons. As mentioned above, oxygen can give rise to destructive free radicals and other harmful species and it seems probable that the production of these agents would increase with increased oxygen pressure. High O_2 pressure may therefore increase the susceptibility to photoinhibition. On the other hand, oxygen may also serve to dissipate energy in photorespiration (e.g., OSMOND and BJÖRKMAN 1972; TOLBERT and RYAN 1976) and in other O_2 consuming reactions (e.g., RADMER et al. 1978), thereby reducing the excess of excitation energy under condition of severely restricted CO_2 supply. These questions are further discussed below.

Interaction with Temperature. There is evidence that low temperature increases the sensitivity to photoinhibition. STEEMAN NIELSEN (1942) found that exposure of the marine alga *Fucus serratus* to a moderately high light intensity resulted in substantial photoinhibition when the alga was kept at 5 °C, but not when it was kept at 20 °C. BJÖRKMAN and HOLMGREN (1963) commented that although no photoinhibition occurred when *Solidago* leaves were exposed to about 700 $\mu E\ m^{-2}\ s^{-1}$ at 22 °C, exposure to the same light intensity at leaf temperatures of 10 °C or lower caused photosynthesis to decline with time, indicating that photoinhibition occurred. In more recent studies of the interaction between the effects of high light levels and low temperature on chilling-sensitive plants of tropical origin (TAYLOR and ROWLEY 1971; TAYLOR and CRAIG 1971; TAYLOR et al. 1972), it was found that severe damage to the photosynthetic apparatus occurred when the plants were exposed to a combination of low temperature and high light levels. No such effects were seen under a combination of moderate temperatures and high light levels, nor under a combination of low temperature and low light levels. Finally, POWLES et al. (1980a) investigated the interactive effects of temperature and quantum flux density on photosynthesis in several chilling-sensitive plants, including *Phaseolus vulgaris*. Exposure of single leaves of this species to 2,000 $\mu E\ m^{-2}\ s^{-1}$ at 6 °C and normal air for 3 h resulted in a severe inhibition of ϕ_i but no inhibition of ϕ_i was observed when the leaf temperature was kept above 12 °C or when the light was kept at a low level.

These results are consistent with the notion that low temperature increases the susceptibility to photoinhibition because it depresses the rate of photosynthesis, thereby reducing the diversion of quanta via this process and thus increasing the excess of excitation energy. However, the casual relationships are far from clear and require much further investigation.

Interaction with CO_2 and O_2 Pressure. Severe restriction of the supply of CO_2 to an illuminated leaf, such as would occur under conditions of stomata closure, inevitably results in a reduction of the CO_2 pressure inside the leaf intercellular spaces and in a drastic reduction of the photosynthetic rate, thus presumably increasing the susceptibility to photoinhibition. There are several early reports that destructive effect of high light levels on the photosynthetic apparatus of algal suspensions may be largely attributed to CO_2 depletion in the medium, and recent studies by POWLES and coworkers show that CO_2 pressure also strongly affects the susceptibility to photoinhibition of higher plant leaves. For example, reduction of the intercellular CO_2 pressure during photoinhibition treatment of shade leaves of *Phaseolus vulgaris* from 100 to 10 μbar increased the extent of the photoinhibition by a factor of 3, and in sun leaves photoinhibition was fully prevented at CO_2 pressures greater than 50 μbar (POWLES and CRITCHLEY 1980).

Under conditions of very low O_2 pressure photorespiration does not occur and complete stomatal closure would thus cause the intercellular CO_2 pressure to fall to zero and CO_2 fixation would come to a complete halt. However, in the presence of normal atmospheric O_2 pressure (210 mbar) photorespiratory CO_2 release prevents the intercellular CO_2 pressure from falling below the CO_2 compensation point (~50 μbar CO_2 at 25 °C in C_3 plants). Experiments in which sun leaves of *Phaseolus vulgaris* and other C_3 species were exposed to high light levels under low O_2 pressure (10 mbar) and CO_2 pressures well below 50 μbar showed that such treatments result in severe photoinhibition. Increasing the O_2 pressure to 210 mbar, thus permitting internal generation of CO_2 by photorespiration, and refixation of this CO_2 by the Calvin-Benson cycle, gave full protection against photoinhibition (POWLES and OSMOND 1978; POWLES et al. 1979). Leaves of C_4 plants also suffered photoinhibition when the supply of CO_2 was blocked, but even very low CO_2 pressures were effective in alleviating photoinhibition and 50 μbar CO_2 provided total protection. In contrast to the situation in C_3 plants, changes in the O_2 pressure during the treatment had little effect on photoinhibition (POWLES et al. 1980b). It seems evident that C_4 plants may dispose of excess energy in a cyclic process in which CO_2 released in the bundle-sheath cells is refixed in the mesophyll cells continually consuming ATP and NADPH. The studies of POWLES and coworkers provide strong evidence that maintenance of some minimum level of photosynthetic carbon metabolism is necessary to prevent photoinhibition in both C_3 and C_4 plants.

It is noteworthy that although high O_2 pressure provides protection against photoinhibition under conditions of moderate temperature and severely restricted CO_2 supply, it aggravates the photoinhibition which occurs in chilling-sensitive plants such as *Phaseolus vulgaris* when they are exposed to a combination of low temperatures and high quantum flux densities (POWLES et al. 1980a).

3.5.4 Comparisons Between Sun and Shade Leaves

There is considerable evidence that the low light-saturated photosynthetic capacity of shade leaves is associated with a high susceptibility to photoinhibition. This applies not only to leaves of obligate shade plants but is also true of leaves of sun plants that have developed under a low light regime. Exposure of intact leaves of shade-grown plants of *Solidago virgaurea* to light intensities equivalent to full sunlight caused a gradual decline in P_s and a 2-h exposure resulted in a severe inhibition of light-limited photosynthesis (BJÖRKMAN and HOLMGREN 1963), but no such effects were observed when leaves from high light-grown sun clones of *Solidago* were subjected to the same treatment. GAUHL (1976) showed that shade and sun leaves of *Solanum dulcamara* exhibited differences in their responses to high light exposure similar to those found in *Solidago*. Extreme sensitivity to photoinhibition was observed in the shade fern *Pteris cretica* (HARIRI and PRIOUL 1978). Exposure of intact fronds to light intensities in excess of approximately 300 $\mu E\ m^{-2}\ s^{-1}$ caused a continuous decrease in photosynthesis with time and if the fronds were maintained at 500–600 $\mu E\ m^{-2}\ s^{-1}$ photosynthesis decreased continuously and was almost completely inhibited after 3 to 4 days. As was also the case in the studies with *Solidago* and *Solanum*, the decline was not due to partial stomatal closure, and no concomitant morphological aberrations could be detected.

POWLES and THORNE (1981) compared the effect of exposure to high quantum flux densities on photosynthesis in shade-grown fronds of the fern *Lastreopsis microsora* and of shade and sun leaves of *Phaseolus vulgaris*. Exposing the fern fronds to 1,800 $\mu E\ m^{-2}\ s^{-1}$ for 2 h reduced ϕ_i to approximately one-half of its original value. With *Phaseolus vulgaris* leaves grown at 4% full daylight intensity, exposure to 2,300 $\mu E\ m^{-2}\ s^{-1}$ (approximately 10 times the intensity required to saturate photosynthesis in these shade-grown leaves), resulted in a decline in P_s that could be detected within 5 to 10 min. As mentioned above, photoinhibition of the sun leaves could be fully prevented by maintaining the CO_2 pressure above the CO_2 compensation point, but this was not the case with the shade leaves.

It may be considered that one of the principal characteristics of sun species is their genetically determined capability gradually to increase the capacity for light-saturated photosynthesis in response to an increased light intensity during growth. As discussed in Section 3.4, such an adjustment includes an increase in the level of carbon metabolism enzymes (such as RuBP carboxylase) and of several carriers of the photosynthetic electron transport chain, as well as increased capacity for diffusive transport of CO_2. Obligate shade plants have a very limited capability for such adjustments and the reduced quantum yield and other symptoms of high light injury that are observed when shade plants are grown in bright light may well be a direct consequence. The difference in tolerance to high light intensities between sun and shade species may also be in part related to certain other mechanisms that provide alternative pathways for nondestructive dissipation of excessive levels of excitation energy or otherwise protect the photosystems from high light injury.

There is so far no experimental evidence that sun plants possess any such special mechanisms not present in shade leaves. BJÖRKMAN et al. (1972b) in a comparison of the inhibiting effect of intense light on chloroplasts from sun and shade plants failed to detect any differences in the resistance to photoinhibition. In these experiments, washed chloroplast fragments, isolated from extreme shade leaves of the shade species *Alocasia macrorrhiza* as well as from shade and sun leaves of the sun species *Atriplex triangularis,* were exposed to intense light in the absence of added electron donors or acceptors. The quantum yield and the light-saturated activity of PS II- as well as of PS I-driven electron transport in the presence of suitable donors and acceptors were determined at different preillumination time intervals. The preillumination time required to effect a given loss in electron transport activity was nearly identical for all chloroplast preparations irrespective of species and growth light regime. It thus seems that the pronounced differences in sensitivity to photoinhibition that exist between the different leaf types when they are illuminated under conditions that permit photosynthesis disappear if the chloroplasts are isolated and illuminated under conditions that preclude photosynthesis.

3.5.5 Ecological Significance

It seems evident that photoinhibition does not normally occur in plants under the light regimes encountered in their natural environments. Presumably, the energy dissipation provided by photosynthetic activity, together with other mechanisms, protects the leaves from the potentially destructive effect of high quantum flux densities. However, photoinhibition does occur when leaves developed in weak light are suddenly exposed to bright light, or when obligate shade species (or ecotypes) are kept under strong light. It may also take place under other stress conditions that impose severe restrictions on photosynthetic activity.

Although tolerance to high light intensities may be related to the ability to dissipate energy via photosynthesis, other alternative mechanisms for the safe dissipation of, or protection against the potentially harmful effects of excessive excitation energy may also play an important role. In the longer term, mechanisms for repair of components that are inactivated or destroyed in photoinhibition and in secondary destructive processes (such as the photooxidative bleaching of chlorophyll and carotenoids) are probably important as well.

It is obvious that the low potential of obligate shade plants to increase their light-saturated photosynthetic capacity in response to an increased quantum flux density during growth results in a poor efficiency of light utilization in sunny habitats. However, the high susceptibility to photoinhibition of shade plants might constitute an even greater selective disadvantage. It should also be kept in mind that the susceptibility to photoinhibition of shade plants is likely to increase when other stresses are present. Such interaction should not be ignored in ecological considerations of adaptation to sun and shade habitats. Although the difference in light regime certainly is the dominant and often a paramount factor, sun and shade habitats inevitably differ to some extent also in other correlated factors. For example, open sunny locations tend to be drier than densely shaded situations. GAUHL's (1979) observations are noteworthy in this connection. Clones of *Solanum dulcamara* collected from certain shaded habitats showed no signs of photoinhibition when grown under a high light regime as long as the water supply was ample, but typical symptoms of photoinhibition became evident when the water potential of the root medium was reduced even slightly. Similar reductions in water potential had no effect on these shade clones under a low light regime, and *Solanum* clones from sunny habitats were unaffected, irrespective of the light regime.

References

Alberte RS, McClure PR, Thornber JP (1976) Photosynthesis in trees. Plant Physiol 58:341–344

Armond PA, Mooney HA (1978) Correlation of photosynthetic unit size and density with photosynthetic capacity. Carnegie Inst Washington Yearb 77:234–237

Ballantine JEM, Forde BJ (1970) The effect of light intensity and temperature on plant growth and chloroplast ultra-structure in soybean. Am J Bot 57:1150–1159

Batalin A (1874) Über die Zerstörung des Chlorophylls in lebenden Organen. Bot Ztg 32:432–439

Belay A, Fogg CE (1979) Photoinhibition of photosynthesis in *Asterionella formosa* (Bacillariophyceae) J Phycol 14:341–347

Berry JA, Björkman O (1980) Photosynthetic response and adaptation to temperature in higher plants. Ann Rev Plant Physiol 31:491–543

Berzborn RJ, Roos P (1980) Abstr 5th Int Cong Photosynthesis p 61

Björkman O (1968a) Carboxydismutase activity in shade-adapted and sun-adapted species of higher plants. Carnegie Inst Washington Yearb 67:487–488

Björkman O (1968b) Further studies on differentiation of photosynthetic properties in sun and shade ecotypes of *Solidago virgaurea* L. Physiol Plant 21:84–89

Björkman O (1973) Comparative studies on photosynthesis in higher plants. In: Giese A (ed) Current topics in photobiology, photochemistry and photophysiology, Vol. 8. Academic Press, New York, pp 1–63

Björkman O, Holmgren P (1963) Adaptability of the photosynthetic apparatus to light intensity in ecotypes from exposed and shaded habitats. Physiol Plant 16:889–914

Björkman O, Ludlow MM, Morrow PA (1972a) Photosynthetic performance of two rainforest species in their native habitat and analysis of their gas exchange. Carnegie Inst Washington Yearb 71:94–102

Björkman O, Boardman NK, Anderson JM, Thorne SW, Goodchild DJ, Pyliotis NA (1972b) Effect of light intensity during growth of *Atriplex patula* on the capacity of photosynthetic reactions, chloroplast components and structure. Carnegie Inst Washington Yearb 71:115–135

Björkman O, Badger M, Armond PA (1978) Thermal acclimation of photosynthesis: effect of growth temperature on photosynthetic characteristics and components of the photosynthetic apparatus in *Nerium oleander*. Carnegie Inst Washington Yearb 77:262–282

Blackman GE (1956) Influence of light and temperature on leaf growth. In: Milthorpe FL (ed) The growth of leaves. Buttersworth, London, pp 151–169

Blackman GE (1968) The application of concept of growth analysis to the assessment of productivity. In: Eckardt F (ed) Functioning of terrestrial ecosystems at the primary productivity level. UNESCO, Paris, pp 243–260

Blackman GE, Black JN (1959) Physiological and ecological studies in the analysis of plant environment. Ann Bot 23:51–63

Blenkinsop PG, Dale JE (1974) The effects of shade treatment and light intensity on ribulose-1,5-diphosphate carboxylase activity and fraction I protein level in the first leaf of barley. J Exp Bot 25:899–914

Boardman NK (1977) Comparative photosynthesis of sun and shade plants. Ann Rev Plant Physiol 28:355–377

Boardman NK, Anderson JM, Thorne SE, Björkman O (1972) Photochemical reactions of chloroplasts and components of the photosynthetic electron transport chain in two rainforest species. Carnegie Inst Washington Yearb 71:107–114

Boardman NK, Anderson JM, Björkman O, Goodchild DJ, Grimme LH, Thorne SW (1974) Chloroplast differentiation in sun and shade plants: Relationship between chlorophyll content, grana formation, photochemical activity and fractionation of the photosystems. Port Acta Biol Ser A 14:13–236

Boardman NK, Björkman O, Anderson JM, Goodchild DJ, Thorne SW (1975) Photosynthetic adaptation of higher plants to light intensity: Relationship between chloroplast structure, composition of the photosystems and photosynthetic rates. In: Avron A (ed) Proc 3rd Intern Congr on Photosynthesis, Elsevier, Amsterdam, pp 809–827

Bowes G, Ogren WL, Hageman RH (1972) Light saturation photosynthesis rate, RuDP carboxylase activity, and specific leaf weight in soybeans grown under different light intensities. Crop Sci 12:77–79

Bunce JA, Patterson DT, Peet MM (1977) Light acclimation during and after leaf expansion in soybean. Plant Physiol 60:255–258

Butler WL (1977) Chlorophyll fluorescence: A probe for electron transfer and energy transfer. In: Trebst A, Avron M (eds) Encyclopedia of Plant Physiology, New Series: Vol. 5 (Photosynthesis I) Springer, Berlin, Heidelberg, New York pp 149–167

Chabot BF, Chabot JF (1977) Effects of light and temperature on leaf anatomy and photosynthesis in *Fragaria vesca*. Oecologia 26:363–377

Chabot BF, Jurik TW, Chabot JF (1979) Influence of instantaneous and integrated light-flux density on leaf anatomy and photosynthesis. Am J Bot 66:940–945

Charles-Edwards DA, Ludwig LJ (1975) The basis of expression of leaf photosynthetic activities. In: Marcelle R (ed) Environmental and biological control of photosynthesis. W Junk, The Hague, pp 37–43

Clough JM, Terri JA, Alberte RS (1979) Photosynthetic adaptation of *Solanum dulcamara* L. to sun and shade environments I. A comparison of sun and shade populations. Oecologia 38:13–21

Collatz GJ (1978) The interaction between photosynthesis and ribulose-P_2 concentration – effect of light, CO_2 and O_2. Carnegie Inst Washington Yearb 77:248–251

Cooper CS (1967) Relative growth of alfalfa and birdsfoot trefoil seedlings under low light intensities. Crop Sci 7:176–178

Craig H (1953) The geochemistry of stable carbon isotopes. Geochim Cosmochim Acta 3:53–92

Crookston RK, Treharne KJ, Ludford P, Ozbun JL (1975) Response of beans to shading. Crop Sci 15:412–416

Doley D (1978) Effects of shade on gas exchange and growth in seedlings of *Eucalyptus grandis* Mill ex Maiden. Aust J Plant Physiol 5:723–738

Egle K (1960) Menge und Verhältnis der Pigmente. In: Ruhland W (ed) Encyclopedia of plant physiol. Springer, Berlin, Göttingen, Heidelberg. Vol. V, pp 492–496

Ehleringer J (1978) Implications of quantum yield differences on the distributions of C_3 and C_4 grasses. Oecologia 31:255–267

Ehleringer J, Björkman O (1977) Quantum yields for CO_2 uptake in C_3 and C_4 plants. Plant Physiol 59:86–90

Ehleringer J, Björkman O (1978) A comparison of photosynthetic characteristics of *Encelia* species possessing glabrous and pubescent leaves. Plant Physiol 62:185–190

Ehleringer J, Björkman O, Mooney HA (1976) Leaf pubescence: effect on absorptance and photosynthesis in a desert shrub. Science 192:376–377

El-Sharkawy M, Hesketh J (1965) Photosynthesis among species in relation to characteristics of leaf anatomy and CO_2 diffusion resistance. Crop Sci 5:517–521

Emerson R, Arnold W (1932a) A separation of the reactions in photosynthesis by means of intermittent light. J Gen Physiol 15:391

Emerson R, Arnold W (1932b) The photochemical reaction in photosynthesis. J Gen Physiol 16:191–205

Evans GD, Hughes AP (1961) Plant growth and the aerial environment. I. Effects of artificial shading on *Impatiens parviflora*. New Phytol 60:150–180

Fleischhacker PH, Senger H (1978) Adaptation of the photosynthetic apparatus of *Scenedesmus obliquus* to strong and weak light conditions. II. Differences in photochemical reactions, the photosynthetic electron transport and photosynthetic units. Physiol Plant 43:43–51

Foyer CH, Hall DO (1980) Oxygen metabolism in the active chloroplast. Trends in Biochemical Sciences 1980:188–191

Friend DJC (1975) Adaptation and adjustment of photosynthetic characteristics of gametophytes and sporophytes of Hawaiian Tree-fern (*Cibotium glaucum*) grown at different irradiances. Photosynthetica 4:48–57

Gabrielsen EK (1948) Effects of different chlorophyll concentrations on photosynthesis in foliage leaves. Physiol Plant 1:5–37

Gabrielsen EK (1960) Beleuchtungsstärke und Photosynthese. In: Ruhland W (ed) Encyclopedia of plant physiology. Springer, Berlin, Göttingen, Heidelberg. Vol. V/2, pp 27–47

Gaffron H, Wohl K (1936) Zur Theorie der Assimilation. Naturwissenschaften 24:81–90

Gauhl E (1969) Differential photosynthetic performance of *Solanum dulcamara* ecotypes from shaded and exposed habitats. Carnegie Inst Washington Yearb 67:482–487

Gauhl E (1970) Leaf factors affecting the rate of light-saturated photosynthesis in ecotypes of *Solanum dulcamara*. Carnegie Inst Washington Yearb 68:633–636

Gauhl E (1976) Photosynthetic response to varying light intensity in ecotypes of *Solanum dulcamara* L. from shaded and exposed environments. Oecologia 22:275–286

Gauhl E (1979) Sun and shade ecotypes of *Solanum dulcamara* L.: Photosynthetic light-dependence characteristics in relation to mild water stress. Oecologia 39:61–70

Goodchild DJ, Björkman O, Pyliotis NA (1972) Chloroplast ultrastructure, leaf anatomy, and content of chlorophyll and soluble protein in rainforest species. Carnegie Inst Washington Yearb 71:102–107

Grahl H, Wild A (1972) Die Variabilität der Photosyntheseeinheit bei Licht- und Schattenpflanzen. Untersuchungen zur Photosynthese von experimentell induzierten Licht- und Schattentypen von *Sinapis alba*. Z Pflanzenphysiol 67:443–453

Grahl H, Wild A (1973) Lichtinduzierte Veränderungen im Photosyntheseapparat von *Sinapis alba*. Ber Dtsch Bot Ges 86:341–349

Grahl H, Wild A (1975) Studies on the content of P-700 and cytochromes in *Sinapis alba* during growth under two different light intensities. In: Marcelle R (ed) Environmental and biological control of photosynthesis, W. Junk, The Hague, pp 107–113

Hanson HC (1917) Leaf-structure as related to environment. Am J Bot 4:533–560

Hariri M, Brangeon JL (1977) Light-induced adaptive responses under greenhouse and controlled conditions in the fern *Pteris cretica* var. *ouvardii*. I. Structural and infrastructural features. Physiol Plant 41:280–288

Hariri M, Prioul JL (1978) Light-induced adaptive responses under greenhouse and controlled conditions in the fern *Pteris cretica* var. *ouvardii*. II. Photosynthetic capacities. Physiol Plant 42:97–102

Harvey G (1980) Seasonal alteration of photosynthetic unit sizes in three herb layer components of a deciduous forest community. Am J Bot 67:293–299

Hatch MD, Slack CR, Bull TA (1969) Light-induced changes in the content of some enzymes of the C_4 dicarboxylic acid pathway of photosynthesis and its effect on other characteristics of photosynthesis. Phytochemistry 8:697–706

Hesselman H (1904) Zur Kenntnis des Pflanzenlebens schwedischer Laubwiesen. Eine physiologisch-biologische und pflanzengeographische Studie. Beih Bot Centralbl 17: 341–460

Hiesey W, Nobs MA, Björkman O (1971) Experimental studies of the nature of species.

V. Biosystematics, genetics and physiological ecology of the Erythranthe section of *Mimulus*, Publication No. 628, Carnegie Inst Wash, Washington, D.C.

Hiroi T, Monsi M (1963) Physiological and ecological analyses of shade tolerance of plants. 3. Effect of shading on growth attributes of *Helianthus annuus*. Bot Mag 76:121–129

Hiroi T, Monsi M (1964) Physiological and ecological analyses of shade tolerance of plants. 4. Effect of shading and distribution of photosynthate in *Helianthus annuus*. Bot Mag 77:1–9

Hitz WD, Steward CR (1980) Oxygen and carbon dioxide effects on the pool size of some photosynthetic and photorespiratory intermediates in soybean (*Glycine max* [L]. Merr.). Plant Physiol 65:442–446

Holman R (1930) On solarization of leaves. Univ Calif Publ Bot 16:139–151

Holmgren P (1968) Leaf factors affecting light-saturated photosynthesis in ecotypes from exposed and shaded habitats cultivated under two light regimes. Physiol Plant 21:676–698

Hughes AP, Cockshull KE (1971) The effects of light intensity and carbon dioxide concentration on the growth of *Chrysanthemum morifolium* cv. Bright Golden Anne. Ann Bot 35:899–914

Jagels R (1970) Photosynthetic apparatus in *Selaginella*. I. Morphology and photosynthesis under different light and temperature regimes. Can J Bot 48:1843–1852

Jones LW, Kok B (1966a) Photoinhibition of chloroplast reactions. I. Kinetics and action spectra. Plant Physiol 41:1037–1043

Jones LW, Kok B (1966b) Photoinhibition of chloroplast reactions. II. Multiple effects. Plant Physiol 41:1044–1049

Jurik TW, Chabot JF, Chabot BF (1979) Ontogeny of photosynthetic performance in *Fragaria virginiana* under changing light regimes. Plant Physiol 63:542–547

Kandler O, Sironval C (1959) Photooxidation processes in normal green *Chlorella* cells. II. Effects on metabolism. Biochim Biophys Acta 33:207–215

Kok B (1956) On the inhibition of photosynthesis by intense light. Biochim Biophys Acta 21:234–244

Ku SB, Edwards GE (1978) Oxygen inhibition of photosynthesis. III. Temperature dependence of quantum yield and its relation to O_2/CO_2 solubility ratio. Planta 140:1–6

Ledig FT, Borman FM, Wenger KF (1970) The distribution of dry matter growth between shoot and roots in Loblolly pine. Bot Gaz 131:349–359

Lewandowska M, Hart JW, Jarvis PG (1976) Photosynthetic electron transport in plants of Sitka spruce subjected to different light environments during growth. Physiol Plant 37:269–274

Lichtenthaler H (1971) The unequal synthesis of the lipophilic plastidquinones in sun and shade leaves of *Fagus silvatica* L. Z. Naturforsch 26b:832–842

Loach K (1970) Shade tolerance in tree seedlings. II. Growth analysis of plants raised under artificial shade. New Phytol 67:273–286

Louwerse W, van der Zweerde W (1977) Photosynthesis, transpiration and leaf morphology of *Phaseolus vulgaris* and *Zea mays* grown at different irradiances in artificial and sunlight. Photosynthetica 11:11–21

Ludlow MM, Wilson GL (1971) Photosynthesis of tropical pasture plants. II. Photosynthesis and illuminance history. Aust J Biol Sci 24:1065–1075

Mache R, Loiseaux S (1973) Light saturation of growth and photosynthesis of the shade plant *Marchantia polymorpha*. Cell Sci 12:391–401

McClendon JH (1962) The relationship between the thickness of deciduous leaves and their maximum photosynthetic rate. Am J Bot 49:320–322

McCree KJ (1972) The action spectrum, absorptance and quantum yield of photosynthesis in crop plants. Agric Meteorol 9:191–216

Medina E (1970) Relationship between nitrogen level, photosynthetic capacity and carboxydismutase activity in *Atriplex patula* leaves. Carnegie Inst Washington Yearb 69:655–662

Medina E (1971) Effect of nitrogen supply and light intensity during growth on the photosynthetic capacity and carboxydismutase activity of leaves of *Atriplex patula* ssp. *hastata*. Carnegie Inst Washington Yearb 70:551–559

Melis A, Brown JS (1980a) Spectrophotometric determination of system I and II reaction centers in different photosynthetic membranes. Proc Natl Acad Sci USA 77:4712–4716

Melis A, Brown JS (1980b) Stoichiometry of system I and II reaction centers in photosynthetic membranes. Carnegie Inst Washington Yearb 79:172–175

Mohanty P, Boyer JS (1976) Chloroplast response to low leaf water potentials. IV. Quantum yield is reduced. Plant Physiol 57:704–709

Montfort CL (1941) Lichtlähmung und Lichtbleichung bei Wasserpflanzen. Grundsätzliches zur physiologischen Gestalt der submersen Blütenpflanzen. Planta 32:121–149

Mooney HA, Björkman O, Collatz GJ (1977) Photosynthetic acclimation to temperature and water stress in the desert shrub *Larrea divaricata*. Carnegie Inst Washington Yearb 76:328–335

Mooney HA, Björkman O, Collatz GJ (1978) Photosynthetic acclimation to temperature in the desert shrub, *Larrea divaricata*. I. Carbon exchange characteristics of intact leaves. Plant Physiol 61:406–410

Mousseau M, Costes F, de Kouchkovski Y (1967) Influence de conditions d'éclairement pendant la croissance sur l'activité photosynthétique de feuilles entière et de chloroplastes isolées. CR Paris D 264:1158–1161

Myerscough PJ, Whitehead FH (1966) Comparative biology of *Tussilago farfara*, *Chamaenerion angustifolium*, *Epilobium montanum* and *Epilobium adenocaulon*. I. General biology and germination. New Phytol 65:192–210

Neales TF, Treharne KJ, Wareing PF (1971) A relationship between net photosynthesis, diffusive resistance and carboxylating enzyme activity in bean leaves. In: Hatch MD, Osmond CB, Slatyer RO (eds) Photosynthesis and photorespiration, Wiley-Interscience, New York, pp 89–96

Nobel PS (1976) Photosynthetic rates of sun versus shade leaves of *Hyptis emoryi* Torr. Plant Physiol 58:218–223

Nobel PS (1977) Internal leaf area and cellular CO_2 resistance: photosynthetic implications of variations with growth conditions and plant species. Physiol Plant 40:137–144

Nobel PS (1980) Leaf anatomy and water-use efficiency. In: Turner NC, Kramer PJ (eds) Adaptation of plants to water and high temperature stress, Wiley-Interscience pp 43–55

Nobel PS, Zaragosa LJ, Smith WK (1975) Relationship between mesophyll surface area, photosynthetic rate, and illumination level during development of leaves of *Plectranthus parviflorus* Henekel. Plant Physiol 55:1067–1070

O'Leary MH (1981) Carbon isotope fractionation in plants. Phytochemistry 20:553–567

Öquist G, Brunes L, Hällgren JE (1981) Photosynthetic efficiency of *Betula verrucosa* acclimated to different light intensities. Plant, Cell Environment (in press)

Osmond CB, Björkman O (1972) Simultaneous measurements of oxygen effects on net photosynthesis and glycolate metabolism in C_3 and C_4 species. Carnegie Inst Washington Yearb 71:141–148

Osmond CB, Björkman O, Anderson DJ (1980) Physiological processes in plant ecology. Springer, Berlin, Heidelberg, New York

Park RB, Pon NG (1963) Chemical composition and substructure of lamellae isolated from spinach chloroplasts. J Mol Biol 6:105–114

Patterson DT, Duke SO (1979) Effect of growth irradiance on the maximum photosynthetic capacity of water hyacinth [*Eichhornia crassipes* (Mart.) Solms]. Plant Cell Physiol 20:177–184

Patterson DT, Bunce JA, Alberte RS, van Volkenburgh E (1977) Photosynthesis in relation to leaf characteristics of cotton from controlled and field environments. Plant Physiol 59:384–387

Patterson DT, Duke SO, Hoagland RE (1978) Effect of irradiance during growth on the adaptive photosynthetic characteristics of velvet leaf and cotton. Plant Physiol 61:402–405

Pearce RB, Lee DR (1969) Photosynthetic and morphological adaptation of alfalfa leaves to light intensity at different stages of maturity. Crop Sci 9:791–794

Powles SB (1979) The role of carbon assimilation and photorespiratory carbon cycling in the avoidance of photoinhibition in intact leaves of C_3 and C_4 plants. PhD thesis, Aust Nat Univ Canberra

Powles SB, Critchley C (1980) The effect of light intensity during growth on photoinhibition of intact attached bean leaflets. Plant Physiol 65:1181–1187

Powles SB, Osmond CB (1978) Inhibition of the capacity and efficiency of photosynthesis in bean leaflets illuminated in the absence of CO_2 at low O_2 concentrations – a protective role for photorespiration. Aust J Plant Physiol 5:619–629

Powles SB, Thorne SW (1981) Effect of high light treatments in inducing photoinhibition of photosynthesis in intact leaves of low-light grown *Phaseolus vulgaris* and *Lastreopsis microsora*. Planta 1981 (in press)

Powles SB, Osmond CB, Thorne SW (1979) Photoinhibition of intact attached leaves of C_3 plants illuminated in the absence of both carbon dioxide and of photorespiration. Plant Physiol 64:982–988

Powles SB, Berry JA, Björkman O (1980a) Interaction between light intensity and chilling temperatures on inhibition of photosynthesis in chilling-sensitive plants. Carnegie Inst Washington Yearb 79:157–160

Powles SB, Chapman KSR, Osmond CB (1980b) Photoinhibition of intact attached leaves of C_4 plants: Dependence of CO_2 and O_2 partial pressures. Aust J Plant Physiol 7:737–747

Prioul JL (1973) Éclairement de croissance et infrastructure des chloroplastes de *Lolium multiflorum* Lam. Relation avec les resistances au transfert de CO_2. Photosynthetica 7:373–381

Rabinowitch EI (1951) The light factor. Photosynthesis II. Wiley-Interscience, New York, pp 964–1191

Radmer RJ, Kok B, Ollinger O (1978) Kinetics and apparent K_m of oxygen cycle under conditions of limiting carbon dioxide fixation. Plant Physiol 61:914–917

Reyss A, Prioul JL (1975) Carbonic anhydrase and carboxylase activities from plants (*Lolium multiflorum*) adapted to different light regimes. Plant Sci Lett 5:189–195

Satoh K (1970a) Mechanism of photoinactivation in photosynthetic system. I. The dark reaction in photoinactivation. Plant Cell Physiol 11:15–27

Satoh K (1970b) Mechanism of photoinactivation in photosynthetic systems. II. The occurence and properties of two different types of photoinactivation. Plant Cell Physiol 11:29–38

Satoh K (1970c) Mechanism of photoinactivation in photosynthetic systems. III. Site and mode of photoinactivation in photosystem I. Plant Cell Physiol 11:187–197

Senger H, Fleischhacker PH (1978) Adaptation of the photosynthetic apparatus of *Scenedesmus obliquus* to strong and weak light conditions. I. Differences in pigments, photosynthetic capacity, quantum yield and dark reactions. Physiol Plant 43:35–42

Singh KD, Gopal B (1973) The effects of photoperiod and light intensity on the growth of some weeds of crop fields. In: Slayter RO (ed) Plant response to climatic factors. UNESCO, Paris, pp 73–75

Singh M, Ogren WL, Widholm JM (1974) Photosynthetic characteristics of several C_3 and C_4 plant species grown under different light intensities. Crop Sci 14:563–566

Skene DS (1974) Chloroplast structure in mature apple leaves grown under different levels of illumination and their response to illumination. Proc R Soc London Ser B 186:75–78

Stahl E (1883) über den Einfluß des sonnigen und schattigen Standortes auf die Ausbildung der Laubblätter. Jena Z Naturwiss 16:162–200

Steemann Nielsen E (1942) Der Mechanismus der Photosynthese. Dansk Bot Ark 11 (2) pp 95

Taylor AO, Craig S (1971) Plants under climatic stress. II. Low temperature, high light effects on chloroplast ultrastructure. Plant Physiol 47:719–725

Taylor AO, Rowley JA (1971) Plants under climatic stress. I. Low temperature, high light effects on photosynthesis. Plant Physiol 47:713–718

Taylor AO, Jesper NM, Christeller JT (1972) Plants under climatic stress. III. Low temperature, high light effects on photosynthetic products. Plant Physiol 47:798–802

Thornber JP (1975) Chlorophyll-proteins: light-harvesting and reaction center components of plants. Ann Rev Plant Physiol 26:127–158

Tolbert NE, Ryan FJ (1976) Glycolate biosynthesis and metabolism during photorespiration. In: Burris RH, Black CC (eds) CO_2 metabolism and plant productivity. University Park Press, pp 141–159

Turrell FM (1936) The area of internal exposed surface of dicotyledon leaves. Am J Bot 23:255–263

Ursprung A (1917) Über die Stärkebildung im Spectrum. Ber Dtsch Bot Ges 35:44–82

von Mohl H (1837) Untersuchungen über die winterliche Färbung der Blätter. Vermischte Schriften botanischen Inhalts. Ludwig Friedrich Fues, Tübingen, pp 375–392

Whitehead FH (1969) Rational of physiological ecology. Intecol Bull (London) 1:34–42

Whitehead FH (1973) The relationship between light intensity and reproductive capacity. In: Slatyer RO (ed) Plant response to climatic factors. UNESCO, Paris, pp 73–75

Wild A (1979) Physiologie der Photosynthese höherer Pflanzen. Die Anpassung an Lichtbedingungen. Ber Dtsch Bot Ges 92:341–364

Wild A, Höhler T (1978) Die Wirkung unterschiedlicher Lichtintensitäten während der Anzucht auf die CO_2-Kompensationslage, die Glykolsäure-Oxidase- und Ribulosebiphosphat-Carboxylase-Aktivitäten bei *Sinapis alba*. Pflanzenphysiol 87:413–428

Wild A, Grahl H, Zickler H-O (1972) Untersuchungen über den Ferredoxingehalt von experimentell induzierten Licht- und Schattentypen von *Sinapis alba*. Z Pflanzenphysiol 68:283–285

Wild A, Ke B, Shaw E (1973) The effect of light intensity during growth of *Sinapis alba* on the electron-transport components. Z Pflanzenphysiol 69:344–350

Wild A, Rühle W, Grahl H (1975) The effect of light intensity during growth of *Sinapis alba* on the electron-transport and the non-cyclic photophosphorylation. In: Marcelle R (ed) Environmental and biological control of photosynthesis. W Junk, The Hague, pp 115–121

Willmot A, Moore PD (1973) Adaptation to light intensity in *Silene alba* and *S. dioica*. Oikos 24:458–464

Willstätter R, Stoll A (1918) Untersuchung über die Assimilation der Kohlensäure. Julius Springer, Berlin

Wilson D, Cooper JP (1967) Assimilation of *Lolium* in relation to leaf mesophyll. Nature 214:989–992

Wilson D, Cooper JP (1969) Effect of light intensity during growth on leaf anatomy and subsequent light-saturated photosynthesis among contrasting *Lolium* genotypes. New Phytol 68:1125–1135

Wilson D, Cooper JP (1970) Effect of selection for mesophyll cell size on growth and assimilation in *Lolium perenne* L. New Phytol 69:233–245

Wong SC (1979) Stomatal behavior in relation to photosynthesis. PhD thesis, Aust Nat Univ, Canberra

Woolhouse HW (1967) Leaf age and mesophyll resistance as factors in the rate of photosynthesis. Hilger J 11:7–12

4 Non-photosynthetic Responses to Light Quality

D.C. MORGAN and HARRY SMITH

CONTENTS

4.1 Introduction

In nature, survival is dependent on the sensitivity with which an organism can perceive its environment. One environmental resource obviously of prime importance to plants is light, its optimum harvest by photosynthesis being essential for the survival of both the individual organism and the species. Photosynthetic optimisation has been rendered possible through the evolution of highly sensitive perception mechanisms by which many different aspects of the continuously variable and always complex radiation environment may be detected. The information gathered from the environment by these mechanisms allows the plant to adapt, or acclimate, to the light conditions by appropriately modulating its metabolism or development. The non-photosynthetic responses to light quality – namely, photomorphogenesis, phototropism and photoperiodism – which form the subject of this chapter are the physiological manifestations of the environmental perception mechanisms.

The origins of our present knowledge and understanding of photomorphogenesis, phototropism and photoperiodism can be traced at least to the early part of this century, but the presently emergent appreciation of the functions of these phenomena in plants growing under natural conditions has awaited the development of reliable instrumentation for the analysis and simulation of natural light spectra. Clearly, there is an abundant literature. The work discussed below is limited to that which has a direct bearing on the natural environment and in which the light environment is accurately defined. As this is, apparently, the first review of non-photosynthetic responses to light quality, it seems essential to set out in detail the spectral variation encountered in nature, and the responses to such variation as it is simulated in the controlled environment. The photoreceptors are given only minor attention here, since the detailed discussion which they deserve will appear in the later Photomorphogenesis volume of the Encyclopedia.

4.2 The Natural Light Environment (cf. also Chaps. 1 and 2, this Vol.)

4.2.1 Daylight

At solar elevations of greater than 10° the global radiation between 400 nm and 800 nm has a characteristically uniform spectrum which is made up from direct sunlight and diffuse skylight. This is daylight. When the solar disk has less than 10° elevation the spectrum of the global radiation between 400 nm and 800 nm is radically altered by atmospheric processes. We have called this period twilight (see below). The spectral distribution of direct sunlight at the Earth's surface (Fig. 4.1a[1]) is a result of attenuation of the direct solar beam in the Earth's atmosphere by ozone absorption, Rayleigh scattering, Mie scattering, and absorption by oxygen and water molecules (GATES 1966; ROBINSON 1966; HENDERSON 1977). Scattering produces diffuse skylight (Fig. 4.1b), which is rich in blue light due to Rayleigh scattering.

A typical spectral distribution for daylight under a clear midsummer sky is shown in Fig. 4.1c. Cloud cover and dust/haze (Mie particles) produce two frequently-encountered variations of this typical spectral distribution. Clouds appear to act as non-selective diffusing filters, which reflect a considerable proportion of direct sunlight. The resulting spectrum shows an increase in the proportion of blue (scattered) light, but very little change at the longer wavelengths (600–800 nm) (Fig. 4.1d) (TAYLOR and KERR 1941; HULL 1954; ROBERTSON 1966; HOLMES and SMITH 1977a). HOLMES and SMITH (1977a) found

1 The spectra for Figs. 1, 3 and 4 must unfortunately be presented in relative units (normalised to 1 at 600 nm), because it is not feasible to convert the published data to absolute photon units (μmol m^{-2} s^{-1}). However, we have included an approximate photon fluence rate per nm at 600 nm in the figure legends

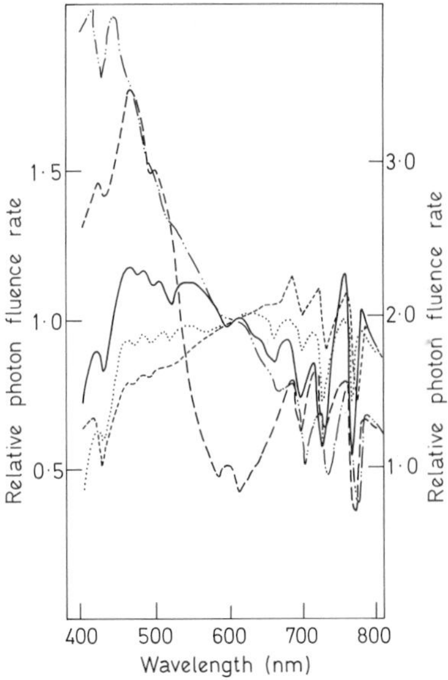

Fig. 4.1. Relative spectral photon distribution for a) direct sunlight (------), b) diffuse skylight (.. — —.. —), c) daylight with clear sky (······), d) daylight with sun obscured by cloud (———), e) twilight (— — —). The twilight spectrum was recorded at 17:15 GMT; sunset was at 17:10 GMT. The curves are normalised to 1.0 at 600 nm (a–d: left ordinate, e: right ordinate). (After HOLMES and SMITH 1977a.) Approximate fluence rates per nm at 600 nm: c) 6 μmol m^{-2} s^{-1} nm^{-1}; d) 1.5 μmol m^{-2} s^{-1} nm^{-1}; e) 0.0007 μmol m^{-2} s^{-1} nm^{-1}

that the effect of clouds on the ratio of red (600 nm) to far-red (730 nm) light was only ca. 5%, the values ranging from 1.03 to 1.22, the mean value on cloudless days being 1.15. The presence of dust or haze in the atmosphere results in a reduction in the proportion of blue light and an increase in the proportion of red light in the daylight spectrum (TAYLOR and KERR 1941; HULL 1954; ROBERTSON 1966).

4.2.2 Twilight

The variations in spectral distribution of daylight are small in comparison with those which occur at solar elevations of less than 10°, i.e., at sunset and sunrise. As solar elevation decreases the pathlength through the atmosphere (air mass) becomes very large leading to enhancement of all the processes of attenuation of the direct solar beam – scattering preferentially attenuates the shorter wavelengths, and refraction, which also becomes important at low solar elevations, preferentially enhances the longer wavelengths. Twilight spectra, therefore, are relatively rich in blue light (from diffuse skylight) and far-red light (from attenuated and refracted direct sunlight), and poor in orange-red light (Fig. 4.1e). This pattern becomes more exaggerated as the solar elevation decreases (ROBERTSON 1966; SHROPSHIRE 1973; MUNZ and MCFARLAND 1973; GOLDBERG and KLEIN 1977; HOLMES and SMITH 1977a). These changes can be detected beneath a forest canopy (TASKER 1977) but they are sensitive to variation in weather, dust and water content of the horizon sky.

Finally, twilight spectra will persist for very different periods of time, depending upon latitude and solar declination.

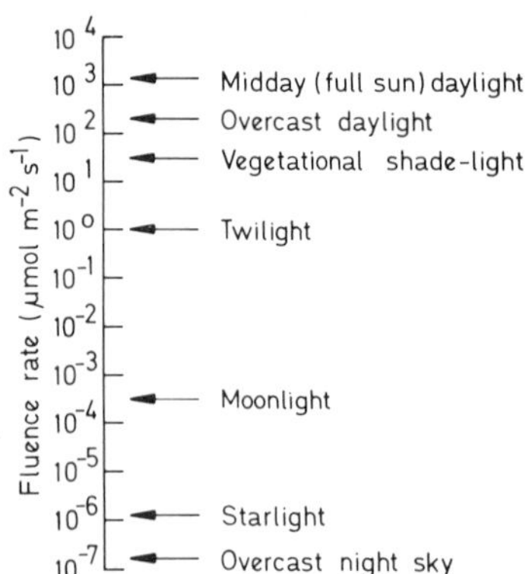

Fig. 4.2. Typical photon fluence rates (400–700 nm) for natural radiation. (Adapted from MUNZ and MCFARLAND 1973)

4.2.3 Moonlight and Starlight

At night the sole sources of light are moonlight and starlight. Moonlight is the reflection of the direct solar beam. Mie scattering and refraction by the particles of the lunar surface preferentially attenuate the shorter wavelengths (KOPAL 1969) with the result that the moonlight spectrum contains a slightly higher proportion of the longer wavelengths than direct sunlight, to which it is otherwise very similar (MUNZ and MCFARLAND 1973). The spectral distribution of individual stars depends upon their colour temperature. Global starlight has a spectrum which is similar to moonlight (MUNZ and MCFARLAND 1973), but with a fluence rate several orders of magnitude lower (Fig. 4.2).

4.2.4 Vegetational Shadelight

This is the light which is encountered within or beneath a vegetation canopy. It has two components: unfiltered daylight (direct sunlight, diffuse skylight, or diffuse light from clouds) which has passed through holes in the canopy, and filtered or attenuated daylight, the spectrum of which has been altered by the canopy by the processes of absorption, reflection and transmission. The spectra for reflection and transmission of green leaves are largely a consequence of absorption by chlorophyll (WOOLLEY 1971).

Figure 4.3a is a spectrum taken from beneath a broadleaf deciduous canopy on a clear sunny day; it shows the typical troughs in the blue and red, the minor peak in the green, and the major band in the far-red. Spectra from beneath coniferous woodland tend to have a more uniform distribution of radiation (Fig. 4.3b); over the 400–700 nm waveband there is only a minor peak in the blue, and the proportion of far-red light tends to be lower (COOMBE 1957; VEZINA and BOULTER 1966; FEDERER and TANNER 1966). Vegetation structure and density have a large effect on the spectral distribution of shadelight. GOODFELLOW and BARKHAM (1974) studied a range of canopy densities for a beechwood under a clear sky with the sun obscured by the vegetation, and found, as the canopy became less dense, and the number of gaps increased,

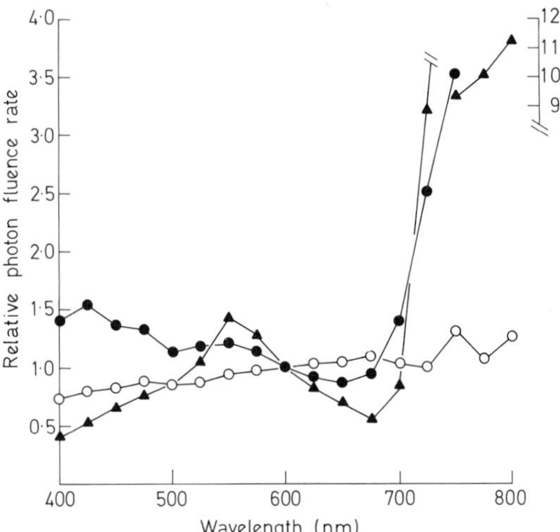

Fig. 4.3. Relative spectral photon distribution for a) broadleaf deciduous – beech – woodland, ▲ (Adapted from TASKER and SMITH 1977); b) needle-leaf coniferous – jack pine – woodland, ●. (Adapted from FEDERER and TANNER 1966); c) a sunfleck from within a wheat canopy, ○. (After HOLMES and SMITH 1977b)
The curves are normalised to 1.0 at 600 nm. Approximate photon fluence rates per nm at 600 nm: a) 0.075 μmol m^{-2} s^{-1} nm^{-1}; b) 0.4 μmol m^{-2} s^{-1} nm^{-1}; c) 4.5 μmol m^{-2} s^{-1} nm^{-1}

that the proportion of diffuse skylight increased, and the spectrum became enriched with blue light. Climatic conditions have only a relatively small effect on the spectrum of vegetational shadelight. In some cases the proportion of blue light under an overcast sky has been observed to decrease (HOLMES and SMITH 1977b; TASKER 1977), whilst in others an increase has been observed (FEDERER and TANNER 1966; VEZINA and BOULTER 1966); all show a consistent increase in the red, and a generally more uniform spectral distribution. The latter may reflect the proportionately greater contribution of non-attenuated diffuse light from the clouds.

The most striking aspect of all these vegetational shadelight spectra is the large difference in attenuation between the far-red and the visible radiation. The degree of shading may, therefore, be described numerically by calculating the ratio of photon fluence rates at two wavelengths, one in the far-red and one in the visible. The ratio of photon fluence rates in 10 nm bandwidths centred at 660 nm and 730 nm has been used (SINCLAIR and LEMON 1973; HOLMES and SMITH 1977b), chiefly because 660 nm and 730 nm are the action maxima of the photoreceptor phytochrome (Sect 4.3.1). This ratio is denoted by the symbol ζ (zeta) in many publications, but in the text of this chapter we use the more descriptive abbreviation R:FR[2]. For closed vegetational canopies the spectral variations reported in the literature are summarised in terms of R:FR in Table 4.1. R:FR is generally lower beneath broadleaf deciduous woodland than beneath coniferous woodland, but the variation is quite considerable. R:FR is always lower in shadelight than in daylight.

Finally, for deciduous canopies the vegetational shadelight spectrum varies with season. TASKER and SMITH (1977) have observed a biphasic change in

2 Abbreviations peculiar to Chapter 4: R:FR, or ζ (zeta), the ratio of photon fluence rate between 655 nm and 665 nm (red light) to that between 725 nm and 735 nm (far-red light); Pfr, the far-red-light-absorbing form of phytochrome; Pr, the red-light-absorbing form of phytochrome; ϕ (phi), the phytochrome photoequilibrium, i.e., Pfr/Pr+Pfr

Table 4.1. Estimation of R:FR for shadelight beneath various vegetation canopies. In daylight R:FR is 1.15

Canopy	Sky conditions	R:FR	Reference
a. Crops			
Wheat	Clear	0.49	HOLMES and SMITH (1977b)
	Overcast	0.59	HOLMES and SMITH (1977b)
Maize	Clear	0.20	YOCUM et al. (1964)
Sugar beet	Partially overcast	0.11–0.41	HOLMES and SMITH (1975)
Tea	Overcast	0.09–0.15	HADFIELD (1974)
b. Broadleaf deciduous woodland			
Beech	Overcast	0.36–0.97	GOODFELLOW and BARKHAM (1974)
	All conditions	0.16–0.64	TASKER and SMITH (1977)
	?	0.0	STOUTJESDIJK (1972a)
Oak	Clear	0.12–0.17	FEDERER and TANNER (1966)
	Hazy	0.32	FEDERER and TANNER (1966)
	All conditions	0.37–0.77	TASKER and SMITH (1977)
Sweet chestnut	Clear	0.12	COOMBE (1957)
Sugar maple	Clear	0.14–0.28	VEZINA and BOULTER (1966)
	Clear	0.08–0.11	FEDERER and TANNER (1966)
	Overcast	0.21	FEDERER and TANNER (1966)
Birch	All conditions	0.56–0.78	TASKER and SMITH (1977)
Mixed	?	0.12	STOUTJESDIJK (1972a)
Hawthorn shrub	?	0.30	STOUTJESDIJK (1972a)
Alder shrub	?	0.35	STOUTJESDIJK (1972a)
c. Coniferous evergreen woodland			
Spruce	Clear	0.33	COOMBE (1957)
	Clear	0.15	FEDERER and TANNER (1966)
	Overcast	0.46	FEDERER and TANNER (1966)
Red pine	Clear	0.47–0.76	VEZINA and BOULTER (1966)
	Clear	0.33	FEDERER and TANNER (1966)
	Partially overcast	0.55	FEDERER and TANNER (1966)
	Hazy	0.61	FEDERER and TANNER (1966)
White pine	Clear	0.25–0.26	FEDERER and TANNER (1966)
	Hazy	0.49	FEDERER and TANNER (1966)
Jack pine	Clear	0.32	FEDERER and TANNER (1966)
	Overcast	0.76	FEDERER and TANNER (1966)
d. Tropical rain forest			
Montane	Bright	0.22–0.30	STOUTJESDIJK (1972b)
	Overcast	0.77	STOUTJESDIJK (1972b)
Lowland	?	0.26	STOUTJESDIJK (1972b)

R:FR values during the leaf phase of beechwood, and, of course, the spectral distribution changes during the autumn before the leaves abscise (FEDERER and TANNER 1966; TASKER and SMITH 1977).

4.2.5 Sunflecks

Sunflecks are a special case of vegetational shadelight in which direct sunlight breaks through a gap in the canopy. They are composed of direct sunlight, sunlight reflected from vegetation, diffuse skylight and vegetation-filtered diffuse skylight. Published spectra show that in the centre of all sunflecks direct sunlight predominates (Fig. 4.3c) (YOCUM et al. 1964; VEZINA and BOULTER 1966; HOLMES and SMITH 1977b), and the fluence rate is similar to that of the incident daylight. However, the spectra can vary considerably across a sunfleck (TASKER 1977), and at the periphery R:FR may be lower than in ambient shade. With increasing gap size the spectral distribution tends towards that of daylight from a clear sky (Fig. 4.1c).

4.2.6 Underwater

In water, downwelling radiation is refracted and absorbed. Refraction leads to the effect of limiting angle – below an optically smooth water surface direct radiation is confined to a cone with a half-angle of 48.6° (WEINBERG 1976). Weak diffuse radiation enters from other angles. The contrast between diffuse and direct radiation diminishes with depth, and, at depth, weak diffuse upwelling radiation can be a significant proportion of the total radiation (SPENCE 1975; WEINBERG 1976).

Absorption is wavelength-dependent, and is determined by the degree and type of turbidity. In very clear water (Fig. 4.4a) – Crater Lake, USA, has a purity approaching that of distilled water – far-red light is rapidly attenuated, and the spectrum of downwelling radiation is predominantly blue. Dissolved organic matter and peat-derived material selectively and strongly attenuate this blue light (Fig. 4.4b). A suspension of algae or other water plants also causes selective attenuation, and this depends upon the cell or colony size, the cell or colony concentration, and the pigment content (KIRK 1976), which, of course, can vary with the season (see Chap. 9, this Vol.). DUBINSKY and BERMAN (1979) followed the spectral changes in downwelling radiation of Lake Kinneret, Israel, over the period of bloom of the alga *Peridinium*. When the bloom had subsided, and the small amount of algae were evenly dispersed, the spectral distribution was similar to that of Crater Lake, with the exception of some absorption of blue light. Three months earlier, when the bloom was at its peak and the algae concentrated at 1 m, the spectral distribution was markedly different (Fig. 4.4c), for attenuation of blue light exceeded that of red light. This effect had been observed before for both freshwater and seawater (SAKAMOTO and HOGUETSU 1963; SPENCE et al. 1971).

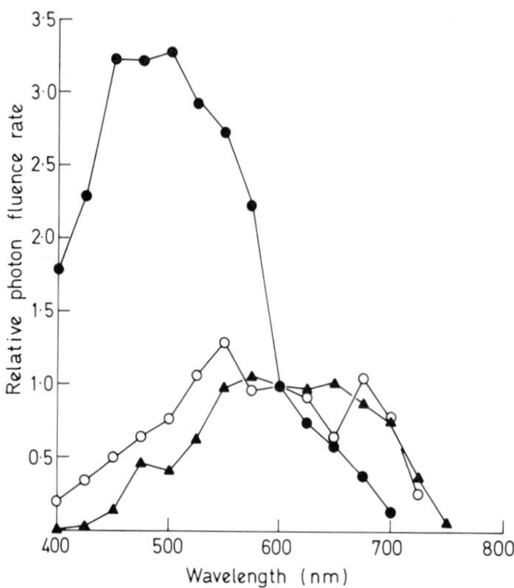

Fig. 4.4. Relative spectral photon distribution for downwelling radiation in a) Crater Lake at 5 m depth, ● (Adapted from TYLER and SMITH 1970); b) Loch Uanagan at 1 m depth, ▲ (Adapted from SPENCE et al. 1971); c) Lake Kinneret at 1 m depth, ○ (Adapted from DUBINSKY and BERMAN 1979.) The curves are normalised to 1.0 at 600 nm. Approximate photon fluence rates per nm at 600 nm: a) 1.6 μmol m^{-2} s^{-1} nm^{-1}; b) 2.4 μmol m^{-2} s^{-1} nm^{-1}; c) 0.20 μmol m^{-2} s^{-1} nm^{-1}

4.3 Putative Photoreceptors

4.3.1 Phytochrome

PRATT (1979) has reviewed the chemistry of this photoreceptor, which is a photochromic protein. The molecule can exist in two interconvertible forms: Pr, which absorbs maximally at 660 nm (red), and Pfr, which absorbs maximally at 730 nm (far-red). The action spectra for conversion of one form to the other overlap (Fig. 4.5a), and thus at any wavelength below 730 nm, or in broadband irradiation, the two forms will be continuously interconverted – a process known as cycling – and will come to a dynamic equilibrium, normally expressed quantitatively as Pfr/Ptotal and symbolised ϕ. In dark-grown tissue several other processes – synthesis of Pr, "destruction" of Pfr, dark reversion from Pfr to Pr – may affect the photoequilibrium or the total amount of phytochrome. Less is known for green plants because their phytochrome cannot be measured spectrophotometrically. Several problems are particularly important: the effect of chlorophyll screening; the effect of cycling; the magnitude of "destruction"; and the effect of the different molecular environment. It is outside the scope of this article to discuss these problems, which will be dealt with in the Photomorphogenesis Volume of this Encyclopedia.

4.3.2 Chlorophyll

MANCINELLI and RABINO (1978) have reviewed the small amount of evidence which suggests that photosynthesis is involved in photomorphogenic responses. The data suggest that in adult green plants photosynthesis may only affect responses through its role of energy transduction. This contribution is important

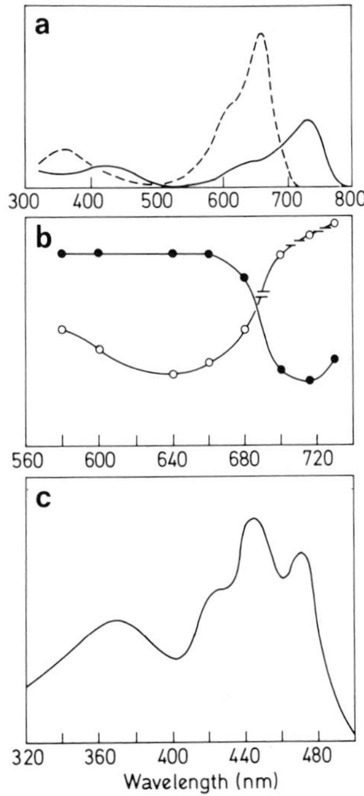

Fig. 4.5 a–c. Action spectra; **a** relative quantum yield for photoconversion of phytochrome; ------ Pr → Pfr, ——— Pfr → Pr (After BUTLER et al. 1964a); **b** relative photophosphorylation activity (quantum requirement per ATP) for ● cyclic photophosphorylation and ○ non-cyclic photophosphorylation (after AVRON and BEN-HAYYIM 1969); **c** the blue-light receptor (After CURRY 1969)

for responses in natural, and simulated natural (broadband) light environments, however, since it may mask or interact with the photomorphogenic response. In younger, de-etiolating, tissue some evidence suggests that photosynthesis may have a true photomorphogenic role (SCHNEIDER and STIMSON 1971, 1972), in which chlorophyll may detect red and far-red light by the balance of cyclic (action maximum in the far-red), and non-cyclic (action maximum in the red) photophosphorylation (Fig. 4.5b).

4.3.3 The Blue-Light Receptor

PRESTI and DELBRÜCK (1978) and GRESSEL (1979) have reviewed the chemistry of this photoreceptor, which absorbs maximally in the blue. The blue-light receptor has an action band between 350 nm and 500 nm, with action maxima at approximately 450 nm and 370 nm and subpeaks or shoulders at about 420 nm and 480 nm (Fig. 4.5c). Carotenoids, a flavin, and a flavin-cytochrome complex have all been proposed as the photoreceptor, but the current consensus, for most responses, favours a flavin.

4.3.4 Detection of Light Quality and Light Quantity

All of the naturally occurring spectra (Sect. 4.2) could be resolved with a combination of two photoreceptors – one for light quantity and one for light quality.

Light quality can only be detected by a photoreceptor, or a complex of two photoreceptors, which is able to absorb two or more wavelengths of light, and compare their relative magnitude. This criterion is fulfilled by phytochrome, and possibly the photosynthetic complex. The blue-light receptor absorbs in the ultraviolet and blue, but these wavelengths do not have a photoreversible effect, and furthermore the ultraviolet content of natural light is small, because of attenuation by atmospheric ozone. The blue-light receptor may therefore only act as a light quantity detector.

4.4 The Responses

4.4.1 Seed Germination

For many species light-sensitised seed germination has been categorised as being positively photoblastic (germination promoted by white light) or negatively photoblastic (germination inhibited by white light) (SMITH 1975 lists some examples). It is misleading strictly to categorise these two responses, however, since they are in reality two different manifestations of the same underlying physiological phenomenon (SMITH 1973). Light sensitivity appears to ensure that germination will occur only when the seed is either buried or exposed, depending upon the physiological state of the photoreceptor. SAUER and STRUIK (1964) found that many more seeds, in soils from a variety of habitats, would germinate if the soil was disturbed in the light, than if it was disturbed in the dark. The seedlings which emerged were all weed/pioneer species, which were absent from the vegetation covering three of the five habitats. WESSON and WAREING (1969a, b) made similar observations, and also found that the seeds of normally light-insensitive species became light-requiring following a period of burial. Even shallow burial will cause inhibition of germination. WOLLEY and STROLLER (1978) found that 1 mm of silty-clay loam, or 1 mm of sand, reduced daylight to 2%–6% of its incident value, whilst at depths greater than 2 mm attenuation was complete. However, in many cases promotion of germination by light is restricted to a rather limited temperature range (e.g., MANCINELLI et al. 1967). The surface of bare ground will be subject to large diurnal fluctuations in temperature (STROLLER and WAX 1973), and there are no definitive data which demonstrate a light-stimulated germination under such field conditions.

Photoblastic seeds are sensitive to light quality, with action maxima in the red and far-red. These two wavebands act antagonistically (BORTHWICK et al. 1952), which suggests that in nature, phytochrome may act to detect exposure to light following burial, and thus trigger germination. The seeds of many species show large variability in their germination behaviour, however, some requiring light, others not. Of those which require light some have a photoperiodic requirement, whilst others merely require a threshold value of fluence rate. Such variation may clearly have adaptive significance, but no specific examples have been reported.

Light-sensitive seed germination is not simply an on/off light/dark triggered process. When red and far-red are given as admixtures of various proportions, germination is sensitive to R:FR over a wide range (CUMMING 1963; KENDRICK and FRANKLAND 1968; FRANKLAND 1976). It has been proposed that this is also an ecologically significant phenomenon, since the daytime R:FR of terrestrial light is related to the depth of vegetational shade (Table 4.1) (CUMMING 1963; BLACK 1969; TAYLORSON and BORTHWICK 1969; VAN DER VEEN 1970; STOUTJESDIJK 1972a; SMITH 1973), and there is stronger evidence for this. Seeds of many species are prevented from germinating, either partially or completely, when they receive sunlight or artificial light which has been filtered through one or two green leaves (BLACK 1969; TAYLORSON and BORTHWICK 1969; VAN DER VEEN 1970). Inhibition of germination by leaf-filtered light is not a result of the low fluence rate, which when given as white light does not inhibit germination (TAYLORSON and BORTHWICK 1969). Also, prolonged irradiation with leaf-filtered light will prevent germination in "Grand Rapids" lettuce at both high temperature, when germination is light-requiring, and low temperature, when seeds will germinate in darkness (BLACK 1969). GORSKI (1975) found that beneath natural vegetation canopies germination was inhibited not only in all of the light-requiring species which were tested, but also in 14 out of the 19 "light-insensitive" species. Similarly, STOUTJESDIJK (1972a) observed that for a number of species a hawthorn canopy inhibited germination; others were unaffected. VAZQUEZ-YANES (1980) found that the germination of two shade-intolerant species was inhibited by the shadelight of a rain forest canopy, but that the germination could be fully promoted if the shadelight was filtered through a red narrow band-pass filter. STOUTJESDIJK (1972a) questioned the possibility that in temperate climates shade-inhibition kept seeds dormant until a canopy gap was formed, since here much germination occurs in the spring before the leaf phase. However, KING (1975) found that the winter annuals *Arenaria serpyllifolia, Veronica arvensis* and *Cerastium holosteoides*, which do germinate during the leaf phase, are almost completely inhibited from germinating by leaf-filtered light, and in nature they are only found on the open ground of ant hills. Of course, such light-sensitive germination does not necessarily explain their natural distribution, although it is highly suggestive.

Finally, the light environment during seed maturation can be detected by the immature seed (SHROPSHIRE 1973), and can affect the subsequent light-sensitive germination response (GUTTERMAN and PORATH 1975). Seeds of *Arabidopsis thaliana* which had matured in incandescent light showed no subsequent dark germination, and required less quanta of red to promote germination than seeds which had matured under fluorescent light (SHROPSHIRE 1973). These light sources are not satisfactory analogues of natural shadelight and daylight (Sect. 4.4.2) and it would be profitable to know whether maturation in better simulations of natural radiation would have a similar effect.

Such "carry-over" effects are probably the result of the phytochrome photoequilibrium established within the seed during maturation which is "fixed" when the seed dehydrates. The relationship between the incident radiation and the photoequilibrium established is not simple, however, for surrounding tissues, even the seed coat itself (MANCINELLI et al. 1967), can change the spectral distribution of light incident within the seed, and Pfr can revert to Pr in conditions of partial hydration which do not allow germination (HSIAO and VIDAVER 1973).

4.4.2 Photomorphogenesis – Red/Far-Red Ratio

There are several methods of simulating the naturally occurring changes in R:FR for studies in the controlled environment of the effects of twilight and vegetational shadelight. The simplest and most unsatisfactory method is to compare the effects of white fluorescent light alone (R:FR 3–6), with those of either admixtures with incandescent light, or incandescent light alone (R:FR 0.7–0.8). In terrestrial nature the highest R:FR is found in daylight (R:FR=1.15); R:FR values in shade and at twilight are lower. Fluorescent light does not, therefore, simulate daylight well, whilst light from tungsten filament lamps is equivalent to only very sparse shade. Furthermore, there is a considerable difference in the fluence rate of blue light between these two sources. These problems can be partially overcome if fluorescent light is used as a background white light source for a constant spectrum and fluence rate of photosynthetically active radiation, and far-red light (>700 nm) is added from either filtered incandescent light (HEATHCOTE et al. 1979) or far-red emitting fluorescent tubes (DEITZER et al. 1979). Full simulation of the spectral photon distributions of natural radiation poses major technical problems, and has not yet been achieved, except for certain very small sources.

Stem Extension. In 1957, DOWNS et al. were the first to demonstrate red/far-red reversible control of stem development in light-grown plants. A 5-min exposure to far-red light, at the end of an 8-h white (fluorescent) light photoperiod, increased internode extension by up to 400% in a range of bean (*Phaseolus vulgaris*) varieties, and in sunflower (*Helianthus annuus*) and morning glory (*Ipomoea hederacea*). When far-red light was immediately followed by 5-min red light, the effect was fully reversed. These observations have been repeated many times in many more species (e.g., SELMAN and AHMED 1962; KASPERBAUER 1971; LECHARNY and JACQUES 1974; VINCE-PRUE 1977). KASPERBAUER (1971) found the effects of end-of-day far-red light-treatment of tobacco in the controlled environment to be similar to the effects of vegetational shade in field conditions. Comparisons of the effect of growth in incandescent light with growth in fluorescent light, with the same amount of photosynthetically active radiation, for the whole of the daily white light period have also suggested that R:FR can markedly affect stem extension (BORTHWICK 1957; COOKE 1969; HOLMES and SMITH 1975, 1977c). In all of these cases the incandescent light elicited a higher stem extension rate. The best evidence that the R:FR of vegetational shade-light can modulate stem extension rate, however, comes from studies in which supplementary far-red light has been added to background white light for the whole of the daily photoperiod. In this situation, stem extension rate is proportional to the phytochrome photoequilibrium (estimated from R:FR) for the range of photoequilibria found in natural shade (Fig. 4.6) (MORGAN and SMITH 1976; HOLMES and SMITH 1977c).

Fig. 4.6. Effect of 21 d simulated shadelight on development in *Chenopodium album*. *Left* control – white fluorescent light; *right* simulated shadelight – white fluorescent light + supplementary far-red light (After MORGAN and SMITH 1976)

It is possible that this is a response to the end-of-day light treatment, and that daytime-simulated vegetational shadelight does not affect development. In this case the significance of the response in nature would be less plausible, because of the reduction in R:FR during twilight (Sect. 4.2.2.). However, MORGAN and SMITH (1978a) found that, in 16-h photoperiods, 80% of the increased elongation was due to daytime irradiation. End-of-day irradiation with shadelight may be of more significance for shorter (8-h) photoperiods (VINCE-PRUE 1977), but, for temperate regions at least, these occur during the period of least-active vegetative growth.

The daytime stem extension response to simulated shade has a systematic relationship to species habitat – species from open habitats react by large increases in stem extension rate, species from woodland/shade habitats react less strongly or not at all (MORGAN and SMITH 1979). These two classes of response may have ecological significance – species from open habitats may overtop an herbaceous vegetation canopy, whilst woodland herbs cannot overtop their canopy. Indeed, an extreme response by a woodland plant may make it more susceptible to fungal attack (VAARTAJA 1962). GRIME (1966) has listed several other characters which distinguish shade-avoiding and shade-tolerating flowering plants.

It may also be of ecological significance that stem extension rate will react to a change in R:FR within 15 min (MORGAN and SMITH 1978b; LOVEYS 1979), since sunflecks can have a similar life-time (WOODS and TURNER 1971). However, the large changes in temperature and humidity which also accompany natural sunflecks may obscure or inhibit a response to light quality. Over the long-term RAJAN et al. (1971) and WARRINGTON et al. (1976) observed R:FR-modulated stem extension for a range of temperature and fluence rate. Unfortunately, the species studied were all crop plants, and the lamp combinations used produced a range of R:FR, most of which is not found in terrestrial nature.

There is some evidence to suggest that the response to the R:FR of simulated shade may change with plant age. McLAREN and SMITH (1978) observed that 5-day-old seedlings of *Rumex obtusifolius* did not show the petiole elongation observed when 35-day-old plants were transferred to simulated shadelight. GABA and BLACK (1979) found that cucumber seedlings, which had been de-etiolated for 30 h in white light, took a further 8 h to respond to red light. In comparison 5-week-old *Chenopodium album* plants (MORGAN and SMITH 1978b) and 5-week-old tomato plants (LOVEYS 1979) responded to supplementary far-red light within 15 min. The possibility that the response changes with the age of light-grown plants requires further investigation.

Apical Dominance. Field observations of many species have shown that plants growing in the open branch profusely, whilst shaded plants may show complete apical dominance (BOGORAD and McILRATH 1960; KASPERBAUER 1971; TUCKER and MANSFIELD 1972). In the controlled environment, R:FR has been found to exert a remarkable degree of control over this response. Short periods of far-red light or incandescent light at the end of the daily white light period will completely suppress axillary bud outgrowth in *Xanthium strumarium* (BOGORAD and McILRATH 1960; TUCKER and MANSFIELD 1972), tobacco (KASPERBAUER

1971), tomato (TUCKER 1976) and *Fuchsia* (VINCE-PRUE 1977). The far-red light only suppresses bud-break; elongation of axillaries is promoted if far-red light is given after outgrowth has started (TUCKER and MANSFIELD 1972).

There have been no reports of apical dominance in realistic shade simulation. In the studies of *Xanthium* discussed above incandescent light was considered to represent shadelight, whereas in fact it corresponds to only very sparse shade, and it may be that an experimental treatment with a R:FR equivalent to daylight would suppress axillary bud outgrowth. *Chenopodium album* and *Polygonum persicaria*, however, only show partial inhibition of outgrowth, even when incandescent light is used during the whole white light period (TUCKER 1972), and in these species more realistic shadelight simulation may produce the full expression of apical dominance which was originally observed in the field. Furthermore, apical dominance is not asserted when *C. album* and *P. persicaria* are grown in either high or low fluence rate fluorescent light, which indicates that, in natural shade, light quantity would probably not be responsible for suppression of axillary bud outgrowth.

It has been suggested that apical dominance may be another morphological adaptation to vegetational shade (TUCKER and MANSFIELD 1972). If axillary bud outgrowth is suppressed by shadelight, the maximum of reserves may be used for rapid stem elongation. It would be of interest to study apical dominance in a shade-tolerant plant in which stem extension is not highly responsive to shadelight.

Although BOGORAD and MCILRATH (1960) found that excision of the three youngest leaves of *Xanthium* prevented the assertion of apical dominance by end-of-day far-red, TUCKER and MANSFIELD (1972) showed that, for plants treated with incandescent light for the whole of the daily white light photoperiod, excision had no effect. This may indicate that the axillary buds are involved in photoperception for this response. Endogenous buds in root cultures of *Convolvulus arvensis* are suppressed from outgrowth by darkness and far-red light, but will elongate in red light or white light (BONNETT 1972). *Convolvulus* produces horizontal roots, and a single plant may colonize an area of greater than 3 m radius in one season. Photoperception by the bud may detect the soil surface or degree of vegetational shade. In rhizome sections of *Agropyron repens* buds break freely in darkness, but are partially suppressed by light (LEAKEY et al. 1978). Incandescent light is more effective than fluorescent light. Following bud break a single shoot asserts dominance. Fluorescent light cannot prevent assertion of dominance, whilst incandescent light can, but only during specific parts of the year. Clearly, these latter observations may have ecological significance, but a more realistic simulation of the natural light environment is required.

Others. Increase of stem extension rate and assertion of apical dominance are only two of the many developmental changes which occur in light-grown plants in response to R:FR decreases (Table 4.2). A particular response may vary between species either in magnitude or direction. In some cases the responsivity is systematically related to habitat, in other cases it is not (MORGAN and SMITH 1979). Reports of plants showing no response at all are rare (MCDONOUGH

Table 4.2. Photomorphogenic changes (in addition to those discussed in the text) shown by plants in response to an increase of the proportion of far-red light in the incident radiation (+ = stimulation, − = inhibition)

Developmental factor	Effect	Species	Light treatment	Reference
Petiole length	1. +	Strawberry	End-of-day far-red	VINCE-PRUE (1976)
	2. +	*Cucurbita pepo*	Simulated shadelight	HOLMES and SMITH (1977c)
Leaf length	1. +	Tobacco	End-of-day far-red	KASPERBAUER and HIATT (1966)
	2. −	Mosses	Simulated shadelight	HODDINOTT and BAIN (1979)
Leaf area	1. −	*Chenopodium album*	Tungsten/fluorescent	HOLMES and SMITH (1975)
	2. −	*Rumex obtusifolius*	Simulated shadelight	MCLAREN and SMITH (1978)
Stem dry weight	1. +	Tomato	Tungsten/fluorescent	HURD (1974)
	2. +	*Cucurbita pepo* (hypocotyl)	Simulated shadelight	HOLMES and SMITH (1977c)
Transition to climbing form	1. −	*Phaseolus*	Nightbreak far-red	KRETSCHMER et al. (1977)
Flowering	1. +	Wheat	Tungsten/fluorescent	FRIEND et al. (1961)
	2. +	Strawberry	Tungsten/fluorescent	COLLINS (1966)
Senescence	1. +	*Marchantia*	End-of-day far-red	DEGREEF and FREDERICQ (1972)
Chlorophyll content	1. −	*Marchantia*	End-of-day far-red	FREDERICQ and DEGREEF (1966)
	2. −	*Rumex obtusifolius*	Simulated shadelight	MCLAREN and SMITH (1978)
	3. +	*Veronica persica*	Simulated shadelight	FITTER and ASHMORE (1974)
Free sugars/organic acids	1. +	Tobacco	End-of-day far-red	KASPERBAUER et al. (1970)
Ethylene production	1. +	Peach	Natural shade	EREZ (1977)

and BROWN 1969), and may result from unrealistic shadelight simulation. Developmental responses to supplementary far-red light have also been observed in liverworts, mosses, and ferns, and it seems likely that the R:FR detection mechanism functions universally in green plants. FRANKLAND and LETENDRE (1978) observed some changes in developmental pattern of *Circaea lutetiana* grown in low fluence rate simulated-shadelight (supplementary far-red) in comparison with plants grown in fluorescent light at the same fluence rate. They found that plants grown in the latter most closely resembled plants grown in natural shade, and concluded that there was no evidence to suggest that in *C. lutetiana* changes in spectral quality are necessary for the plant to detect shading. *C. lutetiana* is a woodland/shade-tolerant plant, and it has been shown that such plants are less responsive to simulated shadelight quality (MORGAN and SMITH 1979). In a study of shade-tolerant and shade-intolerant species of *Veronica* in low fluence rate simulated shade, FITTER and ASHMORE (1974) found that development in the shade-tolerant *V. montana* was unaffected by either the low fluence rate or the supplementary far-red light component of the simulated shade. Development in the shade-intolerant *V. persica* was radically changed, and showed a far-red specific reduction in leaf area. In natural shade, compared with daylight, *V. persica* showed a similar change in developmental pattern.

In the aquatic fern, *Marsilea vestita*, exposure to far-red light followed by return to white light is the most effective means of eliciting the production of aerial fronds (GAUDET 1965). In water far-red light is rapidly attenuated (Sect. 4.2.6). It is possible that the increase in far-red light as submerged fronds approach the surface may be the signal for development of aerial fronds in this species, but more accurate simulations of natural spectra are required to test this hypothesis. BODKIN et al. (1980) have evidence that R:FR can control heterophylly in *Hippuris vulgaris*, an aquatic angiosperm.

The Photoreceptor. Several workers have suggested that it is phytochrome which detects the R:FR spectral changes associated with vegetational shadelight (VEZINA and BOULTER 1966; KASPERBAUER 1971; SINCLAIR and LEMON 1973; SMITH 1973), and the bulk of the evidence now supports this proposal.

The phytochrome photoequilibrium, established in non-green tissue, is most sensitive to that range of R:FR found in nature (Fig. 4.7) (SMITH and HOLMES 1977). The response to daytime supplementary far-red light, for a particular species, is similar to the response to end-of-day far-red light, and the end-of-day response is red/far-red reversible. In *Fuchsia* final internode length is linearly related to the phytochrome photoequilibrium which, in non-green tissue, would be established by the all-night low fluence rate light treatments of mixed red and far-red light (VINCE-PRUE 1977). (However, this response becomes non-linear at photoequilibria of less than 0.3.) Similarly, stem extension rate in *Chenopodium album* is linearly related to the photoequilibrium which would be established in non-green tissue by the daytime simulated shadelight environment (MORGAN and SMITH 1978a). This correlative evidence suggests that phytochrome is controlling the photomorphogenic responses to the R:FR balance of natural spectra. MORGAN et al. (1980) have produced direct evidence that this is so – the increased

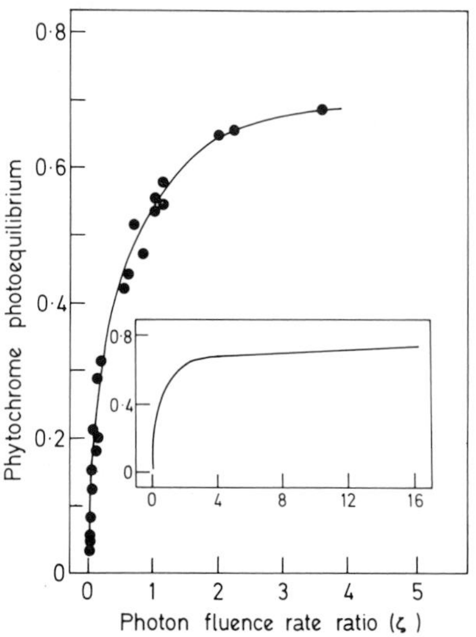

Fig. 4.7. The relationship between the zeta (ζ, R:FR) of the incident radiation and the phytochrome photoequilibrium established in dark-grown *Phaseolus vulgaris* tissue (After Smith and Holmes 1977)

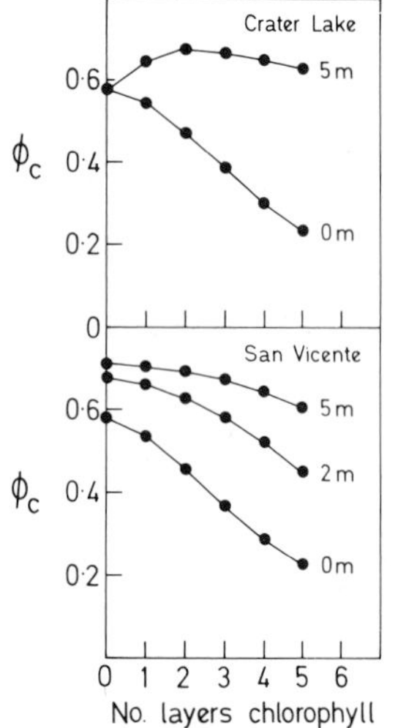

Fig. 4.8. The relationship between the calculated phytochrome photoequilibrium (φ_c), depth, and the number of layers of chlorophyll for downwelling radiation in Crater Lake and San Vicente Lake. φ_c was calculated using the formula from Tasker (1977), the data for phytochrome absorbance from Butler et al. (1964b), and the spectra for the Lakes from Tyler and Smith (1970). The transmission spectrum of chlorophyll was for a solution with a concentration of 6.95×10^{-5} mg mm^{-2}

stem extension rate found in *Sinapis alba* growing in simulated shadelight (white light plus supplementary far-red) can be reversed by adding red light to the spectrum.

For aquatic plants phytochrome may also detect spectral changes in the light environment, but the significance of this is less clear. It is possible to calculate the phytochrome photoequilibrium that would be established in non-green tissue by the underwater spectrum (Fig. 4.8). In the extremely clear water of Crater Lake the difference in photoequilibrium between 0 m and 5 m is insignificant. The effect of a layer of chlorophyll, which absorbs blue light more strongly than red light, depends on depth. At 0 m the far-red light has a dominant effect. At 5 m, where far-red light is absent, the greater loss of blue light causes an increase in photoequilibrium. In the lake at San Vicente,

USA, which is "heavily contaminated with phytoplankton, zooplankton, and plankton decomposition products" (TYLER and SMITH 1970), the photoequilibrium shows a steady decline with depth. The photoequilibrium, therefore, is a complex function of depth, turbidity, and the depth at which turbidity occurs, and, as such, its predictive value as an index of shade or depth would be limited.

4.4.3 Photomorphogenesis – Blue Light Effects

Three photoreceptors – chlorophyll, phytochrome, and the blue-light receptor – may absorb blue light in green plants. Simulation in the controlled environment of natural changes in the fluence rate of blue light will result in a change in photosynthetic output, unless the fluence rate is high enough to saturate photosynthesis, or the fluence rate of red light is adjusted to keep photosynthesis constant. Neither of these possibilities is particularly satisfactory – light sources rich in blue tend to have a low output, and therefore maintaining saturated photosynthesis is technically difficult; adjustment of the red light fluence rate, whilst keeping photosynthesis constant, may sensitize other photoreceptors and mask or interact with the response to blue light. It is difficult, therefore, to distinguish which photoreceptor is involved in a particular photomorphogenic response without resorting to the use of unnatural light treatments, and, since most of the published reports have focused on the photoreceptor, it is difficult to assess the significance of blue light responses in nature. MEIJER (1958, 1959) was able to demonstrate that the relative internode growth in red and blue light could change when different pretreatment lighting conditions were used, by growing "green" plants for long periods in narrow waveband spectral sources. This indicated that two independent photoreceptors were involved. However, in these experiments the light treatments were set at equal energy rather than equal photon fluence rates, and, therefore, a photosynthetic effect cannot be ruled out (see Chap. 2, this Vol.). More recent evidence suggests that two independent photoreceptors operate in young, newly de-etiolated cucumber seedlings. Both red and blue light inhibit hypocotyl elongation in these seedlings, but they have different sites of perception (BLACK and SHUTTLEWORTH 1974) and they produce different response kinetics (GABA and BLACK 1979). The observations of THOMAS and DICKINSON (1979), that hypocotyl length in very young seedlings of lettuce, tomato, and cucumber growing in high fluence rate yellow SOX (low pressure sodium discharge) light is inhibited by as much as 80% by a low fluence rate blue light admixture, is highly suggestive of photoperception by the blue-light photoreceptor. Light from SOX (λ_{max} 590 nm) photoconverts phytochrome with high quantum efficiency to establish maximum photoequilibrium. Blue light will tend to reconvert some of this Pfr to Pr, but the quantum efficiency for this conversion is much lower (Fig. 4.5a), and, with the low fluence rate of the supplementary blue light used, it is not likely to have had any significant effect on phytochrome photoequilibrium. It is similarly unlikely to have had any large effect on phytochrome cycling, or on photosynthesis.

The work discussed above indicates that the blue-light receptor probably functions in light-grown plants, although the plants used in many cases have been very young, and have been de-etiolated for only a short time. Furthermore, the blue light fluence rates used have been rather low, at least in comparison with daylight. For the three species considered by THOMAS and DICKINSON (1979) inhibition had already reached 50%–80% with a fluence rate between 400 nm and 500 nm of only 10 μmol m^{-2} s^{-1}. Although in deep shade blue light can be attenuated to a lower value, such a fluence rate between 400 nm and 500 nm is not atypical for vegetational shade. Clearly then, it is not possible to deduce from these data the significance of the blue-light receptor in nature; older fully de-etiolated plants treated with a range of fluence rate of blue light equivalent to that found in nature must be used to establish this.

4.4.4 Phototropism and Turgor Movements

There are two types of orientation response – growth curvature resulting from an imbalance in extension growth (phototropism), and movements resulting from turgor changes in specific cells. Orientation of the lamina either towards or away from the largest fluence rate would clearly be advantageous in different situations. SHELL and LANG (1976) calculated that photosynthesis could be increased by between 10% and 23% in *Helianthus* because of the phototropic movements by which the leaves track the sun. Observations of dark-grown material in narrow waveband light have provided evidence that orientation, by phototropism at least, is determined by the fluence rate of blue light, following photodetection, probably by a flavin (Sect. 4.3.3; Fig. 4.5c). However, the control of orientation movements in light-grown plants by the light quality of natural or simulated natural spectra is a sadly neglected subject, which must await adequate experimentation before its ecological significance can be assessed.

4.4.5 Photoperiodism

This topic was comprehensively reviewed by VINCE-PRUE in 1975, and is discussed in detail in Chapter 5, this Volume. Although photoperiodic responses occur in light-dark cycles in which the white light must exceed a critical fluence rate, the many experiments with light quality suggest that phytochrome is involved, whilst photosynthesis is probably only important in so far as it provides metabolites.

The possible effects of natural variation of light quality – daylight, vegetational shade, twilight – in photoperiodism have been largely unexplored. The red/far-red spectral shifts which occur at twilight (Sect. 4.2.2) are an astronomical phenomenon, and therefore have the timing precision required for photoperiodic control (WAGNER 1976), but the absolute changes in R:FR are highly variable, and no larger than may be experienced in shadelight during the day (TASKER 1977). It would be difficult, therefore, for a timing mechanism to operate by detecting twilight spectral shifts without the screening effect of a circadian oscillation in sensitivity (HOLMES and WAGNER 1980). Furthermore, numerous experiments in the controlled environment have demonstrated photoperiodic control in the absence of "dawn" and "dusk" spectral shifts. DEITZER et al. (1979) have demonstrated that simulated shadelight (white light plus supplementary far-red light), can shorten the time to onset of floral initiation, and accelerate floral development in comparison with simulated daylight (white light alone) in the facultative long-day plant "Wintex" barley. This stimulation by shadelight was subject to circadian oscillation.

4.4.6 Others

Light quality also influences stomatal aperture (ZEIGER and HEPLER 1977), chloroplast orientation (HAUPT 1972) and "sleep" movements (FONDEVILLE et al. 1966), but again a discussion of its effect in nature is precluded because of the unnatural light treatments used for experimental study.

4.5 Concluding Remarks

The evidence strongly suggests that the perception of light quality is of vital importance to plants growing in nature.

There is now a quite comprehensive description of the natural light environment, although not all of the data are in photon units, and there are some omissions, e.g., natural herbaceous canopies and moonlight. Natural spectra are in a continuous state of flux. The variations are large, but predictable, being determined by factors such as vegetation canopies and solar angle. Plant photoreceptors are most sensitive to precisely those wavelength ranges over which the largest spectral variations occur.

The literature describing the responses of green plants to simulations of natural spectra is not as comprehensive as that which describes the spectra themselves. Modulation of the R:FR balance can sensitize seed germination, and dramatically alter subsequent development. These responses are almost certainly under phytochrome control. For photomorphogenesis in blue light, phototropism, and photoperiodism, however, there has been no demonstration of control by simulated natural spectra. Furthermore, although phototropism and photoperiodism clearly occur in natural situations there has been little or no investigation of the possible role of light quality in these non-photosynthetic responses. Such experiments are particularly difficult since the spectral changes which occur in nature are usually accompanied by other environmental changes. This is more of a philosophical, than an actual, difficulty, however, since as physiologists we must believe that processes and mechanisms recorded under defined conditions have their counterparts in nature. What is difficult to determine, on the other hand, is the relative importance of the several, different, physiological responses to the several different but associated, environmental changes. At this point, the worrying can be left to the ecologists!

References

Avron M, Ben-Hayyim (1969) Interaction between two photochemical systems in photoreactions of isolated chloroplasts. In: Metzner H (ed) Progress in photosynthesis research III. International Union of Biological Sciences, Tübingen

Black M (1969) Light controlled germination of seeds. Symp Soc Exp Biol 23:193–217

Black M, Shuttleworth JE (1974) The role of the cotyledons in the photocontrol of hypocotyl extension in *Cucumis sativa* L. Planta 117:57–66

Bodkin PC, Spence DHN, Weeks DC (1980) Photoreversible control of heterophylly in *Hippuris vulgaris* L. New Phytol 84:533–542

Bogorad L, McIlrath WJ (1960) Effect of light quality on axillary bud development in *Xanthium*. Plant Physiol 35: suppl 32

Bonnett HT (1972) Phytochrome regulation of endogenous bud development in root cultures of *Convolvulus arvensis*. Planta 106:325–330

Borthwick HA (1957) Light effects on tree growth and seed germination. Ohio J Sci 57:357–364

Borthwick HA, Hendricks SB, Parker MW, Toole EH, Toole VK (1952) A reversible photoreaction controlling seed germination. Proc Nat Acad Sci USA 38:662–666

Butler WL, Hendricks SB, Siegelman HW (1964a) Action spectra of phytochrome in vitro. Photochem Photobiol 3:521–538

Butler WL, Siegelman HW, Miller CO (1964b) Denaturation of phytochrome. Biochemistry 3:851–857

Collins WB (1966) Floral initiation in strawberry and some effects of red and far-red irradiation as components of continuous white light. Can J Bot 44:663–668

Cooke IJ (1969) The influence of far-red light on the development of tomato seedlings. J Hortic Sci 44:285–292

Coombe DE (1957) The spectral distribution of shadelight in woodlands. J Ecol 45:823–830

Cumming BG (1963) The dependence of germination on photoperiod, light quality, and temperature in *Chenopodium* spp. Can J Bot 14:1211–1233

Curry GM (1969) Phototropism. In: Wilkins MB (ed) Physiology of plant growth and development. McGraw-Hill, Maidenhead, pp 243–273

De Greef JA, Fredericq H (1972) Enhancement of senescence by far-red light. Planta 104:272–274

Deitzer GF, Hayes R, Jabben M (1979) Kinetics and time dependence of the effect of far-red light on the photoperiodic induction of flowering in Wintex barley. Plant Physiol 64:1015–1021

Downs RJ, Hendricks SB, Borthwick HA (1957) Photoreversible control of elongation in Pinto beans and other plants under normal conditions of growth. Bot Gaz 118:199–208

Dubinsky Z, Berman T (1979) Seasonal changes in the spectral composition of downwelling irradiance in Lake Kinneret (Israel). Limnol Oceanogr 24:652–663

Erez A (1977) The effect of different portions of the sunlight spectrum on ethylene evolution in peach (*Prunus persica*) apices. Physiol Plant 39:285–289

Federer CA, Tanner CB (1966) Spectral distribution of light in the forest. Ecology 47:555–560

Fitter AH, Ashmore CJ (1974) Response of two *Veronica* spp to a simulated woodland light climate. New Phytol 73:997–1001

Fondeville JC, Borthwick HA, Hendricks SB (1966) Leaflet movement of *Mimosa pudica* L. indicative of phytochrome action. Planta 64:357–364

Frankland B (1976) Phytochrome control of seed germination in relation to the light environment. In: Smith H (ed) Light and plant development. Butterworths, London, pp 477–491

Frankland B, Letendre RJ (1978) Phytochrome and effects of shading on growth of woodland plants. Photochem Photobiol 27:223–230

Fredericq H, De Greef JA (1966) Red, far-red photoreversible control of growth and chlorophyll content in light grown thalli of *Marchantia polymorpha* L. Naturwissenschaften 53:337

Friend DJC, Helson VA, Fischer JE (1961) The influence of the ratio of incandescent to fluorescent light on the flowering response of Marquis wheat grown under controlled conditions. Can J Plant Sci 41:418–427

Gaba V, Black M (1979) Two separate photoreceptors control hypocotyl growth in green seedlings. Nature 278:51–54

Gates DM (1966) Spectral distribution of solar radiation at the earth's surface. Science 151:523–529

Gaudet JJ (1965) The effect of various environmental factors on the leaf form of the aquatic fern *Marsilea vestita*. Physiol Plant 18:674–686

Goldberg B, Klein WH (1977) Variations in the spectral distribution of daylight at various geographical locations in the earth's surface. Sol Energy 19:3–13

Goodfellow S, Barkham JP (1974) Spectral transmission curves for a beech (*Fagus sylvatica* L.) canopy. Acta Bot Neerl 23:225–230

Gorski T (1975) Germination of seeds in the shadow of plants. Physiol Plant 34:342–346

Gressel J (1979) Blue light photoreception. Photochem Photobiol 30:749–754

Grime JP (1966) Shade avoidance and shade tolerance in flowering plants. In: Bainbridge R, Evans GC, Rackham O (eds) Light as an ecological factor. Blackwell, Oxford, pp 187–207

Gutterman Y, Porath D (1975) Influences of photoperiodism and light treatments during fruit storage on the phytochrome and on the germination of *Cucumis prophetarum* and *Cucumis sativa* seeds. Oecologia 18:37–43

Hadfield W (1974) Shade in north-east Indian tea plantations. II. Foliar illumination and canopy characteristics. J Appl Ecol 11:179–199

Haupt W (1972) Short-term phenomena controlled by phytochrome. In: Mitrakos K, Shropshire W (eds) Phytochrome. Academic Press, New York pp 349–368

Heathcote L, Bambridge KR, McLaren JS (1979) Specially constructed growth cabinets for simulation of the spectral photon distributions found under natural vegetation canopies. J Exp Bot 30:347–353

Henderson ST (1977) Daylight and its spectrum, 2nd edn. Adam Hilger, Bristol

Hoddinott J, Bain J (1979) The influence of simulated canopy light on the growth of six apocarpous moss species. Can J Bot 57:1236–1242

Holmes MG, Smith H (1975) The function of phytochrome in plants growing in the natural environment. Nature 254:512–514

Holmes MG, Smith H (1977a) The function of phytochrome in the natural environment. I. Characterisation of daylight for studies in photomorphogenesis and photoperiodism. Photochem Photobiol 25:533–538

Holmes MG, Smith H (1977b) The function of phytochrome in the natural environment II. The influence of vegetation canopies on the spectral energy distribution of natural daylight. Photochem Photobiol 25:539–545

Holmes MG, Smith H (1977c) The function of phytochrome in the natural environment. IV. Light quality and plant development. Photochem Photobiol 25:551–557

Holmes MG, Wagner E (1980) A re-evaluation of phytochrome involvement in time measurement in plants. J Theor Biol 83:255–265

Hsiao AL, Vidaver W (1973) Dark reversion of phytochrome in lettuce seeds stored in a water-saturated atmosphere. Plant Physiol 51:459–463

Hull JN (1954) Spectral distribution of radiation from sun and sky. Illum Eng (London) 19:21–28

Hurd RG (1974) The effect of an incandescent supplement on the growth of tomato plants in low light. Ann Bot 38:613–623

Kasperbauer MJ (1971) Spectral distribution of light in a tobacco canopy and effects of end-of-day light quality on growth and development. Plant Physiol 47:775–778

Kasperbauer MJ, Hiatt AJ (1966) Photoreversible control of leaf shape and chlorophyll content in *Nicotiana tabacum*. Tob Sci 10:29–32

Kasperbauer MJ, Tso TC, Sorokin TP (1970) Effects of end-of-day red and far-red radiation on free sugars, organic acids, and amino acids of tobacco. Phytochemistry 9: 2091–2095

Kendrick RE, Frankland B (1968) Kinetics of phytochrome decay in *Amaranthus* seedlings. Planta 82:317–320

King TJ (1975) Inhibition of seed germination under leaf canopies in *Arenaria serpyllifolia, Veronica arvensis,* and *Cerastium holosteoides.* New Phytol 75:87–90

Kirk JTO (1976) A theoretical analysis of the contribution of algal cells to the attenuation of light within natural waters. III Cylindrical and spheroidal cells. New Phytol 77:341–358

Kopal Z (1969) Photometry of scattered moonlight. In: Kopal Z (ed) The moon. D Reider Publ Co, Dordrecht, Holland, pp 357–370

Kretchmer PJ, Ozbun JL, Kaplan SL, Laing DR, Wallace DH (1977) Red and far-red light effects on climbing in *Phaseolus vulgaris* L. Crop Sci 17:797–799

Leakey RRB, Chancellor RJ, Vince-Prue D (1978) Regeneration from rhizome fragments of *Agropyron repens* (L.) Beauv. IV Effects of light on bud dormancy and development of dominance amongst shoots on multi-node fragments. Ann Bot 42:205–212

Lecharny A, Jacques R (1974) Phytochrome et croissance des tiges; variations de l'effet de la lumiere en function du temps et du lieu de photoperception. Physiol Veg 12:721–738

Loveys BR (1979) The influence of light quality on levels of abscisic acid in tomato plants, and evidence for a novel abscisic acid metabolite. Physiol Plant 46:79–84

Mancinelli AL, Rabino I (1978) The "high irradiance responses" of plant photomorphogenesis. Bot Rev 44:129–180

Mancinelli AL, Yanio Z, Smith P (1967) Phytochrome and seed germination: 1. Temperature dependence and relative Pfr levels in the germination of dark germinating tomato seeds. Plant Physiol 42:333–337

McDonough WT, Brown RW (1969) Seedling growth of grasses and forbs under various incandescent-fluorescent wattage ratios. Agron J 61:485–486

McLaren JS, Smith H (1978) Phytochrome control of the growth and development of *Rumex obtusifolius* under simulated canopy light environments. Plant Cell Environ 1:61–67

Meijer G (1958) Influence of light on the elongation of gherkin seedlings. Acta Bot Neerl 7:614–620

Meijer G (1959) Spectral dependence of flowering and elongation. Acta Bot Neerl 8:189–246

Morgan DC, Smith H (1976) Linear relationship between phytochrome photoequilibrium and growth in plants under simulated natural radiation. Nature 262:210–212

Morgan DC, Smith H (1978a) The relationship between phytochrome photoequilibrium and development in light grown *Chenopodium album* L. Planta 142:187–193

Morgan DC, Smith H (1978b) Simulated sunflecks have large, rapid effects on plant stem extension. Nature 273:534–536

Morgan DC, Smith H (1979) A systematic relationship between phytochrome-controlled development and species habitat, for plants grown in simulated natural radiation. Planta 145:253–258

Morgan DC, O'Brien T, Smith H (1980) Rapid photomodulation of stem extension in light-grown *Sinapis alba* L.: Studies on kinetics, site of perception, and photoreceptor. Planta 150:95–101

Munz FW, McFarland WN (1973) The significance of spectral position in the rhodopsins of tropical marine fishes. Vision Res 13:1829–1874

Pratt LH (1979) Phytochrome: Function and properties. In: Smith KC (ed) Photochemical and photobiological reviews 4, Plenum Press, New York pp 59–124

Presti D, Delbrück M (1978) Photoreceptors for biosynthesis, energy storage and vision. Plant Cell Environ 1:81–100

Rajan AK, Betteridge B, Blackman GE (1971) Interrelationships between the nature of the light source, ambient air temperature, and the vegetative growth of different species within growth cabinets. Ann Bot 35:323–343

Robertson GW (1966) The composition of solar and sky spectra available to plants. Ecology 47:640–643

Robinson N (1966) Solar radiation. Elsevier, New York

Sakamoto M, Hogetsu K (1963) Spectral change of light with depth in some lakes and its significance to the photosynthesis of phytoplankton. Plant Cell Physiol 4:187–198

Sauer J, Struik G (1964) A possible ecological relation between soil disturbance, light flash, and seed germination. Ecology 45:884–886

Schneider MJ, Stimson WR (1971) Contributions of photosynthesis and phytochrome to the formation of anthocyanin in turnip seedlings. Plant Physiol 48:312–315
Schneider MJ, Stimson WR (1972) Phytochrome and photosystem 1 interaction in the high-energy photoresponse. Proc Natl Acad Sci USA 69:2150–2154
Selman IW, Ahmed EOS (1962) Some effects of far-red irradiation and gibberellic acid on the growth of tomato plants. Ann Appl Biol 50:479–485
Shell GSG, Lang ARG (1976) Movements of sunflower leaves over a 24 h period. Agric Meteorol 16:161–170
Shropshire W (1973) Photoinduced parental control of seed germination and the spectral quality of solar radiation. Sol Energy 15:99–105
Sinclair TR, Lemon ER (1973) The distribution of 660 and 730 nm radiation in corn canopies. Sol Energy 15:89–97
Smith H (1973) Light quality and germination: Ecological implications. In: Heydecker W (ed) Seed ecology. Butterworths, London, pp 219–231
Smith H (1975) Phytochrome and photomorphogenesis. McGraw-Hill, London
Smith H, Holmes MG (1977) The function of phytochrome in the natural environment. III. Measurement and calculation of phytochrome photoequilibrium. Photochem Photobiol 25:547–550
Spence DHN (1975) Light and plant response in fresh water. In: Evans GC, Bainbridge R, Rackham O (eds) Light as an ecological factor II. Blackwell, Oxford, pp 93–133
Spence DHN, Campbell RM, Chrystal J (1971) Spectral intensity in some Scottish freshwater lochs. Freshwater Biol 1:321–337
Stoutjesdijk Ph (1972a) Spectral transmission curves of some types of leaf canopies with a note on seed germination. Acta Bot Neerl 21:185–191
Stoutjesdijk Ph (1972b) A note on the spectral transmission of light by tropical rainforest. Acta Bot Neerl 21:346–350
Stroller EW, Wax LM (1973) Temperature variations in the surface layers of an agricultural soil. Weed Res 13:273–282
Tasker R (1977) Phytochrome and the radiation environment of woodlands. PhD Thesis, University of Nottingham
Tasker R, Smith H (1977) The function of phytochrome in the natural environment. V. Seasonal changes in the radiant energy quality in woodlands. Photochem Photobiol 26:487–491
Taylor AH, Kerr GA (1941) The distribution of energy in the visible spectrum of daylight. J Opt Soc Am 31:3–8
Taylorson RB, Borthwick HA (1969) Light filtration by foliar canopies; significance for light-controlled weed seed germination. Weed Sci 17:48–51
Thomas B, Dickinson HG (1979) Evidence for two photoreceptors controlling growth in de-etiolated seedlings. Planta 146:545–550
Tucker DJ (1972) The effects of light quality on apical dominance in *Xanthium strumarium*. PhD Thesis, University of Lancaster
Tucker DJ (1976) Effect of far-red light on the hormonal control of side shoot growth in the tomato. Ann Bot 40:1033–1042
Tucker DJ, Mansfield TA (1972) Effects of light quality on apical dominance in *Xanthium strumarium* and the associated changes in endogenous levels of abscisic acid and cytokinins. Planta 102:140–151
Tyler JE, Smith RC (1970) Measurement of spectral irradiance underwater. Gordon and Breach, New York
Vaartaja O (1962) The relationship of fungi to survival of shaded tree seedlings. Ecology 43:547–549
Van der Veen R (1970) The importance of the red-far-red antagonism in photoblastic seeds. Acta Bot Neerl 19:809–812
Vazquez-Yanes C (1980) Light quality and seed germination in *Cecropia obtusifolia* and *Piper auritum* from a tropical rainforest in Mexico. Phyton 38:33–35
Vezina PE, Boulter DWK (1966) The spectral composition of near UV and visible radiation beneath forest canopies. Can J Bot 44:1267–1284
Vince-Prue D (1975) Photoperiodism in plants. McGraw-Hill, London

Vince-Prue D (1976) Photocontrol of petiole elongation in light-grown strawberry plants. Planta 131:109–114

Vince-Prue D (1977) Photocontrol of stem elongation in light-grown plants of *Fuchsia hybrida*. Planta 133:149–156

Wagner E (1976) The nature of periodic time measurement: Energy transduction and phytochrome action in seedlings of *Chenopodium rubrum*. In: Smith H (ed) Light and plant development. Butterworths, London, pp 419–443

Warrington IJ, Mitchell KJ, Halligan G (1976) Comparisons of plant growth under four different lamp combinations and various temperature and irradiance levels. Agric Meteorol 16:231–245

Weinberg S (1976) Submarine daylight and ecology. Mar Biol 37:291–304

Wesson G, Wareing PF (1969a) The role of light in the germination of naturally occurring populations of buried weed seeds. J Exp Bot 20:402–413

Wesson G, Wareing PF (1969b) The induction of light sensitivity in weed seeds by burial. J Exp Bot 20:414–425

Woods DB, Turner NC (1971) Stomatal response to changing light by four tree species of varying shade tolerance. New Phytol 70:77–84

Woolley JT (1971) Reflectance and transmittance of light by leaves. Plant Physiol 47:656–662

Woolley JT, Stroller EW (1978) Light penetration and light-induced seed germination in soil. Plant Physiol 61:597–600

Yocum CS, Allen LH, Lemon ER (1964) Photosynthesis under field conditions: VI. Solar radiation balance and photosynthetic efficiency. Agron J 56:249–253

Zeiger E, Hepler PK (1977) Light and stomatal function: Blue light stimulates swelling of guard cell protoplasts. Science 196:887–889

5 Responses to Photoperiod

F.B. SALISBURY

CONTENTS

5.1 Introduction

Photoperiodism appears via one manifestation or another in many if not most eukaryotic organisms, including plants and animals. In plants, reproduction is often controlled or influenced by response to relative lengths of day and night, and form of a plant (e.g., internode lengths, stem heights, leaf shapes) is virtually always influenced by photoperiod. Other plant responses, including tuber formation and even relative concentrations of various growth regulators and metabolites, are also influenced by photoperiod. Clearly, most if not all plants and animals of virtually any ecosystem are as they are in response to the photoperiods to which they have been exposed. Hence, the complete understanding of ecosystems must include an understanding of the photoperiodic responses of their component organisms.

In spite of this obvious truism, little scientific research has been directed toward understanding effects of photoperiodism on the plants and animals of any given ecosystem. Vast amounts of research have aimed at understanding either the individual responses to photoperiod of various species or cultivars or the physiological mechanism of plant or animal response, often with the goal of improving yields and qualities of crops. Relating such information to ecosystems becomes almost overwhelming. Even individual genotypes exhibit unique photoperiodism responses, so information gained from one species or cultivar can seldom be applied exactly to another, even if the second is closely related. Ultimate understanding of the role of photoperiodism in any natural ecosystem would require detailed knowledge of the photoperiodic responses of the ecosystem's members – knowledge we have scarcely begun to accumulate (see Sect. 5.4.1). Thus, this chapter must be limited to generalities.

The goal is to emphasize those aspects of photoperiodism that are especially relevant to ecology, but almost by definition, few if any areas of photoperiodism research do not have at least *some* potential bearing on plant response to environment. Hence, choice of subjects becomes arbitrary. Purely internal matters, particularly the possible role of a flowering hormone, are not stressed.

Reviews on photoperiodism or related topics include those of HILLMAN (1976), EVANS (1969, 1971, 1975), SCHWABE (1971), VINCE-PRUE (1975), and ZEEVAART (1976). Papers presented at the international Colloquium on the Physiology of Flowering (Anon. 1979) are also relevant. ZEEVAART noted an abundance of reviews before 1976, but little active research on basic mechanisms of photoperiodism; since 1976, no major reviews have appeared. Yet in spite of the paucity of papers on basic mechanisms, recent reports on applied work with diverse species are abundant – so abundant that only about a third could be mentioned here, although most relate at least indirectly to ecology.

5.2 Varieties of Plant Response to Photoperiod

Figure 5.1 shows day-length or photoperiod, the key variable of this chapter, as a function of time during the year and at several latitudes. The figure also shows *rates* of changing daylengths throughout the year, since plants in the field respond as day and/or night lengths *change*. Note that changes at any latitude are most rapid for about three months centered around the spring and autumn equinoxes and are least rapid around the summer and winter solstices.

Actual plant perception of day and/or night length is a perception of changing light levels during twilight, and this perception is strongly modified by spectral quality. These matters are discussed in Section 5.5.

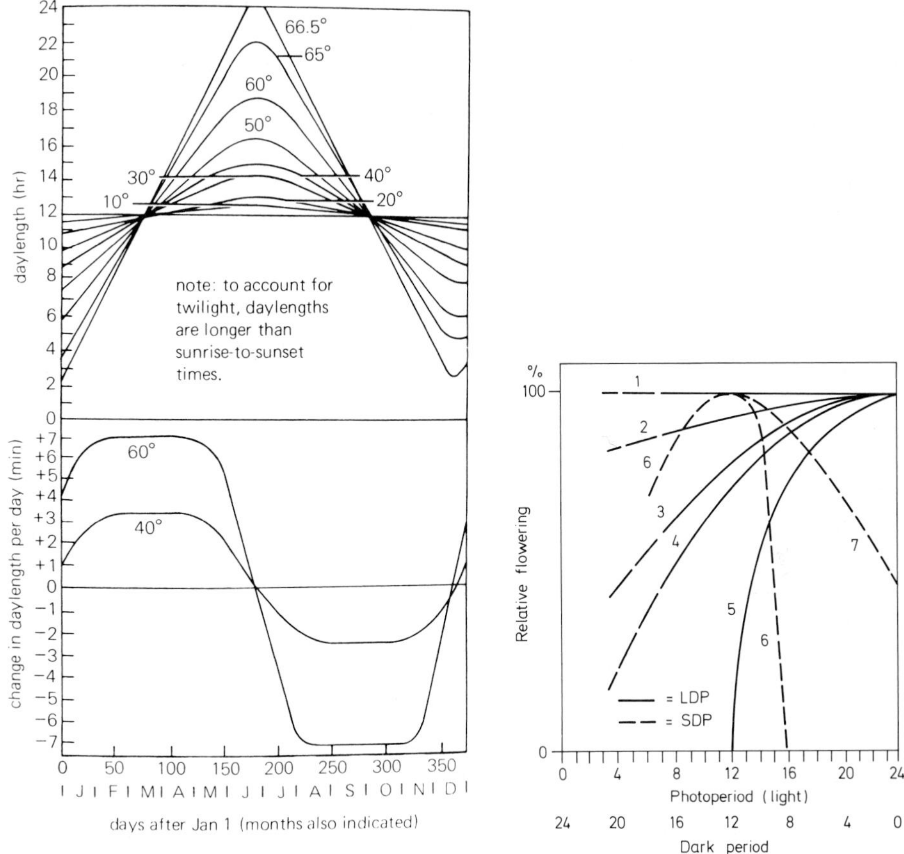

Fig. 5.1 **Fig. 5.2**

Fig. 5.1. *Top* daylength at various latitudes as a function of time during the year. *Bottom* rate of change in daylength at two latitudes as a function of time during the year. (SALISBURY and ROSS 1978)

Fig. 5.2. Diagram illustrating flowering (and other) responses to various daylengths. Each *line* represents a different hypothetical plant. Flowering can be measured in various ways, such as counting the number of flowers on each plant, classifying the size of the buds according to a series of arbitrary stages, or taking the inverse of the number of days until the first flower appears. *1* A truly day-neutral plant, flowering about the same at all daylengths. (With many species, there is little or no flowering when days are unusually short; e.g., from 4 to 6 h long.) *2* Plant slightly but probably insignificantly promoted in its flowering by long days. *3* and *4* Both plants quantitatively promoted by long days (to different degrees), although they flower on any daylength. *5* Qualitative or absolute long-day plant such as henbane (flowers *only* when days are *longer* than about 12 h). *6* Qualitative short-day plant such as cocklebur (flowers *only* when days are *shorter* than about 15.6 h; nights longer than 8.3 h). Note that cocklebur also fails to flower if days are *shorter* than 3 to 5 h (in that sense, it is also a long-day plant). *7* Quantitative short-day plant, flowers on any daylength but better under short days. Note that different species have different critical day and night lengths, not just the 12 and 15.6 h shown for henbane and cocklebur

In the broadest sense, the response to photoperiod is either to increasing daylength (type: long-day), to decreasing daylength (type: short-day), or the response is independent of daylength (type: day-neutral).[1] The important point is that the response is not to *absolute* daylength but to increasing or decreasing daylength. The SDP, *Xanthium strumarium* (cocklebur), for example, flowers when days are *shorter* than about 15.6 h, while the LDP, *Hyoscyamus niger* (black henbane), flowers when days are *longer* than about 12.5 h. Thus both the SDP and the LDP flower when days are between 12.5 and 15.6 h long. A strict DNP would flower at virtually any daylength, although a more typical situation is some *promotion* by SD's or LD's (see Sect. 5.3). These relationships are summarized in Fig. 5.2; they surely apply to other photoperiod responses as well as to flowering, although this has not been tested in every case.

5.2.1 Flowering: The Initiation of Reproductive Development

The most widely studied plant response to photoperiod is that of flowering in angiosperms, the response originally reported by GARNER and ALLARD (1920). Much discussion here concerns work on flower initiation in response to photoperiod, frequently as modified by temperature. It is important to emphasize, however, that many plant responses besides flowering are controlled or influenced by photoperiod. Again, this was well-documented in an early paper of GARNER and ALLARD (1923).

5.2.2 Responses Other Than Flowering

VINCE-PRUE (1975) devoted eight chapters in her book to photoperiod responses other than flower initiation. Each chapter is documented with many literature citations (one with as many references as space allows in this chapter) and with a number of tables summarizing species and responses. SALISBURY and ROSS (1978) summarized much of this information in a brief table adapted to teaching, and SCHWABE (1971) has similar summary tables. Some examples follow, when possible from the more recent literature.

5.2.2.1 Germination or Ability to Germinate

It has long been known that the germination of certain seeds can be controlled or at least influenced by daylength (MAYER and POLJAKOFF-MAYBER 1963; VINCE-PRUE 1975). A few seeds are promoted in their germination by SD's, but promotion by LD's is more common (e.g., rice, two cultivars required 8 h or more of light; BHARGAVA 1975). In general it has been difficult to separate photoperiod effects on germination from other effects of light and from other environmental factors. For example, in rice (BHARGAVA 1975), more cycles were required with

1 Abbreviations: SD('s) = short day(s), LD('s) = long day(s), SDP('s) = short-day plant(s), LDP('s) = long-day plant(s), and DNP('s) = day-neutral plant(s). It is assumed that SDP (for example) is pronounced *short-day plant*; hence *a* SDP rather than *an* SDP

shorter photoperiods, although a defined critical day of 8 h occurred with sufficient cycles.

POURRAT and JACQUES (1975) have shown that the photoperiod to which mother plants are exposed while their seeds are developing may strongly influence later germination of the seeds.

5.2.2.2 Growth and Plant Form: Stem Elongation, Leaf Growth, etc.

The vegetative growth of a plant is extremely sensitive to photoperiod. Typically, stems are longer under LD's, producing a taller, less dense plant. The phenomenon is seldom studied with the care it deserves, often being mentioned only as an aside in a report on flowering response to photoperiod. Furthermore, it is greatly complicated by at least two factors: First, flowering itself often strongly modifies plant form, as meristems become reproductive, producing flowers instead of stems and leaves (e.g., BUTTROSE and SEDGLEY 1978; SHARMA and NANDA 1976). Formation of tubers may have the same effect (e.g., EWING and WAREING 1978; MURTI and SAHA 1975). Nevertheless, effects of photoperiod on vegetative form can often be observed with DNP's, in which case flowering is about the same under both LD's and SD's. Second, plant growth is highly sensitive to many factors besides photoperiod. These include temperature, humidity, soil moisture, mechanical stresses, etc. (e.g., SALISBURY and ROSS 1978). Quantity of light is especially important and difficult to control. Although plants should be given comparable amounts of photosynthetically active radiation with different photoperiods, few studies have taken this precaution (as did MILFORD and LENTON 1976), so the following representative papers must be considered as only indicative of the possibilities, rather than definitive in every case.

By far the most common vegetative growth response to photoperiod is an increased rate of stem elongation under long days (recognized by GARNER and ALLARD 1923). Examples include such monocots as tossa jute (*Corchorus olitorius*; OKUSANYA 1979), tulips (HANKS and REES 1979), and *Cordyline terminalis* (SEIBERT 1976); such herbacious dicots as California poppy (*Eschscholtzia californica*; SHARMA and NANDA 1976), potatoes (HAMMES and NEL 1975; MURTI et al. 1975), cassava (*Manihot esculenta*; LOWE et al. 1976), flax (*Linum usitatissimum*; YANAGISAWA 1975), alfalfa (SATO 1971), various vegetables (SOFFE et al. 1977), and chrysanthemums (LANGTON 1977), and even such woody dicots as citrus species (WARNER et al. 1979) and oaks (IMMEL et al. 1978). Conifer shoot elongation is especially responsive to LD's (e.g., ARNOTT 1979; WHEELER 1979, and references therein). LD's promote runnering in strawberry (GUTTRIDGE 1976). Frequently, elongation in response to LD's occurs best at higher temperatures (e.g., AUNG 1976; WRIGHT 1976), and in the case of tossa jute, the LD elongation occurred only at temperatures of 24 °C or above (BOSE 1974). TINUS (1976) describes interesting interactions between photoperiod and CO_2. LD's and high CO_2 acted synergistically to increase stem height and thickness, leaf number, and dry weight. MACDOWALL (1977) reported that growth of Marquis wheat was dependent only upon the total daily incident radiant energy and was quite independent of photoperiod.

Heins and Wilkins (1979) report that LD's inhibit stem elongation in *Alstroemeria* (especially at low soil temperatures). This is because LD's lead to flowering, and flowering stems of this and many other species elongate less than vegetative ones. Sawhney et al. (1978) report that LD's produce shorter stems with *fewer* flowers in *Impatiens balsamina*.

Morita et al. (1978) found that spring shoot emergence and elongation of *Viburnum awabuki* and *Pinus thunbergii* was promoted by LD's, but without autumn chilling at 3 °C, high temperature had to be applied in spring to observe LD elongation. With fall chilling, however, both species showed a flourishing shoot growth even under SD's. Stem extension has even been observed in several species of intertidal brown algae, although this is probably a response to total light rather than a true photoperiod response (Stromgren 1978).

Leaf number and area are also typically promoted by LD's (e.g., Sato 1979, who examined many aspects of soybean leaf morphology at the macro and the micro levels). Hedley and Harvey (1975) observed an increase in leaf area of *Antirrhinum majus* (snapdragon) not only in response to LD's, but also in response to a brief light interruption of a long night. Milford and Lenton (1976) confirmed that a LD-induced increase in sugarbeet leaf area was a true photoperiodic effect by extending the day with low light levels.

Daylength also influences tillering in grasses. Fairey et al. (1975) observed fewer tillers under LD's applied to barley (a LDP), but Sato (1974) observed more tillers under LD's applied to rice (a SDP).

5.2.2.3 Root Formation and Growth

Effects of photoperiod on root formation have long been studied (Vince-Prue 1975). LD's, for example, significantly increase rooting of Douglas fir cuttings (Bhella and Roberts 1974). The photoperiod experienced by the mother plant also influences ability of cuttings to form roots. Barba and Pokorny (1975), for example, showed that *Rhododendron* cultivars produced cuttings that rooted better when they were grown on LD's. LD's or SD's given to the cuttings themselves did not significantly influence rooting.

Root/shoot ratios are often influenced by photoperiod. Immel et al. (1978), for example, found increased root growth compared to shoot growth under SD's applied to oaks (see also MacDowall 1977).

5.2.2.4 Storage Organ Formation and Development

Because of their agricultural value, much study has been devoted to initiation and development of potato tubers in response to SD's (see Vince-Prue 1975). Many potato cultivars have an absolute SD requirement for tuber initiation, but others have only a facultative requirement (Murti et al. 1975). Because LD's promote shoot elongation and growth, however, tuber yields may be highest under relatively long days (e.g., 15 h; Hammes and Nel 1975; Murti and Saha 1975). Response to length of day is localized in the leaves (Ewing and Wareing 1978) and is truly photoperiodic, as shown by the effectiveness of a night interruption (Murti and Banerjee 1977).

Cassava (*Manihot esculenta*) initiates more root tubers under SD's (Lowe et al. 1976), and the spider plant (*Chlorophytum comosum*) forms stolons in response to a night break with incandescent or pure red light (Heins and Wilkins 1978). Bulb formation in onions is a LD response (Kedar et al. 1975).

5.2.2.5 Flower Development

Vince-Prue (1975) gives examples from the literature for four possible combinations of photoperiod effects on flower initiation and development: photoperiodic requirement for flower initiation but DN for flower development; DN for initiation, photoperiodic requirement for development; same photoperiodic requirement for initiation and flowering; and different photoperiodic requirements for initiation and flowering. The same photoperiodic requirement may be most common (e.g., Aung 1976; Langton 1977; Skrubis and Markakis 1976), but see Shillo (1976) for an example of a DNP with a LD promotion of flower development (*Limonium sinuatum*), and Cockshull (1976) for an opposite response in *Chrysanthemum*.

5.2.2.6 Sex Expression

Sex expression in many species is often strongly influenced by photoperiod, but no simple relationship exists: Either femaleness or maleness can be promoted by either SD's or LD's, depending on species (reviews by Vince-Prue 1975; Schwabe 1971; recent examples include Hugel 1976; Sedgley and Buttrose 1978).

5.2.2.7 Seed Filling: Yield

Many studies have been carried out to investigate effects of photoperiod and other environmental factors on seed filling and thus yield of selected agricultural crops. In a few cases, there seemed to be clear-cut effects of photoperiod on seed development (e.g., in soybeans, Thomas and Raper 1976). The situation is often complicated, however, in that ultimate seed yield is strongly dependent upon the amount of plant available to produce the seeds, and this in turn can be strongly influenced by photoperiod, as we have seen. Patterson et al. (1977) nicely document these relationships with soybeans (see also Allison and Daynard 1979). Hamner (1969) says that soybean cultivars from northern latitudes are so sensitive to photoperiod that they give maximum yields only within a band of latitude about 80 km wide. Cultivars from more southern latitudes are successful over a wider range.

5.2.2.8 Abscission, Dormancy, and Frost Hardiness

Bentley et al. (1975) observed abscission of flower buds of certain varieties of common bean (*Phaseolus vulgaris*) when daylengths were *longer* than 13 to 14 h. They showed that the perception of the daylength occurred in the leaves, while the inhibitory effects were expressed in the buds. It has long been known (see VINCE-PRUE 1975) that leaves of many deciduous woody plants abscise in response to the SD's of autumn, and often leaf abscission is closely correlated with a cessation of bud growth and an induction of dormancy. Willows became dormant in response to SD's, although abscisic acid contents of the buds did not change (ALVIM et al. 1978). YOUNG and HANOVER (1977) prevented onset of dormancy in blue spruce (*Picea pungens*) with days longer than 12 h or with a 2-h night interruption with red or white light. The red-light effect could be reversed with far-red, implicating phytochrome. Closely correlated with abscission and dormancy is a developing frost hardiness in response to SD's (e.g., JOHNSON and HAVIS 1977). CHRISTERSSON (1978) found that SD's promoted frost hardiness in *Pinus sylvestris* and *Picea abies* only when applied at 20 °C but not at 2 °C, at which temperature frost hardiness developed anyway. (Problems of frost resistance are discussed in detail in Chap. 13, of heat resistance in Chap. 14, this Vol.)

5.2.2.9 Senescence

Annuals senesce and die after flowering in the year they germinate. Frequently, the same photoperiodic conditions that promote flowering, flower development, and seed filling, also promote senescence, but this is not always the case. Soybeans, for example, are classical SDP's, but senescence may be hastened by exposure to LD's and strong lighting (e.g., LACHAUD 1976). LD's also promoted senescence in an early flowering genetic line of peas, but senescence was induced by the total hours of light and darkness rather than by the length of the dark period; further, senescence required flower and fruit development as well as LD's (PROEBSTING et al. 1976).

5.2.2.10 Miscellaneous Effects

Typically for such a survey, several responses have been reported that seem to defy classification into the above groups. Some examples: SD induction of flowering in *Xanthium* is accompanied by a transient change of leaf phyllotaxis (ERICKSON and MEICENHEIMER 1977). HERATH and ORMROD (1979) found differences in stomatal density on leaves of winged beans in response to different photoperiods. LAVIGNE et al. (1979) report a strong photoperiod control of essential oils and thus flower fragrance in *Jasminum grandiflorum*. SEIBERT (1976) observed increased stem elongation and anthocyanin pigmentation of *Cordyline terminalis* in response to a night interruption (equivalent to LD's). Note that all these effects and several others can be important in agriculture (where they are most often studied) and are surely important in natural ecosystems (where they have hardly been considered).

5.3 The Classification of Response Types

Research during the past 60 years has greatly expanded our knowledge of the photoperiodic response types described by GARNER and ALLARD (1920).

There are now hundreds to thousands of papers describing flowering and other responses of nearly as many species to photoperiod, temperature, nutrients, etc. In the following discussion, flowering is emphasized, but the same principles must apply to other responses. In a majority of modern papers, other responses are mentioned along with flowering.

Consideration of the ecology of flowering requires a comprehensive classification scheme. Although there are numerous complications of temperature and a few other factors, only a few additions have been made to the three basic flowering types of GARNER and ALLARD (1920): SDP's, LDP's, and DNP's. I developed a preliminary classification nearly 20 years ago (SALISBURY 1963a). Considering the various responses that had been observed, it was possible to suggest that 777 possible response types might eventually be encountered! Published examples were found for 48 of these possible categories. The schema was modified and expanded by VINCE-PRUE (1975). If this table were again brought up to date, adding literature citations, its space would probably exceed the entire space allotment for this review. Thus, Table 5.1 (SALISBURY and ROSS 1978) is abridged from VINCE-PRUE's (1975) classification. SCHWABE (1971) also presents a brief classification. In the remainder of this discussion, we will consider some important responses used in the tables of SALISBURY (1963a) and VINCE-PRUE (1975), noting a few recent papers.

Table 5.1. Twenty-five photoperiodic response types with selected representative species[a]

Short-day plants

1. Qualitative (Absolute) SDP
   Andropogon gerardi Bluestem
   Cattleya trianae Orchid
   Chenopodium polyspermum Goosefoot
   Chenopodium rubrum Red goosefoot
   *Chrysanthemum morifolium** Chrysanthemum
   Glycine max Soybean
   Lemna perpusilla Strain 6746 Duckweed
   Xanthium strumarium Cocklebur
   *Zea mays** Maize or corn
2. SDP at high temperature; quantitative SDP at low temperature
   Fragaria × ananassa Pine strawberry
3. SDP at high temperature; day-neutral at low temperature
   Pharbitis nil E Japanese morning glory
   *Nicotiana tabacum** Maryland mammoth tobacco
4. SDP at low temperature; day-neutral at high temperature
   Cosmos sulphureus cv. Orange flare Cosmos
5. SDP at high temperature; LDP at low temperature
   Euphorbia pulcherrima Poinsettia
6. Quantitative SDP
   Cannabis sativa cv. Kentucky Hemp or marijuana

[a] Mostly from VINCE-PRUE, 1975 and SALISBURY, 1963a (see also SALISBURY and ROSS, 1978). Note that *Chrysanthemum morifolium, Nicotiana tabacum, Gossypium hirsutum, Helianthus annuus*, and *Zea mays* (all marked*) appear in more than one category, illustrating variabilities of varieties or cultivars within species. To conserve space, the lists have been greatly abbreviated, although common categories are included

Table 5.1 (continued)

*Chrysanthemum morifolium** Chrysanthemum
Datura stramonium H (older plants are day-neutral) Jimsonweed datura
*Gossypium hirsutum** Upland cotton
*Helianthus annuus** Sunflower

7. Quantitative SDP, require or accelerated by low-temperature vernalization
Allium cepa Onion
*Chrysanthemum morifolium** Chrysanthemum

8. Quantitative SDP at high temperature, DN at low temperature
Zygocactus truncatus Christmas cactus

Long-day plants

9. Qualitative (Absolute) LDP
Agropyron smithii Bluestem wheatgrass
Agrostis palustris Creeping bentgrass
Arabidopsis thaliana Mouse ear cress
Avena sativa, Spring strains, Oats
Chrysanthemum maximum Pyrenees chrysanthemum
Festuca elatior Meadow fescue
Hyoscyamus niger, Annual strain, Black henbane
Lemna gibba Swollen duckweed
Nicotiana sylvestris Tobacco
Raphanus sativus Radish
Rudbeckia hirta Black-eyed Susan
Sedum spectabile Showy stonecrop

10. LDP, require or accelerated by low-temperature vernalization
Arabidopsis thaliana, Biennial strains, Mouse ear cress
Beta saccharifera Sugar beet
Bromus inermis Smooth bromegrass
Hyoscyamus niger, Biennial strain, Black henbane
Lolium temulentum Darnel ryegrass
Oenothera suaveolens Evening primrose
Triticum aestivum Winter wheat (also other cereals)

11. LDP at low temperature; quantitative LDP at high temperature
Beta vulgaris Common beet

12. LDP at high temperature, day-neutral at low temperature
Cichorium intybus Chicory

13. LDP at low temperature; day-neutral at high temperature
Rudbeckia bicolor Pinewoods coneflower

14. LDP, low temperature vernalization will substitute (at least partly) for the LD requirement
Silene armeria Sweet william silene

15. Quantitative LDP
Hordeum vulgare Spring barley
*Nicotiana tabacum** cv. Havana A Tobacco
Triticum aestivum Spring wheat

16. Quantitative LDP, require or accelerated by low-temperature vernalization
Digitalis purpurea Foxglove
Secale cereale Winter rye

17. Quantitative LDP at high temperature, day-neutral at low temperature
Lactuca sativa Lettuce

Dual-daylength plants

18. Long-short-day plants
Kalanchoe laxiflora Kalanchoe
Cestrum nocturnum (at 23 °C, day-neutral at >24 °C), night-blooming jasmine

Table 5.1 (continued)

19. Short-long-day plants
    Trifolium repens White clover
20. Short-long-day plants; require or accelerated by low-temperature vernalization
    Dactylis glomerata Orchardgrass
    Poa pratensis Kentucky bluegrass
    (in these plants, the SD is required for induction and LD for development of the inflorescence)
21. Short-long-day plants, low temperature substitutes for the SD effect and, after low temperature, plants respond as LDP
    Campanula medium Canterbury bells

Intermediate-day plants

22. Plants flower when days are neither too short nor too long
    Chenopodium album Lambsquarters goosefoot
    Saccharum spontaneum Sugar cane

Ambiphotoperiodic plants

23. Plants quantitatively inhibited by intermediate daylengths
    Madia elegans Tarweed
    Setaria verticillata Hooked bristlegrass

Day-neutral plants

24. Day-neutral plants: These are the plants with least response to daylength for flowering. They flower at about the same time under all daylengths but may be promoted by high or low temperature, or by a temperature alternation
    Cucumis sativus Cucumber
    *Gossypium hirsutum** Upland cotton
    *Helianthus annuus** Sunflower
    Helianthus tuberosus Jerusalem artichoke
    Lunaria annua Dollar plant
    *Nicotiana tabacum** Tobacco
    Phaseolus vulgaris Kidneybean
    Poa annua Annual bluegrass
    *Zea mays** Maize or corn
25. Day-neutral plants; require or accelerated by low-temperature vernalization
    Allium cepa Onion
    Daucus carota Wild carrot

5.3.1 Additional Daylength Response Types

Some species flower only in response to LD's followed by SD's, a situation encountered in late summer and autumn (examples in Table 5.1). Contrasted to these long-short-day plants are the short-long-day plants that flower only when exposed to SD's followed by LD's, as in spring and early summer. A few examples are known of plants that respond only to an intermediate daylength, remaining vegetative when exposed to days that are either too short or too long (intermediate-day plants). Their counterparts, plants that remain vegetative on intermediate daylengths but flower on either SD's or LD's, are also known (called ambiphotoperiodic). A recently reported intermediate-day plant is *Cyperus rotundus*, the purple nutsedge (Williams 1978), which on a world basis is probably the most important weed.

SCHWARTZ and KOLLER (1975) suggest an explanation for the intermediate-day response of *Bouteloua eriopoda*. They interpret the lack of flowering promotion on LD's as a "supra-induction" in which the meristems degenerate in a characteristic manner (nucleoli disintegrate, nuclei shrink, and finally the cells collapse).

5.3.2 Qualitative or Quantitative Responses

Virtually all photoperiodically sensitive species can be further classified as either qualitative (absolute) or quantitative (facultative) in their response to photoperiod. Those that have an absolute requirement for a given daylength (either SD, LD, or one of the other combinations) will remain vegetative for an indefinite time (months to years in available reports) as long as they are not exposed to the floral inductive daylength. *Xanthium strumarium* (cocklebur) is a classical absolute SDP; it can be grown to the status of a small tree in the greenhouse if it is never allowed to experience a dark period longer than about 8.3 h. The more typical response, however, is probably the quantitative one in which flowering is accelerated or number of flowers may be increased by suitable photoperiod, but plants will eventually flower at virtually any normal daylength (e.g., from 8 to 24 h).

Many plants otherwise classified as DN prove to have a quantative photoperiod response. Tomatoes, for example, have been considered classic DNP's, but depending upon cultivar, they may be promoted in their flowering by SD's and/or warm temperatures (AUNG 1976). Chrysanthemums, on the other hand, have traditionally been thought of as SDP's, but LANGTON (1977) found that most of 30 cultivars would flower on both SD's and LD's, although more quickly on SD's. Two cultivars were nearly true DNP's. Among the most important quantitative LDP's are the cereals (e.g., FAIREY et al. 1975). Many if not most species may become more quantitative and less absolute in their response as they get older (chrysanthemum: COCKSHULL 1976; *Antirrhinum majus:* HEDLEY and HARVEY 1975; even *Xanthium,* unpublished).

5.3.3 Photoperiod Interactions with Temperature

In virtually every case where studies have been extensive enough, photoperiodic response has been readily modified by changing temperature. Yet one time-measuring feature, the critical day or critical night (the maximum day length or minimum night length that permits flowering in SDP's, or the minimum day length or maximum night length that allows flowering in LDP's), remains relatively resistant to changing temperatures (reviews in SCHWABE 1971; VINCE-PRUE 1975).

With *Xanthium,* for example, the critical dark period was extended about 20 min by lowering the temperature from 30 to 15 °C (from about 8.75 to 9.1 h, single dark period), although a further drop to 10 °C extended the critical night another 1.3 h (to 10.5 h; SALISBURY 1963a and 1963b). The Q_{10} for the temperature influence on timing between 15 and 30 °C was about 1.02. KREBS and ZIMMER (1978) found a critical night for two *Begonia* species of about 14 h at 14 °C and 10 h at 26 °C. With *Pharbitis* (TAKIMOTO and HAMNER 1964), critical night was increased from about 9.8 h at 25 °C to nearly 12 h at 20 °C, but a further drop of only 2 °C (to 18 °C) extended the critical night to 18 to 20 h! Yet timing as displayed by the time of maximum sensitivity to a light interruption of a dark period was *not* changed over the entire range of temperatures studied. It is conceivable that in *Pharbitis* the marked effects on critical night observed at reduced temperature are really effects on synthesis of a flowering stimulus rather than on the photoperiodism clock. The resistance of the time of maximum sensitivity to a night interruption suggests that actual time measurement in many SD species is relatively insensitive to temperature over a wide range, which is typical of other manifestations of the biological clock (see HILLMAN 1976; VINCE-PRUE 1975).

Flowering of many species is promoted by a brief to prolonged exposure to temperatures close to the freezing point; that is, they can be vernalized (see VINCE-PRUE 1975). In a few cases plants may be virtually DN after vernalization (e.g., celery and *Lunaria biennis*). More commonly, however, vernalization is followed by a qualitative or quantitative response to photoperiod. The winter cereals, for example, have a quantitative requirement for vernalization followed by a quantitative requirement for LD's (with a few exceptions having an absolute requirement for LD's; PIRASTEH and WELSH 1975). A few species are known with a vernalization requirement followed by a SD promotion or requirement (e.g., chrysanthemums, a few grasses; see HABJORG 1978a; SCHWABE 1971). In a few interesting cases, a vernalization requirement can be replaced with a SD treatment (examples in SALISBURY 1963a; VINCE-PRUE 1975). MORITA et al. (1978) report that *growth* (not flowering) of *Viburnum awabuki* is promoted either by low temperatures (analogous to vernalization) or by LD's.

It is common for the photoperiodism response itself (as contrasted to critical day or night) to change with temperature. For example, a plant may have a day-length requirement at one temperature and be DN at another (examples in SCHWABE 1971; VINCE-PRUE 1975). Virtually all possible combinations are known: DN at low temperature, qualitative or quantitative day-length response (SD or LD) at high temperature; DN at high temperature, day-length response at low temperature; and qualitative or quantitative at one temperature, and the opposite at higher or lower temperature. Probably the most common situation is a DN response at one temperature with a quantitative photoperiod response at another.

Some examples from the recent literature include the following: *Jasminum grandiflorum* is an absolute LDP at moderate temperature (28 °C) but becomes quantitative in its LD response at 12 and 17 °C (LAVIGNE et al. 1979). Jute plants have an absolute SD response at both 24 and 32 °C, but SD's are *more effective* at the higher temperatures (BOSE 1974). A similar pattern appears in several mung-bean cultivars (AGGARWAL and POEHLMAN 1977) and in alfalfa (SATO 1971). Floret initiation in maize was independent of photoperiod but promoted at high temperatures (ALLISON and DAYNARD 1979). Different species of *Stylosanthes* (pencilflower) were SDP's, LDP's, and DNP's, but low temperatures promoted flowering in all three types (CAMERON and MANNETJE 1977). *Limonium* is a DNP, but flower initiation is promoted by low temperatures, although flower development is promoted by high temperatures and LD's (SHILLO 1976).

5.3.4 Other Complexities

An example of the interesting complexities that can be observed when sufficient effort goes into study of a species is provided by *Silene armeria* (WELLENSIEK 1969). In this plant, production of flowers may be brought about by exposing the plants to LD's at 20 °C or to SD's at 5° or 32 °C. Induction is also possible at any photoperiod or temperature by treating plants with GA_7.

Another example is provided by *Sinapis alba* seedlings, for many years considered LDP's (BODSON et al. 1977). White mustard flowers after being exposed to a single LD cycle of 16–20 h, even if the irradiance is relatively low (25 J m^{-2} s^{-1}). But only 8 h of light induce flowering (3 to 6 cycles) if irradiance is high enough (96 J m^{-2} s^{-1}). High irradiance also promotes flowering during the first 8 h of a 16-h cycle but inhibits during the second 8 h. Photosynthesis seems to play a role, since CO_2 is essential to observe either high irradiance effect, yet flowering could not be accounted for by enhanced assimilate supply to the shoot apex. The workers conclude: "Thus, although there is no absolute requirement for long days to induce flowering in *S. alba,* light reactions other than photosynthesis probably contribute to photoperiodic induction in this species."

Other environmental factors such as nutrient levels have also been shown to interact with photoperiod in flower induction and other responses (examples in VINCE-PRUE 1975). One of the most important of these from an ecological standpoint is the genetic variability within a species (Sect. 5.5 below).

5.4 Some General Ecological Aspects of Photoperiodism

The fundamental feature of photoperiodism is the measurement of light and dark intervals. The photoperiodism clock is an interval timer, a kind of chronometer or stop watch activated by a light detection system (a photometer, as it were), and functioning as a count-down alarm system. When a light or dark interval has been measured, the clock activates various developmental processes, including initiation of flowers and other processes. In this section, possible roles for such a chronometer in species populations are discussed.

5.4.1 Photoperiodism and Latitude; the Importance of Ecotypes

To a first approximation, a response to photoperiod ties a species to a given latitude, since daylengths at any time during the year are a function of latitude

(Fig. 5.1). Actually, the latitudinal range over which a given species can do well in the wild or as a crop is broader than might be implied by considering such sensitive species as soybeans (see Sect. 5.2.2.7). There are at least two reasons, and the second is probably the most important:

First, flowering and other photoperiod responses can often occur within fairly broad limits at various times of the year without endangering survival of the species. This is well illustrated by photoperiodically sensitive ornamental species that are grown over a wide latitudinal range.

Second, plants in the wild almost always become genetically adapted to the latitudes where they exist. That is, they develop physiological ecotypes based on varying responses to photoperiod. In a series of papers, McMillan (most recent, 1975) has demonstrated a range of critical nights in complexes of *Xanthium strumarium* collected all over the world. In a related study, using various species instead of ecotypes, Habjorg (1978b; see also Arnott 1974) examined several woody ornamentals grown in Scandinavia. Plants collected near Århus (56° North) had a critical day for shoot elongation of about 14 to 16 h; plants collected at Trondheim (63° North), 16–18 h; and at Alta (70° North), 20–24 h. There was also an altitude effect, with longer critical days being observed in plants from higher altitudes, apparently adapting such plants to begin growth later in the spring.

Habjorg (1978a) also studied selections of *Poa pratensis* collected from a wide latitudinal range in Europe. Again, critical days for flowering were closely correlated with latitude, but this time for ecotypes within a single species. Hodgkinson and Quinn (1978) studied *Danthonia caespitosa* in Australia, observing several ecotypes that correlated well with latitude. Cameron and Mannetja (1977) found the same thing with twelve species of *Stylosanthes* in Australia, which included SDP's, LDP's, and DNP's.

Paton (1978) found a weak growth response to photoperiod in selected *Eucalyptus* species but concluded that, in contrast to Northern-Hemisphere woody species, response types were largely unrelated to the latitude or altitude of the seed source. Kotecha et al. (1975) also found no correlation between photoperiodic class and latitude of origin in several sesame (*Sesamum indicum*) cultivars.

In spite of such exceptions, the rule seems to be that photoperiodic ecotypes develop within natural populations and that cultivars produced by controlled breeding exhibit a wide range of photoperiodic types within a species. Indeed, much breeding aims at producing DN cultivars, completely avoiding latitudinal restrictions. This has been largely achieved with wheat (e.g., Hunt 1979) and even soybeans (Hamner 1969).

In any case, it is seldom acceptable to apply published data for a given species unless the specific ecotype or cultivar under question has been studied. Hence, in a natural ecosystem consisting of several dozen plant species, laborious studies would have to be performed to determine the photoperiodic sensitivities and interactions with temperature and other factors of all member species. Then the assembled information would have to be collated into a single comprehensive scheme that might provide some insight into ecosystem functions. Perhaps the near hopelessness of such a task is the reason that it has never been attempted, even in a preliminary way (to the best of my knowledge). Nevertheless, modern phytotrons make the plant studies possible, and computers can accumulate and process vast amounts of information. Perhaps such an endeavor can soon be initiated. Our complete understanding of ecosystems depends upon such studies.

5.4.2 Anticipation of Seasonal Events

It is imperative for perennial plants in temperate latitudes to become frost-hardy before the cold season arrives and to become active again in the spring. Such a pattern would be an eminently "logical" response to photoperiod, allowing a plant to "anticipate" and thus prepare for approaching difficult environmental conditions. Indeed, photoperiod often strongly influences leaf abscission and dormancy, although it often acts as a sort of "back-up" or "fail-safe" system to induction of dormancy in response to decreasing average temperatures. That is, photoperiod may induce dormancy only above certain temperatures (Sects. 5.2.2.8 and 5.2.2.9). In temperate regions during an unusually extended and warm autumn, leaves on those trees that are strongly controlled by photoperiod fall off at about the same date as usual, regardless of the warm temperatures, but trees that are less sensitive to photoperiod but more under the control of temperature keep their leaves much later than usual.

As a rule, frost hardening and dormancy are promoted by SD's (e.g., in Douglas-fir seedlings, MCCREARY et al. 1978), but LAIDLAW and BERRIE (1977) observed hardening of perennial ryegrass seedlings under LD's in West Scotland and suggested that this might prepare plants against early autumn frosts.

5.4.3 Coordination of Phenological Events Within a Population

Some plants in a population might germinate early, gaining a head start over others. Or some might have more favorable light, nutrient, or soil moisture conditions. If flowering depended only on achieving a certain size (as in some DNP's), larger plants might flower earlier than smaller ones. Control by photoperiod, however, insures that all members of a population flower simultaneously and at an appropriate time. If cross-pollination is necessary or at least advantageous, there is a strong selection pressure for such simultaneous flowering. Indeed, this could be one of the main forces that led to the development of photoperiodism in plants and that maintains high sensitivity to photoperiod in many species. The availability of pollinators might be an additional selection factor that could influence actual flowering time for a given species at a given location.

SCHWARTZ and KOLLER (1975) observed a synchronization of flowering in response to an ideal photoperiod (13 h) in populations of *Bouteloua eriopoda*, an apparent intermediate-day plant. Synchronization of flowering has been documented in complex natural ecosystems (e.g., NEWELL and TRAMER 1978), but such situations need to be examined experimentally as they relate to photoperiod.

Synchronization by daylength depends upon an accurate and uniform response to photoperiod among members of a population. With *Xanthium*, uniformity of photoperiod detection increases with increasing number of inductive cycles (SALISBURY 1963b).

5.5 Some Ecological Implications of Photoperiodism Mechanisms

In photoperiodism, a plant must discriminate between day and night (some low photon flux detected as darkness), measure the duration of one or both, and in response control some process such as flowering (noted also in Sect. 5.4). Ecological implications of the internal mechanisms that control the phenological events (e.g., a flowering hormone) are not readily apparent, but the measurement of time is clearly of ecological significance, since the measuring system can be strongly influenced (e.g., rephased) by environment, especially light and temperature. The plant's measurement of light level (irradiance) is also of ecological and physiological interest. For one thing, the plant's response to light level is strongly influenced by the spectral quality of the light, and both level and quality change in natural environments with the seasons (see Fig. 5.1) and as a function of the plant's immediate environment.

Figure 5.3 shows six representative spectra for natural light, extending over seven orders of magnitude from full summer sunlight to light near the end of twilight and from the full moon (see Chaps. 2 and 4, this Vol.). For the sake of better spectral comparison, each curve is presented with its own linear scale (less "flattening" than on a log scale). Figure 5.4 shows representative light levels and red/far-red ratios during twilight. In addition to the rapid drop in light level during dusk (approximately an order of magnitude each ten minutes, depending on latitude and time of year), the most striking feature of Figs. 5.3 and 5.4 is the shift in spectral quality. Compared to direct sunlight, light from a clear blue sky is greatly enriched in blue wavelengths (not surprisingly!). During twilight, red wavelengths increase relative to blue as the light passes through increasing amounts of atmosphere (blue scattered), and far-red increases relative to red. Moonlight usually appears predominantly yellow, since blue wavelengths are reduced, and the eye perceives the full spectrum minus blue as yellow. Note the dips in the spectral curves at 688 and 762 nm; they are absorption bands of O_2, which is highly constant in the atmosphere. The less-noticeable band at 723 nm is caused by H_2O and thus might vary slightly with changing atmospheric humidity. All three of these bands are in the part of the spectrum to which phytochrome is responsive. Changing red/far-red ratios during twilight have been noted by several workers (e.g., HOLMES and SMITH 1977a; MORGAN and SMITH, Chap. 4, this Vol.; SHROPSHIRE 1973), and daylight spectra have often been reported (e.g., the above plus KADMAN-ZAHAVI and EPHRAT 1974; TASKER and SMITH 1977).

Such local factors as clouds, terrain (especially mountains), and air pollution often strongly influence both irradiance and quality of the light environment.

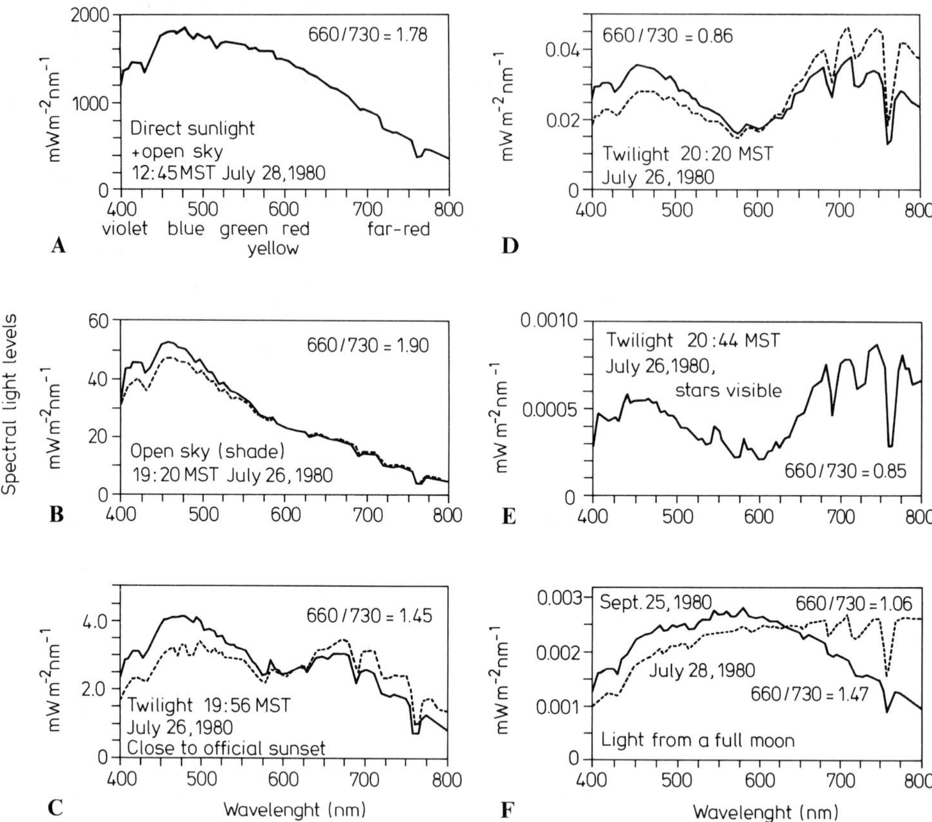

Fig. 5.3. Spectral distributions of natural light energies, including sunlight (**A**), skylight measured at four times before and during twilight (**B** to **E**), and light from the full moon (**F**). Note the greatly different scales of the various curves and that the energy levels at the end of the twilight measurement (**E**) are an order of magnitude lower than light from the full moon. Curves **B**, **C**, and **D** were made while light levels were changing rapidly, each scan requiring 10 min; hence, they were "corrected" by lowering the long-wavelength (red) end where the scans began by an amount proportional to the scans taken 12 min later, and raising the short-wavelength (blue) end in the same way. That is, the curves were "rotated" (by computer) around their center points, down on the red end and up on the blue end. In B, C, and D, solid lines are corrected; dashed lines are raw data. All scans were made with a Model 2900 Auto-Photometer attached to a high-resolution diffraction-grating monochromator (Model 700-31), manufactured by Gamma Scientific, Inc., San Diego, California, U.S.A. Data were obtained with a cosine receptor above the photomultiplier tube (Mamamatsu R928), except for the dashed line for moonlight (**F**), which represents data obtained with a fiber optics probe aimed directly at the moon. Light levels at 660 nm for these and other scans are shown in Fig. 5.4. (See SALISBURY 1981 for these and other data plotted on a logarithmic scale)

From a leaf's viewpoint, the most important changes in irradiance and spectral quality are caused by other leaves – by canopy shading. Far-red wavelengths are greatly enriched in light that has passed through leaves (HOLMES and SMITH 1977b; MORGAN and SMITH, Chap. 4, this Vol.; TASKER and SMITH 1977).

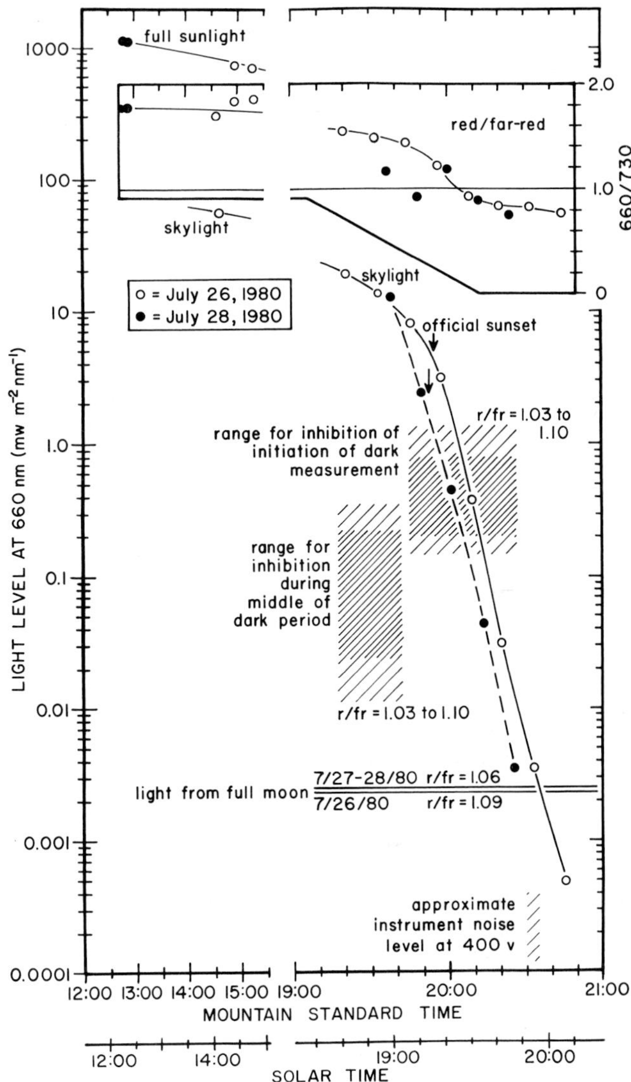

Fig. 5.4. Light levels at 660 nm as a function of time on July 26 and 28, 1980 at Logan, Utah, U.S.A., including light from a nearly full moon. The *cross-hatched areas* represent ranges of light levels that inhibit initiation of dark measurement in *Xanthium* or inhibit flowering when applied for 2 h during the middle of a 16-h dark period. The *inserted graph* shows ratios of light levels at 660 nm to levels at 730 nm for all the sunlight and twilight points; other red/far-red ratios are given as numerals. (SALISBURY 1981)

5.5.1 The Plant's Measurement of Light Level

How dark is dark? Or, when does a plant begin to measure the dark period during twilight, and when do plants respond to moonlight? The light level just bright enough to inhibit flowering of SDP's and promote LDP's is called the threshold irradiance. An example for *Xanthium* is shown in Fig. 5.5. Incan-

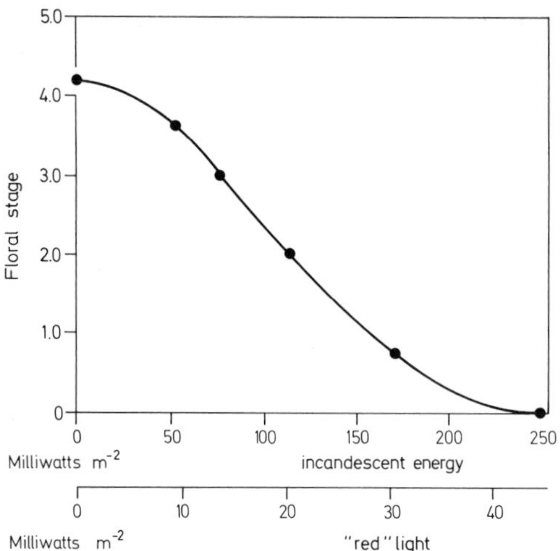

Fig. 5.5. Results of an experiment designed to measure threshold light for *Xanthium* dark-period inhibition. Light from a 7.5-W incandescent lamp was applied continuously during a 16-h "dark" period to plants arranged in concentric circles around the lamp, and degree of flowering (Floral Stage) was evaluated 9 days later. (SALISBURY 1963a)

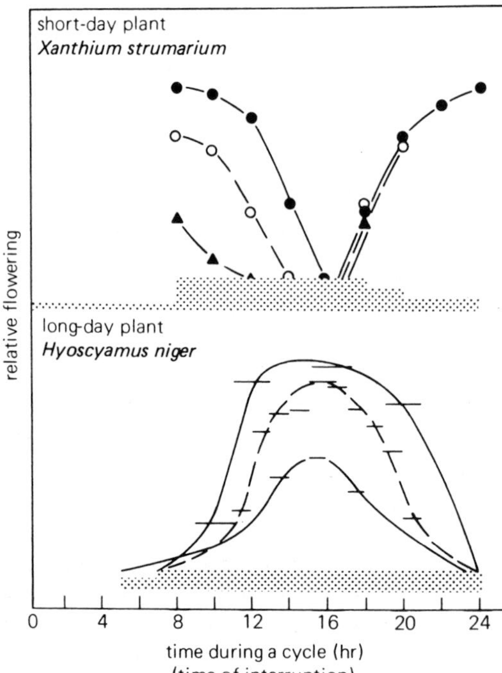

Fig. 5.6. Effects of a light interruption given at various times during dark periods of various lengths (*shaded bars*) on subsequent flowering of a SDP and a LDP. Interruptions with *Xanthium* were 60 s; with *Hyoscyamus*, times are indicated by *length of the data lines*. (Data for *Xanthium* from SALISBURY and BONNER 1956; for *Hyoscyamus* from CLAES and LANG 1947; Figure from SALISBURY and ROSS 1978)

descent light at about 50 mW m^{-2} inhibited only slightly, while light at 200 mW m^{-2} inhibited completely.

At equal total energies, is response to a light interruption of a dark period equivalent, whether it is applied at a low irradiance for an extended time or a higher irradiance for a brief time? That is, does reciprocity hold? PARKER et al. (1946) concluded that reciprocity did hold within a factor of 1.5, during the 5th to 7th hour of a 12-h dark period for *Xanthium* and near the middle of a 14-h dark period for soybeans. Yet reciprocity clearly fails if it is determined by measuring the total light energy required to completely inhibit flowering of a SDP or maximally promote flowering of a LDP. This is because these maximum effects are only possible at certain times during the dark period, regardless of light levels used. That is, the process saturates at certain levels of inhibition or promotion, so additional light energy has no effect. Figure 5.6 shows examples for the SD *Xanthium* and the LD *Hyoscyamus* . Note, that effects depend on length of the dark period as well as time during the dark period. Clearly, any study of light effects must consider the photoperiodism clock.

Studies are further complicated by the fact that, at least in *Xanthium*, light might play as many as three separate roles in photoperiodism:

1. Light during the day prevents the initiation of measurement of the dark period (i.e., "dark measurement").
2. Light may reset the phases of the photoperiodism clock. PAPENFUSS and SALISBURY (1967) showed that a light interruption of an inductive dark period for *Xanthium* increasingly inhibited flowering during the first 6 h of darkness but did not change timing; beginning at about 6 h, timing was rephased by the light interruption (i.e., the time of maximum light sensitivity was suddenly shifted about 10 h).
3. Light at dawn terminates the events of the dark period.

It has been possible with *Xanthium* to separate experimentally the inhibition of dark measurement from inhibition in the middle of a 16-h dark period (SALISBURY 1981). Initiation of dark measurement was delayed for 2 h with incandescent light filtered through a red filter (a mixture of approximately equal amounts of red and far-red light) at a level of 0.2 to 1.0 mW m^{-2} nm^{-1}, whereas inhibition during the middle 2 h of a 16-h dark period required only 0.01 to 0.3 mW m^{-2} nm^{-1} (light measured at 660 nm). This suggests that the first two possible roles of light are indeed different. (Note ranges in Fig. 5.4).

These results (SALISBURY 1981) suggest that *Xanthium* plants shift from their day-time mode of metabolism or pigment balance to their dark mode as light levels at 660 nm (modified by presence of far-red as discussed below) drop from about 1.0 to 0.2 mW m^{-2} nm^{-1}. Figure 5.4 shows light levels at 660 nm during evening twilight passing through the light range over which cockleburs shift from their day mode to their night mode. Although the range needs sharpening by further experimentation, it is apparent that *Xanthium* plants react during twilight almost as if they were controlled by an on/off switch. The shift from the day to the night mode occurs in only about 5.5 to 11 min, although the human eye sees twilight as lasting from 30 to 45 min or more and could further detect changes in light level caused by clouds during the day.

Two decades ago, TAKIMOTO and IKEDA (1961) took a more direct and ecological approach to the study of plant sensitivity during dusk and dawn twilight. They covered plants at various times during both dusk and dawn, comparing their flowering with plants left uncovered and noting the level of twilight perceived by the plants as darkness. They found considerable variation among five SD species: *Oryza sativa* was relatively insensitive to light both in the morning and in the evening; *Glycine max, Perilla frutescens,* and *Pharbitis nil* were relatively insensitive at dusk but more sensitive at dawn; and *Xanthium* was highly sensitive in the evening but less so in the morning. In an earlier paper (TAKIMOTO and IKEDA 1960), they concluded that clouds might well influence photoperiodic time measurement, but perhaps less so for plants that are most sensitive to light during dusk and/or dawn, since the lowest light levels seemed to be less influenced by clouds. KATAYAMA (1964) applied the same methods with *Oryza*, finding that all cultivars studied exhibited fairly high sensitivity to morning light, but only some cultivars were especially sensitive in the evening. Furthermore, clouds extended the night for *Oryza*.

Figure 5.4 also shows the probable range of light sensitivity of *Xanthium* plants during the 7th to the 9th hour of a 16-h inductive dark period (0.01 to 0.3 $mW\ m^{-2}\ nm^{-1}$) as well as maximum light levels at 660 nm for light from the full moon. Note that red/far-red ratios of the light used to determine the sensitivity ranges were similar to red/far-red ratios just after sunset and also of moonlight. Figure 5.4 suggests that maximum levels of moonlight are not high enough to influence flowering in the middle of a dark period, even though sensitivity to light increases by about an order of magnitude at that time compared to dusk. Unfortunately, the experiments (SALISBURY 1981) were not designed to accurately measure the lowest light levels to which *Xanthium* plants might respond in the middle of a dark period, so the conclusion that *Xanthium* plants do not respond to light from the full moon must remain tentative.

The values for moonlight (Figs. 5.3 and 5.4) are close to maximum levels for the summer full moon. Since the moon stays close to the path of the ecliptic in the sky, the full moon is relatively low in the night sky in summer and high in winter (at latitudes several degrees from the equator). The full moon near its zenith was only 30° above the southern horizon on July 26, 1980 (Figs. 5.3 and 5.4). By September 25, and because the moon deviates several degrees from the plane of the ecliptic, it was 50° above the horizon at midnight. BÜNNING and MOSER (1969) have suggested that normal sleep movements of the leaves of many species position leaf blades in a nearly vertical position at night; hence, they might be nearly parallel to rays from the full moon overhead at midnight and thus least sensitive to those rays. But as noted, the moon is *not* overhead in summer, and its intensity is apparently not high enough to influence *Xanthium* photoperiodism anyway.

Photoperiodism (not photosynthesis) in *Xanthium* plants apparently "ignores" daylight changes in light levels (e.g., caused by clouds), shifts from the day to night mode within a few minutes during dusk, and then "ignores" minor fluctuations in light levels during the night.

What pigment is absorbing the light involved in photoperiodism? BORTHWICK et al. (1952) showed that the red (ca. 660 nm) inhibition of *Xanthium* flowering could be reversed by immediately following the irradiation with far-red (ca. 730 nm). Since then, the red effectiveness and the far-red reversal have been demonstrated in numerous cases (e.g., TURNER and KARLANDER 1975, with *Cassia*; see also below). There are exceptions. JULIEN and SOOPRAMANIEN (1975) report that red, orange, and green light were the most inhibitory to inflorescence development in sugar cane (*Saccharum*), while blue, far-red, and ultraviolet acted like darkness. Attempts to reverse the inhibitory effect of red light with far-red light failed.

In most cases, however, it is clear that night interruptions with light act via phytochrome (probably both rephasing and terminating dark processes). Furthermore, we shall see below that effects of end-of-day light quality are often reversible when red follows far-red or vice versa.

5.5.2 The Role of Spectral Quality in Detecting End-of-Day

(see Chap. 4, this Vol.)

Knowledge of phytochrome and its implication in photoperiodism suggests several physiological questions: What change in the pigment signals the plant that night (or morning) has begun? Is it a decrease in P_{fr} to some critical concentration, as P_{fr} is metabolically destroyed or converted in the dark to P_r? This was suggested by BORTHWICK et al. (1952) and has been a major hypothesis ever since. Or does the plant detect some ratio of P_{fr} to P_{total}, as suggested by several workers (e.g., HOLMES and SMITH 1977b; SHROPSHIRE 1973)? Could the plant detect darkness when the rate of cycling between P_{fr} and P_r slows to some critical speed (see JOSE and VINCE-PRUE 1978)? How do these possibilities relate to the photoperiodism clock? Do phytochrome changes in some way account for time measurement? Or does an independent clock in some way control the condition of phytochrome? Or does such a clock control the responses of the plant to the status of phytochrome? Or does reality in some way combine and/or add to these possibilities?

One thing is certain: Changes in the balance between red and far-red wavelengths (as well as others) have often been observed to influence photoperiodism, and such spectral changes occur during twilight and as light passes through leaves (Figs. 5.3 and 5.4; Sect. 5.5). Hence, these physiological questions have strong ecological implications, justifying an examination of some experimental results.

BORTHWICK et al. (1952) tested the concept that a drop in P_{fr} concentration initiates dark measurement. Thirty minutes of high far-red irradiance (sunlight through a Corning red purple ultra filter) were reported to shorten the critical night for *Xanthium* by about 2 h (red light lengthened it by about 30 min). It was suggested that the immediate conversion of P_{fr} to P_r by far-red light eliminated the time required for metabolic dark conversion. Unfortunately, BORTHWICK and coworkers were never able to repeat these results with *Xanthium* (personal communication), nor have they been repeated by others (e.g., SALISBURY 1981), although high light levels were seldom used. Is it possible that the drop in P_{fr} in the dark is normally so rapid (e.g., only 5 to 10 min) that it cannot be detected by experiments designed to study effects of end-of-day far-red treatment on critical night?

Exposure of *Xanthium* plants to temperatures of 10 °C during the first 2 h of darkness delayed initiation of dark measurement (lengthened critical night) by about 0.8 h, but exposure to 10 °C between the 5th and 7th h had no effect on timing (SALISBURY 1963b). Thus, initiation of dark timing is apparently temperature-sensitive, but dark measurement is not. (Temperature insensitivity is typical of circadian clocks; e.g., HILLMAN 1976; VINCE-PRUE 1975.) The temperature-sensitive process that initiates dark measurement could well be a metabolic change in level of P_{fr}.

If darkness is detected either via a drop in P_{fr} concentration or a change in P_{fr}/P_{total}, there should be some mixture of red and far-red wavelengths that would be detected by the plant as darkness. That is, the mixture should either lower the P_{fr} to the critical level or establish the critical ratio. In the null experiments of EVANS and KING (1969; see also KATO 1979), mixtures of red and far-red wavelengths were sought that would have no effect when applied at various times during the dark period. Such innocuous mixtures should indicate P_{fr}/P_r ratios in the plant. Because of the overlapping absorption spectra for P_{fr} and P_r, some P_{fr}/P_r ratios could not be established by mixtures of wavelengths (VINCE-PRUE 1975), but EVANS and KING (1969) did find many mixtures that were detected by the plant at various times as darkness. SALISBURY (1981) found that relatively high levels of far-red were detected by *Xanthium* plants as darkness, both during the first 2 h and from the 7th to the 9th hour of a 16-h dark period. For example, 2 h of far-red plus 8 h of darkness gave virtually the same level of flowering as 10 h of darkness. If the far-red brought about a drop in P_{fr} level more rapid than usual, however, and that immediately initiated dark measurement, then far-red plus darkness should have produced *more* flowering than an equivalent time in darkness (as originally claimed by BORTHWICK et al. 1952).

Numerous experiments have revealed effects of red and/or far-red applied before an inductive (SDP) or inhibitory (LDP) dark period. In spite of confusing results, some patterns seem to be emerging:

Several workers have reported on effects of red and far-red applied at the end-of-day to *Pharbitis nil*, the Japanese morning glory. NAKAYAMA et al. (1960) found that flowering was inhibited by end-of-day far-red and repromoted by red previous to a 16-h inductive dark period – providing the treatment was applied to seedlings! With older plants there was no response to end-of-day red or far-red. Both red and far-red in the middle of the dark period inhibited flowering of seedlings, but only red was inhibitory and reversed by far-red when night interruption was given to mature plants. These authors emphasized that effects of red were opposite at the end-of-day and 4 h later.

KING (1974) confirmed these *Pharbitis* results and showed that far-red near the end-of-day shortened the critical night, provided that night was preceded by 5 min of red. Yet far-red *lengthened* the critical dark period when it was given 9 h before the onset of darkness (again preceded by 5 min of red), clearly implicating the photoperiodism clock.

KING and VINCE-PRUE (1978) applied 24 h of far-red to *Pharbitis* before an inductive dark period, but to obtain flowering, the far-red had to be terminated by a 10-min red exposure. VINCE-PRUE et al. (1978) then measured phytochrome photometrically in *Pharbitis*. Following 24 h of white light and 10 min of red, P_{fr} required 3 h of darkness to drop to zero, at which time total phytochrome had dropped by about half. KING et al. (1978) followed phytochrome levels with both photometric and physiological techniques, which agreed well with each other. P_{fr} dropped within 30 to 90 min and there was some correlation with the critical dark period. Yet they were unable to observe a correlation of phytochrome levels with the time of maximum sensitivity to a light interruption of the inductive dark period. On this basis, they conclude that P_{fr} disappearance does *not* signal the transition from light to darkness. Yet, dark measurement could be initiated by a relatively small drop in P_{fr} or a relatively small change in P_{fr}/P_{total}. Finally, OGAWA and KING (1979) report that the promotion by red at the beginning of an inductive dark period for *Pharbitis* is greatly increased by treatment with benzyladenine (a cytokinin). They suggest that the effect is upon translocation of floral stimulus rather than induction itself.

Even with *Xanthium*, the phytochrome system is implicated during the light period (e.g., SALISBURY 1965), although there are no clear-cut, *Pharbitis*-type promotive or inhibitory effects of red or far-red light. SALISBURY (1981) found that levels of red required to extend the daylight period (or to inhibit when applied in the middle of a dark period) were from 2 to 7 times as high when they were accompanied by approximately equivalent amounts of far-red (red/far-red = ca. 1.0) as when much less far-red was present (red/far-red = 7 to 9).

WILLIAMS and MORGAN (1979) found that, contrary to *Pharbitis*, end-of-day far-red *promoted* flowering of sorghum, another SDP. The effect was red-far-red reversible and synergistic with applied gibberellic acid. There were differences among genotypes, but the far-red promotion was easily observed.

Many workers have applied far-red light or mixtures of different wavelengths for 12 or more hours before a dark period (e.g., VINCE-PRUE et al. 1978; see above). REID and MURFET (1977) found that far-red given as a 16-h photoperiod extension was more effective than red in lowering the node to the first flower in peas. They also found that a light interruption of a dark period was not effective in their system until at least 12 h after the beginning of the previous photoperiod, implicating again the photoperiodism clock.

KADMAN-ZAHAVI and EPHRAT used greenhouses covered with polyvinyl chloride (opaque to ultraviolet) and colored celluloid filters. They reported in 1974 that flowering of some species was promoted by about equal amounts of blue and far-red but inhibited by blue with virtually no far-red. Other species reacted just the opposite. KADMAN-ZAHAVI and EPHRAT (1976a) studied a number of grasses (*Hordeum, Triticum, Setaria, Sorghum, Oryza,* and *Zea*), some LDP's and some SDP's. All reacted the same: flowering and elongation were promoted by blue plus far-red but inhibited by blue alone. Responses were intermediate in orange plus far-red, clear, and neutral shade. Night interruptions of 1.75 min in the middle of a 16-h dark period promoted flowering in the LD barley (*Hordeum*), but only if plants were in the blue plus far-red and not if they were in blue alone. KADMAN-ZAHAVI and EPHRAT (1976b) further observed a promotion of flowering and elongation in the LD *Hyoscyamus niger* by blue plus far-red but inhibition by blue. End-of-day far-red (15 min) also promoted flowering, even on 8-h days (thereby nullifying the LD requirement). End-of-day red delayed flowering in all greenhouses. The authors concluded that day-time P_{fr} levels are important, being high in the blue greenhouse and low in blue plus far-red.

DEITZER et al. (1979) found enhanced flowering of the LD wintex barley by addition of far-red to the main light period, suggesting that the high irradiance response was involved. They approximated the red/far-red ratios of leaf shade or twilight. They could eliminate the usual LD requirement of barley with sufficient far-red during the light period, but their most striking far-red promotions of flowering were observed only when far-red was added to continuous white light. Brief intervals of far-red given at different times revealed a circadian rhythm of response to far-red.

The above experiments, although complex, show a general promotion or inhibition by red or far-red wavelengths applied at the end-of-day or during the day, actual responses depending strongly upon species. Other experiments are more difficult to interpret. For example, GUTTRIDGE (1976) describes complex effects with strawberry, which is a SDP for flowering. Pure red from midnight to dawn inhibited flowering and promoted runnering – but did not have this effect from dusk to midnight. Incandescent light (mixtures of red and far-red) was always effective in inhibiting flowering and promoting runnering plus especially long petioles.

Several experimental results are not easily compatible with the suggestion (JOSE and VINCE-PRUE 1978) that plants detect darkness by responding to a slower rate of cycling between P_{fr} and P_r. For example, mixtures of red and far-red require higher levels of red to achieve the same delay in the initiation of dark measurement than when red is applied alone (SALISBURY 1981). But mixtures should produce *faster* cycling than the lower levels of equally effective pure red. Furthermore, cycling would probably be quite temperature-insensitive, but initiation of dark timing is delayed at low temperatures (SALISBURY 1963b).

What remains is to relate all this to plants in natural or agricultural environments. We have seen that red/far-red ratios during the day and at the end-of-day can influence flowering in various ways, from simple promotion or inhibition to changing of critical day or night; such effects surely occur under natural conditions. Yet it is seldom valid to apply laboratory data to field conditions, since the red/far-red ratios may have been different in the two situations. Nevertheless, there are possible generalizations. For example, a plant growing in the enriched far-red light under a leaf canopy might flower under otherwise unfavorable daylengths, which could insure seed production before the low light levels led to photosynthetic starvation. But what about plants that are *inhibited* by high far-red levels? It is time to devise field experiments to test some of the possibilities.

5.5.3 The Photoperiodism Clock

As is clear from experiments already described, the photoperiodism clock cannot be ignored in any discussion of light effects. An initial suggestion was that changing P_{fr} levels might account for the clock (BORTHWICK et al. 1952). By now the situation is seen quite differently: it is the clock that controls the manner in which a plant responds to phytochrome status. Since phytochrome status is determined by the light environment, the photoperiodism clock takes on ecological significance. Not only does the clock control the plant's response to the light environment, but the environment (especially light but other factors as well) phases and may otherwise influence the clock (e.g., BÜNNING 1978; FRIEND 1975). As a further complication, HOROWITZ and EPEL (1978) observed in etiolated zucchini seedlings that the clock may influence the nature of phytochrome, specifically maximum absorption wavelengths of P_{fr}, which seemed to follow a circadian rhythm.

Several manifestations of plant growth and function are controlled by a biological clock; that is, they exhibit a circadian (approximately 24-h) rhythm, even under constant conditions of light and temperature. Notable among these are the sleep movements of leaves (brief discussion in SALISBURY and ROSS 1978). In the mid 1950's, researchers began to ask whether the photoperiodism clock was similar to the clock that controls circadian rhythms. By the mid 1960's, many papers documented similarities between the two clocks (reviewed by VINCE-PRUE 1975). Yet SALISBURY and DENNEY (1974) found that the two clocks could be separated in *Xanthium*. Neither critical dark period nor the time of maximum sensitivity to a light interruption of an inductive dark period could be predicted from leaf positions. Thus leaf positions in *Xanthium* do not act as a set of "hands" for the photoperiodism clock. These conclusions have been confirmed for *Pharbitis* by BOLLIG (1977) and for *Chenopodium rubrum* by KING (1975, 1979).

One of the most important conclusions from work on the photoperiodism clock is that plant response in photoperiodism is neither to daylength nor to night length, although early experiments emphasized the importance of the dark period. It is now clear from many experiments, including several discussed above (Sect. 5.5.2), that conditions during the day strongly influence such responses as time measurement and light sensitivity during the night.

5.6 A Final Word

Much remains to be learned about photoperiodism in plants. The overwhelming observation is that various species respond in a legion of ways to light and temperature components of the environment. Does this mean that there are dozens to hundreds of different mechanisms of response? Most workers in the field would probably make the statement of faith that a few basic mechanisms are probably involved, these being "tuned" in various ways by genetic modifications. Certainly the very complexity of the responses is itself of high ecological significance; it provides profound insight into the complexity of ecosystems. Attempts to apply the vast body of information about plant responses to photoperiod have usually involved agricultural systems. It may well be time to begin a similar application to natural ecosystems, partially by theory but mostly by field experiments.

References

Aggarwal VD, Poehlman JM (1977) Effects of photoperiod and temperature on flowering in mungbean (*Vigna radiata* (L.) Wilczek). Euphytica 26:207–219

Allison JCS, Daynard TB (1979) Effect of change in time of flowering, induced by altering photoperiod or temperature, on attributes related to yield in maize. Crop Sci 19:1–4

Alvim R, Thomas S, Saunders PF (1978) Seasonal variation in the hormone content of willow. II. Effect of photoperiod on growth and abscisic acid content of trees under field conditions. Plant Physiol 62:779–780

Anonymous (ed) (1979) La physiologie de la floraison. International Colloquium on the Physiology of Flowering. Editions due Centre de la Recherche Scientifique, Paris

Arnott JT (1974) Growth response of white-Engelmann spruce provenances to extended photoperiod using continuous and intermittent light. Can J For Res 4:69–75

Arnott JT (1979) Effect of light intensity during extended photoperiod on growth of amabilis fir, mountain hemlock, and white and Engelmann spruce seedlings. Can J For Res 9:82–89

Aung LH (1976) Effects of photoperiod and temperature on vegetative and reproductive responses of *Lycopersicon esculentum* Mill. J Am Soc Hortic Sci 101:358–360

Barba RC, Pokorny TA (1975) Influence of photoperiod on the propagation of two rhododendron cultivars. J Hortic Sci 50:55–59

Bentley B, Morgan CB, Morgan DG, Saad FA (1975) Plant growth substances and effects of photoperiod on flower bud development in *Phaseolus vulgaris*. Nature 256:121–122

Bhargava SC (1975) Photoperiodicity and seed germination in rice. Indian J Agric Sci 45:447–451

Bhella HS, Roberts AN (1974) The influence of photoperiod and rooting temperature on rooting of Douglas-fir [*Pseudotsuga menziesii* (Mirb.) France]. J Am Soc Hortic Sci 99:551–555

Bodson M, King RW, Evans LT, Bernier G (1977) The role of photosynthesis in flowering of the long-day plant *Sinapis alba*. Aust J Plant Physiol 4:467–478

Bollig I (1977) Different circadian rhythms regulate photoperiodic flowering response and leaf movement in *Pharbitis nil* (L.) Choisy. Planta 135:137–142

Borthwick HA, Hendricks SB, Parker MW (1952) The reaction controlling floral initiation. Proc Natl Acad Sci USA 38:929–934

Bose TK (1974) Effect of temperature and photoperiod on growth, flowering and seed formation in tossa jute. Indian J Agric Sci 44:32–35

Bünning E (1978) Wechselwirkungen zwischen circadianer Rhythmik und Licht. Physiol Veg 16:799–804

Bünning E, Moser I (1969) Interference of moonlight with the photoperiodic measurement of time by plants, and their adaptive reaction. Proc Natl Acad Sci USA 62:1018–1022

Buttrose MS, Sedgley M (1978) Some effects of light intensity, daylength and temperature on growth of fruiting and non-fruiting watermelon (*Citrullus lanatus*). Ann Bot 42:599–608

Cameron DF, Mannetje L (1977) Effects of photoperiod and temperature on flowering ot twelve *Stylosanthes* species. Aust J Exp Agric Anim Husb 17:417–424

Christersson L (1978) The influence of photoperiod and temperature on the development of frost hardiness in seedlings of *Pinus silvestris* and *Picea abies*. Physiol Plant 44:288–294

Claes H, Lang A (1947) Die Wirkung von β-Indolylessigsäure und 2, 3, 5-Trijodbenzoesäure auf die Blütenbildung von *Hyoscyamus niger*. Z Naturforsch 26:56–63

Cockshull KE (1976) Flower and leaf initiation by *Chrysanthemum morifolium* Ramat in long days. J Hortic Sci 51:441–450

Deitzer GF, Hayes R, Jabben M (1979) Kinetics and time dependence of the effect of far-red light on the photoperiodic induction of flowering in wintex barley. Plant Physiol 64:1015–1021

Erickson RO, Meicenheimer RD (1977) Photoperiod induced change in phyllotaxis in *Xanthium*. Am J Bot 64:981–988

Evans LT (ed) (1969) The Induction of Flowering: Some Case Histories. Macmillan of Australia, South Melbourne pp 488

Evans LT (1971) Flower induction and the florigen concept. Ann Rev Plant Physiol 22:365–394
Evans LT (1975) Daylength and the flowering of plants. Menlo Park, CA: Benjamin
Evans LT, King RW (1969) Role of phytochrome in photoperiodic induction of *Pharbitis nil.* Z Pflanzenphysiol 60:277–288
Ewing EE, Wareing PF (1978) Shoot, stolon, and tuber formation on potato (*Solanum tuberosum* L.) cuttings in response to photoperiod. Plant Physiol 61:348–363
Fairey DT, Hunt LA, Stoskoff NC (1975) Day-length influence on reproductive development and tillering in 'Fergus' barley. Can J Bot 53:2770–2775
Friend DJC (1975) Light requirements for photoperiodic sensitivity in cotyledons of dark-grown *Pharbitis nil.* Physiol Plant 35:286–296
Garner WW, Allard HA (1920) Effect of the relative length of day and night and other factors of the environment on growth and reproduction in plants. J Agric Res 18:553–606
Garner WW, Allard HA (1923) Further studies in photoperiodism, the response of the plant to relative length of day and night. J Agric Res 23:871–920
Guttridge CG (1976) Growth, flowering and runnering of strawberry after daylength extension with tungsten and fluorescent lighting. Sci Hortic 4:345–352
Habjorg A (1978a) Climatic control of floral differentiation and development in selected latitudinal and altitudinal ecotypes of *Poa pratensis* L. Norges Landbrukshogskole Meldinger 57(7):1–21
Habjorg A (1978b) Photoperiodic ecotypes in Scandinavian trees and shrubs. Norges Landbrukshogskole Meldinger 57(33):2–20
Hammes PS, Nel PC (1975) The effect of photoperiod on growth and yield of potatoes (*Solanum tuberosum* L.) in controlled environments. Agroplantae 7:7–12
Hamner KC (1969) *Glycine max* (L.) Merrill. In: Evans LT (ed), The induction of flowering. Some case histories. Macmillan of Australia, South Melbourne pp 62–89
Hanks GR, Rees AR (1979) Photoperiod and tulip growth. J Hortic Sci 54:29–46
Hedley CL, Harvey DM (1975) Variation in the photoperiodic control of flowering of two cultivars of *Antirrhinum majus* L. Ann Bot 39:257–263
Heins RD, Wilkins HF (1978) Influence of photoperiod and light quality on stolon formation and flowering of *Chlorophytum comosum* (Thunb.) Jacques. J Am Soc Hortic Sci 103:687–689
Heins RD, Wilkins HF (1979) Effect of soil temperature and photoperiod on vegetative and reproductive growth of *Alstroemeria* 'Regina'. J Am Soc Hortic Sci 104:359–365
Herath HMW, Ormrod DP (1979) Effects of temperature and photoperiod on winged beans (*Psophocarpus tetragonolobus* (L.) D.C.) Ann Bot 43:729–736
Hillman WS (1976) Biological rhythms and physiological timing. Ann Rev Plant Physiol 27:159–179
Hodgkinson KC, Quinn JA (1978) Environmental and genetic control of reproduction in *Danthonia caespitosa* populations. Aust J Bot 26:351–364
Holmes MG, Smith H (1977a) The function of phytochrome in the natural environment. I. Characterization of daylight for studies in photomorphogenesis and photoperiodism. Photochem Photobiol 25:533–538
Holmes MG, Smith H (1977b) The function of phytochrome in the natural environment. II. The influence of vegetation canopies on the spectral energy distribution of natural daylight. Photochem Photobiol 25:539–545
Horowitz BA, Epel BL (1978) Circadian changes in activity of the far-red form of phytochrome: physiological and in vivo-spectrophotometric studies. Plant Sci Lett 13:9–14
Hugel B (1976) Opposite sex-expression of flower primordia of the long-day plant *Lemna gibba* and of the short day plant *Lemna paucicostata* in vitro. (Ger.) Z Pflanzenphysiol 77:395–405
Hunt LA (1979) Photoperiodic responses of winter wheats from different climatic regions. Z Pflanzenzuecht 82:70–80
Immel MJ, Rumsey RL, Carpenter SB (1978) Comparative growth responses of northern red oak and chestnut oak seedlings to varying photoperiods. For Sci 24:554
Johnson JR, Havis JR (1977) Photoperiod and temperature effects on root cold acclimation. J Am Soc Hortic Sci 102:306–308

Jose AM, Vince-Prue D (1978) Phytochrome action: a reappraisal. Photochem Photobiol 27:209–216

Julien MAR, Soopramanien GC (1975) Effect of night break on floral initiation and development of *Saccharum*. Crop Sci 16:625–629

Kadman-Zahavi A, Ephrat E (1974) Opposite response groups of short-day plants to the spectral compositions of the main light period and to end of day red or far-red irradiations. Plant Cell Physiol 15:693–699

Kadman-Zahavi A, Ephrat E (1976a) Development of plants in filtered sunlight. II. Effects of spectral composition, light intensity, daylength and red and far-red irradiations on long- and short-day grasses. Isr J Bot 25:11–23

Kadman-Zahavi A, Ephrat E (1976b) Development of plants in filtered sunlight. III. Interaction of the spectral composition of main-light periods with end-of-day red or far-red irradiations and with red night interruptions in bolting and flowering of *Hyoscyamus niger*. Isr J Bot 25:203–210

Katayama TC (1964) Photoperiodism in the genus *Oryza*. II. Jpn J Bot 18:349–383

Kato A (1979) Effect of interruption of the nyctoperiod with an R/FR-mixture of various ratios of red and far-red light on flowering in *Lemna gibba* G3. Plant Cell Physiol 20:1273–1283

Kedar N, Levy D, Goldschmidt EE (1975) Photoperiodic regulation of bulbing and maturation of Bet Alpha onions (*Allium cepa* L.) under decreasing daylength conditions. J Hortic Sci 50:373–380

King RW (1974) Phytochrome action in the induction of flowering in short-day plants: effect of photoperiod quality. Aust J Plant Physiol 1:445–457

King RW (1975) Multiple circadian rhythms regulate photoperiodic flowering responses in *Chenopodium rubrum*. Can J Bot 53:2631–2638

King RW (1979) Photoperiodic time measurement and effects of temperature on flowering in *Chenopodium rubrum* L. Aust J Plant Physiol 6:417–422

King RW, Vince-Prue D (1978) Light requirement, phytochrome and photoperiodic induction of flowering of *Pharbitis nil* Chois. I. No correlation between photomorphogenetic and photoperiodic effects of light pretreatment. Planta 141:1–7

King RW, Vince-Prue D, Quail PH (1978) Light requirement, phytochrome and photoperiodic induction of flowering of *Pharbitis nil* Chois. III. A comparison of spectrophotometric and physiological assay of phytochrome transformation during induction. Planta 141:15–22

Kotecha AK, Yermanos DM, Shropshire FM (1975) Flowering in cultivars of sesame (*Sesamum indicum*) differing in photoperiodic sensitivity. Econ Bot 29:185–191.

Krebs O, Zimmer K (1978) Flower formation in *Begonia boweri* and an offspring of *Begonia cleopatra*. I. Temperature and critical daylength. Gartenbauwissenschaft 43:87–90

Lachaud S (1976) Influence de l'eclairement et de la photoperiode sur le vieillissement du *Soja biloxi*. CR Soc Biol 171:180–187

Laidlaw AS, Berrie AMM (1977) The relative hardening of roots and shoots and the influence of day-length during hardening in perennial ryegrass. Ann Appl Biol 87:443–450

Langton FA (1977) The response of early-flowering chrysanthemums to daylength. Sci Hortic 7:277–289

Lavigne C, Cosson L, Jacques R, Miginiac E (1979) Effect of day length and temperature on *Jasminum grandiflorum* L. growth, flowering and essential oil chemical composition (FR.) Physiol Veg 17:363–373

Lowe SB, Mahno JD, Hunt LA (1976) The effect of daylength on shoot growth and formation of root tubers in young plants of cassava (*Manihot esculenta* Grantz). Plant Sci Lett 6:57–62

Macdowall FDH (1977) Growth kenetics of Marquis wheat. VII. Dependence on photoperiod and light compensation point in vegetative phase. Can J Bot 55:639–643

Mayer AM, Poljakoff-Mayber A (1963) The Germination of Seeds. Pergamon Press, New York

McCreary DD, Tanaka Y, Lavender DP (1978) Regulation of Douglas-fir seedling growth and hardiness by controlling photoperiod. For Sci 24:142–152

McMillan C (1975) The *Xanthium strumarium* complexes in Australia. Aust J Bot 23:173–192

Milford GEJ, Lenton JR (1976) Effect of photoperiod on growth of sugar beet. Ann Bot 40:1309–1315

Morita M, Iwamoto S, Higuchi H (1978) Interrelated effects of thermo- and photoperiodism on growth and development of ornamental woody plants. III. Modification of growth responses of *Viburnum awabuki* K. Koch and *Pinus thunbergii* Parl. by fall chilling pretreatment. J Jpn Soc Hortic Sci 47:425–430

Murti GSR, Banerjee VN (1977) Effects of light breaks on tuber initiation and growth of the potato. JIPA 4:7–10

Murti GSR, Saha SN (1975) Effect of stage of perception of photoperiodic stimulus and number of short-day cycles on tuber initiation and development of potato. Indian J Plant Physiol 18:184–188

Murti GSR, Saha SN, Banerjee VN (1975) Studies on the photoperiod response of some newly released varieties of potato. Indian J Plant Physiol 18:41–45

Nakayama S, Borthwick HA, Hendricks SB (1960) Failure of photoreversible control of flowering in *Pharbitis nil.* Bot Gaz 121:237–243

Newell SJ, Tramer EJ (1978) Reproductive strategies in herbaceous plant communities during succession. Ecology 59:228–234

Ogawa Y, King RW (1979) Establishment of photoperiodic sensitivity by benzyladenine and a brief red irradiation in dark grown seedlings of *Pharbitis nil* Chois. Plant Cell Physiol 20:115–122

Okusanya OT (1979) Quantitative analysis of the effects of photoperiod, temperature, salinity and soil types on the germination and growth of *Corchorus olitorius.* Oikos 33:444–450

Papenfuss HD, Salisbury FB (1967) Aspects of clock resetting in flowering of *Xanthium.* Plant Physiol 42:1562–1568

Parker, MW, Hendricks SB, Borthwick HA, Scully NJ (1946) Action spectrum for the photoperiodic control of floral initiation of short-day plants. Bot Gaz 108:1–26

Paton DM (1978) Eucalyptus physiology. I. Photoperiodic responses. Aust J Bot 26:633–642

Patterson DT, Peet MM, Bunce JA (1977) Effect of photoperiod and size at flowering on vegetative growth and seed yield of soybean. Agron J 69:631–635

Pirasteh B, Welsh JR (1975) Monosomic analysis of photoperiod response in wheat. Crop Sci 15:503–505

Pourrat Y, Jacques R (1975) The influence of photoperiodic conditions received by the mother plant on morphological and physiological characteristics of *Chenopodium polyspermum* L. seeds. Plant Sci Lett 4:273–279

Proebsting WM, Davies PJ, Marx GA (1976) Photoperiodic control of apical senescence in a genetic line of peas. Plant Physiol 58:800–802

Reid JB, Murfet IC (1977) Flowering in *Pisum*: the effect of light quality on the genotype *lf e Sn Hr.* J Exp Bot 28:1357–1364

Salisbury FB (1963a) The Flowering Process. Pergamon Press, New York

Salisbury FB (1963b) Biological timing and hormone synthesis in flowering of *Xanthium.* Planta 59:518–534

Salisbury FB (1965) Time measurement and the light period in flowering. Planta 66:1–26

Salisbury FB (1981) The twilight effect: initiating dark measurement in photoperiodism of *Xanthium.* Plant Physiol. 67:1230–1238

Salisbury FB, Bonner J (1956) The reactions of the photoinductive dark period. Plant Physiol 31:141–147

Salisbury FB, Denney A (1974) Noncorrelation of leaf movements and photoperiodic clocks in *Xanthium strumarium* L. In: Chronobiology. Scheving LE, Halberg F, Pauly JE (eds). Proc Int Soc Chronobiol, Little Rock, AR, Nov 8–10 1971. Iqaku Shoin, Ltd. pp 679–686

Salisbury FB, Ross CW (1978) Plant Physiology, 2nd edn. Wadsworth Publishing Co, Inc., Belmont, CA

Sato K (1971) Growth and development of alfalfa plant under controlled environment. I. The effects of daylength and temperature on the growth and chemical composition. Proc Crop Sci Soc Jpn 40:120–125

Sato K (1974) Growth responses of rice plant to environmental conditions. III. The effects of photoperiod and temperature on the growth and chemical composition. Proc Crop Sci Soc Jpn 43:401–410

Sato K (1979) The growth responses of soybean plant to photoperiod and temperature. III. The effects of photoperiod and temperature on the development and anatomy of photosynthetic organ (Jap.) Jpn J Crop Sci 48:66–74

Sawhney S, Sawhney N, Nanda KK (1978) Effect of varying lengths of inductive and supplementary non-inductive photoperiods on vegetative growth and flowering of *Impatiens balsamina*. Plant Cell Physiol 10:647–653

Schwabe WW (1971) Physiology of vegetative reproduction and flowering. In: Stewart FC (ed) Plant Physiology – A Treatise, Vol. 6A, Academic Press, New York pp 233–411

Schwartz A, Koller D (1975) Photoperiodic control of shoot-apex morphogenesis in *Bouteloua eriopoda*. Bot Gaz 136:41–49

Sedgley M, Buttrose MS (1978) Some effects of light intensity, daylength and temperature on flowering and pollen tube growth in the watermelon (*Citrullus lanatus*). Ann Bot 42:609–616

Seibert M (1976) The effects of night-lighting on growth and pigmentation in greenhouse-grown tropical ornamentals. HortScience 11:46–47

Sharma R, Nanda KK (1976) Effect of photoperiod on growth and development of *Eschscholtzia californica* Cham. Indian J Plant Physiol 19:202–206

Shillo R (1976) Control of flower initiation and development of statice (*Limonium sinuatum*) by temperature and daylength. Acta Hortic 64:197–203

Shropshire W, Jr (1973) Photoinduced parental control of seed germination and the spectral quality of solar radiation. Sol Energy 15:99–105

Skrubis B, Markakis P (1976) The effect of photoperiodism on the growth and the essential oil of *Ocimum basilicum* (sweet basil). Econ Bot 30:389–394

Soffe RW, Lenton JR, Milford GFJ (1977) Effects of photoperiod on some vegetable species. Ann Appl Biol 85:411–415

Stromgren T (1978) The effect of photoperiod on the length growth of five species of intertidal Fucales. Sarsia 63:155–158

Takimoto A, Hamner K (1964) Effect of temperature and preconditioning on photoperiodic response of *Pharbitis nil*. Plant Physiol 39:1024–1030

Takimoto A, Ikeda K (1960) Studies on the light-controlling flower initiation of *Pharbitis nil*. VI. Effect of natural twilight. Bot Mag Tokyo 73:175–181

Takimoto A, Ikeda K (1961) Effect of twilight on photoperiodic induction in some short day plants. Plant Cell Physiol 2:213–229

Tasker R, Smith H (1977) The function of phytochrome in the natural environment. V. Seasonal changes in radiant energy quality in woodlands. Photochem Photobiol 26:487–491

Thomas JF, Raper CD, Jr (1976) Photoperiodic control of seed filling for soybeans. Crop Sci 16:667–672

Tinus RW (1976) Photoperiod and atmospheric CO_2 level interact to control black walnut (*Juglans nigra* L.) seedling growth (abst.) Plant Physiol 57 (Suppl):106

Turner BC, Karlander EP (1975) Photoperiodic control of floral initiation in sicklepod (*Cassia obtusifolia* L.). Bot Gaz 136:1–4

Vince-Prue D (1975) Photoperiodism in Plants. McGraw-Hill Book Co., London

Vince-Prue D, King RW, Quail PH (1978) Light requirement, phytochrome and photoperiodic induction of flowering of *Pharbitis nil* Chois. II. A critical examination of spectrophotometric assays of phytochrome transformations. Planta 141:9–14

Warner RM, Worku Z, Silva JA (1979) Effect of photoperiod on growth responses of citrus rootstocks. J Am Soc Hortic Sci 104:232–235

Wellensiek SJ (1969) *Silene armeria* L. In: Evans LT (ed) The Induction of Flowering. Some Case Histories. Macmillan of Australia, South Melbourne pp 350–363

Wheeler N (1979) Effect of continuous photoperiod on growth and development of lodgepole pine seedlings and grafts. Can J For Res 9:276–283

Williams EA, Morgan PW (1979) Floral initiation in sorghum hastened by gibberellic acid and far-red light. Planta 145:269–272

Williams RD (1978) Photoperiod effects on the reproductive biology of purple nutsedge (*Cyperus rotundus*). Weed Sci 26:539–542

Wright RD (1976) Temperature and photoperiod on the growth of *Ilex cornuta* Lindl. et Paxt. HortScience 11:44–46

Yanagisawa Y (1975) Effects of combinations of different day-length conditions during the growth of flax plants (*Linum usitatissimum* L.) on their growth and development. II. The effects of transfer from short day to natural long day. Proc Crop Sci Soc Jpn 44:375–381

Young E, Hanover JW (1977) Effects of quality, intensity, and duration of light breaks during a long night on dormancy in blue spruce (*Picea pungens* Engelm.) seedlings. Plant Physiol 60:271–273

Zeevaart JAD (1976) Physiology of flower formation. Ann Rev Plant Physiol 27:321–348

6 Plant Response to Solar Ultraviolet Radiation

M.M. CALDWELL

CONTENTS

6.1 Introduction

In keeping with the tone of these volumes on plant physiological ecology, in this chapter the ecological context of plant response to solar UV radiation will be emphasized rather than UV photophysiology. Several recent reviews can be consulted for more information on physiological mechanisms (JAGGER 1976; WELLMANN 1976; THOMAS 1977; FRAIKIN and RUBIN 1979). A review of KLEIN (1978) is particularly comprehensive.

Although interest in the effects of solar UV radiation on plants arose over a century ago, comparatively little attention had been paid to this limited portion of the solar spectrum until recent suggestions that man's activities might reduce the atmospheric ozone layer, and thus cause an increase of solar UV at the Earth's surface.

This chapter will attempt to develop perspectives about the nature of the solar UV environment, the mechanisms of plant response to this radiation, and the morphological and physiological characteristics of plants which afford tolerance of this radiation. Prospects concerning the consequences of a reduced atmospheric ozone layer will also be offered. The emphasis is clearly directed to terrestrial higher plant life, but reference to phytoplankton will also be made.

6.2 Ultraviolet Bands: Common UV Sources and Potential Chromophores

6.2.1 Common Sources of UV Radiation

Electromagnetic radiation from the sun contains a small proportion of UV (wavelengths shorter than 400 nm) – about 7% of the solar radiation that reaches the Earth's surface. Yet, UV photons are effectively absorbed by important biological macromolecules which result in a variety of photochemical reactions. Thus, the biological significance of UV radiation is proportionately much greater than its small contribution to the total solar energy reaching the Earth.

Because photochemical reactions can be highly dependent on the wavelength of radiation, it is misleading to categorically treat all UV merely in terms of energy. Only absorbed photons can elicit photochemical reactions and the absorption spectra of chromophores, compounds that mediate photochemical reactions in biological systems, can be very wavelength-specific. To lend some perspective, the spectral irradiance from common sources of UV radiation and the absorption spectra of some plant chromophores are shown in Fig. 6.1. Major subdivisions of the UV spectrum are also indicated.

Although there are several possible chromophores that can mediate UV-induced reactions in plants, the most destructive action of UV involves the nucleic acids. The irradiance at 30 cm from a common 15-W germicidal lamp is only 10^{-4} of the energy coming from the midday sun. However, radiation from such a low pressure mercury vapor lamp is almost entirely at 254 nm,

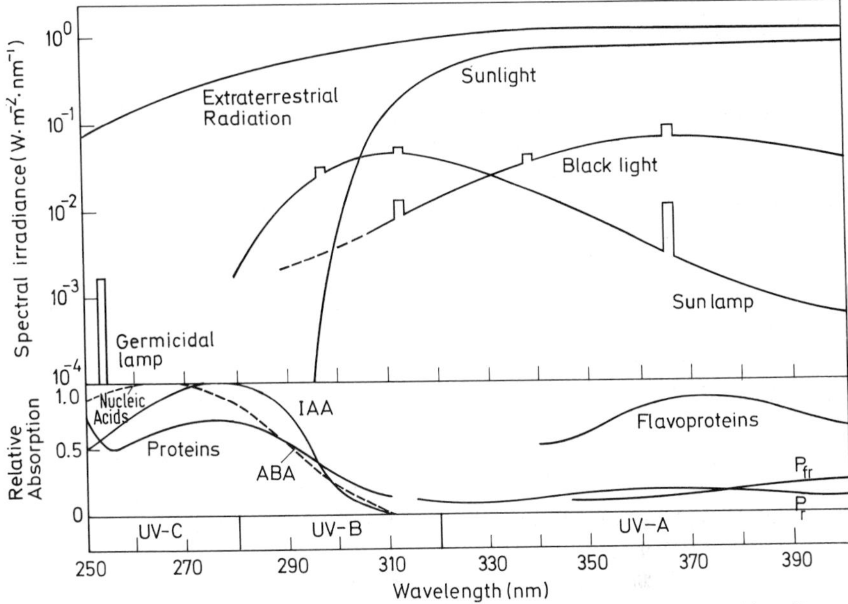

Fig. 6.1. Spectral irradiance at 30 cm from three common UV lamps, solar spectral irradiance before attenuation by the Earth's atmosphere (extraterrestrial), and as would be received at sea level at midday in the summer at temperate latitudes. The absorption spectra of some common potential UV chromophores in plants are also represented: abscisic acid, ABA, is represented by the same curve as for nucleic acids, indoleacetic acid, IAA, the two forms of phytochrome, P_{fr} and P_r, and general absorption spectra for flavoproteins and proteins. Major subdivisions of the ultraviolet spectrum are indicated. As originally defined UV-C is UV radiation less than 280 nm, UV-B is 280 to 315 nm, and UV-A is 315 to 400 nm; however, 320 nm is now more commonly used as a division between UV-B and UV-A

a wavelength where radiation is efficiently absorbed by nucleic acids. For this reason, this small quantity of radiant energy can be very biologically effective – especially for lower forms of plant life.

The common "black light bulb" emits more energy than a germicidal lamp, but still an order of magnitude less than solar radiation. Because most of the energy is at longer wavelengths, radiation from such bulbs is much less destructive. Another common type of lamp is the fluorescent "sun lamp". Although this lamp emits less energy than the black light, it is more biologically destructive because the spectral distribution overlaps more effectively with the tail of the nucleic acid absorption spectrum.

6.2.2 Potential UV Chromophores in Plants

The chromophores shown in Fig. 6.1 are some of the candidates commonly reputed to mediate a variety of UV-elicited phenomena in plants. These biological compounds have key roles in plant cell function or structure, and any alterations of these compounds due to UV might be expected to cause physiological alterations in plants.

The role played by potential chromophores in plant responses to solar UV radiation has not been well established. For example, the quantum yield of some chromophores may be too low for solar UV. Such is apparently the case with abscisic acid, ABA (LINDOO et al. 1979). The specific response of a potential chromophore to solar UV radiation also will be greatly diminished if it absorbs appreciably in the visible spectrum where solar flux is much greater. For example, though phytochrome transitions can be effected by UV radiation (BUTLER et al. 1964), this response would be overwhelmed by the response to red and far-red radiation in daylight.

Another complication in the identification of responsible UV chromophores lies in the distinction between primary and secondary effects of UV radiation. For example, changes in DNA or RNA, which can be readily caused by UV, may result in a series of secondary effects of UV radiation, such as reduced protein synthesis, altered nitrate reductase activity (WRIGHT and MURPHY 1975) or petiolar abscission (KLEIN 1967). The protein, and to a certain extent the lipid components, of plant membranes also effectively absorb UV. Alteration of membrane characteristics may not only result in changes in membrane permeability and ionic balance (WRIGHT and MURPHY 1978), but may be also ultimately responsible for partial inhibition of photosynthesis (MANTAI et al. 1970; BRANDLE et al. 1977) and respiratory changes (ROY and ABBOUD 1978; WRIGHT 1978).

6.2.3 UV Action Spectra: Ecological Utility

Biological responses to UV can be very wavelength-dependent. This wavelength specificity makes it necessary to develop weighting functions to express UV, whether from the sun or from lamps, in biologically meaningful terms. As will be developed in Section 6.3.2, use of different weighting functions can result in radically different conclusions concerning the significance of solar UV radiation intensification due to atmospheric ozone reduction. Biological action spectra normally serve as a basis of these weighting functions.

A generalized equation to describe the use of a biological weighting function is:

$$\text{effective irradiance} = \int I_\lambda E_\lambda \, d\lambda$$

where I_λ is the spectral irradiance, and E_λ is the relative effectiveness of irradiance at wavelength λ to elicit a particular biological response. The limits to the integration are prescribed by the wavelengths where either I_λ or E_λ approach zero. Irradiance can be expressed on either a photon or energy basis.

The weighting function, E_λ, is normally taken as an action spectrum for a particular biological response. Traditionally, biological action spectra have been determined by exposing organisms to monochromatic radiation at the same irradiance and scoring the biological response, or by supplying the amount of monochromatic flux necessary to elicit a certain threshold response. In either case, the emphasis has been to determine the fine structure of the action spectrum with as much spectral resolution as possible in order that potential chromophores might be identified. Seldom have the tails of these action spectra been elucidated, as usually they contribute little to identification of chromophores. Since solar spectral irradiance changes by orders of magnitude within the UV spectrum, tails of action spectra can be quite important even though these may represent greatly diminished biological effectiveness compared to the maxima. Thus, the spectra to be used as weighting functions should ideally include any tails that lie within the wavebands where there is solar flux.

Since biological action spectra are normally determined with monochromatic radiation, the assumption must be made that radiation at different wavelengths does not interact when presented simultaneously. Yet there is evidence that synergisms may be involved when organisms are exposed to polychromatic UV radiation (ELKIND et al. 1978; ELKIND and HAN 1978). Various repair systems activated by light also may complicate spectral assessments. Photoreactivation is a process that repairs a certain type of DNA damage. This mechanism is driven by UV-A and blue light and can greatly diminish the resultant damage caused by UV-C or UV-B. Plant photosynthesis also appears to be much less inhibited by UV-B in the presence of strong visible light than if leaves are only subjected to UV-B radiation. Although it is unlikely that photoreactivation per se is effective in repairing photosynthetic damage, some mechanism is operating to alleviate the inhibition. Thus, repair systems or other interactions that might alter the expression of UV damage in plants diminish the usefulness of biological action spectra determined with monochromatic radiation. To be useful in an ecological context, radiation weighting functions should be verified by exposing organisms to different spectral distributions of polychromatic radiation to determine if the response is consistent with a proposed action spectrum.

Since most of the available plant action spectra for both UV-A and UV-B radiation were depicted in an earlier review (CALDWELL 1971) only a few general characteristics of these spectra will be mentioned.

Though fewer than a dozen plant UV-A action spectra are available, all of these spectra extend into the visible and/or UV-B wavebands. Since solar radiation contains so much more energy in the visible waveband, the importance of UV-A may be overwhelmed for plant responses that are elicited by both visible and UV-A radiation. These action spectra in the UV-A waveband usually do not change rapidly as a function of wavelength, nor is there any apparent common chromophore that mediates these responses; the action spectra differ considerably. Because neither the UV-A action spectra nor the solar spectral irradiance change abruptly in the UV-A spectrum, the use of biological weighting functions is of less ecological utility than for the UV-B spectrum.

Generalizations concerning plant UV-B action spectra must be based on fewer than a dozen available action spectra, all of which have been determined with monochromatic radiation. Furthermore, the tails of these spectra that may lap into the UV-A spectrum have usually not been elucidated. Unlike the UV-A spectrum, UV-B action spectra do have a common tendency in that the biological responsiveness increases rather sharply with decreasing wavelength. Nevertheless, small differences in these action spectra can be important. An example of two ecologically important plant response spectra will be depicted, as this demonstrates how small differences in basically similar action spectra can influence the evaluation of solar UV.

Higher plants that are sensitive to solar UV, and most likely would be affected if the Earth's ozone layer were substantially reduced, exhibit two primary symptoms. Both photosynthetic capacity and leaf expansion are reduced. The latter phenomenon is not solely the result of reduced photosynthates, but apparently involves inhibited or delayed cell division (SISSON and CALDWELL 1976; DICKSON and CALDWELL 1978). Though action spectra are not available for these UV responses, two spectra can be used as likely candidates. For photosynthetic inhibition, the UV action spectrum for inhibition of a partial photosynthetic reaction, dichlorophenolindophenol reduction by isolated spinach chloroplasts (JONES and KOK 1966a), may serve as a model. For inhibition of cell division, a composite action spectrum for DNA damage compiled by SETLOW (1974) may be most appropriate. A plot of these spectra on linear coordinates suggests these spectra are rather similar and that both exhibit the characteristics of most plant UV-B action spectra – namely a decided increase

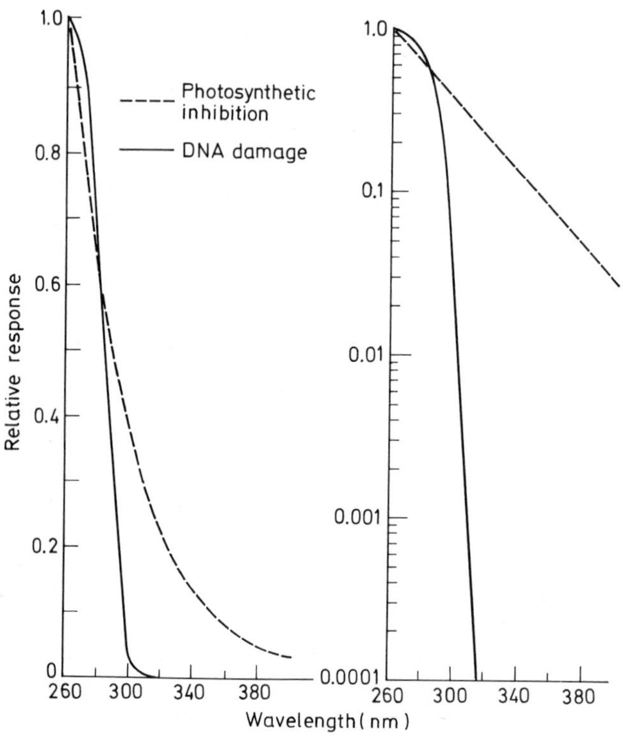

Fig. 6.2. Action spectra for UV damage in microorganisms mediated by DNA as compiled by SETLOW (1974) and for photoinhibition of a partial photosynthetic reaction of isolated chloroplasts (JONES and KOK 1966a), plotted with linear and logarithmic ordinate scales

of response with decreasing wavelength (Fig. 6.2). When plotted with a logarithmic ordinate, the differences in these spectra are more apparent – especially at longer wavelengths. The differences in the tails of these spectra would not be of great consequence were it not for the fact that solar UV spectral irradiance at the Earth's surface increases very steeply with increasing wavelength in this portion of the spectrum. The implications of these action spectra when used as weighting functions will be developed in Section 6.3.2 after first considering the characteristics of solar UV in nature.

6.3 Solar UV Irradiation

6.3.1 Attenuation of UV Radiation in the Atmosphere

Ozone in the Earth's atmosphere is the primary constituent that prevents all of the UV-C, and most of the UV-B radiation from reaching the Earth's surface.

Although ozone is decidedly the most important atmospheric constituent which attenuates solar UV-C and UV-B, other extinction processes operate in the atmosphere. Scattering and absorption by the molecules of the atmosphere and aerosols cause attenuation of solar UV. The magnitude and wavelength dependence of these extinction processes for the UV spectrum are shown in Fig. 6.3. The Rayleigh scattering by molecules of the atmosphere increases decidedly at shorter wavelengths (λ^{-4}), although it is not nearly as wavelength-dependent as ozone absorption. Furthermore, since this is scattering rather than absorption of radiation, a component of the scattered radiation will reach the ground as diffuse skylight. Because Rayleigh scattering is so effective in the UV, as much as 50% to 75% of the total global (direct solar beam plus skylight) that reaches the ground may be in the form of diffuse radiation (BENER 1972). Scattering and absorption by aerosols is largely independent of wavelength and of generally small magnitude.

Solar UV-A radiation is little influenced by ozone, undergoes less scattering, and, therefore, behaves in a reasonably predictable fashion like solar visible flux. It is the UV-B component which is most variable.

Measurements of solar UV-B global irradiance are not easily accomplished and it is not surprising that rather few data sets exist. The variable and abrupt truncation of the solar UV-B spectrum (Fig. 6.1) combined with the highly wavelength-specific biological effectiveness of this radiation (Fig. 6.2) requires that spectral irradiance be measured with sufficient accuracy and spectral resolution to be meaningful. The spectral irradiance can change as much as an order of magnitude within 7 nm in this portion of the spectrum. The most extensive series of measurements were conducted by BENER in Switzerland (BENER 1960, 1963, 1964, 1972).

Because of the difficulty and expense of conducting solar UV spectral irradiance measurements it may be advisable for many applications to calculate

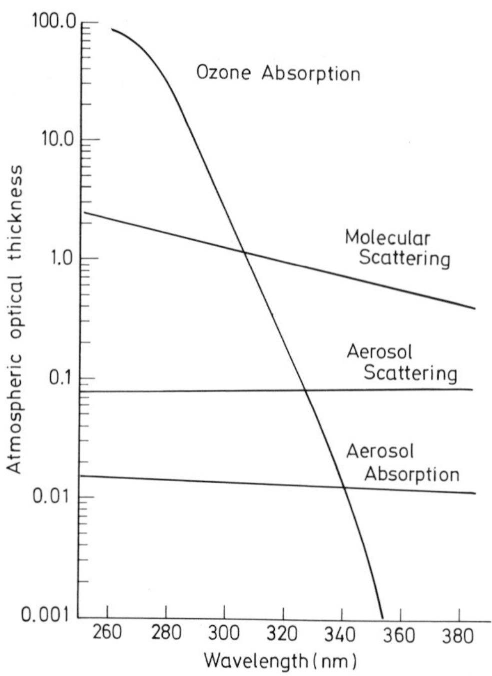

Fig. 6.3. Solar radiation attenuation processes in the atmosphere as a function of wavelength. Each of the four processes that attenuate direct beam solar radiation are represented for a standard atmospheric thickness as would be encountered under cloudless sky conditions with minimal atmospheric aerosols. Ozone absorption is the product of the ozone absorption coefficient from Inn and Tanaka (1953) and a standard atmospheric ozone thickness of 0.300 atm cm as is typical of temperate latitudes during the summer. Scattering by air molecules and aerosols, as well as aerosol absorption, are represented from the data presented by Dave and Halpern (1976)

rather than to measure solar UV irradiance. Although calculation of the direct beam solar UV irradiance presents little difficulty, a deterministic calculation of the diffuse radiation component is expensive and tedious (Dave and Halpern 1976). A useful algorithm developed by Green et al. (1980) effectively approximates the deterministic solutions of Dave and Halpern (1976) and allows calculation of diffuse and direct beam UV spectral irradiance as a function of solar angle, atmospheric conditions, and elevation above sea level. This model also allows prediction of the solar spectral irradiance that would occur at various locations if the ozone layer were reduced.

The first concerns that human activities might interfere with the natural atmospheric ozone equilibrium arose in the early 1970's. These were focused on the injection of nitrogen oxides into the stratosphere by high altitude aircraft – primarily supersonic transports (Johnston 1971). These concerns were founded largely on atmospheric models that were designed to simulate the chemistry and transport dynamics of the atmosphere. As the models were subsequently refined and better measurements of various rate coefficients were made, the importance of nitrogen oxides as catalysts of ozone reduction was considerably diminished and the importance of halogens as ozone catalysts was recognized. Molina and Rowland (1974) proposed that stratospheric photodissociation of chlorofluoromethanes – principally from spray propellants and refrigerants – might constitute a significant source of chlorine which could act as a catalyst of ozone reduction. These compounds are so inert that they are not removed from the lower atmosphere by precipitation. As they eventually migrate to the stratosphere, they are exposed to solar UV-C which causes their photodissociation. The released elemental chlorine can have an appreciable residence time in the stratosphere as a catalyst for ozone destruction. If present global release rates of chlorofluoromethanes are maintained, an ozone reduction of 16% will eventually result. The dynamics of the stratosphere are so slow that a century will be required for an equilibrium ozone reduction to be achieved. If release rates increase 7% per year from 1980 to 2000 and remain constant thereafter, an ozone reduction at equilibrium in excess of 30% is predicted (National Academy of Sciences 1979).

The change of solar UV flux at a particular time and location that would result from ozone reduction can be readily calculated. Solar irradiance at the shorter UV-B wavelengths would be most affected by changes in ozone concentration as shown in Fig. 6.4. At longer wavelengths the ozone absorption coefficient is too small to significantly affect the solar flux, and below approximately 280 nm the ozone absorption coefficient is so large that even if there were a massive 90% ozone reduction, the atmosphere would still be effectively opaque to UV-C radiation.

6.3.2 The Radiation Amplification Effect

A 16% ozone reduction would only result in a 1% increase in total solar UV flux between 280 and 400 nm (Fig. 6.4). This would be of trivial consequence if radiation throughout this waveband were of equal biological effectiveness. However, when the biological weighting functions based on the action spectra shown in Fig. 6.2 are employed, a very different picture emerges. Plots of the products of spectral irradiance and these weighting functions are shown in Fig. 6.4. Since these plots are on linear coordinates, the area under the curves in each case represents the relative change in the integrated biologically effective flux due to ozone reduction. For radiation that might inhibit photosynthesis there would only be a 2% increase in effective flux under these conditions. In contrast, for DNA-damaging radiation, the effective increase would by 47%.

Although these two action spectra are similar in many respects (Fig. 6.2) the differences at longer wavelengths in the tails of these spectra yield strikingly different conclusions concerning the consequences of atmospheric ozone reduction. When UV-B is presented to sensitive higher plants, depression of photosynthesis is one of the most immediate responses (SISSON and CALDWELL 1977; BOGENRIEDER and KLEIN 1977). Photosynthesis of phytoplankton is especially liable to photoinhibition by UV-B radiation (JITTS et al. 1976; SMITH et al. 1980). Despite the sensitivity of the photosynthetic apparatus to UV-B radiation, a change in solar UV-B radiation as a consequence of ozone reduction may be unimportant for photoinhibition of photosynthesis when compared to other phenomena whose action spectra are similar to that for DNA damage. This conclusion can be reached only if inhibition of photosynthesis in intact plants under natural conditions corresponds to the action spectrum described by JONES and KOK (1966a). This is yet to be determined in higher plants; however,

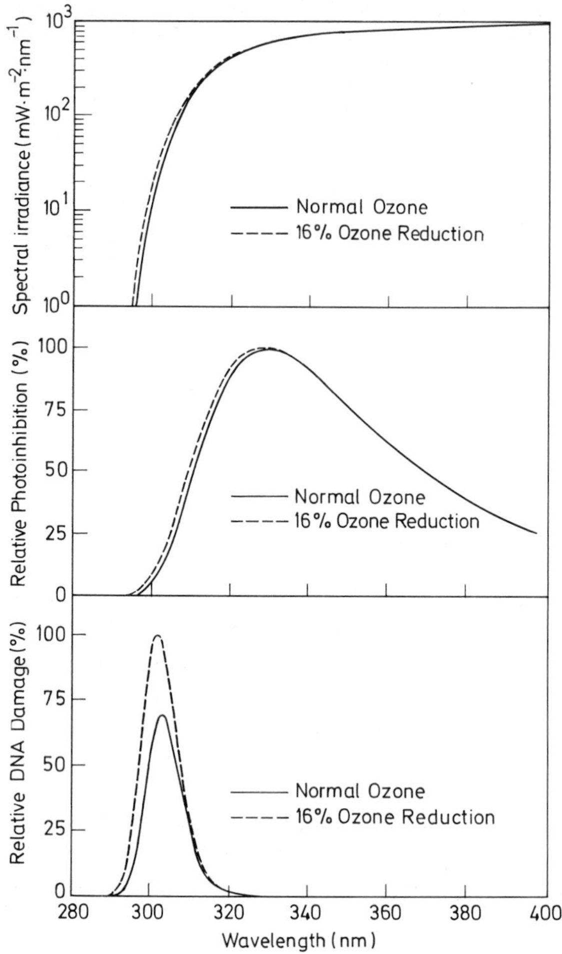

Fig. 6.4. Global spectral irradiance calculated from the model of GREEN et al. (1980) for normal (0.310 atm cm) ozone concentrations at midday (zenith angle 12°) during the summer at sea level and for a 16% reduction of the atmospheric ozone layer *(dashed line, top frame)*. In the *middle* and *bottom frames* are the product curves for global spectral irradiance and the weighting functions based on the action spectra for photoinhibition of isolated chloroplasts (JONES and KOK 1966a) and the DNA damage action spectrum (SETLOW 1974) normalized to peak values. The areas underneath the respective product curves represent the relative biologically effective flux under normal and depleted ozone conditions

it does appear to apply in the case of marine phytoplankton (SMITH et al. 1980).

The radiation amplification factor, RAF, is the relative change of biologically effective UV-B radiation resulting from a unit change in the atmospheric ozone layer. The RAF is a function of the initial quantity of ozone at a particular location, solar angle, and the percentage ozone change itself. It is most useful to express the RAF for the change in daily integrated dose of biologically effective UV. Table 6.1 contains RAF values for 16% ozone reduction at the time of year of maximum solar radiation at three different latitudes. Three action spectra are employed to illustrate the different RAF values. For larger changes in atmospheric ozone, the RAF is proportionately greater; at 40° N, a 30% ozone reduction would translate into 119% increase of DNA-effective radiation – about 4% for each percent of ozone change as opposed to about 3% at a 16% ozone reduction.

Table 6.1. Radiation amplification factors for a 16% ozone reduction using different action spectra. The percentage increase in daily biologically effective UV dose is given for the Julian date of maximum solar radiation at each latitude

Latitude	Date	RAF for 16% ozone reduction Action spectrum		
		DNA-damage SETLOW (1974)	General plant response CALDWELL (1971)	Photoinhibition of photosynthesis JONES and KOK (1966a)
20°	141, 205	46.3	32.0	2.2
40°	172	47.5	35.0	2.3
70°	172	49.4	44.0	2.3

6.3.3 Other Considerations: Atmospheric Pathlength, Elevation Above Sea Level, Cloud Cover and Ground Reflectance

Changes in solar angle, whether the result of time of day, season of year, or latitude, alter the effective optical pathlength through the atmosphere. As the sun sinks to lower altitudes in the sky and the optical thickness of the atmosphere effectively increases, the wavelength-dependent nature of UV-B attenuation results in a truncation of the solar UV spectrum at increasingly longer wavelengths. This extinction occurs at a much more rapid rate than attenuation of visible radiation.

At higher elevations above sea level, the pathlength through the atmosphere is decreased. However, since most of the atmospheric ozone is in the stratosphere above 15 km, even in the highest mountain areas the pathlength through the ozone layer is effectively the same. Nevertheless, reduced optical thickness of aerosols and reduced attenuation by Rayleigh scattering can result in increased solar UV at higher elevations. A few generalizations about the behavior of solar UV at different elevations can be offered based on a few spectral irradiance measurements (BENER 1972; CALDWELL et al. 1980) and the theoretical calculations of DAVE and FURUKAWA (1966). The increase with elevation is most pronounced in the direct beam component of solar UV because of the reduced attenuation at higher elevations. However, the diffuse component undergoes only a modest increase with elevation and in some circumstances might actually decrease because of the reduced atmospheric scattering at higher elevations. There is a small spectral shift toward shorter wavelengths with elevation above sea level because of the wavelength dependence of Rayleigh scattering.

Cloud cover greatly influences solar UV irradiance. If the solar image is obscured by clouds, there is an attenuation of the direct beam solar UV. However, at the shorter UV wavelengths a large proportion of the global solar radiation (40%–75%) is in the diffuse component (BENER 1963, 1972; CALDWELL et al. 1980). Thus, the influence of cloud cover on skylight UV is of considerable significance. The measurements of BENER (1964) indicate that cloud cover can either increase or decrease diffuse UV irradiance. Generally, he reported that low, dark cloud cover would cause a decrease of skylight UV, while high altitude cloud cover would often result in an increase of this diffuse UV. The net effect on global UV radiation is not readily estimated. In any case, clouds should generally attenuate global UV-B to a lesser degree than total solar radiation because the diffuse component constitutes such a large proportion of global UV radiation.

A high ground UV reflectance not only results in a significant upwelling UV flux, but also enhances the downward flux because the reflected UV is scattered in the atmosphere and a portion of this is returned as downward flux. Since scattering in the atmosphere is more effective at shorter wavelengths, this effect is particularly pronounced in the UV-B. BENER (1960) determined that snow cover could result in an increase of downward UV flux of 20% to 60% depending on solar angle. High surface albedos usually would only be found in nature when the ground is covered by snow. Most soil and vegetation surfaces have UV reflectance of less than 10% (BENER 1960; CALDWELL 1971; GAUSMAN et al. 1975). In alpine areas where snow cover often lingers near the summer solstice, plants near snowbanks would be exposed to reflected UV from the snow surface as well as the enhanced downward UV flux from the sky. The effective dose could well be doubled under such circumstances.

6.3.4 Geographic Variation of Solar UV Radiation

There is a pronounced latitudinal gradient of solar UV-B due to greater prevailing solar angles at higher latitudes, as well as a natural latitudinal gradient of atmospheric ozone concentration. Although most of the atmospheric ozone is produced at tropical latitudes, stratospheric circulation patterns cause atmospheric ozone to accumulate at higher latitudes. During the season of maximum solar radiation, atmospheric ozone thickness varies by a factor of 1.5 from the equator to the poles (DUETSCH 1971).

The latitudinal gradient of biologically effective UV-B radiation for the day of maximum solar radiation is shown in Fig. 6.5. (The date of maximum solar radiation at each latitude corresponds to the time when solar zenith angles are minimal.) Above 23° latitude this occurs at the summer solstice, but at latitudes less than 23°, this time is progressively closer to the equinox, March 21 and September 21 (which are the dates of maximum solar radiation at the equator.) This latitudinal gradient of radiation has been calculated by the model of GREEN et al. (1980) using a DNA damage weighting function. Although the radiation amplification effect has been discussed with respect to changes in the atmospheric ozone layer (Sect. 6.3.2), the concept also applies to the natural latitudinal UV-B radiation gradient. As was the case for ozone reduction (Table 6.1), when the DNA-damage spectrum is used as a weighting function, the latitudinal gradient of this effective radiation is much steeper than if other action spectra are employed as weighting functions.

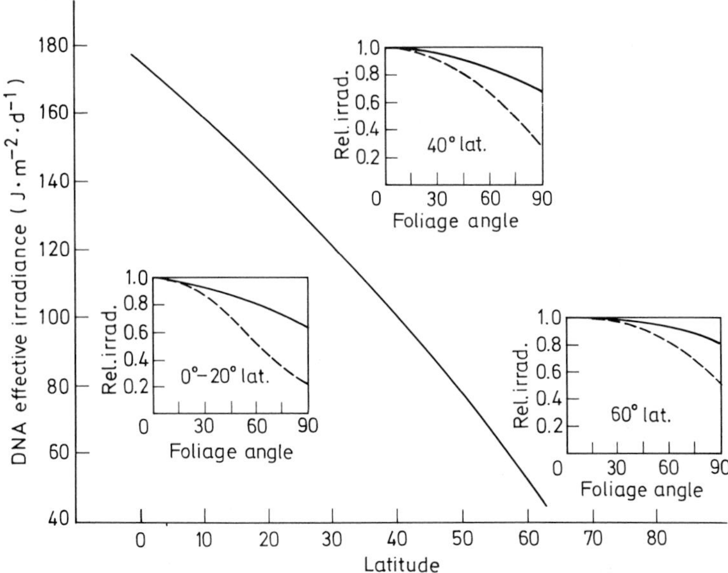

Fig. 6.5. Latitudinal gradient of daily DNA-effective UV-B irradiation for the season of maximum solar radiation at sea level based on the model of GREEN et al. (1980). *Inserts* Relative reduction of DNA-effective irradiation for direct beam *(dashed line)* and global irradiation *(continuous line)* as a function of foliage inclination angle at three different latitudes

6.3.5 Penetrance in Natural Waters

Since the first concerns about atmospheric ozone depletion arose, interest has focused on the fate of aquatic organisms. Plankton generally appear to be much more sensitive to solar UV-B radiation than terrestrial vegetation (JITTS et al. 1976; WORREST et al. 1978; SMITH et al. 1980). Thus, the degree of protection afforded to plankton by UV attenuation in water is of obvious importance. Pure water is quite transparent to UV, but scattering and absorption by dissolved organic matter and other materials in natural waters can alter the penetrance of solar UV considerably.

The diffuse attenuation coefficient for contrasting types of natural waters estimated from the recent measurements of SMITH and BAKER (1979) are shown in Fig. 6.6. The diffuse attenuation coefficient is an optical parameter that relates spectral irradiance just beneath the water surface to the downward spectral irradiance at some depth in the water. Curve A is the coefficient based on their measurements in clear waters of low productivty (0.05 mg chlorophyll a m^{-3}) in the middle of the Gulf of Mexico. This corresponds rather well with earlier measurements of JERLOV (1950) in the eastern Mediterranean Sea and CALKINS (1975) in clear waters near Puerto Rico. The dashed line represents an extrapolation from their measurements which were only reliable to 310 nm.

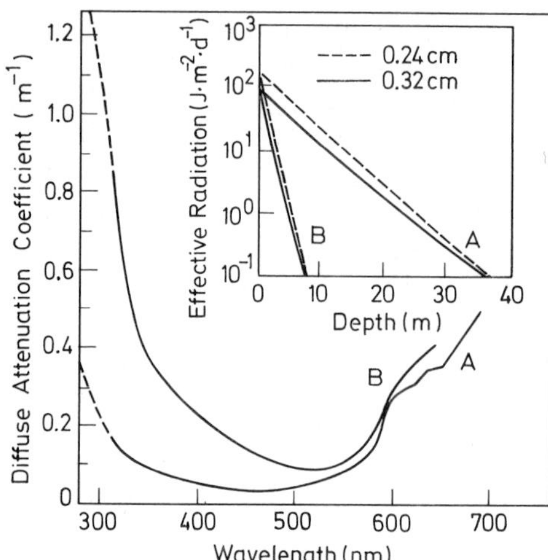

Fig. 6.6. Diffuse attenuation coefficients for clear water of low productivity (Curve *A*) and for waters of moderate productivity (Curve *B*). *Insert* Downward flux of DNA-effective UV-B irradiance in clear and moderately productive waters based on these diffuse attenuation coefficients and solar global irradiance at two atmospheric ozone concentrations. (Adapted from SMITH and BAKER 1979)

The situation is quite different in moderately productive ocean waters with relatively high concentrations of dissolved matter as represented in curve B. This is based on measurements taken near Tampa, Florida, where chlorophyll a concentrations were 0.5 mg m^{-3}. The insert in Fig. 6.6 illustrates the projected penetration of DNA-damaging solar UV-B into clear and moderately productive waters for atmospheric ozone concentrations of 0.32 and 0.24 atm cm. It appears that a protective biological feedback mechanism is operative in the more productive waters. The dissolved organic matter resulting from biological activity may provide a UV-B radiation screen that could be effective even in shallow waters. This may also have occurred in the primeval Earth environment when solar UV-C radiation penetrated freely to the Earth's surface and life was largely confined to the marine environment.

6.4 Response of Plants to UV Irradiation

6.4.1 Nucleic Acids

6.4.1.1 UV Inactivation

The photochemical liability of nucleic acids can be very pronounced when biological systems are exposed to UV-C, and, to a lesser extent, UV-B radiation. The photobiology of nucleic acid inactivation by UV radiation is well advanced, although most of the work has been conducted on bacterial and viral systems. These studies are also primarily restricted to 254-nm radiation because it is conveniently generated by low pressure mercury lamps, it is easy to measure, and is very effectively absorbed by nucleic acids. A number of photochemical lesions result when DNA absorbs UV radiation, the most common involve polymers of pyrimidine bases termed dimers (MURPHY 1975). Not surprisingly, these lesions result in loss of DNA biological activity; for instance, UV-C interferes with the ability of DNA to serve as a template for replication and also causes loss of transforming activity (MURPHY 1975).

Ultraviolet radiation is also efficiently absorbed by RNA, and proteins for that matter, but much higher doses must be absorbed before inactivation occurs. This may be related to the relative cellular abundance of these compounds compared to DNA. There is evidence that many of the same photochemical lesions of DNA can be elicited by UV-B radiation, although radiation in this waveband is less effective (HARM 1979; ROTHMAN and SETLOW 1979). Extrapolations from the use of UV lamps in the laboratory suggest that bacteria and some algae could be quite sensitive to normal solar UV-B (NACHTWEY 1975). The DNA of higher plants appears to be much better shielded than that of bacteria or algae. For example, doses of 254-nm radiation necessary to kill leaves of higher plants appear to be some 4 orders of magnitude greater for the most sensitive higher plant species than for killing UV-resistant species of algae (CLINE and SALISBURY 1966; NACHTWEY 1975). There are no indications that higher plants growing in nature would suffer mortality directly due to solar UV-B radiation – at least in the sporophyte phase of the life cycle.

Pollen may be one stage of the higher plant life cycle where DNA might be vulnerable to solar UV-B radiation. The pollen grain wall can transmit 20% of the incident UV-B radiation (STADLER and UBER 1942). Furthermore, nuclear DNA is limited to a single set of haploid chromosomes and therefore the benefits of redundancy of genetic material are missing. Pollen viability can be reduced under laboratory conditions following exposure of pollen to UV radiation (WERFFT 1951; CHANG and CAMPBELL 1976). WERFFT (1951) also demonstrated that pollen viability could be reduced as much as 90% when exposed to 8 h of normal sunlight. These experiments were conducted at a latitude of 52° N where solar UV-B would not be particularly intense. Unfortunately, she did not selectively filter sunlight to determine which portion of the spectrum was responsible for the damage. Whether there is sufficient exposure of pollen grains to sunlight between release of pollen from the anthers and successful fertilization for significant reduction of viability to take place is an unanswered question.

6.4.1.2 Molecular Repair Systems

Damage incurred by nucleic acids due to absorption of UV radiation can be repaired at the molecular level by a variety of repair systems. These systems have been studied primarily in bacteria but their occurrence has been documented in many representatives of both the plant and animal kingdoms.

Photoreactivation is an enzymatic mechanism driven primarily by UV-A and blue light which effectively splits pyrimidine dimers – the most common lesion in DNA caused by absorption of UV-C. Photoreactivation, PR, has also been demonstrated in a variety of higher plants for phenomena such as UV-C inhibition of chlorophyll synthesis (SCHIFF et al. 1961), epidermal damage (CLINE and SALISBURY 1966), growth inhibition (KLEIN 1963), and synthesis of nitrate reductase (WRIGHT and MURPHY 1975). Photoreactivation is limited to the monomerization of dimers (SETLOW 1966).

There are also repair systems that do not require light for operation. Excision repair involves a series of enzymatic reactions that identify a lesion in a DNA strand, literally excise the lesion from the strand, synthesize a replacement patch using the complementary strand as a template and seal the patch back into the existing strand. Excision repair is not limited to pyrimidine dimers but can repair other nucleic acid lesions. Another type of dark repair system, termed postreplication or recombinational repair, takes place after normal DNA replication. In this case, the undamaged portions of the DNA strand undergo normal replication leaving an appropriate-sized gap in place of the lesion. After this replication has occurred, a patch is synthesized using the redundant information in the cell as a template and the functional DNA strand is restored (SMITH 1978). After several unsuccessful attempts by other workers, HOWLAND (1975) demonstrated that dark repair systems are operative in higher plants.

Most organisms appear to have more than one functional repair system which helps greatly to alleviate UV-induced damage to nucleic acids. The degree to which these repair systems are effective in reducing the stress imposed by solar UV-B radiation on plants in nature is difficult to assess. There are, however, some indications that photoreactivation of UV-B-induced damage in alpine plant leaves may take place in nature (CALDWELL 1968).

Exposure of organisms to UV or visible irradiation before they are subjected to inactivating UV-C radiation can sometimes lessen the effects of the shortwave UV. This phenomenon is known as photoprotection, PP. This is not a direct molecular repair process but rather an indirect modification of cellular metabolism manifested primarily in the delay of cell division (JAGGER 1960; KLEIN 1978). This delay of cell division may allow more time for normal repair processes such as excision repair to operate or it may increase the capacity of these repair systems (see KLEIN 1978). Photoprotection is well known in bacteria but has only been demonstrated once in higher plant tissue cultures (KLEIN 1963).

6.4.1.3 Genetic and Other Nonlethal Effects

Apart from the lethal UV damage caused by unrepaired nucleic acid lesions, nonlethal effects of UV which involve nucleic acids may also be of ecological significance. These may involve genetic alterations that are not immediately lethal or reduced growth due to impaired cell division or cell enlargement.

Acceleration of mutation rates and chromosomal aberrations in plants by UV radiation has been well documented. A likely candidate as a chromophore for these genetic alterations is DNA. Action spectra for mutations and chromosomal aberrations generally correspond to the absorption spectrum of DNA (KNAPP et al. 1939; HOLLAENDER and EMMONS 1941; STADLER and UBER 1942;

KIRBY-SMITH and CRAIG 1957). However, UV absorption by proteins also may be involved in chromosomal breaks (WOLFF 1972). Mistakes in the repair of DNA lesions by certain error-prone repair systems, rather than the original UV-induced DNA lesion, are probably the major cause of mutation (SMITH 1978).

It is difficult to estimate the contribution of solar UV to the normal mutation rates of plants. The tissues in plants that give rise to gametes are probably very well shielded from solar UV, although this has not been explicitly demonstrated. Pollen is the most plausible stage where mutations and chromosomal aberrations in germinal cells might take place.

Nucleic acids may also serve as a chromophore for nonlethal effects of UV radiation such as a reduction in maximum potential leaf expansion. Ultraviolet-B can cause a reduction in leaf enlargement which appears to act primarily through a delay or reduction of leaf cell division (DICKSON and CALDWELL 1978). Although it remains to be demonstrated, this apparent effect on cell division may well be mediated by nucleic acid absorption of UV. In normal solar radiation, this reduction of maximum potential leaf expansion should be quite small for most species. Nevertheless, small reductions of leaf expansion combined with other alterations in growth processes may be reflected as subtle changes in the competitive balance of plant species.

Solar radiation lesions observed in the field on fruits of cantaloupes (*Cucumis melo*) have been attributed to solar UV (LIPTON 1977). Whether such phenomena involve nucleic acid chromophores is not known.

6.4.2 Photosynthesis

It has long been known that UV-C can readily inhibit photosynthesis. ZILL and TOLBERT (1958) demonstrated that partial inhibition of photosynthesis could be detected in wheat leaves at about 10^{-4} of the dose that CLINE and SALISBURY (1966) found necessary to kill leaves of this species.

JONES and KOK (1966a, b) suggested that UV photoinhibition is independent of oxygen, which clearly distinguishes it from photobleaching by intense visible radiation of chlorophyll which is quite oxygen-dependent. In studies with UV-B radiation, chlorophyll concentration was not significantly reduced even though there was an appreciable decline of photosynthesis. Chlorophyll reduction was only evident in plant leaves which were exposed to particularly large doses of UV-B flux (SISSON and CALDWELL 1976; BRANDLE et al. 1977; TERAMURA et al. 1980) or when plants were exposed to UV-B while in an environment of very low visible irradiance (VU et al. 1981).

The site of UV photoinhibition has not been fully elucidated. Destruction of plastoquinone (MANTAI and BISHOP 1967), inhibition of electron transport associated with photosystem II (YAMASHITA and BUTLER 1968; BRANDLE et al. 1977), disruption of thylakoid membranes (MANTAI et al. 1970), or direct inhibition of primary photochemistry at the reaction center of photosystem II and to a lesser extent by inactivation of photosystem I (OKADA et al. 1976) have all been implicated. Partial inhibition of carboxylating enzyme activity has also been demonstrated (VU et al. 1981).

Ultraviolet damage to the photosynthetic system of individual leaves appears to be cumulative through time (SISSON and CALDWELL 1976) and reciprocity applies for reduction of photosynthesis for leaves of a given age. That is, photosynthetic reduction is only a function of total UV-B dose and independent of dose rate even when dose rates were much lower than one would expect in the ambient environment and were presented to individual leaves over a period of forty days (SISSON and CALDWELL 1977). JONES and KOK (1966a) also demonstrated that reciprocity applied in short-term experiments with isolated chloroplasts. If reciprocity is a general phenomenon associated with UV inhibition of photosynthesis, this represents a fundamental difference from damage associated with UV photolesions of nucleic acids which appear to be strikingly dependent on dose rate. For example, mortality of the unicellular alga *Chlamydomonas reinhardi* was demonstrated to be much greater if the same UV-B dose was applied at higher rates (NACHTWEY 1975). Since nucleic acid repair systems are comparatively slow enzymatic processes, the same dose presented at low dose rates allows more complete repair. Thus, it would appear that nucleic acid repair systems do not participate in ameliorating photosynthetic damage. Furthermore, there is no fundamental reason to suggest that nucleic acids would be involved as a chromophore in the inhibition of photosynthesis.

Even though photoreactivation may not participate in reducing the impact of UV on photosynthesis, it has been observed frequently that when plants are subjected to chronic UV-B exposure at comparatively low levels of visible irradiance, as would be commonly experienced in many growth chambers (PAR less than 400 $\mu E\ m^{-2}\ s^{-1}$), greater depressions of photosynthesis result than when the same UV-B flux is presented to the plant with higher visible irradiance (SISSON and CALDWELL 1976; TERAMURA et al. 1980). A similar phenomenon has been apparent in a series of studies designed to evaluate the potential consequences of increased UV-B (as would result from atmospheric ozone reduction) on plant growth. Experiments conducted in growth chambers or greenhouses with low visible irradiance always resulted in much greater plant growth depression due to UV-B irradiation than similar experiments conducted under field conditions with comparable UV-B supplementation (BIGGS et al. 1975; BIGGS and KOSSUTH 1978). Whether this phenomenon stems in part from increased UV sensitivity of leaves which have developed under low light conditions and thus possess many of the morphological and physiological characteristics of shade-adapted leaves, or whether there is an immediate repair or protective process driven by visible irradiance is not clear. In any case, the assessment of potential consequences of ozone layer reduction is greatly complicated since those experiments which are conducted in the field are necessarily much more difficult and less well controlled than those conducted under growth chamber conditions.

Most of the evidence suggests that UV, especially at shorter wavelengths, is most likely to cause some degree of reduced photosynthetic competence. However, there is also some indication that photosynthesis might be stimulated to a small degree by UV, particularly at longer UV wavelengths. Since chlorophyll and accessory photosynthetic pigments absorb in the UV-A, if this radiation reaches mesophyll tissue it can drive the photosynthetic reaction. McCREE and KEENER (1974) demonstrated that removal of the epidermis of *Brassica oleracea* could lead to a 10% increase in photosynthetic rates and this was attributed to UV-A. Recently, DALEY et al. (1978) suggested that ribulose bisphosphate carboxylase could be activated in response to UV-A. The degree to which this might result in enhanced leaf photosynthetic rates is unclear. There has been some suggestion from studies with intact leaf photosynthesis that a slight amount of UV stimulation could take place especially when the leaves are simultaneously exposed to reasonably high flux of visible light (TERAMURA et al. 1980). Still, the weight of the evidence suggests UV to be a generally detrimental factor for plant photosynthesis and that with even low flux levels, damage can accumulate during the ontogeny of individual leaves.

6.4.3 Growth and Development

If net photosynthesis and leaf enlargement potential are reduced, even to a small degree (Sects. 6.4.1.3 and 6.4.2), proportional reductions of plant growth and productivity would be expected (e.g., VU et al. 1981). However, UV radiation may also lead to other irregularities in the growth and development of plants. Disproportionate growth of plant organs such as altered root/shoot ratios (BIGGS and KOSSUTH 1978), depressed flower development (KLEIN et al. 1965), loss of apical dominance (BIGGS et al. 1975), abscission of leaves (KLEIN 1967) and even altered mineral nutrient concentrations (ANDERSEN and KASPERBAUER 1973) have been observed as plant responses to various UV lamp systems. However, it remains to be demonstrated that solar UV will cause any of these changes in nature, although occasionally phenomena such as promotion of flowering have been observed when solar UV has been removed by filtration under field conditions (CALDWELL 1968; KASPERBAUER and LOOMIS 1965).

As described in Section 6.2, several plant growth regulators absorb UV and conceivably could be involved directly as chromophores for alterations of plant growth and development. Alternatively, other UV chromophores could either transfer excitation energy directly to growth regulators or could potentiate still other reactions such as oxidation of key substrates.

The potential effects of UV on metabolic processes and plant development are indeed manifold. However, to be of ecological consequence, several criteria must be met: (1) the primary chromophore or sensitizing pigment must absorb UV in the waveband where solar UV flux occurs in nature, (2) the quantum efficiency of the primary photoreaction must be of sufficient magnitude so that solar UV that reaches these chromophores in plant tissues can be effective, (3) the quantum efficiency of these reactions in the visible spectrum, where there is much more solar flux, must be sufficiently low so that the influence of UV radiation predominates, (4) the photoreaction itself must be of sufficient metabolic consequence so as not to be bypassed or reversed by alternative processes.

Certainly, if solar UV does result in changes in plant growth or developmental patterns, these could be of equal or greater importance than the direct effects on plant photosynthesis or nucleic acids discussed earlier.

6.5 UV Avoidance and Adaptation

6.5.1 Defense

Most of the effects of UV radiation, especially with respect to nucleic acids and photosynthesis, are considered as deleterious. Unprotected plant tissues can be quite susceptible to solar UV radiation that reaches the Earth's surface – even at temperate latitudes. Plants thus must avail themselves of several lines of defense against solar UV in nature. These fall into two principal categories of radiation repair and avoidance; nucleic acid repair systems (Sect. 6.4.1.2) or other repair or damage quenching systems may also be operative. For example, terminal oxidases may be important in suppressing photooxidation, if this is a significant component of UV-mediated damage as has been suggested to be the case for *Chlorella* (PULICH 1974). Apart from nucleic acid repair systems whose efficacy has been vividly documented for bacterial systems, and by extension is presumed to play an important role in higher plants, there is little yet known concerning other possible repair mechanisms. Therefore, the emphasis of this section will be directed to avoidance of solar UV.

6.5.2 Avoidance: Geometrical Considerations

Inclined foliage generally receives less incident direct-beam solar radiation than horizontal foliage. Thus, foliage inclination can represent an effective means of reducing the solar heat load on leaves (e.g., MOONEY et al. 1977) since approximately 90% of the total solar shortwave radiation is in the direct-beam component. However, for UV radiation, and especially UV-B, 40% to 75% of the radiation is in the isotropic diffuse component, which substantially reduces the effectiveness of leaf inclination as an avoidance mechanism for solar UV radiation. To illustrate the magnitude of radiation avoidance due to leaf inclination angle, the relative reduction of incident UV-B radiation load on foliage inclined at different angles from the horizontal is depicted in Fig. 6.5. The reduction of DNA-effective radiation on a daily basis is shown for both the direct beam component and the total global UV-B radiation at three latitudes. These calculations are based on a model developed in earlier studies (CALDWELL et al. 1980). Steeply inclined foliage does receive substantially less direct beam UV-B radiation – especially at lower latitudes. This corresponds with what is expected for the solar heat load on leaves. However, for global UV-B radiation, even vertical foliage receives at least 70% of the daily effective UV-B radiation load.

In plant canopies, UV-B radiation incident on foliage would be greatly attenuated due to the shading by other foliage. This has been very adequately described in the calculations of ALLEN et al. (1975).

6.5.3 Avoidance: Plant Optical Properties

Although foliage orientation affords little potential reduction of incident global UV irradiance, attenuation of UV in outer tissue layers remains a significant avoidance mechanism for sunlit foliage. Since mesophyll tissues of plant leaves must be adequately exposed to visible irradiance, attenuation of solar irradiance in the epidermis must be highly wavelength-selective. This is achieved primarily by selective absorption of UV in the epidermis, which in turn is primarily attributed to phenolic compounds such as flavonoids and flavones (LAUTENSCHLAGER-FLEURY 1955; CALDWELL 1971), although the cuticular waxes of some species also absorb to a certain extent in the UV (WUHRMANN-MEYER and WUHRMANN-MEYER 1941).

Reflection of UV for most species is generally less than 10% (GAUSMAN et al. 1975; CALDWELL 1971), although exceptions exist where leaf UV reflectance can be appreciable (ROBBERECHT et al. 1980; MULROY 1979; CLARK and LISTER 1975). In the few species where leaf UV reflectance is sizeable, the leaves do not selectively reflect UV but also reflect a large proportion of the visible radiation (e.g., MULROY 1979).

Plants in environments where UV-B radiation is particularly intense are notably effective in epidermal UV absorption. Perhaps the largest natural gradient of solar UV-B radiation exists in the arctic-alpine environments extending from sea level in the Arctic to high elevations at low latitudes. The daily DNA-effective UV-B at different locations on this arctic-alpine gradient is shown for the season of year of maximum solar radiation in Fig. 6.7. Epidermal UV transmittance spectra for plant species and the calculated average effective UV-B reaching mesophyll tissues of leaves of species at these various locations

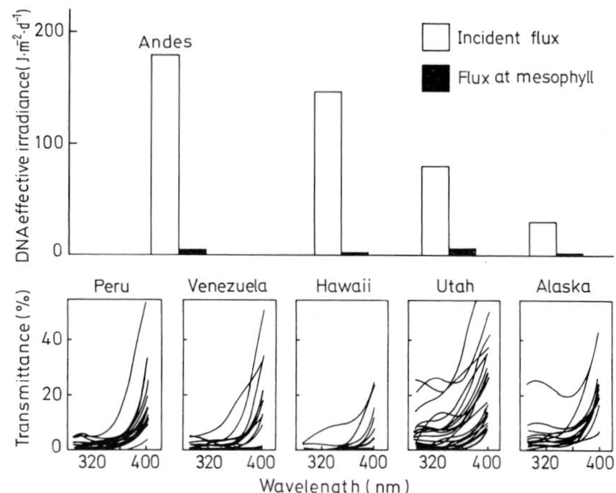

Fig. 6.7. Daily DNA-effective irradiation for the season of maximum solar radiation at four locations along a latitudinal arctic-alpine gradient from measurements of CALDWELL et al. (1980) *(open bars)*, epidermal spectral transmittance of leaves of several species at each of five locations along this latitudinal arctic-alpine gradient, and the calculated average DNA-effective flux of UV-B radiation reaching the mesophyll tissues of horizontal leaves at each location *(solid bars)*. (ROBBERECHT et al. 1980)

on this gradient are also shown. At temperate and arctic latitudes where solar UV-B is less intense, there is greater variability in epidermal transmittance among species (ROBBERECHT et al. 1980). However, in high UV-B radiation areas at equatorial and tropical latitudes, all species exhibit a consistently low epidermal transmittance – an average value for these species is less than 2%. Although this latitudinal solar UV-B gradient from arctic to tropical alpine environments represents more than a sevenfold difference in daily effective UV-B irradiance, the calculated mean effective UV-B reaching mesophyll tissues of tropical alpine plant leaves is not substantially different from that of species at higher latitudes.

6.5.4 The Adaptive Response

Exotics from temperate latitudes which have been introduced into low latitude, high elevation environments also possess very low epidermal UV transmittance, which suggests that some acclimatization may have taken place (ROBBERECHT et al. 1980). Indeed short-term changes of epidermal UV transmittance have been found for some, but not all, species exposed to enhanced UV-B radiation for a few days' time (ROBBERECHT and CALDWELL 1978). This change of absorption has been attributed to pigments in the epidermis. WELLMANN (1974) demonstrated that flavonoid pigment synthesis could be induced by UV-B radiation at intensities that occur in normal daylight. Although this is certainly not the only environmental factor which can influence flavonoid synthesis and concentrations in plant tissues, the UV-B induction of flavonoid synthesis appears to be a very plausible mechanism for short-term UV-B radiation acclimatization. The extent to which this phenomenon accounts for the consistently low epidermal transmittance of both native and exotic species in intense solar UV flux environments is not known. Other pigments may also play an important role in UV absorption. For example, many alkaloids exhibit appreciable UV-B absorption. LEVIN (1976) has presented a convincing negative correlation of alkaloid-bearing plant species with latitude. Species that possess alkaloids are twice as prevalent in tropical floras as in the floras of temperate latitudes. Although the selective advantage of alkaloids has primarily been associated with deterring herbivores, alkaloids in plant leaves may also function as UV radiation filters. Other changes in plant tissues, such as cuticular thickness and mesophyll structure, may also play a role in UV acclimatization.

6.6 The Ecological Significance of UV Radiation

6.6.1 Accommodation of Greater Solar UV Flux

If plants have not been acclimatized to UV radiation, there is sufficient actinic UV flux even at temperate latitudes to inhibit photosynthesis. A particularly vivid example was described by BOGENRIEDER and KLEIN (1977) in which seedlings of *Rumex alpinus* grown under greenhouse conditions exhibited greatly depressed photosynthetic rates after two to three days of exposure to normal solar radiation in UV-transparent cuvettes. Some of the plants were killed after this three-day exposure to solar UV radiation. Plants in the control group exposed to solar radiation in cuvettes with walls which absorb all UV but are equally transparent to visible irradiance were not affected. Another species, *Lactuca sativa,* was not nearly so sensitive when subjected to the same experiment.

Even if plants such as *Rumex alpinus* are extremely sensitive to solar UV, as long as they can acclimatize to the UV environment of their natural habitat it is perhaps not of great importance that they would be extremely sensitive to solar UV if previously grown in a greenhouse environment. Of greater ecological significance is the question of the capacity of plant species to acclimatize to still greater intensities of solar UV radiation than they experience in their natural habitat. This is germane not only to the question of introduction of species to environments of greater UV flux, but also for the more ominous problem of global atmospheric ozone reduction.

The occurrence of temperate latitude exotics at high elevations in the tropics, where DNA-effective UV-B is more than three times that which would occur with a 16% ozone reduction at temperate latitudes (CALDWELL et al. 1980), attests to the acclimatization ability of many species. While one cannot argue with the success of these temperate latitude exotics that do exist in tropical highlands, nothing is known about the degree to which solar UV prevented the establishment of temperate latitude species that do not occur in these high UV flux environments. When plants have been exposed to UV-B lamps in experiments designed to evaluate potential consequences of atmospheric ozone reduction, an impressive variability in UV sensitivity among different plant species has come to light (BIGGS et al. 1975, BIGGS and KOSSUTH 1978). Why some species are more sensitive to and others are remarkably tolerant of very intense UV-B flux is a largely unanswered question.

If acclimatization to high UV flux environments is merely a phenotypic response, the rate at which atmospheric ozone depletion takes place would not matter. If, on the other hand, acclimatization involves genotypic changes for some species, the rate at which the atmospheric ozone layer changes must be considered. About half of the equilibrium ozone reduction is now predicted to occur within the next 30 years (National Academy of Sciences 1979). This may be an insufficient time for genotypic changes to occur for long-lived plant species or those that reproduce primarily asexually.

It would seem to be a rather simple matter to conduct studies to evaluate consequences of ozone reduction using UV-B lamps to simulate the increased radiation. Indeed, many studies of this type have been conducted in the past few years, both in the laboratory and in the field environment (BIGGS et al. 1975; KRIZEK 1975; BIGGS and KOSSUTH 1978; VU et al. 1981; FOX and CALDWELL 1978). Fluorescent sun lamps (see Fig. 6.1) fitted with plastic film filters to remove the shorter-wavelength radiation (SISSON and CALDWELL 1975) have been used in these studies. Although this provides an inexpensive experimental system for UV-B supplementation, the spectral composition does not exactly match that of solar radiation. To equate a certain amount of lamp UV-B supplementation with solar UV-B for a particular ozone reduction, one must resort to comparing the integrated biologically effective UV-B from each. This necessarily involves an assumed action spectrum to be used as a weighting function (Sect. 6.2.3). Depending on the action spectrum employed, the estimates of the equivalent solar UV-B provided by these lamps can vary by at least a factor of two. Other limitations and sources of error are also inherent in this experimental approach (National Academy of Sciences 1979).

Despite the uncertainties involved, the experiments conducted thus far indicate that ozone reduction of the magnitude now contemplated (about 16%) would likely result in reduced yields of some sensitive crop species such as sugarbeets, tomatoes, and maize (National Academy of Sciences 1979). The consequences in nonagricultural ecosystems are much more difficult to predict. The range of UV sensitivity of nonagricultural species overlaps with that of crop species and many species appear to be just as sensitive to this radiation as the most sensitive agricultural plants (BIGGS and KOSSUTH 1978). Still the major consequence of ozone reduction, or for that matter, even the primary influence of solar UV in the present climate may involve changes in competitive balance of species rather than a reduction of primary productivity per se (CALDWELL 1977).

6.6.2 Competitive Balance of Plant Species

Because plant species vary so much in UV sensitivity, imposition of UV stress may be reflected in changes in competitive balance of species. A UV-tolerant species may even exhibit increased growth and productivity if it can avail itself of more resources at the expense of UV-sensitive competitors. Such hypotheses were tested in one experiment using supplementary UV-B radiation in the field (FOX and CALDWELL 1978). Indeed, for most of the pairs of competing species tested, a shift in competitive balance was noted at least at some stage in plant development. Several species did exhibit an increase in aboveground production under this rather substantial UV supplement in the field (equivalent to more than 40% ozone reduction). However, in almost every case of increased production, a decrease in production or density of the competing species was documented. Furthermore, in almost every situation, the combined biomass of the competing species was not significantly less under UV-B radiation enhancement than under control conditions. Since most species encounter interspecific competition at some phase of their life cycle, the effects of UV radiation, or any stress factor for that matter, must be considered in this competitive context. This greatly complicates the assessment of environmental perturbations such as increased solar UV radiation. In some cases, the same species might be either benefited or subject to a competitive disadvantage due to the UV radiation supplement depending on the species with which it is placed in competition (FOX and CALDWELL 1978).

Even without a reduction of the ozone layer, there is evidence that solar UV radiation which now strikes the Earth may be a subtle, though influential, factor in the balance of plant competition. Recently, BOGENRIEDER, BRUZEK, and KILIANI (unpublished data) have demonstrated that the balance between competing species pairs can be changed when UV is filtered out of normal sunlight in Germany at 50° N latitude, a region where solar UV is not particularly intense.

6.7 Conclusions

Solar UV radiation represents but a small fraction of the total solar energy reaching the Earth's surface. Yet, because this radiation is absorbed effectively by nucleic acids and other plant chromophores, there are potential photochemical effects. When considered with other factors of the environmental complex, solar UV radiation in most cases is a rather subtle influence on plant growth and development – at least if plants have been suitably acclimatized. The importance of this environmental factor for sensitive plants may be greater at tropical latitudes, especially at high elevations, or as the atmospheric ozone layer is slowly reduced in the coming decades. This may be reflected in reductions of productivity of certain sensitive crop species and shifts in the competitive balance of species in nonagricultural ecosystems.

The degree of importance that solar UV assumes is very dependent on the appropriate action spectra for plant responses to solar UV radiation and the ensuing radiation amplification factors that magnify the importance of any changes in latitude or atmospheric ozone concentration.

Higher plant species vary considerably in their sensitivity to this radiation but the basis for these differences is not well understood. At the present time it is difficult to predict which species would be most affected if the UV radiation climate changes. It is apparent that plants can acclimatize to UV radiation. Yet, the degree to which changes in plant optical properties or the protection afforded by molecular repair systems can accommodate significant increases of UV radiation is unknown.

While it would be irresponsible at this point to support the notion that reduction of the atmospheric ozone layer necessarily portends grave consequences for terrestrial and aquatic plant life, the effects are potentially insidious.

The effects of increased solar UV radiation would be on a global scale. Because of the long atmospheric time constants, man's activities influence the ozone equilibrium only after a period of decades. Therefore, the benefits of predicting the probable consequences of ozone reduction in time for corrective measures are obvious.

Acknowledgments. A series of helpful discussions with D.S. NACHTWEY contributed substantially to several facets of this review. Some of the computer modeling depicted was sponsored by a contract from the National Aeronautics and Space Administration. Assistance with computer programming by R. RYEL and with preparation of the manuscript by S. FLINT, R. DZUREC, J. RICHARDS, and R. ROBBERECHT is gratefully acknowledged.

References

Allen LH Jr, Gausman HW, Allen WA (1975) Solar ultraviolet radiation in terrestrial plant communities. J Environ Qual 4:285–294

Andersen R, Kasperbauer MJ (1973) Chemical composition of tobacco leaves altered by near-ultraviolet and intensity of visible light. Plant Physiol 51:723–726

Bener P (1960) Investigation on the spectral intensity of ultraviolet sky and sun+sky radiation under different conditions of cloudless weather at 1590 m a.s.l. Contract AF 61(052)-54 Tech Sum Rep No 1 US Air Force

Bener P (1963) The diurnal and annual variations of the spectral intensity of ultraviolet sky and global radiation on cloudless days at Davos, 1590 m a.s.l. Contract AF 61(052)-618 Tech Note No 2 US Air Force

Bener P (1964) Investigation on the influence of clouds on ultraviolet sky radiation. Contract AF 61(052)-618 Tech Note No 3 US Air Force

Bener P (1972) Approximate values of intensity of natural ultraviolet radiation for different amounts of atmospheric ozone. Final Tech Rep Contract No DAJ37-68-C-1017 US Army

Biggs RH, Kossuth SV (1978) Impact of solar UV-B radiation on crop productivity. Final rep of UV-B biological and climate effects research. Terrestrial FY 77. Univ Florida, Gainesville

Biggs RH, Sisson WB, Caldwell MM (1975) Response of higher terrestrial plants to elevated UV-B irradiance. In: Nachtwey DS, Caldwell MM, Biggs RH (eds) Impacts of climatic change on the biosphere, CIAP Monog 5, Part 1: Ultraviolet radiation effects. US Dept Trans, Springfield VA, pp 4–34 to 4–50

Bogenrieder A, Klein R (1977) Die Rolle des UV-Lichtes beim sog. Auspflanzungsschock von Gewächshaussetzlingen. Angew Bot 51:99–107

Brandle JR, Campbell WF, Sisson WB, Caldwell MM (1977) Net photosynthesis, electron transport capacity, and ultrastructure of *Pisum sativum* L. exposed to ultraviolet-B radiation. Plant Physiol 60:165–169

Butler WL, Hendricks SB, Siegelman HW (1964) Action spectra of phytochrome in vitro. Photochem Photobiol 3:521–528

Caldwell MM (1968) Solar ultraviolet radiation as an ecological factor for alpine plants. Ecol Monogr 38:243–268

Caldwell MM (1971) Solar UV irradiation and the growth and development of higher plants. In: Giese AC (ed) Photophysiology. Academic Press, New York, Vol. 6, pp 131–177

Caldwell MM (1977) The effects of solar UV-B radiation (280–315 nm) on higher plants: implications of stratospheric ozone reduction. In: Castellani A (ed) Research in photobiology. Plenum Publishing Corp, New York, pp 597–607

Caldwell MM, Robberecht R, Billings WD (1980) A steep latitudinal gradient of solar ultraviolet-B radiation in the arctic-alpine life zone. Ecology 61:600–611

Calkins J (1975) Measurements of the penetration of solar UV-B into various natural waters. In: Nachtwey DS, Caldwell MM, Biggs RH (eds) Impacts of climatic change on the biosphere, CIAP Monog 5, Part 1: Ultraviolet radiation effects. US Dept Trans, Springfield VA, pp 2–267 to 2–296

Chang DCN, Campbell WF (1976) Responses of *Tradescantia* stamen hairs and pollen to UV-B irradiation. Environ Exper Bot 16:195–199

Clark JB, Lister GR (1975) Photosynthetic action spectra of trees. II. The relationship of cuticle structure to the visible and ultraviolet spectral properties of needles from four coniferous species. Plant Physiol 55:407–413

Cline MG, Salisbury FB (1966) Effects of ultraviolet radiation on the leaves of higher plants. Radiat Bot 6:151–163

Daley LS, Dailey F, Criddle RS (1978) Light activation of ribulosebisphosphate carboxylase. Purification and properties of the enzyme in tobacco. Plant Physiol 62:718–722

Dave JV, Furukawa PM (1966) Scattered radiation in the ozone absorption bands at selected levels of a terrestrial, Rayleigh atmosphere. Meteorol Monogr 7 (29)

Dave JV, Halpern P (1976) Effect of changes in ozone amount on the ultraviolet radiation received at sea level of a model atmosphere. Atmos Environ 10:547–555

Dickson JG, Caldwell MM (1978) Leaf development of *Rumex patientia* L. (Polygonaceae) exposed to UV irradiation (280–320 nm). Am J Bot 65:857–863

Duetsch HU (1971) Photochemistry of atmospheric ozone. Adv Geophys 15:219–322

Elkind MM, Han A (1978) DNA single-strand lesions due to "sunlight" and UV light: a comparison of their induction in Chinese hamster and human cells, and their fate in Chinese hamster cells. Photochem Photobiol 27:717–724

Elkind MM, Han A, Chang-Liu CM (1978) "Sunlight"-induced mammalian cell killing: a comparative study of ultraviolet and near-ultraviolet inactivation. Photochem Photobiol 27:709–715

Fox FM, Caldwell MM (1978) Competitive interaction in plant populations exposed to supplementary ultraviolet-B radiation. Oecologia 36:173–190

Fraikin GY, Rubin LB (1979) Some physiological effects of near-ultraviolet light on microorganisms. Photochem Photobiol 29:185–188

Gausman HW, Rodriguez RR, Escobar DE (1975) Ultraviolet radiation reflectance, transmittance, and absorptance by plant leaf epidermises. Agron J 67:720–724

Green AES, Cross KR, Smith LA (1980) Improved analytic characterization of ultraviolet skylight. Photochem Photobiol 31:59–65

Harm W (1979) Relative effectiveness of the 300–320 nm spectral region of sunlight for the production of primary lethal damage in *E. coli* cells. Mutat Res 60:263–270

Hollaender A, Emmons CW (1941) Wavelength dependence of mutation production in the ultraviolet with special emphasis on fungi. Cold Spring Harbor Symp Quant Biol 9:179–186

Howland GP (1975) Dark-repair of ultraviolet-induced pyrimidine dimers in the DNA of wild carrot protoplasts. Nature 254:160–161

Inn ECY, Tanaka Y (1953) Absorption coefficient of ozone in the ultraviolet and visible regions. J Opt Soc Am 43:870–873

Jagger J (1960) Photoprotection from ultraviolet killing in *Escherichia coli* B. Radiat Res 13:521–539

Jagger J (1976) Effects of near-ultraviolet radiation on microorganisms. Photochem Photobiol 23:451–454

Jerlov NG (1950) Ultra-violet radiation in the sea. Nature 166:111–112

Jitts HR, Morel A, Saijo Y (1976) The relation of oceanic primary production to available photosynthetic irradiance. Aust J Mar Freshwater Res 27:441–454

Johnston H (1971) Reduction of stratospheric ozone by nitrogen oxide catalysts from supersonic transport exhaust. Science 173:517–522

Jones LW, Kok B (1966a) Photoinhibition of chloroplast reactions. I. Kinetics and action spectra. Plant Physiol 41:1037–1043

Jones LW, Kok B (1966b) Photoinhibition of chloroplast reactions. II. Multiple effects. Plant Physiol 41:1044–1049

Kasperbauer MJ, Loomis WE (1965) Inhibition of flowering by natural daylight on an inbred strain of *Melilotus*. Crop Sci 5:193–194

Kirby-Smith JS, Craig DL (1957) The induction of chromosome aberrations in *Tradescantia* by ultraviolet radiation. Genetics 42:176–187

Klein RM (1963) Interaction of ultraviolet and visible radiations on the growth of cell aggregates of *Ginkgo* pollen tissue. Physiol Plant 16:73–81

Klein RM (1967) Effect of ultraviolet radiation on auxin-controlled abscission. Ann NY Acad Sci 144:146–153

Klein RM (1978) Plants and near-ultraviolet radiation. Bot Rev 44:1–127

Klein RM, Edsall PC, Gentile AC (1965) Effects of near ultraviolet and green radiations on plant growth. Plant Physiol 40:903–906

Knapp EA, Reuss A, Risse O, Schreiber H (1939) Quantitative Analyse der mutationsauslösenden Wirkung monochromatischen UV-Lichtes. Naturwissenschaften 27:304

Krizek DT (1975) Influence of ultraviolet radiation on germination and early seedling growth. Physiol Plant 34:182–186

Lautenschlager-Fleury D (1955) Über die Ultraviolettdurchlässigkeit von Blattepidermen. Ber Schweiz Bot Ges 65:343–386

Levin DA (1976) Alkaloid-bearing plants: an ecogeographic perspective. Am Nat 110:261–284

Lindoo SJ, Seeley SD, Caldwell MM (1979) Effects of ultraviolet-B radiation stress on the abscisic acid status of *Rumex patientia* leaves. Physiol Plant 45:67–72

Lipton WJ (1977) Ultraviolet radiation as a factor in solar injury and vein tract browning of cantaloupes. J Am Soc Hortic Sci 102:32–36

Mantai KE, Bishop NI (1967) Studies on the effects of ultraviolet irradiation on photosynthesis and on the 520 nm light-dark difference spectra in green algae and isolated chloroplasts. Biochim Biophys Acta 131:350–356

Mantai KE, Wong J, Bishop NI (1970) Comparison studies on the effects of ultraviolet irradiation on photosynthesis. Biochim Biophys Acta 197:257–266

McCree KJ, Keener ME (1974) Effect of atmospheric turbidity on the photosynthetic rates of leaves. Agric Meteorol 13:349–357

Molina JM, Rowland FS (1974) Stratospheric sink for chlorofluoromethanes: chlorine atom-catalysed destruction of ozone. Nature 249:810–812

Mooney HA, Ehleringer J, Björkman O (1977) The energy balance of leaves of the evergreen desert shrub *Atriplex hymenelytra*. Oecologia 29:301–310

Mulroy TW (1979) Spectral properties of heavily glaucous and nonglaucous leaves of a succulent rosette-plant. Oecologia 38:349–357

Murphy TM (1975) Effects of UV radiation on nucleic acids. In: Nachtwey DS, Caldwell MM, Biggs RH (eds) Impacts of climatic change on the biosphere, CIAP Monog 5, Part 1: Ultraviolet radiation effects. US Dept Transportation, Springfield VA, pp 3–21 to 3–44

Nachtwey DS (1975) Linking photobiological studies at 254 nm with UV-B. In: Nachtwey DS, Caldwell MM, Biggs RH (eds) Impacts of climatic change on the biosphere, CIAP Monog 5, Part 1: Ultraviolet radiation effects. US Dept Transportation, Springfield VA, pp 3–50 to 3–73

National Academy of Sciences (1979) Protection against depletion of stratospheric ozone by chlorofluorocarbons. Washington DC

Okada M, Kitajima M, Butler WL (1976) Inhibition of photosystem I and photosystem II in chloroplasts by UV radiation. Plant Cell Physiol 17:35–43

Pulich WM (1974) Resistance to high oxygen tension, streptonigrin, and ultraviolet irradiation in the green alga *Chlorella sorokiniana* strain ors. J Cell Biol 62:904–907

Robberecht R, Caldwell MM (1978) Leaf epidermal transmittance of ultraviolet radiation and its implications for plant sensitivity to ultraviolet-radiation induced injury. Oecologia 32:277–287

Robberecht R, Caldwell MM, Billings WD (1980) Leaf ultraviolet optical properties along a latitudinal gradient in the arctic-alpine life zone. Ecology 61:612–619

Rothman RH, Setlow RB (1979) An action spectrum for cell killing and pyrimidine dimer formation in Chinese hamster V-79 cells. Photochem Photobiol 29:57–62

Roy RM, Abboud S (1978) The effects of far ultraviolet radiation on respiration and cation accumulation by isolated mitochondria of *Phaseolus vulgaris*. Photochem Photobiol 27:285–288

Schiff JA, Lyman H, Epstein HT (1961) Studies of chloroplast development in *Euglena*. II. Photoreversal of the UV inhibition of green colony formation. Biochim Biophys Acta 50:310–318

Setlow RB (1966) Cyclobutane-type pyrimidine dimers in polynucleotides. Science 153:379–386

Setlow RB (1974) The wavelengths in sunlight effective in producing skin cancer: a theoretical analysis. Proc Natl Acad Sci USA 71:3363–3366

Sisson WB, Caldwell MM (1975) Lamp/filter systems for simulation of solar UV irradiance under reduced atmospheric ozone. Photochem Photobiol 21:453–456

Sisson WB, Caldwell MM (1976) Photosynthesis, dark respiration, and growth of *Rumex patientia* L. exposed to ultraviolet irradiance (288–315 nm) simulating a reduced atmospheric ozone column. Plant Physiol 58:563–568

Sisson WB, Caldwell MM (1977) Atmospheric ozone depletion: reduction of photosynthesis and growth of a sensitive higher plant exposed to enhanced UV-B radiation. J Exp Bot 28:691–705

Smith KC (1978) Multiple pathways of DNA repair in bacteria and their roles in mutagenesis. Photochem Photobiol 28:121–130

Smith RC, Baker KS (1979) Penetration of UV-B and biologically effective dose-rates in natural waters. Photochem Photobiol 29:311–324

Smith RC, Baker KS, Holm-Hansen O, Olson R (1980) Photoinhibition of photosynthesis in natural waters. Photochem Photobiol 31:585–592

Stadler LJ, Uber FM (1942) Genetic effects of ultraviolet radiation in maize. IV. Comparison of monochromatic radiations. Genetics 27:84–118

Teramura AH, Biggs RH, Kossuth SV (1980) Effects of ultraviolet-B irradiance on soybean. II. Interaction between ultraviolet-B and photosynthetically active radiation on net photosynthesis, dark respiration, and transpiration. Plant Physiol 65:483–488

Thomas G (1977) Effects of near ultraviolet light on microorganisms. Photochem Photobiol 26:669–673

Vu CV, Allen LH Jr, Garrard LA (1981) Effects of supplemental UV-B radiation on growth and leaf photosynthetic reactions of soybean (*Glycine max* L.). Physiol Plant (in press)

Wellmann E (1974) Regulation der Flavonoidbiosynthese durch ultraviolettes Licht und Phytochrom in Zellkulturen und Keimlingen von Petersilie (*Petroselinum hortense* Hoffm.). Ber Deutsch Bot Ges 87:267–273

Wellmann E (1976) Specific ultraviolet effects in plant morphogenesis. Photochem Photobiol 24:659–660

Werfft R (1951) Über die Lebensdauer der Pollenkörner in der freien Atmosphäre. Biol Zentr 70:354–367

Wolff S (1972) Chromosome aberrations induced by ultraviolet radiation. In: Photophysiology. Academic Press, New York, Vol. VII, pp 189–205

Worrest RW, Van Dyke H, Thomson BE (1978) Impact of enhanced simulated solar ultraviolet radiation upon a marine community. Photochem Photobiol 27:471–478

Wright LA Jr (1978) The effect of UV radiation on plant membranes. PhD dissertation. Univ. Calif. Davis CA

Wright LA Jr, Murphy TM (1975) Photoreactivation of nitrate reductase production in *Nicotiana tabacum* var. *xanthi*. Biochim Biophys Acta 407:338–346

Wright LA Jr, Murphy TM (1978) Ultraviolet radiation-stimulated efflux of 86-rubidium from cultured tobacco cells. Plant Physiol 61:434–436

Wuhrmann-Meyer K, Wuhrmann-Meyer M (1941) Untersuchungen über die Absorption ultravioletter Strahlen durch Kutikular- und Wachsschichten von Blättern. I. Planta 32:43–50

Yamashita T, Butler WL (1968) Inhibition of chloroplasts by UV-irradiation and heat-treatment. Plant Physiol 43:2037–2040

Zill LP, Tolbert NE (1958) The effects of ionizing and ultraviolet radiations on photosynthesis. Arch Biochem Biophys 76:196–203

7 Responses to Ionizing Radiation

S. Ichikawa

CONTENTS

7.1 Introduction

Various types of ionizing radiations cause diversely different injuries in living organisms, namely, from those at genic or molecular level to those at ecosystem level. These biological responses at different levels result in many cases from some primary radiation effects (sometimes rather simple) on molecules composing living organisms, but magnified through the complicacy and delicacy of life phenomenon and of the structure of organisms, by the self-reproducing function (genetic phenomenon) characteristic of living organisms, and also by the organic interrelation between different organisms. Ionizing radiations to which plant species are exposed include those from space and those emitted by natural or man-made radioactive nuclides. Among them, special attention must be paid to man-made radiation, since many of the man-made radioactive nuclides which have not existed on the Earth (and, of course, which have not been encountered by organisms in their long evolution) are often concentrated heavily in biological tissues and through food chains, and they therefore tend to exhibit much greater genetical, physiological and ecological influences, as will be described in this chapter.

7.2 Ionizing Radiations and Their Effects on Plants

7.2.1 Ionizing Radiations and Their Characteristics

7.2.1.1 Particles and Electromagnetic Radiation

Radiation, flow of energy transmitted through space, can be classified into two categories, particles and electromagnetic radiation, based on their physical nature as shown in Table 7.1. The effects of the particle radiation are largely determined by the amount of their kinetic energy, which is nearly proportional to the square of velocity multiplied by mass.

In the case of electrically charged particles, an electrical interaction occurs between the particle and the orbital electrons of molecules existing along the track of the particle. Through this interaction, a part (Q) of the kinetic energy of the particle is transferred to the orbital electron. When the amount of energy transferred is smaller than the energy required for ionization of the molecule (I), namely, $Q<I$, the orbital electron is shifted to an orbit of higher energy and an electric excitation results. When $Q>I$, however, the orbital electron leaves the electric field of the molecule, becoming a free electron, and the molecule is ionized. Especially, when Q is very much larger than I, the free electron produced possesses a large amount of energy $(Q-I)$, and thus a secondary electron beam (delta rays) is produced. All these electrical interactions occur within a very short time of 10^{-15} s. The effects of the electrically charged particles are thus due to production of excited molecules, positive ions, and free electrons. The tracks of these particles are short, and can only penetrate small amounts of matter, because of the occurrence of the electrical interactions.

Table 7.1. Classification of radiation

Particles	Charged	Light	Electrons, *beta rays*[a]	Ionizing radiation
		Heavy	Protons, beams of ions, *alpha rays*[a]	
	Not charged	High speed	Fast neutrons	
		Slow speed	Thermal neutrons	
Electromagnetic radiation		Short wave	X-rays, *gamma rays*[a]	
		↕	Ultraviolet light	Nonionizing radiation
			Visible light, infrared light	
		Long wave	Microwaves, radiowaves	

[a] Shown in *italics* are those emitted from radioactive nuclei

For example, alpha rays penetrate biological tissues at most 0.1 mm, and beta rays about 1 cm at most.

Neutrons, which are not charged electrically, penetrate much longer than charged particles. Fast neutrons transfer their kinetic energy to nuclei mainly through elastic collisions with nuclei, producing ejected nuclei. In an elastic collision, small nuclei receive higher kinetic energy than large nuclei. Since biological tissues contain large amounts of hydrogen whose nucleus is smallest in mass, the biological effects of fast neutrons are mainly due to production of ejected protons (hydrogen nuclei) which are electrically charged particles. On the other hand, thermal neutrons which have very low kinetic energy are captured by nuclei. The nucleus which captured a neutron is radioactivated, and radioactive nuclei release alpha particles, beta rays, protons, and/or gamma rays. Boron and lithium have very much larger neutron capture cross-sections than other elements, thus their influence in determining biological effects of thermal neutrons is very significant. Since both fast and thermal neutrons eventually produce charged particles (and often gamma rays which also produce charged particles; see below), the final effects of neutrons are similar to those of charged particles.

The flows of photons having both particle and wave characteristics are called electromagnetic radiation. The amount of energy of this type of radiation is inversely proportional to its wave length. Among the electromagnetic radiations (see Table 7.1), X-rays and gamma rays have very short wave length and therefore possess very high energy and large penetrating power. When X- and gamma rays pass closely to nuclei, the energy of photons is absorbed to produce electron beams through either of the following three different processes, namely, photoelectric effect, Compton effect, or electron-pair creation. The photoelectric effect is the process in which the energy of photons is entirely absorbed to produce an ejected electron, while in the Compton effect a part of the energy of photons is absorbed to produce an ejected electron and the other part of energy remains as scattering gamma rays. The third process, the electron-pair creation, is that in which the whole energy of photons is transferred to the creation of a pair of negative and positive electrons and also to kinetic energy of these electrons. Therefore, the effects of X- and gamma rays are eventually those of electrons, namely, of charged particles, and this is the reason why X- and gamma rays are classified as ionizing radiation (see Table 7.1).

The kinetic or electromagnetic energy of such ionizing radiation is mostly very high, being 15 keV to 400 MeV (Table 7.2). The only exception is the kinetic energy of thermal neutrons, which is at most 0.025 eV. Excepting thermal neutrons, the figures shown in Table 7.2 are much higher if compared with the energy range for chemical bindings in molecules composing living organisms (shown in the same table) or with the energy required for chemical activation (1 to 10 eV). Such extremely high energy of ionizing radiations is the most fundamental point to understand the remarkably tremendous effects of ionizing radiations on living organisms. Thermal neutrons, though they have low kinetic energy, radioactivate various nuclei and make them release high-energy ionizing radiations as explained above.

Table 7.2. Energy ranges of various ionizing radiations compared with those for chemical bindings

Radiation	Energy (MeV)	Chemical binding		Energy (eV/molecule)
Electrons	2–20	Covalent bond	H−H	4.5
Beta rays	0.015–5		C−C	2.6–3.5
			N−N	1.6
Protons	5–400		P−P	2.3
Alpha rays	5–10		O−O	1.5
			C−H	3.5–5.2
Fast neutrons	0.3–15		C−N	2.1–3.2
Thermal neutrons	$\cong 2.5 \cdot 10^{-8}$		C=C	6.3
			O=O	5.1
X rays	0.05–1	Hydrogen bond	O−H···O	~0.25
Gamma rays (^{60}Co)	0.662		C−H···O	0.11
			N−H···O	~0.1
Gamma rays (^{137}Cs)	1.17, 1.33		N−H···N	~0.25

7.2.1.2 Units of Radiation Dose and Exposure

The dose of ionizing radiations with diverse physical characteristics can be commonly expressed by the amount of energy absorbed per unit weight of substance. The dose expressed in such a way means absorbed dose, and its unit commonly used is rad. The absorbed dose of 1 rad means the quantity of radiation which results in an energy absorption of 100 erg per 1 g of substance. In 1973, the International Commission on Radiation Units and Measurements recommended the use of a new unit for the absorbed dose, Gy (Gray). The dose of 1 Gy means the absorption of 1 J of radiation energy per 1 kg of substance, and corresponds to 100 rad. The dose rate is the amount of dose per unit time, and thus expressed in $\mathrm{Gy\,h^{-1}}$, $\mathrm{Gy\,s^{-1}}$, etc.

On the other hand, another unit, R (Roentgen), is often used but only for X- and gamma rays. The amount of radiation expressed as 1 R is that producing 1 esu (electrostatic unit) of positive and negative ion pairs in 1 cm^3 of air at NTP (normal temperature and pressure: 0 °C and 760 mm). Therefore, the amount of radiation expressed in R does not mean absorbed dose but indicates exposure. It is possible, however, to convert roughly from R to Gy (or rad) depending on the materials irradiated. In the tissues of living organisms, for example, an exposure of 100 R roughly corresponds to a dose of 0.96 to 0.97 Gy (or 1 R roughly corresponds to 0.96 to 0.97 rad). Exposure rate is the amount of exposure per unit time, as expressed in $\mathrm{R\,h^{-1}}$, $\mathrm{R\,s^{-1}}$, etc.

Both the dose in Gy (or rad) and exposure in R do not represent directly a biologically effective amount of radiation, because each different radiation has specific relative biological efficiency (RBE, see Sect. 7.3.2.2). Therefore, in the case of expressing a quantity of radiation which results in the same biological effect, the unit of dose equivalent, rem, is used. The dose equivalent of 1 rem means the radiation dose which gives the biological effect equivalent to that given with 0.01 Gy or 1 rad of gamma or X-rays (the standard radiation).

7.2.2 Effects of Ionizing Radiations on Plants

7.2.2.1 Radiation Effects Observed at Genic to Ecosystem Levels

When plants are exposed to ionizing radiations, various responses are observed at different levels, namely, from genic, chromosomal, and cellular levels to the levels of tissues, organs, and individuals, and further up to population and ecosystem levels.

The effects of ionizing radiations at the genic level are observed as induction of various mutations. Alterations, deletions, and additions of the bases of DNA or changes in the base sequence may result in mutations, but ionizing radiations chiefly tend to induce base-deletion type mutations. They may alter or modify genetic information, lead to partial or complete loss of genic functions, or may even result in lethality of cells or individual plants.

At the chromosomal level, ionizing radiations break chromosomes. The broken ends of chromosomes may undergo reunion or rejoining, or may be healed, thus the chromosome breakages may result in production of acentric, dicentric, and ring chromosomes. The chromosome breakages may also result in formation of deletions (or deficiencies), duplications, translocations, and inversions of chromosomes. It is possible to classify them into one-hit (or one-break) and two-hit (or two-break) aberrations, and the terminal deletions belong to the former while the interstitial deletions, dicentrics, rings, translocations, and inversions belong to the latter. There are both chromosome and chromatid types in structural aberrations of chromosomes (Fig. 7.1). Ionizing radiations also induce some numerical changes of chromosomes such as aneuploidy, polyploidy, and haploidy. These changes in structure and number of chromosomes are mostly accompanied by alterations of hereditary phenomena and may often result in loss of reproductive integrity of cells or in partial or complete sterility observable at individual plant level.

The effects observed at the cellular level include changes in reproductive ability of cells (loss of reproductive integrity, mitotic delay, lengthened cell cycle time, etc.), changes in cell morphology (giant cells, misshapen cells, etc.), modifications in differentiation (abnormal differentiation, dedifferentiation, gall formation, etc.), and killing of cells. These changes often affect the growth, physiological functions and/or reproductive ability of individual plants, sometimes seriously.

At the levels of tissues and organs, it is observed that ionizing radiations cause the delay and inhibition of development, abnormal development or malformation, partial or complete inhibition of normal function, and abnormal functioning. Any of them, if occurring, inevitably has at least some effects on the growth, physiological functions, and/or reproductive ability of the individual plants.

Radiation effects observed at the individual plant level are partial or complete growth inhibition, physiological disturbances, various morbid phenomena, sterility, and even lethality. These effects are mostly those which appear as the combined results of the radiation effects at each of the levels mentioned above.

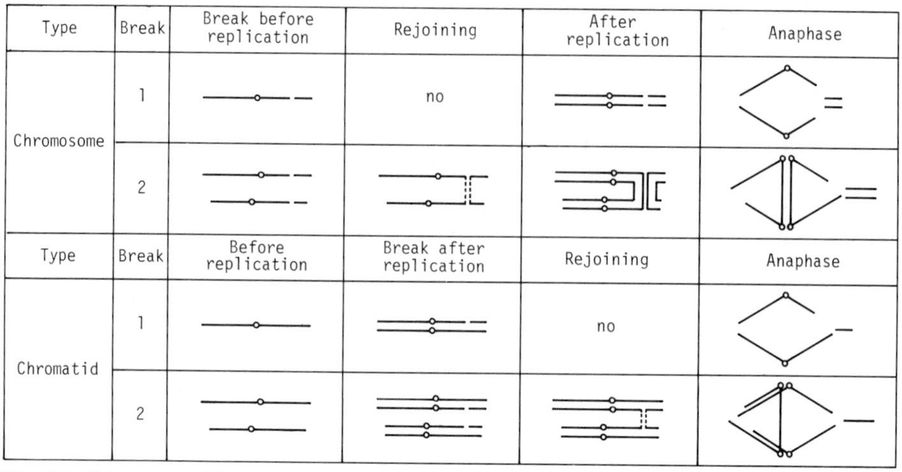

Type	Break	Break before replication	Rejoining	After replication	Anaphase
Chromosome	1		no		
	2				

Type	Break	Before replication	Break after replication	Rejoining	Anaphase
Chromatid	1			no	
	2				

Fig. 7.1. Diagram to illustrate some examples of chromosome- and chromatid-type aberrations

The effects of radiation observable at the level of population are increases of genetic variation and greater frequencies of unadaptive (including lethal) genotypes and karyotypes. Shifts in sex ratio may occur in dioecious plants with sex chromosomes. These radiation effects can break down the balances in gene frequency which have been kept adaptive to the environment, and reduce the reproductive function of the population as a whole.

Concerning the effects of ionizing radiations at the ecosystem level, only a few experimental results are available. However, relative decreases and increases in number were reported to occur between different species in the same ecosystem because of their diversely different radiosensitivities (Woodwell and Sparrow 1965). Such a phenomenon can lead to a breakdown of the balance in the ecosystem.

Occurrences of mutations and chromosomal aberrations are generally termed to be genetic effects of radiations regardless of whether they were found in the somatic or gametic cells of the irradiated plants or in the progeny. On the other hand, all other responses are called somatic injuries in general, although many of them were caused directly or indirectly from damages to genes or chromosomes.

7.2.2.2 Dose-Response Curves and Quantitative Expression of Effects

The relationships between radiation doses (or exposures) and the biological effects are, in many cases, not simple. Mutation frequency in higher plants, for example, often increases nonlinearly with increasing radiation dose in high-dose ranges (Fig. 7.2B), although it ordinarily increases linearly with lower doses (Fig. 7.2A). Two-hit chromosomal aberrations show exponential increases with radiation dose (Fig. 7.2C), and nonlinear relationships with dose are also observed for growth inhibition (Fig. 7.2D) and survival fraction. However, in cases of survival data (and also in many cases of growth inhibition data),

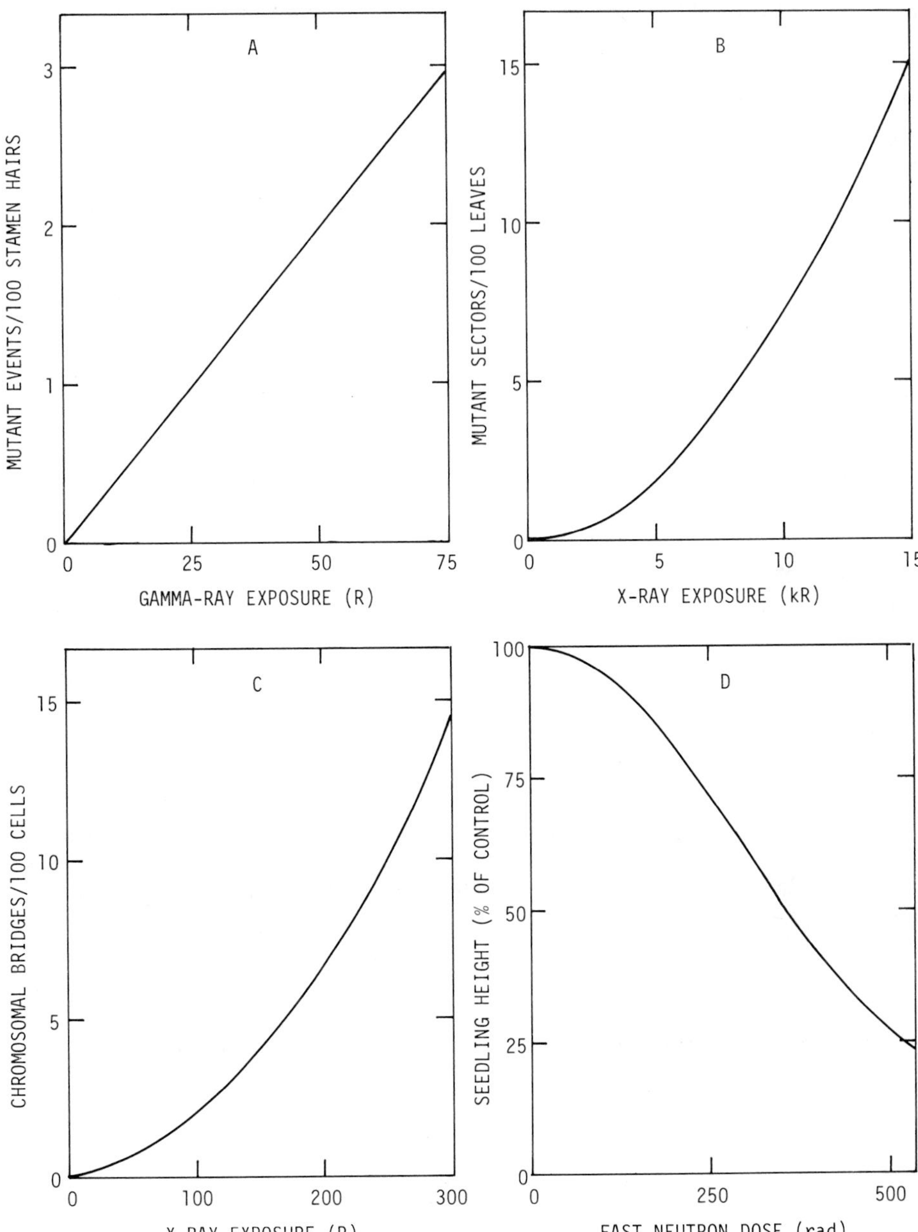

Fig. 7.2. Typical dose-response curves of some biological effects of radiation. **A** Somatic mutation frequency in *Tradescantia* stamen hairs in relatively small-dose range (ICHIKAWA 1972b). **B** Somatic mutation frequency in the third leaves of *al*/+ diploid oat seedlings (NISHIYAMA et al. 1966). **C** Frequency of chromosomal bridges at anaphase in barley root-tip meristem (ICHIKAWA et al. 1965). **D** Growth inhibition of diploid *Triticum* seedlings (ICHIKAWA 1970b)

it is possible to obtain linear dose-response relationships if the data are plotted on a semi-log graph, namely, to see the decrease of the logarithm of survival fraction with increasing dose (see Fig. 7.5).

For comparing the effects of radiation between different species, different treatments, different biological responses, etc., it is necessary to express the relationships between radiation doses and the biological effects quantitatively. One of the commonly used ways of expressing the dose-response relationship is to show how much dose of radiation was required for producing a certain amount of a specific biological effect. The LD_{50} (50% lethal dose) and D_0 (or D_{37}; the dose required for reducing survival fraction to $e^{-1}=0.37$ at the linear portion of a survival curve drawn on a semi-log graph as shown in Fig. 7.5) are such examples. The other way commonly used is to show how great a quantity of a specific biological effect was induced with a certain dose of radiation. The mutation rate per unit dose is an example of this second way of quantitative expression. Occasionally, the above two ways are used together, for example, to show the frequency of chromosomal bridges per unit dose at the dose which induced a certain amount of chromosomal bridges.

7.3 Radiosensitivity and Its Modification

It is well known that different plant species show different sensitivities to ionizing radiation, and even an almost 500-fold difference in radiosensitivity has been reported (SPARROW and WOODWELL 1962). It is also known that the radiosensitivity differs between different varieties, races, strains, etc., even within the same species, and that it also varies considerably between different tissues or organs, and depends on the developmental stage and environmental conditions. On the other hand, even though all botanical conditions are kept constant, different responses to radiation are observed if some physical conditions such as type and intensity (dose rate) of radiation are changed.

7.3.1 Biological Factors Determining or Modifying Radiosensitivity

7.3.1.1 Nuclear and Chromosome Volumes and DNA Content

SPARROW and his co-workers, conducting gamma-ray irradiation of growing plants of many different species, reached a conclusion that there was an inverse relationship between radiation dose required for producing a certain extent of growth inhibition and the volume of nucleus at interphase (SPARROW et al. 1961; SPARROW and EVANS 1961). The relationship meant that species having larger interphase nuclear volumes were more radiosensitive, suffering a certain extent of growth inhibition with smaller doses.

SPARROW's group further found that radiosensitivity was more strongly correlated to a value obtained by dividing the interphase nuclear volume of each species by its somatic chromosome number, the value which they called the interphase chromosome volume or ICV (SPARROW et al. 1965). The ICV was the value representing roughly an average volume of one chromosome at interphase, supposing the nucleus is mostly occupied by chromosomes at interphase. They found an evident inverse correlation of the ICV with the dose required for inhibiting growth or killing plants, that is, the larger the ICV, the higher was the radiosensitivity. Such inverse correlation was similarly observed after chronic and acute irradiation treatments (SPARROW 1965, 1966) as shown in Fig. 7.3, and in both herbaceous and woody plants (SPARROW et al. 1965, 1968a). Also, the correlation of the ICV with radiosensitivity was essentially unchanged when the radiosensitivity was measured in terms of seed setting (YAMAKAWA and SPARROW 1965) or pollen fertility (YAMAKAWA and SPARROW 1966). According to MILLER and SPARROW (1964), the difference in radiosensitivity between

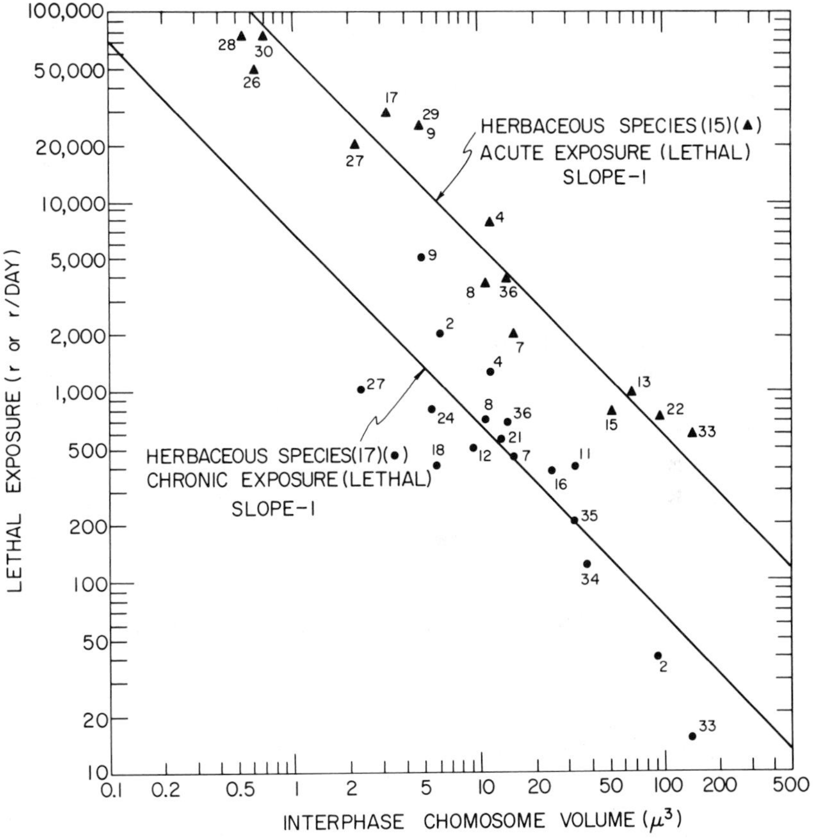

Fig. 7.3. Relation of acute and chronic lethal exposures to interphase chromosome volume in herbaceous species. *2 Anethum graveolens; 4 Aphanostephus skirrobasis; 7 Chlorophytum elatum; 8 Crepis capillaris; 9 Gladiolus* sp.; *11 Haworthia attenuata; 12 Helianthus annuus; 13 Hyacinthus orientalis; 15 Lilium longiflorum; 16 Luzula purpurea; 17 Mentha spicata; 18 Nicotiana glauca; 21 Pisum sativum; 22 Podophyllum peltatum; 24 Ricinus communis; 26 Sedum alfredi* var. *nagasakianum; 27 S. oryzifolium; 28 S. rupifragum; 29 S. ternatum; 30 S. tricarpum; 33 Trillium grandiflorum; 34 Tulbaghia violacea; 35 Vicia faba; 36 Zea mays* (SPARROW 1965, 1966)

different cell types of *Marchantia polymorpha* could be also correlated to their different ICV values.

On the other hand, OSBORNE and LUNDEN (1964, 1965) reported that radiosensitivity was more strongly correlated to nuclear volume than to ICV when radiation treatments were conducted on seeds of 12 different plant species. Their argument does not seem to be strongly negative evidence against the findings by SPARROW's group, since the number of species they employed was not very large, and the variation in chromosome number among those species was also rather small. It should be taken into consideration that many other factors affecting plant responses to radiation often accompany such cases of irradiation of seeds (see Sect. 7.3.1.5).

It was also demonstrated that radiosensitivity was positively correlated to DNA content per nucleus (see SPARROW and EVANS 1961) and more closely to DNA content per chromosome (BAETCKE et al. 1967; SPARROW et al. 1968b). All these findings proved the idea that chromosomes and DNA in chromosomes play the major role in determining radiosensitivity. Thus, it became possible to predict the radiosensitivity of a plant species if the ICV or the DNA content per chromosome of the species was known (see SPARROW et al. 1968a).

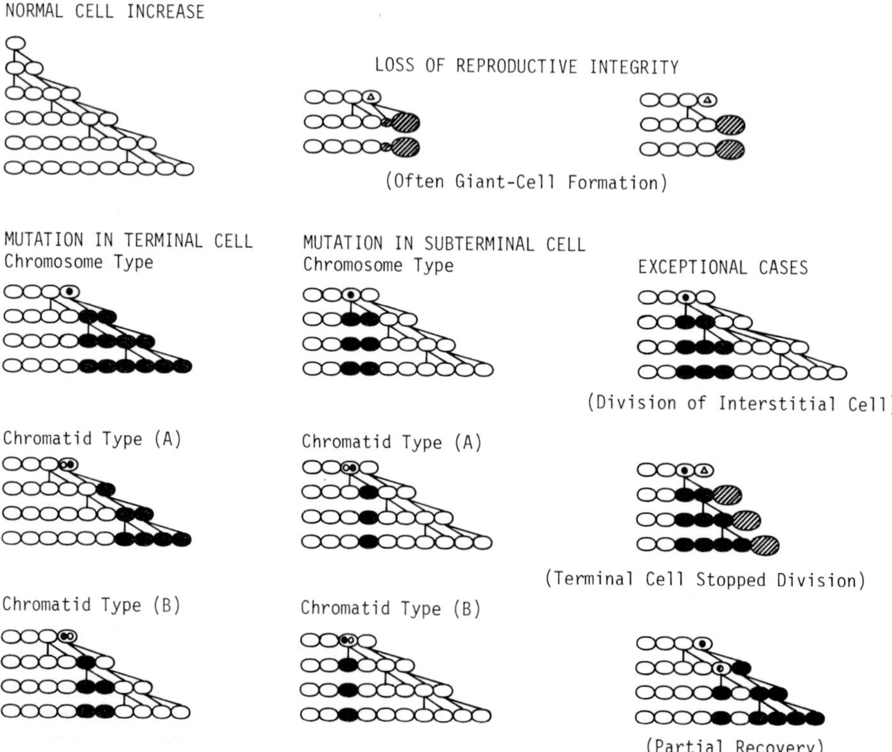

Fig. 7.4. Diagram to illustrate the pattern of cell increase and the appearance of mutations and other radiation effects in the stamen hairs of *Tradescantia*

Employing the stamen hairs of *Tradescantia* species, ICHIKAWA and SPARROW (1967) and ICHIKAWA (1972a) further proved that the correlation of radiosensitivity with ICV evidently stood also at single-cell level. The stamen-hair system of *Tradescantia* can be regarded as an essentially single-meristematic-cell system (ICHIKAWA and SPARROW 1967) rarely found in the plant kingdom, since each of the stamen hairs is composed of a single chain of about 20 to 35 cells and its development (increase in cell number) is largely due to repeated divisions of the terminal cell which is meristematic until a young stamen hair completes its development. Therefore, conducting radiation treatment of young stamen hairs which are undergoing development, the effects on individual meristematic terminal cells can be directly observed in the resultant individual stamen hairs, thus allowing the analysis at single-cell level (ICHIKAWA et al. 1969), as illustrated in Fig. 7.4.

The constant relationship between nuclear parameters and radiosensitivity mentioned above indicated that the amount of energy (eV) absorbed per average chromosome (or per DNA of average chromosome) at the radiation dose giving a certain extent of biological effect (D_0, LD_{50}, 50% growth dose, etc.) is roughly constant for any species regardless of the different target sizes of genetic substances (SPARROW 1965, 1966; SPARROW et al. 1965). This is a very important point for understanding the nature of apparent radiosensitivity.

7.3.1.2 Polyploidy

The first report of differential radiosensitivity due to ploidy level was made by STADLER (1929), who reported that *Avena* and *Triticum* polyploids were more resistant to X-rays than diploids. He considered that the lower sensitivity of polyploids was due to gene reduplication, based especially on much lower frequencies of induced chlorophyll mutants in polyploids than in diploids. Indeed, many studies performed later using *Triticum* species have shown that chlorophyll mutants are induced in diploid species at high frequencies while in tetraploids evidently much less frequently, and in hexaploids only rarely, as reviewed by MACKEY (1959) and ICHIKAWA (1970c). On the other hand, radiosensitivity measured in terms of radiation-induced injuries has been reported to be much higher in diploid *Triticum* species than in the tetra- and hexaploid relatives, but no consistent difference in sensitivity has been found between the two levels of polyploidy (see MACKEY 1959; ICHIKAWA 1970c).

Polyploidy vs. radiosensitivity relationships similar to that in *Triticum*, namely, that resistance to radiation generally increases in polyploids but not always parallel with the levels of ploidy, have been observed also in the polyploid series of *Avena, Aegilops, Oryza, Tradescantia, Nicotiana, Chrysanthemum, Sedum, Solanum, Brassica,* and *Rumex* as reviewed by ICHIKAWA (1970c).

On the other hand, artificially produced autotetraploids have been confirmed to be, with few exceptions, evidently more resistant to radiation than their original diploid forms. For example, the autotetraploids of *Secale cereale, Zea mays, Hordeum vulgare, Avena strigosa, Triticum boeoticum, Beta vulgaris, Raphanus sativus, Astragalus sinicus, Capsicum annuum, Citrullus vulgaris, Brassica oleracea,* and *Zinnia elegans* have been reported to show remarkably higher resistances as compared with their original diploids as reviewed by ICHIKAWA (1970c). Correlated results reported recently are higher sensitivities of haploid protoplasts than the original diploid protoplasts of *Petunia hybrida* and *Datura innoxia* (KRUMBIEGEL 1979).

The remarkably lower radiosensitivity of autotetraploids had been regarded as support for the idea of STADLER (1929) mentioned above. FUJII and MATSUMURA (1959) also considered the genetic redundancy in polyploids to be the main factor which makes the polyploids more radioresistant than diploids. Namely, they argued that the possibility of a radiation-induced genetic damage being compensated or masked must be much higher in artificially produced autotetraploids than in diploids, since genetic substances are completely reduplicated in the autotetraploids. Concerning the fact that naturally existing polyploids (mostly having the nature of allopolyploids) do not always show a clear relationship between ploidy level and radiosensitivity, they considered that the probability of such compensation or masking must be lower in those polyploids as compared with artificially produced autotetraploids, since the genetic substances are not always reduplicated corresponding to the extent of the differentiation accomplished.

However, in 1965, SPARROW and his co-workers threw doubt on such a generally accepted idea with the finding that the data from *Chrysanthemum, Rumex, Sedum, Tradescantia,* and some other genera could be best explained by reduced chromosome volumes in polyploids rather than by genetic redundancy (SPARROW 1965; SPARROW et al. 1965). Especially, it was clearly demonstrated in the studies on the stamen hairs of *Tradescantia* that the radiosensitivities of the polyploid series were strongly correlated with their ICV's, namely, higher polyploids had smaller ICV's and showed lower radiosensitivities (ICHIKAWA and SPARROW 1967; ICHIKAWA 1970c, 1972a). A similar relationship was also found between ICV's and pollen abortion rates in $2x$ to $12x$ *Tradescantia* species (UNDERBRINK et al. 1973a). In the $2x$ to $20x$ species of *Rumex,* also, higher polyploids were found to have generally smaller chromosomes (ICHIKAWA et al. 1971) and polyploids were generally more resistant to radiation than diploids (ICHIKAWA and SPARROW 1966; ICHIKAWA 1970c), though higher than $14x$ polyploids were extremely radiosensitive (see Sect. 7.3.1.3 for such unexpected results). According to MILLER (1970) who compared radiosensitivity between diploid and tetraploid forms of *Tagetes, Zinnia,* and *Petunia,* clear difference in the sensitivity was found only when there was a clear difference in ICV between the di- and tetraploids.

It seems most reasonable to conclude that the principle that the radiosensitivity of each species is determined primarily by the ICV characteristic of each species holds true also in polyploids. In the case of comparison between a diploid and its artificially produced autotetraploid, where the main factor (ICV) is unchanged, the influence of the modifier (genetic redundancy) becomes evident. In the case of comparison of allopolyploids or naturally existing polyploids with their diploid relatives, on the other hand, the higher the extent of differentiation or diploidization, the greater becomes the difference in the main factor, and thus the effect of the main factor tends to become more evident while the influence of the modifier tends to disappear (ICHIKAWA 1970c).

7.3.1.3 Cell Cycle

It is known that radiosensitivity differs between different periods of cell cycle, and also depending on the length of cell cycle time. The mitotic cycle of cells is composed of four different periods, G_1 (post-mitotic or pre-synthetic period), S (DNA-synthetic period), G_2 (pre-mitotic or post-synthetic period), and M (mitotic period). Morphologically, the cells at the G_1 through G_2 periods remain at interphase.

BREWEN (1964) observed more than twice as many chromatid aberrations in the root-tip cells of *Vicia faba* when the cells were irradiated at the G_2 period, in comparison with the cases of irradiation at the G_1 or S periods, and he considered the increased target number to be the main cause of the twofold sensitivity of the G_2 period. The highest sensitivity of the G_2 period was also demonstrated by HORSLEY et al. (1967a, 1967b) in *Oedogonium cardiacum*, an alga species.

On the contrary, HAMPL et al. (1971) reported that the radiosensitivity of *Chlorella pyrenoidosa* was highest in the cells just before entering the S period. DEERING (1968) studied the radiosensitivity of zoospores of *Blastocladiella emersonii*, an aquatic phycomycete, and reported that they were most sensitive just after nuclear division and most resistant just before. The data from *Saccharomyces cerevisiae* obtained by BEAM et al. (1954) also showed the G_2 period to be the most resistant in the cell cycle.

YAMAGUCHI and TATARA (1973) investigated the radiosensitivity vs. mitotic cycle relationship in the first-leaf meristem of barley, and reported the highest sensitivity of the period from late S to early G_2 and also of the late G_1 period for chromosomal aberration production. However, when the sensitivity was determined in terms of mitotic index and loss of dividing ability, the middle G_1 period and the late G_1 to S period, respectively, were most sensitive.

Such conflicting results seem to indicate that different conclusions could be obtained depending on how the radiosensitivity was determined and what endpoint was used for the determination, and also suggest that different biological responses might result from different causes or through different processes.

Concerning the effects of the length of cell cycle time in determining radiosensitivity, SPARROW and EVANS (1961) stated that the species or cell types possessing longer cell cycle times were more sensitive to radiation in general. According to VAN'T HOF and SPARROW (1963), there was an evident positive correlation between nuclear volume and mitotic cycle time, namely, the larger the nuclear volume, the longer was the mitotic cycle time. They explained that the species with a longer mitotic cycle time would absorb a higher dose of radiation per cell generation during a chronic radiation treatment (but not so with an acute treatment), thus the longer mitotic cycle time, besides the larger nuclear volume, must have increased the apparent radiosensitivity. Indeed, even in the same species, increased radiosensitivity resulted after chronic irradiation in the case in which the mitotic cycle time was lengthened (SPARROW et al. 1961, 1971). The extremely high radiosensitivities of $14x$ to $20x$ *Rumex* species (ICHIKAWA and SPARROW 1966; ICHIKAWA 1970c) mentioned in the previous section may be understood in connection with their very long mitotic cycle times.

Table 7.3. The principal nuclear factors determining or modifying plant radiosensitivity listed by EVANS and SPARROW (1961) or found later

	Factors tending to increase sensitivity	Factors tending to decrease sensitivity
1	Large nucleus	Small nucleus
2	High DNA content/nucleus	Low DNA content/nucleus
3	Low chromosome number	High chromosome number
4	Diploid or haploid	Polyploid
5	Large chromosomes	Small chromosomes
6	High DNA content/chromosome	Low DNA content/chromosome
7[a]	S to G_2 (or G_1 to S) period	M to G_1 (or G_2) period
8[b]	Long mitotic cycle time	Short mitotic cycle time
9	Long chromosome arms	Short chromosome arms
10	Acrocentric chromosomes	Metacentric chromosomes
11	Normal centromere	Polycentric or diffused centromere
12	Much heterochromatin	Little heterochromatin
13	Small nucleolus	Large nucleolus

[a] Conflicting results have been reported (see text in Sect. 7.3.1.3)
[b] Of particular importance only under conditions of chronic irradiation

7.3.1.4 Other Nuclear Factors

There are some other nuclear factors known to modify radiosensitivity. These factors are summarized in Table 7.3 (the 9th to 13th factors) together with the major factors (up to the 8th) explained above.

Loss of chromosome segments following radiation-induced breakages of chromosomes is considered to be one of the major causes leading to cell killing or inactivation. If we consider the case in which one break occurs at random with a radiation dose and the broken chromosome segment distal to the centromere is lost without rejoining, the size of the chromosome segment to be lost must be larger on the average with chromosomes possessing longer arms than with those having shorter arms. Likewise, even for the chromosomes of the same length, larger chromosome segments must be lost on the average with acrocentric chromosomes than with metacentric chromosomes. The species having polycentric chromosomes or diffused centromeres on chromosomes must be more tolerant to radiation than the species with normal centromeres because of the much lower possibility of losing broken segments of chromosomes. The high tolerance of *Luzula* species to radiation is known to be a good example of such case (EVANS and SPARROW 1961).

A higher radiosensitivity of heterochromatic regions of chromosomes than euchromatic regions was first found in a moss species, *Sphaerocarpos donnellii*, and similar findings were reported later also in higher plants such as *Cannabis sativa, Impatiens balsamina, I. sultani,* and *Vicia faba* (see EVANS and SPARROW 1961). The nucleolar chromosome arms of *Vicia faba* were found to be involved in chromosome interchange less frequently than expected from their lengths, and the nucleolus was considered to have acted as a physical barrier reducing the possibility of association between nucleolar and other chromosomes (see EVANS and SPARROW 1961). The ratio of nucleolus to nucleus differs considerably among species, being about 20% in *Vicia faba* but much higher in *Crepis capillaris* (EVANS and SPARROW 1961). Therefore, comparing two species having similar nuclear volumes but one possessing a larger ratio of nucleolus to nucleus than the other, the former must have a smaller ICV value, thus being more resistant to radiation than the latter.

Besides those factors listed in Table 7.3, some data have suggested that individual chromosomes may have some genetic factors determining or influencing radiosensitivity of the species. For example, 21 different monosomics of *Triticum aestivum,* each of which is missing one particular chromosome different from each other, were demonstrated to show differential radiosensitivities, but the monosomics of the same homoeologous group tended to exhibit similar sensitivities (TSUNEWAKI 1975).

7.3.1.5 Physiological Factors

Water content, temperature, oxygen concentration, as well as deficiencies of some nutritional elements, are known to be physiological factors of particular importance which greatly modify responses of plants, particularly of seeds, to ionizing radiations.

Remarkably higher radiosensitivity of soaked seeds as compared to dry or dormant seeds first discovered by STADLER (1930) is well known through repeated studies also by other workers. Therefore, it was easily considered that higher contents of water made the seeds more susceptible to radiation. In case of soaked seeds, however, some physiological and developmental changes accompanying germination phenomenon might also be involved. EHRENBERG (1955) changed the water content of barley seeds variously by storing them in air containing different moisture levels prior to radiation treatments, and found the lowest radiosensitivity at 12% to 20% water contents. Similar results, the lowest sensitivity of cereal seeds occurring at about 8% to 15%, were obtained later by other workers. ATAYAN and GABRIELIAN (1978) reported recently a great increase of sensitivity of *Crepis capillaris* seeds at 2.4% water content as compared with seeds of 7.6% and 12.6% water contents. Concerning the higher sensitivity at lower water content, which was an unexpected result from earlier experiments with soaked seeds, a slower decay of radiation-induced free radicals at lower water content was demonstrated to be the most likely cause of the higher radiosensitivity, employing the technique of measuring electron spin resonance (ESR) (EHRENBERG 1959).

The temperature during irradiation treatment also modifies the effects of radiation. The effect of temperature has been most clearly observed in the case of irradiation of growing plants. In *Tradescantia,* SPARROW et al. (1961, 1971) reported that higher frequencies of somatic mutations were induced at 8 or 12.5 °C than at 20 or 21 °C. A similar temperature effect was also reported in rice plants in the range of 10 to 30 °C (YAMAGATA and FUJIMOTO 1970). At least for the temperature effect to modify responses of growing plants to radiation treatments, some repair mechanisms which are more effective at higher temperature are very likely involved, because TAKAHASHI and ICHIKAWA (1976) demonstrated clearly that the spontaneous somatic mutation frequency in *Tradescantia* stamen hairs was also higher at lower temperature.

The presence or absence or the concentration of oxygen during irradiation treatment is also known to modify plant responses to radiation. Irradiation of soaked barley seeds in air or in oxygen resulted in remarkable increases of radiation effects, whereas no such difference was observed in case of irradiation of dry seeds (see CALDECOTT and NORTH 1961). Similar results were obtained by STEIN and SPARROW (1966) who irradiated the growing *Kalanchoe* plants in nitrogen, carbon dioxide, air, and pressurized oxygen-enriched air. They found an enhanced growth inhibition in the plants irradiated in air as compared with those treated without oxygen, and more enhanced radiation effects by the treatment in the oxygen-enriched air. UNDERBRINK et al. (1975) reported an evident oxygen enhancement for somatic mutations in X-ray- and neutron-irradiated stamen hairs of *Tradescantia*.

Cultivation of plants under phosphorus-deficient conditions was reported to enhance radiosensitivity in *Antirrhinum majus, Oryza sativa*, and *Avena strigosa* (see NISHIYAMA and AMANO 1963). The deficiency of calcium was also reported to increase the sensitivity of chromosomes to radiation in *Tradescantia paludosa* (STEFFENSEN 1957). In case of thermal neutron irradiation, even a slight increase of boron content increases plant radiosensitivity greatly (see ICHIKAWA 1975), since the neutron capture cross-section of ^{10}B is extremely large, as described above (see Sect. 7.2.1.1).

As a phenomenon specific to irradiation of dry seeds, a conspicuous effect modifying radiosensitivity termed "storage effect" is well known. According to KONZAK et al. (1961), the longer the post-irradiation period before allowing germination, the greater extent of radiation effects observed, and the storage effect varied greatly depending on water content of irradiated seeds, temperature, and presence or absence of oxygen during the storage period.

Presence or absence of oxygen at the time of soaking irradiated seeds also modifies greatly the radiation injuries. Post-irradiation soaking of seeds under the presence of oxygen resulted in enhanced radiation injuries (CALDECOTT and NORTH 1961; KONZAK et al. 1961), and this phenomenon was termed "oxygen effect". The oxygen effect is also modified by other factors. For example, the oxygen effect almost disappeared when the irradiated seeds were once soaked under the absence of oxygen, then redried, and stored before soaking again to allow germination (KONZAK et al. 1961). According to HABER and RANDOLPH (1967), ESR signals (see above) decreased very much by a similar once-soaking and redrying treatment of gamma-ray irradiated lettuce (*Lactuca sativa*) seeds. Their finding indicated that radiation-induced free radicals decayed rapidly by the once-soaking treatment. It suggests that the oxygen effect is the result of interaction of free radicals with oxygen, and the storage effect mentioned above is also understood as the long-term effect of the free radicals.

On the contrary, presence of oxygen after irradiation acts as a protective factor from radiation effects in case of irradiation of soaked seeds or growing plants. WOLFF and LUIPPOLD (1958) reported a remarkable increase of chromosomal aberrations in *Vicia faba* root tips under anoxic condition after irradiation. This phenomenon suggested that oxygen must be necessary for supplying energy for the rejoining of broken ends of chromosomes. Support for this idea has been obtained from the experiments using 2,4-dinitrophenol (DNP), which suppresses oxidative phosphorylation, during and/or after irradiation (WOLFF and LUIPPOLD 1955; ICHIKAWA et al. 1965).

7.3.1.6 Repair Mechanisms

Living cells possess some kinds of inherent repair mechanisms of radiation injuries. The existence of such repair mechanisms has been recognized especially through dose fractionation experiments and also based on dose-rate effect. Fractionation of dose and chronic irradiation, in many cases, resulted in fewer radiation injuries than acute single irradiation with the same total dose (see Sect. 7.3.2.1), suggesting occurrence of recovery between fractionated doses or during chronic irradiation.

Knowledge of the repair mechanisms has been obtained predominantly from microorganisms, and the molecular bases of some repair mechanisms of UV-induced DNA damages (photoreactivation, excision repair, and recombinational repair) have been clarified at least partially. However, the repair mechanisms of ionizing radiation-induced DNA damages have not yet been made clear even with microorganisms, excepting the understanding that there are both recombinational and rejoining repairs, the former being common in the case of UV and the latter being specific for ionizing radiation.

Only little is known about the repair mechanisms in higher plants. In the 1960's, the photoreactivation function of UV-induced damage (see Chap. 6, this Vol.) was shown to be present in plant cells (see IKENAGA et al. 1974). On the other hand, for the existence of the excision repair of UV damages in plant cells, HOWLAND (1975) and SOYFER and CIEMINIS (1977) recently found positive evidence from carrot protoplasts and grass pea (*Lathyrus sativa*). The rejoining repair of ionizing radiation-induced DNA single-stranded breaks has been also recently proved to be present in carrot protoplasts (HOWLAND et al. 1975) and barley embryos (YAMAGUCHI et al. 1975). For DNA double-stranded breaks by ionizing radiation, evidence of rejoining repair has been obtained only from yeast (RESNICK 1975) and *Physarum polycephalum*, a Myxomycete species (BREWER 1979). More evidences of repair mechanisms in higher plants have been accumulating rapidly (see VELEMINSKY and GICHNER 1978).

RILEY and MILLER (1966) and YAMAGUCHI (1974) proved the higher radiosensitivity of desynaptic and partially asynaptic strains of barley and rice plants, respectively, and their findings suggested the existence of a specific type of recombinational repair (probably specific for higher plants) in which homologous chromosome must be utilized as the template.

Lower spontaneous mutation frequency at higher temperature in *Tradescantia* (TAKAHASHI and ICHIKAWA 1976) and a similar relationship for radiation-induced mutations in *Tradescantia* (SPARROW et al. 1961, 1971) and in rice plants (YAMAGATA and FUJIMOTO 1970) suggest that a repair mechanism which is more effective at higher temperature, thus presumably an enzymatic one, is very likely involved. Highly mutable clones of *Tradescantia* described by SPARROW and SPARROW (1976) and TAKAHASHI and ICHIKAWA (1976) seem to be those resulting from mutations related to such repair mechanism.

Besides the above repair mechanisms at DNA level, the meristem of higher plants possesses some cell kinetic recovery mechanisms. Loss of reproductive ability of some cells in a meristem is often compensated by repeated divisions of other cells in the meristem (ICHIKAWA and IKUSHIMA 1967), or severe injury of meristematic tissue is compensated by reorganizing the meristem from more radioresistant cells, such as those of the quiescent center (CLOWES and HALL 1963), or even from a small number of surviving cells (YAMAKAWA and SEKIGUCHI 1968).

7.3.2 Physical Factors and Radiosensitivity

7.3.2.1 Dose Rate and Fractionated Dose

Greater biological efficiency of higher dose (exposure) rates has been described over many years for a variety of biological endpoints in plants such as growth inhibition, lethality, sterility, etc., as reviewed by NAUMAN et al. (1975) and ICHIKAWA et al. (1978), and the phenomenon is called "dose-rate effect". The dose-rate effect has been also clearly observed for the loss of reproductive integrity at single-cell level in *Tradescantia* stamen hairs (ICHIKAWA et al. 1978) as shown in Fig. 7.5.

Some results contradictory to the above have been also reported (see MATSUMURA 1965). It should be noted, however, that all such results contrary to the expectation from the above dose-rate effect are those obtained from experiments of irradiation of dry seeds in which much less repair or recovery must have taken place. As explained in an earlier section (7.3.1.5), storage effect plays an important role in modifying the responses of irradiated dry seeds. Thus contradictory results from dry seeds are considered to arise by concealment of the dose-rate effect (which must be weak with dry seeds) by the storage effect (MATSUMURA 1965).

The effects of dose rate and of fractionation of exposure have been most clearly seen for the induction of two-break-type chromosomal aberrations, the induction frequency being evidently higher at higher dose rates and also with single exposure than with fractionated exposure of the same total dose (see ICHIKAWA et al. 1965), an example of which is shown in Fig. 7.6. It is considered that higher dose rates and single exposure induce larger numbers of chromosome breaks within a short, limited time than do lower dose rates and fractionated exposure, thus the chance of yielding chromosomal exchanges increases with the former. At lower dose rates or with fractionated exposure, such a chance decreases because of restitution and healing of broken chromosome ends. Even in the cases of fractionated exposure, however, when respiratory inhibitors were given during the interval period between the first and second exposures to suppress restitution and healing, effects comparable or close to a single exposure of the same total dose were observed (see ICHIKAWA et al. 1965). An anoxic condition was also reported to keep the broken ends open, suppressing restitution and healing (WOLFF and LUIPPOLD 1958). From the dose fractionation experiments, it has been proved that there are two kinds of chromosome rejoining,

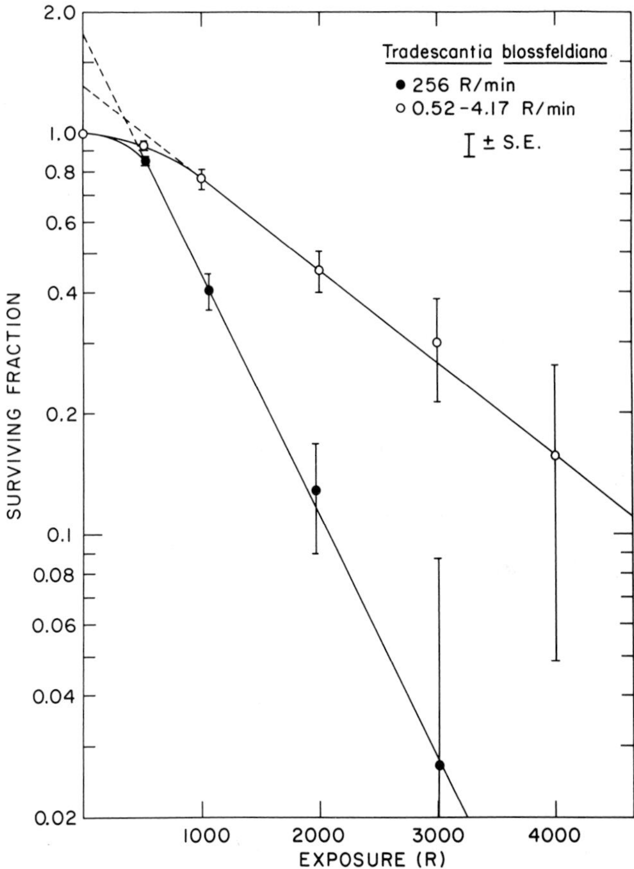

Fig. 7.5. Survival curves for stamen hair cells of *Tradescantia blossfeldiana* exposed at two different exposure rates (ICHIKAWA et al. 1978)

one occurring within a short period of time and the other after a longer time (WOLFF and LUIPPOLD 1956; ICHIKAWA et al. 1965), as shown in Fig. 7.6.

For somatic mutation frequency in plants determined by the specific loci method, dose-rate effects have been often observed as shown in Fig. 7.7. Greater efficiency in inducing somatic mutations with higher dose rates has been repeatedly confirmed by many workers as reviewed by NAUMAN et al. (1975) and ICHIKAWA et al. (1978), and lowered somatic mutation frequencies by fractionation of exposure have been also reported (DAVIES and WALL 1960; NISHIYAMA et al. 1966). These findings suggest that at least a part (probably the majority with higher doses) of mutations induced by ionizing radiations are due to small interstitial deletions of chromosome segments where the specific marker genes are located. This has been supported by the exponential dose-response curves in a relatively high dose range (SPARROW et al. 1961, 1972; NISHIYAMA et al. 1964), responses to DNP (NISHIYAMA et al. 1966), and the RBE values (see Sect. 7.3.2.2) comparable with those for chromosomal aberrations (MATSUMURA

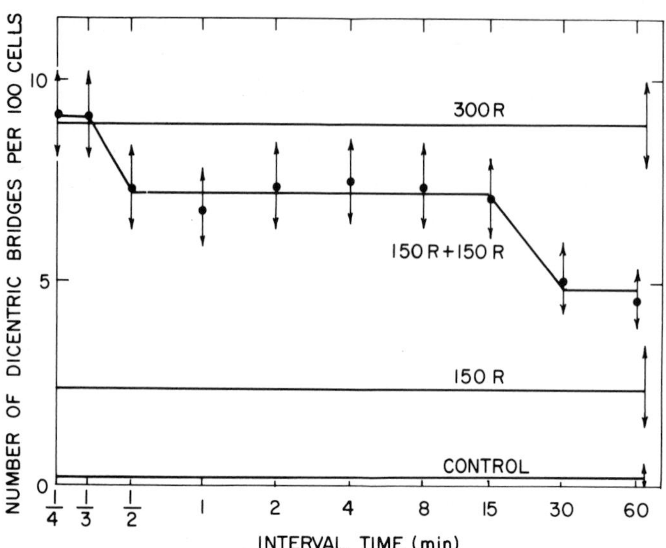

Fig. 7.6. Effects of fractionations of X-ray exposure with intervals of 1/4 to 60 min on the yield of dicentric chromosome bridges in barley root-tip meristems. *Arrows* show fiducial intervals at 5% level (ICHIKAWA et al. 1965)

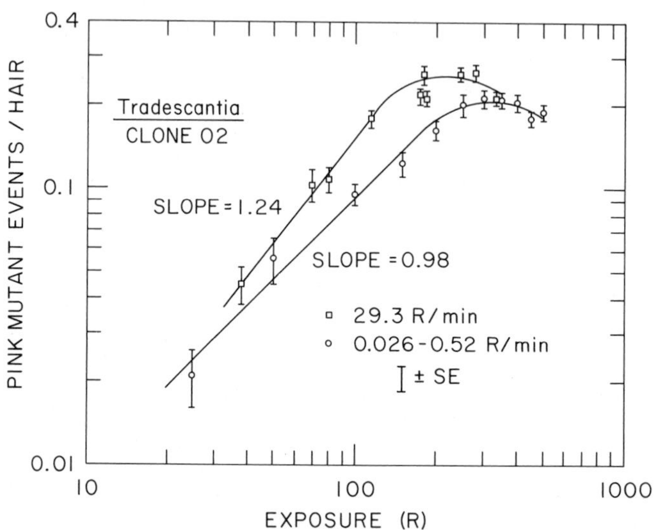

Fig. 7.7. Mutation-response curves for *Tradescantia* clone 02 following exposure to gamma rays delivered at high and low exposure rates (ICHIKAWA et al. 1978)

et al. 1963). It is important, however, that the dose-response curve of somatic mutations in *Tradescantia* stamen hairs has been proved to be linear in the ranges of smaller doses (ICHIKAWA 1971; SPARROW et al. 1972; ICHIKAWA and TAKAHASHI 1977) or at lower dose rates (NAYAR and SPARROW 1967; ICHIKAWA and SPARROW 1968; ICHIKAWA et al. 1978). The linear relationship indicates that predominantly one-hit events occur in cases of irradiation at low dose rates and with smaller doses (ICHIKAWA et al. 1978).

Table 7.4. The RBE (relative biological efficiency) values of various ionizing radiations determined in plants for mutation induction

Radiation	Material	Mutation type	RBE	Reference
14 MeV n	*Triticum* (2*x*) dry seeds	In progeny	10–15	MATSUMURA (1966)
14 MeV n	*Avena* (2*x*) dry seeds	Somatic	24	IKUSHIMA (1972a)
14 MeV n	*Avena* (2*x*) soaked seeds	Somatic	1.7–4.0	IKUSHIMA (1972a)
14 MeV n	Maize dry seeds	Somatic	47–51	SMITH (1967)
14 MeV n	*Arabidopsis* dry seeds	Somatic	15	FUJII (1964)
14 MeV n	*Tradescantia* stamen hairs	Somatic	4.8–5.3	ICHIKAWA (1970a)
4.7 MeV n	Rice dry seeds	In progeny	19	MATSUMURA and MABUCHI (1965)
4.7 MeV n	*Triticum* (2*x*) dry seeds	In progeny	40	MATSUMURA (1966)
1.8 MeV n	Maize dry seeds	Somatic	47–69	SMITH et al. (1964), SMITH (1967)
1.5 MeV n	*Triticum* (2,4*x*) dry seeds	In progeny	25	MATSUMURA (1966)
1.5 MeV n	*Arabidopsis* dry seeds	Somatic	16	FUJII (1964)
1.5 MeV n	Maize dry seeds	Somatic	72	SMITH et al. (1964)
1.25 MeV n	Maize dry seeds	Somatic	64–79	SMITH (1967)
1.0 MeV n	Maize dry seeds	Somatic	71	SMITH et al. (1964)
0.65 MeV n	Maize dry seeds	Somatic	88	SMITH et al. (1964)
0.65 MeV n	*Tradescantia* stamen hairs	Somatic	40	DAVIES and BATEMAN (1963)
0.43 MeV n	Maize dry seeds	Somatic	82–102	SMITH et al. (1964), SMITH (1967)
0.43 MeV n	*Tradescantia* stamen hairs	Somatic	20–24	UNDERBRINK et al. (1971)
0.43 MeV n	*Tradescantia* stamen hairs	Somatic	40–45	KAPPAS et al. (1972)
0.08 MeV n	*Tradescantia* stamen hairs	Somatic	16	UNDERBRINK et al. (1971)
28 GeV p	Maize dry seeds	Somatic	3.5–5.2	SMITH et al. (1974)
$\alpha + {}^7$Li	*Triticum* (2,4*x*) soaked seeds	In progeny	4.7–9.2	IKUSHIMA (1972b)
$\alpha + {}^7$Li	*Triticum* (2,4*x*) soaked seeds	In progeny	29	MATSUMURA et al. (1963)
He ion	*Arabidopsis* dry seeds	Somatic	10	FUJII et al. (1966)
He ion	*Arabidopsis* dry seeds	Somatic	21	HIRONO et al. (1970)
Li ion	*Arabidopsis* dry seeds	Somatic	16	HIRONO et al. (1970)
C ion	*Arabidopsis* dry seeds	Somatic	12	HIRONO et al. (1970)
C ion	*Arabidopsis* dry seeds	Somatic	35	FUJII et al. (1966)
N ion	*Tradescantia* stamen hairs	Somatic	3–14	UNDERBRINK et al. (1973b)
O ion	*Arabidopsis* dry seeds	Somatic	11	HIRONO et al. (1970)
Ne ion	*Arabidopsis* dry seeds	Somatic	1.3	HIRONO et al. (1970)
Ar ion	*Arabidopsis* dry seeds	Somatic	1.6	HIRONO et al. (1970)
Ar ion	*Arabidopsis* dry seeds	Somatic	5	FUJII et al. (1966)

7.3.2.2 Relative Biological Efficiencies of Different Radiations

Qualitatively and quantitatively different biological effects have been observed with different ionizing radiations. The quantitative differences are more readily seen, and the absorbed dose required for producing a certain biological effect differs greatly between different radiations. To express such differences in biological efficiency of radiation, a ratio of absorbed doses of a radiation and of the standard radiation required for producing the same biological effect is taken, and is called relative biological efficiency (RBE, or relative biological effectiveness). Usually, 250 kV X-rays or ^{60}Co (or ^{137}Cs) gamma rays are used as the standard radiation. For example, if 0.01 Gy or 1 rad of a radiation gave the same biological effect as 0.1 Gy or 10 rad of gamma rays, the RBE value of the radiation is 10.

The RBE depends on the linear energy transfer (LET) coefficient of the radiation, becoming maximum when the LET is around 40 to 80 keV/μm. For example, fast neutrons having energy of 2 to 0.4 MeV have LET values of the above range, and thus exhibit the maximum RBE value. Such a relationship is clearly seen in Table 7.4 in which the RBE values determined for various radiations in plant materials are listed.

Some of the RBE values obtained in plants are extremely high as seen in Table 7.4, and they contrast with rather smaller RBE values of at most 10 from animals and microorganisms. However, it is important to note that these extremely high RBE values were mostly obtained from irradiation of dry seeds. Much-reduced RBE values have been obtained in cases of irradiation of soaked seeds or active tissues with higher water contents (ICHIKAWA 1970a, 1970b, 1975; IKUSHIMA 1972a). In the case of irradiation of dry seeds, the RBE values were lowered by increasing the water content higher than 13% or by decreasing it lower than 13% (CONGER and CARABIA 1972). The RBE values are also a function of radiation doses applied, as demonstrated by UNDERBRINK and SPARROW (1974).

7.4 Long-Term Effects of Natural and Man-Made Radiations

7.4.1 Genetic Effects of Low-Level Radiation

The genetic effect of radiation at low levels has been relatively well studied with higher plants rather than with animals and microorganisms. This is solely because two botanical systems, *Tradescantia* stamen hairs heterozygous for flower color and cereal pollen grains possessing suitable starch characters, have been proved to be most excellent test system for such studies.

In particular, the characteristics of the *Tradescantia* stamen-hair system, that is, the capability of detecting all pink mutant cells easily without being concealed by other cells as well as the ease of handling a great number of samples, have proved to be especially suitable for studying the genetic effects of low-level radiation (see UNDERBRINK et al. 1973c).

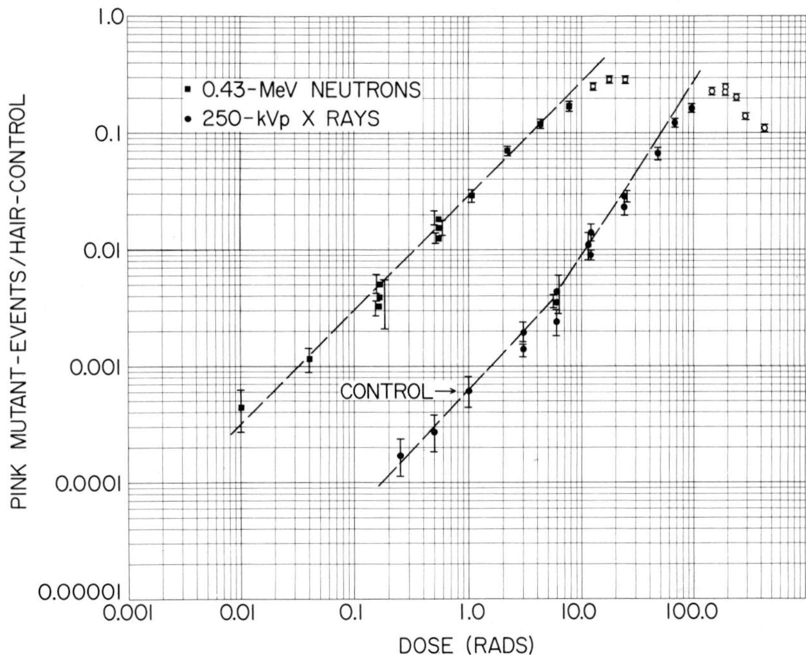

Fig. 7.8. Neutron and X-ray dose-response curves for pink mutant events in the stamen hairs of *Tradescantia* clone 02 in small-dose ranges (Sparrow et al. 1972)

Ichikawa (1971, 1972b, unpublished), performing gamma-ray irradiation experiments of *Tradescantia* at low levels, demonstrated that the somatic pink mutation frequency in the stamen hairs kept a linear relationship with chronic gamma-ray exposure down to 2.1 R and with the scattering radiation exposure down to 0.72 R. Sparrow et al. (1972) further demonstrated that the somatic pink mutation frequency in *Tradescantia* stamen hairs increased linearly with increasing acute X-ray dose in the extremely small-dose range of 0.25 to 6 rad as shown in Fig. 7.8. The mutation frequency with 0.43 MeV fast neutrons was linear down to 0.01 rad. However, increase of mutations in the X-ray dose range over 6 rad was exponential at the relatively high dose rates they applied (Fig. 7.8). Similar dose-mutation relationships at low or relatively low radiation levels have been reported repeatedly (see Ichikawa and Takahashi 1977).

The pollen grains of some cereals are also suitable for studying mutation frequency at low radiation levels, since the mutations of their starch characters can be easily detected by staining with iodine-potassium iodide, and since a very large number of pollen grains can be collected. De Nettancourt et al. (1977) demonstrated the mutation responses at low radiation levels with this pollen system, and argued that mutation induction per unit dose was higher in the range of smaller doses.

7.4.2 Genetic Effects of Natural Radiation

The dose (equivalent) of natural background radiation varies generally from about 70 to 130 mrem yr^{-1} and averages about 100 mrem yr^{-1} for human bodies, of which about 30 and 40 mrem yr^{-1} are external doses from space (cosmic rays) and the earth (from naturally existing radioactive nuclides), respectively. The remaining about 30 mrem yr^{-1} is an internal dose due to uptake of naturally existing radioactive nuclides (predominantly ^{40}K). The dose of cosmic rays increases with elevation, and the radiation dose from the earth differs considerably depending on geological characteristics. The internal dose for plants also varies depending on geological factors, being different from animals with high mobility and especially in contrast to the case of human beings whose foods come from various domestic and world-wide sources, thus resulting in much smaller differences in internal dose.

Some regions in the world show rather higher natural background radiation levels. Such regions are known in Kerala in India, Minas Gerais in Brazil, and Colorado in USA, and the external radiation levels at those regions are as high as 10 to 400 times the average level (70 mrem yr^{-1}).

The genetic effects of such natural background radiations are much better-known for higher plants than for any other organisms. This is mainly due to the immobile nature of plants and also because of the existence of excellent test systems in the plant kingdom. MERICLE and MERICLE (1965) carried out experiments at a region in Colorado, USA, where natural radiation levels were 0.10 to 0.25 mR h^{-1}, and found increases in somatic mutation frequency in *Tradescantia* petals and stamen hairs with only 84 mR total exposure. NAYAR et al. (1970) also investigated the genetic effects of exposure at 0.08 to 1.3 mR h^{-1} from monazite sand (containing ^{232}Th) from Kerala, India, and reported increased mutation frequencies in *Tradescantia* stamen hairs. They also found that contribution of internal exposure in inducing mutations was larger than that of external radiations. Increases of cytological abnormalities were reported by GOPAL-AYENGAR et al. (1970) for five out of seven species examined which grew in a region in Kerala where external radiation level ranged from 0.7 to 1.1 mR h^{-1}, as compared with the same species exposed at 0.02 mR h^{-1}.

7.4.3 Genetic Effects of Increasing Man-Made Radiation

Nuclear explosions have been repeated a great number of times since 1945 in air, water, and underground, scattering into the global environment a large quantity of man-made radioactive nuclides, many of which have not previously existed on the Earth. On the other hand, so-called peaceful uses of nuclear energy have been promoted on a huge scale in the last two decades, producing also a huge amount of man-made radioactive nuclides (although almost all of them are said to be isolated from the environment, a small part is released to the environment constantly and some other part eventually). Uses of radioisotopes in various ways and forms have also increasingly expanded.

Many of such various man-made radioactive nuclides released are long-lived, thus their releases inevitably result in higher environmental radiation level. More important is that some of the man-made radioactive nuclides are concentrated greatly in biological tissues. For example, ^{131}I is known to be concentrated from air into plant tissues with an extremely high concentration factor of $3.5–10 \cdot 10^6$ (MARTER 1963; SOLDAT 1963). Such a high concentration is never observed with natural radioactive nuclides which have been present on the Earth throughout the long evolutionary courses of living organisms, but is specifically conspicuous with man-made radioactive nuclides which no organisms have ever encountered. Therefore, in case of man-made radioactive nuclides, internal exposure becomes more important than external exposure. Nevertheless, most of the environmental monitoring performed near nuclear facilities or on the occasions of nuclear explosions have been measuring only radiation doses in the air which correspond to external exposures for living organisms.

ICHIKAWA and NAGATA (1975) reported significantly increased somatic mutation frequencies in the stamen hairs of *Tradescantia* placed near a nuclear power station. Similar data have been obtained repeatedly from the surrounding areas of other nuclear facilities (ICHIKAWA 1981). At any of the areas, the increases in radiation dose monitored in the air were very small, being at most 9 mR yr^{-1}, but the increases of mutation frequency observed corresponded to those induced by much higher exposures (estimated to be about 150 to 300 mrem). This discrepancy may be understood, at least in part, to be indicative of the importance of internal exposure due to incorporation and concentration of man-made radioactive nuclides.

The data which supported the above consideration were obtained recently by ICHIKAWA and NAGASHIMA (1979), who investigated the somatic mutation frequency in the stamen hairs of *Tradescantia* grown for 76 days in a soil sample taken from the Bikini Island where a hydrogen bomb explosion test had been conducted in 1954. The soil sample contained 186 ± 9 pCi$\cdot g^{-1}$ ^{137}Cs, 1.69 ± 0.12 pCi$\cdot g^{-1}$ ^{60}Co, and some other nuclides. Significantly high mutation frequency was observed for those plants as compared to the controls, and the mutation frequency observed corresponded to that induced with about 500 mrem. On the other hand, the external radiation exposures of the inflorescences measured averaged only 62 mR, though the internal exposures were estimated to be much higher, being about 250 to 500 mrem. The importance of the internal exposures from ^3H and from ^{131}I at low levels was also recently demonstrated by NAUMAN et al. (1979) and TANO and YAMAGUCHI (1979), respectively, in *Tradescantia* stamen hairs.

The genetic effects of low-level man-made radiations have to be studied more intensively, not only by using *Tradescantia* stamen hairs but also by means of pollen grain testing.

References

Atayan RR, Gabrielian JY (1978) The influence of postradiation moisture alteration on biological after-effect in *Crepis* seeds. Environ Exp Bot 18:9–17

Baetcke KP, Sparrow AH, Nauman CH, Schwemmer SS (1967) The relationship of DNA content to nuclear and chromosome volumes and to radiosensitivity (LD_{50}). Proc Nat Acad Sci USA 58:533–540

Beam CA, Mortimer RK, Wolfe R, Tobias C (1954) The relation of radioresistance to budding in *Saccharomyces cerevisiae*. Arch Biochem Biophys 49:100–122

Brewen JG (1964) Studies on the frequencies of chromatid aberrations induced by X-rays at different times of the cell cycle of *Vicia faba*. Genetics 50:101–107

Brewer EN (1979) Repair of radiation-induced DNA double-strand breaks in isolated nuclei of *Physarum polycephalum*. Radiat Res 79:368–376

Caldecott RS, North DT (1961) Factors modifying the radiosensitivity of seeds and the theoretical significance of the acute irradiation of successive generations. In: Mutation and plant breeding. NAS-NRC, Washington DC, pp 365–404

Clowes FAL, Hall EJ (1963) The quiescent centre in root meristem of *Vicia faba* and its behaviour after acute X-irradiation and chronic gamma-irradiation. Radiat Bot 3:45–53

Conger BV, Carabia JV (1972) Modification of the effectiveness of fission neutrons versus ^{60}Co gamma radiation in barley seeds by oxygen and seed water content. Radiat Bot 12:411–420

Davies DR, Bateman JL (1963) A high relative biological efficiency of 650-keV neutrons and 250-kVp X-rays in somatic mutation induction. Nature 200:485–486

Davies DR, Wall ET (1960) Induced mutations at the V^{by} locus of *Trifolium repens*. I. Effects of acute, chronic and fractionated doses of gamma radiation on induction of somatic mutations. Heredity 15:1–15

Deering RA (1968) Radiation studies of *Blastocladiella emersonii*. Radiat Res 34:87–109

De Nettancourt D, Eriksson G, Lindegren D, Puite K (1977) Effects of low doses by different types of radiation on the waxy locus in barley and maize. Hereditas 85:89–100

Ehrenberg L (1955) Factors influencing radiation induced lethality, sterility, and mutations in barley. Hereditas 41:123–146

Ehrenberg L (1959) Radiobiological mechanisms of genetic effects: a review of some current lines of research. Radiat Res Suppl 1:102–123

Evans HJ, Sparrow AH (1961) Nuclear factors affecting radiosensitivity. II. Dependence on nuclear and chromosome structure and organization. Brookhaven Symp Biol 14:101–127

Fujii T (1964) Radiation effects on *Arabidopsis thaliana*. I. Comparative efficiencies of X-rays, fission and 14 MeV neutrons in somatic mutation. Jpn J Genet 39:91–101

Fujii T, Matsumura S (1959) Radiosensitivity in plants. III. Experiments with several polyploid plants. Jpn J Breed 9:245–252

Fujii T, Ikenaga M, Lyman JT (1966) Radiation effects on *Arabidopsis thaliana*. II. Killing and mutagenic efficiencies of heavy ionizing particles. Radiat Bot 6:297–306

Gopal-Ayengar AR, Nayar GG, George KP, Mistry KB (1970) Biological effects of high background radioactivity: studies on plants growing in the monazite-bearing areas of Kerala coast and adjoining regions. Indian J Exp Biol 8:313–318

Haber AH, Randolph ML (1967) Gamma-ray-induced ESR signals in lettuce: evidence for seed-hydration-resistant and -sensitive free radicals. Radiat Bot 7:17–28

Hampl W, Altmann H, Biebl R (1971) Unterschiede und Beeinflussung der Erholungskapazität von Chlorellazellen in verschiedenen Stadien des Zellzyklus. Radiat Bot 11:201–207

Hirono Y, Smith HH, Lyman JT, Thompson KH, Baum JW (1970) Relative biological effectiveness of heavy ions in producing mutations, tumors, and growth inhibitions in the crucifer plant, *Arabidopsis*. Radiat Res 44:204–223

Horsley RL, Fucikovsky LA, Banerjee SN (1967a) Studies on radiosensitivity during the cell cycle in *Oedogonium cardiacum*. Radiat Bot 7:241–246

Horsley RL, Banerjee SN, Banerjee M (1967b) Analysis of lethal responses in *Oedogonium cardiacum* irradiated at different cell stages. Radiat Bot 7:465–476

Howland GP (1975) Dark-repair of ultraviolet-induced pyrimidine dimers in the DNA of wild carrot protoplasts. Nature 254:160

Howland GP, Hart RW, Yette ML (1975) Repair of DNA strand breaks after gamma-irradiation of protoplasts isolated from cultured wild carrot cells. Mutat Res 27:81–87

Ichikawa S (1970a) Relative biological efficiency of 14.1 MeV fast neutrons and ^{137}Cs gamma rays in the stamen hairs of *Tradescantia reflexa* Rafin. Jpn J Genet 45:205–216

Ichikawa S (1970b) Relative biological efficiency of 14.1 MeV fast neutrons and ^{137}Cs gamma rays in the pre-soaked seeds of *Triticum boeoticum* Boiss. and its autotetraploid. Jpn J Genet 45:217–224

Ichikawa S (1970c) Polyploidy and radiosensitivity in higher plants. Gamma Field Symp 9:1–17

Ichikawa S (1971) Somatic mutation rate at low levels of chronic gamma-ray exposures in *Tradescantia* stamen hairs. Jpn J Genet 46:371–381

Ichikawa S (1972a) Radiosensitivity of a triploid clone of *Tradescantia* determined in its stamen hairs. Radiat Bot 12:179–189

Ichikawa S (1972b) Somatic mutation rate in *Tradescantia* stamen hairs at low radiation levels: finding of low doubling doses of mutations. Jpn J Genet 47:411–421

Ichikawa S (1975) The biological effects of heavy particles from the nuclear reaction of ^{10}B(n, α)^{7}Li in diploid wheat seeds containing widely different ^{10}B contents. Ann Rep Res React Inst Kyoto Univ 8:32–43

Ichikawa S (1981) *In situ* monitoring with *Tradescantia* around nuclear power plants. Environ Health Perspect 37:145–164

Ichikawa S, Ikushima T (1967) A developmental study of diploid oats by means of radiation-induced somatic mutations. Radiat Bot 7:205–215

Ichikawa S, Nagashima C (1979) Changes in somatic mutation frequency in the stamen hairs of *Tradescantia* grown in the soil samples from Bikini Island (in Jpn). Jpn J Genet 54:436

Ichikawa S, Nagata M (1975) Increase of mutation rate in *Tradescantia* at the surrounding area of the Hamaoka nuclear power plant (in Jpn). Kagaku 45:417–426

Ichikawa S, Sparrow AH (1966) Polyploidy and radiosensitivity in the genus *Rumex*. Genetics 54:341

Ichikawa S, Sparrow AH (1967) Radiation-induced loss of reproductive integrity in the stamen hairs of a polyploid series of *Tradescantia* species. Radiat Bot 7:429–441

Ichikawa S, Sparrow AH (1968) The use of induced somatic mutations to study cell division rates in irradiated stamen hairs of *Tradescantia virginiana* L. Jpn J Genet 43:57–63

Ichikawa S, Takahashi CS (1977) Somatic mutation frequencies in the stamen hairs of stable and mutable clones of *Tradescantia* after acute gamma-ray treatments with small doses. Mutat Res 45:195–204

Ichikawa S, Ikushima T, Nishiyama I (1965) Two kinds of chromosome rejoinings in X-rayed two-rowed barley. Radiat Bot 5:513–523

Ichikawa S, Sparrow AH, Thompson KH (1969) Morphologically abnormal cells, somatic mutations and loss of reproductive integrity in irradiated *Tradescantia* stamen hairs. Radiat Bot 9:195–211

Ichikawa S, Sparrow AH, Frankton C, Nauman AF, Smith EB, Pond V (1971) Chromosome number, volume and nuclear volume relationships in a polyploid series ($2x$-$20x$) of the genus *Rumex*. Can J Genet Cytol 13:842–863

Ichikawa S, Nauman CH, Sparrow AH, Takahashi CS (1978) Influence of radiation exposure rate on somatic mutation frequency and loss of reproductive integrity in *Tradescantia* stamen hairs. Mutat Res 52:171–180

Ikenaga M, Kondo S, Fujii T (1974) Action spectrum for enzymatic photoreactivation in maize. Photochem Photobiol 19:109–113

Ikushima T (1972a) Relative biological effectiveness of 14.1 MeV neutrons for somatic mutations induced in the diploid oat seeds presoaked in water. Jpn J Genet 47:265–275

Ikushima T (1972b) Relative biological effectiveness of heavy particles from the reaction ^{10}B(n, α)^{7}Li for cytogenetic effects in a ploidy series of the genus *Triticum*. Jpn J Genet 47:401–410

Kappas A, Sparrow AH, Nawrocky MM (1972) Relative biological effectiveness (RBE)

of 0.43-MeV neutrons and 250-kVp X-rays for somatic aberrations in *Tradescantia subacaulis* Bush. Radiat Bot 12:271–281
Konzak CF, Nilan RA, Harle JR, Heiner RE (1961) Control of factors affecting the responses of plants to mutagens. Brookhaven Symp Biol 14:128–157
Krumbiegel G (1979) Responses of haploid and diploid protoplast from *Datura innoxia* Mill. and *Petunia hybrida* L. to treatment with X-rays and a chemical mutagen. Environ Exp Bot 19:99–103
MacKey J (1959) Mutagenic responses in *Triticum* at different levels of ploidy. In: Proceedings of the first international wheat genetics symposium, Winnipeg. pp 89–111
Marter WL (1963) Radioiodine release incident at the Savannah River plant. Health Phys 9:1105–1109
Matsumura S (1965) Relation between radiation effects and dose rate of X- and γ-rays in cereals. Jpn J Genet 40, Suppl: 1–11
Matsumura S (1966) Radiation genetics in wheat. IX. Differences in effects of gamma-rays and 14 MeV, fission and fast neutrons from Po-Be. Radiat Bot 6:275–283
Matsumura S, Mabuchi T (1965) Differences in effects of γ-rays and fast neutrons from Po-Be source on paddy rice. Seiken Ziho 17:37–39
Matsumura S, Kondo S, Mabuchi T (1963) Radiation genetics in wheat. VIII. The RBE of heavy particles from $^{10}B(n, \alpha)^{7}Li$ reaction for cytogenetic effects in einkorn wheat. Radiat Bot 3:29–40
Mericle LW, Mericle RP (1965) Biological discrimination of differences in natural background radiation level. Radiat Bot 5:475–492
Miller MW (1970) The radiosensitivity of three pairs of diploid and tetraploid plant species: correlation between nuclear and chromosomal volume, roentgen exposure and energy absorption per chromosome. Radiat Bot 10:273–279
Miller MW, Sparrow AH (1964) Relationship between nuclear volume and radiosensitivity of different cell types in gemmae of *Marchantia polymorpha* L. Nature 204:596–597
Nauman CH, Underbrink AG, Sparrow AH (1975) Influence of radiation dose rate on somatic mutation induction in *Tradescantia* stamen hairs. Radiat Res 62:79–96
Nauman CH, Klotz PJ, Schairer LA (1979) Uptake of tritiated 1,2-dibromoethane by *Tradescantia* floral tissues: relation to induced mutation frequency in stamen hair cells. Environ Exp Bot 19:201–215
Nayar GG, Sparrow AH (1967) Radiation-induced somatic mutations and the loss of reproductive integrity in *Tradescantia* stamen hairs. Radiat Bot 7:257–267
Nayar GG, George KP, Gopal-Ayengar AR (1970) On the biological effects of high background radioactivity: studies on *Tradescantia* grown in radioactive monazite sand. Radiat Bot 10:287–292
Nishiyama I, Amano E (1963) Radiobiological studies in plants. VIII. Radiosensitivity of plants cultured under phosphorus deficient conditions. Jpn J Genet 38:61–68
Nishiyama I, Ichikawa S, Amano E (1964) Radiobiological studies in plants. X. Mutation rate induce by ionizing radiations at the *al* locus of sand oats. Radiat Bot 4:503–516
Nishiyama I, Ikushima T, Ichikawa S (1966) Radiobiological studies in plants. XI. Further studies on somatic mutations induced by X-rays at the *al* locus of diploid oats. Radiat Bot 6:211–218
Osborne TS, Lunden AO (1964) Seed radiosensitivity: a new constant? Science 145:710–711
Osborne TS, Lunden AO (1965) Prediction of seed radiosensitivity from embryo structure. Radiat Bot 5 Suppl: 131–150
Resnick MA (1975) The repair of double-stranded breaks in chromosomal DNA of yeast. In: Hanawalt PC, Setlow RB (eds) Molecular mechanisms for repair of DNA. Plenum, New York, pp 549–556
Riley R, Miller TE (1966) The differential sensitivity of desynaptic and normal genotypes of barley to X-rays. Mutat Res 3:355–359
Smith HH (1967) Relative biological effectiveness of different types of ionizing radiations: cytogenetic effects in maize. Radiat Res Suppl 7:190–195
Smith HH, Bateman JL, Quastler H, Rossi HH (1964) RBE of monoenergetic fast neutrons: cytogenetic effects in maize. In: Biological effects of neutron and proton irradiation, 2. IAEA, Vienna, pp 233–248

Smith HH, Woodley RG, Maschke A, Combatti NC, McNulty P (1974) Relative cytogenetic effectiveness of 28 GeV protons. Radiat Res 57:59–66

Soldat JK (1963) The relation between ^{131}I concentrations in various environmental samples. Health Phys 9:1167–1171

Soyfer VN, Cieminis KGK (1977) Excision of thymine dimers from the DNA of UV-irradiated plant seedlings. Environ Exp Bot 17:135–143

Sparrow AH (1965) Comparison of tolerances of higher plant species to acute and chronic exposures of ionizing radiation. Jpn J Genet 40 Suppl:12–37

Sparrow AH (1966) Research uses of gamma field and related radiation facilities at Brookhaven National Laboratory. Radiat Bot 6:377–405

Sparrow AH, Evans HJ (1961) Nuclear factors affecting radiosensitivity. I. The influence of nuclear size and structure, chromosome complement, and DNA content. Brookhaven Symp Biol 14:76–100

Sparrow AH, Sparrow RC (1976) Spontaneous somatic mutation frequencies for flower color in several *Tradescantia* species and hybrids. Environ Exp Bot 16:23–43

Sparrow AH, Woodwell GM (1962) Prediction of the sensitivity of plants to chronic gamma irradiation. Radiat Bot 2:9–26

Sparrow AH, Cuany RL, Miksche JP, Schairer LA (1961) Some factors affecting the responses of plants to acute and chronic radiation exposures. Radiat Bot 1:10–34

Sparrow AH, Sparrow RC, Thompson KH, Schairer LA (1965) The use of nuclear and chromosomal variables in determining and predicting radiosensitivities. Radiat Bot 5 Suppl:101–132

Sparrow AH, Rogers AF, Schwemmer SS (1968a) Radiosensitivity studies with woody plants. I. Acute gamma irradiation survival data for 28 species and prediction for 190 species. Radiat Bot 8:149–186

Sparrow AH, Baetcke KP, Shaver DL, Pond V (1968b) The relationship of mutation rate per roentgen to DNA content per chromosome and to interphase chromosome volume. Genetics 59:65–78

Sparrow AH, Schairer LA, Nawrocky MM, Sautkulis RC (1971) Effects of low temperature and low level chronic gamma radiation on somatic mutation rates in *Tradescantia*. Radiat Res 47:273–274

Sparrow AH, Underbrink AG, Rossi HH (1972) Mutations induced in *Tradescantia* by small doses of X-rays and neutrons: analysis of dose-response curves. Science 176:916–918

Stadler LJ (1929) Chromosome number and the mutation rate in *Avena* and *Triticum*. Proc Nat Acad Sci USA 15:876–881

Stadler LJ (1930) Some genetic effects of X rays in plants. J Heredity 21:3–19

Steffensen D (1957) Effects of various cation inbalances on the frequency of X-ray-induced chromosomal aberrations in *Tradescantia*. Genetics 42:239–252

Stein OL, Sparrow AH (1966) The effect of acute irradiation in air, N_2 and CO_2 on the growth of the shoot apex and internodes of *Kalanchoe* cv "Brilliant Star". Radiat Bot 6:187–201

Takahashi CS, Ichikawa S (1976) Variation of spontaneous mutation frequency in *Tradescantia* stamen hairs under natural and controlled environmental conditions. Environ Exp Bot 16:287–293

Tano S, Yamaguchi H (1979) Effects of low-dose irradiation from ^{131}I on the induction of somatic mutations in *Tradescantia*. Radiat Res 80:549–555

Tsunewaki K (1975) Radiological study of wheat monosomics. III. The differential survival rate of 21 monosomics and the disomic to gamma-ray irradiation. Bull Inst Chem Res Kyoto Univ 53:30–42

Underbrink AG, Sparrow AH (1974) The influence of experimental end points, dose, dose rate, neutron energy, nitrogen ions, hypoxia, chromosome volume and ploidy level on RBE in *Tradescantia* stamen hairs and pollen. In: Symposium on biological effects of neutron irradiation. IAEA, Vienna, pp 185–214

Underbrink AG, Sparrow RC, Sparrow AH, Rossi HH (1971) Relative biological effectiveness of 0.43-MeV and lower energy neutrons on somatic aberrations and hair-length in *Tradescantia* stamen hairs. Int J Radiat Biol 19:215–228

Underbrink AG, Sparrow AH, Pond V, Takahashi CS, Kappas A (1973a) Radiation-induced pollen abortion in several Commelinaceous taxa: its relation to chromosomal parameters. Radiat Bot 13:215–227

Underbrink AG, Schairer LA, Sparrow AH (1973b) The biophysical properties of 3.9 GeV nitrogen ions. V. Determinations of the relative biological effectiveness for somatic mutations in *Tradescantia*. Radiat Res 55:437–446

Underbrink AG, Schairer LA, Sparrow AH (1973c) *Tradescantia* stamen hairs: a radiobiological test system applicable to chemical mutagenesis. In: Hollaender A (ed) Chemical mutagens; principles and methods for their detection, Vol. III. Plenum, New York, pp 171–207

Underbrink AG, Sparrow AH, Sautkulis D, Milles RE (1975) Oxygen enhancement ratios (OER's) for somatic mutations in *Tradescantia* stamen hairs. Radiat Bot 15:161–168

Van't Hof J, Sparrow AH (1963) A relationship between DNA content, nuclear volume and minimum mitotic cycle time. Proc Nat Acad Sci USA 49:897–902

Veleminsky J, Gichner T (1978) DNA repair in mutagen-injured higher plants. Mutat Res 55:71–84

Wolff S, Luippold HE (1955) Metabolism and chromosome break rejoining. Science 122:231–232

Wolff S, Luippold HE (1956) The production of two chemically different types of chromosomal breaks by ionizing radiations. Proc Nat Acad Sci USA 42:510–514

Wolff S, Luippold HE (1958) Modification of chromosomal aberration yield by postirradiation treatment. Genetics 43:493–502

Woodwell GM, Sparrow AH (1965) Effects of ionizing radiation on ecological systems. In: Woodwell GM (ed) Ecological effects of nuclear war. Brookhaven Nat Lab, Upton, pp 20–28

Yamagata H, Fujimoto M (1970) Effects of temperature on the induction of somatic mutation by acute gamma radiation exposures in rice plants. Bull Inst Chem Res Kyoto Univ 48:72–77

Yamaguchi H (1974) Mutations with gamma irradiation on dormant seeds of a partially asynaptic strain in rice. Jpn J Genet 49:81–85

Yamaguchi H, Tatara A (1973) A correlation between growth inhibition of the first leaf and mitotic delay after irradiation of soaking barley seeds. Jpn J Breed 23:221–230

Yamaguchi H, Tatara A, Naito T (1975) Unscheduled DNA synthesis induced in barley seeds by gamma rays and 4-nitroquinoline 1-oxide. Jpn J Genet 50:307–318

Yamakawa K, Sekiguchi F (1968) Radiation-induced internal disbudding as a tool for enlarging mutation sectors. Gamma Field Symp 7:19–39

Yamakawa K, Sparrow AH (1965) Correlation of interphase chromosome volume and reduction of viable seed set by chronic irradiation of 21 cultivated plants during reproductive stages. Radiat Bot 5:557–566

Yamakawa K, Sparrow AH (1966) The correlation of interphase chromosome volume with pollen abortion induced by chronic gamma irradiation. Radiat Bot 6:21–38

8 The Aquatic Environment

W.N. WHEELER and M. NEUSHUL

CONTENTS

8.1 Introduction

The plant scientist who seeks to evaluate quantitatively the various environmental factors that influence plant life in aquatic habitats faces a formidable task. For those interested in a more extensive discussion, the marine aquatic environment is introduced in KINNE's *Marine Ecology* (1970, 1971, 1972) vol 1, while HUTCHINSON's *Treatise on Limnology* (1957, 1967) vols 1, 2 covers the freshwater habitat. We will concentrate here on a description of macronutrients, water motion, and irradiance in both fresh and salt water. For a treatment of such factors as salinity, dissolved gases, and micronutrients, the reader is directed to such works as WETZEL (1975), RILEY and SKIRROW (1975) and HILL (1963).

The energy driving the aquatic environment is external. The ultimate source is the sun which heats the water while generating winds which move the water. Irradiance influences the plant through photosynthesis and photomorphogenesis. Water motion affects the transport of mass, heat, and momentum on both a very small (boundary layer) scale and on a very large (planetary) scale.

It is important to describe fluctuations in the energy sources with time since these fluctuations will influence the behavior of plants. It is also important to consider the size spectrum of the plants themselves. In the aquatic environment, plants range in size from nanoplankton and algal spores around 1×10^{-5} m to the giant kelp at 50 m, which is a span of almost 7 orders of magnitude.

8.2 Water Motion

8.2.1 Turbulence

Movement in natural bodies of water is predominantly turbulent. Turbulence is a property of fluid motion that consists of irregular fluctuating movements which are superimposed on the general pattern of flow. These variations in the current velocity are three-dimensional and may be thought of as packets of randomly moving fluid called eddies. These randomly fluctuating vortices can also be thought of as groups of waves covering a large spectrum of wavelengths corresponding to eddy diameters.

Since turbulent fluctuations are the result of interactions with outside energy sources, the eddy diameter depends on the "wavelength" of the energy source. Another property of turbulence is the dissipation of its energy through viscous shear stresses which increase the internal energy of the fluid at the expense of its kinetic energy. The result is a transformation of the energy into ever-decreasing wavelengths. These two properties together define the two ends of a spectrum of diameters with frequencies on the order of days^{-1} (tidal influence) to ms^{-1} and smaller (Fig. 8.1). Of most interest to oceanographers and limnologists are the medium- and long-length frequencies. The medium-length frequences (min^{-1} to h^{-1}) have been correlated with phytoplankton patchiness (THERRIAULT et al. 1978) and implicated in modeling studies (KEMP and MITCH 1979) designed to establish turbulence levels which allow the coexistence of a number of phytoplankton species in the same "niche" or patch as first hypothesized by HUTCHINSON (1961). Long-period fluctuations are due to climatic variability caused primarily by local variation in solar radiation which produces sea surface temperature anomalies (CUSHING and DICKSON 1976; LASKER 1978) and variability in current movements due to fluctuations in winds and climate (LASKER 1978).

Turbulent fluctuations with much smaller frequencies are found in the region of the benthos (bottom-dwelling organisms) where interactions between the water, the bottom, the animals, and the plants have dissipated many of the larger frequencies (GUST and SCHRAMM 1974). These smaller frequencies are important in terms of mass transport and drag which are considered later (Sects. 8.2.4, 8.2.5).

The spatial scale on which turbulence takes place can be directly measured through the diffusive properties of turbulent flow. Here, rate of transport of a substance is defined by how far it moves by eddy diffusion in a given direction

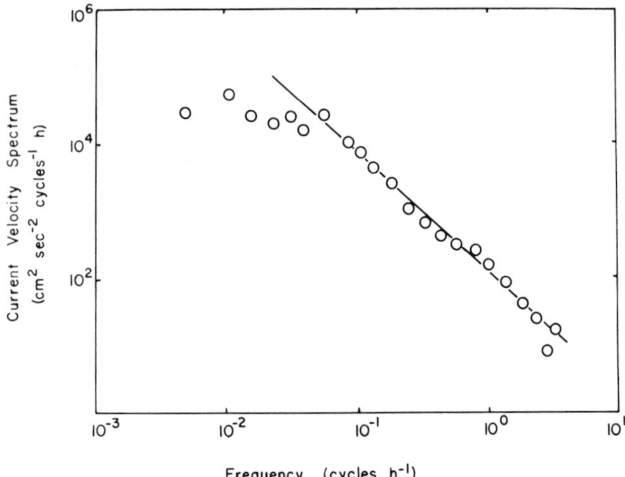

Fig. 8.1. Turbulent power spectral distribution. The spectrum was calculated from the variance of current measurements over varied time scales in St. Margarets Bay, Canada. (After THERRIAULT et al. 1978)

through time. The rate is proportional to the eddy diffusion coefficient. The coefficient is determined through the physical transport of labeled molecules, radionuclides of naturally occurring elements such as Rd, Sr, and Rn, or through macronutrients such as phosphorus.

Although three-dimensional, the scale and intensity of the vertical component of the turbulent eddy diffusion often differ greatly from the horizontal component (BOWDEN 1964), being generally 1–4 orders of magnitude smaller than the corresponding vertical component (BOWDEN 1964; IMBODEN and EMERSON 1978). Values of the coefficient from the Greifensee in Switzerland vary from 0.05 to 0.2 $cm^2 s^{-1}$ in the vertical direction to 100–1,000 $cm^2 s^{-1}$ in the horizontal direction. For comparison, still water in a bottle has a coefficient near 0.25 $cm^2 s^{-1}$.

8.2.2 Mixing

The stability of a water column is a function of two interacting components: (1) the work against gravity produced by the density gradient in the water and (2) the energy supplied to the eddies from the shearing flow (MORTIMER 1974). The Richardson number (Ri) is a nondimensional number reflecting the ratio between these components. Thus,

$$Ri = g(d\rho/dz)/\rho(dU/dz)^2$$

where g is the acceleration due to gravity, $d\rho/dz$ is the density gradient, and ρ is the density and $(dU/dz)^2$ is the energy component in terms of velocity

(U). When the ratio falls below 0.25, the flow becomes unstable and mixing occurs. With a given density gradient, an increase in external energy applied to the system will decrease the ratio with a tendency toward mixing.

Solar radiation warms the surface water, creating a temperature gradient and concomitantly a density gradient. Wind increases the kinetic energy of the system through shear stresses and the production of breaking waves. In temperate regions in the winter, solar irradiance is weak, and its effects in producing a density gradient are easily overcome by winter winds which mix the water (in the ocean to a depth of between 50 and 300 m) and cool the mixed water through convection. In the summer, the reverse occurs, a much higher solar irradiance creates a steep density gradient which the weaker summer winds cannot overcome. This sets up a seasonal discontinuity layer or thermocline.

There is, then, a seasonal variation in the formation and breakdown of the thermocline typical of high and middle latitudes in the ocean and freshwater lakes (MONIN et al. 1977; HUTCHINSON 1957). The depth of this layer increases in winter and decreases in summer. Variability in this pattern with respect to freshwater lakes led HUTCHINSON and LÖFFLER (1956) to a now widely held lake-classification scheme based on the degree of mixing.

Within the mixed layer, phytoplankton populations are for the most part homogeneously distributed (BOUGIS 1976). When the depth of this layer rapidly becomes thinner relative to the phytoplankton compensation depth (photosynthesis = respiration), phytoplankton blooms often occur (SVERDRUP 1953). This is the case in spring, when the mixed layer becomes shallower, and solar irradiance increases, phytoplankton blooms should and do occur (BOUGIS 1976). The sedimentation rate of phytoplankton is also greatly influenced by the depth of the mixed layer (see Sect. 8.2.4).

The mixing processes in the ocean are generally related to large-scale winds which move large bodies of water. Upwelling is one such form of mixing. Upwelling takes place when seasonal winds blow toward the equator along western coasts (California, Peru, and Benguala currents). The stress of the wind on the water combined with Coriolis forces move the water away from the shore. The water is replaced by water which "upwells" from deeper water (50–300 m; PICARD 1963). Away from the coasts, in oceanic habitats, the interactions of currents moving in opposite directions may lead to large-scale upwelling in the open ocean, for example along the equator in the eastern tropical Pacific.

8.2.3 Boundary Layers

The exchange of mass (dissolved gases, nutrients, etc.) as well as heat and momentum between the water and plant surfaces occurs on a very short-length scale. The exchange takes place through a thin layer of fluid adjacent to the surface which is produced as frictional forces act to slow down fluid motion in that region (cf. Chap. 15, this Vol.).

The thickness of this layer is dependent upon chracteristics of the fluid, the flow, and the surface. Because the boundary layer is a region where viscous forces predominate over inertial forces, it is reasonable to assume that the characteristics of the boundary layer are determined by the ratio of these forces. This ratio is known as the Reynolds number (Re),

$$\mathrm{Re} = \rho l/\eta\ U$$

where η is the viscosity and l is the characteristic dimension (e.g., diameter or length) of the surface. Although experimentally determined for only such simple shapes as flat plates, rods, and spheres, it is possible in a general way to guess at the flow characteristics over a surface by calculating the Reynolds number. Experiments (SCHLICHTING 1968) have shown that Reynolds number less than 10^4 for a flat plate (with a zero pressure gradient) indicate laminar flow and boundary layers. Reynolds numbers greater than 10^4 indicate turbulent flows and boundary layers. The transition between laminar and turbulent is, however, critically dependent on the surface roughness (SCHLICHTING 1968). It is not surprising, then, to find that the rugose nature of the blades of the giant kelp *Macrocystis pyrifera* produce turbulent flow with a Reynolds number less than 10^3 (WHEELER 1980).

The surface mentioned above can be either a plant or a part of the substrate. When the latter, it is called the benthic boundary layer. Because of their size, some plants, juveniles, and plant propagules inhabit this region. NEUSHUL (1972) has called this community the boundary layer community. Propagules must sink (sediment) through this region, settle, and attach (SAWADA et al. 1972; CHARTERS et al. 1973). Nutrients which are regenerated in the sediments (Sects. 8.2.6, 8.2.7) must diffuse through the benthic boundary layer to the waters above.

8.2.4 Drag

At very low Reynolds numbers (<1), a laminar boundary layer persists over the entire surface of a plant (substrate) without separation. The frictional force of the moving fluid against the surface (drag) is, in this case, entirely due to skin-friction (viscous drag).

Stokes' law can be modified to calculate the intrinsic terminal sinking rate (U_t) of particles the size and shape of phytoplankton or plant propagules due to viscous drag. Here,

$$U_t = (2/9)\ g r^2 (\Delta\rho)/\eta\,\Theta$$

where g is the acceleration due to gravity, r is the radius of the particle and Θ is a shape factor (HUTCHINSON 1967). Physiological state (fat accumulation and the ionic composition within the cell vacuole) and the change in physiological state with age play large roles in the modification of ($\Delta\rho$) and thus the intrinsic sinking rate (SMAYDA 1970). For particulates, surface coatings, electrolytic forces, and dissolved organic substances may reduce drag – giving sinking rates an order of magnitude greater than Stoksian predictions (CHASE 1979). Aggregations

Table 8.1. Intrinsic sinking rates of a number of marine plants and spores

Species	Velocity $m\,s^{-1} \times 10^{-6}$	Reference
Agardhiella tenera	116	COON et al. (1972)
Callophyllis flabellulata	15	COON et al. (1972)
Cryptopleura violacea	97	COON et al. (1972)
Gelidium robustum	43	COON et al. (1972)
Myriogramme spectabilis	59	COON et al. (1972)
Anabaena flos-aquae (straight form)	3.9–4.5	BOOKER and WALSBY (1979)
Anabaena flos-aquae (helical form)	4.5–8.1	BOOKER and WALSBY (1979)
Asterionella japonica	5	Cited in SMAYDA (1970)
Chaetoceros curvisetus	12–29	Cited in SMAYDA (1970)
Coccolithus huxleyi	15	Cited in SMAYDA (1970)
Coscinodiscus concinnus	708	Cited in SMAYDA (1970)
Dunaliella tertiolecta	2	Cited in SMAYDA (1970)
Leptocylindricus danicus	0.9–5	Cited in SMAYDA (1970)
Monochrysis lutheri	5	Cited in SMAYDA (1970)
Noctiluca miliaris	300	Cited in SMAYDA (1970)
Phaeodactylum tricornutum	0.02–0.05	Cited in SMAYDA (1970)

of particulates, propagules, or plankton may also alter the intrinsic sinking rate (SMAYDA 1970; REYNOLDS 1979; BOOKER and WALSBY 1979). The intrinsic sinking rate may or may not be related to the effective (actual) sinking rate (Table 8.1). REYNOLDS (1979) found good agreement between Stoksian predictions and effective sinking rates of *Lycopodium* spores only through a highly stratified lake enclosure. Turbulence within the mixed layer slowed the effective sinking rate. According to REYNOLDS (1979) the greater the mixed depth, the slower the effective sinking rate or,

$$dP/dt = -U_t/z_m$$

where dP/dt is the loss in population (spores or phytoplankton) with time, and z_m is the depth of the mixed layer.

As Reynolds numbers increase beyond a few hundred, flow separation occurs, and the drag becomes proportional to the square of the velocity and the cross-sectional area (pressure drag). Thus, the drag force (F_d) can be expressed by

$$F_d = (C_d/2)\, U^2 A.$$

Here, A is the cross-sectional area of the plant normal to the flow and C_d is the drag coefficient. Using measurements of the drag of the kelps, *Eisenia* (CHARTERS et al. 1969) and *Nereocystis* (KOEHL and WAINWRIGHT 1977), it is possible to estimate the stipe breakage velocities. This gives an indication of the forces that can be found in the near shore and intertidal regions. Where these plants occur, *Eisenia* would have a breakage velocity of $9\ m\,s^{-1}$ and *Nereocystis* $3\ m\,s^{-1}$. These plants are not adapted to very rough exposed intertidal areas where waves can generate calculated velocities of $14\ m\,s^{-1}$ (JONES and

Table 8.2. Hydrodynamic properties of some marine macrophytes

Species	Breaking stress ($MN\ m^{-2}$)	Modulus of elasticity ($MN\ m^{-2}$)	Reference
Ascophyllum nodosum	1.48	14.1	DELF (1932)
Eisenia arborea		12.5–24.1	CHARTERS et al. (1969)
Fucus serratus	4.21	65.0	DELF (1932)
Halydris siliquosa		53.4	DELF (1932)
Laminaria digitata	0.93	12.7	DELF (1932)
Nereocystis luetkeana	3.64	13–30	KOEHL and WAINWRIGHT (1977)

DEMETROPOULOS 1968). Mechanical adaptations such as high moduli of elasticity (DELF 1932; Table 8.2) and a tendency to clump or otherwise become modified in form enables these plants to survive in subtidal areas where although 14 m s^{-1} velocities are rare, breakage point velocities do occur with relatively high frequency.

8.2.5 Mass Transport

On the other end of the water velocity scale, stagnation, and small water velocities can create diffusion stresses. Under these conditions, Fick's first law governs the rate of diffusion of metabolites to and from plant surfaces. It can be described as

$$J_j = D_j(\Delta C_j)/\Delta X$$

where J_j is the flux of the diffusing molecule, j, D_j is the diffusion coefficient of species j, (ΔC_j) is the concentration gradient and ΔX is the diffusion boundary layer thickness, which is somewhat smaller than the physical boundary layer thickness.

Fick's law can be simplistically modified to work under more realistic conditions in nature. The diffusion coefficient D_j is replaced by another D_j which is the eddy diffusion coefficient (or turbulent diffusion coefficient).

When the flux, J_j, is smaller than the plant's ability to take up species j, the plant is said to be under a diffusion or mass transport stress. Phytoplankton, although small in size, do experience mass transport stresses. MUNK and RILEY (1952) first called attention to the fact that various-shaped phytoplankters would sink with variable speeds, influencing their mass transport abilities. Since then, the problem of mass transport (carbon, nitrogen, phosphorus, and other metabolic molecules) to and from phytoplankton cells has been well documented (GAVIS 1976).

Larger aquatic plants can also face severe mass transport stresses under low water motion conditions. WHEELER (1980) has demonstrated that the giant kelp, *Macrocystis,* encounters such stresses when currents over the fronds are less than 6 cm s^{-1}. WESTLAKE (1967) has shown such stresses for river plants in flows less than 1 cm s^{-1}, while LOCK and JOHN (1979) found 5 cm s^{-1} a critical speed for river periphyton phosphate uptake.

8.2.6 Nitrogen

Inorganic macronutrients are distributed throughout the aquatic environment by the movement of water. However, the transformation of these molecules within the aquatic environment is the result of biological cycling (see Chapt. 2, Vol. 120). The period and amplitude of these cycles varies depending on the scale being considered. For instance, on a global scale, STEVENSON (1972) has estimated that 1.7×10^{20} g of nitrogen (N) are present on Earth. Of this, 97.6% belongs to the lithosphere, slightly less than 2.3% to the atmosphere, and the remainder to the hydrosphere and biosphere. The approximately 3.9×10^{18} g N in the atmosphere are in the elemental form, which is not directly available to plants. Only a few prokaryotes can fix N_2. In spite of this fact, organisms fix 2.2–2.3×10^{14} g of N yr^{-1} (SÖDERLUND and SVENSSON 1976) of which somewhere between 0.09 and 37% is fixed in aquatic environment.

N-fixation by blue-green algae and bacteria occurs in anaerobic environments, either within the plants themselves (heterocysts) or in anaerobic areas surrounding the plants (CARPENTER and PRICE 1976). Values range from 65.7 g N m^{-2} yr^{-1} (WIEBE et al. 1975) for blue-green algal mats on Eniwetok Atoll to 7×10^{-4} g N m^{-2} yr^{-1} in the Sargasso Sea (CARPENTER and MCCARTHY 1975).

Although large quantities of N are fixed, this accounts for only 3–10% of the N utilized by plants. Recycling accounts for the other 90–97% made available for plant production (SÖDERLUND and SVENSSON 1976). Thus, N turnover rates can be important indicators of plant production. On a global scale, N has been estimated to turn over in between 1 to 20 days.

Within freshwater and coastal communities most of the regeneration takes place in the sediments. Regeneration rates from sediments vary, the N being turned over in some small ponds at the rate of 67% d^{-1} (SUGIYAMA and KAWAI 1978) to 0.7% d^{-1} in the Baltic (HALLBERG et al. 1976).

Nitrogen contributions from zooplankton vary, being dependent on the amounts already present in the water. In an estuary near Beaufort, North Carolina, U.S.A., ammonium excretion by zooplankton accounted for 16% of the utilizable N (SMITH 1978). In oligotrophic (nutrient-poor) environments, excretion can account for as much as 90% (JAWED 1973).

Within the ocean most of the regeneration takes place in the top 200 m of the water column. Fecal pellets or marine snow produced in the upper layer are micro-habitats for bacterial regeneration of N within the upper layers. These micro-habitats can produce micro-patches of available N where traditional methodology would show no nutrients. Such micro-scale patches are immediately exploited by the phytoplankton, contributing to phytoplankton patchiness (MCCARTHY and GOLDMAN 1979).

Inorganic nitrogen is made available through fixation, mineralization, and excretion. It is removed from the environment through denitrification, assimilation, and sedimentation. Denitrification is a microbial process. Rates range from 4.6 μg N m^{-2} yr^{-1} in a Danish fjord (OREN and BLACKBURN 1979) to 8–16 mg N m^{-2} d^{-1} in a polluted river (NAKAJIMA 1979).

Plants assimilate inorganic N primarily as nitrate (NO_3^-). SÖDERLUND and SVENSSON (1976) and BOUGIS (1976) estimate that 12 and 30 g N m^{-2} yr^{-1} respectively are assimilated from the ocean alone. Ammonium may also be sorbed by the sediments and not released to the interstitial water under anoxic conditions (HALLBERG et al. 1976; ROSENFELD 1979).

Laboratory studies and field correlations indicate that N is the most limiting nutrient (DUGDALE 1967; RYTHER and DUNSTAN 1971; EPPLEY et al. 1979) in the coastal marine environment. This is, perhaps, the reason that some marine macrophytes have been shown to store N (CHAPMAN and CRAIGIE 1977) and others have been found in symbiotic relationships with N fixers (HANSON 1977).

8.2.7 Phosphorus

In freshwater communities, phosphorus (P) rather than N is generally limiting. In contrast to N which is primarily in the atmosphere, inorganic P is found primarily in the lithosphere. In natural waters, P occurs in solution in both inorganic (soluble reactive P: SRP) and organic forms (soluble unreactive P: SUP), as well as adsorbed to organic, colloidal, and inorganic particles. Although SRP is for the most part orthophosphate, SRP levels can be as high as six times the orthophosphate levels (LEAN and CHARLTON 1976).

SUP can be transformed to SRP either by the plants themselves through hydrolysis via alkaline phosphatase or via photodegradation with UV radiation (FRANCKO and HEATH 1979; MORSE and COOK 1978). Phosphorus in sediments is present mainly as apatite or in association with free cations (usually iron and calcium). Under anoxic conditions phosphates are released from sediments (MORTIMER 1942; MARTENS et al. 1978).

Under aerobic conditions, the sediment may act as a sink for P. Phytoplankton and macrophytes are also a temporary sink for P. They assimilate to a large degree only orthophosphate, although in special circumstances other forms are also assimilated (KUHL 1974). Phosphorus assimilation can be in quantities far greater than amounts immediately required by plants. The excess P is either stored or excreted as unreactive polyphosphates.

The turn-around time between assimilation and excretion and reassimilation can be on the order of seconds for some phytoplankton (POMEROY 1960). Because of its fast turnover rate and general scarcity in freshwaters (HUTCHINSON 1957, 1967), it is generally held that P is the most limiting nutrient in freshwaters although inorganic carbon has also been implicated (SCHINDLER et al. 1972). Adding (loading) phosphates to freshwaters usually results in increased biomass production (SCHINDLER 1971). Phosphate loading of natural waters occurs through introduction of man-made detergents, fertilizers, and sewage. This eutrophication process results in tremendous blooms of phytoplankton and macrophytes (see SCHINDLER 1971). Thus, changes in physical factors can bring about large-scale responses from the associated flora.

8.2.8 Plant Response

Because the interactions between water and plant determine survival and influence plant function, it is not surprising to find relationships between the type of water motion in a given habitat and plant morphology. Thus, phytoplankton shape (SMAYDA 1970) and blue-green algal colony shape (BOOKER and WALSBY 1979) have been correlated with intrinsic sinking velocity. The blades of the giant kelp, *Macrocystis*, produce turbulent boundary layers in nonturbulent water flow, which, by decreasing the effective boundary layer thickness, enhances mass transport (WHEELER 1980). High and low water motion morphologies are common in macrophyte habitats (NORTON 1969; GERARD and MANN 1979), as well as different morphologies for submerged and emerged leaves (see SCULTHORPE 1967). Depending on the size of the plant, different morphologies, even in the same location, are a response to different hydrodynamic habitats. NEUSHUL (1972) has classified marine macrophytes according to their hydrodynamic environment, and ALEYEV (1976) has classified the plankton according to size and hydrodynamic environment.

8.3 Irradiance

Solar radiation is the source of energy utilized by plants in the manufacture of complex organic substances. The degree of availability of this energy directly controls the amount of organic matter synthesized. The energy that is not utilized by photosynthesis contributes to the potential and kinetic energy of the aquatic environment.

This radiant energy is carried by electromagnetic waves, and may be described in terms of its spectral composition. Visible light is the radiant energy within the wavelength range from 350 to 750 nm. The range utilized by plants for photosynthesis, photosynthetically active radiation (PAR), is restricted to the range from 350 to 700 nm. (For detailed discussion of photosynthetically active radiation see Chapt. 2, this Vol.)

8.3.1 Definitions and Units

The quantity of energy transferred by radiation (in Joules, J) is represented by the symbol Q. Radiant flux (F), dQ/dt, is defined as the time rate of flow of radiant energy (measured in watts, W). The irradiance, E(z) at a depth z is the total radiant flux per unit area incident on an element of surface (the flux from the hemisphere covering the element's surface). $E_d(z)$ is the downwelling (downward) irradiance (flux incident per unit area measured on a horizontally oriented surface facing upward). The upwelling (upward) irradiance $E_u(z)$ is the flux per unit area measured on the downward facing side of the surface.

While the SCOR working group (TYLER 1974) has recommended the use of the above definitions for oceanographers, photobiologists have defined the energy incident on a given surface area as energy fluence (F; $J\,m^{-2}$). The time rate of flow of the energy fluence is the energy fluence rate and has the units of $W\,m^{-2}$. These terms can also be defined in terms of quanta, e.g., photon fluence or photon fluence rate (RUPERT 1974), as discussed in detail in Chapter 2 of this Volume.

Rough conversions can be made between the radiometric and quantum aspects of irradiance. MOREL and SMITH (1974) found the ratio, Q/E (total quanta/total energy) remained constant above water at 2.77×10^{18} quanta $s^{-1}\,W^{-1}$, regardless of sun elevation (above 22°) and meteorological conditions. Below the surface the Q/E ratio exhibited greater variability ($\pm 10\%$), but was predictably dependent on the optical properties of the water. The average Q/E for natural waters is $2.5 \pm 0.25 \times 10^{18}$ quanta $s^{-1}\,W^{-1}$. For Lake Kinneret, DUBINSKY and BERMAN (1979) further divided this into three water types characterized by the plant pigment in them (abundant chlorophyll, low chlorophyll, and abundant peridinin, a red pigment of dinoflagellates) with respective Q/E's of 2.7–2.8, 2.5–2.6 and 2.96×10^{18} quanta $s^{-1}\,W^{-1}$. For waters with high gilvin (yellow substance, see Sect. 8.3.2) content, SPENCE et al. (1971) found 3.07×10^{18} quanta $s^{-1}\,W^{-1}$.

8.3.2 Absorption

Water itself absorbs radiant energy throughout the visible spectrum, absorption being the strongest in the red wavelengths. Seasalts have little influence (MOREL 1974) and so the absorption of light in pure seawater, as well as in distilled water, is minimal in the blue and maximal in the red region. Light is absorbed in aquatic habitats by pigmented phytoplankton, particulate matter (JERLOV 1976) and dissolved substances (KIRK 1976b). In many inland and coastal waters a yellow substance called gilvin or gelbstoff can be present in large quantities, influencing the spectral irradiance immensely (KIRK 1976b). Gilvin is thought to be composed of humic substances that are found in waters run-off from land or phenolic compounds released from algae (SIEBURTH and JENSEN 1970).

8.3.3 Scattering

Scattering does not in itself attenuate light, but increases the optical pathlength followed by the light, thereby increasing absorption. The scattering of wavelengths by water and by small particulates is generally dependent upon the reciprocal of the fourth power of the wavelength. However, phytoplankton and larger particulates scatter light almost independently of wavelength (see discussion in JERLOV 1976).

In the clearest natural waters scattering plays a minor role, but there are minor differences between fresh and salt water. The scattering coefficient, which is 30% higher in seawater (35–39‰) appears to be related to the presence of dissolved ions in seawater (MOREL 1974). Particulates in clear natural waters are present only in small quantities (0.02–0.17 mg m^{-3}; JERLOV 1976). Phytoplankton are therefore the major contributor to scattering and absorption. Light absorbed by phytoplankton and gilvin virtually eliminates blue light from deeper natural waters. The red end of the spectrum is absorbed by the water itself, creating a situation where, in deeper waters, the irradiance spectrum is nearly monochromatic (JERLOV 1976).

8.3.4 Attenuation Coefficient, K

The attenuation of radiant energy in natural water has been found to obey the Bouguer-Beer law:

$$dE = -K\,E\,dz$$

where K is the attenuation coefficient. If K is constant with depth (assuming a mixed water column) then, the irradiance at depth z [E(z)] can be defined as

$$E(z) = E(o)\,e^{(-Kz)} \text{ or } K_d(z) = [\ln E(z) - \ln E(o)]/z$$

where E(o) is the energy incident on the surface of the water and K_d is the diffuse attenuation coefficient measured in the vertical direction. Another way to measure attenuation is through beam transmission. Here the energy of a beam of light is measured before and after being transmitted through a fixed distance of water,

$$T = e^{(-Cz)}.$$

C is a total of the absorption of the water, particulates, and dissolved substances but only a portion of the scattered energy. That energy which is scattered away from the beam's path is not measured. The beam transmittance coefficient, C, and the K_d can be related by including the forward scattering.

8.3.4.1 Spectral Variation in K_d

The range of water conditions produced by variable amounts of particulates and gilvin has led JERLOV to propose a system to classify a body of water according to its transmission or attenuation spectrum (JERLOV 1976, 1977; Fig. 8.2). In this system, there are five types of oceanic water characterized by transmission windows in the blue with corresponding high transmission coefficients approaching those of pure water (Fig. 8.2). These are types I, Ia, Ib, II, and III. For areas with increasing gilvin and particulates, he defined nine gradations of coastal water, from the most transparent at 1 to the least at 9 (Fig. 8.2). There appears to be little difference between fresh and salt water. The clearest water measured to date is that of Crater Lake (K_d=0.037 m^{-1};

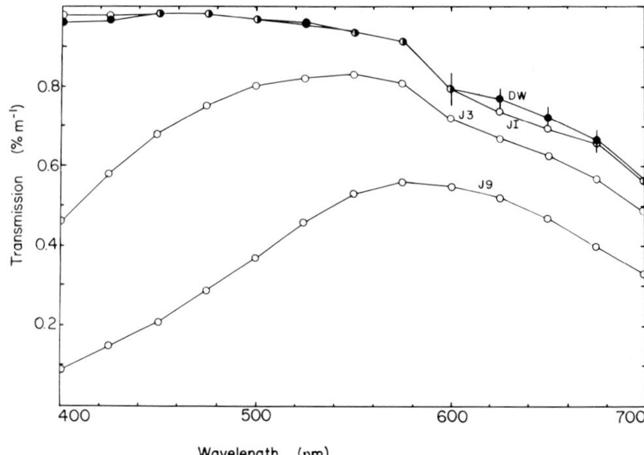

Fig. 8.2. Transmission spectra for a variety of marine water types and distilled water. *DW* distilled water; *JI* clearest oceanic water; *J3, J9* two types of coastal water. *Vertical bars* represent the range of values measured by different authors. (After JERLOV 1976)

SMITH et al. 1973) and the least clear is that from an East African lake (K_{540} = 100 m^{-1}; MELACK and KILHAM 1974).

8.3.4.2 Temporal Variation in K_d

The temporal variation in the K_d can be quite large. Both the K_d and the spectral distribution of underwater irradiance are dependent on changes in the pathlength of the light as the sun's angle changes, and on phytoplankton patchiness and wind-related processes (CLARKE 1938; KAIN 1971; KIRK 1977; HARGER 1979; LÜNING and DRING 1979). Runoff during storms in winter, as well as wind-related mixing, can drastically change the near-shore underwater light climate. HARGER (1979) has demonstrated that the submarine irradiance in a kelp forest (15 m deep) can change by orders of magnitude within 1 day. These variations can and in many cases do exceed the range observed seasonally. In the winter months the irradiance was correlated more with hydrographic conditions than with atmospheric ones as was the case in summer. LÜNING (1971) related loss of water clarity to winds over the Beaufort scale 6 (10.8–13.8 m s^{-1}). Changes in tidal height with the above factors caused variations as much as 250 times in the daily irradiance near the Isle of Man (KAIN 1971). BINDLOSS (1976) and DUBINSKY and BERMAN (1979) in seasonal studies on freshwater lakes found that phytoplankton blooms were the most important factors determining underwater irradiance. In more open waters where hydrographic conditions are much more uniform, seasonal changes in submarine irradiance seems less noticeable (POOLE and ATKINS 1929), although little data exists for this region.

8.3.4.3 Spatial Variation in Irradiance

Irradiance levels can change drastically with the microhabitat of an organism. Sandy bottoms reflect more irradiance (BRAKEL 1979). Large boulders, crevasses or reef structure in general, and vegetation can greatly modify the light impinging on a given surface area (ERNST 1957; FORSTNER and RÜTZLER 1970).

Vegetation can change not only the irradiance, but the spectrum as well. Phytoplankton in concentrations greater than 10 mg m^{-3} can significantly alter the underwater irradiance and spectra (TALLING 1960). Size and shape of the phytoplankter also play a significant role in their light-absorbing characteristics (KIRK 1976a). Canopies of the giant kelp, *Macrocystis* can cut the irradiance by 2 orders of magnitude, depending on the thickness of the surface canopy (NEUSHUL 1971), much in the same way that irradiance in forests is attenuated (TASKER and SMITH 1977). However, in the marine environment, pigments other than chlorophyll a can be dominant. Fucoxanthin, a brown pigment, absorbs primarily in the blue-green, while phycoerythrin and phycocyanin absorb mainly in the green and blue-green region respectively. Little work is available dealing with this effect on undergrowth algae.

8.3.5 Plant Response

The distribution of red and blue light in the aquatic environment may be more important than previously considered. Research on phytochrome and blue light responses may apply to aquatic plants. LÜNING and DRING (1975) have noticed blue light photoperiodic responses in some brown algae, while MÜLLER and CLAUSS (1976) have demonstrated other photomorphogenetic responses in brown algae. So far, these effects have been found to be limited to the Phaeophyta, but investigations with other divisions are continuing.

The levels of irradiance necessary to induce photomorphogenetic effects, or to sustain growth have been generally considered to lie in the former case below the 1% level of surface irradiance and the latter case above the 1% level. However, recent research (LÜNING and DRING 1979) has shown that many species of algae have growth compensation points between 0.01–0.05%. Red algal crusts have been found in the sea at depths representing 0.01% surface irradiance calculated over a year's time. Kelps such as *Laminaria hyperborea* have been shown exist to depths representing 0.5% in many places around the world. Because their pigmentation is complementary to most green coastal waters, red algae have been postulated to have evolved as a deep water division (ENGELMANN 1883). In contrast, however, red algae do equally as well in the high intertidal zone. Although the pigmentation of an alga confers an advantage in the absorption and conversion of incident energy of specific wavelengths, recent research has shown that factors such as morphology (RAMUS 1978) and internal organization of the photosynthetic unit (RAMUS et al. 1977; PRÉZELIN 1976) play much larger roles. Responses to light in aquatic plants are discussed in detail in the following Chapter of this Volume.

8.4 Concluding Remarks

We have attempted in this chapter to provide a dynamic view of the aquatic environment. To emphasize this, we have considered the manifold effects of water motion on the aquatic plant community. The dynamic aspect of the environment can be thought of as being periodic – in the sense that the energy sources of the aquatic environment manifest themselves in cycles. The cycles or period of the energy input vary on a temporal scale from milliseconds to centuries, and spatially from microns to planetary in scope. An understanding of the vegetation and its productivity necessitates, then, an understanding of the periodic behavior of the energy source. In this review we have tried to define a number of the more important cycles influencing the ecosystem and associated flora.

As botanists, we can see certain periodic phenomena (variability) within the plant ecosystem (variability in population dynamics, growth rates, productivity, etc.) for which we cannot identify any single physical factor. This is the result of plant behavior. Aquatic plants and plants in general have been shown to be integrators of environmental variables (Evans 1972) and further, to respond to changes in environmental factors with a certain lag phase (Doty 1971; Cushing and Dickson 1976). Thus, storage of essential nutrients by many of the aquatic plants, as well as the effect of physical factors in influencing the distribution of plant propagules, determine the behavior of an individual or a plant population at a later date. Although it is important to identify the driving factors, it is almost impossible to significantly correlate plant behavior with fluctuating environmental variables (Evans 1972). An understanding of the variability within the system, however, can only come about from an elucidation of the driving function(s).

The aquatic botanist should be aware then, of the problem of scale. Spatial, temporal, and energy scales interact to define a given set of circumstances to which the plant must respond. The nature of this response is the subject of intensive research.

Acknowledgment. The authors would like to thank W. Anikouchine, A.C. Charters, D. Coon, S. MacIntyre, M. Polne, and J. Woessner for critically reading the manuscript.

References

Aleyev YG (1976) Biohydrodynamics and ecology of life forms of pelagial. Int Rev Ges Hydrobiol 61:137–147

Bindloss ME (1976) The light climate of Loch Leven, a shallow Scottish lake, in relation to primary production by phytoplankton. Freshwater Biol 6:501–518

Bowden KF (1964) Turbulence. Oceanogr Mar Biol Ann Rev 2:11–30

Bougis P (1976) Marine plankton ecology. Elsevier/North-Holland Amsterdam, New York

Booker MJ, Walsby AE (1979) The relative form resistance of straight and helical blue-green algal filaments. Br Phycol J 14:141–150

Brakel WH (1979) Small-scale spatial variation in light available to coral reef benthos: quantum irradiance measurements from a Jamaican reef. Bull Mar Sci 29:406–413

Carpenter EJ, McCarthy JJ (1975) Nitrogen fixation and uptake of combined nitrogeneous nutrients by *Oscillatoria (Trichodesmium) thiebautii* in the western Sargasso Sea. Limnol Oceanogr 20:389–401
Carpenter EJ, Price CC (1976) Marine *Oscillatoria (Trichodesmium)*: Explanation for aerobic nitrogen fixation without heterocysts. Science 191:1278–1280
Chapman ARO, Craigie JS (1977) Seasonal growth in *Laminaria longicruris*: relations with dissolved inorganic nutrients and internal reserves of nitrogen. Mar Biol 40:197–205
Charters AC, Neushul M, Barilotti C (1969) The functional morphology of *Eisenia arborea*. Proc Int Seaweed Symp 6:89–105
Charters AC, Neushul M, Coon D (1973) The effect of water motion on algal spore adhesion. Limnol Oceanogr 18:884–896
Chase RRP (1979) Settling behaviour of natural aquatic particles. Limnol Oceanogr 24:417–426
Clarke GL (1938) Seasonal changes in the intensity of submarine illumination off Woods Hole. Ecology 19:89–106
Coon DA, Neushul M, Charters AC (1972) The settling behavior of marine algal spores. Proc Int Seaweed Symp 7:237–242
Cushing DH, Dickson RR (1976) The biological response in the sea to climatic changes. Adv Mar Biol 14:1–122
Delf EM (1932) Experiments with the stipes of *Fucus* and *Laminaria*. J Exp Biol 9:300–313
Doty MS (1971) Antecedent event influence on benthic marine algal standing crops in Hawaii. J Exp Mar Biol Ecol 6:161–166
Dubinsky Z, Berman T (1979) Seasonal changes in the spectral composition of downwelling irradiance in Lake Kinneret (Israel). Limnol Oceanogr 24:652–663
Dugdale RC (1967) Nutrient limitation in the sea: dynamics, identification and significance. Limnol Oceanogr 12:685–695
Engelmann TW (1883) Farbe und Assimilation. Bot Ztg 41:18–29
Eppley RW, Renger EH, Harrison WG (1979) Nitrate and phytoplankton production in southern California coastal waters. Limnol Oceanogr 24:483–494
Ernst J (1957) Studien über die Seichtwasser-Vegetation der Sorrentiner Küste. Pubbl Stn Zool Napoli 30: (suppl) 470–518
Evans GC (1972) The quantitative analysis of plant growth. Univ of California Press, Los Angeles, Berkeley
Forstner H, Rützler K (1970) Measurements of the micro-climate in littoral marine habitats. Oceanogr Mar Biol Ann Rev 8:225–249
Francko DA, Heath RT (1979) Functionally distinct classes of complex phosphorus compounds in lake water. Limnol Oceanogr 24:463–473
Gavis J (1976) Munk and Riley revisited: nutrient diffusion transport and rates of phytoplankton growth. J Mar Res 34:161–179
Gerard V, Mann KH (1979) Growth and production of *Laminaria longicruris* (Phaeophyta) populations exposed to different intensities of water movement. J Phycol 15:33–41
Gust G, Schramm W (1974) Erfassung der Strömungsstruktur im Substratbereich mariner Biotope. Mar Biol 26:365–367
Hallberg RO, Bågander LE, Engvall A-G (1976) Dynamics of phosphorus, sulfur and nitrogen at the sediment-water interface. In: Nriagu JO (ed) Environmental biogeochemistry Vol 2, pp 295–308
Hanson RB (1977) Pelagic *Sargassum* community metabolism: carbon and nitrogen. J Exp Mar Biol Ecol 29:107–118
Harger BWW (1979) Coastal oceanography and hard substrate ecology in a Californian kelp bed. PhD thesis, Univ of California, Santa Barbara
Hill MN (ed) (1963) The sea. Vol I, II, Wiley and Sons, New York
Hutchinson GE (1957) A treatise on limnology. Vol. I. Geography, physics and chemistry. Wiley and Sons, New York
Hutchinson GE (1961) The paradox of the plankton. Am Nat 95:137–145
Hutchinson GE (1967) A treatise on limnology. Vol. II. Introduction to lake biology and the limnoplankton. Wiley and Sons, New York

Hutchinson GE, Löffler H (1956) The thermal classification of lakes. Proc Nat Acad Sci Wash DC 42:84–86

Imboden DM, Emerson S (1978) Natural radon and phosphorus as limnologic tracers: horizontal and vertical eddy diffusion in Greifensee. Limnol Oceanogr 23:77–90

Jawed M (1973) Ammonia excretion by zooplankton and its significance to primary productivity during summer. Mar Biol 23:115–120

Jerlov NG (1976) Marine optics. Elsevier North-Holland 2nd edn, Amsterdam, New York

Jerlov NG (1977) Classification of seawater in terms of quanta irradiance. J Cons 37:281–287

Jones WE, Demetropoulos A (1968) Exposure to wave action: measurements of an important ecological parameter on rocky shores on Anglesy. J Exp Mar Biol Ecol 2:46–63

Kain JM (1971) Continuous recording of underwater light in relation to *Laminaria* distribution. In: Crisp DJ (ed) Fourth European Marine Biology Symposium, Cambridge Univ Press, pp 335–346

Kemp WM, Mitsch WJ (1979) Turbulence and phytoplankton diversity: a general model of the "Paradox of Plankton". Ecol Modelling 7:201–222

Kinne O (ed) (1970) Marine Ecology. Vol I, Environmental factors, pt 1. Wiley and Sons, New York

Kinne O (ed) (1971) Marine Ecology. Vol I, Environmental factors, pt 2. Wiley and Sons, New York

Kinne O (ed) (1972) Marine Ecology. Vol I, Environmental factors, pt 3. Wiley and Sons, New York

Kirk JTO (1976a) A theoretical analysis of the contribution of algal cells to the attenuation of light within natural waters. III. Cylindrical and spheroidal cells. New Phytol 77:341–358

Kirk JTO (1976b) Yellow substance (Gelbstoff) and its contribution to the attenuation of photosynthetically active radiation in some inland and coastal south-eastern Australian waters. Aust J Mar Freshwater Res 27:61–71

Kirk JTO (1977) Use of a quanta meter to measure attenuation and underwater reflectance of photosynthetically active radiation in some inland and coastal south-eastern Australian waters. Aust J Mar Freshwater Res 28:9–21

Koehl MAR, Wainwright SA (1977) Mechanical adaptations of a giant kelp. Limnol Oceanogr 22:1067–1071

Kuhl A (1974) Phosphorus. In: Stewart WDP (ed) Algal physiology and biochemistry. Univ of California Press, Berkeley, Los Angeles, pp 636–654

Lasker R (1978) Ocean variability and its biological effects – Regional review – Northeast Pacific. Rapp PV Reun Cons Int Explor Mer 173:168–181

Lean DRS, Charlton MN (1976) A study of phosphorus kinetics in a lake ecosystem. In: Nriagu JO (ed) Environ Biogeochem Vol 2, pp 283–294

Lock MA, John PH (1979) The effect of flow patterns on uptake of phosphorus by river periphyton. Limnol Oceanogr 24:376–383

Lüning K (1971) Seasonal growth of *Laminaria hyperborea* under recorded light conditions near Helgoland. In: Crisp DJ (ed) Fourth European Marine Biological Symposium, Cambridge Univ Press, pp 347–361

Lüning K, Dring MJ (1975) A photoperiodic response mediated by blue light in the brown alga *Scytosiphon lomentaria*. Planta 125:25–32

Lüning K, Dring MJ (1979) Continuous underwater light measurement near Helgoland (North Sea) and its significance for characteristic light limits in the sublittoral region. Helgoländer Wiss Meeresunters 32:403–422

Martens CS, Berner RA, Rosenfeld JK (1978) Interstitial water chemistry of anoxic Long Island Sound sediments. 2. Nutrient regeneration and phosphate removal. Limnol Oceanogr 23:605–617

McCarthy JJ, Goldman JC (1979) Nitrogenous nutrition of marine phytoplankton in nutrient-depleted waters. Science 203:670–672

Melack JM, Kilham P (1974) Photosynthetic rates of phytoplankton in East African alkaline, saline lakes. Limnol Oceanogr 19:743–755

Monin AS, Kamenkovich M, Kort VG (1977) Variability of the oceans. Wiley and Sons, New York
Morel A (1974) Optical properties of pure water and pure sea water. In: Jerlov NG, Steeman Nielson E (eds) Optical aspects of oceanography. Academic Press, New York, London, San Francisco, pp 1–24
Morel A, Smith RC (1974) Relation between total quanta and total energy for aquatic photosynthesis. Limnol Oceanogr 19:591–600
Morse JW, Cook N (1978) The distribution and form of phosphorus in North Atlantic Ocean deep-sea and contintental slope sediments. Limnol Oceanogr 23:825–830
Mortimer CH (1942) The exchange of dissolved substances between mud and water in lakes III and IV. Summary and references. J Ecol 30:147–201
Mortimer CH (1974) Lake hydrodynamics. Mitt Int Verein Limnol 20:124–197
Müller S, Clauss H (1976) Aspects of photomorphogenesis in the brown alga *Dictyota dichotoma*. Z Pflanzenphysiol 78:461–465
Munk WH, Riley GA (1952) Absorption of nutrients by aquatic plants. J Mar Res 11:215–240
Nakajima T (1979) Denitrification by the sessile microbial community of a polluted river. Hydrobiologia 66:57–64
Neushul M (1971) Submarine illumination in *Macrocystis* beds. Nova Hedwigia 32:241–254
Neushul M (1972) Functional interpretation of benthic marine algal morphology. In: Abbott IA, Kurogi M (eds) Contributions to the systematics of benthic marine algae of the North Pacific. Jap Soc Phycol, Kobe, pp 47–74
Norton T (1969) Growth form and environment in *Sacchoriza polyschides*. J Mar Biol Ass UK 49:1025–1045
Oren A, Blackburn TH (1979) Estimation of sediment denitrification rates at in situ nitrate concentrations. Appl Environ Microbiol 37:174–176
Picard GL (1963) Descriptive physical oceanography. Pergamon Press Ltd, Oxford
Pomeroy LR (1960) Residence time of dissolved phosphate in natural waters. Science 131:1731–1732
Poole HH, Atkins WRG (1929) Photo-electric measurements of submarine illumination throughout the year. J Mar Biol Ass UK 16:297–324
Prézelin BB (1976) The role of peridinin-chlorophyll a-proteins in the photosynthetic light adaptation of the marine dinoflagellate, *Glenodinium sp*. Planta 130:225–233
Ramus J (1978) Seaweed anatomy and photosynthetic performance: the ecological significance of light guides, heterogeneous absorption and multiple scatter. J Phycol 14:352–362
Ramus J, Lemons F, Zimmerman C (1977) Adaptation of light harvesting pigments to downwelling light and the consequent photosynthetic performance of the eulittoral rockweeds *Ascophyllum nodosum* and *Fucus vesiculosus*. Mar Biol 42:293–303
Reynolds CS (1979) Seston sedimentation, experiments with *Lycopodium* spores in a closed system. Freshwater Biol 9:55–76
Rosenfeld JK (1979) Ammonium absorption in nearshore anoxic sediments. Limnol Oceanogr 24:356–364
Riley JP, Skirrow G (eds) (1975) Chemical oceanography, 2nd edn. Academic Press, New York
Rupert CS (1974) Dosimetric concepts in photobiology. Photochem Photobiol 20:203–212
Ryther J, Dunstan WM (1971) Nitrogen, phosphorus and eutrophication in the coastal marine environment. Science 171:1008–1013
Sawada T, Koga S, Uchiyama S (1972) Some observations on carpospore adherence in *Polysiphonia japonica* Harvey. Sci Bull Fac Agr Kyushu Univ 26:223–226
Schindler DW (1971) Carbon, nitrogen and phosphorus and the eutrophication of freshwater lakes. J Phycol 7:321–329
Schindler DW, Brunskill GJ, Emerson S, Broecker WS, Peng T-H (1972) Atmospheric carbon dioxide: its role in maintaining phytoplankton standing crops. Science 177:1192–1194
Schlichting H (1968) Boundary-layer theory. 6th edn. McGraw-Hill, New York
Sculthorpe D (1967) The biology of aquatic vascular plants. St Martin's Press, New York
Sieburth J McN, Jensen A (1970) Production and transformation of extracellular organic

matter from littoral marine algae: A resume. In: Wood DW (ed) Organic matter in natural waters. Univ of Alaska, Inst Mar Sci Publ no 1, pp 203–223

Smayda T (1970) The suspension and sinking of phytoplankton in the sea. Oceanogr Mar Biol Ann Rev 8:353–414

Smith RC, Tyler JE, Goldman CR (1973) Optical properties and color of Lake Tahoe and Crater Lake. Limnol Oceanogr 18:189–199

Smith SL (1978) The role of zooplankton in the nitrogen dynamics of a shallow estuary. Estuarine Coastal Mar Sci 7:555–565

Söderlund R, Svensson BH (1976) The global nitrogen cycle. In: Svensson BH, Söderlund R (eds) Nitrogen, phosphorus and sulfur-global cycles. SCOPE Report 7. Ecol Bull (Stockholm) pp 23–73

Spence DH, Campbell RM, Chrystal J (1971) Spectral intensity in some Scottish freshwater lochs. Freshwater Biol 1:321–337

Stevenson FJ (1972) Nitrogen: element and geochemistry. In: Fairbridge RW (ed) The encylopedia of geochemistry and environmental science. Van Nostrand Reinhold Co, New York, pp 795–801

Sugiyama M, Kawai A (1978) Microbiological studies on the nitrogen cycle in aquatic environments – IV. Metabolic rate of ammonium nitrogen in freshwater regions. Bull Jap Soc Sci Fish 44:351–355

Sverdrup HL (1953) On conditions for the vernal blooming of phytoplankton. J Cons 18:287–295

Talling JF (1960) Self-shading effects of a planktonic diatom. Wetter Leben 12:235–242

Tasker R, Smith H (1977) The function of phytochrome in the natural environment – V. Seasonal changes in radiant energy quality in woodlands. Photochem Photobiol 26:487–491

Therriault J-C, Lawrence DJ, Platt T (1978) Spatial variability of phytoplankton turnover in relation to physical processes in a coastal environment. Limnol Oceanogr 23:900–911

Tyler JE (1974) Photosynthetic radiant energy. Recommendations SCOR working group 15 Sci Com Oceanic Res Proc 10:30–42

Westlake DF (1967) Some effects of low-velocity currents on the metabolism of aquatic macrophytes. J Exp Bot 18:187–205

Wetzel RG (1975) Limnology. WB Saunders, Philadelphia, London, Toronto

Wheeler WN (1980) The effect of boundary layer transport on the fixation of carbon by the giant kelp *Macrocystis pyrifera*. Mar Biol 56:103–110

Wiebe WJ, Johannes RE, Webb KL (1975) Nitrogen fixation in a coral reef community. Science 188:257–259

9 Responses to Light in Aquatic Plants

S.W. JEFFREY

CONTENTS

9.1 Introduction: Characteristics of Aquatic Plants in Relation to Light

Aquatic environments contain a wider range of light environments than land environments. Even at a few meters' depth, the absorbing and scattering properties of different waters yield a profoundly altered light spectrum and reduced intensity, compared to that just above the water's surface (see also the previous Chapter of this Volume). Aquatic plants have evolved a number of light-harvesting pigment systems for trapping those portions of the visible spectrum available in different water types and depths. In contrast, terrestrial plants which receive the full visible spectrum of white light normally utilize only one light-harvesting pigment system (cf. Chap. 3, this Vol.).

Aquatic environments include shallow freshwater ponds, lakes and streams, fresh and saline inland seas, brackish-water marshes, swamps and estuaries, and a wide variety of marine habitats. These extend from inshore coastal regions, surf zones, and fringing reefs to the wide spaces of the open sea and the abyssal deeps. Inland waters cover less than 1% of the Earth's surface, with an average photic depth of about 10 m (LIKENS 1975). The oceans extend over more than 70% of the Earth's surface and have a maximum photic depth of about 200 m.

9.1.1 Diversity of Aquatic Plant Systems

Aquatic ecosystems, both freshwater and marine, support an immense diversity of plants and plant communities. At least 2,100 genera and more than 27,000 species of macroscopic and microscopic aquatic plants are known (SCAGEL et al. 1969; BOLD and WYNNE 1978).

Some characteristics of aquatic plants are listed in Table 9.1. The macroalgae or seaweeds occur in three algal Divisions. The Rhodophyta (red algae) comprise some 4,000 species. Most are benthic macroscopic plants, although unicellular forms are known (e.g. *Porphyridium, Rhodella*). Red algae are found from the spray zone down to depths as great as 175 m (DAWSON 1966; LÜNING and DRING 1979). The Phaeophyta (brown seaweeds), with about 1,500 species, include the largest of all the algae, the kelps. Members of the Chlorophyta, which includes the green seaweeds, are of less importance in the sea compared to their overwhelming dominance in freshwaters. The Charophyta (stoneworts) are a small group of green algae found in fresh or brackish waters. The Cyanophyta (blue-green algae or cyanobacteria), although mostly microscopic, form macroscopic mats, skeins, or tufts in shallow habitats (WHITTON 1973; FOGG 1973). Minute unicellular forms are widely distributed in the sea (WATERBURY et al. 1979). Useful introductions to the biology and ecology of all these groups and recent bibliographies are given by DAWSON (1966), ROSCOWSKI and PARKER (1971), BOLD and WYNNE (1978) and CLAYTON and KING (1981).

The microscopic unicellular algae which make up the floating plant populations (phytoplankton) of the open sea and inland waters range from about 2 to 200 μm, and occur in many algal Divisions (Table 9.1). The silica-walled Bacillariophyta (diatoms) account for 5,500–10,000 species (HOSTETTER and STOERMER 1971). The Dinophyta (dinoflagellates) are second to the diatoms in species diversity and biomass, with about 1,200 species (COX and ZINGMARK 1971; TAYLOR 1976). Other unicellular groups (Table 9.1) are not as numerous or morphologically diverse, but they can dominate the phytoplankton in particular seasons or habitats (JEFFREY and VESK, 1981a). Introductions to the unicellular algae are given by ROSCOWSKI and PARKER (1971), DODGE (1973), ROUND (1973), LASKIN and LECHEVALIER (1978), JEFFREY and VESK (1981a) and JEFFREY (1981).

The distribution of aquatic plants in marine and freshwater habitats is shown in Table 9.1. The red and brown seaweeds and the unicellular golden-brown coccolithophorids are almost exclusively marine, whereas the macroscopic stoneworts and microscopic eustigmatophytes and xanthophytes occur predominant-

Table 9.1. Aquatic plants and their major light-harvesting pigments

Taxon	Common name	Predominant form		Predominant habitat		Chlorophylls				Major accessory light-harvesting pigments	
		Unicellular	Macroscopic plants	Marine	Freshwater	a	b	c_1	c_2	Biliproteins	Carotenoids
Prokaryotes											
Cyanophyta	Blue-green algae (cyanobacteria)	+		+	+	+				+	
Prochlorophyta	Prochloron	+		+		+	+				
Eukaryotes											
Rhodophyta	Red algae	+	+	+		+				+	
Cryptophyta	Cryptomonads	+		+	+	+			+	+	
Dinophyta	Dinoflagellates	+		+	+	+			+		Peridinin
Chrysophyta											
Chrysophyceae	Golden brown flagellates; Silicoflagellates	+		+	+	+		+	+		Fucoxanthin
Rhaphidophyceae (Chloromonadophyceae)	Chloromonads	+		+	+	+		+	+		Fucoxanthin[a]
Prymnesiophyta (Haptophyta)	Golden brown flagellates with haptonema; coccolithophorids	+		+		+		+	+		Fucoxanthin
Eustigmatophyta	Eustigmatophytes	+			+	+					
Bacillariophyta	Diatoms	+		+	+	+		+	+		Fucoxanthin
Phaeophyta	Brown algae		+	+		+		+[b]	+[b]		Fucoxanthin
Xanthophyta	Yellow-green algae	+			+	+					
Euglenophyta	Euglenoids	+		+	+	+	+				
Chlorophyta											
Chlorophyceae	Green algae	+	+	+	+	+	+				Siphonaxanthin[a]
Prasinophyceae	"Scaly" green flagellates	+		+	+	+	+				Siphonaxanthin[a]
Charophyceae	Stoneworts		+		+	+	+				
Angiosperms	Seagrasses;		+	+		+	+				
	Freshwater macrophytes		+		+	+	+				

[a] Some species only [b] Trace amounts only

ly in fresh or brackish waters. All other Divisions are well represented in both marine and freshwaters.

Many freshwater angiosperms occur in lakes and streams. The seagrasses (Zosteraceae) represent a small group of terrestrial angiosperms that have colonized the sea and are now totally aquatic (Bonotto 1976; King 1981a). About 50 species are recognized in 12 genera. They are remarkable in flowering and pollinating under water. Seagrasses are found on tropical and temperate shores, often in embayments and sheltered estuaries. Mangroves and marsh grasses represent another small group of angiosperms rooted within tidal reach, but they maintain a predominantly terrestrial habit (Bonotto 1976; King 1981b).

9.1.2 Global Patterns of Productivity of Aquatic Plants

Because of the relatively small volume of inland waters, freshwater plant production contributes only a few percent to the world total (Table 9.2). Marine systems, covering a vast area, contribute significantly, equivalent to half the terrestrial plant production. The coastal kelp forests have productivities said to rival those of tropical rain forests, generally considered the world's most productive plant ecosystem (Mann 1973; Lieth 1975). Of the five major marine habitats considered in Table 9.2, the microscopic plants of the open sea are the most productive, contributing 74% of the marine production or 24% of the global total. The importance of aquatic plants in global plant productivity is thus obvious, although these plants and their ecological responses are not yet well known to terrestrial plant physiologists. (For detailed discussion of phytoplankton productivity in aquatic ecosystems see Chapt. 10, Vol. 12D).

Table 9.2. Importance of aquatic plants in global plant productivity. (After Whittaker and Likens 1975)

Plant community	Total net primary production (dry matter)			Chlorophyll	
	Mean ($g\ m^{-2}\ yr^{-1}$)	Total ($10^9\ t\ yr^{-1}$)	% of total	Total (10^6 t)	% of total
Aquatic plants – freshwater					
Swamps and marsh	3,000	6.0		6.0	
Lake and stream	400	0.8		0.5	
Aquatic plants – marine					
Open ocean	125	41.5		10.0	
Upwelling zones	500	0.2		0.1	
Continental shelf	360	9.6		5.3	
Algal beds and reefs	2,500	1.6		1.2	
Estuaries (excluding marsh)	1,500	2.1		1.4	
Total aquatic freshwater plants		6.8	3.9	6.5	2.7
Total aquatic marine plants		55.0	31.9	18.0	7.4
Total terrestrial plants		110.7	64.2	219.5	89.9
Global total		172.5	100.0	244.0	100.0

9.1.3 Changes in Spectral Irradiance in Different Water Types

Aquatic plants living in clear shallow waters (e.g., seagrasses and freshwater plants) and plants that are only partly submerged (mangroves and marsh grasses) receive the same full visible spectrum of the sun's radiation as terrestrial plants. Their responses to light are discussed in Chapter 3, this Volume. Plants living deeper than about 0.5 m receive smaller portions of the visible spectrum, often at significantly reduced intensities.

As discussed in Chapter 8, this Volume, the sun's electromagnetic energy is selectively absorbed and scattered as it penetrates the upper layers of any body of water (fresh or marine) in processes that are strongly wavelength-dependent. All infra-red and visible red light (>750 nm) is absorbed by the water itself in the first half meter (JERLOV 1976), reducing the solar energy available to one half. Attenuation of light with depth in clear waters results in the rapid disappearance of red and yellow light with blue and blue-green

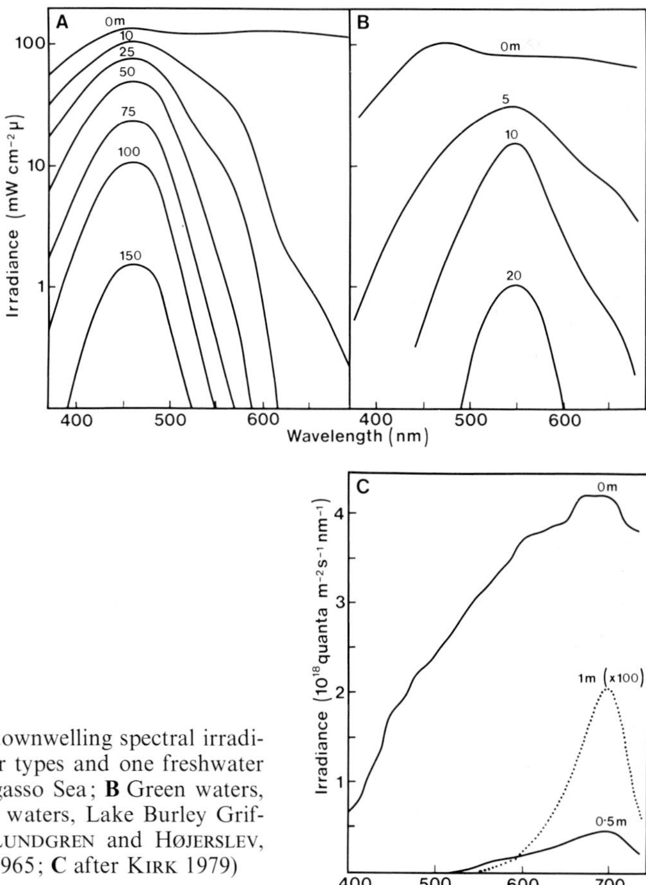

Fig. 9.1. Distribution of downwelling spectral irradiance in two oceanic water types and one freshwater lake. **A** Blue waters, Sargasso Sea; **B** Green waters, Baltic Sea; **C** Orange-red waters, Lake Burley Griffin, Australia. (**A** after LUNDGREN and HØJERSLEV, 1971; **B** after AHLQUIST 1965; **C** after KIRK 1979)

light (maximum about 480 nm) penetrating furthest (to about 200 m depth in clear "desert blue" ocean waters, Fig. 9.1A).

In coastal waters and upwelling regions, the optical properties of seawater are further modified by increased concentrations of phytoplankton and suspended particles of inorganic or organic origin. Dissolved yellow substances ("Gelbstoff", KALLE 1966) from land runoff, detritus, and marine humic materials modify the light field by strongly absorbing in the violet and blue-green region (KALLE 1966; BROWN 1977). This shifts the underwater irradiance from blue-green (480 nm) to green and yellow wavelengths (around 550 nm) and can reduce the depth of the euphotic zone to a few meters (JERLOV 1976). The downwelling spectral irradiance of a typical green ocean (e.g. the Baltic Sea) is shown in Fig. 9.1B, which has a 1% surface light depth of 20 m and maximum light penetration around 550 nm.

Studies of ocean color are becoming increasingly important, not only for tracing and characterizing water masses and their phytoplankton content by remote sensing (MOREL and PRIEUR 1977; SMITH and BAKER 1978a, b), but in using optical methods to measure the in situ quantum efficiency of phytoplankton photosynthesis (JERLOV and STEEMANN NIELSEN 1974; MOREL and SMITH 1974; TYLER 1975; HØJERSLEV et al. 1977; HØJERSLEV 1978; MOREL 1978). In shallow euphotic zones, the phytoplankton, which absorb throughout the visible spectrum according to their pigment composition (see Sect. 9.2.2), are often the major determinant of their own light environment (TALLING 1971; LORENZEN 1972). Instrumentation for light measurements is becoming increasingly precise, but now accurate knowledge of both the spectral characteristics of natural phytoplankton populations and the ability of their pigment systems to adapt to changing light regimes are needed for the interpretation of irradiance measurements (MOREL 1978).

The optical properties of some moderately deep freshwater lakes closely resemble marine green waters. The 20-m-deep Lake Kinneret (Israel) normally has a light penetration maximum around 525–575 nm depending on the phytoplankton biomass and species composition (DUBINSKY and BERMAN 1976, 1979). In other shallow lakes the available photosynthetic radiation is limited to the orange-red region of the spectrum (KIRK 1976, 1979). These lakes contain a large concentration of Gelbstoff and suspended mineral particles, which absorb almost the entire blue and blue-green wavelengths within a few centimeters of the surface. Figure 9.1C shows the downwelling spectral irradiance of one such lake, in which the only radiation left at 1.0 m is in the region 600–700 nm. The absorption of red light by the water itself is exceeded by the absorption of violet and blue-green light by the dissolved and suspended materials.

Thus, in marine and freshwaters, great variations occur in the spectral composition of underwater irradiance. Predominantly blue, blue-green, green, or yellow to red photic zones all exist in different waters. Water movements can transport planktonic algae rapidly from one light regime to another. Flexibility of the photosynthetic apparatus to changes in both light quality and intensity are thus essential to plant survival in aquatic systems.

9.1.4 Vertical Distribution of Benthic and Planktonic Algae Within Photic Zones

The zonation of benthic macrophytes (seaweeds) with depth is not as distinct as was once thought. Algal groups have their own light preferences, and different species, even in the same genus, often show marked preferences. Generally the Chlorophyta grow in shallow water, and both the Phaeophyta and Rhodophyta extend from shallow to deeper waters. However, some green algae (e.g. *Palmoclathrus*, WOMERSLEY 1971) occur only between 10 and 50 m, and many Rhodophyta occur near low tide level or even intertidally (e.g. some coralline algae).

LÜNING and DRING (1979) show that the lower limit of the kelps (Phaeophyta) is at the 0.5% to 1% surface light depth, whereas the zone for other seaweeds (reds, brown, and greens) can extend to the 0.05% to 0.1% surface light depth (Fig. 9.2). These light intensities are 10 to 20 times less than the generally

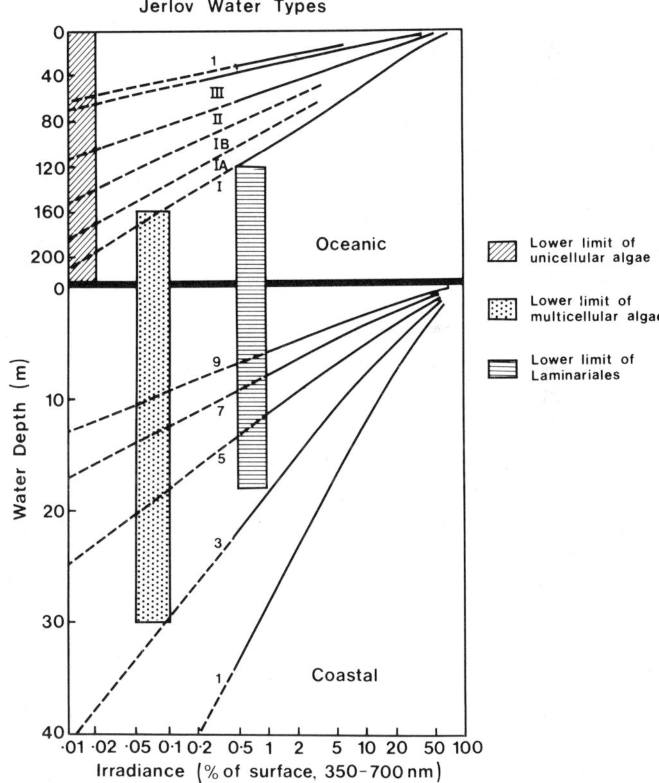

Fig. 9.2. Percentage of surface downwelling irradiance (350–700 nm) as a function of depth and Jerlov water types. *Solid lines* according to JERLOV (1970); *broken lines* projections of JERLOV's curves into the lower percentage ranges according to LÜNING and DRING (1979). Critical light levels: 0.5%–1% of surface irradiance for Laminariales; 0.05%–0.1% for multicellular marine algae; 0.01%–0.05% for unicellular planktonic algae; 1% surface irradiance classically regarded as lower limit of euphotic zone, (STEEMANN NIELSEN 1975)

accepted 1% surface irradiance set as the limit of the euphotic zone (STEEMANN NIELSEN 1975).

The depth distribution of phytoplankton also extends below the 1% surface light level. The micro-algae often form deep sub-surface maxima even within well-mixed water columns. Differences in chlorophyll biomass and depth of the euphotic zone of different waters are shown in Fig. 9.3. The deepest chlorophyll maxima (90 to 130 m depth) are found in low chlorophyll biomass waters

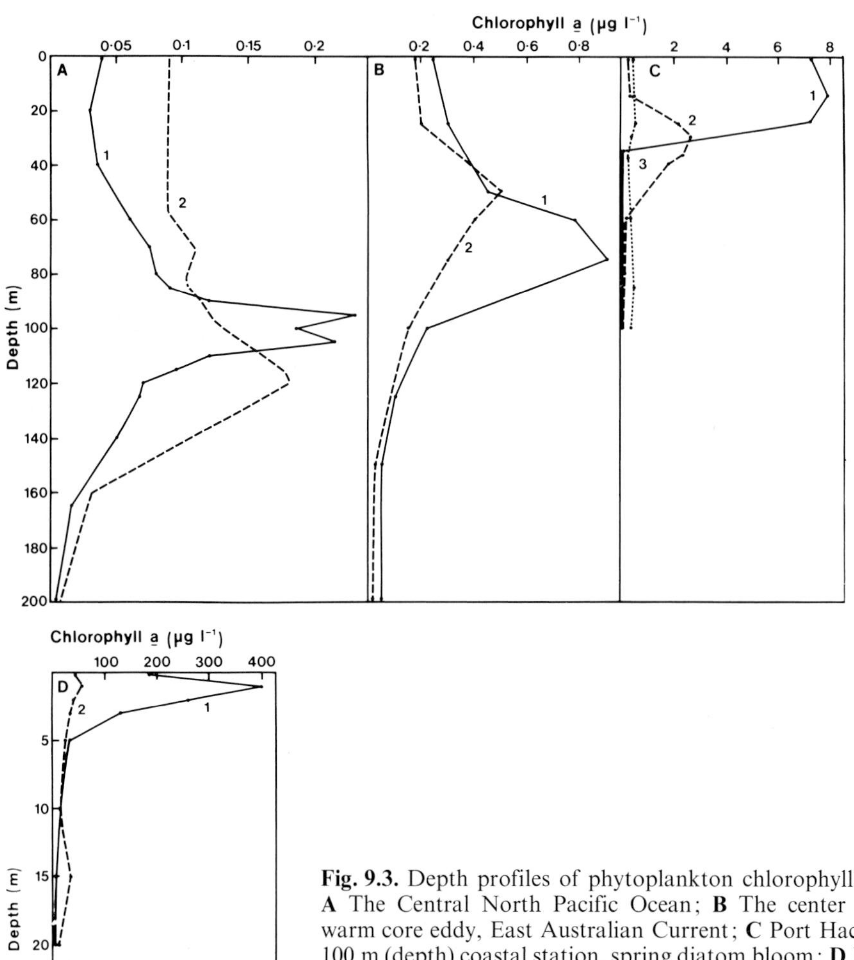

Fig. 9.3. Depth profiles of phytoplankton chlorophyll a in **A** The Central North Pacific Ocean; **B** The center of a warm core eddy, East Australian Current; **C** Port Hacking 100 m (depth) coastal station, spring diatom bloom; **D** Lake Kinneret, dinoflagellate bloom. Curves *1, 2* in **A, B** represent typical but unrelated chlorophyll profiles; curves *1, 2* and *3* in **C** represent events at the beginning (curve *1*), middle, 7 days later (curve *2*) and end, 13 days later (curve *3*) of the diatom bloom; curves *1* and *2* in **D** represent a bloom of dinoflagellates concentrated by vertical migration in the surface layers (curve *1*), and 15 days later (curve *2*). **A** after SHULENBERGER (1978) (curve *1*), KIEFER et al. (1976) (curve *2*); **B** after JEFFREY and HALLEGRAEFF (1980); **C** after JEFFREY (1974); **D** after DUBINSKY and BERMAN (1979)

like the Central North Pacific Gyre (Fig. 9.3A). More enriched waters of warm-core eddies of the East Australian Current show shallower maxima at 50 to 80 m depth (Fig. 9.3B). Coastal phytoplankton blooms with higher chlorophyll biomass show chlorophyll maxima at about 15 to 30 m depth (Fig. 9.3C), whereas lakes with extremely high chlorophyll biomass (up to 400 $\mu g\ l^{-1}$; Fig. 9.3D) may have shallow maxima near the surface (1 to 2 m depth).

The phytoplankton certainly have mechanisms for active depth selection. Motile dinoflagellates can undergo significant vertical migrations even through temperature discontinuities (EPPLEY et al. 1968; KAMYKOWSKI and ZENTARA 1977; BLASCO 1978). Positive buoyancy occurs in some nonmotile species (centric diatoms and nonmotile dinoflagellates) by a combination of metabolic (ionic), nutritional and light gradient control (ANDERSON and SWEENEY 1977; KAHN and SWIFT 1978). While phytoplankton can be homogeneously distributed by water mixing, or can be concentrated by hydrodynamic features (fronts, thermoclines, and discontinuity layers), many species actively select preferred depths depending on light color and intensity.

9.2 Light-Harvesting Pigment Systems in Algae

9.2.1 Distribution of Pigment Systems

The algae collectively have pigment systems that can harvest the entire range of wavelengths of the visible spectrum (350 to 700 nm). These pigments use light in two ways – as a source of energy for photosynthesis and as a means of interpreting photosignals from the environment. In the latter category the most important sensor pigment known is the blue chromoprotein, phytochrome, which controls many aspects of higher plant development (MITRAKOS and SHROPSHIRE 1972; cf. Chap. 4, this Vol.). These photo-receptor pigments are discussed in Sections 9.3 and 9.4.

The major photosynthetic light-harvesting pigments, known from classic studies of photosynthetic action spectra (e.g. HAXO 1960; HALLDAL 1970; KAGEYAMA et al. 1977), and their distribution in the algal taxa are shown in Table 9.1. Chlorophyll a is universally present. Biliproteins are restricted to the more primitive groups (blue-green and red algae, and cryptomonads). Chlorophyll c_2 is present in all the chlorophyll c-containing algae (JEFFREY 1976; JEFFREY et al. 1975). Chlorophyll c_1 is found together with c_2 in the fucoxanthin-containing brown algal classes, and both c_1 and c_2 occur in trace quantities in the yellow-green algae. The photosynthetically active carotenoids – fucoxanthin and peridinin – are found only in the brown algal line, whereas siphonaxanthin has a similar function in some green algae. Chlorophyll b occurs in the green algae and higher plants, and in members of the newly erected Division Prochlorophyta (LEWIN 1976).

Some exceptions to the above generalizations are appearing as more algae are examined with modern techniques (JEFFREY 1980). These include atypical distributions of chlorophylls c_1 and c_2, modifications to the structure of fucoxanthin in some species and the finding of the dinoflagellate carotenoid peridinin in some red algae!

9.2.2 In Vivo Absorption Spectra of Algal Classes

Although photosynthetic action spectra are required to distinguish active from inactive pigments, in vivo absorption spectra can suggest which algal groups might be best suited to the various marine light environments. The in vivo spectra of a diatom and a dinoflagellate (Fig. 9.4A) show a relatively high absorption in the blue to green region (450 to 550 nm) due to chlorophyll c and the carotenoids fucoxanthin or peridinin respectively. In contrast, marine green algae (Fig. 9.4A) have a low absorbance in the green to yellow region (520 to 600 nm), except for species which contain the 540 nm-absorbing carotenoid siphonaxanthin (Kageyama et al. 1977; Yokohama et al. 1977). Marine

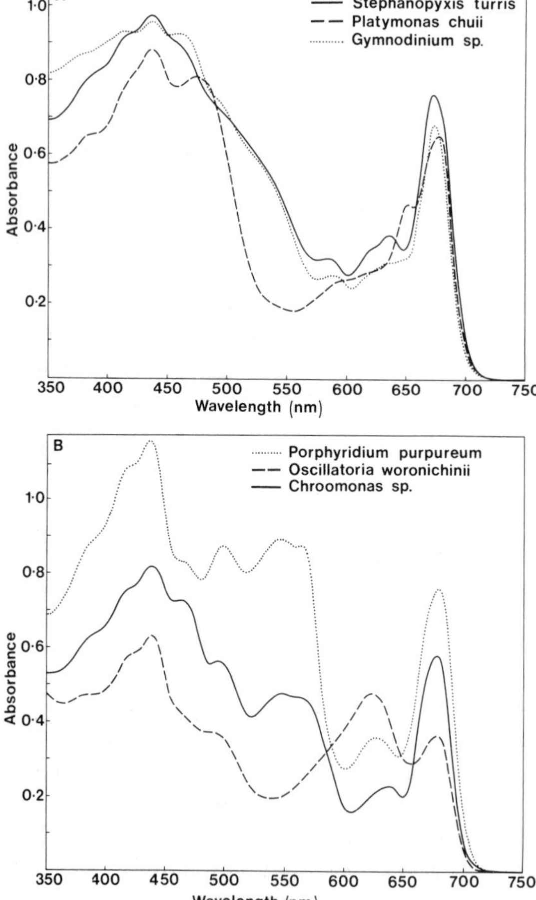

Fig. 9.4. In vivo absorption spectra of algal divisions: **A** A diatom, *Stephanopyxis turris,* a dinoflagellate, *Gymnodinium* sp., and a green alga, *Platymonas chuii.* **B** Three biliprotein-containing algae: a red alga, *Porphyridium purpureum,* a blue-green alga, *Oscillatoria woronichinii* and a cryptomonad, *Chroomonas* sp. (Jeffrey 1980)

green algae have more chlorophyll b than higher plants (JEFFREY 1968a; WOMERSLEY 1971; KEAST and GRANT 1976; NAKAMURA et al. 1976; WOOD 1979), with chlorophyll a:b ratios ranging from 1.02 to 2.00, and 2.8 to 3.4, respectively. The biliprotein-containing algae show enhanced absorption in the green to yellow region (500 to 600 nm) if phycoerythrin is dominant, or in the orange-red region (600 to 650 nm) if phycocyanin is dominant (Fig. 9.4B). Thus some algae are better equipped to harvest blue-green light (brown algal groups), green light (browns, reds, cryptomonads and siphonaxanthin-containing green algae), orange light (blue-green algae), and red and blue light (green algae).

9.2.3 Chlorophylls

Three types of conjugated molecules are responsible for the in vivo absorption spectra and act as light-harvesting pigments – chlorophylls, carotenoids, and biliproteins. Chlorophylls a and b are well known from higher plant studies (VERNON and SEELY 1966). They are magnesium-containing conjugated tetrapyrroles with one ring (IV) completely reduced, contain a cyclopentanone ring conjoint with ring III, and have a propionic acid residue esterified to the C_{20} alcohol phytol at position C-7 on ring IV. Chlorophyll b differs from chlorophyll a by replacement of the methyl group at position C-3 on ring II with an aldehyde group, which changes both its spectral properties and its biological function.

Chlorophyll c, which is the least well-known chlorophyll, is a mixture of two spectrally distinct components c_1 and c_2 (JEFFREY 1968b, 1969, 1972). These compounds are porphyrin rather than chlorin derivatives with ring IV unsaturated like a porphin, but with a cyclopentanone ring and central magnesium atom like a chlorophyll (DOUGHERTY et al. 1966, 1971; WASLEY et al. 1970; STRAIN et al. 1971; BUDZIKIEWICZ and TARAZ 1971; and Fig. 9.5). An acrylic acid side chain, with a free and very acidic carboxyl group which is not esterified, replaces the propionic acid side chain of chlorophylls a and b. Chlorophylls c_1 and c_2 differ only in alkyl groups attached to the porphyrin ring system – c_1 having an ethyl group and c_2 having a vinyl group at the C-4 position on ring II (Fig. 9.5). The absorption spectra of chlorophylls c_1 and c_2 in organic solvents are of the magnesium-porphyrin type (Fig. 9.5), exhibiting a three-banded spectrum with a very small absorption peak in the red region, at shorter wavelengths (629 nm) than that of chlorophyll b (645 nm) and chlorophyll a (660 nm). They exhibit a much stronger absorption in the Soret region (JEFFREY 1972) than chlorophylls a and b (STRAIN et al. 1963), due to differences in the conjugation pathway of the ring system of porphyrins compared to chlorins. The position of the absorption maxima of chlorophylls c_1 and c_2 are almost identical, but the band intensities differ markedly due to the different substituents attached to the ring system (JEFFREY 1980). The fluorescence spectra of chlorophylls c_1 and c_2 extend well into the region of chlorophyll a absorption (JEFFREY 1972), making these pigments good candidates for an energy-transferring function in photosynthesis (MACCOLL and BERNS 1978).

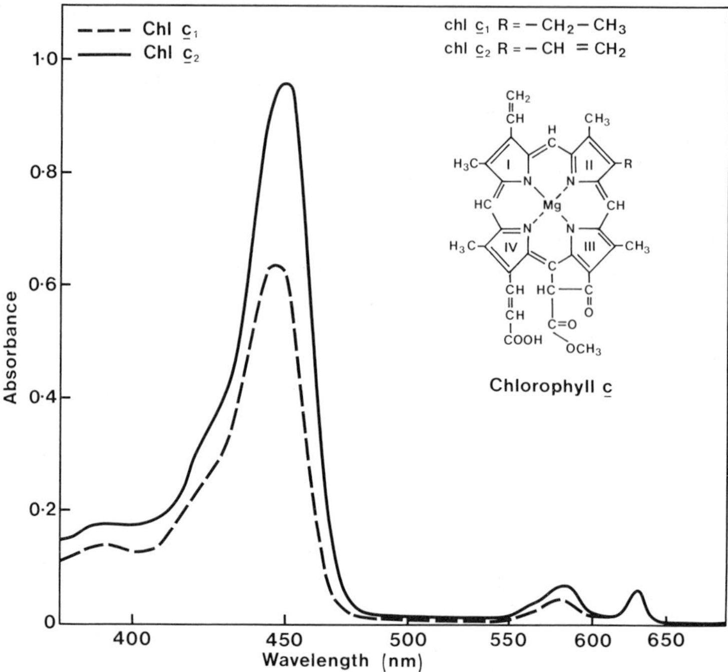

Fig. 9.5. Absorption spectra of chlorophylls c_1 and c_2 from *Sargassum flavicans* in diethyl ether (concentration 1.6 μg ml^{-1}; 4 cm cells; JEFFREY 1969)

9.2.4 Carotenoids

About four or five major carotenoids are found in each algal class, and these may be supplemented by up to a dozen minor or occasional carotenoids of unknown function. Carotenoids are usually yellow, orange, or red isoprenoid, polyene pigments. Xanthophylls are highly oxygenated derivatives of the parent hydrocarbon carotene. The simplest carotenoids are found in blue-green and red algae, and during the course of evolution more than 15 structural modifications to the basic skeleton have been added (LIAAEN-JENSEN 1977, 1978, 1979). Of about 60 algal carotenoids identified chemically, a light-harvesting function has been definitely established for only three, fucoxanthin, peridinin, and siphonaxanthin. A fourth, β-carotene, is a possible candidate for a light-harvesting function in photosystem I in higher plants and brown seaweeds (THRASH et al. 1979; BARRETT and ANDERSON 1980). Five other carotenoids are known to be involved in epoxide cycle reactions, the function of which is still unclear (JEFFREY 1980).

The absorption spectra and chemical structures of the four light-harvesting carotenoids are shown elsewhere (JEFFREY 1980). In organic solvents the absorption maxima of these carotenoids lie between 450 and 470 nm, but when complexed with protein in vivo, the absorption maxima show a red shift of 20 to 90 nm. This extends their useful absorbing range well into the green spectral region (up to 550 nm; Fig. 9.4A).

9.2.5 Biliproteins

The three most primitive algal groups – reds, blue-greens and cryptomonads – contain biliproteins as light-harvesting accessory pigments. There are two main types, phycoerythrins and phycocyanins. Each biliprotein consists of a protein moiety with the chromophores, open-chain tetrapyrroles, covalently attached. Only two types of native chromophores (phycoerythrobilin and the closely related phycocyanobilin) are known, but the wide variety of spectrally different biliproteins which have been isolated from red, blue-green and cryptomonad algae is thought to be due to specific chromophore attachments and a varied protein environment rather than to a multiplicity of chromophores (GANTT 1975; GYSI and CHAPMAN 1981).

Biliproteins absorb in the 540 to 650 nm region, 540 to 565 nm for the red phycoerythrins, 610 to 640 nm for the blue phycocyanins, and about 650 nm for the blue allophycocyanins. Both phycocyanins and allophycocyanins always occur together in red and blue-green algae, and many species contain only these two components. Phycoerythrin seems more dispensable, since some blue-green algae can adapt to function without it. The absorption and fluorescence properties of these biliproteins overlap sufficiently to allow energy to be transferred with high efficiency in the sequence phycoerythrin → phycocyanin → allophycocyanin → chlorophyll a. In the cryptomonads allophycocyanin is absent; its function may be replaced by chlorophyll c_2 (MACCOLL and BERNS 1978).

9.2.6 Chlorophyll-Protein Complexes

In contrast to the biliproteins, which are located in densely packed phycobilisome particles on the outer surface of the thylakoids in red and blue-green algae (GANTT 1975), but within the intrathylakoid space in the cryptomonads (GANTT et al. 1971; LICHTLÉ et al. 1980), the chlorophylls and carotenoids are located within the thylakoid system itself, complexed to proteins, and organized in functional photosynthetic units. These chlorophyll–protein complexes can be released from the thylakoids by treatment with detergents and subsequently separated using acrylamide gel electrophoresis. Many chlorophyll–protein complexes with light-harvesting and reaction center properties have been isolated from higher plants by this technique (reviewed by THORNBER and BARBER 1979).

Light-harvesting and reaction center pigment proteins have also been isolated from several algal groups (reviewed by JEFFREY 1980). The well-characterized water-soluble pigment complex, peridinin-chlorophyll a-protein from dinoflagellates (Fig. 9.6A), contains either 4 peridinins and 1 chlorophyll a per protein or 9 peridinins and 2 chlorophyll a's per protein (PRÉZELIN and HAXO 1976). No chlorophyll c_2 is present. This pigment-protein probably represents a major light-harvesting complex of dinoflagellates.

The major light-harvesting complex of brown algae is an orange-brown chlorophyll a/c_2-fucoxanthin-protein (BARRETT and ANDERSON 1980; Fig. 9.6B). A P-700-chlorophyll a-β-carotene protein has also been isolated from brown

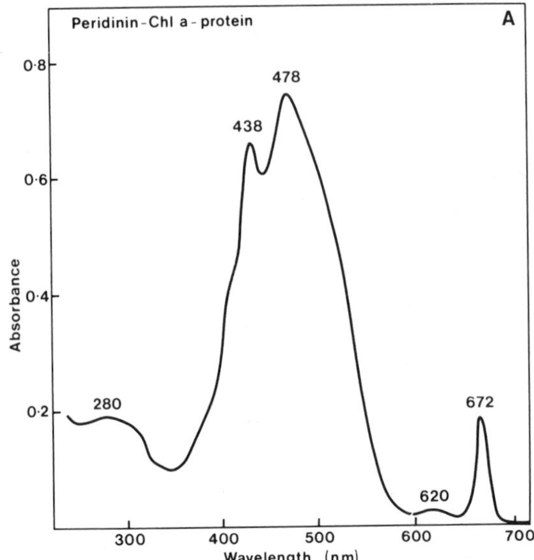

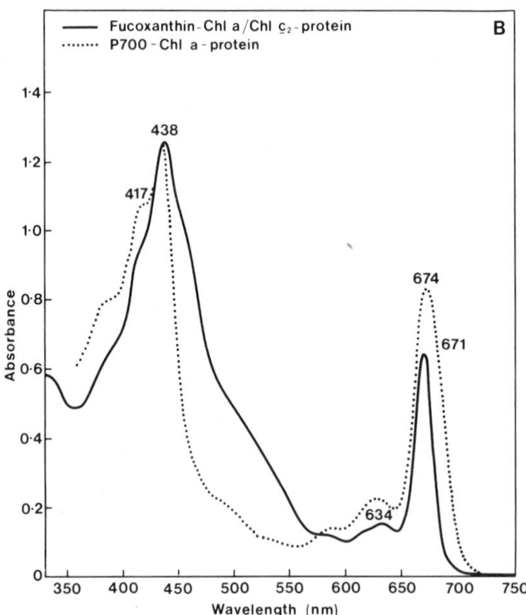

Fig. 9.6. Absorption spectra of light-harvesting pigment proteins from a dinoflagellate and a brown seaweed. **A** Peridinin-chlorophyll a–protein complex from *Glenodinium* (dinoflagellate) (PRÉZELIN and HAXO 1976); **B** P-700-chlorophyll a-β-carotene–protein complex, and the major light-harvesting chlorophyll a/c_2-fucoxanthin-protein complex, from the brown seaweed *Acrocarpia paniculata*. (BARRETT and ANDERSON 1980)

algae (Fig. 9.6B). This complex occurs in all higher plants and algae that have been examined. Fluorescence excitation spectra of these and other complexes from brown algae clearly implicate chlorophyll c, fucoxanthin and β-carotene, but not violaxanthin, in a light-harvesting role. This latter pigment instead plays a key role in the epoxide cycle (JEFFREY 1980).

9.3 Light Regulation of the Photosynthetic Apparatus in Aquatic Plants

The previous section reviewed the photosynthetic pigments of aquatic plants that harvest light energy in restricted photic environments. In this section, several light-induced responses which optimize the photosynthetic apparatus to natural light fields are considered.

In terrestrial plants three major photoresponses are recognized in chloroplast formation – (1) the photoconversion of protochlorophyll(ide) into chlorophyll(ide), (2) the formation of the physiologically active far-red form of phytochrome which controls chloroplast structural development, chlorophyll formation, photophosphorylation and the level of Calvin cycle enzymes (MOHR 1977) and (3) blue light-induced responses (PRESTI and DELBRÜCK 1978; VOSKRESENSKAYA 1979) probably mediated by a flavin noncovalently bound to protein (KASEMIR 1979). In aquatic plants two types of light quality control of the photosynthetic apparatus are known – (1) red and green light regulation of biliprotein synthesis in red, blue-green and possibly cryptomonad algae (BOGORAD 1975; VESK and JEFFREY 1977) and (2) blue-green light regulation of the thylakoid system in unicellular members of brown and green marine algal groups (JEFFREY and VESK 1977; VESK and JEFFREY 1977). Many white light photosynthetic responses have also been documented for algae in the marine and oceanographic literature, but since the major proportion of photosynthetic activity in the sea is not conducted in white light, these effects may have only limited ecological significance.

9.3.1 Complementary Chromatic Adaptation

This phenomenon is the process by which red and blue-green algae alter their rates of biliprotein synthesis in response to changes in the color of the prevailing light (BOGORAD 1975). Green light (maximum 540 to 560 nm) stimulates the synthesis of green light-absorbing phycoerythrin, whereas red light (maximum 640 to 660 nm) stimulates the synthesis of red light-absorbing phycocyanin, with only minor changes in the chlorophyll or carotenoid content of the cells (FUJITA and HATTORI 1960; DIAKOFF and SCHEIBE 1973). The transition occurs around 590 nm. Chromatic adaptation occurs in a variety of phycoerythrin-containing red and blue-green algae (DE MARSAC 1977), and involves de novo synthesis of biliprotein (BENNETT and BOGORAD 1973). Phycoerythrin synthesis in chromatically adapting cells occurs within 45 to 90 min of transfer from red to green light in processes which can be blocked by the RNA synthesis inhibitor rifamycin (GENDEL et al. 1979). The evidence suggests that chromatic adaptation involves gene regulation at the transcriptional level.

The photoreceptors for chromatic adaptation (adaptochromes), which are thought to be hemoproteins or biliproteins (BOGORAD 1975) have not yet been positively identified. Photoreversible pigments (phycochromes) from blue-green algae (SCHEIBE 1972; BJÖRN and BJÖRN 1976) show red-green shifts, and may have adaptochrome-like functions. OHAD et al. (1980) have recently shown that allophycocyanin has identical photoresponses to the phycochromes, which suggests that allophycocyanin may be the true identity of the phycochromes. Whether allophycocyanin is also an elusive adaptochrome needs further evidence.

Chromatic adaptation thus allows the light-harvesting capacity of both red and blue-green algae to be optimized in green (Fig. 9.1 B) or orange to red light (Fig. 9.1 C) in marine and freshwaters. This has been demonstrated in situ in red seaweeds by RAMUS et al. (1976a, b).

9.3.2 Blue-Green Light Responses in Unicellular Algae

Until recently it was not known if the blue-green light environment of the open ocean exerted any photocontrol on the photosynthetic apparatus of unicellular planktonic algae. Recent studies, however, have shown that low intensity blue-green light (maximum 480 nm) does exert a profound effect on the pigment content, chloroplast ultrastructure and photosynthetic capacity of six classes of unicellular algae (JEFFREY and VESK 1977; VESK and JEFFREY 1977). The pigment content increased from 17% to 506% in blue-green light, being most marked in diatoms (both centric and pennate), with lesser effects in the motile taxa (dinoflagellates, green flagellates, cryptomonads, chrysomonads, and prymnesiophytes, Table 9.3). There was no significant change in the ratio of the light-harvesting carotenoids fucoxanthin or peridinin to chlorophyll a in eight species tested (VESK and JEFFREY 1977), and only a small increase in the ratio chlorophyll c:chlorophyll a. However, in the cryptomonad *Chroomonas* sp., a significant increase in the relative proportions of phycoerythrin to chlorophyll a occurred, which suggested that, in this species only, a blue-green light chromatic adaptation effect was operating.

The structural basis for these pigment changes was an increase in the number of chloroplasts (Fig. 9.7A, B), and an increase in the thylakoid system of the chloroplasts (Fig. 9.7C, D). In *Stephanopyxis turris* grown on white light, the chloroplasts (Fig. 9.7C) contained 3-thylakoid bands surrounded by girdle lamellae beneath which lay a single ring genophore. A large central pyrenoid was present. In cells grown on blue-green light, the chloroplasts (Fig. 9.7D) were packed with 3-thylakoid bands and had multiple fibrillar (DNA) areas beneath loops of girdle lamellae which were associated with the thylakoid stacks. The pyrenoid contained dense paracrystalline areas. Similar ultrastructural changes were also observed in four other diatoms, a dinoflagellate, and a cryptomonad (VESK and JEFFREY 1977).

Blue-green light also increased the content of polar lipid, protein, RNA, and DNA and increased photosynthetic carbon fixation (WALLEN and GEEN 1971a, b; SHIMURA and FUJITA 1975; JEFFREY and VESK 1977). The latter may be associated with the ultrastructural changes seen in the pyrenoid since this is the site of ribulosebisphosphate carboxylase activity (HOLDSWORTH 1971).

These results suggest that complementary chromatic adaptation (blue-green light increasing the concentration of blue-green light-absorbing carotenoids)

Table 9.3. Chlorophyll content of algae grown in 400 μW cm^{-2} white or blue-green light (JEFFREY and VESK 1981b)

Algae	Chlorophyll a+c (μg 10^{-6} cells)		% Increase with blue-green
	White	Blue-green	
Diatoms			
Amphiprora sp.	8.95	13.98	56
Asterionella glacialis	3.0	3.6	20
Biddulphia aurita	5.2	11.6	123
Cylindrotheca closterium	1.6	3.1	94
Ditylum brightwellii	27.93	65.38	134
Gyrosigma sp.	15.2	19.9	31
Lauderia annulata	8.9	15.7	76
Nitzschia closterium	1.6	2.9	81
Phaeodactylum tricornutum	0.16	0.39	146
Rhizosolenia setigera	9.96	60.36	506
Skeletonema costatum	1.81	2.31	28
Stephanopyxis turris	25.9	52.1	101
Thalassiosira eccentrica	106.6	150.0	41
Thalassiosira gravida	28.6	27.9	−2
Thalassiosira rotula	16.0	16.7	4
Dinoflagellates			
Amphidinium carterae	6.5	8.8	35
Gymnodinium sp.	4.6	9.1	98
Prorocentrum micans	56.6	67.3	19
Golden-brown flagellates (prymnesiophytes, chrysomonads)			
Cricosphera carterae	5.1	6.1	20
Isochrysis galbana	0.2	0.5	150
Olisthodiscus luteus	11.8	13.8	17
Cryptomonads			
Chroomonas sp.	1.32	2.05	55
Marine green flagellates			
Dunaliella tertiolecta[a]	0.94	1.31	39

[a] (chlorophyll a+b)

probably does not occur at low light intensities in those algae with carotenoids and chlorophyll c as accessory light-harvesting pigments. Instead, blue-green light regulates the photosynthetic apparatus by increasing the total amount of pigments without changing their proportions, increasing the photosynthetic capacity, and reducing the onset of chlorophyll degradation in older cells (JEFFREY and HALLEGRAEFF unpublished). Blue-green light may thus increase the viability of phytoplankton at depth in the sea, by "switching on" the synthesis of the light-harvesting pigment apparatus for more efficient photon capture.

Many photoreceptors, primary reactions, metabolic processes, and physiological responses influenced by blue light are known in higher plants, green algae,

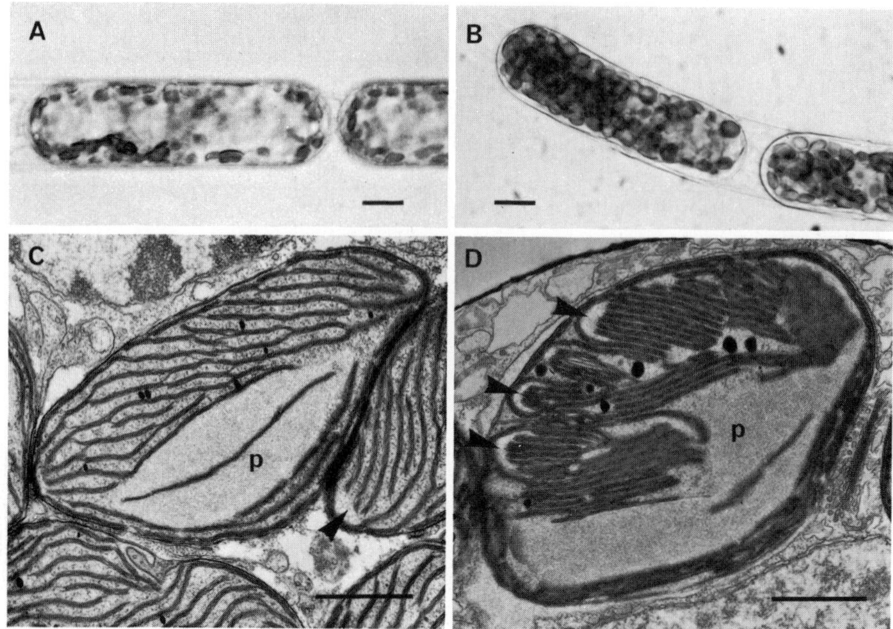

Fig. 9.7. A, B Light micrographs of the diatom *Stephanopyxis turris* grown in low intensity white light (**A**) and blue-green light (**B**) (400 $\mu W\,cm^{-2}$). Note increased number of chloroplasts in cell in blue-green light (*Bar* = 10 µm). **C, D** electron micrographs of chloroplasts of *S. turris* from white light (**C**) and blue-green light (**D**) cultures. Note extensively developed thylakoid system, dense pyrenoid (*P*) and multiple fibrillar (DNA) regions (*arrow heads*) lying beneath girdle lamellae, in the blue light chloroplast (*Bar* = 1 µm) (JEFFREY 1980)

fungi, and yeasts (PRESTI and DELBRÜCK 1978; VOSKRESENSKAYA 1979; SENGER 1980). SENGER and BRIGGS (1980) conclude that "there is no one distinct blue light system analogous to the phytochrome system at the red end of the spectrum. Blue light systems, predominant in lower organisms (yeasts, fungi), seem to represent early evolutionary phases in photoregulation". The blue-green light responses in unicellular marine algae reviewed here are no doubt a direct response to the ancient blue-green light aquatic environment where the first ancestral prokaryotes originated (OLSON 1978). These blue light mechanisms became less important (although some were retained) with the colonization of the high intensity white light environment on land. In this case evolution of the phytochrome system switched environmental control mainly to the red–far-red spectral region.

9.3.3 White Light Responses in Algae

Although the full visible spectrum of white light is found only in a few specialized marine and freshwater habitats (e.g. the sea surface, intertidal regions, polar ice fields and shallow ponds and tide pools), most studies of aquatic photosynthesis and algal metabolism have utilized white light regimes almost exclusively (e.g. BEARDALL and MORRIS 1976; JØRGENSEN 1977). In view of the red, green, and blue-green light algal responses (Sects. 9.3.1, 9.3.2) and current knowledge of marine light fields (Sect. 9.1.3), it is urgent to reassess white light models of photoadaptive responses, and relate these to the performance of aquatic plants in specific underwater light fields as is being done for brown seaweeds by LÜNING (1980) and phytoplankton by VESK and JEFFREY (1977).

Photosynthetic responses to white light in bloom-forming marine and freshwater dinoflagellates have recently received attention (PRÉZELIN 1976; PRÉZELIN and ALBERTE 1978; PRÉZELIN and SWEENEY 1978, 1979). In some species low irradiances of white light altered the size of the photosynthetic units (PSU) by increasing the amount of light-harvesting peridinin-chlorophyll a–proteins while keeping the PSU number constant (see also Chap. 3, this Vol.). Other species adapted by increasing the number of PSU's while keeping the size constant. Pigment responses were completed within 3 to 18 h.

Rapid turnover of chlorophyll in white light has also been found in diatoms (RIPER et al. 1979). Chlorophyll turnover times varied from 3 to 10 h for chlorophyll a and from 7 to 26 h for chlorophyll c, appreciably faster than chlorophyll turnover in higher plants. This suggests that both diatoms and dinoflagellates have the capacity for rapid chlorophyll synthesis already noted in the discussion of blue-green light responses. However, investigators should now realise that the "white light ocean" does not exist, and study marine plants under more natural green, blue-green and yellow-green irradiances similar to specific underwater light fields.

9.4 Light Control of Physiological Processes in Aquatic Plants

Light controls a range of physiological processes in aquatic plants in addition to those connected with photosynthesis. In these types of photoregulation, the light absorbed by the photoreceptor(s) acts as a signal rather than the energy source (PRESTI and DELBRÜCK 1978). These responses include blue light induction of fertility in female gametophytes of kelps, light-oriented chloroplast movements, phototaxis and photokinesis in various unicellular algae, red–far-red effects in kelps and unicellular algae, and photobiological responses connected with circadian rhythms (EHRET and WILLE 1970; HERMAN and SWEENEY 1975; SWEENEY et al. 1979). Other photoperiodic responses include light/dark interconversions of epoxidic xanthophylls in many algal classes (HAGER 1975).

9.4.1 Photomorphogenesis

Blue light of relatively high quantum flux is necessary for the development of fertility in both male and female gametophytes of several species of *Laminaria* (kelps) (LÜNING and DRING 1975; LÜNING 1980). The action spectrum for the induction of fertility has a major peak at 430 to 450 nm with no response above 500 nm, and thus differs from the action spectra for photosynthesis which show peaks at red, green, and blue wavelengths (LÜNING and DRING 1975). The action spectrum for the blue light response resembles those of other blue light responses in bacteria, fungi, plants, and animals (PRESTI and DELBRÜCK 1978) for which the photoreceptor riboflavin has been suggested. No differential effects of blue and red light on growth of *Laminaria* were detected, and lower temperatures reduced the quantum requirement for the blue light response. In an elegant study LÜNING (1980) proposes that in early spring, underwater irradiance and its blue light component increase suddenly due to calm weather at the study site (Helgoland), and both male and female gametophytes can mature rapidly. In locations where blue light is limited (e.g. beneath algal canopies) fertile gametophytes are not produced, but vegetative growth can continue by utilizing the available green light from fucoxanthin-activated photosynthesis.

9.4.2 Chloroplast Movements

Light-induced chloroplast movements are well known for freshwater algae, mosses, and higher plants (see review by HAUPT and SCHÖNBOHM 1970, and Volume 7, this Series). Control is usually achieved by a blue light photoreceptor system, although in the green alga *Mougeotia,* chloroplast movements are regulated by phytochrome. The freshwater xanthophyte *Vaucheria* has been extensively investigated in this regard (see review by BRIGGS and BLATT 1980). Positive photoresponse is characterized by aggregation of organelles in the light, by the formation of a cortical fiber reticulum induced by blue light (BLATT and BRIGGS 1980). Maximum response is at 480 nm, with wavelengths longer than 530 nm inactive.

Light-induced chloroplast movements have also been found in intertidal brown seaweeds (NULTSCH and PFAU 1979). In the low intensity arrangement, chloroplasts lie beneath the cell walls facing the light. In high light intensity they move parallel to the light direction and lie against the side walls. No such chloroplast movements were found in other species of intertidal red or green algae, except *Ulva lactuca,* whose chloroplast displacements followed circadian rhythms. Action spectra for these brown algal responses have not yet been reported. The results suggest that the brown algal pigment system may need more protection from high light intensity in the intertidal region than those of the green and red algae.

9.4.3 Phototaxis

Phototaxis involves locomotive movements in which organisms move in response to light. Such movements are well-known for both marine and freshwater phytoflagellates – e.g., euglenoids, dinoflagellates, cryptomonads, and green flagellates, and for those pennate diatoms that possess a raphe system (see reviews by NULTSCH 1974; HARPER 1977). Action spectra for phototaxis resemble either the blue light spectrum of riboflavin (e.g. *Euglena,* CHECCUCCI et al. 1976; the diatom, *Nitzschia*, NULTSCH 1971), the spectrum of phycoerythrin (the cryptomonad *Cryptomonas* sp., WATANABE and FURUYA 1974), or particular biliproteins (red and blue-green algae, NULTSCH et al. 1979). The action spectra for phototaxis often show a close similarity to parts of the photosynthetic action spectrum, and NULTSCH et al. (1979) suggest that photokinetic or phototaxic effects are due to additional ATP supply from noncyclic or pseudocyclic photophosphorylation to the motor apparatus. The occurrence of eyespot structures containing carotenoids in close proximity to both the flagellar apparatus and the photoreceptor of many unicellular algae (DODGE 1973) suggests the importance of shadow-casting pigment masses in photic orientation.

9.4.4 Red–Far-Red Effects in Algae

That phytochrome may be active in aquatic systems has been inferred from physiological studies of only a few species of algae (DRING 1970; DUNCAN and FOREMAN 1980). The small number of reports may be due to the assumption that red and far-red light do not penetrate far enough in aquatic habitats to stimulate physiological responses. However, DRING (1970) and DUNCAN and FOREMAN (1980) calculated that in shallow coastal areas where plants such as kelps grow upward to the surface, the energy of the available red radiation (660 nm), and the red–far-red ratios could be sufficient to activate the phytochrome system. Red–far-red physiological responses recorded in algae include the far-red light stimulation of stipe elongation in kelps and its inhibition by red light (DUNCAN and FOREMAN 1980) and far-red light inhibition of cell division in planktonic algae and its reversal by red light (LIPPS 1973). Phytochrome has so far been extracted from only two algal species (TAYLOR and BONNER 1967; VAN DER VELDE et al. 1978). Obviously, the significance of phytochrome in marine systems has yet to be established. (For detailed discussion of red-far-red effects in algae with respect to locomotion and intracellular movements see Vol. 7 of this Series.)

9.5 Conclusions

The most fundamental response that aquatic plants have made to their light environment is the evolution of an array of light-harvesting pigment proteins capable of trapping the particular portions of the visible spectrum available. Photosynthesis in light-limited environments (e.g. the sea) has characteristics which are basically different from those of white light-saturated terrestrial environments. For example, inspection of Table 9.2 reveals the remarkable fact that 7.4% of the world's chlorophyll (in marine plants) fixes 31.9% of the world's carbon, whereas 89.9% of the world's chlorophyll (in terrestrial plants) fixes only 64.2% of the world's carbon. This suggests a higher efficiency for marine photosynthesis (photosynthetic productivity of aquatic ecosystems will be discussed in Chap. 10, Vol. 12D). These figures however may be biased since only the blue bands of chlorophyll would be used in the sea, whereas in terrestrial habitats both red and blue bands are active. Also in marine photosynthesis the carotenoids, fucoxanthin and peridinin, and the accessory chlorophylls b and c, would make additional light-harvesting contributions. New studies on chlorophyll turnover and on highly active chlorophyllases in unicellular marine plants (JEFFREY and HALLEGRAEFF in preparation) point to a potential for fast adaptation of the photosynthetic apparatus to changing light fields, not found in (or required by) terrestrial plants. Furthermore, the role of carotenoids differs in marine and terrestrial plants. From a predominantly light-harvesting role in the sea, they assume a photo-protective function on land. They may also have other functions still incompletely understood in both environments (e.g. the epoxide cycle, see JEFFREY 1980). The question of the two chlorophylls c is also unresolved. These widely distributed pigments in most marine plant groups undoubtedly have a light-harvesting function, but why are two closely related forms required for those taxa that contain the fucoxanthin pigment system? Further, are there any reasons, apart from photoinhibition, for the greater efficiencies of marine photosynthesis found below rather than at the surface, of the photic zone?

This chapter reviewing functional responses to light by plants in aquatic systems reveals large areas of ignorance. Many exciting questions challenge investigators of the deep-blue sea.

Acknowledgments. I wish to thank Dr. R.J. King for helpful discussions on the aquatic macrophyte sections, and Dr. P. Hindley for constructive editorial help.

References

Ahlquist CD (1965) Strålningsenergins (från sol och himmel) (fördelning i. N. Östersion, Ålandshav och S. bottenhavet; unpublished)

Anderson LWJ, Sweeney BM (1977) Diel changes in sedimentation characteristics of *Ditylum brightwelli*: changes in cellular lipid and effects of respiratory inhibitors and ion-transport modifiers. Limnol Oceanogr 22:539–552

Barrett J, Anderson JM (1980) The P700-chlorophyll *a*-protein complex and two major light-harvesting complexes of *Acrocarpia paniculata* and other brown seaweeds. Biochim Biophys Acta 590:309–323

Beardall J, Morris I (1976) The concept of light intensity adaptation in marine phytoplankton: some experiments with *Phaeodactylum tricornutum*. Mar Biol 37:377–387

Bennett A, Bogorad L (1973) Complementary chromatic adaptation in a filamentous blue-green alga. J Cell Biol 58:419–435

Björn GS, Björn LO (1976) Photochromatic pigments from blue-green algae: phycochromes A, B and C. Physiol Plant 36:297–304

Blasco D (1978) Observations on the diel migration of marine dinoflagellates off the Baja California coast. Mar Biol 46:41–47

Blatt MR, Briggs WR (1980) Blue-light-induced cortical fibre reticulation concomitant with chloroplast aggregation in the alga *Vaucheria sessilis*. Planta 147:355–362

Bogorad L (1975) Phycobiliproteins and complementary chromatic adaptation. Ann Rev Plant Physiol 26:369–401

Bold HC, Wynne MJ (1978) Introduction to the algae, structure and reproduction. Prentice Hall Inc. Englewood Cliffs, New Jersey

Bonotto S (1976) Cultivation of plants: multicellular. In: Kinne O (ed) Marine ecology: A comprehensive, integrated treatise of life in oceans and coastal waters. Vol III Cultivation Part I. Wiley and Sons, New York pp 467–529

Briggs WR, Blatt MR (1980) Blue light responses in the siphonous alga *Vaucheria*. In: Senger H (ed) The Blue Light Syndrome. Springer-Verlag, Berlin, Heidelberg, New York

Brown M (1977) Transmission spectroscopy examinations of natural waters. C. Ultraviolet spectral characteristics of the transition from terrestrial humus to marine yellow substance. Estuarine ε Coastal Mar Sci 5:309–317

Budzikiewicz H, Taraz K (1971) Chlorophyll *c*. Tetrahedron 27:1447–1460

Checcucci A, Colombetti G, Ferrara R, Lenci F (1976) Action spectra for photoaccumulation of green and colourless *Euglena*: evidence for identification of receptor pigments. Photochem Photobiol 23:51–54

Clayton M, King RJ (eds) (1981) Marine botany: an Australasian perspective. Longman Cheshire, Melbourne 468 pp

Cox ER, Zingmark RG (1971) Bibliography of the Pyrrophyta. In: Roscowski JR, Parker BC (eds) Selected papers in phycology. Dept of Botany, Univ of Nebraska, Lincoln, pp 803–808

Dawson EY (1966) Marine Botany: an Introduction. Holt, Rinehart and Winston, New York

De Marsac NT (1977) Occurrence and nature of chromatic adaptation in Cyanobacteria. J Bacteriol 130:82–91

Diakoff S, Scheibe J (1973) Action spectra for chromatic adaptation in *Tolypothrix tenuis*. Plant Physiol 51:382–385

Dodge JD (1973) The fine structure of algal cells. Academic Press, New York

Dougherty RC, Strain HH, Svec WA, Uphaus RA, Katz JJ (1966) Structure of chlorophyll *c*. J Am Chem Soc 88:5037–5038

Dougherty RC, Strain HH, Svec WA, Uphaus RA, Katz JJ (1971) The structure, properties and distribution of chlorophyll *c*. J Am Chem Soc 92:2826–2833

Dring MJ (1970) Photoperiodic effects in microorganisms. In: Halldal P (ed) Photobiology of microorganisms. Wiley-Interscience, London, pp 345–368

Dubinsky Z, Berman T (1976) Light utilization efficiencies of phytoplankton in Lake Kinneret (Sea of Galilee). Limnol Oceanogr 21:226–230

Dubinsky Z, Berman T (1979) Seasonal changes in the spectral composition of upwelling irradiance in Lake Kinneret (Israel). Limnol Oceanogr 24:652–663

Duncan MJ, Foreman RE (1980) Phytochrome-mediated stipe elongation in the kelp *Nereocystis* (Phaeophyceae). J Phycol 16:138–142

Ehret CF, Wille JJ (1970) The photobiology of circadian rhythms in protozoa and other eukaryotic microorganisms. In: Halldal P (ed) Photobiology of microorganisms. Wiley-Interscience, London, pp 369–416

Eppley RW, Holm-Hansen O, Strickland JDH (1968) Some observations on the vertical migration of dinoflagellates. J Phycol 4:333–340

Fogg GE (1973) Physiology and ecology of marine blue-green algae. In: Carr NG, Whitton BA (eds) The biology of blue-green algae. Blackwell, Oxford, pp 368–378

Fujita Y, Hattori A (1960) Effect of chromatic light on phycobilin formation in a blue-green alga, *Tolypothrix tenuis*. Plant Cell Physiol 1:293–303

Gantt E (1975) Phycobilisomes: light-harvesting pigment complexes. Bio Science 25:781–788

Gantt E, Edwards MR, Provasoli L (1971) Chloroplast structure of the Cryptophyceae. Evidence for phycobiliproteins within intrathylakoid spaces. J Cell Biol 48:280–290

Gendel S, Ohad I, Bogorad L (1979) Control of phycoerythrin synthesis during chromatic adaptation. Plant Physiol 64:786–790

Gysi JR, Chapman DJ (1981) Phycobiliproteins and phycobilins. In: Black CC, Mitsui A (eds) Handbook of Biosolar Resources. Vol I, Fundamental Principles, Part 1. CRC Press, Florida (in press)

Hager A (1975) Die reversiblen lichtabhängigen Xanthophyllumwandlungen im Chloroplasten. Ber Dtsch Bot Ges 88:27–44

Halldal P (1970) The photosynthetic apparatus of microalgae and its adaptation to environmental factors. In: Halldal P (ed) Photobiology of microorganisms. Wiley-Interscience, London, pp 17–55

Harper MA (1977) Movements. In: Werner D (ed) The Biology of diatoms. Blackwell, Oxford, pp 224–249

Haupt W, Schönbohm E (1970) Light-oriented chloroplast movements. In: Halldal P (ed) Photobiology of microorganisms. Wiley-Interscience, London, pp 283–307

Haxo FT (1960) The wavelength dependence of photosynthesis and the role of the accessory pigments. In: Allen MB (ed) Comparative biochemistry of photoreactive systems. Academic Press, New York, pp 339–359

Herman EM, Sweeney BM (1975) Circadian rhythm of chloroplast ultrastructure in *Gonyaulax polyhedra*, concentric organization around a central cluster of ribosomes. J Ultrastruct Res 50:347–354

Højerslev NK (1978) Daylight measurements appropriate for photosynthetic studies in natural sea waters. J Cons Int Explor Mer 38:131–146

Højerslev NK, Jerlov NG, Kullengerg G (1977) Colour of the ocean as an indicator in photosynthesis studies. J Cons Int Explor Mer 37:313–316

Holdsworth RH (1971) The isolation and partial characterization of the pyrenoid protein of *Eremosphaera viridis*. J Cell Biol 51:499–513

Hostetter HP, Stoermer EF (1971) Bibliography on the Bacillariophyceae. In: Roscowski JR, Parker BC (eds) Selected papers in phycology. Dept of Botany, Univ of Nebraska, Lincoln, pp 784–796

Jeffrey SW (1968a) Quantitative thin layer chromatography of chlorophylls and carotenoids from marine algae. Biochim Biophys Acta 162:271–285

Jeffrey SW (1968b) Two spectrally distinct components in preparations of chlorophyll *c*. Nature 220:1032–1033

Jeffrey SW (1969) Properties of two spectrally different components in chlorophyll *c* preparations. Biochim Biophys Acta 177:456–467

Jeffrey SW (1972) Preparation and some properties of crystalline chlorophyll c_1 and c_2 from marine algae. Biochim Biophys Acta 279:15–33

Jeffrey SW (1974) Profiles of photosynthetic pigments in the ocean using thin layer chromatography. Mar Biol 26:101–110

Jeffrey SW (1976) The occurrence of chlorophyll c_1 and c_2 in algae. J Phycol 12:349–354

Jeffrey SW (1980) Algal pigment systems. In: Falkowski PG (ed) Primary productivity in the sea. Brookhaven Symp Biol 31, Plenum Press, New York, pp 33–58

Jeffrey SW (1981) Phytoplankton ecology, with particular reference to the Australasian region. In: Clayton MN, King RJ (eds) Marine botany: an Australasian perspective. Longman Cheshire, Melbourne, pp 241–291

Jeffrey SW, Hallegraeff GM (1980) Studies of phytoplankton species and photosynthetic pigments in a warm core eddy of the East Australian Current. II. A note on pigment methodology. Mar Ecol Progr Ser 3:295–301

Jeffrey SW, Vesk M (1977) Effect of blue-green light on photosynthetic pigments and chloroplast structure in the marine diatom *Stephanopyxis turris*. J Phycol 13:271–279

Jeffrey SW, Vesk M (1981a) The phytoplankton – systematics, morphology and ultrastructure. In: Clayton MN, King RJ (eds) Marine botany: an Australasian perspective. Longman Cheshire, Melbourne pp 138–179

Jeffrey SW, Vesk M (1981b) Blue-green light effects in marine micro-algae: enhanced thylakoid and chlorophyll synthesis. In: Akoyunoglou G (ed) Proc Fifth Int Congr Photosyn 5: (in press)

Jeffrey SW, Sielicki M, Haxo FT (1975) Chloroplast pigment patterns in dinoflagellates. J Phycol 11:374–384

Jerlov NG (1970) Light – general introduction. In: Kinne O (ed) Marine Ecology. Vol I, part I. Wiley and Sons, pp 95–102

Jerlov NG (1976) Marine Optics. Elsevier Oceanogr Ser 14, Elsevier, Amsterdam Oxford New York

Jerlov NG (1977) Classification of seawater in terms of quanta irradiance. J Cons Int Explor Mer 37:281–287

Jerlov NG, Steemann Nielsen E (eds) (1974) Optical aspects of oceanography, Academic Press, London

Jørgensen EG (1977) Photosynthesis. In: Werner D (ed) The biology of diatoms. Blackwell, Oxford, pp 150–168

Kageyama A, Yokohama Y, Shimura S, Ikawa T (1977) An efficient excitation energy transfer from a carotenoid, siphonaxanthin to chlorophyll *a* observed in a deep-water species of chlorophycean seaweed. Plant Cell Physiol 18:477–480

Kahn N, Swift E (1978) Positive buoyancy through ionic control in the non-motile marine dinoflagellate *Pyrocystis noctiluca* Murray ex Schuett. Limnol Oceanogr 23:649–658

Kalle K (1966) The problem of Gelbstoff in the sea. Oceanogr Mar Biol Ann Rev 4:91–104

Kamykowski D, Zentara S (1977) The diurnal vertical migration of motile phytoplankton through temperature gradients. Limnol Oceanogr 22:148–151

Kasemir H (1979) Control of chloroplast formation by light. Cell Biol Int Rep 3:197–214

Keast JF, Grant BR (1976) Chlorophyll *a*:*b* ratios in some siphonous green algae in relation to species and environment. J Phycol 12:328–331

Kiefer DA, Olson RJ, Holm-Hansen O (1976) Another look at the nitrite and chlorophyll maxima in the central North Pacific. Deep Sea Res 23:1199–1208

King RJ (1981a) Marine angiosperms: Seagrasses In: Clayton MN, King RJ (eds) Marine botany: an Australasian perspective. Longman Cheshire, Melborne, p 200–210

King RJ (1981b) Mangroves and saltmarsh plants. In: Clayton MN, King RJ (eds) Marine botany: an Australasian perspective. Longman Cheshire, Melbourne, p 308–328

Kirk JTO (1976) Yellow substance (Gelbstoff) and its contribution to the attenuation of photosynthetically active radiation in some inland and coastal South-Eastern Australian waters. Aust J Mar Freshwater Res 27:61–71

Kirk JTO (1979) Spectral distribution of photosynthetically active radiation in some South-eastern Australian waters. Aust J Mar Freshwater Res 30:81–91

Laskin AI, Lechevalier HA (1978) Handbook of Microbiology: Vol. II, Fungi, Algae, Protozoa, and Viruses, 2nd edn. CRC Press, Inc, Florida USA

Lewin RA (1976) Prochlorophyta as a proposed new division of algae. Nature 261:697–698

Liaaen-Jensen S (1977) Algal carotenoids and chemosystematics. In: Faulkner DJ, Fenical WH (eds) Marine Natural Products Chemistry, Plenum Press, New York, pp 239–259

Liaaen-Jensen S (1978) Marine carotenoids. In: Scheuer PJ (ed) Marine natural products, Vol II Academic Press, New York pp 1–73

Liaaen-Jensen S (1979) Carotenoids – a chemosystematic approach. Pure Appl Chem 51:661–675

Lichtlé C, Jupin H, Duval JC (1980) Energy transfers from Photosystem II to Photosystem I in *Cryptomonas rufescens* (Cryptophyceae). Biochim Biophys Acta 591:104–112

Lieth H (1975) Primary production of the major vegetation units of the world. In: Lieth H, Whittaker RH (eds) Primary productivity of the biosphere. Ecological Studies 14 Springer, Berlin, Heidelberg, New York pp 203–215

Likens GE (1975) Primary production of inland aquatic ecosystems. In: Lieth H, Whittaker RH (eds) Primary productivity of the biosphere. Ecological Studies 14 Springer, Berlin, Heidelberg, New York pp 185–202

Lipps MJ (1973) The determination of far-red effect on marine phytoplankton. J Phycol 9:237–242

Lorenzen CJ (1972) Extinction of light in the ocean by phytoplankton. J Cons Int Explor Mer 34:262–267

Ludgren B, Højerslev NK (1971) Daylight measurements in the Sargasso Sea. Results from the "Dana" expedition January-April 1966. Univ Copenhagen, Inst Phys Oceanogr Rep 14:1–44

Lüning K (1980) Critical levels of light and temperature regulating the gametogenesis of three *Laminaria* species (Phaeophyceae). J Phycol 16:1–15

Lüning K, Dring MJ (1975) Reproduction, growth and photosynthesis of gametophytes of *Laminaria saccharina* grown in blue and red light. Mar Biol 29:195–200

Lüning K, Dring MJ (1979) Continuous underwater light measurement near Helgoland

(North Sea) and its significance for characteristic light limits in the sublittoral region. Helgol Wiss Meeresunters 32:403–424

MacColl R, Berns DS (1978) Energy transfer studies on cryptomonad biliproteins. Photochem Photobiol 27:343–349

Mann KH (1973) Seaweeds: their productivity and strategy for growth. Science 182:975–981

Mitrakos K, Shropshire W (eds) (1972) Phytochrome. Academic Press, London

Mohr H (1977) Phytochrome and chloroplast development. Endeavour New Series 1:107–114

Morel A (1978) Available, usable and stored radiant energy in relation to marine photosynthesis. Deep Sea Res 25:673–688

Morel A, Prieur L (1977) Analysis of variations in ocean colour. Limnol Oceanogr 22:709–722

Morel A, Smith RC (1974) Relation between total quanta and total energy for aquatic photosynthesis. Limnol Oceanogr 19:591–600

Nakamura K, Ogawa T, Shibata K (1976) Chlorophyll and peptide compositions in the two photosystems of marine green algae. Biochim Biophys Acta 423:227–236

Nultsch W (1971) Phototactic and photokinetic action spectra of the diatom *Nitzschia communis*. Photochem Photobiol 14:705–712

Nultsch W (1974) Movements. In: Stewart WDP (ed) Algal physiology and biochemistry. Blackwell, Oxford, pp 864–893

Nultsch W, Pfau J (1979) Occurrence and biological role of light-induced chromatophore displacements in seaweeds. Mar Biol 51:77–82

Nultsch W, Schuchart H, Dillenburger M (1979) Photomovement of the red alga *Porphyridium cruentum* (Ag.) Naegeli I. Photokinesis. Arch Microbiol 122:207–212

Ohad I, Schneider HAW, Gendel S, Bogorad L (1980) Light-induced changes in allophycocyanin. Plant Physiol 65:6–12

Olson JM (1978) Precambrian evolution of photosynthetic and respiratory organisms. In: Hecht MK, Steere WC, Wallace B (eds) Evolutionary biology. Vol II, Plenum Press, New York pp 1–37

Presti D, Delbrück M (1978) Photoreceptors for biosynthesis, energy storage and vision. Plant Cell Environ 1:81–100

Prézelin BB (1976) The role of peridinin-chlorophyll *a*-proteins in the photosynthetic light adaptation of the marine dinoflagellate, *Glenodinium* sp. Planta 130:225–233

Prézelin BB, Alberte RS (1978) Relationships of photosynthetic characteristics and the organization of chlorophyll in marine dinoflagellates. Proc Natl Acad Sci 75:1801–1805

Prézelin BB, Haxo FT (1976) Purification and characterization of peridinin-chlorophyll *a*-proteins from the marine dinoflagellate *Glenodinium* sp. and *Gonyaulax polyedra*. Planta 128:133–141

Prézelin BB, Sweeney BM (1978) Photoadaptation of photosynthesis in *Gonyaulax polyedra*. Mar Biol 48:27–35

Prézelin BB, Sweeney BM (1979) Photoadaptation of photosynthesis in two bloom-forming dinoflagellates. In: Taylor DL, Seliger HH (eds) Toxic dinoflagellate blooms. Elsevier North Holland, Amsterdam, New York pp 101–106

Ramus J, Beale SI, Mauzerall D (1976a) Correlation of changes in pigment content with photosynthetic capacity of seaweeds as a function of water depth. Mar Biol 37:231–238

Ramus J, Beale SI, Mauzerall D, Howard KL (1976b) Changes in photosynthetic pigment concentration in seaweeds as a function of water depth. Mar Biol 37:223–229

Riper DM, Owens TG, Falkowski PG (1979) Chlorophyll turnover in *Skeletonema costatum*, a marine plankton diatom. Plant Physiol 64:49–54

Roscowski JR, Parker BC (1971) (eds) Selected papers in phycology. Dept Botany, Univ of Nebraska, Lincoln, Nebraska

Round FE (1973) The biology of the algae. 2nd edn. Edward Arnold, London

Scagel RF, Bandoni RJ, Rouse GE, Schofield WB, Stein JR, Taylor TMC (1969) Plant diversity: an evolutionary approach. Wadsworth Pub Co Inc Belmont, California

Scheibe J (1972) Photoreversible pigment: occurrence in a blue-green alga. Science 176:1037–1039

Senger H (ed) (1980) The Blue Light Syndrome. Proc Int Conf on the Effect of Blue

Light on Plants and Microorganisms, Marburg, W. Germany, Springer-Verlag, Berlin, Heidelberg, New York

Senger H, Briggs WR (1980) The blue light receptor(s): primary reactions and subsequent metabolic changes In: Senger H (ed) Effect of blue light on plants and microorganisms, Springer-Verlag, Berlin, Heidelberg, New York

Shimura S, Fujita Y (1975) Changes in the activity of fucoxanthin-excited photosynthesis in the marine diatom *Phaeodactylum tricornutum* grown under different culture conditions. Mar Biol 33:185–194

Shulenberger E (1978) The deep chlorophyll maximum and mesoscale environmental heterogeneity in the western half of the North Pacific central gyre. Deep Sea Res 25:1193–1208

Smith RC, Baker KS (1978a) The bio-optical state of ocean waters and remote sensing. Limnol Oceanogr 23:247–259

Smith RC, Baker KS (1978b) Optical classification of natural waters. Limnol Oceanogr 23:260–267

Steemann Nielsen E (1975) Marine photosynthesis. Elsevier Oceanogr Ser 13. Elsevier, Amsterdam, Oxford, New York

Strain HH, Thomas MR, Katz JJ (1963) Spectral absorption properties of ordinary and deuterated chlorophylls *a* and *b*. Biochim Biophys Acta 75:306–311

Strain HH, Cope BT, McDonald GN, Svec WA, Katz JJ (1971) Chlorophylls c_1 and c_2. Phytochem 10:1109–1114

Sweeney BM, Prézelin BB, Wong D, Govindjee (1979) *In vivo* chlorophyll *a* fluorescence transients and the circadian rhythm of photosynthesis in *Gonyaulax polyedra*. Photochem Photobiol 30:309–311

Talling JF (1971) The underwater light climate as a controlling factor in the production ecology of freshwater phytoplankton. Mitt Int Ver Theor Angew Limnol 19:214–243

Taylor AO, Bonner BA (1967) Isolation of phytochrome from the alga *Mesotaenium* and the liverwort *Sphaerocarpos*. Plant Physiol 42:762–766

Taylor FJR (1976) Dinoflagellates from the International Indian Ocean Expedition. Biblio Bot 132:1–234

Thornber JP, Barber J (1979) Photosynthetic pigments and models for their organization *in vivo*. In: Barber J (ed) Photosynthesis in relation to model systems. Elsevier, North Holland, Amsterdam, New York pp 27–70

Thornber JP, Markwell JP, Reinman S (1979) Plant chlorophyll-protein complexes: recent advances. Photochem Photobiol 29:1205–1216

Thrash RJ, Fang HL-B, Leroi GE (1979) On the role of forbidden low-lying excited states of light-harvesting carotenoids in energy transfer in photosynthesis. Photochem Photobiol 29:1049–1050

Tyler JE (1975) The *in situ* quantum efficiency of natural phytoplankton populations. Limnol Oceanogr 20:976–980

Van der Velde HH, Henrika Wagner AM (1978) The detection of phytochrome in the red alga *Acrochaetium daviesii*. Plant Sci Lett 11:145–149

Venrick EL, McGowan JA, Mantyla AW (1973) Deep maxima of photosynthetic chlorophyll in the Pacific Ocean. Fish Bull 71:41–52

Vernon LP, Seely GR (eds) (1966) The chlorophylls. Academic Press, New York

Vesk M, Jeffrey SW (1977) Effect of blue-green light on photosynthetic pigments and chloroplast structure in unicellular marine algae from six classes. J Phycol 13:280–288

Voskresenskaya NP (1979) Effect of light quality on carbon metabolism. In: Gibbs M, Latzko E (eds) Photosynthesis II – Photosynthetic carbon metabolism and related processes. Encyclopedia of plant physiology, new series 6. Springer-Verlag, Berlin, Heidelberg, New York pp 174–180

Wallen DG, Geen CH (1971a) Light quality in relation to growth, photosynthetic rates and carbon metabolism in two species of marine planktonic algae. Mar Biol 10:34–43

Wallen DG, Geen GH (1971b) Light quality and concentration of proteins, RNA, DNA and photosynthetic pigments in two species of marine planktonic algae. Mar Biol 10:44–51

Wasley JFW, Scott WT, Holt AS (1970) Chlorophyllides *c*. Canad J Biochem 48:376–383

Watanabe M, Furuya M (1974) Action spectrum of phototaxis in a cryptomonad alga, *Cryptomonas* sp. Plant Cell Physiol 15:413–420

Waterbury JB, Watson SW, Guillard RRL, Brand LE (1979) Widespread occurrence of a unicellular marine planktonic cyanobacterium. Nature 277:293–294

Whittaker RH, Likens GE (1975) The biosphere and man. In: Lieth H, Whittaker RH (eds) Primary productivity of the biosphere. Ecological Studies 14. Springer, Berlin, Heidelberg, New York pp 305–328

Whitton BA (1973) Freshwater plankton. In: Carr NG, Whitton BA (eds) The biology of blue-green algae. Blackwell, Oxford, pp 353–367

Womersley HBS (1971) *Palmoclathrus*, a new deep water genus of Chlorophyta. Phycologia 10:229–233

Wood AM (1979) Chlorophyll *a*:*b* ratios in marine planktonic algae. J Phycol 15:330–332

Yokohama Y, Kageyama A, Ikawa T, Shimura S (1977) A carotenoid characteristic of chlorophycean seaweeds living in deep coastal waters. Bot Marina 20:433–436

10 Responses of Macrophytes to Temperature [1]

J.A. BERRY and J.K. RAISON

CONTENTS

1 CIW Publication number 747

10.1 Introduction

Temperature is a major factor determining the natural distribution of plants, and the success and timing of agricultural crops. Habitats occupied by plants show dramatic differences in temperature during the period of active growth, ranging from just above freezing in polar or alpine areas to over 50 °C in the hottest deserts. Moreover, in many habitats the same individual plant is subjected to wide seasonal variations in temperature and even diurnal temperature fluctuations may be considerable.

We know that the primary sites affected by temperature are at a molecular level, such as the structure of substances, the position of chemical equilibria, and the rate of chemical transformations. These molecular events are all affected by temperature in ways which are now (at least in general terms) fairly well understood. These primary effects are, however, very difficult to interpret in the context of plant sciences where the significant observations are the effect of temperature on the rate of a physiological processes, plant growth, or the timing of some developmental event such as seed germination, flowering, or the breaking of dormancy. Ecological phenomena such as the distribution of plants in various natural communities or the competition among species are even more difficult to interpret in relation to the primary effects of temperature. In this chapter we attempt to develop some links between the effects of temperature at a molecular level and the impact of temperature at other levels. Our plan is to provide a brief treatment of temperature at successively more abstract levels, beginning at the plant environment interface and working back toward the molecular events.

10.2 Tissue Temperature

Even though temperature is the most commonly measured environmental variable, the thermal regime of plants is often not adequately characterized. The temperature most frequently used to characterize an environment is that of the air in a standard meteorological enclosure. Plant tissue temperature may differ quite significantly from that of the air. Leaves being exposed to radiant heat exchange may experience somewhat larger extremes of temperature than the air. Roots, owing to the large thermal mass of their substrate, are buffered from the diurnal extremes of temperature and are nearer the mean daily temperature. Other plant parts may fall between these extremes.

For the most part, plant temperature is determined by purely physical interactions with the environment (see Chapt. 1, this Vol.), however, direct metabolic control of plant tissue temperature does occur in some plants. For example, the heat produced by rapid uncoupled respiration of stored carbohydrate causes the inflorescence of *Symplocarpus foetidus* (eastern skunk cabbage) to be significantly warmer than air temperature during a period of rapid development in late winter. Over this time the respiratory production of heat appears to be regulated in such a way as to compensate for changes in ambient temperature and to maintain tissue temperature at 20 to 30 °C (KNUTSON 1974). In other members of the Arum family, respiratory heat production may serve to volatilize odoriferous substances which attract pollinators (MEEUSE 1975). Such metabolic thermoregulation requires that the tissue have a small surface to volume ratio and access to a copious supply of respiratory substrates. Only a few plant organs, such as rapidly developing shoot apices, might thus be expected to thermoregulate by metabolic means, and this phenomenon is comparatively rare (for review see MEEUSE 1975).

Purely physical heat exchange processes may nevertheless result in substantial deviations between leaf temperature and that of the air as measured in a standard meteorological enclosure, and plants occupying the same physical environment may experience quite different tissue temperatures. For example, in a hot desert environment stem succulents such as *Carnegia gigantea* may reach temperatures 10 to 15 °C above the air temperature (NOBEL 1978), while rapidly transpiring leaves of desert ephemerals may maintain leaf temperatures below air temperature (LANGE 1959; PEARCY et al. 1974; see also Chapt. 14, this Vol.).

In many instances the leaf temperatures may be more favorable for physiological activity than the air temperature. For example, desert species with leaf temperatures cooler than a hot ambient temperature (SMITH 1978); also alpine plants (GATES et al. 1954) or winter annuals with leaf temperatures above a cool ambient air temperature (REGEHR and BAZZAZ 1976). Morphological characteristics which may enable plants to maintain a more favorable temperature may represent an evolutionary alternative to physiological adaptations. Such modifications include the ability to form reflective leaf surfaces (MOONEY et al. 1977; EHLERINGER 1980), the light-seeking or light-avoiding leaf movements (EHLERINGER and FORSETH 1980) and rosette or cushion growth forms. In other instances, differences between tissue and air temperature may accentuate an unfavorable temperature regime.

In considering temperature as an ecological factor it is also important to note that other factors may co-vary with temperature in a given habitat. Seasonal changes in air temperature in temperate latitudes are linked with changes in day length, solar elevation, and precipitation regimes. Temperature may be only one of a catena of environmental factors.

10.3 Productivity in Relation to Temperature

Interest in the effect of temperature on the development of crops has been a prominent concern of plant scientists for at least 250 years. In 1735 DE RÉAUMUR developed the summation index for expressing the combined influence of time and temperature on the development of plants. By totaling the mean daily temperature for each day of the months of April, May, and June, and comparing this sum to the date that plants reached a certain stage of development, he concluded that each species required a certain amount of heat (degree-days) to reach a certain stage of development. This index was the first quantitative effort to relate year-to-year variation in climate and the plant development (see WANG 1960). This concept is still used, with some modifications to estimate the time required for seed germination or maturation of crops under cool temperatures (BIERHUIZEN 1973).

One basic flaw of this approach was made clear in the 1860's with SACHS' measurements of the temperature response for growth of different species (see Fig. 3.3 in JOHNSON et al. 1974). He demonstrated that growth was most rapid at an optimum temperature and that it declined at temperatures above and below this optimum until high and low temperature thresholds for growth were reached. These studies were apparently the first to demonstrate the concept of physiological optima and limits. LIVINGSTON and SHREVE (1921), in their effort to relate the vegetation types to climate, devised a physiological summation index based upon empirical measurements of the temperature dependence for the growth of *Zea mays* which took into account limitations to growth at both low and high temperature. They acknowledged, however, that the empirical curve which they used was not adequate, since they were certain that considerable differences existed in the temperature responses of individual species and for a species at different stages of development. However, they lacked the experimental procedures to pursue this point. An index (ROBERTSON 1973) for estimating the development of wheat takes into account curvilinear dependence of growth on maximum and minimum daily temperature and the corresponding dependence of growth on light and the photoperiod for five successive developmental intervals. While this index is very useful for predicting the development of this crop and for predicting the geographic limits for the cultivation of this crop, it is specific.

10.3.1 Primary Productivity

Differences among species in their response to temperature were highlighted in studies (BJÖRKMAN et al. 1974a) of growth and survival of plants native to contrasting thermal regimes. Two experimental gardens were used in this study; one was located at Bodega Head along the coast of Northern California; the other was located on the floor of Death Valley, California. Maximum daily temperatures during summer in the Bodega Head site seldom exceeded 20 °C, while those in Death Valley usually exceeded 40 °C and occasionally exceeded 50 °C. Plants native to several environments were transplanted to these gardens. (In the Death Valley garden this was done during the spring while temperatures were moderate.) Although plants in the Death Valley garden were provided with abundant water, the growth of many of these species was severely limited by the high temperatures which occurred during the summer months. In contrast, some species which failed in the Death Valley garden thrived in the cool coastal garden, but temperatures in this garden were obviously too cool for other species. The responses in the two gardens were scored into three general categories (Table 10.1): (1) Those unable to survive, (2) those which survived but made only slow growth, and (3) those which maintained rapid growth. Plants which maintained rapid growth in the coastal garden

Table 10.1. Summer performance of several species of *Atriplex* and *Tidestromia* in transplant gardens located in hot desert (Death Valley) and cool maritime regions (Bodega Head). (BJÖRKMAN et al. 1974a)

Death Valley Garden		
Unable to survive	Slow growth	Rapid growth
A. glabriuscula *A. triangularis* *A. sabulosa* *A. rosea*	*A. hymenelytra* *A. lentiformis*	*T. oblongifolia*
Bodega Head Garden		
Unable to survive	Slow growth	Rapid growth
T. oblongifolia	*A. hymenelytra* *A. lentiformis*	*A. glabriuscula* *A. triangularis* *A. sabulosa* *A. rosea*

were unable to survive in the Death Valley garden, and the only species, *Tidestromia oblongifolia,* which made rapid growth during summer in the Death Valley garden, was unable to survive in the coastal garden. The adaptations which apparently permit efficient performance of plants at one temperature extreme appear to be detrimental at the opposite temperature extreme.

The growth responses of these plants may be related to the thermal regimes of the native habitats of these species. *Tidestromia oblongifolia,* the only species which made rapid growth at high temperature is a native to Death Valley, and is closely related to members of the summer annual flora of the Sonoran Desert. These summer annuals characteristically make their major growth after summer rain storms (SHREVE and WIGGINS 1964). In the cool winter season of Death Valley, *T. oblongifolia* remains dormant.

The species unable to survive in the Death Valley garden but capable of rapid growth in the Bodega Head garden were annuals native either to coastal regions (*Atriplex sabulosa* or *A. glabriuscula*) or to the warm interior valleys of California (*A. rosea* or *A. triangularis*). These species characteristically germinate after winter and spring rain storms. Many extend their growth into the warm summer months but were unable to tolerate the extreme heat of Death Valley. Those species which maintained slow growth in both gardens were perennial species native to Death Valley (*A. hymenelytra, Larrea divaricata* and *A. lentiformis*). These species are active throughout the year in their native habitat, and since the temperatures at the coastal garden were similar to those in Death Valley during winter, it is not surprising that these species could sustain growth in either garden.

Experiments conducted in controlled environment chambers with thermal regimes similar to those of the two gardens but with high atmospheric humidity, abundant water, and nutrients (BJÖRKMAN et al. 1974b) confirmed that the growth responses observed in the field were primarily conditioned by temperature. Quantitative analysis of growth over a range of temperatures showed marked differences between these species in their responses to temperature (Fig. 10.1). These differences in growth response correspond very well to differences in the response of photosynthesis of normal healthy leaves of these species to temperature (Fig. 10.2). This comparison is complicated, however, because the photosynthetic response to temperature may change after prolonged exposure to a given thermal regime. In this example *A. sabulosa* and *A. glabriuscula* exposed to 45 °C or *T. oblongifolia* exposed to 16 °C for several days suffered damage to their photosynthetic capacity (BJÖRKMAN et al. 1975). This accounts for the somewhat stronger inhibition of growth of cool climate species at high temperature and of *T. oblongifolia* at low temperature in Fig. 10.1 as compared to short-term photosynthetic responses to temperature in Fig. 10.2.

Studies of the kinetics and partitioning of growth as a function of temperature (BJÖRKMAN et al. 1974b; OSMOND et al. 1980) show that differences between species in the amount of new growth which is directed to leaves, stems, roots, and flowers account for some differences in growth rate; however, the overall pattern of the response to temperature is dominated by the response of photosynthesis to temperature. Large genetically determined differences in the constitution of the basic growth-related physiological processes apparently underlie the different growth responses of these species. PEARCY (1976, 1977) examined populations of *A. lentiformis*, a perennial species which occurs both on the coast of Southern California (where the climate is much warmer than that at Bodega Head but

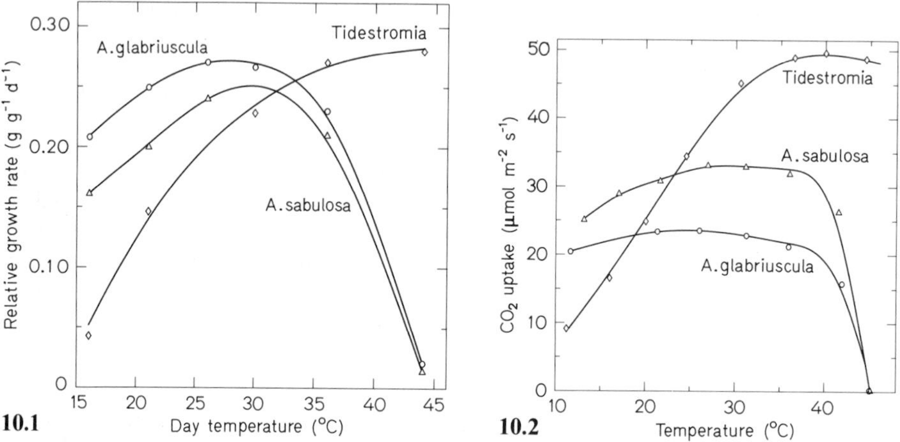

Fig. 10.1. The daily relative growth rate as a function of temperature in controlled environment cabinets for seedlings of *Atriplex glabriuscula, A. sabulosa*, and *Tidestromia oblongifolia*. The light period was 14 h per day with 2,016 μmol quanta (0.4–0.7 μ) cm^{-2} d^{-1}. Night temperature was two-thirds of the day temperature in each regime. (BJÖRKMAN et al. 1974b)

Fig. 10.2. The temperature dependence of photosynthetic CO_2 uptake for leaves of *Atriplex glabriuscula* (C_3), *A. sabulosa* (C_4) and *Tidestromia oblongifolia* (C_4) at a high irradiance (1,600 μmol quanta m^{-2} s^{-1}) and normal CO_2 and O_2 concentrations (320 μbar and 210 mbar, respectively). Leaves of the *Atriplex* species were from plants grown at 16 °C/11 °C and *Tidestromia* was grown at 40 °C/30 °C, day/night temperature. (Redrawn from BJÖRKMAN et al. 1975 and OSMOND et al. 1980)

still maritime) and in Death Valley. He found that cloned plants from these separate populations showed similar growth rates in controlled environment studies at low and moderate temperatures (Table 10.2); however, the growth rate of the cloned plants from the desert populations was superior at hot 43 °C/35 °C (day/night) growth temperature. This difference corresponds to a somewhat poorer performance of plants from the coastal population in the Death Valley experimental garden during summer than the desert populations (BJÖRKMAN et al. 1974a). PEARCY also measured photosynthetic rates of leaves of the two clones at their respective growth temperatures (Table 10.2), and he attributed the poorer growth performance of the coastal clone at high temperature to the lower rate of photosynthesis. These differences in photosynthetic response were apparently related to differences in the long-term effects of temperature upon the photosynthetic characteristics of leaves of these clones (Fig. 10.3). Leaves of either clone had similar intrinsic responses to temperature over the whole range of temperatures when the plants were grown at moderate temperature (23 °C/18 °C, day/night). The temperature response of leaves of the desert clone when grown at hot temperatures (43 °C/30 °C) showed a shift in its temperature response (Fig. 10.3) such that its rate of photosynthesis at the growth temperatures was improved over that of a leaf which had developed at the 23 °C/18 °C regime (ratio = 1.26). The coastal clone, on the other hand, was adversely affected by growth at high temperature and the rate of photosynthesis at 43 °C of leaves from plants grown at 43 °C/30 °C was *lower* than that of leaves which had grown at moderate temperature (ratio = 0.66).

Growth of the desert clone at high temperature clearly results in a beneficial phenotypic modification. Environmentally induced phenotypic modifications that may be interpreted as being adaptive are termed "acclimations" (BERRY and BJÖRKMAN 1980). The coastal clone apparently lacks the ability to acclimate to high growth temperature. These studies illustrate the role of physiological acclimations in permitting a plant to grow over a range of thermal regimes such as may occur in a single site over the course of a year. Further it is demonstrated that the ability to acclimate is most likely genetically determined. The ability of the desert clone to acclimate to high growth temperature would be of selective advantage in its native desert environment, while the inability of the coastal clone to acclimate to high growth temperature would probably be of little selective disadvantage in its native coastal environment.

Table 10.2. The effect of temperature on the relative growth rates and light saturated photosynthetic rates at the growth temperature of cloned plants of *Atriplex lentiformis* taken from a coastal and a desert population of this species. (PEARCY 1976)

Growth regime °C (day/night)	Relative growth rate ($g\ g^{-1}\ day^{-1}$)		Photosynthetic rate ($\mu mol\ m^{-2} s^{-1}$)	
	Coastal	Desert	Coastal	Desert
23/18	0.138 ± 0.010	0.138 ± 0.004	26.2	27.2
33/25	0.155 ± 0.015	0.155 ± 0.007	–	–
43/30	0.114 ± 0.006	0.136 ± 0.006	13.2	30.4

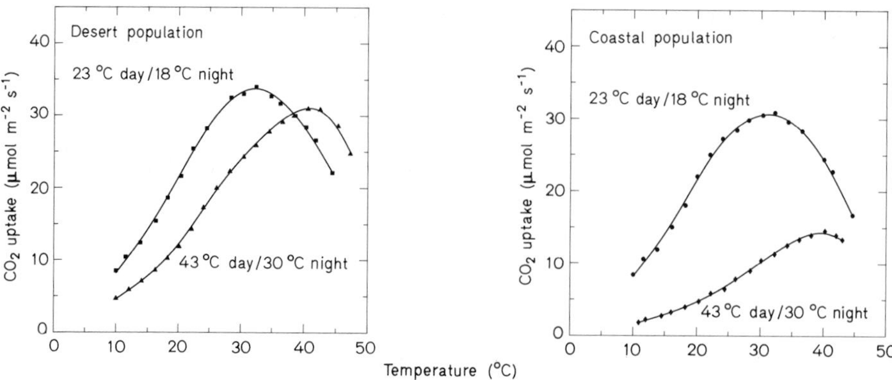

Fig. 10.3. The temperature dependence of photosynthetic CO_2 uptake by leaves of cloned plants of *Atriplex lentiformis* from a coastal and a desert population of this species as influenced by growth temperature. All measurements were made at high light (1,700–1,800 μmol quanta m^{-2} s^{-1}) and normal atmospheric CO_2 and O_2 concentrations. (Redrawn from PEARCY 1977)

MOONEY (1980) draws attention to the importance of seasonal patterns of physiological activity in environments such as Death Valley which have large seasonal changes of temperature. The thermal regime of this continental desert differs as much between seasons (Fig. 10.4) as do widely separated geographical locations. Measurements of photosynthetic activity of plants growing under natural conditions in Death Valley were conducted at four times during a single year. Leaves of evergreen perennial species such as *Larrea divaricata* and *Atriplex hymenelytra* were present and photosynthetically active in all seasons of the year. Other species were present for only a portion of the year. The perennial species (like *Atriplex lentiformis* above) possess a broad temperature tolerance and have considerable capacity to acclimate to different growth temperature (MOONEY et al. 1977; MOONEY 1980). In contrast, species of winter annuals such as *Camissonia claviformis* or the summer-active *Tidestromia oblongifolia* have physiological characteristics which permit each to maintain high rates of photosynthesis in the conditions of its respective growth season (MOONEY et al. 1976; BJÖRKMAN et al. 1971).

The phenomenon of separate or partially separate growing seasons is common in many temperate regions. SHREVE and WIGGINS (1964) divide the ephemeral species of the Sonoran Desert into those which grow in response to winter rains (winter annuals) and those which grow in response to summer rains (summer annuals). Appropriate control of seed germination is an important requirement for such seasonal displacement of activity (WENT 1948, 1949; KOLLER 1972; BEATLEY 1974; Sect. 10.6.7). The Great Plains region of North America has two phenologically separate grass floras; those which are active in spring, and those which are active in mid summer (DODD 1968). The physiological responses of warm season and cool season pasture grasses are reviewed by EVANS et al. (1964). In temperate climates seasonal displacement of growth may result in differences in temperature that correspond to rather large geo-

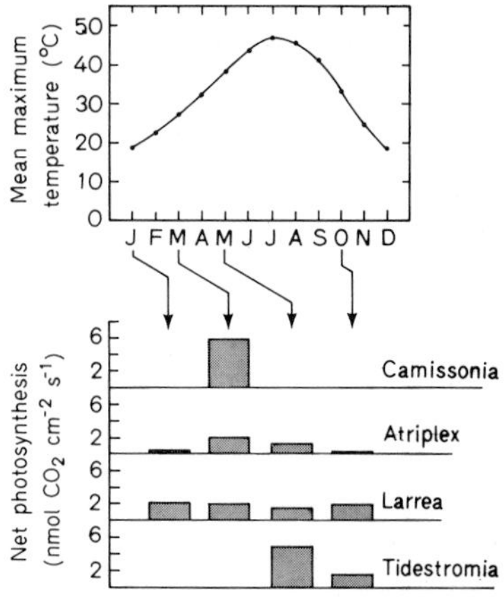

Fig. 10.4. The seasonal maximum photosynthetic capacity for four species native to Death Valley. *Camissonia claviformis* is a winter annual. *Tidestromia oblongifolia* is a summer active herbaceous perennial and *Atriplex hymenelytra* and *Larrea divaricata* are evergreen perennials. Mean maximum daily temperatures are for Furnace Creek, Death Valley. (MOONEY 1980)

graphical separations in the same season. (For detailed discussion of productivity of different ecosystems see Chapts. 5–10, Vol. 12D).

10.3.2 Above and Belowground Productivity

The proportion of primary productivity which is used for above and belowground biomass has a very important impact upon productivity. Not only is belowground biomass concealed and generally of little or no agricultural use, but the growth of roots occurs as an alternative to the growth of new photosynthetic capacity. This in turn may feed back to affect the rate of exponential growth leading to very large cumulative differences in growth (MONSI 1968).

Many natural and agricultural communities divert a significant portion of their growth to belowground growth. Communities of *Atriplex confertifolia* of the Great Basin region of the North America are reported to have sevenfold greater root than shoot biomass (CALDWELL and CAMP 1974). In general, regions which may have the combination of low root temperatures and low nutrient availability tend to have higher root to shoot ratios (OSMOND et al. 1980). DAVIDSON (1969a) conducted experiments in which soil temperature was manipulated while the shoots were exposed to a common shoot environment. Subtropical pasture species produced maximum shoot growth and minimum root to shoot ratio when soil temperature was 35 °C. At lower root temperatures there was an increase in the root to shoot ratio and a decrease in the total shoot biomass accumulated at the completion of the 3-month growth period. Temperate species also showed a tendency for root/shoot to increase at low temperature; however the root/shoot ratio was a minimum at 20–30 °C and also tended to increase at high root temperature. DAVIDSON (1969a) proposed that root/shoot ratio was a minimum at a temperature which was optimal for root function of a particular species. The root/shoot ratio increased at temperatures above or below this optimum resulting in lower rates of shoot growth at these nonoptimal root temperatures. It is postulated that these changes in allocation are required to maintain a balance between root and shoot functions such that the absorption of water, uptake of nutrients, and assimilation of CO_2 occurs at ratio appropriate for the synthesis of new plant biomass (Sect. 10.6.4). Experiments by DAVIDSON (1969b) and others (see THORNLEY 1977) emphasize the interaction of nutrient status of the soil and the effect of temperature upon root/shoot partitioning. When nutrient status is high, the root system may not change as much with temperature (CHAPIN 1980).

10.3.3 Chilling-Sensitive and -Resistant Species

Many important crop plants (e.g., corn, cotton, rice, soybeans, tomatoes, etc.), and presumably many more native species fail to grow or may be damaged with prolonged exposure to low temperature in the range of 20° to about 0 °C (see Chapt. 12, this Vol.). Germination, growth, and reproduction may be sharply restricted at low temperature. This physiological injury to plants is commonly referred to as "chilling injury" (for reviews see LYONS 1973; LYONS et al. 1979). Interest in this phenomenon has been stimulated by efforts to extend the cultivation of plants of agricultural or ornamental use into areas not previously occupied by these species, and by damaging effects of refrigeration to fruit and vegetable products of chilling-sensitive plants. It is a common observation that most species which are sensitive to chilling injury are from tropical or subtropical regions, while plants from temperate regions are much less sensitive (if at all) to low temperatures. Native species have (for the most part) not been examined for chilling responses; however, it seems likely that the responses of tropical and temperate crop plants are representative of plants from these regions. The sensitivity to chilling damage may be an important factor limiting the spread of tropical taxa into temperate climates. This difference in sensitivity does not appear to be simply a quantitative shift in the temperature responses of temperate species to a lower temperature range. Many temperate species appear to be able to tolerate tissue temperatures at least as high as may be tolerated by tropical species. It seems likely that the distinction is related to qualitative differences in the ability of these plants to respond to changes in temperature; however, very little evidence can be put forward to support this.

10.4 Temperature as a Factor in Biogeography

A number of studies have examined the genetic basis of differentiation of species into ecotypes or ecological races along elevational or latitudinal gradients (for reviews see HIESEY and MILNER 1965; COOPER 1963). Attempts to identify crucial physiological differences which distinguish these ecotypes have, however, been less successful. Subtle differences in the seasonal patterns of activity and dormancy coupled with compensating differences in microsite preference that tend to minimize differences in tissue temperature (as opposed to habitat temperature) appear to underlie elevational ecotypes of *Mimulus* (HIESEY et al. 1971).

For the most part physiological differences between plants are difficult to specify without conducting extensive physiological studies of the plant response to temperature and the effect of growth temperature upon these responses. Obviously we do not possess the necessary detailed physiological information to conduct a satisfactory analysis of the interplay between environment, physiological type, and genetics that underlies the dynamics of natural vegetation.

With the discovery of the C_4 pathway it became possible to divide plants on the basis of a significant qualitative difference in physiology (see Chapt. 15, Vol. 12B). Comparative studies of the physiology of C_3 and C_4 plants have shown large differences in the response to temperature and water relations leading to the speculation that C_4 plants would be favored in hot, arid environments (BJÖRKMAN and BERRY 1973).

Studies of the biogeography of C_4 and C_3 plants have tended to support this generalization. STOWE and TEERI (1978) found from several environmental parameters that high pan evaporation correlated best with the abundance of C_4 species of dicots, while warm minimum night temperature correlated best with the abundance of C_4 grass species in North America (TEERI and STOWE 1976). DOLINER and JOLLIFFE (1979) conclude that C_4 species of California tend to be associated with warm and arid climates. These studies have not, however, yielded the expected clear-cut picture of species replacement according to a gradient in a single environmental factor.

Several things have complicated this analysis. First because C_4 and C_3 plants from many families and of differing morphological forms have been considered, the climatic parameters used may bear variable relationships to the actual growth conditions experienced by each plant in its habitat. Second, all C_3 and C_4 plants, including many which occupy rather specialized edaphic habitats (such as saline environments), were included. Also, the analyses have focused upon temperate environments with large seasonal changes in growing conditions. Species of quite different physiological characteristics might co-exist in the same habitat with each having a competitive edge in a different season (MOONEY 1980).

Two recent studies (TIESZEN et al. 1979; RUNDEL 1980) seem to have avoided some of these problems. First, they focus upon elevational gradients along tropical mountain ranges which present seasonally more stable environmental conditions. Second, they focus upon members of the grass family. This family has the advantage that both photosynthetic pathways are widely represented. Species of either type are morphologically similar, and replacement of one photosynthetic type with another results in minimal changes in the structure of the canopy. Grassland as a vegetation type extends over wide environmental gradients; however, the basic structure of the community remains analogous whether it is composed of C_3 or C_4 species.

Ascending the slopes of Mt. Kenya (TIESZEN et al. 1979) or the Hawaiian volcanoes (RUNDEL 1980) there is a complete replacement of C_4 species which dominate at low elevations with C_3 species which dominate at high elevations (Fig. 10.5). This replacement occurs over a narrow elevational band of about

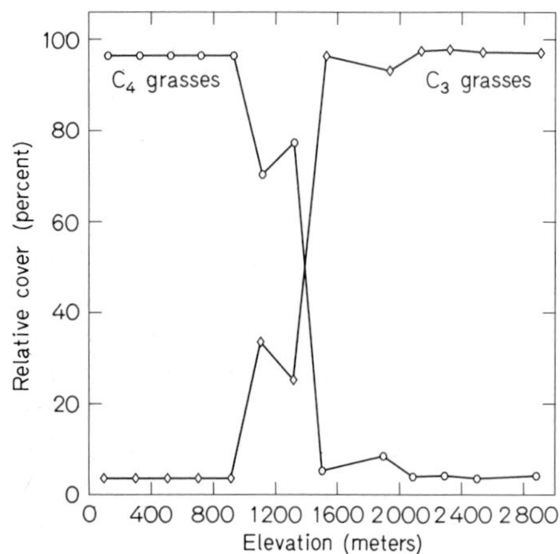

Fig. 10.5. Relative ground coverage by C_3 and C_4 species of grasses along an elevational gradient in Hawaii Volcanoes National Park. (Redrawn from RUNDEL 1980)

600 m corresponding to a cline of about 3 °C. RUNDEL (1980) estimates that mean maximum daily temperature at the midpoint for this replacement is about 22 °C for both the Hawaiian and the African gradients. He also derives a similar estimate for the midpoint temperature from the study of the distribution of C_4 plants in Java by HOFSTRA et al. (1972).

While it is natural to interpret an elevational gradient as a temperature gradient, this interpretation is complicated by changes in other factors with elevation. These may include obvious factors such as changes in water relations or subtle differences such as changes in the herbivores which feed upon the grasslands. The fact that similar replacements are seen along geographically different gradients suggests that temperature plays a major role in determining the success of C_3 and C_4 species. Furthermore, patterns of vegetation change seen in these gradients are unlike those seen along gradients in aridity (MOONEY et al. 1974).

The vegetational change with elevation (temperature) suggests that C_3 plants are able to out-compete C_4 plants at cool temperatures, and C_4 plants are able to out-compete C_3 plants at warm temperatures. While the midpoint occurs at a maximum daily temperature of about 22 °C, it is important to note that this is the air or meteorological temperature. The temperature of the leaf tissue may differ somewhat from this value (Sect. 10.2).

Two reasons that C_3 plants might be expected to be superior at low temperature have been suggested. First, many C_4 plants are unable to tolerate low temperatures, which has led to the postulate that some step in C_4 photosynthesis may be intrinsically sensitive to low temperature. The success of some C_4 plants at very low temperatures (*Atriplex confertifolia,* CALDWELL et al. 1977a, b; *A. sabulosa,* BJÖRKMAN et al. 1975; *Spartina townsendii* LONG et al. 1975) seems to discredit this idea. Also C_4 species were noted at elevations as high as 4,000 m in open xeric sites along both of these equitorial elevational gradients (TIESZEN et al. 1979; RUNDEL 1980). A second proposal is based upon the observation of EHLERINGER and BJÖRKMAN (1977) that the efficiency of light utilization in photosynthesis by C_3 plants increases at low temperature while that of C_4 plants remains constant (Fig. 10.6, page 301). At 15 °C a C_3 plant has a quantum yield 35% higher than that of a C_4 species at that temperature; however, at 35 °C the situation is reversed. Since well-developed grass canopies absorb about 95% of the incident light, and since light use for net CO_2 uptake by a grass canopy may approach the limit set by the quantum yield (LOOMIS et al. 1971), it follows that the species with a higher quantum yield would be able to make more growth with the light it is able to absorb. It seems likely therefore that C_3 species would be competitively favored over an otherwise identical C_4 species at the low temperatures prevailing at high elevations.

The decrease in the quantum yield of C_3 plants with increasing temperatures while that of C_4 plants remains constant (Fig. 10.6) is probably sufficient to explain the dominance of C_4 plants at the higher temperatures at the lower elevations of these gradients. One apparent problem is the midpoint temperature. RUNDEL (1980) noted that the midpoint of the transition between photosynthetic types was near 22 °C rather than 30 °C as predicted by EHLERINGER (1978). A factor not taken into account by EHLERINGER (1978) is the role of the intercellular CO_2 concentration of a C_3 species upon the temperature at which the C_3 quantum yield would be equivalent to a C_4 species. The predicted crossover at 30 °C assumes an intercellular CO_2 concentration of 320 μbar. A lower CO_2 concentration would correspond to a lower temperature for the crossover (Fig. 10.6). No studies have examined the influence of lower atmospheric pressure on the quantum yield of C_3 species; however simulations (POWLES and BERRY, unpublished) do not indicate a large effect of elevation of the relative responses of C_3 and C_4 species.

Other factors in addition to the superior quantum yield may contribute to the success of C_4 species at warmer temperatures. WILSON and FORD (1971) showed that the nitrogen content of C_4 grass species fell while that for C_3 grasses stayed constant as the temperature for growth increased (Table 10.3). This presumably reflects a lower requirement for nitrogen in order to fulfill the photosynthetic functions of the C_4 leaf (BROWN 1978). Given a lower requirement for nitrogen it may be possible for C_4 plants to allocate more new growth to shoots than to roots (Sect. 10.3.1). This by itself would lead to a superior growth rate and would further enhance the benefits of an improved quantum yield. Another factor which may be significant in considering root–shoot interactions is the lower water requirement for C_4 plants. Under comparable conditions C_4 plants generally require about half the amount of water as do C_3 plants per unit of dry matter gain (FISCHER and TURNER 1978). These factors would also tend to favor C_4 species at high temperature.

Other data which lends support to the interpretations that C_3 plants are favored at low temperature and C_4 plants at high temperature comes from an analysis of annual species which grow in the Sonoran Desert either during winter or summer. EHLERINGER (1978) used an empirical model of photosynthesis to simulate the productivity of C_3 and C_4 plants in typical winter and summer conditions for Tucson, Arizona. The simulation indicated that C_3 species should be more productive in the cool winter growing season and C_4 species in the hot summer growing season. In fact the summer annuals are predominantly C_4 and the winter annuals are predominantly C_3 (MULROY and RUNDEL 1978).

Table 10.3. The nitrogen content of leaves of C_3 and C_4 grasses as influenced by temperature. (Data of Wilson and Ford 1971)

Growth regime °C (day/night)	Nitrogen (% of dry wt.)	
	C_3	C_4
15.6/10	5.1	5.2
21.1/15.6	5.1	4.5
26.7/21.1	5.1	2.1
32.2/26.7	5.3	2.4

While the exact physiological basis for the replacement of C_3 with C_4 species is not yet fully understood, it is clear that temperature plays a major role. Further experimentation which attempts to bridge the gap between studies of the quantum yield for photosynthesis of single leaves on the one hand and the competitive success of whole plants within a grass canopy on the other would seem highly appropriate. Nevertheless, the biogeographical and seasonal separations discussed above are probably exemplary of the interplay between the thermal regime of a site and the physiological characteristics of the plants that naturally occupy the site. By necessity other studies of physiological differentiation have been restricted to much narrower ecological perspectives.

10.5 Temperature Effects on Limiting Resources for Growth

Plant growth and development is generally considered according to the law of limiting factors as being limited by the rate of the slowest step in the growth process. According to this dogma an effect of temperature would not be expected when some other environmental factor is strongly limiting for growth. The observation (Fig. 10.6) that the rate of photosynthesis under strictly rate-limiting light intensities is strongly temperature-dependent appears to be a contradiction of the law of limiting factors. This effect can be understood by recognizing that the rate may be limited by the rate of the *slowest* step (in this example light absorption), and by the *efficiency* of that step (in this example the quantum yield). The differential effect of temperature upon the distribution of C_3 and C_4 plants discussed in the previous section may be viewed as an example of the role of physiological efficiency in ecological systems. Temperature also affects the efficiency with which water and nutrients are used for plant growth.

10.5.1 Water

As temperature is increased, the driving function for transpiration of water, the vapor pressure gradient between the intercellular air spaces and the ambient atmosphere ΔW, increases exponentially. For example, ΔW in a cool maritime environment may be 5 mbar while that in a hot desert environment in summer may exceed 100 mbar (MOONEY et al. 1975); the difference in ΔW from winter to summer may be severalfold. If all other factors are held constant, the rate of transpiration increases in direct proportion to the vapor pressure gradient.

Water loss must be tolerated for CO_2 to be taken up by the cells of the leaf, and the efficiency of this exchange (the water use efficiency, see Chapts. 7 and 17, Vol. 12B) is inversely proportional to ΔW (see FISCHER and TURNER 1978). The amount of water which must be exchanged in order to fix a given amount of CO_2 at different temperatures may be calculated. If we assume that the relative stomatal limitation remains constant (i.e., the intercellular CO_2 remains constant, BERRY and DOWNTON 1981), and that the relative humidity is constant as temperature is changed, then the calculations indicate that the water required to fix a given amount of CO_2 increases with temperature according to a Q_{10} of about 1.8. Thus, a given water resource would permit only 55% as much CO_2 uptake at a temperature 10 °C higher. While C_4 species would use less water than C_3 species, the relative effect of temperature increase would be the same for either. The net CO_2 uptake, and thus, the net productivity which could be sustained given a limiting water input is thus lowered as temperature is increased.

10.5.2 Nutrients

Nitrogen and other nutrients are required in plant growth as structural components or cofactors to the many essential functions of the plant. On purely theoretical grounds, since the rates of enzymatic reactions are strongly temperature-dependent, one might expect that a given amount of nutrients could be more effective (i.e., catalyze more growth-related reactions) at a moderate than at a low (and perhaps also at high) temperature. Alternatively, more plant protein might be required to sustain the same rate of productivity at a low than at a moderate temperature. The lower nitrogen requirement of C_4 species grown at high temperature (Table 10.3) is an example of such an effect.

10.6 Physiological and Biochemical Responses to Temperature

Raw materials for growth are the result of photosynthesis, ion uptake, and water absorption. These materials are transported and partitioned to the various growing points of the plant. Respiration is required to enable active uptake and transport processes; to drive the primary biosynthetic reactions forming proteins, nucleic acids, lipids, and carbohydrates needed for growth, and to maintain the essential life processes of the plant. Normal growth is dependent upon the integrated functioning of these processes at an appropriate rate. For example a specific ratio between carbon gain in photosynthesis and nutrient uptake by the roots is required to meet the synthetic requirements for growth. While storage of photosynthate or nutrients may buffer short-term imbalances, these inputs must be balanced over the long term. The rates of photosynthesis and ion uptake may change with temperature. The capacity of the various growing points to act as sinks for photosynthate and nutrients will change, as will the losses due to respiration, with temperature. Also the mechanisms which translocate and partition materials within the plant may respond to temperature. Finally there are effects of temperature on developmental processes such as germination and flowering.

10.6.1 Thermodynamic Considerations

The rates of physiological processes must ultimately be determined by the physical relationships which govern the rates of chemical reactions. In the biological system, however, many more steps are involved, and we must also consider the thermodynamic constraints on the stability of the primary catalysts, the proteins and membranes which enable biological reactions. These considerations make the thermodynamic analysis of living systems immensely more complex than single chemical reactions. Nevertheless the principles derived from the study of less complex systems are useful in analyzing biological phenomena.

10.6.1.1 Temperature Coefficients

The Arrhenius theory provides a basis for analyzing the effect of temperature on the rate of reactions. According to this theory the rate of a reaction should change approximately exponentially with the inverse of the absolute temperature [1/T (K)]; thus a plot of the logarithm of the rate of reactions against the reciprocal of the absolute temperature (an Arrhenius plot) should yield a straight line of slope E_a/R where R is the universal gas constant, 8.314 J mol^{-1} $°K^{-1}$, and E_a is an activation energy (J mol^{-1}). The relation is interpreted in terms of the energy required to activate the reactants according to the equation

$$k = Ae^{-E_a/RT} \tag{10.1}$$

where k is the rate constant and A is a pre-exponential term which need not be temperature-independent. It should be noted that Eq. (10.1) is a simplification. The concept of activation energy has evolved to more sophisticated concepts that deal with the theory of absolute reaction rates; for a review see JOHNSON et al. (1974). It is the complexity of biological systems, however, rather than the lack of mechanistic sophistication which limits the application of these concepts in biology. An activation energy can only be defined in a simple system. Even a reaction catalyzed by a single enzyme may occur by several steps and physiological processes are of course the result of many linked enzymatic reactions. While physiological processes may approximate the Arrhenius relationship over certain temperature regions it is a large extrapolation to interpret this "apparent activation energy" in a strict chemical context. The temperature response of enzymatic reactions may vary depending upon the substrate concentrations (Sect. 10.6.1.2). Also it is possible that the rate over a given temperature range may be controlled by a combination of factors. For example JOHNSON et al. (1974) cite data on the temperature-dependence of the α-rhythm of the occipital cortex of human subjects. While the data appear to fit the Arrhenius relationship, few would be inclined to interpret these data in terms of a single limiting reaction. Nevertheless, the E_a derived from an Arrhenius plot may be a useful coefficient to describe the temperature response. We will use the term "apparent activation energy" to distinguish such measurements from those of the true activation energy conducted under controlled biochemical conditions.

The Q_{10} offers an important advantage for analysis of the temperature dependence of complex physiological processes. Unlike the activation energy, the Q_{10} does not imply a mechanistic explanation. According to the general usage the Q_{10} is defined as the ratio of the rate at one temperature to that at a temperature 10° lower. If the Q_{10} remains constant with temperature then the rate at any temperature can be calculated from the rate at a known temperature from the expression

$$k_2 = k_1 Q_{10}^{\left(\frac{T_2 - T_1}{10}\right)} \tag{10.2}$$

where k_1 and k_2 are the rates and T_1 and T_2 are the corresponding temperatures. Most values for Q_{10} reported in the literature are based upon only two data points, however Q_{10} may be determined graphically or by a linear regression from a plot of log rate (k) vs. temperature (T, °C) which should yield a straight line of

$$\log k = \frac{T \log Q_{10}}{10} + C \tag{10.3}$$

where the slope is $^1/_{10}$ log Q_{10} and C is the intercept at 0 °C. Since the Q_{10} function assumes an exponential relation to temperature (C) and the Arrhenius relation assumes an exponential relation to the inverse of the absolute temperature (1/T, K), the two parameters are fundamentally different. Over the narrow range of temperatures normally considered in plant physiology, however, data which can fit one plot will probably also fit the other plot equally well.

The following expression may be derived to relate an apparent activation energy to Q_{10}.

$$-E_a = \frac{RT(T+10)}{10} \ln Q_{10} \tag{10.4}$$

Where $R = 8.314\ \mathrm{J\ mol^{-1}\ °K^{-1}}$ and T = 20 °C, the following numerical conversion may be applied

$$E_a = -74 \ln Q_{10} (\text{K J mol}^{-1}) \tag{10.5}$$

The constant changes about 10% between 5 and 40 °C.

Aside from the descriptive value, these temperature coefficients have been used to infer a mechanism for physiological processes. For example, comparison of the apparent E_a or Q_{10} measured in vivo, and that of possible rate-limiting steps determined in vitro may be useful in inferring the identity of the rate-limiting step in vivo. Thus, a Q_{10} of 1.3 to 1.5 for a process such as water transport might be interpreted as suggesting limitation by a diffusive process, since such physical processes have a low temperature coefficient relative to most biochemical and chemical reactions which have Q_{10}'s near 2 (see NOBEL 1974a). Processes with Q_{10}'s much greater than 2 are occasionally observed with biological systems. These may be attributed to phenomena such as protein denaturation (LEVITT 1980).

The physiological conditions under which a reaction occurs in vivo may have a strong influence on its temperature dependence. For example, most enzymatic reactions have a lower Q_{10} when the reaction is provided with rate-limiting rather than rate-saturating concentrations of substrate. Some caution is thus appropriate when comparing the temperature dependence of reactions in vivo and in vitro.

10.6.1.2 The Effect of Substrate Concentration

The temperature dependence of enzymatic reactions may be dramatically influenced by the concentration of substrate available to an enzyme. While it is well known that temperature has a strong effect on the maximum capacity of enzymatic reactions under rate-saturating substrate concentration (V_{max}), it is less generally known that temperature also effects the K_m of enzymes for their substrates. The effect of this can be illustrated by considering the rate of an enyzmatic reaction under strictly rate-limiting substrate concentration (when the substrate concentration is small in comparison to the K_m). Under these conditions the rate is approximately

$$V \simeq V_{max} \frac{C}{K_m} \tag{10.6}$$

The temperature dependence under these conditions would be the quotient of the separate dependencies of the K_m and the V_{max}. BADGER and COLLATZ (1976) observed that the V_{max} of RuBP carboxylase reaction increases with a Q_{10} of 2.2 (see Fig. 10.10 below); however the K_m also increases with a Q_{10} of 2.2. From these data it may be predicted that the rate of the RuBP carboxylase reaction would be independent of temperature $Q_{10} = 1.0$ at very low CO_2 concentration and that this would increase to a Q_{10} of 2.2 at rate-saturating CO_2 concentration. It is noteworthy that KU and EDWARDS (1977) observed that the initial slope of the CO_2 response curve of wheat leaves was independent of temperature.

It should be noted that the temperature dependence of the K_m for an enzymatic reaction in not always monotonic as shown above. For example, HOCHACHKA and SOMERO (1973) present data for the temperature dependence of the K_m of several enzymes which have a distinct temperature minimum for the K_m. HAVSTEEN (1973) discusses the kinetic basis for effects of temperature on the K_m of enzyme-catalyzed reactions. HOCHACHKA and SOMERO (1973) place considerable emphasis upon the temperature dependence of the K_m as a possible basis for temperature compensation of metabolic activity in ectothermic animals exposed to short–term variations of temperature. This would occur if the substrate concentration was low in comparison to the K_m as shown in Eq. (10.6) above. The existence of a temperature minimum does not seem to be a necessary aspect of their hypothesis, although it has been observed with some enzymes of plant origin (SIMON 1979; TEERI 1980).

10.6.1.3 Breaks in the Temperature Coefficient

Arrhenius plots are often observed to possess "breaks" or abrupt changes in the value of the temperature coefficient (see also Chapt. 12, this Vol.). These breaks might result from a change in the limiting reaction in a catenary series, or they could result from effects of temperature upon the catalytic activity of the same system. There are abundant examples of breaks occurring in the temperature response of reactions catalyzed by highly purified enzymes. For example, the temperature coefficient of the carboxylation reaction of RuBP carboxylase increases from a Q_{10} of 2.2 to 3.3 at temperatures below 15 °C (see Fig. 10.10 below). In some instances a break in the rate of membrane-associated processes has been associated with the beginning of a phase change in the lipids of that membrane (Sect. 10.6.3.1). Several workers (KAVANAU 1950; LYONS and RAISON 1970; JOHNSON et al. 1974 and WOLFE 1978) suggest that breaks in the temperature coefficient of a single process may be related to molecular conformation changes which effect the catalytic properties of an enzyme.

Changes in the temperature coefficient for physiological processes such as respiration, ion uptake, or even growth are also observed. In some instances a break at a physiological level may be coincident with a break seen at a biochemical of biophysical level. For example, a sharp increase in the temperature coefficient for growth of *Vigna radiata* occurs at about the same temperature as the beginning of a phase transition of membrane lipids of that species (RAISON and CHAPMAN 1976). In other instances no break is seen at a physiological level yet a membrane phase transition is detected or vice versa. A number of studies have attempted to use the presence or absence of breaks in Arrhenius plots of physiological processes to argue for or against change in membrane lipid structure. Such inferences have little meaning without supporting studies of membrane lipids.

A change in the temperature coefficient for a physiological process nevertheless indicates a significant change in the factors controlling the rate of the process over a fairly short range of temperature. This may provide some clue to the identity of the biochemical change that is limiting the rate in vivo.

10.6.1.4 The Temperature Optimum

Most physiological processes reach a peak rate at an optimum temperature. These responses are generally fully reversible over a temperature range extending from well below the optimum to several degrees above the optimum temperature. The increase in rate at temperatures below the optimum may be attributed to the Arrhenius function; however, the progressive decline in the temperature dependence as the temperature approaches the optimum and the eventual change in the sign of the temperature dependence above the temperature optimum are not explained by this concept alone. In some instances temperature stimulation of competing reactions may occur. For example, stimulation of respiration diminishes net photosynthetic activity; however, this effect is of only minor importance when compared to the direct effect of temperature on photosynthesis (Sect. 10.6.2). It seems necessary to postulate some form of reversible inhibition of a rate-controlling step with increasing temperature in order to explain the change in the temperature coefficient with temperature. JOHNSON and coworkers (for a review see JOHNSON et al. 1974) proposed a model to explain the effect of temperature upon the luminescence of bacteria. This model seems to be of general use in interpreting the temperature optimum of other physiological processes. They proposed that an essential enzyme may exist in active (N_a) and inactive (N_i) conformations according to a temperature-dependent equilibrium.

$$N_a \rightleftharpoons N_i$$

As temperature is increased the equilibrium shifts to favor the inactive form. The effect of temperature on the equilibrium constant (κ) may be expressed in terms of the thermodynamics of the conformational change

$$\kappa = e^{-(\Delta H - T\Delta S)/RT} \tag{10.7}$$

The concentration of the active conformation (N_a) may be expressed in terms of the total enzyme concentration (N_0) and the equilibrium constant, κ

$$N_a = N_0 \frac{1}{1+\kappa} \tag{10.8}$$

The rate of the reaction as a function of temperature is the product of the Arrhenius expression [Eq. 10.1], and the concentration of active conformers of the enzyme obtained by substituting [Eq. 10.7] for κ in Eq. 10.8.

$$k = A \frac{1}{1+\kappa} e^{-E_a/RT} \tag{10.9}$$

This expression can be adjusted to fit most physiological temperature responses, and the model it summarizes seems plausible. Values of ΔH and ΔS required to empirically fit data are similar to values obtained in studies of conformational changes of proteins (Sect. 10.6.1.5). Few studies, however, have attempted to identify the site of inhibition in the physiological processes of plants, nor have studies examined the basis of the proposed conformational changes. SHARPE and DE MICHELLE (1977) extended this approach to consider reversible denaturation equilibria at both high and low temperatures.

10.6.1.5 Thermal Stability of Macromolecules

Several recent reviews have dealt with the thermodynamics of assembly and denaturation of biological macromolecules (TANFORD 1980; EDELHOCK and OSBORNE 1976; LAUFER 1975). These reviews emphasize the importance of hydrophobic interactions in establishing and maintaining native biological structures. The association of hydrophobic amino acid side chains or fatty acid chains with like substances and minimizing water contact with these substances permits greater entropy of the system and is a major factor leading to a negative free energy maintaining the conformation of the native state. There is general agreement that the actual changes in protein structure which occur with denaturation (either at high or low temperature) is an unfolding of portions of the peptide chains which both increase the entropy of the protein and decrease the entropy of water (BRANDT 1967). The energetics of such changes are inherently very temperature-dependent; thus, the energetically favored conformation will vary as a function of temperature.

Studies of the influence of temperature upon the interactions between hydrophobic substances and water (TANFORD 1980) indicate that water is likely to exert the greatest stabilizing effect at temperatures near the normal physiological range (20–25 °C). At temperatures above and below this the strength of hydrophobic interactions most likely declines. This decline in the stabilizing influence of water appears to explain both the nearly universal denaturation of proteins which occurs at high temperature and the less common but nonetheless important denaturation of some proteins at low temperatures.

Transformation of proteins between conformational states has been shown to have rather large enthalpy and entropy contributions (typically $\Delta H = 200{,}000\ J\ mol^{-1}$, $\Delta S = 700$ entropy units mol^{-1}). An interesting feature of this is that the enthalpy and entropy terms nearly compensate for one another at normal physiological temperatures (i.e., $\Delta H \simeq T\Delta S$, and the net free energy change is zero at a physiological temperature, typically 300 °K). LUMRY and RAJENDER (1970) review the generality of this coincidence and conclude that it is most likely a property of water and its interaction with hydrophobic moieties which result in this compensation and that the phenomenon of enthalpy-entropy compensation in protein reactions may be mechanistically associated with conformational changes of the protein. LAUFER (1975) pointed out that any reversible reaction in a living system must have a free energy change near zero under physiological conditions. The self-assembly of protein subunits and biomembranes are such reactions, and the compensation noted above may be essential for assembly. This is an interesting idea in that it could explain why it has apparently been impossible for organisms to evolve macromolecular structures which are stable over very wide temperature ranges. The parallel between this thermodynamic argument of LAUFER and the concept of conformational flexibility of proteins advanced by ALEXANDROV (1977) is notable. In a mechanistic sense changes ΔH or ΔS of an unfolding reaction would change the compensation temperature of that reaction, and the temperature limits for the "native" form of the protein. This would seem, as suggested by LUMRY and RAJENDER (1970), to open the way to a chemical treatment of protein evolution.

There has been comparatively much less work on the effect of temperature on membrane structure. This may be because the concepts of how membranes are organized, and the biophysical mechanisms which maintain this organization have changed dramatically in the past few years. Most workers now subscribe to some version of the fluid-mosaic model of SINGER and NICOLSON (1972) to describe and explain membrane structure and stability (for reviews see SINGER 1974; JAIN and WHITE 1977). According to this model the membrane consists of a bilayer of polar lipid molecules which are oriented back-to-back with their polar head groups exposed to water and their nonpolar acyl chains associated to form a hydrophobic core. Proteins may be partially imbedded in or span the bilayer and are mostly free to diffuse within the plane of the membrane. It is thought that intrinsic membrane proteins have a high proportion of hydrophobic amino acid side chains on the surfaces which associate with the lipid acyl chains and that surfaces which associate with the polar head groups of the lipids and the aqueous phase are composed of polar side chains. The forces which cause this association are thus characterized as hydrophobic (TANFORD 1980), and are in concept like the interactions of forces which are thought to maintain the native conformation of protein molecules. As pointed out previously, the balance of these forces can be changed by variations in temperature.

When a membrane becomes denatured it is most likely the protein components of the membrane which change; the lipid bilayer structure, being composed of small robust molecules, is not likely to be damaged. It is not clear whether the proteins unfold, as is the case with soluble proteins, or if they dissociate or aggregate within the membrane when they become denatured. The thermal stability of membranes is not only a function of the properties of water and protein but of the interaction between the proteins and the lipid. As mentioned previously, lipids might undergo a thermal phase transition from a liquid crystalline to a gel state. With a membrane composed of a single lipid this would occur over a very sharp temperature interval. Biological membranes are composed of complex mixtures of lipids and the phase transition actually occurs over a wide temperature range. Within this range some membrane lipids phase separate to give two co-existing phases in different domains of the bilayer. This is a function of lipid composition. The lipid phase separation can affect properties of the bilayer (BLOCK et al. 1976; NOBEL 1974b), and the extent of the interaction between membrane lipids and proteins. Freeze-fracture studies of membranes show that membrane proteins might be excluded from regions of the membrane in which the lipid is solidified (HACKENBROCK 1976). The protein molecules may either accumulate in the remaining fluid phase or be excluded from the membrane (see ARMOND and STAEHELIN 1979).

There is also evidence that the relative fluidity of the membrane may influence the interaction of proteins with the membrane. MCMURCHIE and RAISON (1979) showed that the apparent E_a of succinate oxidase activity increased as the fluidity of the mitochondrial membrane lipids increased. This indicates that membrane lipid fluidity can directly modulate membrane function. RAISON and BERRY (1979) have shown that lipid fluidity is also an important factor in determining the stability of membrane proteins to high temperature (Sects. 10.6.2.2; 10.6.8.2). Their results indicate that denaturation occurs at a critical lipid fluidity, and that thermal stability of plant membranes is modified by altering the lipids and, as a consequence, fluidity. Apparently lipid fluidity or some parameter which correlates with it affects the balance between the opposing forces which maintain or disrupt the native structure of some membrane proteins.

10.6.1.6 Strategies of Adaptation and Acclimation

Changes in temperature represent an important disruptive influence on metabolic functions. Not only is the overall rate of metabolism affected, but the balance between individual steps in metabolic pathways (or even complete pathways) is likely to be affected. The metabolic systems of plants must be well buffered against the influences of temperature changes in order to cope with the temperature variations which they encounter in their native habitats. In addition, plants from contrasting natural environments appear to have metabolic characteristics that compensate in part for the differences in tissue temperature. HOCHACHKA and SOMERO (1973) suggest some general strategies whereby changes at a molecular level may permit an organism to cope with the stresses imposed by temperature. (1) There may be qualitative changes in the properties of key constituents of the metabolic apparatus, such as new enzymes or different lipid mixtures, that have different temperature response characteristics. (2) There may be quantitative changes in the amount of specific constituents which compensate for effects of temperature on the rate of the corresponding reaction. (3) There may be immediate responses of the existing metabolic systems which control or minimize the perturbation caused by a change of temperature. These general strategies are not mutually exclusive. In the following sections we will examine the temperature dependence of some of the major physiological processes, and attempt to identify some of the specific mechanisms which appear to adapt species to contrasting thermal regimes or permit a given individual to acclimate to a changing thermal regime.

10.6.2 Photosynthesis

As the primary source of carbon and energy, photosynthesis plays a dominant role in the logistics of growth. In addition, photosynthesis is among the most responsive of plant processes to temperature. Comparisons of the response of growth (Fig. 10.1) and photosynthesis (Fig. 10.2) to temperature illustrate that species-dependent differences in the photosynthetic responses to temperature are closely linked to the response of plant growth to temperature. Also the ability of species to acclimate to contrasting growth conditions is associated with the ability to acclimate at the level of photosynthesis (Fig. 10.3).

A great many studies have examined the photosynthetic responses to temperature of species native to or grown in various environmental regimes (for a review see BERRY and BJÖRKMAN 1980). Diversity among species in their temperature responses may be characterized according to their temperature optimum, according to their photosynthetic activities over specific temperature ranges, or according to their tolerance to extremes of high or low temperature. The temperature optimum observed with rate-saturating light intensity is the most obvious index of the thermal response. Species from warm habitats usually reach their optimum at a somewhat higher temperature than species from cool habitats, and the temperature optimum may shift if the species is capable of acclimating to a changing growth temperature. Changes in the optimum temperature do not, however, fully compensate for changes in the growth temperature. With several species the optimum has been observed to change by about 1 °C for each 3 °C change in the growth temperature (for references see BERRY and BJÖRKMAN 1980). Species adapted to very cool arctic and alpine environments seldom have temperature optima below 20 °C, leading some workers to suggest that plants may not have fully adapted to such cold environments. Nevertheless, the photosynthetic capacity of cold-adapted species may be quite high at low temperatures characteristic of the native habitat, and the rate of CO_2 uptake at 4 °C may be severalfold higher than the rate of a species from a more mesic environment. The temperature optima of cool-adapted species may, however, be only slightly lower than that species which do poorly at low temperature. The temperature optimum of C_3 species from hot habitats likewise may not be much higher than that of species from cooler habitats. Enhanced photosynthetic capacity at high temperature and, especially, an enhanced capacity to tolerate high leaf temperatures without damage are important features adapting such species to hot natural habitats. The temperature optimum taken by itself may thus underestimate the extent of physiological adaptation or acclimation to different thermal regimes.

10.6.2.1 The Temperature Response Curve

Photosynthesis is an approximately linear sequence of reactions. CO_2 which diffuses into the leaf is fixed and reduced to carbohydrate by enzymatic reactions which occur in the stroma of the chloroplast. The CO_2 fixation reactions are in turn linked to light-driven electron transport reactions occurring on the chloroplast membranes. The response of net CO_2 uptake to temperature is the summation of multiple effects of temperature upon individual steps of the photosynthetic process, and this response interacts strongly with other environmental factors – especially light intensity and CO_2 concentration. Consistent differences are also seen between the responses of C_3 and C_4 species.

The light intensity required to saturate CO_2 uptake is lower at low temperatures; thus, a constant light intensity may be rate-saturating over a portion of the temperature curve and be rate-limiting at higher temperatures. The resultant response curves are, thus, truncated. At light intensity sufficiently high to saturate CO_2 uptake at all temperatures, responses as shown in Figs. 10.2, 10.3 and 10.7 are obtained. Species-dependent differences are sharply apparent

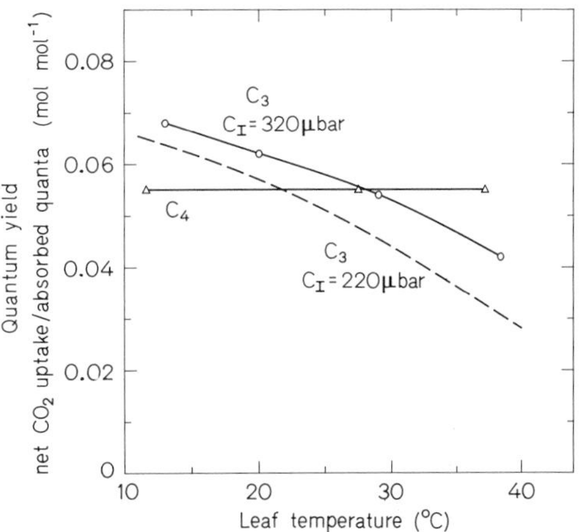

Fig. 10.6. The effect of temperature on the quantum yield for light-limited photosynthesis by leaves of C_3 and C_4 plants. *Data points* are from EHLERINGER and BJÖRKMAN (1977). *Dashed line* indicates the results of a simulation according to the model of BERRY and FARQUHAR (1978) assuming a concentration of 220 µbar CO_2 in the chloroplast and normal atmospheric O_2 (210 mbar)

in these responses. At normal CO_2 concentration, C_4 species reach higher peak rates and have a higher temperature optimum. This difference is more pronounced with the high temperature-adapted species. When the comparisons are conducted at elevated CO_2 concentrations the responses of the C_3 and C_4 species become similar. These studies permit some resolution of the role of carbon metabolism and other factors on the photosynthetic temperature response. At high external concentration of CO_2 responses of C_3 and C_4 species from comparable habitats are similar. Presumably this reflects the temperature dependence of some primary reactions of photosynthesis common to both C_3 and C_4 species. At normal CO_2 concentrations there is a substantial depression of photosynthetic performance of C_3 species. The relative extent of this inhibition at a given temperature is similar for both C_3 species, and it increases continuously with temperature (Fig. 10.8), although the absolute magnitude declines above the temperature optimum. These responses indicate two separate effects of temperature on photosynthesis; one common to both C_3 and C_4 plants, and the other affecting only C_3 plants when they are at normal CO_2 and O_2 concentrations. There is good reason to expect that these separate temperature effects are: (a) upon the capacity of the photosynthetic membranes for electron transport, and (b) upon the efficiency of the reactions of carbon metabolism as affected by photorespiration.

The temperature response for whole chain electron transport by isolated chloroplast membranes of *Larrea divaricata* (Fig. 10.9) is remarkably similar to the response of CO_2 uptake by leaves of heat-adapted species (Fig. 10.7). NOLAN and SMILLIE (1976) report that the temperature optimum for electron transport of *Hordeum vulgare* (barley) chloroplasts is about 32 °C, as would be required to explain the response of the cool climate species of Fig. 10.7. ARMOND et al. (1978a, b) demonstrated corresponding changes in the temperature response of electron transport reactions and of whole leaf photosynthesis upon acclimation to different growth temperatures. These results provide strong support for the notion that an effect of temperature on the chloroplast membranes determines the temperature optimum for the whole leaf when photorespi-

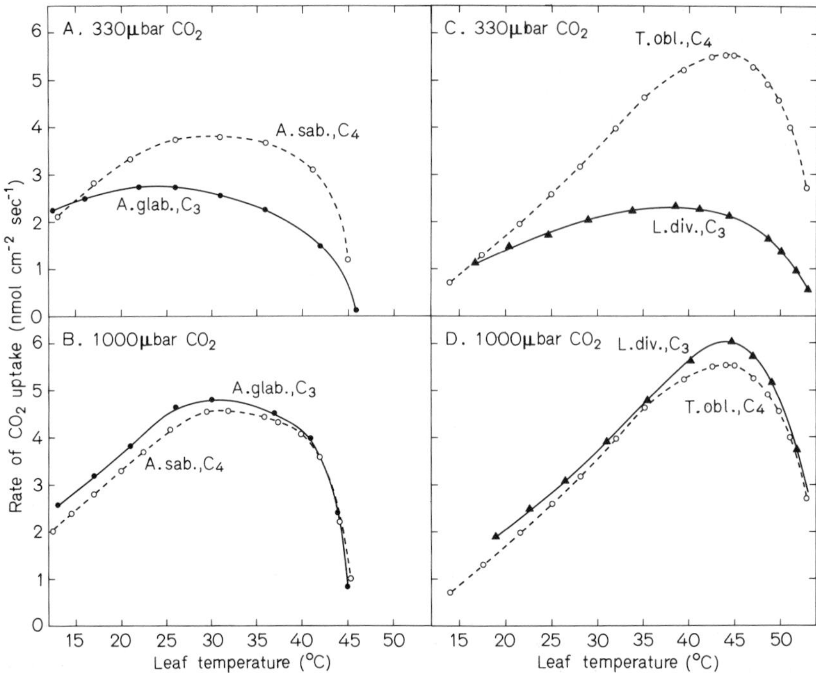

Fig. 10.7. The temperature dependence of net CO_2 uptake at 330 μbar (*top*) and 1,000 μbar CO_2 (*bottom*) of cool-temperature-adapted *Atriplex glabriuscula* (C_3) and *A. sabulosa* (C_4) and of high-temperature-adapted *Larrea divaricata* (C_3) and *Tidestromia oblongifolia* (C_4). The cool-adapted species were grown at 16 °C, the heat-adapted at 45 °C maximum daily temperature. Measurements were conducted at high light intensity and normal O_2 concentration (21%). (OSMOND et al. 1980)

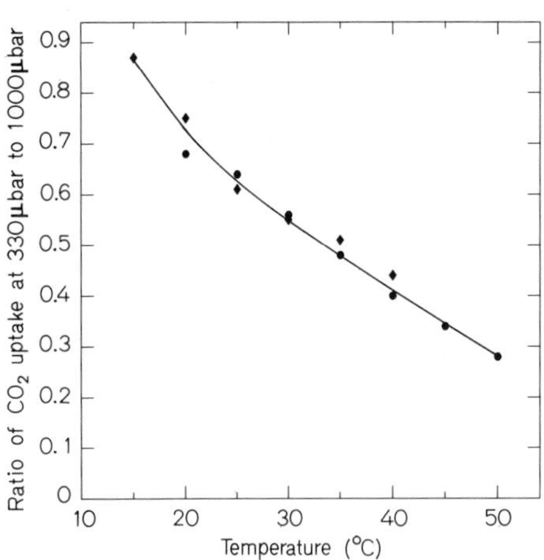

Fig. 10.8. The effect of temperature on the rate of net CO_2 uptake at normal atmospheric (330 μbar) CO_2 relative to that at elevated (1,000 μbar) CO_2 concentrations. Data are for the C_3 species [*A. glabriuscula* (◆) and *L. divaricata* (●)] of Fig. 10.7. The response is similar for cool- and heat-adapted C_3 species, and resembles the response of the quantum yield (Fig. 10.6) of C_3 plants to temperature

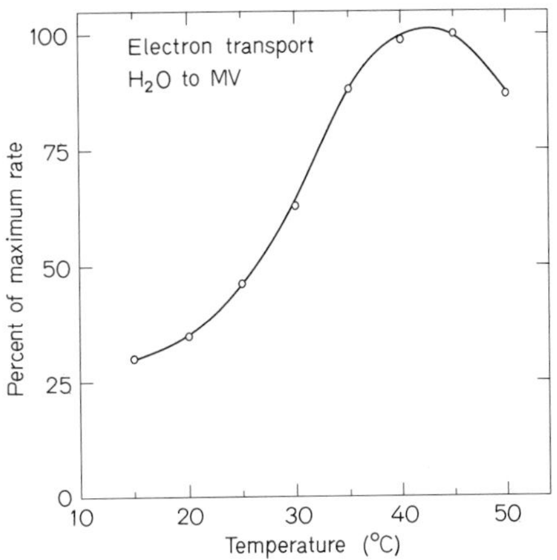

Fig. 10.9. The temperature dependence of whole chain electron transport by chloroplasts isolated from *Larrea divaricata* leaves grown at 45 °C maximum daily temperature. Rates were measured by oxygen exchange using methyl viologen as the electron acceptor. The maximum rate was 1,040 μmol O_2 mg chl^{-1} h^{-1}. (Redrawn from ARMOND et al. 1978)

ration is inhibited. The basis on this effect on the membrane has not been identified; however differences in the fluidity and acyl fatty acid composition of membrane lipids correlate with differences in the upper temperature limit for stability of the chloroplast membrane to heat (Sect. 10.6.8.2). It is likely that this limitation plays a similar role in C_3 and C_4 species. Parameters of Eq. (10.9) have been adjusted to fit experimental measurements of the temperature dependence of electron transport, and these were used to summarize these effects of temperature in a kinetic model of CO_2 assimilation by C_3 plants (FARQUHAR et al. 1979).

At strongly rate-limiting light intensity the response to temperature is dominated by an effect of temperature on the efficiency of light utilization. The quantum yield for CO_2 uptake of C_3 species when measured at normal CO_2 and O_2 concentrations is progressively inhibited by increasing the temperature (Fig. 10.6). Stomatal factors may modify this response slightly by influencing the concentration of CO_2 available to the photosynthetic cells within the leaf; however, there is little evidence to suggest that there are species-dependent differences between C_3 plants in the response of their quantum yield to temperature, except for the range of temperatures over which a species is able to maintain a normal quantum yield (BJÖRKMAN 1975). Leaves of C_3 species provided with high CO_2 or low O_2 concentration during measurement do not show the temperature-dependent inhibition of the quantum yield, and C_4 species lack this temperature dependence even at normal atmospheric concentrations of O_2 and CO_2 (EHLERINGER and BJÖRKMAN 1977).

The effect of temperature on the quantum yield was attributed to a stimulation of photorespiration by increasing temperature (EHLERINGER and BJÖRKMAN 1977). It is now widely accepted that the balance between photorespiration and photosynthesis is regulated by the bifunctional enzyme RuBP carboxylase/oxygenase (LORIMER 1981). The carboxylation reaction (RuBP CO_2ase) results in CO_2 fixation, and supports the photosynthetic carbon reduction pathway. The oxygenation reaction (RuBP O_2ase) results in oxygen uptake, and supports the photorespiratory carbon oxidation pathway (for a review see LORIMER 1978).

The photorespiratory pathway results in a net release of 1/2 CO_2 for each RuBP O_2ase reaction. A convenient parameter to specify the relative rates of these two pathways is the ratio of RuBP O_2ase to RuBP CO_2ase reactions (ϕ). There are ϕ oxygenations per carboxylation.

The branching of photosynthetic carbon metabolism to either the carboxylation or oxygenation of RuBP effects the net uptake of CO_2 per RuBP consumed, and the energy input (as ATP or NADPH) required for each net uptake of CO_2. These effects may be quantitatively related to the branching ratio by expressions presented in Farquhar et al. (1979). Since one CO_2 is released in photorespiration for every two oxygenations, net CO_2 uptake $A=(1-\phi/2)\times$ (the rate of carboxylation of RuBP). The energy costs for regeneration of RuBP from the products of carboxylation or oxygenation are 3 and 3.5 ATP respectively (Berry and Farquhar 1978), hence the ratio $A/ATP=(1-\phi/2)/(3+3.5\phi)$. An analogous expression for the efficiency of NADPH use in CO_2 fixation is $A/NADPH=(1-\phi/2)/(2+2\phi)$, since 2 NADPH are required to regenerate RuBP in either case. These indices of efficiency decrease as ϕ increases. Factors which influence the ratio ϕ are the concentrations of O_2 and CO_2 and effects of temperature upon the kinetic parameters of RuBP carboxylase/oxygenase. Studies of temperature dependence of these kinetic constants (Laing et al. 1974; Badger and Collatz 1977) provide a quantitative basis for understanding the influence of temperature. An Arrhenius plot of the temperature dependence of the V_{max} and K_m of these functions (Fig. 10.10) illustrates the strong temperature dependence of all of these parameters. The key differential effect of temperature is upon the K_m terms. The K_m (CO_2) increases more strongly with temperature than does the $K_m(O_2)$; as a result, at constant substrate concentration, the rate of the oxygenase reaction increases relative to that of the carboxylase reaction. Using a simplified kinetic model of net CO_2 uptake, Berry and Farquhar (1978) showed that the temperature dependence of the kinetic constants predicted a temperature response of the quantum yield which closely matched experimental observations. This effect of temperature on photorespiration is most clearly identified in the response of the quantum yield to temperature. However, as shown above (Fig. 10.8), this stimulation of photorespiration with increasing temperature has a strong effect on net CO_2 uptake of C_3 plants given normal CO_2 and O_2 concentrations, even at rate-saturating light intensities.

Thus far we have considered rate limitation over temperatures approaching and exceeding the temperature optimum. We have assumed that the rate of CO_2 uptake is principally limited by the capacity of the primary energy conversion reactions of the chloroplast membranes. This assumption is based upon comparative studies which show similar responses to temperature by intact leaves and isolated chloroplast membranes and upon studies which show that the major enzymes of carbon metabolism are stable (and presumably fully functional) at temperatures above those which result in complete destruction of sensitive components of the chloroplast membranes (Berry and Björkman 1980).

While the above assumptions appear to be valid at temperatures near or above the optimum, other factors not discussed above may contribute to the temperature dependence at suboptimal temperatures. As temperature is lowered

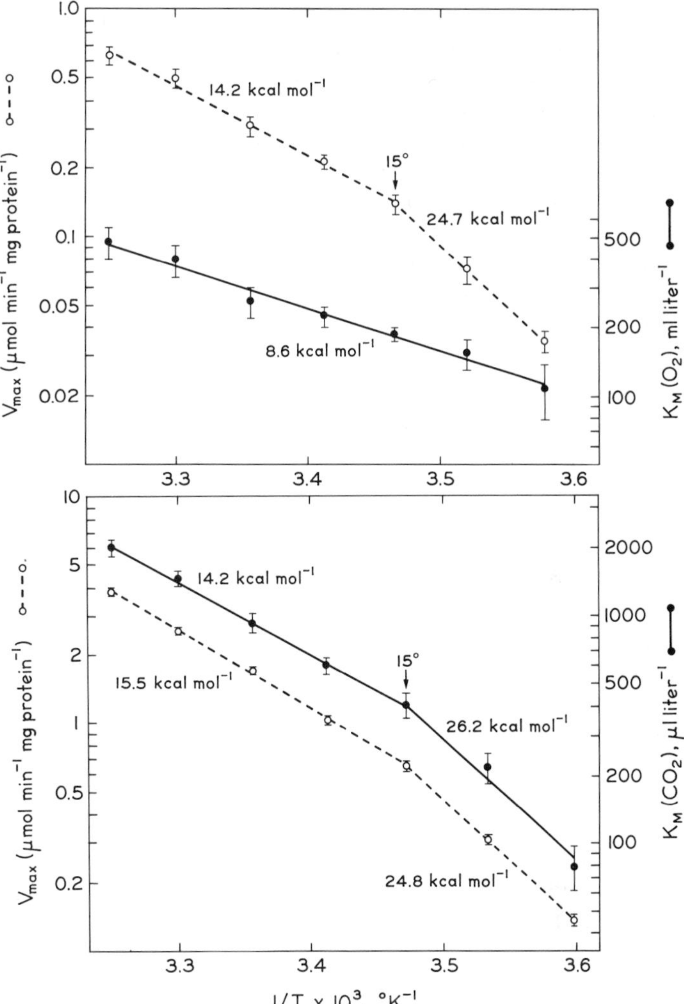

Fig. 10.10. The response to temperature of the V_{max} and K_m of RuBP oxygenase (*top*) and RuBP carboxylase (*bottom*). The kinetic constants were determined by statistical analysis of studies of the rate vs. substrate concentration at each temperature. Concentrations of O_2 and CO_2 are expressed as the gas phase concentration at equilibrium considering the effects of temperature on the pK_a for H_2CO_3 and the solubility of CO_2 and O_2 in water. Activation energies are given in kcal mol^{-1}. There are 4.18 J per cal. (BADGER and COLLATZ 1977)

below the optimum the capacities of all reactions decline according to their individual temperature coefficients. Depending upon the relative capacities of these steps and the relative differences in their temperature coefficients, one or several steps may be limiting over specific temperature regions. Comparative studies have demonstrated a very strong correlation between the activity of RuBP carboxylase in extracts of leaves and the photosynthetic rate of those leaves at suboptimal temperatures; however, the capacity of isolated chloroplast membranes for whole-chain electron transport when compared on an equivalent basis is also similar to that for CO_2 fixation (see Chap. 3, this Vol.). It is very likely that these capacities are equivalent at a point on the temperature response curve. The process which has the strongest temperature dependence would become limiting at temperatures below the equivalence point, and that with the lowest temperature coefficient would be limiting above the equivalence point. A few studies have examined the temperature dependence of partial reactions of photosynthesis at low temperature.

At low temperature the rate of electron transport may respond with a Q_{10} of 2 or higher; however, as temperature is increased the dependence becomes less steep (usually in steps, e.g., SHNEYOUR et al. 1973; RAISON 1974; MURATA and FORK 1976; NOLAN and SMILLIE 1976). Sharp increases in the temperature dependence of electron transport at low temperature have been associated with the onset of phase separation of chloroplast membrane lipids. A strong inhibition of electron transport at low temperature may restrict CO_2 uptake, further low temperature in combination with light may result in damage to the primary photosynthetic reaction (POWLES et al. 1980), inhibition of chlorophyll synthesis (MCWILLIAM and NAYLOR 1967), and bleaching of already developed leaves (TAYLOR and ROWLEY 1971; SLACK et al. 1974).

The capacity of carbon metabolism may be limiting at low temperature. PEARCY (1977) demonstrated that the temperature dependence for net CO_2 uptake by *Atriplex lentiformis*, a C_4 species, at strongly limiting temperature was the same as that of the V_{max} for RuBP carboxylase, and differences among leaves acclimated to contrasting growth temperature in the rate of photosynthesis at low temperature correlated with differences in RuBP carboxylase activity. Because of the CO_2-concentrating mechanism of C_4 plants, the RuBP carboxylase reaction of C_4 species is most likely presented with nearly rate-saturating CO_2 concentration, while in C_3 species the reaction occurs at normal ambient CO_2 concentration. As noted above (Sect. 10.6.1.2) the temperature dependence of this reaction would be much steeper at rate-saturating than at rate-limiting CO_2 concentration. Thus, RuBP carboxylase is more likely to become limiting at low temperatures in C_4 than in C_3 species. Studies with a C_3 species, *Nerium oleander*, indicate that net CO_2 uptake of leaves which developed at a high temperature (45 °C) and was measured at a low temperature (20 °C) was limited by the activity of fructose-1,6-bisphosphate phosphatase (BJÖRKMAN et al. 1978; BJÖRKMAN and BADGER 1979).

10.6.2.2 Adaptation and Acclimation of Photosynthesis

Plants native to (or grown in) thermally contrasting habitats may possess physiological differences which either improve photosynthetic performance or extend the capacity of the leaf to tolerate temperatures of the habitat. These physiological adjustments may be genotypically fixed (in which case there are termed adaptations) or environmentally induced acclimation responses. The adjustments to one extreme of temperature usually result in poorer performance at a contrasting temperature. This suggests that there may be some mechanistic link such that improved performance at one extreme necessarily leads to poorer performance at the other extreme.

Lipids play a role in determining the sensitivity of chloroplast membranes to thermal denaturation and possibly also the sensitivity to low temperatures. Changing the growth temperature of fully expanded leaves of *N. oleander* leaves from low (20 °C) to high (45 °C) growth temperature or vice versa caused rapid upward or downward adjustments in the apparent thermal stability of chloroplast membranes (RAISON et al. 1980) and in the temperature optimum for photosynthesis (BJÖRKMAN et al. 1980). These changes in the chloroplast membrane performance correlated with changes in the fluidity (RAISON and BERRY 1979), and the fatty acid composition of the chloroplast membrane lipids (ROBERTS and BERRY 1980). Lipid fluidity was also correlated with membrane stability in other species where the variation in stability was induced by genetic differences or differences in growth temperature (RAISON and BERRY 1979).

Associated with the changes in lipid properties at high temperature are corresponding changes in the temperature of onset of phase separation. As noted above, this event may be associated with a strong increase in the temperature dependence of chloroplast membrane reactions at temperatures below this point. The phase separation temperature also provides an index of changes in the composition of the membrane.

Since the activity of rate-limiting enzymes of carbon metabolism may also play a role in adapting or acclimating to low temperature, it might be expected that leaves of plants adapted to lower temperature would have more total protein than leaves adapted to a higher temperature. BJÖRKMAN et al. (1976) found that leaves of *Atriplex sabulosa,* a low temperature-adapted C_4 species had 6x the RuBP carboxylase, 2x the soluble protein and 1.2x the total protein per unit dry weight as did leaves of *Tidestromia oblongifolia,* a high temperature-adapted C_4 species. With the C_3 species, *Nerium oleander,* there is only a minor change in total protein or RuBP carboxylase with acclimation to high or low growth temperature; however, the fructosebisphosphate phosphatase of the high temperature acclimated leaf was only 40% of that of a low-temperature-adapted leaf (BJÖRKMAN et al. 1978). These studies emphasize the role of specific rather than general changes in protein levels with temperature.

10.6.3 Respiration

Within a temperature range of 10 to 30 °C the apparent E_a for respiratory activity is about the same for all plants (FORWARD 1960). Variations in the rate of respiration at a given temperature have, however, been noted in plants from arctic regions compared to plants from more temperate and sub-tropical regions (FORWARD 1960). This variation is also apparent in the same species adapted to different temperatures. In some instances acclimation of plants to growth at high temperature has been associated with a decline in the respiratory rate at a constant temperature (PEARCY 1977); however, this has not been observed with other species (MOONEY et al. 1977). Stimulation of respiratory capacity at a given temperature by growth at cool temperature has been demonstrated by a number of workers (BILLINGS 1971; BJÖRKMAN and HOLMGREN 1961; CHAPIN 1974). BJÖRKMAN (1975) compared the high temperature sensitivity of respiratory activity of leaves of higher plants native to contrasting thermal regimes. Respiration of leaves from the cool climate *Atriplex sabulosa* was inhibited at temperatures above 47 °C while similar inhibition did not occur until 55 °C with *Tidestromia oblongifolia* from a hot desert habitat. CAREY and BERRY (1978) showed that the temperature optimum for respiration of barley roots grown at 28 °C was about 5 °C higher than that of roots grown at 10 °C. KINBACHER et al. (1967) demonstrated that plants of *Phaseolus acutifolius* and *P. vulgaris* plants hardened by exposure to high temperature were more resistant to damage to respiration by exposure to high temperatures than the corresponding un-hardened plants.

These studies indicate differences in the capacity for respiration, the temperature optimum for respiration, and the temperature tolerance of respiration of plants adapted or acclimated to contrasting thermal regimes. These differences exactly parallel differences seen at the level of photosynthesis. It should be noted, however, that respiration is considerably more tolerant of high temperature than are photosynthetic reactions of the same plant (BJÖRKMAN 1975).

The oxidation of carbohydrates to CO_2 and water by plants involves reactions of the glycolytic pathway, the tricarboxylic acid pathway and the pentose phosphate pathway. The mitochondrion plays a central role in linking decarboxylation reactions of the above pathways to oxygen uptake and oxidative phosphorylation. In normal tissue the reactions of carbon metabolism are integrated with the oxidative reactions of the mitochondrion by direct metabolic coupling or by mechanisms that control enzyme activity.

KLICKOFF (1968) showed that increases in specific respiratory activity of *Sitanion hystrix* plants along an elevational gradient correlated with differences in the respiratory capacity of mitochondria isolated from those plants. Specific respiratory activity of mitochondria isolated from plants collected at 3,150 m in the Sierra Nevada was 284 compared to 75 $\mu l\ O_2\ h^{-1}\ mg\ N^{-1}$ for plants collected at 1,250 m.

The series of electron transfer reactions involved in mitochondrial oxidation of substrates are catalyzed by membrane-associated enzymes located on the inner mitochondrial membrane. Given appropriate concentrations of an oxidizable substrate and ADP to support oxidative phosphorylation, the rate of oxygen uptake by mitochondria appears to be limited by the activity of the lipid-associated dehydrogenase enzymes of the inner mitochondrial membrane. Coupled respiratory activity of isolated mitochondria in general increases with temperature with an apparent E_a for succinate oxidation of 40 to 60 KJ mol^{-1} up to a maximum after which irreversible denaturation occurs and the measured rate of oxidation declines. With chloroplasts this decline at high temperatures has been associated with differences in lipid fluidity of the chloroplast membrane (Sect. 10.6.2.2). The possibility of a similar mechanism controlling the high temperature stability of mitochondrial membranes has not been examined. MCMURCHIE and RAISON (1979) showed that the apparent E_a of succinate oxidase activity of liver mitochondria from different sources correlated with differences in the fluidity of the lipids of the mitochondrial membrane. Organisms which had a more fluid membrane had a higher apparent E_a at temperatures where the membrane lipids were in the fluid state.

With mitochondria derived from tissue of tropical plants, Arrhenius-type plots of oxidative activity (Fig. 10.11) show a marked increase in slope at low temperatures (below about 10 to 15 °C). This increase in slope represents an increase in apparent E_a for oxidation of from 40 to 120 KJ mol^{-1} (LYONS and RAISON 1970). Measurements with both spin labels (RAISON et al. 1971; RAISON et al. 1980) and fluorescent probes (PIKE et al. 1979) show that the membrane lipids of the mitochondria and chloroplasts of many tropical plants begin to phase separate (Sect. 10.6.1.5) at temperatures coincident with that of the increase in apparent E_a for reactions catalysed by membrane-associated enzymes (a comprehensive tabulation of the available data is proved in an Appendix to LYONS et al. 1979). In addition, the increase in the apparent E_a of oxidation at low temperature can be abolished if the membrane lipids are perturbed by treatment with detergent (RAISON et al. 1971). Based on these findings it was proposed that the increase in apparent E_a for respiratory activity of mitochondria from tropical plants at low temperature was a consequence of changes in membrane lipid order rather than an intrinsic property of the enzymes (RAISON et al. 1971). The characteristics such as the phase separation temperature and fluidity of the lipids which are incorporated into the mitochondrial membrane appear to play an important role in affecting the temperature dependence of these membrane-associated reactions.

10.6.3.1 Respiration and Low Temperature Stress

A very good correlation has been established showing that the apparent E_a for respiratory activity of mitochondria from plants susceptible to chilling (Fig. 10.11) increases below about 10 °C (see Table in Ref. LYONS et al. 1979). The apparent E_a for respiration of tissue from these plants however may or may not show a similar increase at low temperature. Mitochondria from fruit of tomato (*Lycopersicon esculentum*) show the increase in apparent E_a below about 12 °C, the temperature below which symptoms of chilling injury become evident (LYONS and RAISON 1970). Roots of young tomato plants also show an increase

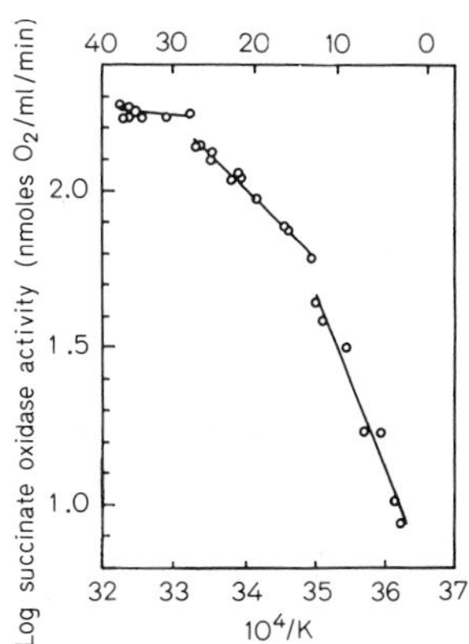

Fig. 10.11. Changes in the succinate oxidase activity of maize root mitochondria as a function of temperature. The mitochondria were isolated from 6-day-old roots by the method of RAISON and CHAPMAN (1976). Measurements were made on four preparations of mitochondria each at six different temperatures. This composite plot was made by normalizing the rates of each preparation at 25 °C. The *straight lines* were fitted by regression analysis. The E_a was 42 kJ mol^{-1} between 12 and 27 °C, and it was 104 kJ mol^{-1} below 12 °C. (RAISON et al. 1979)

in the apparent E_a of respiration from 36 to 96 K J mol^{-1} below 12 °C even when the stems and leaves are maintained at room temperature (VALLEJOS and BREIDENBACH, unpublished data). The apparent E_a for respiration of pollen grains from tomato however is constant from 20 to 0 °C (PATTERSON et al. 1979). Inconsistencies of this type have also been observed with imbibed seeds. Germination of mung bean (*Vigna radiata*) seeds below 15 °C is almost completely inhibited (SIMON et al. 1976) and oxidative activity of mitochondria from the hypocotyl of developing mung beans shows an increase in apparent E_a below this temperature (RAISON and CHAPMAN 1976). However, the apparent E_a for the respiration of mung bean seeds during imbibition and for the growth rate of roots about 1 cm in length was constant between 30 and 5 °C (SIMON et al. 1976). These discrepancies are not necessarily unexpected since the respiratory rate of tissues is often significantly lower than the maximum capacity of the respiratory enzymes of that tissue (Sect. 10.6.3.2).

Plants which show a marked decrease in respiratory activity at temperatures below about 10 °C do not necessarily suffer a decrease in phosphorylation efficiency but they do have a reduced rate of ATP formation (LYONS and RAISON 1970). This could lead to a decrease in the available ATP and hence a decrease in synthetic reactions essential in maintaining cellular integrity. Changes in the ratio of ATP:[ADP]+[AMP] have been measured in cotton seedlings (*Gossypium hirsutum*) after exposure to 3 to 5 °C for two days and compared to nonchilled control seedlings. There was a decrease in total nucleotide levels from 1,328 to 487 nmol g dry $weight^{-1}$ as well as a decrease in the nucleotide ratio by a factor of 1.8 in leaves and 2.6 in roots (STEWART and GUINN 1971). The loss of nucleotide and the increase in nucleosides in the leaves in particular indicate that low temperature also accentuates the dephosphorylation of nucleotides (STEWART and GUINN 1971). The total effect is a marked reduction in the level of high-energy compounds and a corresponding reduction in synthetic activity.

It seems likely that the damaging effect of low-temperature exposure is not the direct result of the lowering of the respiration rate itself, but secondary events, such as the accumulation of toxic concentrations of glycolytic products, may be linked to the effect of temperature on mitochondrial activity. The accumulation of α-ketoglutarate, pyruvate, ethanol, and acetaldehyde in banana tissue maintained at 6 °C for 15 days (MURATA 1969) is a good example of imbalances in metabolism induced by low temperature. This type of imbalance could develop because of the proportionately larger decrease in mitochondrial oxidative activity below about 12 °C (E_a=120 K J mol^{-1}) compared to the decrease in the rate of glycolysis which proceeds with an E_a of 38 K J mol^{-1} over the range of 25 to 0 °C. Thus between 12 and 6 °C the capacity for oxidation by mitochondria would have decreased about 1.6-fold relative to that of glycolysis. Further, the decrease in ATP levels could lead to a stimulation of glycolytic activity. Accumulation of fermentation products has been noted in a variety of plant tissues maintained under low oxygen tension or under nitrogen (FORWARD 1965), and is attributed to an increase in glycolytic activity induced by a reduction in the level of ATP under anaerobic conditions (Pasteur effect). The accumulation of glycolytic products in tissue of banana and other tropical plants at low temperature is similar to that induced by anaerobiosis. The low temperature depression of mitochondrial oxidative activity would lower the level of ATP, perhaps resulting in a stimulation of glycolysis. Accumulation of products of glycolysis might result if the rate of glycolysis exceeded the capacity of the mitochondria to oxidize the glycolytic products. Its seems likely that the observed accumulation is the result of a loss of normal control of the balance between glycolysis and mitochondrial oxidation. This may be a consequence of changes in the capacity of steps in the respiratory sequence, depletion of nucleotide pools, or an alteration in the ion balance between mitochondria and the cytoplasm because of changes in permeability of the membrane at temperatures below the phase transition (Sect. 10.6.1.5).

The imbalance in carbon metabolism is accentuated with the duration of the low-temperature stress and in the short term is reversible. This has been shown with snap beans stored at temperatures below 10 °C for various periods and then returned to 15 °C for measuring the rate of respiration (WATADA and MORRIS 1966). The increased rate of respiration after storage at low temperature is a measure of the amount of oxidizable substrate accumulated during the low-temperature stress. Respiration increased from a control value of 90 to 150 mg $CO_2{}^{-1}$ h^{-1} kg^{-1} after the beans were held for 4 days at 2.5 °C and the rise was attributed to the accumulation of glycolytic products formed during the low-temperature storage but oxidized when the tissue was returned to 15 °C. The increased rate of oxidation observed after the beans were stored for 4 days at 2.5 °C diminished over the next 4 days to the control rate as the accumulated products were oxidized. This shows that the increased rate of respiration was reversible under these conditions and therefore not caused by injury to the tissue. The increase in respiratory activity was even greater if the beans were stored at 0.5 °C, but under these conditions, when returned to 15 °C after exposure for a few days, the rate of respiration increased rather than decreased. This increase in respiration is undoubtedly a response to loss of cellular integrity resulting from damage induced by the low-temperature stress, and indicates with time the imbalances and cellular damage became irreversible.

The increase in respiratory activity induced by loss of cellular integrity (wound response) is an important factor in interpreting the effect of low temperature on respiration of tissue. It has been shown that the apparent E_a for the respiration of cucumber fruit (data of EAKS and MORRIS 1956) increases below 10 °C, if respiration is measured within 24 h of tissue being placed at the various temperatures (LYONS and RAISON 1970). After 10 days of storage, the apparent E_a for respiration was constant between 30 and 0 °C and the relative increase in respiratory activity at temperatures below 10 °C was attributed to loss of cellular integrity and a response of the tissue to wounding (LYONS and RAISON 1970).

10.6.3.2 Respiration and Growth

Respiration was at one time viewed by plant scientists interested in quantitative analysis of crop growth as a "hole in the black box" through which carbon gained in photosynthesis escaped from the plant. Respiration, however, is not an undesirable process, as it provides the energy for ion uptake, osmotic work, and the primary biosynthetic reactions associated with growth and maintenance of plant cells. BEEVERS (1970) argued that respiratory losses by a tissue were most likely a reflection of the demand of that tissue for ATP and reduced pyridine nucleotides, rather than the capacity of the tissue to catalyze respiratory reactions. For example, adding an uncoupling agent such as dinitrophenol at an appropriate concentration to a tissue frequently stimulates the respiratory rate to a level well above the apparent maximum. The addition of an uncoupler abolishes the requirement for ADP, and releases respiration from controls imposed by the internal capacity of the tissue to convert ATP to ADP, or NADH to NAD^+. In normal tissue the regeneration of these limiting metabolites is most probably linked to essential biosynthetic and maintenance activities of the cells. The concept thus emerged of a respiratory machinery governed by the pace of those reactions which consume its products.

MCCREE (1970) noted that a substantial portion of the respiration of white clover was proportioned to the daily net carbon gain by photosynthesis, while the remainder was proportional to the plant mass. This division of respiration led to the concept of a portion of respiration associated with growth (and directly linked to photosynthesis) and another portion linked to maintenance of the life functions of the plant (THORNLEY 1970). It is estimated that synthesis of new plant material from photosynthate occurs with an efficiency of about 75% with the remainder being used in the biosynthetic reactions (MCCREE 1976, PENNING DEVRIES 1975a), and maintenance respiration results in losses of 0.6 to 2% per day on a carbon basis (PENNING DEVRIES 1975b).

While total respiration increases with increasing temperature, little is known of the separate effects of temperature on these components of respiration. It may be significant in consideration of respiratory losses in higher plants, that with bacteria the yield (for substrate conversion to new growth) is independent of temperature, while "maintenance respiration" is strongly dependent upon temperature (TOPEWALA and SINCLAIR 1971). If this is so for higher plants, then a significant portion of total plant respiration may be proportional to new growth and independent of temperature (excluding the temperature dependence of growth itself). Temperature-dependent losses would be restricted to that portion of respiration associated with maintenance. The apparent temperature dependence of respiration might therefore lead to significant overestimates of the impact of respiration on net carbon balance.

10.6.4 Root Function as Influenced by Temperature

It was noted above (Sect. 10.3.1) that temperature exerted a strong influence on the partitioning of new growth to the roots and the shoots. In experiments where root temperature was varied while shoot temperature was held constant and near optimal, there was a distinct "optimum root temperature" where shoot growth was maximal and the root/shoot ratio was minimal (DAVIDSON 1969a). For species of temperate origin the optimum was between 20 and

30 °C. Growth was reduced at low temperature, but growth occurred at near freezing root temperatures (see also NIELSEN et al. 1960a, b). Subtropical species typically have a higher optimum root temperature and growth is often totally inhibited at root temperatures well above freezing (often near 10 °C; DAVIDSON 1969a; WALKER 1969; DAVIS and LINGLE 1961; HARSSEMA 1977). While high root temperatures also inhibit root growth, most work has emphasized the response to low temperature since low soil temperatures in the spring are often an important limitation for crop growth. For example, with maize the total seedling dry weight increased an average of 20% for each °C soil temperature increment up to 26 °C (an apparent Q_{10} of 3.6). Nutritional status of the leaves, leaf length, and leaf numbers, as well as stem length and root numbers showed a similar root temperature dependence (WALKER 1969). Increases in shoot growth with increasing root temperature must in part be related to the greater portion of new growth allocated to roots at low temperature. In studies with tomato seedlings, HARSSEMA (1977) noted that the decrease in dry matter production resulting from root temperature below 20 °C or above 30 °C was primarily due to a decrease in the specific leaf area [the leaf area per g of plant weight (EVANS 1972)], net assimilation rate of the leaves was not affected by root temperature. DAVIDSON (1969a) postulated that these differences in partitioning might be interpreted as compensatory responses that maintain an appropriate balance between root and shoot functions as temperature is changed. For example, if ion or water uptake per unit root mass was inhibited by a decrease in temperature, the plant might compensate by increasing the root mass. If nutrient availability is altered at constant temperature, compensating changes in the root/shoot ratio are generally observed (DAVIDSON 1969b; THORNLEY 1977; CHAPIN 1980). Inhibition of ion uptake by low temperature could thus account for corresponding increases in root/shoot ratio at low temperature. Studies have also correlated inhibition of water uptake at low temperature and an immediate inhibition of leaf growth (KLEINENDORST and BROUWER 1970; WATTS 1971, 1972). This may be related to a loss of turgor of the leaf cells. Osmotic adjustment may permit the cells to regain turgor provided the rate of water uptake by the roots keeps pace with water loss via transpiration.

Studies of the temperature dependence of ion uptake by roots are, however, few. CAREY and BERRY (1978) studied rubidium uptake as a tracer for potassium uptake. Over the temperature range 10 to 30 °C the Q_{10} was 1.5 to 2 for uptake by excised roots of maize and barley; at temperatures below about 10 °C uptake was strongly inhibited ($Q_{10}=5$ to 8). No explanation for this effect was provided; however, a similar inhibition in vivo would have a dramatic effect upon the plant's ability to acquire essential nutrients and water at low temperature. CHAPIN (1974) examined interspecific differences in the capacity of plants to absorb phosphate ion at low temperature (5 °C). Species from high latitudes had much greater uptake capacity (either at rate-limiting or rate-saturating phosphate concentration) than did species from (warmer) lower latitudes.

CLARKSON (1976) studied xylem exudation of decapitated seedlings of barley and rye as a function of root temperature. A sharp increase in the temperature coefficient for exudation was observed at about 10 °C with either species when the plants were grown at 20 °C. The rate of exudation was increased and the temperature of this break shifted to a lower temperature when rye seedlings were preconditioned for several days at 8 °C. CLARKSON et al. (1980) demonstrated that this change was associated with an increase in the ratio of the linolenic to linoleic acids (18:3 to 18:2) in the phospholipids of the root tissues. These studies and those of KUIPER (1964, 1970, 1974) suggest a role of membrane lipids in regulation of root function by temperature. CLARKSON et al. (1980) suggest that changes in the number of ion transport proteins may be important determinants of changes in the rate of exudation with acclimation to low temperature.

Root resistance to water flow can increase sharply below a critical temperature, and the critical temperature can vary depending upon the extent of cold acclimation (KUIPER 1964; MARKHART et al. 1979). The studies of MARKHART et al. (1979) are of particular interest since these workers used hydrostatic pressure to force water through the root system, thus minimizing the possible influence of ion uptake as a driving force. The slope of Arrhenius plots of water flow in soybeans showed a sharp increase at about 14 °C for roots grown at 28 °C/23 °C, and this break was lowered to about 9 °C with roots grown at 17 °C/11 °C. In both instances the Q_{10} increased from 1.4 to about 13, indicating a very strong inhibition by low temperature. In contrast, similar studies using roots of the cold-insensitive broccoli gave a constant Q_{10} for water flow (between 7 and 25 °C). With broccoli roots, however, the Q_{10} varied, depending on the growth temperature regime of plants. For cold-adapted plants, 11 °C day/night, the Q_{10} was 1.9 while for the warm-adapted plants, 28 °C/23 °C, it was 3.7. The abrupt change in water conductivity of the soybean roots compared to the absence of such a change in broccoli roots was attributed to a phase change in the membrane lipids of the soybean and the absence of such changes in the cool-tolerant broccoli (MARKHART et al. 1979). This is a general feature of water uptake by tropical and subtropical plants, and the idea that membrane lipid transitions are involved seems to be a reasonable hypothesis. It remains unclear how membrane properties might be related to water transport.

The pathway of water from the soil through the roots to the xylem must eventually pass through the cytoplasm of the cells of the endodermis because the walls of these cells are impervious to the apoplastic flow of water. The plasma membrane, the cytoplasm of the endodermal cell or the plasmadesmata linking these cells to adjacent cells could be barriers to water entry into the xylem. Although the permeability of lipid bilayers to water may be much lower in the gelled that in the liquid crystalline state (BLOCK et al. 1976), it is unlikely that the membrane undergoes a sharp change of phase which would dramatically affect water permeability (Sect. 10.6.1.5). Also, STOUT et al. (1977), estimate that the cell membrane presents very little resistance to water movement. Possible effects of low temperature on the properties of the cytoplasm of the endodermal cells or the functional integrity of the plasmodesmata linking these cells to the xylem should also be considered.

These studies do not identify the mechanism whereby low temperature inhibits ion and water uptake; however, there is little doubt that root functions may be strongly inhibited at low temperature. The sensitivity of different species to low-temperature inhibition differ and at least some species show the capacity to acclimate to low root temperature.

10.6.5 Temperature and Translocation

A great many studies have examined the effect of local temperature treatment of a segment of a petiole or stem on the rate of translocation across that region (for reviews see; GEIGER and SOVONICK 1975; MINCHIN and TROUGHTON 1980). In general, the rate of translocation is not strongly dependent upon temperature, and the Q_{10} for the rate of translocation is similar to that for the change in viscosity of phloem exudate (GIAQUINTA and GEIGER 1973). With many tropical species, however, a sharp decrease in the rate of translocation is observed at about 10 °C (GIAQUINTA and GEIGER 1973; WEBB 1967, 1970, 1971), while temperate species are usually unaffected or become inhibited only at temperatures near 0 °C (GIAQUINTA and GEIGER 1973; CHAMBERLAIN and SPANNER 1979; WARDLAW 1974).

The investigations of GIAQUINTA and GEIGER (1973) are of particular importance in considering the mechanism of this inhibition by low temperature. Arrhenius-type plots of translocation rate as a function of temperature show a change in the apparent E_a below 0 °C for sugar beet and below 10 °C for the bean. This increase in apparent E_a below about 10 °C is typical for many physiological processes of tropical plants which are sensitive to chilling (nonfreezing) temperatures (RAISON 1974). The apparent E_a for respiration of petiole tissue, streaming velocity of cytoplasmic material of leaf hair cells and the viscosity of sieve tube exudate was, however, constant for both plants between 0 and 25 °C (GIAQUINTA and GEIGER 1973). Since the temperature response of these events was not coincident with the response of the rate of translocation and mass transfer, it was concluded that inhibition of translocation at low temperature does not result from inhibition of metabolic processes or changes in the physical properties of the cytoplasm. The lack of a stoichiometric relationship among petiolar sucrose transport, petiolar respiration, and ATP turnover rates as a function of temperature (COULSON et al. 1972) support this view. Cytological investigation of sieve tubes from petioles of bean, treated at 0 °C, showed that the pores were plugged with cytoplasmic material which was continuous with the cytoplasmic material lining the cell walls (GIAQUINTA and GEIGER 1973). The sieve tube pores of tissue maintained at 25 °C were free of plugs and it was concluded that the inhibition to translocation induced by low temperature is a result of cytoplasmic changes induced by a phase change in the membrane lipids which results in occlusion of the pores. No measurements were made of the response of the sieve tube membrane lipids to temperature to determine if a phase change occurred, but such changes have been observed in membranes of bean at about 10 °C (SHNEYOUR et al. 1973; PIKE and BERRY 1980). Alterations in the structure of cytoplasm at nonfreezing temperatures have been noted with a number of species (PATTERSON et al. 1979) and these have been attributed to a reversible disruption of the microtubular elements that are part of the cytoskeleton of eukaryotic cells. Such changes, however, can be induced by changes in the ionic balance of the cytoplasm as well as by low temperature (OLMSTED and BORISY 1973). Membranes are more permeable to ions at temperatures below the phase change, so the changes observed in the cytoplasm of sieve tubes of bean could be due to the direct effect of temperature on cytoskeletal structure or to an indirect effect of a change in the ionic balance resulting from a change in membrane lipid structure.

Despite the fundamental differences in the mechanisms, phloem transport of assimilates and xylem transport of water respond in very similar ways to low temperatures. The similar strong inhibition of both water uptake and translocation observed with tropical species could be related to effects of low temperature on the properties of the cytoplasm or cell membranes of the endodermal or sieve cells respectively.

10.6.6 Growth

The capacity of the growing points to generate new growth is a function of temperature; however, it is often difficult to determine whether the capacity for growth or the availability of substrates for growing cells is the primary limitation for growth. Studies of the temperature dependence of development of the primary organs of germinating seedlings from stored reserves provide a good system to study the temperature dependence of growth itself since the supply of substrates should be nonlimiting. Elongation of the mung bean hypocotyl (RAISON and CHAPMAN 1976) or the primary leaf of barley (SMILLIE 1976) provide examples of the strong temperature dependence of growth, presumably at rate-saturating substrate concentrations. Mung bean, like many other chilling-sensitive species, failed to grow at temperatures below about 10 °C, while growth in barley continued normally to at least 2 °C. At the lowest temperatures permitting growth the elongation rate increased with a very steep exponential slope (Q_{10} near 10 for barley and over 100 for mung bean). Similar very large Q_{10}'s for seedling growth are reported by MCWILLIAM et al. (1979). This region of very steep temperature dependence extends from 0 to about 10 °C for barley, and from 10 to about 18 °C for mung bean. At higher temperatures the rate continues to increase with temperature (although the Q_{10} is less than 1.5), until the temperature optimum is reached. At superoptimal temperatures growth falls abruptly, with growth of barley being completely inhibited at 36 °C (SMILLIE 1976). These studies indicate the importance of very strong inhibition of growth processes at both high and low temperature, and suggest that growth is not strongly temperature-dependent at intermediate temperatures.

SMILLIE (1976) compared the temperature dependence of elongation and the accumulation of chlorophyll in etiolated barley leaves upon an appropriate induction treatment, and concluded that the rate of growth and chloroplast development responded to temperature according to very similar patterns. Since the greening process and leaf elongation are not interdependent, SMILLIE (1976) suggested that some mechanism(s) common to both processes are affected by high and low temperature.

FEIERABEND and MIKUS (1977) also showed strong inhibition of greening of barley, oats, peas, and wheat seedlings grown at 28 vs. 22 °C, while a similar inhibition of maize was not seen until 34 °C. These workers showed that high temperature caused an almost complete loss of the 23 and 16S component of chloroplast ribosomal RNA. For rye, oats, wheat, barley, and pea grown above 32 °C the loss of chloroplast rRNA results in a lack of accumulation of chlorophyll and inhibition of the synthesis of enzymes dependent on the plastid ribosomes such as RuBP carboxylase (FEIERABEND and MIKUS 1977). Synthesis of cytoplasmic enzymes, dependent on 80S ribosomes, was not inhibited by this treatment. Immunological tests show that in rye, synthesis of the small subunits of RuBP carboxylase continues on the 80S ribosomes during growth at temperatures which are inhibitory to protein synthesis by the chloroplast system (FEIERABEND and WILDNER 1978). Synthesis of other chloroplast-membrane proteins, dependent on 70S ribosomes, such as the DCCD-binding protein of the ATP-ase complex, cytochromes f and b-559, would also be inhibited at high temperatures. High temperature can also accelerate the breakdown of polyribosomes in pear fruit (KU and ROMANI 1970). Somewhat higher temperatures may lead to inhibition of the cytoplasmic protein synthesis system.

Disruption of biosynthetic processes may also occur at low temperature. Inhibition of chlorophyll formation in etiolated seedlings of maize was noted at 16 °C compared with seedlings grown at 28 °C, and inhibition was greater at high than at low light intensity (McWILLIAM and NAYLOR 1967). Under the same conditions the chlorophyll content of wheat was little affected by either low temperature or high light intensity (McWILLIAM and NAYLOR 1967). Chlorophyll production by etiolated cucumber (*Cucumis sativus*) leaves also ceased abruptly below 9 and above 42 °C (SMILLIE 1976). With barley grown at 2 °C, the leaves of most plants were green, although the rate of chlorophyll production was lower when compared to plants grown at 22 °C (SMILLIE 1976; SMILLIE et al. 1978). Light-mediated damage to the chloroplast membrane must play a role in the failure of tropical species to effect net accumulation of chlorophyll at low temperatures (McWILLIAM and NAYLOR 1967; BAGNALL and WOLFE 1978; McWILLIAM et al. 1979); however, SLACK et al. (1974) note a depletion of chloroplast ribosomes and specific enzymes in leaf tissue of chilling-sensitive species subjected to a brief cold stress, suggesting that some primary events in the processes leading to protein synthesis are also disrupted at low temperature.

The effect of chilling temperature on the rate of amino acid incorporation into protein has been determined using plants from both temperate and tropical regions. For intact wheat seedlings, the Q_{10} for incorporation of [^{14}C]-leucine into the total protein fraction is constant over the temperature range of 4 to 25° ($Q_{10}=1.7$ to 2.2) for plants acclimated to temperatures from 4 to 36 °C (WEIDNER and ZIEMENS 1975). A constant E_a (71 kJ mol^{-1}; $Q_{10}=2.6$) has also been noted for the in vitro incorporation of [^{14}C]-leucine by a cell-free system from mitochondria of potato (*Solanum tuberosum*), a temperate plant (TOWERS et al. 1973). In contrast, the temperature response of similar preparations from tropical plants showed a break in the apparent E_a at low temperature. For maize (*Zea mays*) root mitochondria the E_a increased from 38 to 146 kJ mol^{-1}

at 12 °C, and for sweet potato (*Ipomoea batatas*) root from 65 to 119 kJ mol^{-1}, also at 12 °C (TOWERS et al. 1973).

The increase in E_a for protein synthesis in these species corresponded to the temperature at which a change in the ordering of mitochondrial membrane lipids of these species was detected, leading TOWERS et al. (1973) to suggest that a functional relationship between ribosomes and the inner mitochondrial membrane may be important for ribosome function. BERNSTAM (1978) in his review of temperature effects upon protein synthesis of animal and microbial cells makes a strong argument for an influence of membrane properties on events involved in protein synthesis. The experiments discussed here are consistent with there being a disruption of crucial interactions at high temperature (perhaps as a result of excessive membrane fluidity) and at low temperature with tropical species (perhaps as a result of the beginning of a phase separation of the membrane lipids). Differences in the heat stability of the chloroplast and the cytoplasmic systems for protein synthesis might reflect a greater fluidity of the chloroplast than other intracellular membranes, or differences in properties of the corresponding ribosomes, tRNA's, or activating enzymes.

These strong responses of protein synthesis at high and low temperature may be related to the effects of temperature on growth and chloroplast development. The absence of growth at temperatures below about 10 °C and the very strong temperature dependence of growth at low temperatures with tropical species could well be related to a low-temperature block of protein synthesis (although other responses to chilling temperatures may be equally plausible). Also, the strong inhibition of growth at high temperatures is very likely a result of an inhibition of biosynthesis at these temperatures. The strong temperature dependence of growth observed at low temperature with chilling-resistant or -sensitive species is, however, much greater than that observed for protein synthesis either in vivo or in vitro with such species ($Q_{10}=10$ to 100 vs. 2.6 to 8 for growth and protein synthesis respectively). Biosynthetic processes such as protein synthesis are most likely autocatalytic, and as a result the temperature dependence on a sustained basis may be much stronger than that for the rate over a short time interval.

Pretreatment of wheat seedlings at temperatures from 4 to 36 °C resulted in shifts in the temperature optimum for [^{14}C]-leucine incorporation into protein and marked increases in the rate of protein synthesis at low temperatures when the leaves were acclimated to low temperature (WEIDNER and ZIEMENS 1975). A cold hardy cultivar of winter wheat maintained a higher rate of protein synthesis at low temperature when cold-hardened than did a less hardy cultivar (ROCHAT and THERRIEN 1975a, b). By analogy to the acclimation responses of photosynthesis, we might speculate that shifts in the temperature optimum for protein synthesis may reflect changes with acclimation in the fluidity of the membranes to which the ribosomes bind. Further, the changes in the capacity for protein synthesis may be related to quantitative adjustments of the amount of specific components catalyzing protein synthesis during acclimation. Studies have shown changes in the fatty acid composition of membrane lipids during acclimation (Sect. 10.6.8.2), which should lead to a more fluid membrane, and increases in the content of ribosomes of leaves acclimated to low temperature have been noted (SIMINOVITCH et al. 1968; GUSTA and WEISNER 1972). HUNTER and ROSE (1972) note a large increase in the amount of ribosomal RNA in yeast cells cultured at low temperature, and they suggest that this increase may tend to compensate for the decline in the rate of protein synthesis per unit rRNA as temperature is reduced. BIXBY and BROWN (1975) report changes with cold hardening in the properties of ribosomes of black locust *Robinia pseudoacacia* which may result from qualitative changes in the protein components of the ribosome. Changes in the free amino acid pools are observed when plants are grown at low temperature. Proline, for example, increases in leaves of plants grown at low temperature (CHU et al. 1978), apparently a direct effect of low temperature rather than a result of an effect of a change in water relations. In cotton, growth at 10 °C favors formation of glutamate and aspartate in preference to the corresponding amide derivatives present when plants are grown at 23 °C (DUKE et al. 1978). Further studies which examine the temperature response and acclimation of protein synthesis would seem to be central to gaining a better understanding of long-term responses to temperature.

10.6.7 Developmental Switches

A great deal of work has examined temperature as one of several environmental factors controlling the initiation of developmental events such as seed germination, flowering, or dormancy. Yet very little of this work has been aimed at understanding the basic mechanisms of the temperature effects. We consider here some effects of temperature on these general features of developmental events in plants.

10.6.7.1 Imbibition of Seeds

Seed germination is typically best in a rather narrow temperature range which is well within the ultimate high or low temperature limits for growth. Thus, germination of many kinds of seeds is best in the 15 to 30 °C range and is greatly reduced at higher or lower temperature. Correlations between germination and other properties of seeds show that many kinds of seeds tend to leak endogenous sugars and amino acids upon imbibition in water at limiting temperatures. Some seeds appear to be irreversibly damaged during the first few hours of imbibition at chilling temperatures (BRAMLAGE et al. 1978; CHRISTIANSEN 1968; SIMON 1976), suggesting there is some temperature-sensitive event associated with the imbibition process. During the first minutes of imbibition substances leak from seeds into the water phase and the leakage rate declines with time (SIMON 1974; CHRISTIANSEN 1968). This leakage is not merely a mobilization of water-soluble substances adhering to the seed coat, since the same initial leakage and the decline in leakage rate occurs if the imbibed seed is dried and again imbibed. Substances leaking from the seed include amino acids, sugars, organic acids, and phosphates (see SIMON 1974). It appears that these diffuse from within the cells, and the diffusion rate declines as the tissue becomes hydrated.

Leakage is also observed when imbibed seeds are exposed to high temperatures (Simon 1974). Hendricks and Taylorson (1976) demonstrated a correlation between a decline in the percentage germination of seeds at temperatures above about 30 °C and an increase in leakage of endogenous amino acids from seeds held at a constant temperature. With several species leakage and inhibition of germination became significant at temperatures above 30 to 35 °C. In two species which showed enhanced germination in this temperature range there was no enhanced leakage (Hendricks and Taylorson 1976). Leakage appears to be a threshold phenomenon increasing sharply above 30 to 35 °C in sensitive species.

Not only does the leaching of endogenous components such as sugars deplete the reserves available for subsequent growth, but leakage may also lead to fungal attack on the seed (Simon 1974). Thus, the competence of seeds exposed to these conditions may be substantially reduced. For example, germination of cotton seeds at a permissive temperature after imbibition for 48 h at 5 °C was reduced by 40%, and all the seedlings which did develop had serious root abnormalities (Christiansen 1968). Seeds imbibed for several hours at a permissive temperature and then exposed to chilling temperatures (4 °C for 48 h) were affected less severely than seeds imbibed at chilling temperatures. Solute leakage from soybean seeds (*Glycine max*) in the first 10 to 20 min of imbibition, for example, is greater at temperatures below than above 12 °C (Bramlage et al. 1978).

Solute leakage during imbibition can be explained by the need for water in order for membrane lipids to form a bilayer (Simon 1974). No calorimetric, water-ice transition is observed in lipid-water mixtures containing less than 20% water (Ladbrooke and Chapman 1969). Without free water phospholipids in a liquid-crystalline phase do not form a lamellar structure (Luzzati and Husson 1962). Without this there would be no semi-permeable barriers and solutes would leak from cells by simple diffusion. Electronmicrographs of dry (15% water) seeds of soybean (*Glycine max*) show the cells have a plasma membrane but there are regions of cellular disorganization (Webster and Leopold 1977). After imbibition at 25 °C for 20 min the membrane was continuous (Webster and Leopold 1977). The leakage observed in the initial stages of seed imbibition is thus explicable in terms of an incomplete plasma membrane in the dry seed.

The formation of the lipid bilayer is most likely a temperature-dependent process. If the temperature were such that some of the membrane lipids were in the gel phase, formation of a continuous bilayer might not be possible. Alternatively the bilayer formed at low temperature might be functionally imperfect. The membrane lipids of seeds of chilling-sensitive plants have not been investigated. Membrane lipids from other tissues (roots and leaves) of such plants have some gel phase lipids below about 10 °C (Raison 1974; Pike and Berry 1980). The increased leakage of endogenous substances could thus be explained if the seed membranes are similar to those of other tissues of these species (see Lyons et al. 1979). Simon (1976) argues against this view on the grounds that if some membrane lipids are in a gel phase at chilling temperatures, then the Q_{10} for membrane-associated enzyme functions (e.g., mitochondrial respiration), should show a marked increase at chilling temperatures. The problem with this argument is discussed above (Sect. 10.6.1.3).

HENDRICKS and TAYLORSON (1976, 1979) attribute the enhanced leakage of amino acids which occurs at temperatures above 30 °C to a temperature-dependent change in the state of the membranes of the seed. The temperature of this change is too high to be the gel to liquid crystalline transition as currently understood (PIKE et al. 1979). The temperature threshold for leakage is similar to that observed for inhibition of protein synthesis and other membrane-associated reactions at high temperature. This might be related to some change in protein components of the membrane or perhaps in structural order of the lipid. (A change in the temperature coefficient of motion of probes in plant membrane lipids is observed with spin labeled probes at temperatures of 28 to 33 °C, RAISON et al. 1979.)

10.6.7.2 Initiation of Germination

Even under ideal conditions of moisture and temperature some seeds will not germinate until they experience a particular temperature regime while fully imbibed. The temperature regime can be either low-temperature, referred to as stratification, high-temperature, or alternating high and low temperature.

The most common form of temperature treatment is stratification where seeds require exposure to temperatures of 0 to 5 °C for some months before they can be germinated at a high temperature. While the conditions for stratification are similar to those of seeds which fall in autumn and overwinter in the moist leaf-litter of the forest floor, the temperature requirements and time of exposure vary considerably (CROCKER and BARTON 1953). Stratification induces changes in seed metabolism which affect growth. For cherry seeds maintained at 5 °C for up to 16 weeks, increases were noted in the length, the number of cells, and the dry weight of the axis, while for seeds held at 25 °C none of these parameters increased (POLLOCK and OLNEY 1959). It is not clear what effect these changes in metabolism have and how they overcome the factors preventing growth of the embryo. Seedling development in peach seeds which normally requires stratification can be induced if the embryo is excised and the cotyledons removed (FLEMION and PROBER 1960). Seeds induced to germinate without cotyledons were usually deformed but if the seedlings are subjected to chilling temperatures normal growth occurs (BARTON and CROCKER 1948).

It is not clear what role low temperature plays in breaking seed dormancy. Some observations on changes in the level of growth substances would suggest that low temperature alters the balance between growth inhibitors and growth promoters in favor of growth promoters which, under favorable conditions, allows growth (MAYER and POLJAKOFF-MAYBER 1974). Decreases in the level of abscisic acid during stratification has been noted for *Fraxinus americana, Juglans regia*, and *Corylus avellana*, while an increase in the levels of gibberellic acid was found in hazel seed after 6 and 12 weeks of chilling (FRANKLAND

and Wareing 1966). In some seeds the need for chilling can be replaced by treating the seeds with gibberellic acid (Mayer and Poljakoff-Mayber 1974).

Exposure to high temperature before germinating at a low temperature enhances the percentage germination of many seeds. This treatment is effective if applied over a short or long period, and with some seeds exposure to light is a necessary condition during the high-temperature treatment. Mayer and Poljakoff-Mayber (1974) cite examples of desert seeds where germination is increased by increasing storage temperature from 20 to 50 °C but these seeds were killed when exposed to 75 °C. Alternating between high and low temperatures on either a long-term (seasonal or diurnal) or short-term (minutes) basis can also promote germination. Temperature shifts on a seasonal basis such as low-temperature shift during autumn and winter are effective with some species; *Nicotiana tabacum* is a good example. When maintained in the dark at 20 or at 30 °C there is negligible germination but almost 100% germination if the temperature fluctuates between 20 and 30 °C on a diurnal basis (Toole et al. 1955).

Some mechanistic insight into effects of temperature on initiation are provided by Hendricks and Taylorson (1978) in their studies of the interaction of light and temperature in controlling germination of *Amaranthus retroflexus* seeds. Exposure of seeds of this species to a light treatment sufficient to convert only a small fraction (less than 2%) of the phytochrome of the seeds from the red (P_r) to the far-red absorbing form (P_{fr}) resulted in negligible germination (less than 5%) when seeds were held at a constant temperature of 40 °C. However, if the photoconversion was followed by a treatment at a temperature below 32 °C for several minutes, the germination was enhanced substantially when the seeds were returned to 40 °C. Low-temperature treatments of only a few minutes were effective and after 64 min at 15 °C germination was greater than 80%. If the phototransformation followed the low-temperature treatment the response of the seeds was not different from seeds which had not received the low-temperature treatment. Approximately 30-fold less light was required to achieve a maximum stimulation of germination when the light was followed by a low-temperature treatment than when it was not. This dramatic interaction with temperature was interpreted by Hendricks and Taylorson (1979) as indicating a requirement for an interaction of phytochrome in the active (P_{fr}) form with some sort of an organization center. This center is apparently less stable at temperatures above 32 °C. This response is remarkably similar to effects of temperature on initiation of sites of DNA replication in some temperature-sensitive mutants of *E. coli*. These mutants fail to initiate new DNA replication sites at a permissive temperature for growth but initiation can occur if the temperature is lowered for a brief interval. This requirement for low temperature was interpreted according to a model in which replication initiated at a particular site on the cell membrane (Fralick and Lark 1973). It has been postulated, but not clearly established, that the active form of phytochrome needs to be associated with a membrane to express its full effect (Mackenzie et al. 1975; for review see Marmé 1977). The sudden increase in germination from below 5% to over 80%, when the temperature is lowered to below 30 °C (Hendricks and Taylorson 1978), suggests that the membrane lipids must be in a partially ordered state for effective interaction with the active form of phytochrome. This view is supported by the observation that the efficiency of the stimulation induced by the P_{fr} and the low-temperature shift is reduced considerably if the imbibing solution contains anesthetics which induce disorder in the membrane (HENDRICKS and TAYLORSON 1978).

While these data point to the involvement of membrane lipid structure in an event leading to initiation, the question of the interaction of active phytochrome with membranes and the effect of temperature on the structure of seed membrane lipids needs to be clarified.

10.6.7.3 Flowering

There is a considerable literature dealing with temperature as one of several factors affecting flowering. Many articles document changes induced in flowering by high or low temperature after providing an appropriate photoperiod for induction (for a review see EVANS 1969). Flowering of many tropical or subtropical species is reduced when the plants experience cool night temperatures, whereas species from temperate regions are generally not affected. For example, flowering of soybean, *Glycine max*, given short days is severely limited by exposure to 13 °C during dark periods (PARKER and BOTHWICK 1939). BOTHWICK et al. (1941) showed that cooling the leaf petiole was effective in inhibiting induction. This treatment most probably blocks translocation from the leaf of a substance which stimulates flowering.

Sex expression of flowers is also influenced by temperature, but again there is virtually no information on the mechanism of this effect and the diversity of the observations makes it difficult to form some conceptual view of how temperature might act in this regard. In general, high temperature, within the range of normal growth, promotes maleness, while low temperature favors femaleness (HESLOP-HARRISON 1972). This effect of temperature is not confined to any particular time during floral initiation. For hemp, a shift to low temperature in the early stages of flower development will revert "genetic male" plants into female phenotypes. If the same temperature shift occurs later in development, the sex expression of individual floral buds alters, resulting in monoecious plants (FRANKEL and GALUN 1977). In addition to influencing sex expression, high and low temperatures can affect components of the flower bud. Such effects are economically important because they can result in malformed and unmarketable fruit. For example, if floral differentiation of dwarf Cavendish bananas occurs at temperatures below about 12 °C, the female flowers lack one or more carpellar leaves and the resulting fruit is unacceptable on the commercial market (FAHN et al. 1961). Deformed and over-developed ovaries have also been observed in flowers of pepper (*Capsicum annuum*) which attain anthesis at night temperatures below about 8 °C (RYLSKI 1973). Most of the examples of an alteration in sex expression occur in tropical plants exposed for brief periods to chilling temperatures (see FRANKEL and GALUN 1977; MEYER 1966) and suggest that the changes are a manifestation of the injury which develops in these plants at low temperatures. High temperature also effects flower development. Malformation of anthers and pistils occurs in barley following very hot and dry periods in early summer (GREGORY and PURVIS 1947).

Table 10.4. The temperature threshold for inhibition of physiological functions and of various components of the leaves of *Atriplex sabulosa* (C_4) and *Tidestromia oblongifolia* (C_4). All heat treatments were conducted for 10 min with intact leaves under illumination of 1,000 μmol quanta $m^{-2} s^{-1}$. (BJÖRKMAN and BADGER 1979 and BJÖRKMAN et al. 1976, 1980)

Activity	Temperature at 10% inhibition	
	A. sab.	*T. obl.*
Whole leaf functions		
Net photosynthesis	43	51
Respiration	50	55
Retention of ionic substances	52	56
Membrane reactions		
Photosystem I	>55	>55
Photosystem II	42	49
Soluble enzymes		
RuBP carboxylase	49	56
PEP carboxylase	48	54
NAD malate dehydrogenase	54	56
NAD glyceraldehyde-3-P dehydrogenase	51	56
NADP reductase	55	55
3-PGA kinase	51	51
Adenylate kinase	47	49
FBP aldolase	49	55
Phosphoglucomutase	51	53
Phosphohexose isomerase	52	55
Light-activated enzymes		
Ru5P-kinase	44	52
NADP-glyceraldehyde-3-P dehydrogenase	42	51
NADP-malate dehydrogenase	–	51

10.6.8 Molecular Changes

10.6.8.1 Proteins

There have been a number of studies which examine protein differences among species or ecotypes from different thermal regimes. There may be functionally differentiated enzymes from different genotypes, different isozymes of the same genotype, or there may be secondary modification to the same enzyme. Differences in electrophoretic mobility, in heat or cold stability, and in kinetic parameters have been studied (see GRAHAM et al. 1979; HUNER and MACDOWALL 1979a, b; TEERI 1980; SIMON 1979). The problem here is that while very interesting molecular differences have been identified, these are difficult to relate to functional differences at a physiological level. In this regard the studies of BJÖRKMAN et al. (1976, 1980) are exemplary. In these studies the stability of enzymes involved in photosynthetic carbon metabolism were compared for species native to or adapted to contrasting thermal regimes (Table 10.4). The effect of heat treatment under physiologically relevant conditions on the stability of the enzymes was compared to the effect of comparable heat treatments on the photosynthetic capacity of intact leaves. While differences in the stability of all enzymes were observed when species of different thermal tolerance were compared, the comparisons of effects on the biochemical and physiological levels permitted these workers to resolve those components which were most sensitive to heat treatment, and to compare their sensitivity to that of the complete process. Photosynthesis was more labile to high temperature damage than other physiological processes such as respiration or the retention of cellular contents. Most soluble enzymes were stable in vivo to temperatures above that required to inactivate photosynthesis. The inhibition at a physiological level correlated best with the loss of photosystem II electron transport capacity of the chloroplast membranes. Some enzymes that are light activated lost activity at about the same temperature as photosystem II, but these could be reactivated in vitro. It was suggested that the latter effect was a secondary result of the loss of photosystem II activity. In these studies the component most sensitive to high temperature inactivation appeared to be associated with membranes. Carefully correlated biochemical and physiological studies are needed to resolve the significance of differences seen in plant proteins.

10.6.8.2 Membrane Lipids

Changes in the fatty acid composition of membrane lipids with temperature have been noted with a variety of tissues of different plants. In general the abundance of unsaturated fatty acids tends to increase with growth at low temperature for a given species. However, comparisons of the fatty acid composition of the membrane lipids do not provide a good index for predicting the thermal tolerance of different species (for a review see Chap. 12, this Vol.). Bulk properties of the lipid mixtures provide another approach to assessing the functional significance of differences in membrane lipids.

Physical properties which can readily be measured are the fluidity and phase separation temperature. Spin-labeled or fluorescent probes may be used to assess the motion of the fatty acid chains of polar lipids in liposomes prepared from membrane polar lipids. For example, the relaxation time for a spin-labeled probe in chloroplast membrane polar lipids is shown (Fig. 10.12). The relaxation time (τ) is greater when the probe is in lipids from *Nerium oleander* chloroplasts of plants grown at 45 as compared to 20 °C. This is an index of fluidity; the lower the value of τ the more fluid the lipids. Thus, at any temperature in the range of Fig. 10.12 the chloroplast lipids of the plant grown at high temperature were less fluid than the lipids of the plant grown at low temperature. The lipids of the high-temperature-grown plant reach a given fluidity at about 10 °C higher temperature than the low-temperature-grown plant.

The polarization of fluorescence from the probe trans-parinaric acid provides a good index of the presence of solid phase lipids in a preparation. The appearance of only a few percent solid lipid will cause a sharp increase in the polarization of fluorescence. As shown in Fig. 10.13 the polarization of fluorescence from liposomes of lipids from the high-temperature-grown *Nerium oleander* increases steeply at temperatures below 7 °C, indicating that solid lipid is starting to phase-separate at this temperature, while the lipid from low-temperature-grown plants does not show a corresponding increase in polarization until about −3 °C. These studies illustrate substantial differences in the properties of lipids obtained from the same clone of *N. oleander* grown at contrasting temperatures. It should be noted that these studies were conducted with isolated lipids rather than with intact membranes of this species. While the latter approach would be preferable, the procedures presently available can not be used with the intact membrane, because these contain a number of substances which interfere with the determination of the phase properties. At present we must assume that measurements conducted with extracted lipids in vitro reflect the properties of those lipids in vivo.

These lipid differences may be compared with functional differences in the temperature response of photosynthesis. The temperature optimum for CO_2 uptake of high-temperature-grown plants is several degrees higher than of corresponding leaves of low-temperature-grown plants (BJÖRKMAN et al. 1980), and the thermal stability of the pigment–protein complexes of the chloroplast membrane is greater in chloroplast of the high-temperature-grown plant (RAISON et al. 1980). The temperature responses of fully expanded leaves changed over the course of several days upon a transfer from one growth regime to another, and such leaves assumed responses similar to leaves which had developed in that growth regime. The time course for this acclimation response was compared to that of the lipid changes (RAISON and BERRY 1979), and these studies lend further support to the correlation between fluidity and function.

It would be of very great interest to know how these properties, the fluidity, and the phase separation temperature vary among plants. The only comparative

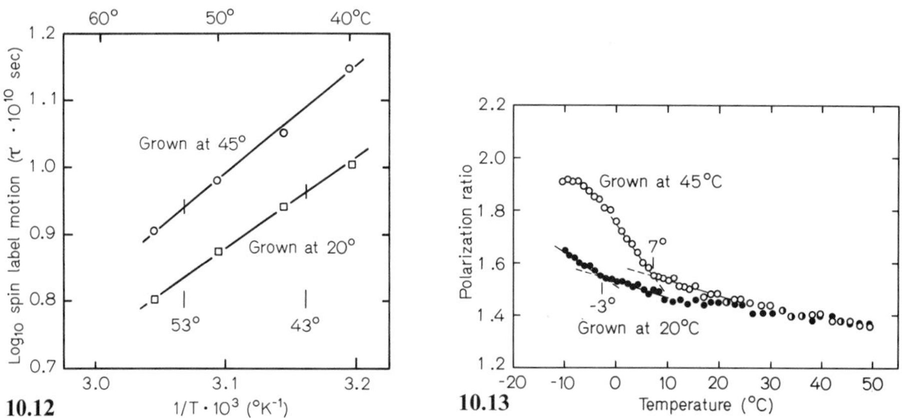

Fig. 10.12. The motion of the spin label derivative of methylstearate, 3-oxazolidinyloxy-2-(10-carbmethoxydecyl)-2-hexyl-4,4-dimethyl, incorporated into liposomes of chlopoplast polar lipids from plants of *N. oleander* grown at 20 °C/15 °C of 45 °C/32 °C. The *tick* marks at 43 and 53 °C indicate the temperatures at which the pigment system of the chloroplast becomes unstable in vivo. (RAISON and BERRY 1979)

Fig. 10.13. Trans-parinaric acid fluorescence polarization of phospholipid vesicles from leaves of *N. oleander* leaves grown at 45 °C/32 °C or 20 °C/15 °C. (PIKE and BERRY 1979)

study which has examined differences in the fluidity of the lipids of chloroplast membranes from various species (RAISON and BERRY 1979) indicates a fairly limited variation among species, but the temperature at which denaturation of the chloroplast membrane was observed was highly correlated with the differences in fluidity. RAISON et al. (1979b) and PIKE and BERRY (1980) have compared the phase separation temperatures of lipid preparations from species having different thermal preferences. Species from alpine areas or which grow during cool seasons of the year had lipid which began to phase-separate at temperature near or below 0 °C, while species from tropical regions or which grow during hot seasons of the year had lipids which began to phase-separate at higher temperatures (usually >10 °C). Evergreen perennial species from temperate desert habitats showed a change in phase-separation temperature with season. A cline in the lipid phase-separation temperature with latitude was also reported (RAISON et al. 1974). The phase-separation temperature in these instances was near or below the minimum temperature likely to occur in the native habitat during the period of active growth.

These differences in bulk properties of membrane lipids are most probably related to the decrease in the abundance of unsaturated fatty acids in the membrane lipids of plants acclimated to high as compared to low growth temperature (PEARCY 1978). However, ROBERTS and BERRY (1980) noted that independent biosynthetic mechanisms appeared to regulate the unsaturation of fatty acids in the galactolipid and phospholipid fractions of the chloroplast membrane, and they concluded from a comparison of changes in lipid fluidity and fatty acid composition during acclimation of membranes of *N. oleander* to a new growth temperature that other factors in addition to the general level of lipid unsaturation must affect the membrane lipid fluidity. THOMPSON (1979) and MAZLIAK (1979) consider the control mechanisms which may result in modification of the extent of unsaturation of fatty acids. A hypothesis put forward by THOMPSON is particularly interesting. He proposes that the enzymes which catalyze the desaturation reactions are themselves membrane-associated enzymes, and that their activity may be modulated by the physical properties of the membrane. According to his proposal, if the membrane were too rigid, the enzyme would become more active, and if the membrane were too fluid the enzyme would be inactivated. In this way the properties of the enzyme would regulate the fluidity of the membrane. While this is an attractive postulate, the details of the mechanisms which regulate lipid properties in higher plants remain to be elucidated.

10.6.8.3 Small Molecules

The pH and ionic concentration of the cytoplasm must have a large influence on the stability of the membranes and proteins of the tissue, and the ability of the tissue to maintain a favorable cytoplasmic environment must be crucial to maintenance of functional macromolecules. In addition, a great many substances have been shown to enhance the stability of membranes and proteins to high or low temperature denaturation (for a review see HEBER and SANTARIUS 1973). These include sugars, sugar alcohols, and D_2O, and these are thought to increase the structure of water and thus tend to stabilize the hydrophobic interactions that maintain native protein and membrane structures (Sect. 10.6.1.5). Some of these may accumulate in the cytoplasm and are thought to play a role in protecting the cell from high or low temperature damage (Chap. 12 this Vol.). These substances are also known to accumulate in plant cells under water stress. SEEMANN et al. (1980) report a correlation between increases in the osmotic potential and increases in the thermal tolerance of chloroplast membranes in vivo as desert winter annuals develop water stress in the late spring. The increased tolerance may in part be related to the increased concentration of osmotically active small molecules.

10.7 Concluding Remarks

For the most part the mechanisms adapting plants to different thermal regimes may be viewed as compensating mechanisms which permit plants to buffer the effects of these different thermal regimes on their metabolic systems. A major concern is the maintenance of appropriate rates of reaction and appropriate balance between reactions as temperature changes. Much remains to be learned concerning the biochemical details of these mechanisms.

Many functional differences have been identified in comparisons of plants adapted or acclimated to different thermal regimes. In most instances the fundamental mechanisms which underlie these differences have not been identified; however, some clear examples of quantitative and qualitative changes in specific plant constituents have been related to the functional performance of the plant. Quantitative increases in the levels of enzymes such as RuBP carboxylase (PEARCY 1977), fructose bisphosphate phosphatase (BJÖRKMAN et al. 1980) appear to enable plants acclimated to low growth temperature to maintain a higher photosynthetic rate at low temperature. Many other quantitative differences seem likely. For example, enhanced protein synthesis by cold-hardened cereals may be related to an increase in the amount of ribosomal RNA (Sect. 10.6.6); changes in ion uptake may be related to the formation of more or less membrane transport proteins (CLARKSON et al. 1980), etc. In general, increases in the quantity of some specific components seem to be a common response to low-temperature limitation of physiological processes.

As temperatures approach the optimum, the importance of differences the susceptibility of specific activities to inhibition at high temperature becomes evident. Such differences are most likely based upon qualitative differences, either of proteins or (in the case of membrane reactions) in the lipids associated with these proteins. The most sensitive of the reactions of photosynthesis to high temperature are associated with membranes (Sect. 10.6.2.1). It may be postulated that high temperature sensitivity in respiration, protein synthesis, and developmental initiation may also be associated with membranes. On the basis of these observations it seems reasonable to propose that membrane structures may be unusually sensitive to disruption at high temperature.

We have also emphasized the importance of chilling sensitivity of plants native to tropical and subtropical regions and the possible role of membrane lipid phase properties in determining sensitivity to this stress (Sect. 10.6.3.1).

Qualitative changes in lipid properties may play a central role in determining the sensitivity to both low and high temperature. Many observations are consistent with this idea. However, proteins are the functional entities of a membrane, and the abundance of temperature-sensitive mutations (many of which must be specific protein alterations) in microbial systems speaks for a role of differences in the properties of proteins in determining temperature sensitivity. Given no change in the properties of the proteins, changes in lipid properties with temperature may modify the temperature response. That we are more aware of the role of lipids may reflect the greater ease with which these changes at a lipid level may be identified.

Adaptive and acclimative responses have been considered primarily from a short-term perspective. Over the long term the primary biosynthetic systems assume a much more important role. Changes in the quantity of enzymes or the synthesis of new enzyme forms in response to a change in temperature is entirely dependent upon the existence of mechanisms of lipid and protein synthesis that are functional under the new conditions. The available information is perhaps too limited to generalize, but the primary biosynthetic reactions appear to be more sensitive to extremes of temperature than are the major growth processes such as photosynthesis and nutrient uptake. It would seem worthwhile to examine the possibility that the capacity of plants to adapt to new thermal regimes is ultimately limited by the tolerances of biosynthetic processes.

References

Alexandrov V Ya (1977) Cells, molecules and temperature, vol 21 Ecol Stud Springer, Berlin, Heidelberg, New York, pp 122–242

Armond PA, Staehelin LA (1979) Lateral and vertical displacement of integral membrane proteins during lipid phase transition in *Anacystis nidulans*. Proc Natl Acad Sci USA 76:1901–1905

Armond PA, Badger MR, Björkman O (1978a) Characteristics of the photosynthetic apparatus developed under different thermal regimes. In: Akoyunoglou G, JH Argysoudi-Akoyunoglou (eds) Chloroplast development. Proc Int Symp Chloroplast Dev. Elsevier North Holland Biomedical Press, Amsterdam, New York, pp 857–862

Armond PA, Schreiber U, Björkman O (1978b) Photosynthetic acclimation to temperature in the desert shrub *Larrea divaricata*. II. Light-harvesting efficiency and electron transport. Plant Physiol 61:411–415

Badger MR, Collatz GJ (1977) Studies on the kinetic mechanism of ribulose-1,5-bisphophate carboxylase and oxygenase reactions, with particular reference to the effect of temperature on kinetic parameters. Carnegie Inst Washington Yearb 76:355–361

Bagnall DJ, Wolfe JA (1978) Chilling-sensitivity in plants. Do the activation energies of growth processes show abrupt change at a critical temperature? J Exp Bot 29:1231–1242

Barton LV, Crocker W (1948) Twenty years of seed research. Faber and Faber, London, 148 pp

Beatle JC (1974) Phenological events and their environmental triggers in Mojave desert ecosystems. Ecology 55:856–863

Beevers H (1969) Respiration in plants and its regulation. In: Setlik I (ed) Prediction and measurement of photosynthetic productivity. Pudoc, Wageningen, pp 209–214
Bernstam VA (1978) Heat effects on protein synthesis. Ann Rev Plant Physiol 29:25–46
Berry JA, Björkman O (1980) Photosynthetic response and adaptation to temperature in higher plants. Annu Rev Plant Physiol 31:491–543
Berry JA, Downton WJS (1981) Environmental regulation of photosynthesis. In: Govindjee (ed) Photosynthesis vol II: CO_2 assimilation and plant productivity. New York: Academic Press, pp. in press
Berry J, Farquhar G (1978) The CO_2 concentrating function of C_4 photosynthesis. In: Hall DO, Goombs J, Goodwin TW (eds) Photosynthesis 77, Proc 4th Int Congr Photosynthesis. Biochemical Society, London, pp 119–131
Bierhuizen JF (1973) The effect of temperature on plant growth development and yield. In: Slatyer RO (ed) Plant response to climatic factors. Unesco, Paris, pp 89–98
Billings WD, Godfrey PJ, Chabot BF, Bourque DP (1971) Metabolic acclimation to temperature in artic and alpine ecotypes of *Oxyria digyna*. Arct Alp Res 3:277–289
Bixby JA, Brown GN (1975) Ribosomal changes during induction of cold hardiness in black locust seedlings. Plant Physiol 56:617–621
Björkman O (1975) Photosynthetic response of plants from contrasting thermal environments. Thermal stability of the photosynthetic apparatus in intact leaves. Carnegie Inst Washington Yearb 74:748–751
Björkman O, Badger MR (1979) Time course of thermal acclimation of the photosynthetic apparatus in *Nerium oleander*. Carnegie Inst Washington Yearb 78:145–148
Björkman O, Berry J (1973) High-efficiency photosynthesis. Sci Am 229:80–93
Björkman O, Holmgren P (1961) Climatic ecotypes of higher plants. Leaf respiration in different populations of *Solidago virgaurea*. Ann R Ag Coll Sweden 27:297–304
Björkman O, Pearcy RW, Harrison AT, Mooney HA (1972) Photosynthetic adaptation to high temperatures: a field study in Death Valley, California. Science 175:786–789
Björkman O, Nobs M, Mooney H, Troughton J, Berry J, Nicholson F, Ward W (1974a) Growth responses of plants from habitats with contrasting thermal environments: Transplant studies in the Death Valley and the Bodega Head experimental gardens. Carnegie Inst Washington Yearb 73:748–757
Björkman O, Mahall B, Nobs M, Ward W, Nicholson F, Mooney H (1974b) Growth response of plants from habitats with contrasting thermal environments: An analysis of the temperature dependence under controlled conditions. Carnegie Inst Washington Yearb 73:757–767
Björkman O, Mooney HA, Ehleringer J (1975) Photosynthetic responses of plants from habitats with contrasting thermal environments: Comparison of photosynthetic characteristics of intact plants. Carnegie Inst Washington Yearb 74:743–748
Björkman O, Boynton J, Berry J (1976) Comparison of the heat stability of photosynthesis, chloroplast membrane reactions, photosynthetic enzymes and soluble protein in leaves of heat-adapted and cold adapted C_4 species. Carnegie Inst Washington Yearb 75:400–407
Björkman O, Badger MR, Armond PA (1978) Thermal acclimation of photosynthesis: Effect of growth temperature on photosynthetic characteristics and components of the photosynthetic apparatus in *Nerium oleander* Carnegie Inst Washington Yearb 77:262–276
Björkman O, Badger MR, Armond PA (1980) Response and adaptation of photosynthesis to high temperatures. In: Turner NC, Kramer PJ (eds) Adaptation of plants to water and high temperature stress. Wiley-Interscience, New York, pp 233–249
Block MC, van Deenen LLM, de Gier J (1976) Effect of the gel to liquid crystalline transition on the osmotic behavior of phosphatidylcholine liposomes. Biochim Biophys Acta 433:1–12
Bothwick HA, Parker MW, Heinze PH (1941) Influence of localized low temperature on Biloxi soybean during photoperiodic induction. Bot Gaz 102:792–800
Bramlage WJ, Leopold AC, Parrish DJ (1978) Chilling stress to soybeans during imbibition. Plant Physiol 61:525–529

Brandt JF (1967) Heat effects on proteins and enzymes. In: Rose AH (ed) Thermobiology. Academic Press, London, New York, pp 25–72

Brown RH (1978) A difference in the N use efficiency in C_3 and C_4 plants and its implication in adaptation and evolution. Crop Sci 18:93–98

Caldwell MM, Camp LB (1974) Below ground productivity of two cool desert communities. Oecologia 17:123–130

Caldwell MM, Osmond CB, Nott DL (1977a) C_4 Pathway Photosynthesis at low temperature in cold-tolerant *Atriplex* species. Plant Physiol 60:157–164

Caldwell MM, White RS, Moore RT, Camp LB (1977b) Carbon balance, productivity, and water use of coldwinter desert shrub communities dominated by C_3 and C_4 species. Oecologia 29:275–300

Carey RW, Berry JA (1978) Effects of low temperature on respiration and uptake of rubidium ions by excised barley and corn roots. Plant Physiol 61:858–860

Chamberlin IS, Spanner DC (1978) The effect of low temperature on the phloem transport of radioactive assimilates in the stolon of *Saxifraga sarmentosa* L. Plant Cell Environ 1:285–290

Chapin FS (1974) Phosphate absorption capacity and acclimation potential along a latitudinal gradient. Science 183:521–523

Chapin FS (1980) The mineral nutrition of wild plants. Annu Rev Ecol Syst 11:233–260

Christiansen MN (1968) Induction and prevention of chilling injury to radicle tips in imbibing cottonseed. Plant Physiol 43:743–746

Chu TM, Jusaitius M, Aspenall D, Paleg LG (1978) Accumulation of free proline at low temperature. Plant Physiol 43:254–260

Clarkson DT (1976) The influence of temperature on the exudation of xylem sap from detatched root systems of rye (*Secale cereale*) and barley (*Hordeum vulgare*). Planta 132:297–304

Clarkson DT, Hall KC, Roberts JKM (1980) Phospholipid composition and fatty acid desaturation in the roots of rye during acclimatization of low temperature. Positional analysis of fatty acids. Planta 149:464–471

Collatz GJ (1978) The interaction between photosynthesis and ribulose-P_2 concentration – Effects of light, CO_2 and O_2. Carnegie Inst Washington Yearb 77:248–251

Cooper JP (1963) Species and population differences in climatic respose. In: Evans LT (ed) Environmental control of plant growth. Academic Press, London, New York, pp 381–404

Coulson CL, Christy AL, Cataldo DA, Swanson CA (1972) Carbohydrate translocation in sugar beet petioles in relation to petiolar respiration and adenosine-5′-triphosphate. Plant Physiol 49:919–923

Crocker W, Barton LV (1953) Physiology of seeds. An introduction to the experimental study of seed and germination problems. Chronica Botanica Co, Waltham Mass

Davidson RL (1969a) Effects of soil nutrients and moisture on root/shoot ratio in *Lolium perenne* and *Trifolium repens*. Ann Bot 33:571–577

Davidson RL (1969b) Effect of root/leaf temperature differentials on root/shoot ratios in some pasture grasses and clover. Ann Bot 33:561–569

Davis RM, Lingle JC (1961) Basis of shoot response to root temperature in tomato. Plant Physiol 36:153–162

Dodd JD (1968) Grassland Associations in North America. In: Gould FW (ed) Grass systematics. McGraw Hill, New York, pp 324–338

Doliner LH, Jolliffe PA (1979) Ecological evidence concerning the adaptive significance of the C_4 dicarboxylic acid pathway of photosynthesis. Oecologia 38:23–34

Downes RW (1970) Differences in transpiration rates between tropical and temperate grasses under controlled conditions. Planta 88:261–273

Duke SH, Schrader LE, Miller MG, Niece RL (1978) Low temperature effects on soybean (*Glycine max* [L] Merr. cv. Wells) free amino acid pools during germination. Plant Physiol 62:642–647

Eaks IL, Morris LL (1956) Respiration of cucumber fruits associated with physiological injury at chilling temperature. Plant Physiol 31:308–314

Edelhock H, Osborne JC (1976) The thermodymanic basis of the stability of proteins, nucleic acids and membranes. Adv Protein Chem 30:183–250

Ehleringer JR (1978) Implications of quantum yield differences on the distribution of C_3 and C_4 grasses. Oecologia 31:255–267

Ehleringer JR (1980) Leaf morphology and reflectance in relation to water and temperature stress. In: Turner NC, Kramer PJ (eds) Adaptation of plants to water and high temperature stress. Wiley-Interscience, New York, pp 295–308

Ehleringer JR, Björkman O (1977) Quantum yields for CO_2 uptake in C_3 and C_4 plants. Dependence on temperature, CO_2 and O_2 concentration. Plant Physiol 59:86–90

Ehleringer J, Forseth I (1980) Solar tracking by plants. Science 210:1094–1098

Evans GC (1972) The quantitative analysis of plant growth. Univ of Calif Press, Berkeley, 734 pp

Evans LT (1969) The nature of flower induction. In: Evans LT (ed) The induction of flowering. MacMillan, Melbourne, pp 457–480

Evans LT, Wardlaw IF, Williams CN (1964) Environmental control of growth. In: Barnard C (ed) Grasses and grasslands. MacMillan, London, pp 102–125

Fahn A, Klarman-Kislev N, Zin D (1961) The abnormal flower and fruit of May flowering, dwarf Cavandish bananas. Bot Gaz 123:116–125

Farquhar G, von Caemmerer S, Berry JA (1980) A biochemical model of photosynthetic CO_2 assimilation in leaves of C_3 species. Planta 149:78–90

Feierabend J, Mikus M (1977) Occurence of high temperature sensitivity of chloroplast ribosome formation in several higher plants. Plant Physiol 59:863–867

Feierabend J, Wildner G (1978) Formation of the large subunit in the absence of the small subunit of ribulose 1,5-bisphosphate carboxylase in 70s ribosome-deficient rye leaves. Arch Biochim Biophys 186:283–291

Fischer RA, Turner NC (1978) Plant productivity in the arid and semiarid zones. Annu Rev Plant Physiol 29:277–317

Flemion F, Prober PL (1960) Production of peach seedlings from unchilled seeds. I. Effects of nutrients in the absence of cotyledonary tissue. Contrib Boyce Thompson Inst 20:409–419

Forseth I, Ehleringer JR (1980) Solar tracking responses to drought in a desert annual. Oecologia 44:159–163

Forward DF (1960) Effects of temperature on respiration. In: Ruhland W (ed) Encyclopedia of plant physiology, vol XII (2). Springer, Berlin, Heidelberg, New York, pp 234–258

Forward DF (1965) The respiration of bulky organs. In: Steward FC (ed) Plant physiology – A treatise, vol IVA. Academic Press, London, New York, pp 311–376

Fralick JA, Lark KG (1973) Evidence for the involvement of unsaturated fatty acids in initiating chromosome replication in *Escherichia coli*. J Mol Biol 80:459–475

Frankel R, Galun E (1977) Pollination mechanisms, reproduction and plant breeding. Springer, Berlin, Heidelberg, New York, pp 145–147

Frankland B, Wareing PF (1966) Hormonal regulation of seed dormancy in hazel (*Corylus avellana* L.) and beech (*Fagus sylvatica* L.). J Exp Bot 17:596

Gates DM, Heisey WM, Milner HW, Nobs MA (1964) Temperature of *Mimulus* leaves in natural environments and in a controlled chamber. Carnegie Inst Washington Yearb 63:418–430

Geiger DR, Sovonick SA (1975) Effects of temperature, anoxia, and other metabolic inhibitors on translocation. In: Pirson A, Zimmermann MH (eds) Phloem transport, vol I. Encyclopedia of plant physiology. Springer, Berlin, Heidelberg, New York, pp 256–286

Giaquinta RT, Geiger DR (1973) Mechanism of inhibition of translocation by localized chilling. Plant Physiol 51:372–377

Graham D, Hockley G, Patterson BD (1979) Temperature effects on phosphoenol pyruvate carboxylase from chilling-sensitive and chilling-resistant plants. In: Lyons JM, Graham D, Raison JK (eds) Low temperature stress in crop plants: The role of the membrane. Academic Press, London, New York, pp 453–461

Gregory FG, Purvis ON (1947) Abnormal flower development in barley involving sex reversal. Nature (London) 160:221–222

Gusta LV, Weiser CJ (1972) Nucleic acid and protein changes in relation to cold acclimation and freezing injury of Korean boxwood leaves. Plant Physiol 49:91–96

Hackenbrock CR (1976) Molecular organization and the fluid nature of the mitochondrial energy transducing membrane. In: Abrahamssen S, Pascher I (eds) Structure of biological membranes. Plenum Press, New York, London, pp 199–234

Harssema H (1977) Root temperature and growth of young tomato plants. Meded Landbouwhogesch. Wageningen 77 (19):1–85

Havsteen B (1973) The thermal dependence of the Michaelis Constant. In: Precht H, Christophersen J, Hensel H, Larcher W (eds) Temperature and life. Springer, Berlin, Heidelberg, New York, pp 310–314

Heber U, Santarius KA (1973) Cell death by cold and heat, and resistance to extreme temperatures. Mechanisms of hardening and dehardening. In: Precht H, Christophersen J, Hensel H, Larcher W (eds) Temperature and life. Springer, Berlin, Heidelberg, New York, pp 232–262

Hendricks SB, Taylorson RB (1976) Variation in germination and amino acid leakage of seeds with temperature related to membrane phase changes. Plant Physiol 58:7–11

Hendricks SB, Taylorson RB (1978) Dependence of phytochrome action in seeds on membrane organization. Plant Physiol 61:17–19

Hendricks SB, Taylorson RB (1979) Dependence of thermal responses of seeds on membrane transitions. Proc Natl Acad Sci USA 76:778–781

Heslop-Harrison J (1959) The experimental modification of sex expression in flowering plants. Biol Rev 32:38–90

Heslop-Harrison J (1972) Sexuality in angiosperms. In: Steward FC (ed). Plant physiology – A treatise, vol VI (c). Academic Press, London, New York, pp 133–289

Hiesey WM, Milner HW (1965) Physiology of ecological races and species. Annu Rev Plant Physiol 16:527–540

Hiesey WM, Nobs MA, Björkman O (1971) Experimental studies on the nature of species. V. Biosystematics, genetics and physiological ecology of the Erythranthe section of *Mimulus*. Carnegie Inst Washington Publ 628:

Hochachka PW, Somero G (1973) Strategies of biochemical adaptation. WB Saunders, Philadelphia

Hofstra G, Aksornkose S, Atmowidjojo S, Banaag JF, Santos-Sastrohoetomo RA, Thus LTN (1972) A study of the occurrence of plants with a low CO_2 compensation point in different habitats in the tropics. Ann Biogr 5:143–157

Huner NPA, MacDowall FDH (1979a) Changes in the net charge and subunit properties of ribulose carboxylase-oxygenase during cold hardening of Puma rye. Can J Biochem 57:155–164

Huner NPA, MacDowall FDH (1979b) The effects of low temperature acclimation of winter rye on catalytic properties of its ribulose bisphosphate carboxylase-oxygenase. Can J Biochem 57:1036–1041

Hunter K, Rose AH (1972) Influence of growth temperature on the composition and physiology of microorganisms. J Appl Chem Biotechnol 22:527–540

Jain MK, White HB (1977) Long range order in biomembranes. Adv Lipid Res 15:1–60

Johnson FH, Eyring H, Stover BJ (1974) The theory of rate processes in biology and medicine. Wiley-Interscience, New York

Kavanau JL (1950) Enzyme kinetics and the rate of biological processes. J Gen Physiol 34:193–209

Kinbacher EJ, Sullivan CY, Knull HR (1967) Thermal stability of malic dehydrogenase from heat hardened *Phaseolus acutifolius* cv. Terpary Bluff. Crop Sci 7:148–151

Kleinendorst AK, Brouwer R (1970) The effect of temperature of the root medium and of the growing point of the shoot an growth, water content and sugar content of maize leaves. Neth J Agric Sci 18:140–148

Klikoff LG (1968) Temperature dependence of mitochondrial oxidative rates of several plant species of the Sierra Nevada. Bot Gaz 129:227–230

Knutson RM (1974) Heat production and temperature regulation in eastern skunk cabbage. Science 186:745–747

Koller D (1972) Environmental control of seed germination. In: Kozlowski TT (ed) Seed biology, vol II. Academic Press, London, New York, pp 1–101
Ku LL, Romani RJ (1970) The ribosomes of pear fruit. Plant Physiol 45:401–407
Ku SB, Edwards GE (1977) Oxygen inhibition of photosynthesis II. Kinetic characteristics as affected by temperature. Plant Physiol 59:991–999
Kuiper PJC (1964) Water uptake of higher plants as effected by root temperature. Meded Landbouwhogesch. Wageningen 63:1–11
Kuiper PJC (1970) Lipids in alfalfa leaves in relationship to cold hardiness. Plant Physiol 45:684–686
Kuiper PJC (1974) Role of lipids in water and ion transport. In: Recent advances in chemistry and biochemistry of plant lipids. Proc Phytochem Soc 12:359–386
Ladbrooke BD, Chapman D (1969) Thermal analysis of lipids, proteins and biological membranes. A review and summary of some recent studies. Chem Phys Lipids 3:304–356
Laing WA, Orgen WL, Hageman RH (1974) Regulation of soybean net photosynthetic CO_2 fixation by the interaction of CO_2, O_2, and ribulose-1,5-diphosphate carboxylase. Plant Physiol 54:678–685
Lange OL (1959) Untersuchungen über Wärmehaushalt und Hitzeresistenz mauretanischer Wüsten- und Savannenpflanzen. Flora 147:595–651
Laufer M (1975) Entropy driven processes in biology; Polymerization of tobacco mosaic virus protein and similar reactions. Springer, Berlin, Heidelberg, New York
Levitt J (1980) Response of plants to environmental stresses. Academic Press, London, New York, pp 347–349
Livingston BE, Shereve F (1921) The distribution of vegetation in the United States, as related to climatic conditions. Carnegie Inst Washington Publ 284:201–216
Long SP, Incoll LD, Woolhouse HW (1975) C_4 photosynthesis from cool temperate regions, with particular reference to *Spartina townsendii.* Nature (London) 257:622–624
Loomis RS, Williams WA, Hall AE (1971) Agricultural productivity. Annu Rev Plant Physiol 22:431–468
Lorimer GH (1981) The carboxylation and oxygenation of ribulose-1,5-bisphosphate: The primary events in photosynthesis and photorespiration. Annu Rev Plant Physiol 32:349–83
Lorimer GH, Woo KC, Berry JA, Osmond CB (1978) The C_2 photorespiratory carbon oxidation cycle in leaves of higher plants: pathway and consequences. In: Hall DO, Coombs J, Goodwin TW (eds) Photosynthesis 77, Proc 4th Int Congr Photosynthesis. Biochemical Society, London, pp 311–322
Lumry R, Rajender S (1970) Enthalpy-entropy compensation in water solution of proteins and small molecules: An ubiquitous property of water. Biopolymers 9:1125–1227
Luzzati V, Husson F (1962) The structure of the liquid-crystalline phases of the lipid-water system. J Cell Biol 12:207–219
Lyons JM (1973) Chilling injury in plants. Annu Rev Plant Physiol 24:445–466
Lyons JM, Raison JK (1970) Oxidative activity of mitochondria isolated from plant tissues sensitive and resistant to chilling injury. Plant Physiol 45:386–389
Lyons JM, Graham D, Raison JK (1979) Low temperature stress in crop plants: The role of the membrane. Academic Press, London, New York, 565 pp
Mackenzie JM Jr, Coleman RA, Briggs WR, Pratt LH (1975) Reversible redistribution of phytochrome within cells upon conversion to its physiologically active form. Proc Natl Acad Sci USA 73:799–803
Markhart AH, Fiscus EL, Naylor AW, Kramer PJ (1979) Effect of temperature on water and ion transport in soybean and broccoli systems. Plant Physiol 64:83–87
Marmé D (1977) Phytochrome: Membranes as possible sites of primary action. Annu Rev Plant Physiol 28:173–193
Mayer AW, Poljakoff-Mayber (1974) The germination of seeds. Pergamon Press, Oxford, 192 pp
Mazliak P (1979) Temperature regulation of plant fatty acyl desaturatases. In: Lyons JM, Graham D, Raison JK (eds) Low temperature stress in crop plants: The role of the membrane. Academic Press, London, New York, pp 391–404
McCree KJ (1970) An equation for the rate of respiration of white clover plants grown

under controlled conditions. In: Setlik I (ed) Prediction and measurement of photosynthetic productivity. Pudoc, Wageningen, pp 221–229

McCree KJ (1976) The role of dark respiration in the carbon economy of a plant. In: Black CC, Burris RH (eds) CO_2 metabolism and plant productivity. Univ Park Press, Baltimore, pp 177–184

McMurchie EJ, Raison JK (1979) Membrane lipid fluidity and its effect on the activation energy of membrane associated enzymes. Biochim Biophys Acta 554:3654–374

McWilliam JR, Naylor AW (1967) Temperature and plant adaptation I. Interaction of temperature and light in the synthesis of chlorophyll in corn. Plant Physiol 42:1711–1715

McWilliam JR, Munokaran W, Kipnis T (1979) Adaptation to chilling stress in sorghum. In: Lyons JM, Graham D, Raison JK (eds) Low temperature stress in crop plants: The role of the membrane. Academic Press, London, New York, pp 491–505

Meeuse BJD (1975) Thermogenic respiration in aroids. Annu Rev Plant Physiol 25:117–126

Minchin PEH, Troughton JH (1980) Quantitative interpretation of phloem translocation data. Plant Physiol 31:191–215

Monsi N (1968) Mathematical models of plant communities. In: Eckhardt F (ed) Functioning of the terrestrial ecosystem at the primary production level. Unesco, Paris, pp 131–149

Mooney HA (1980) Seasonality and gradients in the study of stress adaptation. In: Turner NC, Kramer PJ (eds) Adaptation of plants to water and high temperature stress. Wiley-Interscience, New York, pp 279–294

Mooney HA, Troughton JH, Berry JA (1974) Arid climates and photosynthetic systems. Carnegie Inst Washington Yearb 73:793–805

Mooney HA, Björkman O, Berry JA (1975) Photosynthetic adaptation to high temperature. In: Hadley NF (ed) Environmental physiology of desert organisms. Halsted Press, New York, pp 138–151

Mooney HA, Ehleringer J, Berry J (1976) High photosynthetic capacity of a winter annual in Death Valley. Science 194:322–325

Mooney HA, Ehleringer JR, Björkman O (1977) The energy balance of leaves of the evergreen desert shrub *Atriplex hymenelytra*. Oecologia 29:301–310

Mooney HA, Björkman O, Collatz GJ (1978) Photosynthetic acclimation to high temperature in the desert shrub, *Larrea divaricata* I. Carbon dioxide exchange characteristics of intact leaves. Plant Physiol 61:406–410

Mulroy TW, Rundel PW (1977) Annual plants: adaptations to desert environments. Bioscience 27:109–114

Murata N, Fork DC (1976) Temperature dependence of the light-induced spectral shifts of carotenoids in *Cyanidium caldarium* and higher plant leaves. Evidence for the effect of the physical phase of chloroplast membrane lipids on the permeability of the membrane to ions. Biochim Biophys Acta 461:365–378

Murata T (1969) Physiological and biochemical studies of chilling injury in bananas. Physiol Plant 22:401–411

Nielsen KF, Halstead RL, MacLean AF (1960a) Effects of soil temperature on the growth and chemical composition of lucerne. Proc 8th Int Grassl Congr, pp 287–292

Nielsen KF, Halstead RL, MacLean AF (1960b) The influence of soil temperature on the growth and mineral composition of oats. Can J Soil Sci 40:255–263

Nielsen KF, Halstead RL, MacLean AJ (1961) The influence of soil temperature on the growth and mineral composition of corn, bromegrass and potatoes. Proc Soil Sci Soc Am 25:369–372

Nobel PS (1974a) Introduction to biophysical plant physiology. Freeman, San Francisco

Nobel PS (1974b) Temperature dependence of the permeability of chloroplasts from chilling-sensitive and chilling-resistant plants. Planta 115:369–372

Nobel PS (1978) Surface temperatures of cacti – Influences of environmental and morphological factors. Ecology 59:986–996

Nolan WG, Smillie RM (1976) Multi-temperature effects on Hill reaction activity of barley chloroplasts. Biochem Biophys Acta 440:461–475

Olmsted JB, Borisy GG (1973) Microtubules. Annu Rev Biochem 42:507–533

Osmond CB, Björkman O, Anderson DJ (1980) Physiological processes in plant ecology: Toward a synthesis with *Atriplex*. Springer, Berlin, Heidelberg, New York

Parker WM, Bothwick HA (1939) Effect of variation in temperature during photoperiodic induction upon initiation of flower primordia in Biloxi soybeans. Bot Gaz 101:145–148

Patterson BD, Graham D, Paull R (1979) Adaptation to chilling: Survival, germination, respiration, and protoplasmic dynamics. In: Lyons JM, Graham D, Raison JK (eds) Low temperature stress in crop plants: The role of the membrane. Academic Press, London, New York, pp 25–36

Pearcy RW (1976) Temperature effects on growth and CO_2 exchange rates of *Atriplex lentiformis*. Oecologia 26:245–255

Pearcy RW (1977) Acclimation of photosynthetic and respiratory CO_2 to growth temperatures in *Atriplex lentiformis* (Torr.) Wats. Plant Physiol 59:795–799

Pearcy RW (1978) Effect of growth temperature on the fatty acid composition of the leaf lipids in *Atriplex lentiformis* (Torr) Wats. Plant Physiol 61:484–486

Pearcy RW, Berry JA, Bartholomew B (1974) Field photosynthetic performance and leaf temperatures of *Phragmites communis* under summer conditions in Death Valley, California. Photosynthetica 8:104–108

Penning de Vries FWT (1975a) Use of assimilates in higher plants. In: Cooper JP (ed) Photosynthesis and productivity in different environments. Cambridge Univ Press, London

Penning de Vries FWT (1975b) The cost of maintenance processes in plant cells. Ann Bot 39:77–92

Pike CS, Berry JA (1979) Phase separation temperatures of phospholipids from warm and cool climate plants. Carnegie Inst Washington Yearb 78:163–168

Pike CS, Berry JA (1980) Membrane phospholipid phase separations in plants adapted to or acclimated to different thermal regimes. Plant Physiol 66:238–241

Pike CS, Berry JA, Raison JK (1979) Fluorescence polarization studies of membrane phospholipid phase separations in warm season and cool season plants. In: Lyons JM, Graham D, Raison JK (eds) Low temperature stress in crop plants: The role of the membrane. Academic Press, London, New York, pp 305–318

Pollock BM, Olney HO (1959) Studies of the rest period. I. Growth, translocation and respiratory changes in the embryonic organs of the after ripening cherry seed. Plant Physiol 34:131–142

Powles SB, Berry JA, Björkman O (1980) Interaction between light intensity and chilling temperature on inhibition of photosynthesis in chilling-sensitive plants. Carnegie Inst Washington Yearb 79:157–160

Raison JK (1974) A biochemical explanation of low-temperature stress in tropical and sub-tropical plants. In: Bieleski RL, Ferguson AR, Cresswell MN (eds) Mechanisms of regulation of plant growth. R Soc N Z Bull 12:487–497

Raison JK, Berry JA (1979) Viscotropic denaturation of chloroplast membranes and acclimation to temperature by adjustment of lipid viscosity. Carnegie Inst Washington Yearb 78:149–152

Raison JK, Chapman EA (1976) Membrane phase changes in chilling-sensitive *Vigna radiata* and their significance to growth. Aust J Plant Physiol 3:291–299

Raison JK, Lyons JM, Mehlhorn RJ, Keith AD (1971) Temperature-induced changes in mitochondrial membranes detected by spin labeling. J Biol Chem 246:4036–4040

Raison JK, Chapman EA, Wright LC, Jacobs SWL (1979) Membrane lipid transitions. Their correlation with climatic distribution of plants. In: Lyons JM, Graham DJ, Raison JK (eds) Low temperature stress in crop plants: The role of the membrane. Academic Press, London, New York, pp 177–186

Raison JK, Berry JA, Armond PA, Pike CS (1980) Membrane properties in relation to the adaptation of plants to high and low temperature stress. In: Turner NC, Kramer PJ (eds) Adaptations of plants to water and high temperature stress. Wiley-Interscience, New York, pp 261–273

Regehr DL, Bazazz FA (1976) Low temperature photosynthesis in successional winter annuals. Ecology 57:1297–1303

Roberts JKM, Berry JA (1980) The changes in thylakoid acyl lipid composition of *Nerium oleander* accompanying acclimation to high temperature. Carnegie Inst Washington Yearb 79:147–150

Robertson GW (1973) Development of simplified agroclimatic procedures for assessing temperature effects on crop development. In: Slatyer RO (ed) Plant response to environmental factors. Unesco, Paris, pp 327–344

Rochat E, Therrien HP (1975a) Study of the proteins of resistant, Kharkov, and non-resistant, Selkirk, wheats during cold hardening. I. Soluble proteins. Can J Bot 53: 2411–2416

Rochat E, Therrien HP (1975b) Study of the proteins of resistant, Kharkov, and non-resistant, Selkirk, wheats during cold hardening. II. Soluble proteins and proteins of the chloroplasts and membranes. Can J Bot 53:2417–2424

Rundel PW (1980) The ecological distribution of C_4 and C_3 grasses in the Hawaiian Islands. Oecologia 45:354–359

Rylskki I (1973) The effect of night temperature on shape and size of sweet pepper (*Capsicum annuum*). J Am Soc Hortic Sci 98:149–152

Seemann JR, Berry JA, Downton WJS (1980) Seasonal changes in high-temperature acclimation of desert winter annuals. Carnegie Inst Washington Yearb 79:141–143

Sharpe PJ, de Michele DW (1977) Reaction kinetics of poikilotherm development. J Theor Biol 64:649–670

Shreve F, Wiggins IL (1964) Vegetation and flora of the Sonoran Desert, vol I. Stanford Univ Press, Stanford California, pp 127–142

Shneyour A, Raison JK, Smillie RM (1973) The effect of temperature on the rate of photosynthetic electron transfer in chloroplasts of chilling-sensitive and chilling-resistant plants. Biochim Biophys Acta 292:152–161

Siminovitch D, Rheaume B, Pomeroy K, Le Page M (1968) Phospholipid, protein and nucleic acid increases in protoplasm and membrane structures associated with development of extreme freezing resistance in black locust tree cells. Cryobiology 5:202–225

Simon EW (1974) Phospholipids and plant membrane permeability. New Phytol 73:377–420

Simon EW, Minchin A, McMenamin MM, Smith JM (1976) The low temperature limit for seed germination. New Phytol 77:301–311

Simon JP (1979) Adaptation and acclimation of higher plants at the enzyme level: temperature-dependent substrate binding ability of NAD malate dehydrogenase in four populations of *Lathyrus japonicus* (Leguminosae). Plant Sci Lett 14:113–120

Singer SJ (1974) The molecular organization of membranes. Annu Rev Biochem 44:805–833

Singer SJ, Nicolson GL (1972) The fluid mosaic model of the structure of the cell membranes. Science 175:720–731

Slack CR, Roughan PG, Bassett HCM (1974) Selective inhibition of mesophyll chloroplast development in some C_4 pathway species by low night temperature. Planta 118: 57–73

Smillie RM (1976) Temperature control of chloroplast development. In: Bücher T (ed) Genetics and biogenesis of chloroplasts and mitochondria. Elsevier, North Holland, pp 103–110

Smillie RM, Critchley C, Bain JM, Nott R (1978) Effect of growth temperature on chloroplast structure and activity in barley. Plant Physiol 62:191–196

Smith WK (1978) Temperature of desert plants: Another perspective on the adaptability of leaf size. Science 201:614–616

Sofield I, Evans LT, Wardlaw IF (1977) The effect of temperature and light on grain filling in wheat. In: Bieleski RL, Furguson MM, Cresswell MN (eds) The mechanisms of regulation of plant growth. R Soc N Z Bull 12:909–915

Stewart J McD, Guinn G (1971) Chilling injury and nucleotide changes in young cotton plants. Plant Physiol 48:166–170

Stout DG, Cotts RM, Steponkus PL (1977) The diffusional water permeability of *Elodea* leaf cells as measured by nuclear magnetic resonance. Can J Bot 55:1623–1631

Stowe LG, Teeri JA (1978) The geographic distribution of C_4 species of the dicotyledonae in relation to climate. Am Nat 112:609–623

Tanford C (1980) The hydrophobic effect: formation of micells and biological membranes. Wiley-Interscience, New York

Teeri JA (1980) Adaptation of kinetic properties of enzymes to temperature variablity. In: Turner NC, Kramer PK (eds) Adaptation of plants to water and high temperature stress. John Wiley and Sons, New York, pp 251–260

Teeri JA, Stowe LG (1976) Climatic patterns and the distribution of C_4 grasses in North America. Oecologia 23:1–12

Thompson GA (1979) Molecular control of membrane fluidity. In: Lyons JM, Graham D, Raison JK (eds) Low temperature stress in crop plants: The role of the membrane. Academic Press, London, New York, pp 347–364

Thornley JH (1970) Respiration growth and maintenance in plants. Nature (London) 227:304–305

Thornley JHM (1977) Root:shoot interactions. In: Jennings DH (ed) Integration of activity in the higher plant. SEB Symposium, vol 31. Cambridge Univ Press, Cambridge, pp 367–387

Tieszen LL, Senyimba MM, Imbamba SK, Troughton JH (1979) The Distribution of C_3 and C_4 grasses and carbon isotope discrimination along an altitudinal and moisture gradient in Kenya. Oecologia 37:337–350

Toole EH, Toole VK, Borthwick HA, Hendricks SB (1955) Interaction of temperature and light in germination of seeds. Plant Physiol 30:473–478

Topewalla H, Sinclair CG (1971) Temperature relationships in continuous culture. Biotechnol Bioeng 13:795–813

Towers NR, Kellerman GM, Raison JK, Linnane AW (1973) The biogenesis of mitochondria. Effects of temperature-induced phase changes in membranes on protein synthesis by mitochondria. Biochim Biophys Acta 299:153–161

Tylor AO, Rowley YA (1971) Plants under climatic stress I. Low temperature, high light effects on photosynthesis. Plant Physiol 47:713–718

Walker JM (1969) One-degree increments in soil temperatures affects maize seedling behavior. Soil Sci Soc Am Proc 33:729–736

Wang JY (1960) A critique of the heat unit approach to plant response studies. Ecology 41:785–790

Wardlaw IF (1974) Temperature control of translocation. In: Bieleski RL, Ferguson AR, Cresswell MM (eds) Mechanisms of regulation of plant growth. R Soc N Z Bull 12:533–538

Watada AE, Morris LL (1966) Effect of chilling and nonchilling temperatures on snap bean fruits. Proc Am Hort Soc 89:368–374

Watts WR (1971) Role of temperature in the regulation of leaf extension. Nature (London) 229:46–47

Watts WR (1972a) Leaf extension in *Zea mays*. II. Leaf extension in response to independent variation of the temperature of the apical meristem of the air around the leaves, and of the root-zone. J Exp Bot 23:713–721

Watts WR (1972b) Leaf extension in *Zea mays*. I. Leaf extension and water potential in relation to root zone and air temperatures. J Exp Bot 23:704–712

Webb JA (1967) Translocation of sugars in *Cucurbita melopepo*. IV. Effects of temperature change. Plant Physiol 42:881–885

Webb JA (1970) The translocation of sugars in *Cucurbita melopepo*. V. The effect of leaf blade temperature on assimilation and transport. Can J Bot 48:935–942

Webb JA (1971) Translocation of sugars in *Cucurbita melopepo*. VI. Reversible low temperature inhibition of carbon-14 movement and cold acclimation of phloem tissue. Can J Bot 49:717–733

Webster BD, Leopold AC (1977) The ultrastructure of dry and imbibed cotyledons of soybean. Am J Bot 64:1286–1293

Weidner M, Ziemens C (1975) Preadaptation of protein synthesis in wheat seedlings to high temperature. Plant Physiol 56:590–594

Went FW (1948) Ecology of desert plants. I. Observations on germination in the Joshua Tree National Monument, California. Ecology 29:242–253

Went FW (1949) Ecology of desert plants. II. The effect of rain and temperature on germination and growth. Ecology 30:1–13

Wilson JR, Ford CW (1971) Temperature influences of the growth, digestibility, and carbohydrate composition of two tropical grasses, *Panicum maximum* and *Setaria sphacelata* and two cultivars of the temperate grass *Lolium*. Aust J Agric Res 22:563–571

Wolfe J (1978) Chilling injury in plants – the role of membrane lipid fluidity. Plant Cell Environ 1:241–247

11 Responses of Microorganisms to Temperature

M. Aragno

CONTENTS

11.1 Introduction

Temperature is one of the main environmental factors governing microbial life. Growth, as well as other biological events like fruiting, sporulation, spore germination, motility, and survival are tightly related to temperature, or to temperature changes. Thus, no wonder that microorganisms were most often chosen as models for studies on the effects of temperature on biological processes. Some microorganisms can live in saline environments below 0 °C, whereas others not only resist, but also actively develop in boiling springs near 100 °C. The most heat-resisting living things are bacterial endospores.

Excellent reviews on peculiar aspects of this subject have appeared recently. They deal with the physiological and molecular bases of thermophily (AMELUNXEN and MURDOCK 1978; ESSER 1979; JOHNSON 1979; LANGWORTHY et al. 1979; LJUNGDAHL and SHEROD 1976; REID 1976; STENESH 1976; WELKER 1976; ZUBER 1976, 1979; etc.); ecology, taxonomy, and evolution of thermophiles (BROCK 1978; CASTENHOLZ 1979; TANSEY and BROCK 1978); ecology, physiology, and biochemistry of life at low temperature (BAROSS and MORITA 1978; HERBERT and BHAKOO 1979; INNISS and INGRAHAM 1978; MORITA 1975). Therefore, the scope of the present chapter is not to be a thorough review; rather, we would like to give here an introduction to a broad spectrum of ecophysiological problems regarding microorganisms and temperature.

Heat-killing of microbial cells will not be considered here; we shall focus here on responses of processes of life to temperature. Heat resistance of microorganisms is discussed in Chapters 14 and 12, this Volume.

11.2 Relations of Microorganisms to Temperature

Growth is certainly the best integration of most of the metabolic activities of a microorganism; hence it will be chosen most conveniently to test the effect of temperature on the organism as a whole. But it is by no means the only temperature-sensitive event to be considered. For example, morphogenetical changes, like fruit-body formation, sporulation, and spore germination, may be temperature-dependent in a different way from vegetative growth.

While studying responses of microorganisms to temperature, it is often difficult to assess whether temperature has a direct effect on the processes studied, or whether its role is an indirect one, as temperature can alter other physicochemical characteristics of the environment. Therefore, one must take care when interpreting the results.

11.2.1 Growth

11.2.1.1 Growth Response to Temperature

The most intensively used models of microbial growth generally apply to unicellular organisms which divide by scissiparity, and whose cells age only in response to external factors, such as exhaustion of a nutrient, effect of an inhibitor or changes of the physicochemical environment. Such a model will be discussed here, but one has to keep in mind that it cannot apply without caution to other types of microbial growth, i.e., mycelial growth or budding.

Provided that the cells are fully adapted to the culture medium used, that concentration of all nutrients is saturating the transport systems of the cells, and that no change affecting the whole cell activity occurs in the medium, the doubling time of the biomass in a culture will be constant at any time. Such growth follows exponential kinetics, which may be represented by:

$$N = N_0\, e^{\mu t}$$

where N is the biomass per volume unit, N_0 is the biomass at the beginning of the measurement, t is the time, and μ is the growth rate constant. The semilogarithmic plot of biomass increase vs. time gives a linear relationship:

$$\ln \frac{N}{N_0} = \mu t$$

where the slope is precisely the growth rate μ.

Such growth will be considered as balanced, if not only the biomass, but also all other components of the cell population (e.g., number of cells, protein and nucleic acid contents, enzyme activities) increase at the same rate.

In bacterial cultures, an exponential growth may be readily achieved (sometimes after a lag period enabling the cells to adapt to the medium) and may be maintained for a certain time before growth slows down, due to nutrient exhaustion or to self-inhibitory effects, and finally stops.

Growth rate μ varies with temperature. A typical relation is given in Fig. 11.1 for a mesophilic bacterium, *Escherichia coli.* Such a relation allows us to determine the *cardinal temperatures* which are as a rule constant for a given strain:

- the *minimal temperature* under which no growth occurs
- the *optimal temperature* at which the growth rate μ is maximum
- the *maximal temperature* above which no growth occurs.

The optimal temperature, is as a rule, closer to the maximal than to the minimal temperature.

The rate of a chemical reaction is a function of temperature; it follows the Arrhenius relationship:

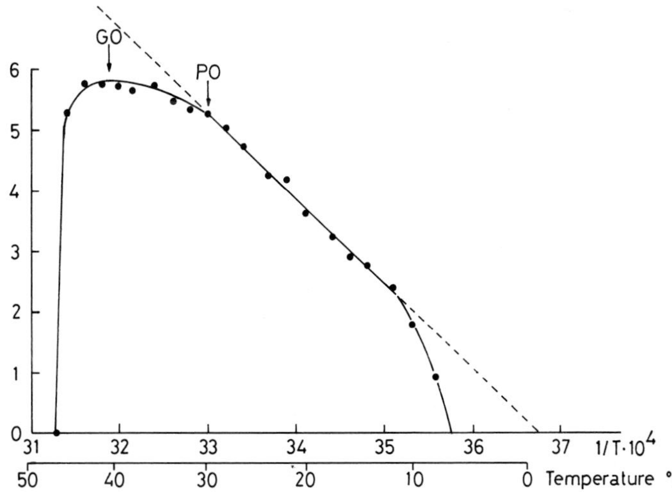

Fig. 11.1. Relationship between growth rate (μ) and temperature in *E. coli,* Arrhenius' plot. (After INGRAHAM 1958.) *GO* growth optimum; *PO* physiological optimum; *dotted line* represents the theoretical Arrhenius' relation

$$\ln v = \frac{-E}{RT} + C$$

where v is the reaction velocity, E is the activation energy of the reaction, R the gas constant, T the absolute temperature (in K), and C a constant. Therefore a plot of the log of reaction velocity vs. T^{-1} gives a straight line. A similar plot of growth rate vs. temperature is given in Fig. 11.1. It appears that, at a defined interval, bacterial growth approximatively follows the Arrhenius law, whereas near the lower and upper limits of growth, the curve falls abruptly to zero. In the interval where a linear relationship occurs it will be possible to define an apparent activation energy of bacterial growth.

There exists confusion in the terminology used to characterize the types of thermal relations of microorganisms. Several parameters may be considered: the cardinal temperatures (minimal, optimal, and maximal), the differences between minimal and maximal temperatures, as well as the activation energy. These parameters do not define clear-cut categories. Rather, there exists among microorganisms a continuum of temperature relations, in which categories are defined more or less arbitrarily. Hence, for convenience, we shall consider here three temperature ranges:
- *psychrobiotic* below 20 °C
- *mesobiotic* between 20 and 45 °C
- *thermobiotic* over 45 °C

Obligate psychrophiles grow readily near 0 °C, but not over 20 °C, whereas *facultative psychrophiles*, although readily growing near 0 °C, can grow (and often have their optimum) above 20 °C.

Mesophiles have a temperature optimum in the mesobiotic range, and do not grow at temperatures near 0 °C.

Thermophiles have temperature optima in the thermobiotic range. *Facultative thermophiles* have a minimal temperature in the mesobiotic or even psychrobiotic range, whereas *obligate thermophiles* grow only in the thermobiotic range. We can use the term *high thermophile* for an organism with temperature optimum at 75 °C or higher, and we would reserve *extreme thermophile* to organisms living in boiling or near-boiling water.

Thermoduric will be used for organisms with relatively heat-resistant cells, without regard on their ability or not to grow at high temperatures (cf. Chap. 12, this Vol.). We prefer to avoid here the use of terms such as psychrotroph (for facultative psychrophile, MORITA 1975) and caldoactive (for high thermophile, WILLIAMS 1975), because their etymology may be a source of confusion.

Eurythermal will be used for organisms with a broad temperature range, whereas *stenothermal* refers to organisms growing only within a narrow temperature interval.

Generally, categories related to the activation energy were not taken into account. In fact, some correlation exists between activation energy and optimum temperature, the former being often lower in psychrophiles than in mesophiles (INGRAHAM 1958; INNISS and INGRAHAM 1978).

11.2.1.2 Factors Determining the Temperature Limits

As stated above, growth obeys the Arrhenius law only in a defined temperature interval. Above and below, the relationship between growth rate and temperature is altered and eventually falls abruptly to zero, a few degrees above or below this interval. The possible causes for these alterations will be discussed here.

Factors Determining the Upper Limit of Growth. The lack of growth above a determined temperature is evidently due to changes of at least one structure or function which is indispensable to the growth process. Heat-sensitive structures in the cell may be membranes (particularly the physical state of their lipids), enzymes and other proteins, the protein-synthesizing apparatus, e.g., nucleic acids and ribosomes (cf. Chap. 12, this Vol.).

It was often presumed that the melting temperature of membrane lipids might determine the upper temperature limit for growth. In return, fatty acid composition of membrane lipids may be altered by changes of growth temperature, thus enabling an adaptation to occur (MARR and INGRAHAM 1962; CRONAN and GELMAN 1975). In *Escherichia coli*, an increase in temperature goes together with a simultaneous increase in the saturated fatty acids fraction and a decrease in the unsaturated fatty acids fraction (Table 11.1). Increasing the proportion of higher melting fatty acids would allow preservation of the physical state of the lipids in the membrane (the so-called "fluidity"). The range in which these modifications occur could determine the temperature limits of growth, but there is little definitive information about generalization of such a mechanism, for example in thermophiles.

Within the growth temperature range, conformation of every essential protein in the cell must allow it to function; life at high temperature would require therefore a corresponding thermostability of the whole protein population. Hence, genetical adaptation to thermophily would imply simultaneous mutations

Table 11.1. Effect of growth temperature on fatty acid composition of *Escherichia coli* ML 30 grown in glucose-minimal medium and harvested during exponential growth. (MARR and INGRAHAM 1962)

Fatty acid (%)	Temperature (°C)							
	10	15	20	25	30	35	40	43
Saturated								
Myristic	3·9	3·8	4·1	3·8	4·1	4·7	6·1	7·7
Palmitic	18·2	21·9	25·4	27·6	28·9	31·7	37·1	48·0
Methylene hexadecanoic	1·3	1·1	1·5	3·1	3·4	4·8	3·2	11·6
Methylene octadecanoic	0	0	0	0	0	0	0	3·7
Unsaturated								
Hexadecenoic	26·0	25·3	24·4	23·2	23·3	23·3	28·0	9·2
Octadecenoic	37·9	35·4	34·2	35·5	30·3	24·6	20·8	12·2

enhancing thermostability of many proteins, which appears as a highly unprobable event. In fact, we must recall that:

1. In mesophiles, many enzymes are active at temperatures higher than the maximum growth temperature. Hence, it might be that one or a small number of enzymes would determine this limit.
2. In thermophiles, enzymes were isolated which are denaturated in vitro at growth temperatures. Their thermostability must be therefore enhanced in vivo; in some cases (Ohta 1967; Amelunxen and Murdock 1978) it was shown that Ca^{2+} ions act as stabilizers.
3. The molecular basis for thermostability of proteins is poorly understood. From comparisons between thermophilic enzymes and their mesophilic counterparts, it appears that no unique or gross difference in structure would explain the higher thermal stability of the former molecules. Such enzymes have similar helical, β-structure and hydrophobic content. The enhanced thermostability appears thus to be related to only a few key substituents in the primary structure.

Therefore, the number of differences implied in the alteration of temperature maximum might be much smaller than could be expected.

It was claimed that nucleic acids from thermophiles had a higher T_m than those from mesophiles. In the genus *Bacillus*, for instance, a certain correlation can be found between the guanine+cytosine content of DNA (related to T_m) and the temperature maximum of growth (Fig. 11.2). Yet, such a correlation does not appear if one considers bacteria from unrelated taxonomic groups.

A correlation seems to exist, too, between the denaturation temperature of ribosomes and the maximum growth temperature (Pace and Campbell 1967). The whole ribosome appears more heat-stable than the isolated rRNA, so that the structural arrangement between r-proteins and rRNA must act as a stabilizing factor. Denaturation temperature of ribosomes from psychrophiles is much higher than the upper limit for growth of the same bacteria (Inniss and Ingraham 1978). Nevertheless, protein synthesis ceases at temperatures much closer to the maximum growth temperature. In vitro protein synthesis experiments with ribosomes and supernatant fractions from psychrophilic Bacilli (*B. insolitus* and *B. psychrophilus*) showed inactivation of ribosomes by a treatment of 10 min at 30 °C, whereas in the same conditions the supernatant remained active (Bobier et al. 1972).

An interesting postulate was made by Allen (1953), who attempted to explain thermophily by a special metabolic state, in which rapid denaturation was compensated by an enhanced rate of protein synthesis; yet experimental evidence supporting this hypothesis is weak.

Factors Determining the Lower Limit of Growth. According to Arrhenius' relation, growth should continue at low temperatures, although at reduced rates, until the medium freezes. It is the case with psychrophiles, but in other bacteria, growth ceases at a temperature which may be considerably higher than the freezing point of the medium. Thus, it is not the growth at low temperature, but the inability of meso- and thermophiles to grow at such temperatures which requires explanation, as pointed out by Foter and Rahn (1936).

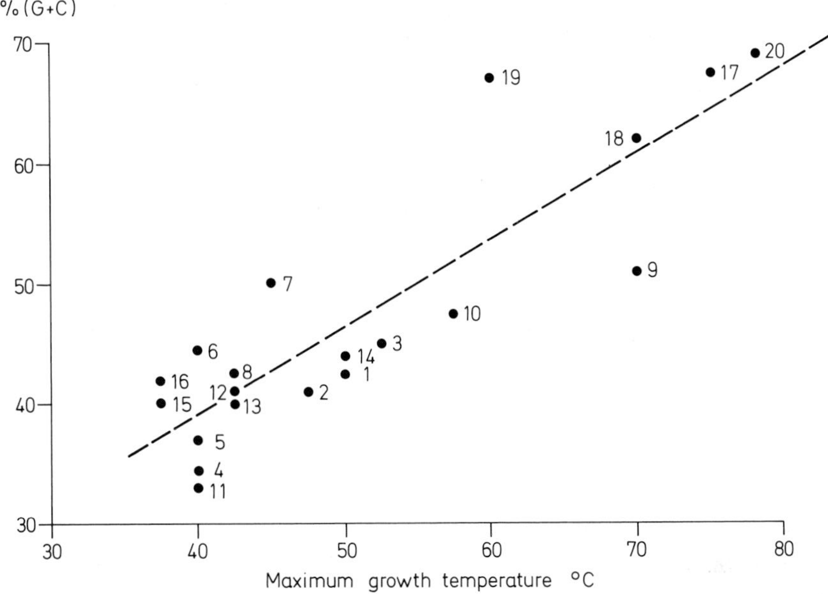

Fig. 11.2. Relationship between guanine (*G*)+cytosine (*C*) content of the whole DNA and maximum growth temperature of different *Bacillus* species.

1 Bacillus subtilis
2 B. pumilus
3 B. licheniformis
4 B. cereus
5 B. megaterium
6 B. polymixa
7 B. macerans
8 B. circulans
9 B. stearothermophilus
10 B. coagulans
11 B. alvei
12 B. firmus
13 B. laterosporus
14 B. brevis
15 B. sphaericus
16 B. pasteurii
17 B. schlegelii
18 B. acidocaldarius
19 B. thermoruber
20 B. thermocatenulatus

Sources of data: *1–16* Buchanan and Gibbons (1974); *17* Schenk and Aragno (1979); *18* Darland and Brock (1971); *19* Aragozzini et al. (1976); *20* Golovacheva et al. (1975). Values presented are the average maximal temperatures and G+C contents given by the authors. The *dotted line* represents the linear regression calculated for these values. Correlation coefficient is 0.875

As stated before, microbial cells respond to temperature by changes in the fatty acid composition of the membrane lipids: the lower the growth temperature, the higher the proportion of low-melting, unsaturated fatty acids and the lower the proportion of high-melting, saturated ones. It is thus possible that, under certain conditions, the minimum growth temperature occurs when the requirement in unsaturated fatty acids allowing the maintenance of membrane fluidity exceeds the cell's biosynthetic capabilities. But it was also shown that *Escherichia coli* cells with 37 °C membranes could grow at much lower temperatures (Shaw and Ingraham 1965; Gelman and Cronan 1972).

Mutants of mesophilic bacteria with decreased minimum temperature were rarely isolated. Isolation failed completely with enteric bacteria, whereas some success was obtained with pseudomonads. Such results would imply that, in the former, a number of mutations are required to decrease minimum temperature; on the other hand, most pseudomonads are facultative psychrophiles, so that the number of mutations necessary to gain psychrophily might be expected to be small in mesophilic strains of this group (INNISS and INGRAHAM 1978).

On the contrary, mutants with *increased* minimum temperature are easily isolated, and occur with approximately the same frequency as mutants with *decreased* maximum temperature (INNISS and INGRAHAM 1978). As a rule, such mutations are restricted to genes which encode proteins requiring very precise conformation to function, particularly allosteric enzymes and ribosomal proteins. Such cold-sensitive conditional mutants show an increase not only of the minimal temperature, but also of the apparent activation energy in the mid-range of temperature. Instead, the optimum and maximum temperatures are unaffected.

Studies on cold-sensitive, histidine-deficient conditional mutants (O'DONOVAN and INGRAHAM 1965) showed, for instance, that all such mutants produced an altered form of phosphoribosyl-ATP-phosphorylase, the first enzyme in the histidine pathway. The mutant enzyme was 100- to almost 1,000-fold more sensitive to feedback inhibition by histidine. Moreover, sensitivity to histidine with or without mutation is 10-fold higher at 20 °C than at 37 °C. The authors assume that, in the case of the mutant, the level of histidine at 20 °C will be too low to allow protein synthesis, whereas it would be still sufficient at 37 °C.

Cold-sensitive mutants unable to synthesize ribosomes at low temperature were often isolated, but there is no evidence that similar defects occur in wild-type mesophiles. In return, polysome formation seems to be altered in mesophiles below their minimum growth temperature. If initiated before the cells are placed at low temperature, translation will be completed at sub-minimal temperature, whereas initiation of synthesis of new peptidic chains will not occur at low temperature. Such phenomena were observed in *E. coli* (DAS and GOLDSTEIN 1968; TAI et al. 1973) and in taxonomically unrelated organisms such as *Azotobacter vinelandii* (OPPENHEIM et al. 1968) and *Bacillus stearothermophilus* (ALGRANATI et al. 1969). Yet a correspondance between the temperature under which polysome formation is repressed and the minimal growth temperature has not always been established.

Care should be taken when extrapolating results obtained with mutants to wild-type strains. These experiments reveal possible mechanisms preventing growth below a defined temperature but, in the case of wild-type mesophiles, the question remains mostly unanswered.

11.2.1.3 The Optimal Temperature: a Compromise

In the laboratory, microorganisms are often cultivated at their optimal temperature, which is considered as giving an ideal environment to the cells.

The optimal temperature is much closer to the maximal than to the minimal one (Fig. 11.1). Moreover, at this temperature, the response of growth to temperature variations is completely altered. In fact, at the optimum, the growth rate should be even higher according to Arrhenius' law. The actual growth rate is therefore a compromise between activatory and inhibitory effects of temperature. Cells grown at optimal temperature are therefore already partially inhibited through temperature.

These considerations are of importance in physiological work, particularly in studies of bioenergetic processes. Hence, one should consider a "physiological optimum", that is the highest temperature where Arrhenius' plot is still linear (see Fig. 11.1); this optimum would be some degrees below the growth optimum.

11.2.1.4 Adaptation to a New Temperature Range

As stated before, increasing the maximum growth temperature, or lowering the minimal temperature by mutations is as a rule a highly inprobable event, which would imply the simultaneous modification of a number of genes. On the contrary, mutations lowering the maximum growth temperature, or increasing the minimal temperature, are much more likely to occur: they may imply only one mutational site, i.e., making an essential protein more heat- or cold-sensitive.

Another type of adaptation was described in aerobic sporeformers. Obligately thermophilic strains of *Bacillus stearothermophilus* did not grow when transferred directly to 37 from 55 °C cultures (JUNG et al. 1974; HABERSTICH and ZUBER 1974). By cultivating these strains at an intermediate temperature (46 °C), and provided that a very rich complex medium (brain heart infusion broth) was used, they could then be adapted to grow at 37 °C. A similar passage through an intermediate temperature was required to re-adapt the mesophilic variants to thermophilic growth. On the contrary, *Bacillus caldotenax* could be adapted directly from 55 to 37 °C and vice versa, without subculturing at an intermediate temperature. There were striking differences in the amount of some enzymes in cells grown at 55 and at 37 °C. Moreover, some enzymes were thermostable in bacteria cultivated at 55 °C and thermolabile when isolated from cells grown at 37 °C. It is not yet known if thermostable and thermolabile isoenzymes are coded by different genes operative at different temperatures, or if only one gene is involved in their synthesis, the differences being acquired during a subsequent step of protein synthesis, or even after enzyme completion, by enzymatic modification. Moreover, such phenomena cannot be generalized, since other Bacilli, such as the facultatively thermophilic *B. coagulans* KU (AMELUNXEN and MURDOCK 1978), behave in a completely different way.

Using DNA from the strict thermophile *B. caldolyticus* as a donor, LINDSAY and CREASER (1975) succeeded in transforming a mesophilic strain of *B. subtilis* into a thermophilic one, growing above 70 °C. Number and function of the genes involved in such a transformation are not known, and confirmation of these results is still expected.

11.2.1.5 Limitation of Temperature Range by a "Facultative" Pathway

Most microorganisms are versatile with regard to their nutritive capabilities, i.e., they may use a variety of carbon- and nitrogen-sources. Hence, one has to distinguish between fundamental processes, such as DNA replication and transcription, RNA translation, ribosome formation and function, ATP synthesis, etc., and "facultative" processes, such as transport and catabolism of a given carbon-source, utilization of a given nitrogen-source (e.g., nitrogen fixation), CO_2 fixation in facultative autotrophs, etc. In most cases, the key enzymes of these latter pathways are inducible, and they are only synthesized when the related nutrients are provided. It is therefore possible that the temperature limits of such a pathway are narrower than those required for the function of fundamental processes of the cell. A good example is given by nitrogen fixation in *Klebsiella pneumoniae* (HENNECKE and SHANMUGAM 1979). The bacterium grows fast with ammonium salts or an organic source of nitrogen at 37 °C and more. Yet, at such temperatures, growth with N_2 or NO_3^- as sole N-sources is prevented. The nitrogen-fixation enzymes are not inhibited by temperature and their activity (in cells previously grown at 30 °C) is even optimal at 39 °C. In this case, temperature was shown to affect the expression of "nif" genes, i.e., genes encoding the components of the nitrogen-fixation system.

Therefore, in some cases, the cardinal temperatures of a given organism may vary according to the nutrients provided.

11.2.2 Induction of Spore Germination in Bacteria and Fungi

Some bacteria and most fungi produce resting cells able to withstand unfavorable environmental conditions (i.e., drought, winter, lack of adequate substrate) and to ensure dispersal of the species. Most are called spores, although their origin may be completely different, like *Bacillus* endospores, arthrospores of streptomycetes, sporangiospores and zygospores of zygomycetes, conidia of molds and meiospores of higher fungi (asco- and basidiospores). In many cases, spores are constitutively dormant (SUSSMAN and HALVORSON 1966), that is, they require a "trigger" to initiate further germination and outgrowth. Chemicals, as well as physical treatments, may act as triggers; temperature shocks, either heat or cold, are often required to break dormancy. (Germination of spores will be discussed in more detail in Chap. 13, Vol. 12B.)

11.2.2.1 Heat-Induced Spore Germination

A heat-shock is in many cases a means of breaking dormancy in spores. This is particularly the case with bacterial endospores (in the genera *Bacillus* and *Clostridium*) and with coprophilous fungi. In these latter, heat activation has a peculiar ecological significance, since spores are kept from germinating as long as they remain on the vegetation. After they have been swallowed by herbivorous, warm-blooded animals, they are submitted to a heating period during the passing through the gastrointestinal tract. Germination then follows immediately the excretion of feces; such fungi are among the first invaders of this peculiar substrate. It must be emphasized that, not only heat, but also chemicals or enzymatic triggers could activate such spores. *Neurospora* ascospores are activated as well by furfuraldehyde or phenyl-ethyl-alcohol (SUSSMAN 1976) as by a heat shock.

The mechanisms by which a heat treatment induces spore germination are poorly understood. In general, heat activation is a reversible phenomenon, and dormancy is restored after a certain delay if the spores are not placed in conditions suitable for germination.

In bacteria, heat induction of spore germination is restricted to the endospores produced by Bacillaceae. Most studies deal with spores from a few species of *Bacillus* (*B. subtilis, megaterium, cereus*); all three belong to the so-called "group I" Bacilli forming relatively thin-walled spores which do not swell the sporangium at maturity. Nevertheless, a limited number of studies deal with *Clostridium* endospores.

It is out of the scope of this paper to review in detail all the research dealing with bacterial endospores activation. This question was thorougly reviewed by KEYNAN and EVENCHIK (1969). We present here only briefly some of the properties of heat activation, and some of the possible mechanisms of this phenomenon.

Contrary to cold activation or post-maturation by aging, heat activation is a reversible phenomenon. In *B. cereus*, deactivation was complete at 28 °C after 96 h; it was much slower at 4 °C, whereas no deactivation occurred at −20 °C (KEYNAN et al. 1964).

Efficiency of heat activation depends on the medium composition and physicochemical characteristics; activation may be optimized by the simultaneous presence of other "germinants", such as L-alanine. Heat activation depends on the presence of water. Lyophilized spores, or spores suspended in high concentrations of glycerol are not activated by heat (BEERS 1958).

The time needed to activate a given population of spores depends on the temperature used. Activation at relatively low temperature requires much more time than activation at higher temperature.

The requirements for heat activation vary greatly from species to species. Whereas spores from mesophilic bacteria are activated in a few minutes at 60 °C, those from *Bacillus stearothermophilus* and other thermophilic Bacilli may require temperatures as high as 105–115 °C.

Many different hypotheses were advanced to explain heat activation, and activation of bacterial endospores in general. In fact, it would be impossible to understand the mechanisms of breaking dormancy without understanding dormancy itself. Among the proposed mechanisms of heat activation of bacterial endospores are included:

- breaking of disulfide bonds in structural proteins, particularly in the spore coat. Then, deactivation could be explained by the reoxidation of free −SH groups with formation of disulfide bonds (GOULD and HITCHINS 1963)
- release of free dipicolinic acid, which is a potent chelating agent and would inactivate a hypothetical inhibitory metal
- inactivation of an inhibitory protein
- activation of a cortex-lytic system, the cortex acting here either as a permeability barrier or as an inhibitor of germination by the pressure it exerts on the protoplast
- change in permeability of some membranes, etc.

In fungi, many hypotheses were advanced, and it is probable that different mechanisms occur in different fungi (see ANDERSON 1978, for a review). In *Neurospora* (SUSSMAN 1976) the effect of a heat shock seems to involve primarily the metabolism of trehalose. The disaccharide trehalose is the main carbon source used during the germination process, whereas dormant spores metabolize mainly lipids. The former is nevertheless present as a reserve in dormant spores, as well as trehalase, the α-glucosidase responsible for its hydrolysis to glucose. But whereas the sugar is located in the cytoplasm, the enzyme is associated with the innermost ascospore wall. Enzyme and substrate are consequently separated; it is supposed that a heat shock increases membrane permeability, allowing trehalose to diffuse out and to be hydrolyzed to glucose.

Another possible mechanism of heat activation implies the presence of a heat-sensitive protein acting as a repressor or inhibitor of a key enzyme of a metabolic pathway linked to the germination process, e.g., membrane-bound ATP-ase in mitochondria. A heat shock would destroy the protein, thus enabling oxidative phosphorylation to occur, which in turn promotes germination.

In *Phycomyces blakesleeanus* sporangiospores, trehalase is temporarily activated by heat (VAN ASSCHE et al. 1972).

In conclusion, little is known about the mechanisms of heat activation. In many cases, high temperature is supposed to act primarily on membranes, but it is not clear whether proteins or lipids are mainly involved.

11.2.2.2 Cold-Induced Spore Germination

In many fungi, particularly mycorrhizal Agaricales and phytopathogens, spores require a treatment at low temperature before germination can occur. It was recently shown that endospores of *Thermoactinomyces vulgaris*, a thermophilic mycelial bacterium abundant in compost and self-heated hay, behaved in a similar manner: germination did not take place unless spores were activated at 20 °C, thus far below the minimal growth temperature. After only 1 h at 20 °C, these spores required a heat shock (10–20 min at 100 °C) before optimal germination occurred, whereas after 24 h at 20 °C, they were activated without needing a further heat treatment (KALAKOUTSKII and AGRE 1976). Contrary to heat activation, cold-induced activation is not reversible, that is, once activated, spores cannot restore their dormancy.

Ecologically significant is the fact that induction by low temperature occurs in fungi whose life cycle is related to higher plants. Dormancy prevents germination of spores in the autumn, when they are generally produced, whereas passage through the winter will activate them, enabling them then to germinate in the spring, together with outgrowth of vegetation. Cold-induced spore germination parallels seeds and buds vernalization.

Mechanisms of cold-induced germination are so far poorly understood.

11.2.3 Developmental Changes Induced by Temperature in Fungi

Apart from its effect on growth rate and from its role as a "trigger" of spore germination, temperature may affect morphogenesis, as it induces developmental

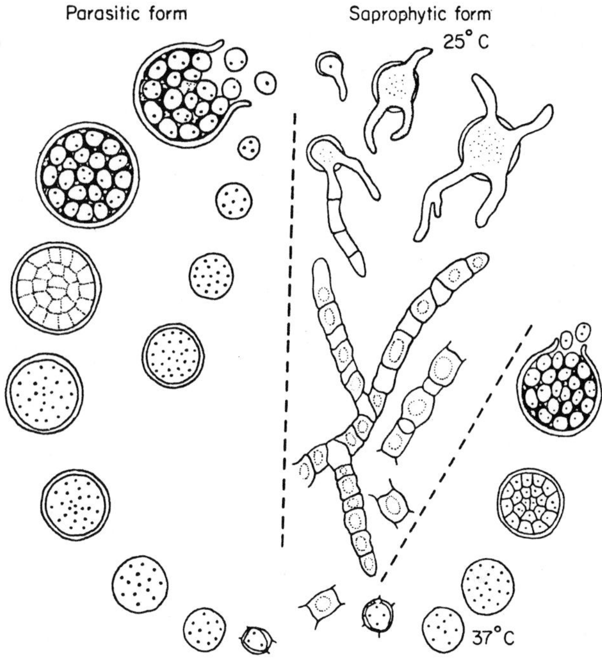

Fig. 11.3. Life cycle and dimorphism in *Coccidioides immitis*. *Left* parasitic form (in vivo at 37 °C); *right* saprophytic forms, at 25 and 37 °C. (WILSON and PLUNKETT 1965)

changes in many fungi; effects on cell growth pattern, as well as on the induction of the spore-bearing apparatus, were observed (ANDERSON 1978).

11.2.3.1 Spore Outgrowth and Vegetative Development

Morphologically, spore outgrowth generally shows two phases; at first, growth is isotropic: it concerns the whole cell area; it is the so-called swelling phase; then, growth becomes restricted to one or a few limited areas of the cell, initiating thus an apical (anisotropic) growth pattern, and eventually generating a mycelium. It was often observed that germination of spores does occur slightly above the maximal growth temperature; in such cases, the anisotropic growth phase is inhibited (ANDERSON 1978; ARAGNO 1973) and an enlarged spherical cell is formed. Its development either stops at this stage or proceeds further to an altered developmental pattern. In *Coccidioides immitis*, at 37 °C, the enlarged cell eventually forms endospores. This fungus is the agent of "St Joaquin Valley Fever", a lung disease of man endemic in south-western USA. The cycle: endospores→enlarged spores is similar in vivo and in vitro at 37 °C, whereas in vitro cultures at 25 °C give rise to mycelium and conidia (Fig. 11.3). In other cases, swollen spores transform to sclerotic cells which divide eventually by septation (*Phialophora* sp.), or give rise to buds, initiating therefore a yeast-like growth (*Histoplasma capsulatum*).

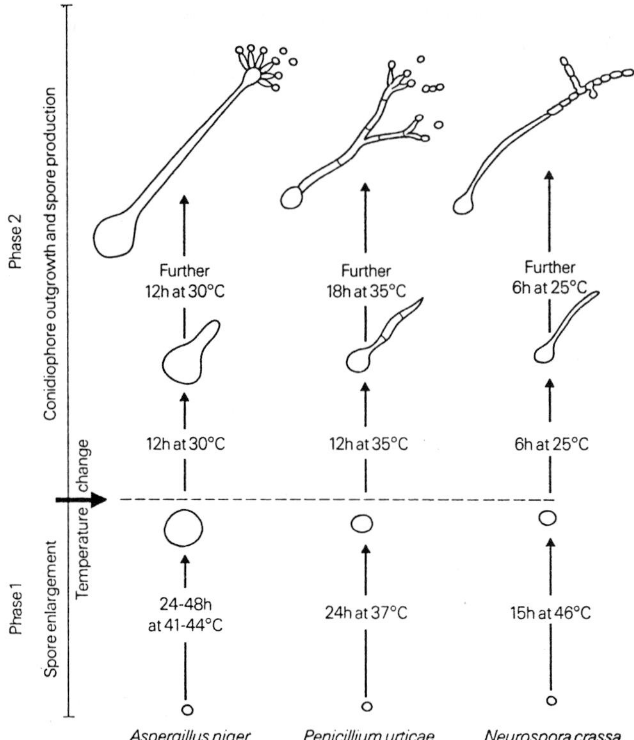

Fig. 11.4. Induction of microcycle conidiation by temperature in three species of molds. *Phase 2* in *P. urticae* also involves nitrogen limitation. (ANDERSON 1978)

Transfer of swollen conidia to temperatures allowing apical growth sometimes leads to microcycle conidiation: in *Aspergillus niger*, for instance, the enlarged cells produced by isotropic outgrowth of conidia at 41–44 °C directly produce a sporangiophore when returned to 30 °C (ANDERSON and SMITH 1971) (Fig. 11.4).

Increase of temperature during mycelial growth may give rise to similar effects: in various fungi, apical growth is substituted by isotropic growth, with production of swollen hyphal segments. In some cases, budding occurs, giving rise to a yeast-like phase.

The mechanism of such biphasic growth patterns remains largely unexplained, although in some cases (*Paracoccidioides brasiliensis*) a modification of cell wall composition accompanies the transition of mycelial to yeast-like cells: the former contain mostly β-1,3-glucans, which give the hyphal walls a more rigid fibrous structure than the walls of the yeast-like phase cells containing mainly short-chained α-1,4-glucans (KANETSUNA et al. 1972).

11.2.3.2 Effect on Sporulation

Temperature limits are frequently different for vegetative growth and for the formation of one or another type of spore-bearing structure. In some cases, the temperature interval in which completion of the spore-bearing apparatus occurs is much narrower than the interval allowing vegetative growth. In *Neurospora crassa*, macroconidia are readily formed at 37–40 °C, whereas fertile perithecia do not appear above 25 °C. At 30 °C, only protoperithecia are formed, and even ascogonia are absent at 35 °C (Hirsch 1954; Ojha and Turian 1968). On the contrary, in *Eurotium herbariorum* and *Ceratostomella fimbriata*, ascospores are produced optimally at 25–30 °C, whereas conidial stages develop only at lower temperatures (Hawker 1957).

Rhytisma acerinum is the agent of tar spot in maple leaves. Like many other phytopathogenic ascomycetes, it develops parasitically during the warm season, forming asexual reproductive structures. Then the fungus remains alive on dead leaves and develops saprophytically, forming apothecia that mature during the winter and sporulate in the spring, when the young leaves appear. Along with the trophic change from parasitism to saprophytism, passage through the cold season seems necessary for complete development of apothecia; indeed, if the fungus on fallen leaves is transferred in the autumn to moist chambers at laboratory temperature, apothecia do not mature.

11.3 Life at High Temperature

11.3.1 Habitats of Thermophiles

It would seem obvious that thermophilic microorganisms live in warm environments. In fact thermophilic sporeformers were often isolated from environments not especially submitted to high temperatures (Allen 1953). Recently, we succeeded in isolating a sporeforming hydrogen bacterium, *Bacillus schlegelii* (Aragno 1978; Schenk and Aragno 1979). The bacterium grows with an optimum temperature of 70 °C, whereas no visible growth occurs in laboratory conditions below 40 °C. So far, the only source of this bacterium is the upper layer of mud from freshwater lakes and ponds, the temperature of which does not exceed 10 °C in the summer. This bacterium sets a problem often encountered in the study of microbial ecology: are the microorganisms isolated from a defined environment active in it, or are they present in a survival form? The fact that most, if not all, thermophiles isolated from cold environments were sporeformers would support the latter hypothesis, but further experimental and field work is still required to clear this problem.

Habitats with elevated temperature, either permanently or occasionally, are frequent in the biosphere. Of course, matter from erupting volcanoes and overheated "dry" steam fumaroles with temperatures of hundreds and more degrees is sterile. Warm environments where life occurs may be grouped in four categories:

- geothermically heated environments, such as hot springs and ponds and heated soils (e.g., solfataras)
- sun-heated substrates (rocks, soil, mud, shallow waters)

- organic materials heated by the energy dissipated by aerobic decomposers, such as self-heated hay, compost, manure, coal refuse piles, and so on
- environments where temperature is increased as a consequence of human activity, such as hot water heaters and tubings, cooling waters from industrial processes, etc. ...

Of these environments, most are subjected to temperature variations of great amplitude, many of them being heated only occasionally. This happens in sun-heated substrates, whereas self-heated organic materials are subjected first to a temperature increase from ambient to 60–70 °C, and once rapidly metabolizable substrates have been exhausted, to a temperature decrease. In such cases, the hot state is only a transient one, during which thermophilic life is possible. In such environments, facultative thermophiles or thermophiles with resting stages will be at an advantage.

On the other hand, geothermically heated environments seem to be as a rule more stable, although geothermic areas are often subjected to dramatic changes in a small scale of time, as pointed out by Brock (1978). In such environments, organisms whose maintenance depends strictly on the stability of environmental conditions (such as stenothermic, strict thermophiles without resting stages) are more likely to occur.

Other characteristics of the hot environments will strongly influence their bacterial flora, e.g., the supply in carbon and energy sources. Many hot springs are relatively poor in organic matter, but many of them contain significant

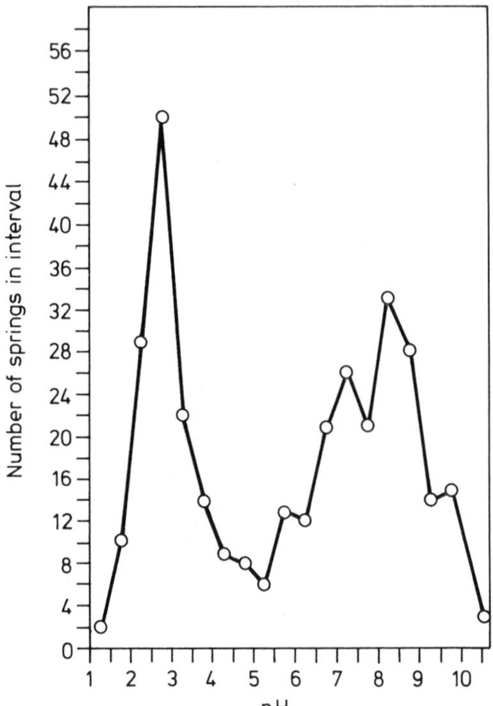

Fig. 11.5. Distribution of thermal springs as a function of pH. (Brock 1978)

amounts of reduced sulfur compounds. It is therefore not surprising that sulfur-oxidizing chemolithoautotrophs exist in springs at near-boiling temperatures.

Among other characteristics of thermal environments, the pH is, along with temperature, a major ecological factor. pH of thermal springs in the world varies from less than 1.5 to 10.5, but the distribution is not even between these limits (Fig. 11.5, BROCK 1978). In fact most springs are either acidic, in the pH range of 2–4, or neutral to slightly alkaline, in the pH range of 7–9. It is therefore evident that quite different microbial floras will predominate in acidic and neutral to alkaline springs.

The occurrence of thermophilic microorganisms in man-made heated environments will be of importance, since some of them (e.g., the sulfate reducers or cellulose hydrolyzers) may cause severe biodeteriorations.

11.3.2 Thermophilic Microorganisms

11.3.2.1 Prokaryotes

As for other environmental extremes, the organisms found at the highest temperatures where life can occur are prokaryotes. In other terms, only prokaryotes are high thermophiles, as defined above. There are important differences, how-

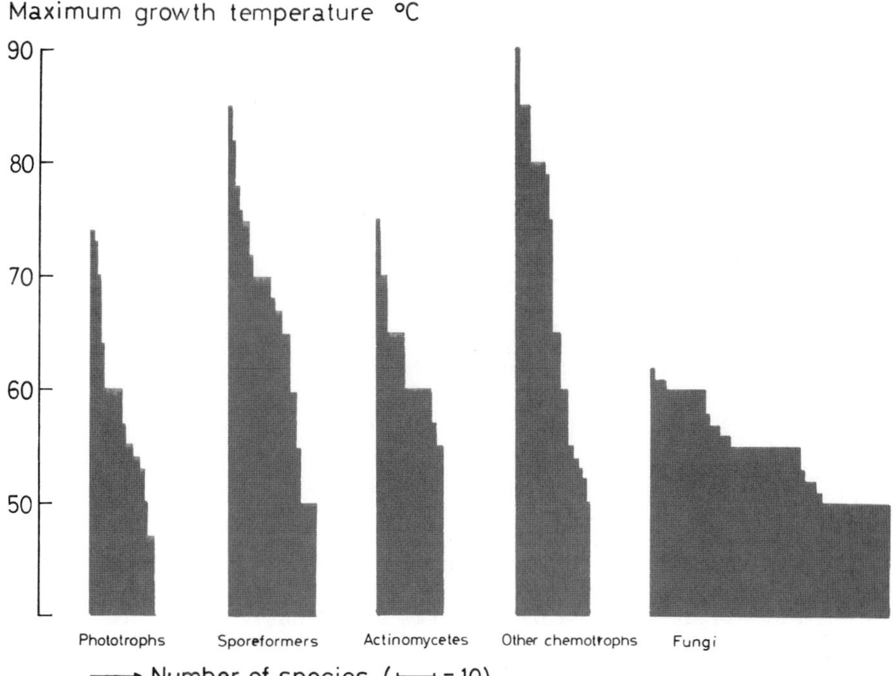

Fig. 11.6. Temperature limits reached by different groups of prokaryotes and fungi. Diagrammatic representation. (Data from BROCK 1978)

ever, with regard to the upper temperature limit reached by different groups of prokaryotes (Fig. 11.6).

Phototrophic Bacteria. Although a great selective pressure should have operated to select photolithoautotrophs in hot springs, no photosynthetic organism develops above 70–72 °C (BROCK 1978). There seems thus to be an intrinsic limit for photosynthesis; as CO_2 fixation occurs at much higher temperatures in chemolithoautotrophs (*Sulfolobus*), photosynthetic membranes are most likely the structure which cannot function at a higher temperature (Chap. 12, this Vol.). The highest thermophiles among phototrophs are a Cyanobacterium, *Synechococcus lividus* and a green nonsulfur bacterium, *Chloroflexus aurantiacus*. Both species are neutrophilic. Another cyanobacterium, *Mastigocladus laminosus*, a N_2-fixing filamentous Nostocale, grows at temperatures up to 60–64 °C.

Chloroflexus and *Synechococcus* were often found associated in the Yellowstone springs, forming stratified mats which may be considered as modern equivalents of Precambrian stromatolites. Other types of stromatolite-like structures, of columnar shape, occur at temperatures below 60 °C. They consist mainly of cyanobacteria of the genus *Phormidium*, and look very much like the Precambrian stromatolites known as "Conophytons".

Endospore-Formers. Prior to the extensive work by BROCK and co-workers on microbial thermophilic life in the Yellowstone thermal areas, most of the thermophiles described were sporeformers, the "classical" one being *Bacillus stearothermophilus*, a species name covering probably a wide range of strains deserving further taxonomic revision. The highest thermophiles among Bacilli are two strains of *B. caldolyticus*, YT-P and YT-G (HEINEN 1971; HEINEN and HEINEN 1972). Other thermophilic Bacilli include strains assumed to be thermophilic variants of well-known mesophilic species, such as *B. subtilis* and *B. sphaericus* (HOLLAUS and KLAUSHOFER 1970), as well as an acidophilic species, *B. acidocaldarius* (DARLAND and BROCK 1971), *B. thermocatenulatus* (GOLOVACHEVA et al. 1975), a red-pigmented species named *B. thermoruber* (ARAGOZZINI et al. 1976), the above-mentioned hydrogen-oxidizing autotroph *B. schlegelii* (SCHENK and ARAGNO 1979), etc. Some anaerobic thermophilic sporeformers were also described, among which cellulolytic *Clostridia* and the sulfate-reducer *Desulfotomaculum nigrificans*. Both types may be relevant in biodeterioration processes.

Actinomycetes. A number of these filamentous, Gram-positive bacteria are thermophiles, occurring mainly in self-heating aerobic decomposition processes. So far, no highly thermophilic actinomycete, growing at 75 °C and over, was described (Fig. 11.6).

Other Chemotrophs. Although most of the early reports on thermophilic microorganisms deal with gram-positive bacteria like *Bacillus* and Actinomycetes, Gram-staining of samples from thermal areas never showed Gram-positive cells (BROCK 1978). In fact, the highest thermophiles ever isolated were Gram-negative, asporogenous bacteria. The genus *Thermus*, of uncertain affiliation, was created by BROCK and FREEZE (1969) for highly thermophilic heterotrophic bacteria. Several species were then described with temperature maxima in culture around 80–85 °C. Extreme thermophiles were also found among sulfur-oxidizing chemolithoautotrophs. A striking event during the research of T.D. BROCK's group on thermophiles was the discovery of *Sulfolobus* (BROCK et al. 1972), the "most thermophilic autotroph available in pure culture", as presented by BROCK himself (it grows at temperatures up to 85–90 °C). It was isolated from acidic springs, with pH between 1.5 and 4.0. The cell wall is devoid of murein and the organism proved, by rRNA-homology, to be more related to methanogens and halobacteria than to any other microbial group. Such organisms are presently considered to belong to a distinct kingdom, the "Archaebacteria", having evolved independently of the other prokaryotes since very early times (WOESE et al. 1978). Among other sulfur oxidizers, we would like to mention *Thiothrix thioparus*, a filamentous denitrifying facultative autotroph growing at neutral pH (CALDWELL et al. 1976). The isolation of a thermophilic, sulfur-oxidizing sporeformer, *"Thiobacillus thermophilica"* (sic) (EGOROVA and DERYUGINA 1963) should be confirmed and its taxonomic position established.

Thermophilic, hydrogen-oxidizing nonsporeformers have also been described, such as *Pseudomonas thermophila, P. hydrogenothermophila, Flavobacterium autothermophilum* (ARAGNO and SCHLEGEL 1980). All these isolates grow in the lower part of the thermobiotic range (45–55 °C). Although molecular hydrogen may be present in thermal areas, so far no systematic research has been made on the occurrence of hydrogen bacteria in such environments.

Methane digestion of organic matter at 60–65 °C is possible and might even be the most suitable solution for the anaerobic treatment of municipal waste (PFEFFER 1977). A thermophilic methanogen, *Methanobacterium thermoautotrophicum* (ZEIKUS and WOLFE 1971), is able to grow autotrophically up to 75 °C in mineral medium gassed with H_2+CO_2. In optimal conditions, its doubling time is very low (3 h), making this species especially well-suited for experimental studies on methanogenesis.

A thermophilic, mycoplasma-like organism, *Thermoplasma acidophilum* was isolated by DARLAND et al. (1970) and BELLY et al. (1973) from coal refuse piles. pH of the samples ranged between 1.17 and 5.21, and temperature was between 32 and 80 °C but they have not been isolated from other thermal habitats. Thermoplasmas appear only superficially to resemble other mycoplasma. The unique structure of their membrane lipids, with ether linkages to glycerol, relate them to *Halobacterium* and *Sulfolobus*, and thus to the Archaebacteria (WOESE et al. 1978).

Life in Boiling Waters. An upper temperature limit for life has not yet been determined. Although isolation and growth of thermophiles in laboratory conditions never succeeded reproducibly above 85 °C, microorganisms do occur and live in boiling hot springs up to 101 °C. The problems encountered when growing organisms above 85 °C are perhaps only technical ones (BROCK 1978).

Several types of boiling water organisms were described (BROCK et al. 1971; BROCK 1978). Some were filamentous forms, other were rod-shaped; all were Gram-negative. Some types occurred in sulfide-containing springs and were shown to require sulfide for the uptake of organic molecules. Other types,

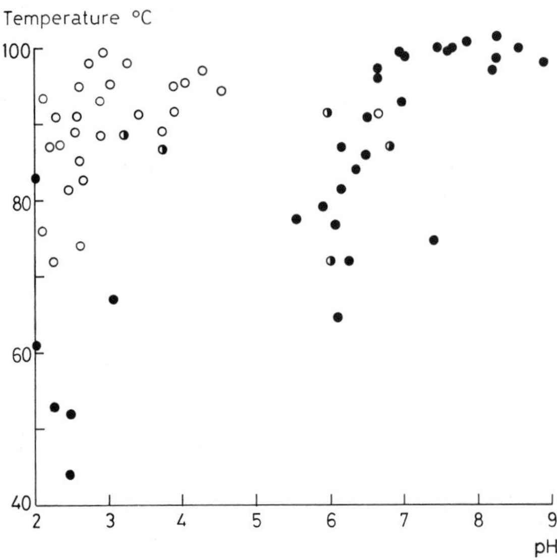

Fig. 11.7. Distribution of hot springs in New Zealand as a function of pH and temperature (Data from BROCK 1978). ○: Springs showing no bacterial development, ●: Springs where bacteria development was observed, ◑ : Springs where occurrence of bacteria was doubtful

from springs with very low sulfide content, were on the contrary inhibited by sulfide. Development of these bacteria was obtained on cover slips and their generation time in situ was estimated to be a few hours, in general. Some types were studied by electronmicroscopy, and revealed unusual cell walls, apparently lacking a typical peptidoglycan layer (BROCK et al. 1971). It is therefore possible that such organisms also belong to Archaebacteria, but experimental evidence is still lacking.

Bacteria living above 90 °C were only found in springs with pH above 6.5, whereas acid boiling springs appeared sterile (Fig. 11.7). Seemingly most, if not all, bacteria observed in acid springs over 70 °C were *Sulfolobus*. Hence, environments with a combination of low pH and high temperature appear very harmful and only *Sulfolobus*, an Archaebacterium with ether- rather than ester-linked hydrophobic chains in its phospholipids, can adapt to it.

Environments with liquid water above 100 °C exist in submarine hot springs, and should be studied with respect to the occurrence of living forms.

Microbial Groups Without Thermophilic Representatives. It is obviously risky to affirm that a defined microbial type or function is absent from a given environment. The microorganisms which we are able to characterize are in general those which we are able to cultivate in pure culture conditions and in sufficient masses. Yet organisms with more subtle requirements are certainly very abundant in natural sites, and bacterial species recognized so far could well be the "emerged part of the iceberg". Thus, the following paragraphs will certainly be outdated in a few years.

So far, the following groups appear to lack true thermophilic representatives: Enterobacteriaceae; spirilla; acetic acid bacteria (*Acetobacter, Gluconobacter*); Micrococci; nitrifying chemolithoautotrophs; Myxobacteriales and Cytophagales; Coryneform bacteria, Mycobacteria and Nocardiae; Mycoplasmas (with one exception, *Thermoplasma*, which does not appear related to the other mycoplasmas, see above). Other groups have few thermophilic representatives, all being low thermophiles, i.e., phototrophic purple bacteria, Pseudomonads, lactic acid bacteria, Spirochetes, and methane-oxidizing bacteria.

Some microbial functions, too, seem to be absent from very hot environments, particularly methane oxidation, nitrification, and nitrogen fixation. The highest known thermophile among nitrogen fixers is *Mastigocladus*, a Nostocale, with temperature maximum around 63–64 °C. It is somewhat surprising, since the low oxygen solubility in hot waters should provide protection to the nitrogenase complex. Perhaps this function has not been studied enough in thermal environments. CO_2 fixation exists (e.g., in *Sulfolobus*) but the occurrence of the ribulose-bisphosphate cycle has not yet been demonstrated in very high thermophiles.

11.3.2.2 Eukaryotes

Only prokaryotes live in hot waters, above 60–62 °C. This seems to be an intrinsic limit for eukaryotic life. Most eukaryotes able to grow between 50 and 60 °C are fungi, the others being only one algal species and some protozoa. Pluricellular animals and plants have still lower temperature maxima. As stated by BROCK (1978), the inability of eukaryotes to grow above 60 °C might reside in the impossibility to form organellar membranes stable and functional at a higher temperature. On the other hand, the much greater dimension of the eukaryotic genome could have rendered the evolution toward high thermophily a far too improbable event.

Fungi. According to the terminology defined at the beginning of this chapter, all thermophilic fungi are facultative thermophiles, their minimal growth temperature rarely exceeding 30 °C, whereas optima range from 35 to 50 °C, and maxima from 50 to 60–62 °C (Fig. 11.6). Their typical habitats are accumulations of self-heating decaying organic matter, such as composts, hay, or manure; they are believed to be significant contributors to the self-heating of their substrate. A review dealing mainly with thermophilic fungi appeared recently (TANSEY and BROCK 1978). These authors list 65 species, with maximum temperatures ranging from 50 to 60 °C (Fig. 11.6). They include Zygomycetes, many Ascomycetes and Deuteromycetes, but only two Basidiomycetes. Surprisingly, no aquatic fungus with motile cells (e.g., Chytrids, Oomycetes) is known so far to occur at such temperatures. Perhaps this is due to the lack of research in that direction? Attempts to isolate such fungi should be made in slightly acidic or neutral waters at 50–55 °C, using organic "traps".

Algae. Only one eukaryotic alga is known to grow at temperatures over 50 °C: *Cyanidium caldarium*, whose temperature limit is between 55 and 60 °C. Among other algae, one species of *Mougeotia* has been found up to 47 °C (STOCKNER 1967). Diatoms were reported in hot environments, but there is no evidence that they were actually living at such temperatures. The maximum temperature for diatoms in culture is 43–44 °C, while they are often dominant in springs between 30 and 40 °C.

Cyanidium caldarium is a unique organism. No other photosynthetic organism is able to grow in acidic environments at temperatures above 40 °C. It can grow at pH values as low as 0 (ALLEN 1959). It is generally considered a Rhodophyte, but it does not resemble any other. It rather looks like a *Chlorella* that has lost its chloroplast and gained an endosymbiotic cyanobacterium.

Other Eukaryotic Microorganisms. Some protozoa were reported which live at temperatures up to 55–58 °C (TANSEY and BROCK 1978). DALLINGER (1887) reported he adapted three species of flagellates to grow up to 70 °C by slowly increasing their incubation temperature for almost seven years! Such experiments have never been repeated so far.

We have not found any report on thermophilic Myxomycetes or Acrasiales.

Pluricellular organisms do not reach temperature maxima as high as unicellular ones. It is often difficult to assess a maximal temperature for higher organisms. Among animals, some insects and ostracods are known to live between 45 and 50 °C. Similar maxima are reached by some mosses, whereas no fish lives above 38 °C.

11.4 Life at Low Temperature

Most of the biosphere is exposed, either occasionally or permanently, to temperatures near the freezing point of water. Schematically, three categories of organisms may be considered in such environments:

- Mesophilic or thermophilic organisms, which do not grow at temperatures near 0 °C, but remain alive in a "dormant" state. This implies that, at least during some periods of a year cycle, the temperature will rise above their minimum temperature, allowing them to develop.
- Facultative psychrophiles able to grow at low temperatures, but whose temperature optima and maxima are in the mesobiotic range. Such eurythermal organisms will be particularly well adapted to environments subjected to important temperature variations
- Obligate psychrophiles, that is organisms with optimum and maximum temperatures in the psychrobiotic range (below 20 °C). Such stenothermal organisms will only survive in environments keeping a relatively stable, low temperature.

In fact, and perhaps due to the abundance of cold environments on Earth, temperatures near 0 °C do not appear as an extreme condition for life. Hence, in contrast with hot environments, a great variety of organisms may be present as long as liquid water is available. In fact, the limit for active life is the occurrence of freezing at 0 °C or below, depending on the concentration of solutes in the water. Liquid water is a peremptory condition for the life processes; moreover freezing of the cell content may be harmful or fatal. Survival at freezing is a complex phenomenon. Microorganisms are intrinsically more or less resistant; they may produce dormant spores or cysts which tolerate freezing better than vegetative cells. Moreover, the physiological state of the cell at the moment of freezing, as well as the temperature and the nature of the medium in which it occurs, may greatly affect the sensitivity of cells to freezing and/or thawing. In laboratory conditions, as rapid freezing as possible, at low temperature and in a protective medium, will ensure an optimal survival. This is of peculiar importance for the conservation of strains in liquid nitrogen (−193 °C) or during their preparation for lyophilization. Ecological aspects of life at low temperature were recently reviewed by BAROSS and MORITA (1978). Problems of low temperature resistance of plants are discussed in Chapters 12 and 13, this Volume.

11.4.1 Low Temperature Habitats

Soils in the temperate zone are exposed to important temperature changes during the year, particularly in their superficial layers which are also the most significant, biologically. In such media, obligately psychrophilic organisms are not likely to occur, since temperatures even a few degrees above the maximum temperature are frequently lethal. In contrast, facultative psychrophiles are able to grow over most of the temperature range occurring in soils, and seem particularly well adapted to these habitats. This is also true for surface waters, shallow ponds sediments and rock and plant surfaces (litho- and phyllosphere).

A peculiar category of environments subjected to broad temperature variations is the surface of snow and ice, where numerous alternations of freezing and thawing occur, hence providing harsh conditions that probably limit the number of organisms living in such habitats.

At the opposite are environments subjected to only minor temperature changes during the course of a year cycle, as can be typically found in waters and sediments of seas and lakes, below the thermocline. For example, Fig. 11.8 shows temperature variations in a small lake in Switzerland (about 500 × 100 m in surface, 10 m in depth) during a year cycle. Whereas surface temperature varies from freezing point to about 23 °C, bottom waters and sediments are maintained within a narrow temperature range (4 to 8 °C): it is hence likely that, since the last glaciation, the bottom of this lake was never submitted to freezing or to temperatures above the psychrobiotic range. Such an habitat appears a priori to be adequate for psychrophiles, and we actually succeeded in isolating strict psychrophiles from it. Other environments, such as caves, rock splits, and deep soils may share similar temperature characteristics. Thus, soils and waters in the Antarctic, which were intensively studied, are by no means the only environments where true psychrophiles do occur! In fact, in these latter habitats, the occurrence of microbial life seems to be restricted by the lack of nutrients as well as by the low temperatures.

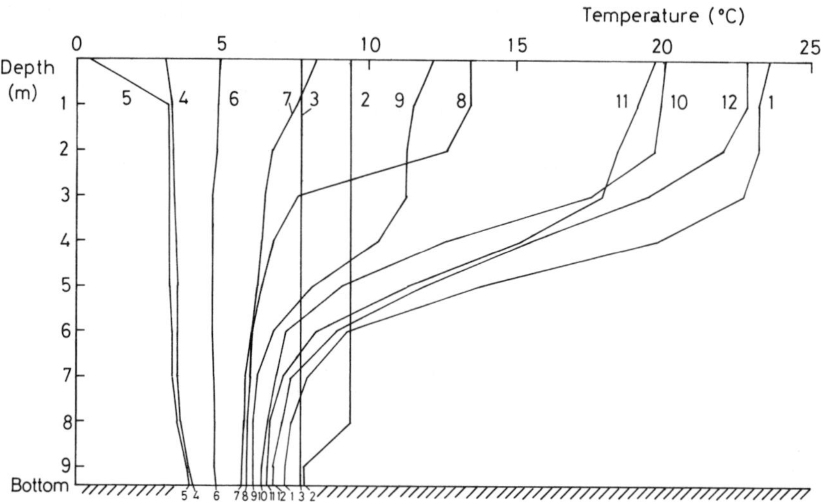

Fig. 11.8. Temperature profiles in a small lake (Le Loclat, near St-Blaise, canton Neuchâtel, Switzerland) during one year cycle (original)

1 23. 8.1973	*5* 20.12.1973	*9* 9.5.1974
2 31.10.1973	*6* 14. 2.1974	*10* 6.6.1974
3 14.11.1973	*7* 21. 3.1974	*11* 5.7.1974
4 4.12.1973	*8* 10. 4.1974	*12* 2.8.1974

Man-made cold environments should be at least mentioned here. Many problems related to conservation of food at low temperature arise from the occurrence of contaminating facultatively psychrophilic microorganisms.

11.4.2 Microorganisms Living at Low Temperature

11.4.2.1 Prokaryotes

Surprisingly, true psychrophilic bacteria were rarely isolated up to some years ago, and their existence was even questioned. In fact, they appear to be numerous and abundant, and only their slow growth rates and the difficulties encountered in manipulating and maintaining the live strains seem to have been the limiting factor of our present knowledge. Table 11.2 gives some examples of recently isolated strains or species of psychrophilic bacteria.

11.4.2.2 Eukaryotes

Eukaryotes living at low temperature are certainly less frequent than prokaryotes; nevertheless, they are often spectacular and were already observed in past centuries. We shall focus here on a few characteristic examples.

Fungi. Herpotrichia nigra causes a brown felt blight on conifers in mountain regions continuously covered with snow during the cold season (GÄUMANN 1951). In nature, the fungus develops exclusively on branches covered with

Table 11.2. Some examples of obligately psychrophilic bacteria. (References: see BAROSS and MORITA 1978, p. 48)

Organism	Origin	Optimum (°C)	Maximum (°C)
Cytophaga psychrophila	Fish	15–20	20
Vibrio marinus	Seawater	15	20
Clostridium spp.	Sea sediment	8·2–10·7	16·3–18·3
Vibrio spp.	Seawater	7–10	13–16
Vibrio AP-2-24	Seawater	4	9

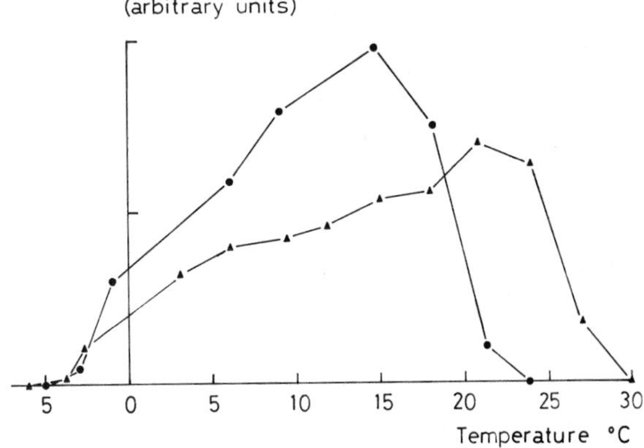

Fig. 11.9. Growth of two psychrophilic fungi as a function of temperature. Growth is expressed as colony diameter in Petri dishes. (GÄUMANN 1951.)
●: *Herpotrichia nigra,*
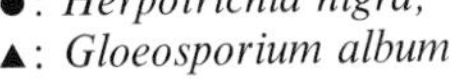
▲: *Gloeosporium album*

snow, at temperatures near 0 °C, although it grows in laboratory cultures with an optimum at 15 °C and a maximum around 25 °C. Interestingly, it shows low minimal temperature (−5 °C) and the fact that, at 0 °C, the growth rate amounts to as much as one third of that reached at optimal temperature (Fig. 11.9). Moreover, it requires a high degree of relative humidity, and it is completely inhibited below 90% R.H. This requirement is actually the limiting ecological factor, and stable conditions of high relative humidity are only found under the snow. Whereas it might physiologically grow in the mesophilic range, *H. nigra* is therefore an apparent, or ecological, strict psychrophile.

Gloeosporium album causes an anthracnose of apples and pears, known as "bitter rot". Its cardinal temperatures, in the laboratory, are: minimum −6 °C, optimum +21 °C, maximum +30 °C (Fig. 11.9). This allows the fungus to continue its development even during storage of apples at low temperature.

Sarcoscypha coccinea is a pretty red cup fungus growing on fallen twigs of various leafy trees. The ascocarps appear exclusively from December to March and may often be found (although not exclusively) under the snow. Vegetative mycelium grows well in laboratory cultures in the mesophilic range; it is therefore likely (although experimental evidence is still lacking) that low temperature is specifically required for the initiation of fruit body formation.

Fig. 11.10. *Diderma alpinum,* a nival Myxomycete. *Left* typical habitat, near melting snow; *arrows* show two groups of fruiting bodies. *Right* a group of fruiting bodies, on dead grass. Chasseral (Swiss Jura, 1500 m altitude), May, 1980. Photograph by the author

Myxomycetes. Fruit bodies of some species of Myxomycetes are often found in the spring in mountain regions, around melting snow areas (Fig. 11.10) (MEYLAN 1914, 1931). They include species of the genera *Diderma, Lamproderma, Lepidoderma,* and *Physarum.* Plasmodia develop under the melting snow, whereas fruiting occurs as soon as the organisms become uncovered. Hence, as with *Herpotrichia nigra,* a requirement for a high moisture along with a low minimal temperature might explain the peculiar ecological niche of these organisms, but experimental work is still needed.

Algae. In high mountains, during the summer, the snow surface is often covered with a pink or red color; this is caused by a bloom of unicellular algae, known as the "snow algae". Most have a high content of carotinoids, which are likely to exert a protective effect against photooxidation. They are many species of snow algae, belonging to the Chlorophyta (*Chlainomonas, Chlamydomonas, Chloromonas, Cylindrocystis, Raphidonema, Stichococcus*), to the Chrysophyta (*Chromulina, Ochromonas*) and to the Cryptophyta (*Cryptomonas*). Most fit our definition of psychrophiles, having optimal and maximal temperatures below 20 °C. HOHAM (1975) considers as true snow algae only those which do not grow above 10 °C. *Chloromonas pichinchae* is perhaps the most psychrophilic organism known so far, having a temperature optimum at about 1 °C, whereas cells show structural abnormalities above 5 °C (HOHAM 1975). Snow algae occur only when air temperature remains permanently above 0 °C for a period of time, causing the snow to melt. Once the bloom has occurred the algae appear to be more or less resistant to alternations of freezing and thawing.

11.5 Conclusion and Evolutionary Considerations

Some microorganisms have adapted to use temperature changes as signals governing morphogenetical events. It is particularly true in coprophilous fungi whose spore germination requires a pretreatment at high temperature, corresponding to the passage through digestive tracts of warm-blooded animals; in contrast, many phytopathogens in the temperate zone require a period at low temperature in order to break dormancy, therefore allowing their life cycles and outgrowth of vegetation to be synchronous.

Higher plants, e.g. trees, must adapt to short-term variations in their physicochemical environment, whereas populations change only in response to long-term variations such as climate changes, soil evolution, and so on. In contrast, the composition of microbial populations may change very rapidly in response to short-term variations in their environment. Such events as, for example, death of an insect in the soil will be followed by a succession of bacterial populations in the microenvironment of the corpse; such a succession might be compared with that of plant populations, e.g., on soil surface uncovered by glaciers; the former takes place in a few days, whereas the latter may need centuries before a climax is reached. Thus, most important in microorganisms will be the diversity of responses to environmental conditions, rather than the adaptive capabilities of a given species. This will be particularly true with temperature, a good example being the succession of populations during the processes of aerobic composting of hay or manure accompanied by self-heating phenomena. Even though in vitro pure cultures often show a broad temperature range for growth, in natural sites the temperature limits might possibly be much narrower. A good example is given by *Herpotrichia nigra* (see Sect. 11.4.2.2); *Escherichia coli,* for its part, grows in the laboratory between 8 and 46 °C (Fig. 11.1), whereas in nature it develops only in the intestinal tract of warm-blooded animals, around 37 °C. In that sense, *E. coli* would be among the worst organisms for a study on physiological responses to temperature!

Along with changes induced in microbes by temperature, the microbial diversity resulting from phylogenetical adaptation, as well as the nature of habitats and ecology, must be considered before unterstanding the responses of the microbial flora to temperature.

Microbial life exists in the biosphere as long as water occurs as a liquid phase. Only waters with a combination of both sub-boiling temperatures and low pH appear sterile. The lower temperature limit does not appear as an extreme environment, and life is possible at about 0 °C for a wide variety of microorganisms and functions. The most psychrophilic organisms are even eukaryotes, the "snow algae". In contrast, hot environments appear much more selective. In the 60–100 °C interval, only prokaryotes and Archaebacteria were observed. The number of functions encountered diminishes with increase of temperature. Photosynthesis occurs only below 70–72 °C, and nitrogen fixers are known so far only below 64 °C. The upper limit for eukaryotic life is about 60–62 °C. The ability of a given group to develop forms able to live at high temperatures seems to be inversely related to its degree of organization, as well at the supracellular as at the subcellular levels.

A mostly unanswered question is that of the origin and evolution of organisms with regard to their adaptation to temperature. Two evolutionary mechanisms in prokaryotes may be considered: the Darwinian evolution, by mutations and selection, and the interspecific exchange of genetic material. This latter consists of the transfer of entire genes or groups of genes, e.g., by means of plasmids, and affects mainly the "facultative" functions. Hence, no wonder that such functions may have a different response to temperature than the fundamental processes of the cell (see Sect. 11.2.1.5).

It is generally thought that thermophilic organisms have evolved from mesophilic ancestors (CASTENHOLZ 1979). Yet recent geochemical evidence suggests that temperature at the Earth surface was relatively high at the time when life is thought to have appeared. Isotopic ratios in nodules from Precambrian rocks lead to the conclusion that average temperature during their formation was 20–30 °C $1{,}2 \times 10^9$ years ago, 37 °C 2×10^9 years ago, and 65–70 °C 3×10^9 years ago (KNAUTH and EPSTEIN 1976; KAPLAN 1978). Other facts would support the hypothesis of a thermophilic origin of life. The probability of mutations lowering the thermal resistance of proteins is much higher than that of mutations enhancing thermostability. Among Archaebacteria (a group which has been evolving separately since very early times), a relatively high proportion of thermophiles is found (*Sulfolobus, Methanobacterium thermoautotrophicum, Thermoplasma,* and possibly boiling water organisms). Perhaps such organisms are relics that subsisted only in extreme environments where there was no competition from the prokaryotes. So far, the only *Archaebacteria* known to live in nonextreme environments are methanogens, the unique metabolic capabilities of which allow them to occupy an ecological niche free from all other organisms.

After cooling of the Earth surface, the pressure was great to select mesophilic and finally psychrophilic organisms, and would have allowed new processes and structures to appear, such as photosynthesis, eukaryotic cells, and pluricellular organization. Of course, such a hypothesis does not exclude a secondary adaptation to thermophily in mesophilic groups (e.g., fungi). If the mesophilic proteins have thermophilic ancestors, such an adaptation is all the more probable, as their temperature sensitivity would be due only to one or a few "weak points", as actually observed. Occurrence of mutations conferring thermostability is therefore less improbable.

Of course, the above considerations are purely speculative and intend only to leave open the debate concerned with the question of a possible thermophilic origin of life. About four billion years have gone by since these early events. Direct records from these times are very scarce, and perhaps one will never know the truth about the origin of life.

References

Algranati ID, Gonzales NS, Bade EG (1969) Physiological role of 70S ribosomes in bacteria. Proc Nat Acad Sci USA 62:574–580

Allen MB (1953) The thermophilic aerobic sporeforming bacteria. Bacteriol Rev 17:125–173

Allen MB (1959) Studies with *Cyanidium caldarium*, an anomalously pigmented chlorophyte. Arch Mikrobiol 32:270–277

Amelunxen RE, Murdock AL (1978) Microbial life at high temperatures: mechanisms and molecular aspects. In: Kushner DJ (ed) Microbial life in extreme environments. Academic Press, London, pp 217–278

Anderson JG (1978) Temperature-induced fungal development. In: Smith JE, Berry DR (eds) The filamentous fungi 3: Developmental mycology. Edward Arnold, London, pp 358–375

Anderson JG, Smith JE (1971) The production of conidiophores and conidia by newly germinated conidia of *Aspergillus niger* (Microcycle conidiation). J Gen Microbiol 69:185–197

Aragno M (1973) Etude de la germination des pycnidiospores de *Coniella diplodiella* (Speg.) Pet. et Syd., agent du coître de la vigne. I. Conditions de la germination. Bull Soc Bot Suisse 83:223–251

Aragno M (1978) Enrichment, isolation and preliminary characterization of a thermophilic, endospore-forming hydrogen bacterium. FEMS Microbiol Letters 3:13–15

Aragno M, Schlegel HG (1981) The hydrogen-oxidizing bacteria. In: Starr MP, Stolp H, Trüper HG, Balows A, Schlegel HG (eds) The procaryotes: a handbook on habitats, isolation and identification of bacteria. Springer, Berlin, Heidelberg, New York (in press)

Aragozzini F, Toppino P, Manachini PL, Craveri R (1976) Fatty acid composition of *Bacillus thermoruber*. Ann Microbiol Enzimol 26:9–13

Assche JA van, Carlier AR, Dekeersmaeker HH (1972) Trehalase activity in dormant and activated spores of *Phycomyces blakesleeanus*. Planta 103:327–333

Baross JA, Morita RY (1978) Microbial life at low temperature: ecological aspects. In: Kushner DJ (ed) Microbial life in extreme environments. Academic Press, London, pp 9–71

Beers RJ (1958) Effect of moisture activity on germination. In: Halvorson HO (ed) Spores. Burgess Publ, Minneapolis, Minn, USA, p 45

Belly RT, Bohlool BB, Brock TD (1973) The genus *Thermoplasma*. Ann NY Acad Sci 225:94–107

Bobier SR, Ferroni GD, Inniss WE (1972) Protein synthesis by the psychrophiles *Bacillus psychrophilus* and *Bacillus insolitus*. Can J Microbiol 18:1837–1843

Brock TD (1978) Thermophilic microorganisms and life at high temperatures. Springer, Berlin, Heidelberg, New York

Brock TD, Freeze H (1969) *Thermus aquaticus* gen. n. and sp. n., a nonsporulating extreme thermophile. J Bacteriol 98:289–297

Brock TD, Brock ML, Bott TL, Edwards MR (1971) Microbial life at 90° C: the sulfur bacteria of Boulder Spring. J Bacteriol 107:303–314

Brock TD, Brock KM, Belly RT, Weiss RL (1972) *Sulfolobus*: a new genus of sulfuroxidizing bacteria living at low pH and high temperature. Arch Mikrobiol 84:54–68

Buchanan RE, Gibbons NE (eds) (1974) Bergey's manual of determinative bacteriology, 8th edn. Williams & Wilkins, Baltimore

Caldwell DE, Caldwell SJ, Laycock JP (1976) *Thermothrix thioparus* gen. et sp. nov. a facultatively anaerobic facultative chemolithotroph living at neutral pH and high temperature. Can J Microbiol 22:1509–1517

Castenholz RW (1979) Evolution and ecology of thermophilic microorganisms. In: Shilo M (ed) Strategies of microbial life in extreme environments. Chemie, Weinheim, pp 373–392

Cronan JE, Gelman EP (1975) Physical properties of membrane lipids: biological relevance and regulation. Bacteriol Rev 39:232–256

Dallinger WH (1887) The president's address. JR Microsc Soc Ser 2:185–199

Darland G, Brock TD (1971) *Bacillus acidocaldarius* sp. nov., an acidophilic thermophilic spore-forming bacterium. J Gen Microbiol 67:9–15

Darland G, Brock TD, Samsonoff W, Conti SF (1970) A thermophilic, acidophilic Mycoplasma isolated from a coal refuse pile. Science 170:1416–1418

Das HK, Goldstein A (1968) Limited capacity for protein synthesis at zero degrees centigrade in *Escherichia coli*. J Mol Biol 31:209–226

Egorova AA, Deryugina Z (1963) The spore-forming thermophile thiobacterium *Thiobacillus thermophilica* Imschenetskii nov. sp. Mikrobiologija 32:437–446

Esser AF (1979) Physical chemistry of thermostable membranes. In: Shilo M (ed) Strategies of microbial life in extreme environments. Chemie, Weinheim, pp 433–454

Foter MJ, Rahn O (1936) Growth and fermentation of bacteria near their minimum temperature. J Bacteriol 32:485–499

Gäumann E (1951) Pflanzliche Infektionslehre. Birkhäuser, Bâle

Gelman EP, Cronan JE (1972) Mutants of *Escherichia coli* deficient in the synthesis of *cis*-vaccinic acid. J Bacteriol 112:381–387

Golovacheva RS, Loginova LG, Salikhov TA, Kolesnikov AA, Zaitseva GN (1975) A new thermophilic species, *Bacillus thermocatenulatus* nov. sp. Mikrobiologija 44:265–266

Gould GW, Hitchins AD (1963) Sensitivation of bacterial spores to lysozyme and to hydrogen peroxyde with agent which rupture disulfid bonds. J Gen Microbiol 33:413

Haberstich HV, Zuber H (1974) Thermoadaptation of enzymes in thermophilic and mesophilic cultures of *Bacillus stearothermophilus* and *Bacillus caldotenax*. Arch Mikrobiol 98:275–287

Hawker LE (1957) The physiology of reproduction in fungi. Cambridge Univ Press, London

Heinen W (1971) Growth conditions and temperature-dependent substrate specificity of two extremely thermophilic bacteria. Arch Mikrobiol 76:2–17

Heinen UB, Heinen W (1972) Characteristics and properties of a caldo-active bacterium producing extracellular enzymes and two related strains. Arch Mikrobiol 82:1–23

Hennecke H, Shanmugam KT (1979) Temperature control of nitrogen fixation in *Klebsiella pneumoniae*. Arch Microbiol 123:259–265

Herbert RA, Bhakoo M (1979) Microbial growth at low temperatures. In: Russel AD, Fuller R (eds) Cold-tolerant microbes in spoilage and the environment. The Society for Applied Bacteriology, Technical Series 13, Academic Press, London

Hirsch HM (1954) Environmental factors influencing the differentiation of protoperithecia and their relation to tyrosinase and melanin formation in *Neurospora crassa*. Physiol Plant 7:72–97

Hoham RW (1975) Optimum temperatures and temperature ranges for growth of snow algae. Arctic Alp Res 7:13–24

Hollaus F, Klaushofer H (1970) Taxonomische Untersuchungen an hochthermophilen Bazillen-Stämmen aus Zuckerfabriksäften. Publ Fac Sci Univ JG Purkyne Brno 47:99–105

Ingraham JL (1958) Growth of psychrophilic bacteria. J Bacteriol 76:75–80

Inniss WE, Ingraham JL (1978) Microbial life at low temperatures: mechanisms and molecular aspects. In: Kushner DJ (ed) Microbial life in extreme environments. Academic Press, London, pp 73–104

Johnson EJ (1979) Thermophile genetics and the genetic determinants of thermophily. In: Shilo M (ed) Strategies of microbial life in extreme environments. Chemie, Weinheim, pp 471–487

Jung L, Jost R, Stoll E, Zuber H (1974) Metabolic differences in *Bacillus stearothermophilus* grown at 55 °C and 37 °C. Arch Microbiol 95:125–138

Kalakoutskii LV, Agre NS (1976) Comparative aspects of development and differentiation in Actinomycetes. Bacteriol Rev 40:469–524

Kanetsuna F, Carbonell LM, Azuma I, Yamura Y (1972) Biochemical studies on the thermal dimorphism of *Paracoccidioides brasiliensis*. J Bacteriol 110:208–218

Kaplan RW (1978) Der Ursprung des Lebens. Thieme, Stuttgart.

Keynan A, Evenchik Z (1969) Activation. In: Gould GW, Hurst A (eds) The bacterial spore. Academic Press, London, pp 359–396

Keynan A, Evenchik Z, Halvorson HO, Hastings JW (1964) Activation of bacterial endospores. J Bacteriol 88:313

Knauth LP, Epstein S (1976) Hydrogen and oxygen ratios in nodular and bedded cherts. Geochim Cosmochim Acta 40:1095

Langworthy TA, Brock TD, Castenholz RW, Esser AF, Johnson EJ, Oshima T, Tsuboi M, Zeikus JG, Zuber H (1979) Life at high temperatures. Group report. In: Shilo M (ed) Strategies of microbial life in extreme environments. Chemie, Weinheim, pp 489–502

Lindsay JA, Creaser EH (1975) Enzyme thermostability is a transformable property between *Bacillus* spp. Nature 255:650–652

Ljungdahl LG, Sherod D (1976) Proteins from thermophilic microorganisms. In: Heinrich MR (ed) Extreme environments. Academic Press, New York, pp 147–187

Marr AG, Ingraham JL (1962) Effect in temperature on the composition of fatty acids in *Escherichia coli*. J Bacteriol 84:1260–1267

Meylan C (1914) Remarques sur quelques espèces nivales de Myxomycètes. Bull Soc Vaudoise Sci Nat 50:1–14

Meylan C (1931) Les espèces nivales du genre *Lamproderma*. Bull Soc Vaudoise Sci Nat 57:147–149

Morita RY (1975) Psychrophilic bacteria. Bacteriol Rev 39:146–167

O'Donovan GA, Ingraham JL (1965) Cold-sensitive mutants of *Escherichia coli* resulting from increased feed-back inhibition. Proc Nat Acad Sci USA 54:451–457

Ohta Y (1967) Thermostable protease from thermophilic bacteria. II. Studies on the stability of the protease. J Biol Chem 242:509–515

Ojha MN, Turian G (1968) Thermostimulation of conidiation and succinic oxidative metabolism of *Neurospora crassa*. Arch Mikrobiol 63:232–241

Oppenheim J, Scheinbuks J, Biava C, Marcus L (1968) Polyribosomes in *Azotobacter vinelandii*. I. Isolation, characterization and distribution of ribosomes, polyribosomes and subunits in logarithmically growing *Azotobacter*. Biochim Biophys Acta 161:386–401

Pace B, Campbell LL (1967) Correlation of maximal growth temperature and ribosome heat stability. Proc Nat Acad Sci USA 57:1109–1116

Pfeffer JT (1977) Methane from urban wastes – process requirements. In: Schlegel HG, Barnea J (eds) Microbial energy conversion. Pergamon Press, Oxford, pp 139–155

Reid BR (1976) Temperature effects on transfer RNA. In: Heinrich MR (ed) Extreme environments. Academic Press, New York, pp 103–117

Schenk A, Aragno M (1979) *Bacillus schlegelii*, a new species of thermophilic, facultatively chemolithoautotrophic bacterium oxidizing molecular hydrogen. J Gen Microbiol 115:333–341

Shaw MK, Ingraham JL (1965) Fatty acid composition of *Escherichia coli* as a possible controlling factor of the minimal growth temperature. J Bacteriol 90:141–146

Stenesh J (1976) Information transfer in thermophilic bacteria. In: Heinrich MR (ed) Extreme environments. Academic Press, New York, pp 85–102

Stockner JG (1967) Observation of thermophilic algal communities in Mount Rainier and Yellowstone National Parks. Limnol Oceanogr 12:13–17

Sussman AS (1976) Activators of fungal spore germination. In: Weber DJ, Hess WM (eds) The fungal spore. Wiley and Sons, New York

Sussman AS, Halvorson HO (1966) Spores: their dormancy and germination. Harper and Row, New York

Tai PC, Wallace BJ, Herzog EL, Davis BD (1973) Properties of initiation-free polysomes of *Escherichia coli*. Biochemistry 12:609–615

Tansey MR, Brock TD (1978) Microbial life at high temperatures: ecological aspects. In: Kushner DJ (ed) Microbial life in extreme environments. Academic Press, London, pp 159–216

Welker NE (1976) Effect of temperature on membrane proteins. In: Heinrich MR (ed) Extreme environments. Academic Press, New York, pp 229–254

Williams RAD (1975) Caldoactive and thermophilic bacteria and their thermostable proteins. Sci Prog Oxf 62:373–393

Wilson JW, Plunkett OA (1965) The fungus diseases of man. Univ California Press, Berkeley

Woese CR, Magrum LJ, Fox GE (1978) Archaebacteria. J Mol Evol 11:245–252

Zeikus JG, Wolfe RS (1971) *Methanobacterium thermoautotrophicum* sp. n., an anaerobic, autotrophic, extreme thermophile. J Bacteriol 109:707–713

Zuber H (ed) (1976) Enzymes and proteins from thermophilic microorganisms. Experientia suppl 26, Birkhäuser, Bâle

Zuber H (1979) Structure and function of enzymes from thermophilic microorganisms. In: Shilo M (ed) Strategies of microbial life in extreme environments. Chemie, Weinheim, pp 393–415

12 Responses to Extreme Temperatures. Cellular and Sub-Cellular Bases

P.L. STEPONKUS

CONTENTS

12.1 Introduction

Environmental stresses result in various cellular stresses, both physical and chemical, which result in cellular lesions – injuries or changes which result in impairment or loss of either a structural or metabolic function. Characteristics which allow a plant to endure a broader thermal range are both constitutive and facultative. The former include stable modifications in structure or function accumulated through natural selection of random mutations or through systematic assembly by plant breeding; the latter, which depend on phylogenetic adaptation a priori, are transient phenotypic modifications elicited in response to various environmental stimuli. Adaptation and acclimation may preclude or mitigate the cellular stresses or alter the sensitivity of the cellular components. Consideration of the physiological, biochemical, and molecular aspects of adaptation and acclimation requires the determination of both the repercussions of the stresses on the cellular environment and what constitutes injury at the organismal, cellular, and molecular level (cf. Chaps. 10 and 11, this Vol.).

12.2 High Temperature Injury

While the effects of temperature on specific plant processes are reasonably characterized around the optimum range, the direct question of how high temperature kills a plant (KURTZ 1961) remains unanswered. A chronological development can be found in the reviews of BĚLEHRÁDEK 1935; WENT 1953; LANGRIDGE 1963; ALEXANDROV 1964; TROSHIN 1967; LANGRIDGE and MCWILLIAM 1967; ALEXANDROV et al. 1970; LEVITT 1972; LARCHER 1973; HEBER and SANTARIUS 1973; ALEXANDROV 1977; BJÖRKMAN et al. 1980; BERRY and BJÖRKMAN 1980. Two types of high temperature stress are generally considered: chronic exposures (days and weeks) to temperatures which just exceed the optimum and acute exposures (minutes and hours) to relatively extreme temperatures. The nature of injury resulting from the two types of thermal stress may be vastly different (e.g. LEITSCH 1916) and should be considered in proposing a mechanism of injury.

12.2.1 Cellular Manifestations

Cytological changes observable upon acute exposures to high temperature include coagulation of the protoplasm, cytolysis, nuclear changes, and altered mitosis (BĚLEHRÁDEK 1935); inhibition of protoplasmic streaming, increased protoplasmic viscosity, and loss of membrane semipermeability (ALEXANDROV 1967). ALEXANDROV (1967) suggests that the various manifestations occur in a succession, protoplasmic streaming being the most sensitive.

Alterations in metabolism affect photosynthesis (WENT 1953; KETELLAPPER 1963; LANGRIDGE and MCWILLIAM 1967; BJÖRKMAN 1975) and respiration

(ALEXANDROV 1967; KINBACHER and SULLIVAN 1967), with photosynthesis being more sensitive (ALEXANDROV 1967; BJÖRKMAN 1975; BERRY et al. 1975), which may lead to respiratory substrate depletions (LEVITT 1972). Both oxidative phosphorylation (SEMICHATOVA et al. 1967; GERONIMO and BEEVERS 1964; BEEVERS and HANSON 1964) and photophosphorylation (EMMETT and WALKER 1969, 1973; SANTARIUS 1973, 1974) are decreased at high temperatures, with noncyclic photophosphorylation more sensitive than cyclic photophosphorylation (SANTARIUS 1975). Heat inactivation of the Hill reaction is commonly reported (EMMETT and WALKER 1969, 1973; MUKOHATA et al. 1973; SANTARIUS 1973, 1975; BERRY et al. 1975; PEARCY et al. 1977; NOLAN and SMILLIE 1976, 1977; CRITCHLEY et al. 1978) with photosystem II more heat-sensitive than photosystem I (SANTARIUS 1975). Light significantly decreases the thermostability of the Hill reaction (AGEEVA 1977). Electron transport reactions are less sensitive than photophosphorylation (SANTARIUS 1973), but seasonal differences in relative sensitivity have been noted (SANTARIUS 1974). Additionally, integrity of the chloroplast envelope appears less affected by heat than photochemical reactions (KRAUSE and SANTARIUS 1975).

Numerous biochemical alterations have been observed and include reductions in protein content (COLE 1969; RAMESHWAR 1971), chlorophyll content (FRIEND et al. 1962; ENGELBRECHT and MOTHES 1960, 1964; COLE 1969; RAMESHWAR 1971; FEIERABEND and MIKUS 1977), and nucleic acid content (COLE 1969). High temperature-induced chlorosis has been attributed to the photodestruction of chlorophyll (MCWILLIAM and NAYLOR 1967) while a deficiency of chloroplast rRNA suggests that a reduction in chloroplast ribosomes occurs (FEIERABEND and SCHRADER-REICHHARDT 1976; SCHAFERS and FEIERABEND 1976; FEIERABEND and MIKUS 1977).

12.2.2 Mechanism of Injury

Thermal denaturation of proteins has long been suggested as the principal cause of high temperature injury (ALEXANDROV 1967, 1977; ALEXANDROV et al. 1970). In recent years, however, attention has been diverted from the denaturation of proteins to the thermal disruption of membrane integrity as a primary step in high temperature injury (SANTARIUS 1973, 1975; BJÖRKMAN 1975; BERRY et al. 1975; SCHREIBER and BERRY 1977; SANTARIUS and MÜLLER 1979; BJÖRKMAN et al. 1980). Thermal inhibition of photosynthesis occurs at temperatures where inhibition of membrane-based processes occurs, and substantially below the temperatures required for thermal denaturation of the many enzymes associated with carbon metabolism (BJÖRKMAN et al. 1980). The Hill reaction, photosystem II activity, and photophosphorylation are most vulnerable (EMMETT and WALKER 1973; MUKOHATA et al. 1973; SANTARIUS 1973, 1975; KRAUSE and SANTARIUS 1975; BERRY et al. 1975; PEARCY et al. 1977; SCHREIBER and BERRY 1977; BJÖRKMAN et al. 1978).

Observations that heat injury is manifested as a rise in light-induced chlorophyll fluorescence (BERRY et al. 1975; KRAUSE and SANTARIUS 1975; SCHREIBER and BERRY 1977; PEARCY et al. 1977; SCHREIBER and ARMOND 1978; SANTARIUS and MÜLLER 1979), as well as abnormal behavior in other chloroplast pigments and components (SMILLIE 1979; SMILLIE and NOTT 1979b), further indicate that degeneration of chloroplast membranes occurs very rapidly upon exposure to high temperatures. Thus, uncoupling of photophosphorylation and inhibition of photoreductive capacity may only be symptoms of a more immediate effect of high temperatures on membrane structure (MUKOHATA et al. 1973) – possibly a breakdown in a light-induced membrane potential (SANTARIUS and MÜLLER 1979).

Reduction in photochemical activities of chloroplasts subjected to high temperatures is associated with an accumulation of free fatty acids (EMMETT and WALKER 1969), and uncoupling of photophosphorylation by free fatty acids is well established (MCCARTY and JAGENDORF 1965). Whether the appearance of free fatty acids is due to the activation of lipases (MOLOTKOVSKY and ZHESTKOVA 1965) or physical alterations related to intrinsic properties of the lipid components and their assembly in membranes (HEBER and SANTARIUS 1973) remains to be demonstrated. The significance of free fatty acids is difficult to assess, however, as many studies include bovine serum albumin, which mitigates the effects of free fatty acids, in the assay medium.

While acute exposures to high temperatures can lead to protein denaturation or membrane disruption, the cause(s) of injury resulting from chronic exposures is unsettled. Other more subtle repercussions of high temperature on cellular activities have been proposed (LANGRIDGE 1963) and include: (a) decreased availability of gases, (b) accelerated breakdown of metabolites, and (c) metabolic rate imbalances. Rate imbalances resulting from exposures to high temperatures were first considered by MACDOUGAL in 1920 (KURTZ 1958). BONNER (1957) and KURTZ (1958, 1961) reiterated this proposal in suggesting that the rate of growth may be limited by the velocity of a single reaction and that there is a temperature-induced deficiency of an essential metabolite, i.e., a "climatic lesion". While there are numerous examples of such climatic or temperature-induced lesions in microorganisms (LANGRIDGE 1963; BROCK 1967; cf. also Chap. 11, this Vol.) there are relatively few reports relating to higher plants. For instance, thermal inactivation of growth of pea epicotyls is prevented by adenine (GALSTON and HAND 1949) or vitamin B mixtures or ribosides (KETELLAPPER 1963), and varietal differences in heat tolerance of peas may be associated with the ability to synthesize adenine at higher temperatures (HIGHKIN 1957). Various vitamins and riboside mixtures are also effective in other plants (KETELLAPPER 1963; LANGRIDGE 1965; LANGRIDGE and GRIFFING 1959).

As the visual manifestations of thermal inactivation of growth are similar to those of senescence, and since retardation of senescence by adenine derivatives and cytokinins is well-documented (LEOPOLD and KRIEDEMANN 1975), the involvement of adenine derivatives in high temperature injury may be more than fortuitous. To wit, acute exposures to high temperatures (50 °C) accelerate senescence of tobacco leaves and can be prevented by applications of kinetin (ENGELBRECHT and MOTHES 1960, 1964); kinetin can prevent heat injury to wheat roots (SKOGQVIST and FRIES 1970); high temperatures cause a reduction in cytokinin content (COLE and STEPONKUS 1968; ITAI and BENZIONI 1974). The fact that roots are a major source of cytokinins (LEOPOLD and KRIEDEMANN

1975) allows for extension of the concept of climatic lesions. Visual symptoms of high temperature injury manifested in the shoot may result from a thermal lesion in the roots, one of which is in cytokinin synthesis. Support for such a proposal is rather direct, i.e., maintaining optimum root temperatures (17 °C) precludes injury to shoots that are exposed to normally lethal temperatures (40 °C) (Steponkus et al. 1970). Furthermore, thermal inactivation of shoot growth in intact plants was significantly reduced by applications of cytokinins to root systems subjected to injurious temperatures (Rameshwar and Steponkus 1969), with zeatin and zeatin riboside being the most effective (Rameshwar 1971).

12.2.3 Adaptive Mechanisms

12.2.3.1 Constitutive Characteristics

Many constitutive characteristics influence the cellular thermal environment in response to the ambient environment and collectively contribute to differences in heat resistance. Such characteristics range from morphological characteristics which moderate the absorption of radiant energy to cellular characteristics which allow the plant to cope with high temperatures. The former characteristics have been considered by Levitt (1972) and Larcher (1973) (cf. also Chap. 14, this Vol.), and only the latter characteristics will be considered.

Differences in thermostability of various enzymes in species, ecotypes, or cultivars of contrasting thermal tolerance have been reported. Thermal stability of malic dehydrogenase is much greater in an ecotype of *Typha latifolia* native to a hot climate than in an ecotype native to a cool climate (McNaughton 1965, 1966). In *Pisum sativum*, five isozymes of malic dehydrogenase can be identified with the majority of the activity attributable to two isozymes which exhibit pronounced differences in thermostability (Steponkus unpublished). In shoots, the thermostable form accounts for the majority of the total activity, while the thermolabile form accounts for the majority of activity in root extracts with discernable differences between cultivars of contrasting thermal tolerance. Contrastingly, Kinbacher (1970) has reported that while thermostability of malic dehydrogenase varied among three bean cultivars, the ranking was not consistent with the inferred heat tolerance based on retention of semipermeability. Differences in thermostability of several soluble photosynthetic enzymes are observed in species from contrasting thermal habitats; but for most, the thermostability is considerably greater than the thermostability of photosynthesis (Björkman et al. 1978, 1980). In contrast, thermostability of photosystem II electron transport and noncyclic photophosphorylation varies significantly in species of contrasting thermal tolerance and coincides with thermostability of photosynthesis. Membrane characteristics responsible for increased thermostability, however, have not been elucidated. In species of widely divergent thermal tolerance, no differences in membrane characteristics are discernable by electron spin resonance (ESR) techniques (Raison and Berry 1978).

HIGHKIN (1957) has suggested that differences in heat tolerance of pea cultivars may be associated with the ability to synthesize adenine at higher temperatures. While a reduction in cytokinin content occurs upon exposure to high temperatures, neither qualitative nor quantitative differences could be discerned in cultivars of contrasting thermal tolerance (COLE 1969). While impairment of cytokinin metabolism in the roots may be a primary site of thermal inactivation of plant growth (STEPONKUS et al. 1970), cultivar differences in high temperature sensitivity may reflect differing sensitivity to the resultant cytokinin deficiency (COLE 1969).

12.2.3.2 Facultative Characteristics

The thermal tolerance of plants, while characteristic for a species, is not constant (LANGE 1967) and varies with the stage of development, age of the organ in question, and exhibits both seasonal and diurnal variation (LARCHER 1973), for detailed discussion see Chapter 14, this Volume. Seasonal fluctuations in heat hardiness are usually characterized by two peaks in resistance which occur in the winter and summer with maximal resistance in winter (LANGE 1967). Increased thermal tolerance can be elicited by cultivation at temperatures for several weeks or months (LANGE 1967; PEARCY et al. 1977), but species differ in their capacity for thermal acclimation (BJÖRKMAN et al. 1980). In species capable of thermal acclimation, an increased thermostability of both the thylakoid membranes and an unidentified extra-thylakoid soluble protein is observed (BJÖRKMAN et al. 1980). While PEARCY et al. (1977) suggest that changes in membranes, in particular the lipids, are important in determining thermal tolerance, no substantial changes in membrane characteristics are discernable by ESR techniques (RAISON and BERRY 1978) but some alterations may be inferred (RAISON et al. 1980). Also, no changes in bulk lipid, fatty acid, or polypeptide composition of thylakoids could be discerned in heat-hardened spinach (SANTARUS and MÜLLER 1979).

In a wide variety of species increased thermal tolerance can be elicited by brief exposures (s to h) to high temperatures (45 to 50 °C) (YARWOOD 1961, 1967; SCHROEDER 1963, 1967; ALEXANDROV 1964; FELDMAN et al. 1967; OLIENIKOVA 1967; KINBACHER and SULLIVAN 1967; ZAVADSKAYA 1967; SANTARIUS and MÜLLER 1979). The cellular and/or molecular basis for the resultant increases in heat hardiness remains obscure. These acute treatments are frequently referred to as "thermal shocks", and ALEXANDROV (1967) has viewed the resulting increased heat resistance as a response to injurious but sublethal effects of heating. Thus, reputed increases in heat resistance based on respiration measurements following thermal shocks (KINBACHER and SULLIVAN 1967) which do not consider accelerated oxygen uptake resulting from uncoupling (SANTARIUS 1963; SEMICHATOVA et al. 1967) may only be manifestations of sublethal injury. FELDMAN et al. (1967) contend that heat hardening results in protein stabilization by either a change in protein structure, the production of protective compounds (sugars and amino acids) which stabilize the proteins, or an acceleration of protein synthesis and repair. LEVITT (1972) interprets several reports as consistent with this proposal, i.e., urease from heat-hardened cucumbers was more thermostable than that isolated from nonhardened plants (FELDMAN 1968) and that heat hardening increased the thermostability of malic dehydrogenase (KINBACHER et al. 1967; KINBACHER 1970) and fraction I protein (SULLIVAN and KINBACHER 1967). In such reports, however, total activity is frequently less in "heat-hardened" plants; the temperature at which activity declines is quite similar in both control and "heat-hardened" plants; and the rate of decline is only slightly decreased. It is possible that the initial heat shock influenced only the most sensitive fraction of the population and may not reflect increased thermostability of the remainder.

Various compounds can minimize high temperature injury (ALEXANDROV et al. 1970; HEBER and SANTARIUS 1973) and ALEXANDROV (1977) suggests that various cellular ligands reduce the conformational flexibility of proteins and thus increase stability. A mechanism of protective action of sugars at high temperatures (MOLOTKOVSKY and ZHESTKOVA 1965), however, has been questioned (ZAVADSKAYA 1967). Since sugars delay inactivation of electron transport and photophosphorylation, SANTARIUS (1973) contends that heat resistance is increased by both protective compounds and direct membrane alterations. Protective compounds are considered responsible for the mid-winter maximum in heat resistance which coincides with maximal cold-hardiness, while membrane alterations are considered responsible for the mid-summer peak in heat resistance. Recently, however, it was concluded that sugars play no protective role in heat hardiness elicited by short-term acclimation (4 to 6 h at 35 °C) (SANTARIUS and MÜLLER 1979).

12.2.4 Conclusions

High temperature injury results in a spectrum of functional and structural aberrations whose incidence and significance to growth and survival depend on the intensity and duration of the thermal stress (LARCHER 1973). Thus, the mechanism of injury will vary accordingly: under some conditions injury may be attributable to metabolic dysfunction resulting from either rate imbalances or to the inactivation of extremely thermolabile enzymes; under other conditions, membrane disruption may result; under extreme temperatures, thermal denaturation of various cellular constituents may occur. Tolerance of high temperatures, which varies greatly among plant species, exhibits seasonal fluctuations. The cellular and molecular characteristics responsible for such differences, however, remain to be elucidated.

12.3 Chilling Injury

Species of tropical and subtropical origin exhibit the least tolerance to low temperatures and are sensitive to temperatures in the range of 0 to 20 °C (cf. Chap. 13, this Vol.). The term "chilling injury" has been used to distinguish this type of injury from that resulting from freezing of the tissue. Visual symptoms of chilling injury are numerous and diverse and represent the integration of a spectrum of metabolic dysfunctions and structural alterations in a manner unique to the organ and species under consideration. A major concern is whether the extreme diversity in visual symptoms represents the direct and unique intervention of temperature with individual cellular processes and constituents or whether there is one primary thermal perturbation common to all of the diverse symptoms. The former option requires the transduction of the thermal influence at numerous and diverse loci while the latter assumes only one such transduction.

12.3.1 Cellular Manifestations

Investigations of cellular manifestations of chilling injury are usually rather productive endeavors – as in sensitive species in a state of degeneration there are few cellular activities which are not influenced. For example, protoplasmic streaming rapidly ceases (LEWIS 1956) or is greatly reduced (PATTERSON and GRAHAM 1977) at temperatures around 10 to 12 °C in chilling-sensitive species and is associated with altered cytoplasmic morphology such as thinning of transvacuolar strands and the formation of numerous small vesicles (DAS et al. 1966) which occur within seconds (PATTERSON et al. 1980).

The influence of chilling on respiratory activities is extensively documented (LYONS 1973). In intact harvested organs which have been chilled, respiration rates increase. The anomalous increase in respiration has been attributed to the uncoupling of oxidative phosphorylation (LEWIS and WORKMAN 1964; CREECIA and BRAMLAGE 1971), and declines in respiratory activity and respiratory control values (MINAMIKAWA et al. 1961) are observed in mitochondria isolated from injured tissues. When mitochondria are isolated from healthy tissues and subjected to chilling in vitro, oxidative activity declines markedly at 12 °C while phosphorylative efficiency is not impaired (LYONS and RAISON 1970).

Impaired photoreductive capacity is commonly observed (MARGULIES and JAGENDORF 1960; MARGULIES 1972; SHNEYOUR et al. 1973; RAISON 1974; MURATA et al. 1975; NOLAN and SMILLIE 1976, 1977; INOUE 1978) and has been suggested as a basis for determining chilling sensitivity (SMILLIE and NOTT 1979a). The decreased photoreductive capacity can, however, be effected at different loci depending on whether chilling occurs in the presence or absence of light or in situ vs. in vitro (GARBER 1977, 1980). Hill activity can be restored by incubating previously chilled leaves in the light at 20 °C, but illumination of isolated chloroplasts is ineffective (KUMM and FRENCH 1945; MARGULIES and JAGENDORF 1960; KANIUGA et al. 1978). Light vs. dark chilling also influences the sensitivity of photophosphorylation (MARGULIES and JAGENDORF 1960) and the decrease in PMS-dependent photophosphorylation is preceded by decreases in proton uptake, osmotic responsiveness, Ca^{2+} ATPase activity, or chlorophyll content (GARBER 1977, 1980). Impaired photosynthetic capacity is frequently associated with chlorosis or other visible lesions (FARIS 1926) and may extend to severe bleaching as a result of chilling at high light intensities. C_4 monocot species are especially sensitive (MCWILLIAM and NAYLOR 1967;

TAYLOR et al. 1974), but similar effects are observed in dicots (VAN HASSELT 1972, 1974; GARBER 1977, 1980). While the temperature/light interaction has prompted the contention that there is no single injurious chilling temperature (BAGNALL and WOLFE 1978), the sequence of abnormal photosynthetic behavior would suggest that photo-bleaching occurs after other photosynthetic activities are impaired (TAYLOR et al. 1974). VAN HASSELT (1974) has suggested that photo-oxidation of chlorophyll is due to the generation of a chlorophyll triplet state when electron transport is impaired.

Alterations in membrane permeability are extensively documented (SIMON 1974) and leakage of cellular constituents from deteriorating tissues is widely observed. More fundamental is the question of whether thermal perturbation of membrane permeability is a cause or symptom of chilling injury (SIMON 1974); abrupt changes in the osmometric behavior of chloroplasts of chilling-sensitive species have been observed (NOBEL 1974).

Degeneration of cell ultrastructure occurs after exposure to relatively long periods of chilling (MOLINE 1976; ILKER et al. 1976), but is also apparent after short periods of chilling. Within 2 h at 5 °C, discontinuities in thylakoid, plastid, and mitochondrial membranes of tomato cotyledons are apparent (ILKER et al. 1980). Partial dilation and microvesiculation of the rough endoplasmic reticulum (NIKI et al. 1978) and changes in the distribution of intramembraneous particles on tonoplast fracture faces (YOSHIDA et al. 1980) are among the first manifestations in a chilling-sensitive culture of *Cornus stolonifera* cells. While chilling injury is generally considered as the result of metabolic dysfunction (LYONS 1973), these reports suggest that gross physical disruption of cellular architecture also occurs quite rapidly.

12.3.2 Mechanism of Injury

Within the vast and diverse array of metabolic dysfunctions and visual symptoms, LYONS and RAISON and coworkers (LYONS and BREIDENBACH 1979; LYONS et al. 1980b) have proposed that: (1) a reversible change in the physical state of the membrane is the primary response to low temperatures in that only above a critical transition temperature are the membranes sufficiently fluid to maintain normal physiological activities; (2) physiological dysfunctions and visible symptoms occur as secondary and time-dependent processes, and these responses include altered function of membrane-associated enzymes, reduced energy supply, loss of cellular compartmentation, ion leakage and other events disruptive of metabolism; and (3) the nature and severity of the symptoms is a function of the temperature extreme and duration of exposure and is characteristic for the morphological and physiological condition of the plant tissue under consideration.

The proposal has evolved from the following observations: (1) swelling characteristics of mitochondria indicated that those isolated from chilling-sensitive tissues were less flexible at low temperatures than were those isolated from chilling-resistant tissues and this difference was associated with a higher proportion of unsaturated fatty acids in mitochondrial membrane lipids of chilling-resistant tissues (LYONS et al. 1964); (2) Arrhenius plots of succinate oxidation by mitochondria isolated from chilling-resistant tissues appeared linear, while anomalies ("breaks" or "discontinuities") were observed in plots of activity of chilling-sensitive tissues (LYONS and RAISON 1970); (3) Alterations in the physical state (fluidity or molecular ordering) of mitochondrial membranes, as inferred from ESR studies of spin labels infused into the membranes, occurred at temperatures which were coincident with temperatures at which anomalies in respiratory activity were observed (RAISON et al. 1971 b). These events occurred immediately in isolated membrane systems at temperatures at which chilling injury was manifested in intact organs and plants, suggesting that a very primary aspect of chilling injury was being addressed (LYONS 1973). Subsequently, RAISON (1973 b) presented a correlation of anomalies in Arrhenius plots of photoreductive capacity in chloroplasts of chilling-sensitive tissues (SHNEYOUR et al. 1973) and the temperature at which an alteration in the physical state of chloroplasts was inferred. A similar correlation, however, was not found in glyoxysomes in that while alterations in the membranes at low temperatures were inferred from ESR studies (WADE et al. 1974), gluconeogenic glyoxysomal enzymes did not exhibit anomalous thermal behavior (BREIDENBACH et al. 1974). The results of numerous studies of ESR and fluorescence probes and respiratory and photoreactive activities as influenced by temperature have been compiled by QUINN and WILLIAMS (1978).

Since its inception (LYONS et al. 1964), the proposal has undergone continuous refinement to accommodate additional evidence and, more importantly, in response to the evolution in interpretation of various techniques used to probe membrane structure and function. While the proposal is supported by considerable and diverse data, there are, however, several reports which contest some of its basic assumptions and inferences, and others which conflict with its basic tenets. Some of the criticisms are directed to early interpretations and are resolvable; some inconsistencies remain to be resolved. Nonetheless, when taken as a working model and conceptual framework upon which to query the cellular and molecular bases for chilling sensitivity, it has served its purpose well. LYONS and BREIDENBACH (1979) have provided extensive documentation relative to the questions of whether low temperatures impose a change in the physical state of cellular membranes and whether this change controls physiological events. The documentation provided, although admittedly correlative and not always without exceptions, can, however, only be viewed as being suggestive of the questions. The interpretations depend on the validity of several fundamental assumptions, and there are other questions which should be addressed.

12.3.2.1 Is it Appropriate to Use Arrhenius Plots to Study Chilling Injury?

The conceptual framework of the LYONS and RAISON hypothesis has evolved from the use of Arrhenius plots to depict the influence of temperature on various biological activities and physical measurements of membrane characteristics. The existence of "breaks" or "discontinuities" in such plots, the temperatures at which they occur, and interpretation of why they occur are cornerstones of the hypothesis; each has been the subject of debate. The question of whether "breaks" or "discontinuities" exist has been of concern for many years (BĚLEHRÁDEK 1953, 1957; KUMAMOTO et al. 1971) and recently reasserted (BAGNALL and WOLFE 1978; WOLFE and BAGNALL 1980a). Whether it is appropriate to fit the data with a series of straight lines or a smooth curve is a major conflict. In very few plots are "discontinuities" (distinct vertical breaks) manifested; more often, "breaks" are inferred when a change in slope of two intersecting lines can be discerned. In general, biological responses exhibit the most pronounced "breaks", while plots of physical probe data are less dramatic. There is the possibility that some of the "breaks" are attributable to artifacts resulting from inappropriate experimental conditions. Both biological (RAISON et al. 1977) and physical probe studies (CANNON et al. 1975; QUINN and WILLIAMS 1978) are vulnerable.

NOLAN and SMILLIE (1976) have suggested that "breaks" per se are not necessarily correlated with chilling sensitivity but that it may be the magnitude of change which distinguishes chilling-sensitive from chilling-resistant species. Thus, it is argued that a smooth curved Arrhenius plot may be approximated by a series of straight lines for a pragmatic diagnostic approach to determine where the change in slope is most pronounced. The decision as to which tangents to select, however, may be somewhat arbitrary and lead to disputable inferences. While it has been suggested that Arrhenius plots be subjected to statistical derivation (WOLFE and BAGNALL 1980b; Epilog in LYONS et al. 1980a), this may do little to resolve the problem, since in some cases where this was done for straight line fits the temperatures at which the breaks were inferred are even more suspect. WOLFE and BAGNALL (1980b) argue that no objective algorithm for selecting the appropriate tangents has been proposed and that the "breaks" are in part determined by the design of the experiment and other procedural aspects. When Arrhenius plots are used in an interpretative approach, the inference that such breaks are indicative of an increase in the activation energies of enzymes due to a phase transition (KUMAMOTO et al. 1971; RAISON 1973a, b) has been strongly criticized (BAGNALL and WOLFE 1978; WOLFE 1978). This aspect is elaborated in the preceding Chapters of this Volume but arguments both for and against this interpretation remain quite speculative.

12.3.2.2 What is the Nature of the Change in the Physical State of Cellular Membranes Elicited by Low Temperatures?

In early reports, the reversible change in the membrane induced by low temperatures was envisaged as a physical change from a liquid crystalline to a solid gel phase in the bulk membrane lipids at some sharply defined temperature (LYONS and RAISON 1970; LYONS 1973). Single breaks in Arrhenius plots of both ESR and biological responses were considered as manifestations of this change. Such an interpretation can no longer be considered appropriate. In biological membranes, major thermotropic phase transitions are much broader, frequently spanning 20 to 30 °C, due to the heterogeneity of the constitutive membrane components and their interactions (QUINN and WILLIAMS 1978). While membrane fluidity is highly influenced by the component fatty acids (acyl chain length, degree of unsaturation and branching), other factors such as polar head group composition and sterols can also markedly influence fluidity. Also, proteins (LETELLIER et al. 1977) and neutral lipids can dramatically alter the thermotropic phase transition (MCKERSIE and THOMPSON 1979). The nature of the transition and the temperature at which it occurs will also depend on the salt and ion concentration of the aqueous phase and the transverse and lateral pressure in the bilayer (WOLFE 1978, 1980).

Subsequent reports which discerned two (and sometimes more) breaks from ESR data (MILLER et al. 1974; RAISON and CHAPMAN 1976; RAISON et al. 1977) suggested that the temperature-induced transition corresponds to a gradual change from a liquid crystalline to a gel phase (RAISON 1974), hence a mixture of fluid and gel was envisaged in the range between the two points (RAISON and CHAPMAN 1976). While plant membranes may exist in a mixed phase at physiological temperatures (MCKERSIE and THOMPSON 1977), there is no basis for the assumption that the two anomalies discerned by ESR reflect the start and termination of such a mixed phase (QUINN and WILLIAMS 1978). Also, such an inference is not consistent with that inferred from fluorescence studies using trans-parinaric acid as a probe (PIKE et al. 1980). Changes in fluorescence intensity and polarization ratio which occurred at 9 and 10 °C, respectively, in maize were interpreted as the initial appearance of detectable solids and not the end of a mixed-phase region as inferred from ESR probes (RAISON and CHAPMAN 1976). Note, however, that only the phospholipids were studied, as pigments in the total polar lipid extracts interfered with the fluorescence studies.

It may be concluded that "breaks" inferred from ESR studies at temperatures in the range of 0 to 20 °C do not reflect major thermotropic transitions in bulk lipids. Indeed, MCMURCHIE (1980) observed a broad but small exothermic transition with an extremely low enthalpy value in calorimetric measurements of tomato membranes, and indicated that less than 1% of the membrane lipids was involved in the phase transition. Whether "breaks" in Arrhenius plots of ESR data are manifestations of minor thermotropic transitions in the membrane lipids (RAISON and MCMURCHIE 1974) or result from erroneous assumptions (CANNON et al. 1975) remains to be resolved.

12.3.2.3 Does Membrane Composition Distinguish Chilling-Sensitive Species from Chilling-Resistant Species?

In initial interpretations, chilling-sensitive and chilling-resistant species were characterized by differences in lipid fatty acid composition (LYONS et al. 1964). Subsequently, the possibility that small changes in fatty acid composition could have a large influence on the temperature at which a lipid mixture solidifies was presented (LYONS and ASMUNDSON 1965). In view of this, the futility of any attempt to discern a causal relationship between fatty acid composition and chilling sensitivity on the basis of bulk compositional analyses should have been apparent. There were, however, many subsequent reports regarding bulk compositional analyses of fatty acids and their relation to both chilling injury (QUINN and WILLIAMS 1978) and freezing injury (STEPONKUS 1978). Some of these reports were not consistent with the concept that chilling sensitivity was related to the extent of fatty acid unsaturation (YAMAKI and URITANI 1972; WILSON 1976, 1980; PATTERSON et al. 1978, 1980). While fatty acid unsaturation can dramatically influence membrane fluidity, differences in chilling-sensitive plants may be due to other characteristics such as polar head group and sterol composition, differences in discrete membrane domains, unique lipid-lipid domains (QUINN and WILLIAMS 1978; LYONS et al. 1980b) or due to unique lipid-lipid interactions (MCMURCHIE 1980). Such possibilities will not be resolved by bulk lipid fatty acid analysis. Therefore, neither support for or evidence against the proposal will be found in bulk compositional analyses.

12.3.2.4 Are the "Breaks" Observed in the Range of 0 to 15 °C Unique to Chilling-Sensitive Plants?

In the early reports, "breaks" in the thermal behavior of membranes from chilling-sensitive plants were consistently observed while such "breaks" were absent in chilling-resistant tissues (RAISON et al. 1971a; RAISON 1973a, b, 1974; WADE et al. 1974). The uniqueness of such responses to chilling-sensitive tissues was questioned when MILLER et al. (1974) reported similar responses in mitochondria isolated from cultivars of both spring and winter wheat. RAISON et al. (1977) suggested several procedural problems associated with the particular ESR probe used in this study. When three other spin probes were used, however, two "breaks" were observed, albeit at lower temperatures. Alterations in the fluidity of barley chloroplasts at 9 an 24 °C were inferred from ESR measurements (NOLAN and SMILLIE 1976). PIKE et al. (1980) have also observed two "breaks" in fluorescence measurements of trans-parinaric acid in both chilling-sensitive and chilling-resistant species, but they differ in the temperature range at which they occur. Other observations of anomalous behavior in chilling-resistant tissues have been inferred by ESR studies (MCGLASSON and RAISON 1973; TORRES-PEREIRA et al. 1974) and by other fluorescent probes (BOROCHOV et al. 1978).

More problematic were "breaks" in Arrhenius plots of biological activities of chilling-resistant plants (POMEROY and ANDREWS 1975). RAISON et al. (1977) have suggested, however, that this atypical response in respiratory activities was attributable to an indirect effect of low temperature on the assay system; and they could not detect anomalous behavior in wheat mitochondrial respiratory activities. NOLAN and SMILLIE (1976, 1977) have inferred "breaks" in Arrhenius plots of photoreduction of DCIP by chloroplasts of both chilling-sensitive and chilling-resistant species, however, they only occurred in the presence of an uncoupler, methylamine. There is also a poor correlation between the temperature at which breaks in photoreductive activities occur and inferred differences in chilling sensitivity of several *Passiflora* species (CRITCHLEY et al. 1978). INOUE (1978) inferred a single "break" in the Arrhenius plot of DCIP photoreductive activity in chloroplasts of a chilling-resistant species, spinach, at −9 °C; whereas two "breaks" (at −9 and +11 °C) were inferred in plants grown at low temperatures (0 to 10 °C). In neither case was an uncoupler added, but the extent of coupling was not ascertained either.

While "breaks" may be inferred in Arrhenius plots of growth of chilling-sensitive species (RAISON and CHAPMAN 1976), they have also been inferred in Arrhenius plots of growth of a chilling-resistant species (SMILLIE 1976). Although a "break" appears at 12 °C in plots of various parameters of early seedling growth in both chilling-sensitive (sorghum) and chilling-resistant species (barley), the departure from linearity is not as pronounced in the resistant species (MCWILLIAM et al. 1980). SMILLIE (1976) has argued that the effects of low temperature on growth and development must be distinguished from chilling injury, and suggested the latter may be due to a change in membrane permeability. MCMURCHIE (1980), however, found little relationship between chilling sensitivity and thermal influences on Mg^{2+}-ATPase or K^{+}-stimulated Mg^{2+}-ATPase activity.

12.3.2.5 How is Control of Biological Activities Effected?

An early inference of the hypothesis was that the slope of the Arrhenius plots of biological activity was proportional to the activation energy of membrane-bound enzymes (RAISON 1973a, b), and the changes in slope were envisaged as consequences of conformational changes in membrane proteins resulting from a bulk phase change (KUMAMOTO et al. 1971). WOLFE (1978) has discussed several theoretical means by which conformational changes in membrane proteins may be effected. Even in the reports which contest the uniqueness of thermal anomalies in biological activity (NOLAN and SMILLIE 1976, 1977), changes in activation energies are inferred. Several authors (WOLFE 1978, 1980) contend that the slope of an Arrhenius plot is not necessarily proportional to an activation energy; and other factors contributing to the slope of Arrhenius plots are partitioning of an enzyme between active and inactive states and relative solubility of an enzyme in different domains. The importance of membrane lateral compressibility to physiological processes has been considered (LINDEN et al. 1973; WOLFE 1980). WOLFE (BAGNALL and WOLFE 1978) contends, however, that changes in membrane lipid mobility would not be expected to cause abrupt changes in Arrhenius plots at either the enzyme or whole plant level.

Further elaboration of the proposal suggests that secondary physiological dysfunctions and visible symptoms result from altered functions of membrane-associated enzymes, reduced energy supply, and altered membrane permeability (LYONS 1973; LYONS and BREIDENBACH 1979). General metabolic imbalances are usually invoked as the means by which alterations in membrane-associated proteins are manifested as injury, but direct experimental confirmation is difficult due to the myriad of events. In some cases, it is difficult to reconcile the immediacy of injury with a metabolic-based argument especially where ultrastructural documentation is provided (ILKER et al. 1980; YOSHIDA et al. 1980). The rather gross and indiscriminate degeneration of cell structure and function could, however, result from some omnipotent toxic compound which is either metabolically generated or physically released from a sequestered location (YOSHIDA et al. 1980).

While several compounds have been considered potentially injurious (LYONS 1973), free fatty acids have received little direct attention in the area of chilling injury. This is rather surprising since many processes which are extremely sensitive to chilling, i.e., reduced photoreductive and phosphorylative capacity, are severely curtailed by low concentrations (10^{-5} to 10^{-6} M) (KROGMANN and JAGENDORF 1959) of free fatty acids (MCCARTY and JAGENDORF 1965). (See KANIUGA et al. 1978 for extensive documentation of uncoupling of both oxidative and photophosphorylation.) ANDERSON et al. (1974) have shown that galactolipase is closely associated with chloroplasts, and lipase treatment of chloroplasts results in the rapid inhibition of electron flow. Up to 50 to 60% of the acyl lipids may be removed, however, without affecting the rate of electron flow provided bovine serum albumin (BSA), a fatty acid scavenger, is present. The involvement of free fatty acids in chilling injury is difficult to deduce from the reported experiments, because in nearly every report of organelle studies, BSA was incorporated in the isolation and/or assay media. The concentrations vary widely and whether the influence of free fatty acids was totally or only partially precluded cannot be determined. Recently, KANIUGA and MICHALSKI (1978) found a fivefold increase in the amount of free fatty acids in leaves of various chilling-sensitive species stored in the dark for 3 days at 0 °C, but no increase was observed in spinach leaves. While the increase may be a consequence of injury, investigation of a causal role is warranted.

12.3.3 Conclusions

The Lyons and Raison hypothesis has been challenged on several grounds. Early interpretations and inferences, i.e., bulk lipid phase transitions, unsaturated fatty acid composition, and alterations in activation energies made the hypothesis especially vulnerable. Although the molecular nature of thermally induced alterations in cellular membranes and the specific manner in which they result in injury remain to be resolved, the most fundamental tenet that thermal influences on the physical properties of cellular membranes can significantly alter biological activities remains as the most attractive explanation for chilling injury.

12.4 Freezing Injury

In recent years there have been several comprehensive reviews on cold hardiness (MAZUR 1969, 1970, 1977a, b; ALDEN and HERMANN 1971; LEVITT 1972; HEBER and SANTARIUS 1973; BURKE et al. 1976; STEPONKUS 1978, 1979; STEPONKUS and WIEST 1978, 1980; LI and SAKAI 1978). This section will focus on the physical and chemical events that occur during freezing, the repercussions on cellular structure and function, and the manner in which cold acclimation may influence these events. Only when these factors are elucidated can aspects of cold acclimation and genotypic diversity in cold hardiness be fully understood. Currently, cellular and molecular lesions are poorly characterized relative to the physicochemical aspects of the freezing process. This deficiency has precluded the formulation of a durable mechanism of freezing injury.

12.4.1 The Freezing Process

The physicochemical events that occur during freezing have been elaborated by MAZUR (1969, 1970), HEBER and SANTARIUS (1973), and STEPONKUS (STEPONKUS 1978; STEPONKUS and WIEST 1980). Initially, ice nucleation occurs in the extracellular solution and results in a disequilibrium in the chemical potential of water in the intracellular solution relative to the chemical potential of the partially frozen extracellular solution with the chemical potential of water molecules in ice a direct function of temperature. Thermodynamic equilibration is achieved either by cellular dehydration and continued extracellular ice formation or by intracellular ice formation. The manner of equilibration is influenced by the rate at which the cell is cooled and the minimum temperature to which it is exposed relative to the efflux of water from the cell.

The amount of water which must be removed to achieve thermodynamic equilibration is a function of the initial osmolality of the intracellular solution. It is often generalized that the rate at which this amount of water is removed is influenced by the water permeability of the plasma membrane and the surface area available for efflux relative to the cooling rate and minimum temperature imposed. However, since intracellular ice formation in isolated cells occurs at freezing rates nearly 60 times faster than those required to cause intracellular ice formation in intact tissues (LEVITT 1972), other factors associated with tissue organization also influence the mass transfer of water. The interaction of these factors will determine the magnitude of disequilibrium incurred and will be manifested as supercooling of the intracellular solution. While the incidence of intracellular ice formation is greatly influenced by the extent of supercooling, it is also influenced by the temperature to which the cells are frozen (DOWGERT and STEPONKUS 1979). Since the probability of intracellular ice formation increases at temperatures lower than -10 °C, regardless of the extent of supercooling (STEPONKUS and WIEST 1980), the incidence of intracellular ice formation is not a simple function of the extent of supercooling.

12.4.2 Repercussions of Freezing on the Cellular Environment

The primary repercussions of the freezing process on the cellular environment include the obvious decrease in temperature, the presence of ice crystals and dehydration of the cell (MAZUR 1969; HEBER and SANTARIUS 1973; STEPONKUS 1978, 1980; STEPONKUS and WIEST 1980). Generally, decreases in temperature or the presence of ice crystals are not the primary cause of injury (HEBER and SANTARIUS 1973); rather, the process of cell dehydration is the most disruptive and injurious repercussion of the freezing process (MAZUR 1970). Cellular dehydration, however, results in a multitude of effects and it is within this array that there is a great divergence in hypotheses on the mechanism of freezing damage (MAZUR 1969, 1970, 1977a, b; HEBER and SANTARIUS 1973; MERYMAN 1974; MERYMAN et al. 1977; STEPONKUS 1978; STEPONKUS and WIEST 1980, for reviews). Several secondary stresses will result from cellular dehydration and will occur in a sequential fashion when their incidence is considered as a function of intensity (STEPONKUS 1980). These secondary stresses include a loss of turgor potential; an increase in solute concentration; reductions in cell volume, surface area, and surface linear dimension; and a small decrease in water activity. Although volume decreases asymptotically, solute concentration increases linearly with decreases in temperature. Therefore the assumption that effects of dehydration can be separated from other effects at low temperatures (RAJASHEKAR et al. 1980) is erroneous.

12.4.3 Manifestations of Freezing Injury

Membrane disruption, as evidenced by a flaccid, water-soaked appearance, is a common manifestation of lethal freezing injury in plant tissues; and nearly all of the methods commonly used to evaluate tissue viability following a freeze-thaw cycle (plasmolytic techniques, vital staining, solute leakage, and reduction of triphenyl tetrazolium chloride) are based on the retention of a semipermeable plasma membrane. The rapidity with which lethal damage is manifested also suggests that injury is not due to metabolic dysfunction. Even in cases where sublethal exposures are experienced (PALTA and LI 1978), damage to the plasma membrane is incurred.

As early as 1912, MAXIMOV suggested that freezing damage was the result of freeze-induced removal of water from the surface of the plasma membrane. Further insight into the nature of freezing damage and cold acclimation emerged from a series of classic papers by SCARTH, LEVITT, and SIMINOVITCH (LEVITT and SCARTH 1936a, b; SCARTH and LEVITT 1937; SIMINOVITCH and SCARTH 1938; SIMINOVITCH and LEVITT 1941; SCARTH et al. 1940; LEVITT and SIMINOVITCH 1940; SCARTH 1941). They demonstrated that both intracellular ice formation and cellular dehydration (extracellular ice formation) result in damage to the plasma membrane, albeit for different reasons. While disruption of the plasma membrane occurs at the moment of intracellular ice formation, damage resulting from cellular dehydration is most frequently manifested upon thawing during deplasmolysis of the cell.

Consideration of specific membrane lesions resulting from freezing were prompted by studies of chloroplast and mitochondrial membranes (HEBER and SANTARIUS 1964; HEBER 1967). Numerous papers by HEBER and coworkers (HEBER and SANTARIUS 1973) provided considerable insight into the effects of freezing on the function of chloroplast membranes. Photophosphorylation was shown to be the most labile process, and uncoupling was attributed to altered permeability. GARBER and STEPONKUS (1976a) demonstrated that the situation was considerably more complex. Following a slow freeze-thaw cycle, there were three lesions in light-induced proton uptake by isolated thylakoids: loss of plastocyanin, loss of coupling factor (CF_1) and loss of osmotic responsiveness. Subsequently, VOLGER et al. (1978) have verified the release of CF_1 from thylakoid membranes.

Recently, a spectrum of lesions in the plasma membrane, as indicated by altered osmometric behavior of isolated protoplasts during a freeze-thaw cycle, has been reported (STEPONKUS et al. 1979a). Intracellular ice formation in cells retaining a large amount of water will result in the immediate physical disruption of the plasma membrane. Cellular dehydration may result in different lesions depending on the freeze-thaw protocol. One form of injury commonly observed in nonacclimated protoplasts cooled to relatively warm subzero temperatures (−5 °C) and rewarmed is an expansion-induced lysis during warming – a phenomenon qualitatively described by SIMINOVITCH and LEVITT (1941). This lesion is the result of a contraction-induced alteration in plasma membrane resilience that decreases the maximum critical surface area of the plasma membrane (WIEST and STEPONKUS 1978a; STEPONKUS and WIEST 1978, 1980). Two other forms of injury occur in cells subjected to freeze-induced dehydration at temperatures below −5 °C. Following cooling, cells may exhibit a total loss of osmotic responsiveness and remain in the contracted state during warming – suggestive of a complete loss of semipermeability of the plasma membrane. Alternatively, cells may exhibit altered osmometric behavior during warming. Such cells expand but to a lesser extent at any given osmolality than normal cells. While the cells remain physically intact, this manifestation is suggestive of a partial loss of semipermeability or leakiness of the plasma membrane. The incidence of these various lesions is influenced by the minimum temperature imposed and duration of the freezing cycle and varies with the degree of hardiness of the tissue. Such observations suggest that attempts to determine *the* moment of injury (RAJASHEKAR et al. 1980) can only be considered for the particular tissues and freezing conditions employed.

12.4.4 Mechanism of Injury

Although numerous theories on the mechanism of freezing injury have been presented (MAZUR 1969, 1970, 1977a, b; MERYMAN 1974; MERYMAN et al. 1977; LEVITT 1972; HEBER and SANTARIUS 1973; STEPONKUS and WIEST 1980, for reviews), none has successfully escaped major criticism. Those hypotheses which claim the broadest universality are the most susceptible. As the freezing process results in a multitude of cellular stresses which can be viewed as a sequential series of stress barriers (MAZUR 1977b; STEPONKUS and WIEST 1980), it is reasonable to envisage different cellular lesions resulting from different stresses. In view of this, then no single mechanism of freezing injury would be capable of explaining freezing injury in tissues in different states of acclimation, or in contrasting species of widely divergent degrees of hardiness, or in cell suspensions frozen under different conditions. LOVELOCK (1953) considered that the myriad of events attendant on dehydration may present individual and vastly different causes of injury.

MAZUR (1977b) has proposed that between the time the cells first encounter extracellular ice nucleation and the time they are returned to post-thaw conditions, they meet a sequence of events – any one of which is potentially lethal. He suggests that solute concentration is the first such event and is followed by cell contraction. While extracellular solute concentration precedes cellular dehydration, cell volume decreases asymptotically with decreases in temperature while solute concentration increases linearly. Thus, in terms of intensity, stresses resulting from cellular contraction reach peak intensity at relatively warm subzero temperatures before solute concentrations achieve appreciably high levels. Whether reductions in cell volume result in a potentially lethal membrane lesion before a lesion resulting from increased electrolytes has been argued (MERYMAN 1968, 1974; MERYMAN et al. 1977; MAZUR 1977b). Under prescribed conditions, survival of protoplasts subjected to a freeze-thaw cycle can be quantitatively predicted on the basis of cellular volumetric changes (WIEST and STEPONKUS 1978a) which suggests that lesions resulting from volumetric changes take precedence over other potentially lethal lesions resulting from other stresses.

12.4.4.1 Reduction in Cell Volume

MERYMAN (1968, 1971) has proposed the "minimum critical volume hypothesis" to account for freeze-induced injury in a wide spectrum of biological cells. While MAZUR (1977b) has criticized the hypothesis, he has employed its basic tenet that cells possess a minimum critical volume in proposing a mechanism of injury. WIEST and STEPONKUS (1979) have questioned this basic premise of the hypothesis, i.e., anomalous osmometric behavior, and have rejected the application of this hypothesis to plant cells (STEPONKUS and WIEST 1980). While a reduction in cell volume is the appropriate cell parameter when considering stresses associated with freeze-induced cellular dehydration, it is more appropriate to consider reductions in cell surface area (i.e., altered spatial relations in the plasma membrane) when considering cellular lesions. During freeze-induced contraction a reduction in the plasma membrane surface area results in an alteration of the resilience of the plasma membrane which limits the expansion potential during expansion upon warming (WIEST and STEPONKUS 1978a). This contraction-induced alteration is a continuous (or nonresolvably discrete) function of contraction rather than the result of some minimum critical volume. Loss of membrane material has been observed by labeling the plasma membrane with Concanavalin A-fluorescein (WIEST and STEPONKUS 1978b).

12.4.4.2 Solute Concentration

Freezing injury has long been attributed to the increased concentration of cellular solutes, especially electrolytes (MAXIMOV 1929). Several hypotheses invoke solute concentration as a major factor contributing to freezing injury, but like those based solely on volume reduction, they are unable to accommodate injury which occurs under varied conditions (MAZUR 1977b). Effects of electrolyte concentration on the freeze-induced inactivation of light-induced proton uptake in chloroplast thylakoids have been considered (HEBER and SANTARIUS 1973; STEPONKUS et al. 1977; STEPONKUS 1979). The demonstration that CF_1 is released from the thylakoid membrane following freezing (GARBER and STEPONKUS 1976a) provides a system in which to study the influence of freezing stresses on membrane structure and function with respect to a single protein which has both a structural and functional role. While exposure of thylakoids to high osmolalities of NaCl results in inactivation of Ca^{2+}-ATPase activity (STEPONKUS et al. 1977) and suggests that salt concentration during freezing may be a primary stress, the extent of CF_1 release is not quantitatively similar in the two treatments. Other factors which influence the fractional volume of solution that is frozen modify the rate or extent of CF_1 inactivation (STEPONKUS et al. 1977; STEPONKUS 1979).

12.4.5 Cold Acclimation

As cellular membranes are the primary site of freezing injury, it follows that cold acclimation must involve cellular alterations that allow the membranes to survive lower freezing temperatures. As freezing results in a sequence of potentially lethal cellular stresses and resultant lesions, it follows that cold acclimation is the integration of many diverse cellular alterations. Such alterations may be in the membranes themselves, in the cellular environment, or a combination of both (STEPONKUS 1971). Changes in the cellular environment may either alter the freezing stresses or they may result in direct protection of the membrane. Both possibilities were identified by CHANDLER (1913) and have been invoked in recent times (HEBER and SANTARIUS 1973; LEVITT 1972; MAZUR 1969, 1970; OLIEN 1967, 1977). Similarly, changes in the membrane may either alter the resultant cellular stresses (SCARTH and LEVITT 1937) or they may result in an increased tolerance to the cellular stresses (STEPONKUS et al. 1979b).

12.4.5.1 Mitigation of Cellular Stresses

The series of physicochemical events that follow extracellular ice nucleation and which can be precluded or mitigated include intracellular ice formation, extracellular ice formation, and the secondary stresses of cellular volumetric changes, solute concentration and the accumulation of toxic solutes resulting from cellular dehydration.

Intracellular Ice Formation. SIMINOVITCH and SCARTH (1938) observed that lethal intracellular ice formation occurred at a slower freezing rate in nonacclimated tissues than in acclimated tissues, and it is generally suggested that cold acclimation is associated with a decrease in the incidence of intracellular ice formation (LEVITT 1972; OLIEN 1967). Several factors interact in determining the probability of intracellular ice formation and may be influenced by cold acclimation.

Increased cellular solute concentration is one of the most universal manifestations of cold acclimation (ALDEN and HERMANN 1971; LEVITT 1972). The decreased solute potential of the intracellular solution would serve to depress the freezing point and decrease the amount of water which would have to be removed from the cell in order to achieve thermodynamic equilibrium at subzero temperatures. Thus, at any time during cooling, the thermodynamic disequilibrium between the intra- and extracellular solutions would be minimized; less supercooling of the intracellular solution would diminish the probability of intracellular ice formation.

It has been proposed that cold acclimation results in an increase in membrane water permeability permitting rapid movement of water to extracellular sites of nucleation (LEVITT and SCARTH 1936b), and this has been invoked to explain the observation that intracellular ice formation occurs at slower freezing rates in nonacclimated than in acclimated tissues (SIMINOVITCH and SCARTH 1938; LEVITT 1972). STOUT et al. (1977) indicated that there may not be a direct cause-and-effect relationship between the two observations. Calculations of the rate of ice formation during freezing at a relatively high rate, viewed in relation to membrane water permeability, suggested that water permeability of the plasma membrane was not rate-limiting in nonacclimated tissues. Although LEVITT (1978) has argued against this interpretation, he failed to acknowledge that individual cells can tolerate cooling rates that are 60 times greater than intact tissues (LEVITT 1972). Furthermore, in protoplasts isolated from nonacclimated tissue, the probability of intracellular ice formation remains minimal at rates up to 180 °C h^{-1} (DOWGERT and STEPONKUS 1979). This would speak very directly to the question of whether limited plasma membrane water permeability is primarily responsible for the increased incidence of intracellular ice formation in nonacclimated tissues that occurs at much slower rates in intact tissues.

Since the plasma membrane serves as an effective barrier to nucleation of the intracellular solution, avoidance of intracellular ice formation may result from a plasma membrane alteration which increases its effectiveness as such a barrier. STOUT et al. (1977) suggest that the resistance to water efflux may be controlled by heat transfer mechanisms that are influenced by freezing rates. As there may be extracellular factors which alter the rate of freezing (OLIEN 1967), the incidence of intracellular factors may be altered accordingly. Thus, there are several ways to explain the observation that acclimated cells survive faster cooling rates than do nonacclimated cells.

Extracellular Ice Formation. Whereas the previous section has addressed alterations which influence the location of ice formation, it is suggested that alterations in the patterns and types of extracellular ice formation may also occur (OLIEN 1967). Various cell wall mucilages can affect the structure of ice masses in the vicinity of cell walls with smaller and more imperfect crystals observed in hardier tissues (OLIEN 1967). Also, the arabo-xylan polymers can influence the location and rate of extracellular ice propagation within tissues (OLIEN 1974). Whether the polymers are specifically associated with the cold acclimation process or whether they are constitutive characteristics has been discussed (STEPONKUS 1978).

Cellular Dehydration. The most common cellular alteration that occurs during cold acclimation is an increase in osmotically active substances which decreases the solute potential of the intracellular solution. Hence less water will have to be removed from the cell to achieve thermodynamic equilibrium and the extent of cellular dehydration and extracellular ice formation is reduced accordingly (CHANDLER 1913; MAXIMOV 1929). Specifically, a doubling of the internal concentration will decrease the extent of cellular dehydration by 50% at any given subzero temperature. Thus, the secondary stresses associated with cellular volumetric changes, i.e., changes in cell surface area or linear spatial arrangement of cellular membranes, would be effectively mitigated.

Toxic Solute Concentrations. While increases in solute concentration during cold acclimation will mitigate volumetric decreases at subzero temperatures, the final solute concentration will be unaffected. If, however, injury results from the concentration of specific toxic solutes, then an increase in cellular solutes may confer protection by nonspecific, colligative dilution of the toxic compounds (LOVELOCK 1957; MAZUR 1969; HEBER and SANTARIUS 1973). The requirement that the solute whose concentration is increased during acclimation be a neutral or protective substance is not even necessary. Moderate increases in the concentration of several solutes, which individually may be toxic at the high concentrations incurred at subzero temperatures, may colligatively serve to preclude the attainment of a critical concentration of any one species. SANTARIUS (1973) considers that sugars may also confer protection by a direct interaction with cellular membranes, perhaps as envisaged by STEPONKUS (1971). Observations that on a molar basis certain sugars are more effective in conferring protection against freezing damage to isolated chloroplast thylakoids (HEBER and SANTARIUS 1973; STEPONKUS et al. 1977) are the basis for this supposition. The differential cryoprotection appears, however, to be due to nonideal activity at the extremely high concentrations experienced during freezing (LINEBERGER and STEPONKUS 1980). When cryoprotection is analyzed as a function of the mole fraction of NaCl to which the thylakoids are exposed during freezing, protection of cyclic photophosphorylation and its component reactions is not dependent upon the chemical identity of the protective sugar. HEBER (1968) has also indicated that mitigation of toxic substances may also be effected by cryoprotective proteins in a manner other than a colligative reduction of toxic compounds. Proteins isolated from hardy tissues can protect isolated thylakoid membranes against freezing injury (HEBER and ERNST 1967; HEBER 1968, 1970). The mechanism of this protection remains to be elucidated.

12.4.5.2 Increased Tolerance of Cellular Stresses

In addition to factors which mitigate the intensity of cellular stresses, cold acclimation may also involve changes in cellular membranes so that susceptibility to the cellular stresses is decreased. In 1937, SCARTH and LEVITT suggested that cold acclimation may render the plasma membrane more tolerant of freezing stresses, i.e., more resistant to dehydration and less easily ruptured by deplasmolysis or tension (SCARTH et al. 1940; SIMINOVITCH and LEVITT 1941). Recently, STEPONKUS et al. (1979b) have reported that cold acclimation increases the tolerance of the plasma membrane to contraction-expansion-induced stresses. The tolerable surface area increment increases from 1,000 μm^2 in nonacclimated rye protoplasts to 3,000 μm^2 in acclimated protoplasts. In nonacclimated protoplasts this value is exceeded when the cells are frozen below -2 °C and subsequently thawed, whereas in acclimated protoplasts the corresponding value approaches the surface area change which would obtain if all of the osmotically active water were removed. The nature of membrane changes following cold acclimation remains undetermined, but change in membrane ultrastructure have been directly observed by freeze-fracture electron microscopy (GARBER and STEPONKUS 1976b) or inferred from altered functional characteristics (WIEST and STEPONKUS 1977; STOUT et al. 1978).

12.4.6 Conclusions

Freezing of plant tissues produces a sequence of cellular stresses which result in various membrane lesions depending on the freezing protocol and the extent of acclimation of the tissue. Cold acclimation results in both mitigation of the cellular stresses and decreased sensitivity of the membranes to the cellular stresses. Neither freezing injury nor cold acclimation can be ascribed to any singular factor under all conditions or at different stages of acclimation.

References

Ageeva OG (1977) Effects of light on thermostability of Hill reaction in pea and spinach chloroplasts. Photosynthetica 11:1–4

Alden J, Hermann RK (1971) Aspects of the cold-hardiness mechanism in plants. Bot Rev 37:37–142

Alexandrov VY (1964) Cytophysiological and cytoecological investigations of heat resistance of plant cells toward the action of high and low temperature. Q Rev Biol 39:35–77

Alexandrov VY (1967) A study of the changes in resistance of plant cells to the action of various agents in the light of cytoecological considerations. In: Troshin AS (ed) The Cell and Environmental Temperature. Pergamon Press, New York, pp. 142–151

Alexandrov VY (1977) Cells, molecules and temperature. Conformational flexibility of macromolecules and ecological adaptation. Ecological Studies, Vol. 21, Springer, Berlin, Heidelberg, New York

Alexandrov VY, Lomagin AG, Feldman NL (1970) The responsive increase in thermostability of plant cells. Protoplasma 69:417–458

Anderson MM, McCarty RE, Zimmer EA (1974) The role of galactolipids in spinach chloroplast lamellar membranes. I. Partial purification of a bean leaf galactolipid lipase and its action on subchloroplast particles. Plant Physiol 53:699–704

Bagnall DJ, Wolfe JA (1978) Chilling sensitivity in plants: Do activation energies of growth processes show an abrupt change at a critical temperature? J Exp Bot 29:1331–1342

Beevers L, Hanson JB (1964) Oxidative phosphorylation by root and shoot mitochondria from corn seedlings as affected by temperature. Crop Sci 4:549–550

Bělehrádek J (1935) Temperature and Living Matter. Protoplasma Monographien. Borntraeger, Berlin, 8:50–54

Bělehrádek J (1957) Physiological aspects of heat and cold. Ann Rev Physiol 19:59–82

Berry J, Björkman O (1980) Photosynthetic response and adaptation to temperature in higher plants. Ann Rev Plant Physiol 31:491–543

Berry JA, Fork DC, Garrison S (1975) Mechanistic studies of thermal damage to leaves. Carnegie Inst Washington Yearb 74:751–759

Björkman O (1975) Thermal stability of the photosynthetic apparatus in intact leaves. Carnegie Inst Washington Yearb 74:748–751

Björkman O, Badger M, Armond PA (1978) Thermal acclimation of photosynthesis: effect of growth temperature on photosynthetic characteristics and components of the photosynthetic apparatus in *Nerium oleander*. Carnegie Inst Washington Yearb 77:262–276

Björkman O, Badger M, Armond P (1980) Response and adaptation of photosynthesis to high temperatures. In: Turner NC, Kramer PJ (eds) Adaptation of plants to water and high temperature stress. Wiley-Interscience, New York, pp 233–249

Bonner J (1957) The chemical cure of climatic lesions. Eng Sci 20:28–30

Borochov A, Halevy AH, Borochov H, Shinitzky M (1978) Microviscosity of plasmalemmas in rose petals as affected by age and environmental factors. Plant Physiol 61:812–815

Breidenbach RW, Wade NL, Lyons JM (1974) Effect of chilling temperatures on the activities of glyoxysomal and mitochondrial enzymes from castor bean seedlings. Plant Physiol 54:324–327

Brock TD (1967) Life at high temperatures. Science 158:1012–1019

Burke MJ, Gusta LV, Quamme HA, Weiser CJ, Li PH (1976) Freezing and injury in plants. Ann Rev Plant Physiol 27:507–528

Cannon B, Polnaszek CF, Butler KW, Eriksson LEG, Smith ICP (1975) The fluidity and organization of mitochondrial membrane lipids of the brown adipose tissue of cold-adapted rats and hamsters as determined by nitroxide spin probes. Arch Biochem Biophys 167:505–518

Chandler WH (1913) The killing of plant tissue by low temperature. Mo Agr Expt Stn Bull 8:141–309

Cole FD (1969) Thermal inactivation of plant growth. Ph. D. Thesis, Univ Arizona, Tucson, 83 pp

Cole FD, Steponkus PL (1968) Some effects of high temperature on plant growth. Plant Physiol 43:S-32

Creecia RP, Bramlage WJ (1971) Reversibility of chilling injury to corn seedlings. Plant Physiol 47:389–392

Critchley C, Smillie RM, Patterson BD (1978) Effect of temperature on photoreductive activity of chloroplasts from passionfruit species of different chilling sensitivity. Aust J Plant Physiol 5:433–438

Das RM, Hildebrandt AC, Riker AJ (1966) Cinephotography of temperature effects on cytoplasmic streaming, nucleolar activity and mitosis in single tobacco cells in microculture. Am J Bot 53:253–259

Dowgert MF, Steponkus PL (1979) Responses of cereal protoplasts to a freezethaw cycle: I. Dehydration *vs* intracellular ice formation. Plant Physiol 63:S36

Emmett JM, Walker DA (1969) Thermal uncoupling in chloroplasts. Biochim Biophys Acta 180:424–425

Emmett JM, Walker DA (1973) Thermal uncoupling in chloroplasts. Inhibition of photophosphorylation without depression of light-induced pH change. Arch Biochem Biophys 157:106–113

Engelbrecht L, Mothes K (1960) Kinetin als Faktor der Hitzeresistenz. Ber Dtsch Bot Ges 73:246–257

Engelbrecht L, Mothes K (1964) Weitere Untersuchungen zur experimentellen Beeinflussung der Hitzewirkung bei Blättern von *Nicotiana rustica*. Flora (Jena) 154:279–298

Faris JA (1926) Cold chlorosis of sugar cane. Phytopathology 16:885–891

Feierabend J, Mikus M (1977) Occurrence of a high temperature sensitivity of chloroplast ribosome formation in several higher plants. Plant Physiol 59:863–867

Feierabend J, Schrader-Reichhardt U (1976) Biochemical differentiation of plastids and other organelles in rye leaves with a high-temperature-induced deficiency of plastid ribosomes. Planta 129:133–145

Feldman NL (1968) The effects of heat hardening on the heat resistance of some enzymes from plant leaves. Planta 78:213–225

Feldman NL, Alexandrov VY, Zavadskaya IG, Kislyuk IM, Lomagin AG, Lyutova MI, Jaskuliev (1967) Heat hardening of plant cells under natural and experimental conditions. In: Troshin AS (ed) The cell and environmental temperature. Pergamon Press, New York, pp 152–160

Friend DJ, Helson VA, Fisher JE (1962) The rate of dry weight accumulation in Marguis wheat as affected by temperature and light intensity. Can J Bot 40:939–955

Galston AW, Hand ME (1949) Adenine as a growth factor for etiolated peas and its relation to the thermal inactivation of growth. Arch Biochem 22:434–443

Garber MP (1977) Effect of light and chilling temperatures on chilling-sensitive and chilling-resistant plants. Pretreatment of cucumber and spinach thylakoids in vivo and in vitro. Plant Physiol 59:981–985

Garber MP (1980) Low temperature response of chloroplast thylakoids. In: Lyons JM, Raison JK, Graham D (eds) Low temperature stress in crop plants: the role of the membrane, Academic Press, New York, pp 203–214

Garber MP, Steponkus PL (1976a) Alterations in chloroplast thylakoids during an *in vitro* freeze-thaw cycle. Plant Physiol 57:673–680

Garber MP, Steponkus PL (1976b) Alterations in chloroplast thylakoids during cold acclimation. Plant Physiol 57:681–686

Geronimo J, Beevers H (1964) Effects of aging and temperature on respiratory metabolism of green leaves. Plant Physiol 39:786–793

Heber U (1967) Freezing injury and uncoupling of phosphorylation from electron transport in chloroplasts. Plant Physiol 42:1343–1350

Heber U (1968) Freezing injury in relation to loss of enzyme activities and protection against freezing. Cryobiology 5:188–201

Heber U (1970) Proteins capable of protecting chloroplast membranes against freezing. In: Ciba Found Symp on The Frozen Cell. Churchill, London, pp 175–188

Heber U, Ernst R (1967) A biochemical approach to the problem of frost injury and frost hardiness. In: Asahina E (ed) Cellular injury and resistance in freezing organisms. Volume II, Inst Low Temp Sci Sapporo, Japan, pp 63–77

Heber U, Santarius KA (1964) Loss of adenosine triphosphate synthesis caused by freezing and its relationship to frost hardiness problems. Plant Physiol 39:712–719

Heber U, Santarius KA (1973) Cell death by cold and heat and resistance to extreme temperatures. Mechanisms of hardening and dehardening. In: Precht H, Christopherson J, Hensel H, Larcher W (eds) Temperature and life. Springer Berlin, Heidelberg, New York, pp 232–263

Highkin HR (1957) The relationship between temperature resistance and purine and pyrimidine composition in peas. Plant Physiol 32:S1

Ilker R, Waring AJ, Lyons JM, Breidenbach RW (1976) The cytological responses of tomato-seedling cotyledons to chilling and the influence of membrane modifications upon these responses. Protoplasma 90:229–252

Ilker R, Breidenbach RW, Lyons JM (1980) Sequence of ultrastructural changes in tomato cotyledons during short periods of chilling. In: Lyons JM, Raison JK, Graham D (eds) Low temperature stress in crop plants: the role of the membrane. Academic Press, New York, pp 97–113

Inoue H (1978) Break points in Arrhenius plots of the Hill reaction of spinach chloroplast fragments in the temperature range from −25 to 25 °C. Plant Cell Physiol 19:355–363

Itai C, Benzioni A (1974) Regulation of plant response to high temperature. In: Bieleski RL, Ferguson AR, Cresswell MM (eds) Mechanisms of regulation of plant growth. Roy Soc N Z Bull No 12, pp 477–482

Kaniuga A, Michalski W (1978) Photosynthetic apparatus in chilling-sensitive plants. II. Changes in free fatty acid composition and photoperoxidation in chloroplasts following cold storage and illumination of leaves in relation to Hill activity. Planta 140:129–136

Kaniuga A, Sochanowicz B, Zabek J, Krzystyniak K (1978) Photosynthetic apparatus in chilling-sensitive plants. I. Reactivation of Hill reaction activity inhibited on the cold and dark storage of detached leaves and intact plants. Planta 140:121–128

Ketellapper HJ (1963) Temperature-induced chemical defects in higher plants. Plant Physiol 38:175–179

Kinbacher EJ (1970) Relative thermal stability of malic dehydrogenase from heat-hardened and unhardened *Phaseolus* sp. Crop Sci 10:181–184

Kinbacher EJ, Sullivan CY (1967) Effect of high temperature on the respiration rate of *Phaseolus* sp. Proc Am Soc Hortic Sci 90:163–168

Kinbacher EJ, Sullivan CY, Knull HR (1967) Thermal stability of malic dehydrogenase from heat-hardened *Phaseolus acutifolius* 'Tepary Buff'. Crop Sci 7:148–151

Krause GH, Santarius KA (1975) Relative thermostability of the chloroplast envelope. Planta 127:285–299

Krogmann DW, Jagendorf AT (1959) Inhibition of the Hill reaction by fatty acids and metal chelating agents. Arch Biochem Biophys 80:421–430

Kumamoto J, Raison JK, Lyons JM (1971) Temperature 'breaks' in Arrhenius plots: a thermodynamic consequence of a phase change. J Theor Biol 31:47–51

Kumm J, French CS (1945) The evolution of oxygen from suspensions of chloroplasts; the activity of various species and the effects of previous illumination of the leaves. Am J Bot 32:291–295

Kurtz EB (1958) Chemical basis for adaptation in plants. Science 128:1115–1117

Kurtz EB (1961) A chemical basis for the adaptation of plants. In: Shields LM, Gardner JL (eds) Bioecology of the arid and semiarid lands of the Southwest. New Mexico Highlands Univ Bull 212, pp 23–26

Lange OL (1967) Investigations on the variability of heat resistance in plants. In: Troshin AS (ed) The cell and environmental temperature, Pergamon Press, New York, pp 131–141

Langridge J (1963) Biochemical aspects of temperature response. Ann Rev Plant Physiol 14:441–462

Langridge J (1965) Temperature-sensitive, vitamin-requiring mutants of *Arabidopsis thaliana*. Aust J Biol Sci 18:311–321

Langridge J, Griffing B (1959) A study of high temperature lesions *Arabadopsis thaliana*. Aust J Biol Sci 12:117–135

Langridge J, McWilliam JR (1967) Heat responses of higher plants. In: Rose AH (ed) Thermobiology. Academic Press, New York, pp 231–292

Larcher W (1973) Limiting temperatures for life functions. In: Precht H, Christopherson J, Hensel H, Larcher W (eds) Temperature and life. Springer, Berlin, Heidelberg, New York, pp 195–231

Leitsch I (1916) Some experiments on the influence of temperature on the rate of growth of *Pisum sativum*. Ann Bot 30:25–46

Leopold AC, Kriedemann PE (1975) Plant growth and development. 2nd edn. McGraw-Hill, New York

Letellier L, Moudden H, Shechter E (1977) Lipid and protein segregation in *Escherichia coli* membrane: morphological and structural study of different cytoplasmic membrane fractions. Proc Nat Acad Sci USA 74:452–456

Levitt J (1972) Responses of plants to environmental stresses. Academic Press, New York

Levitt J (1978) An overview of freezing injury and survival, and its interrelationships to other stresses. In: Li PH, Sakai A (eds) Plant cold hardiness and freezing stress. Academic Press, New York, pp 3–15

Levitt J, Scarth GW (1936a) Frost hardening studies with living cells. I. Osmotic and bound water changes in relation to frost resistance and the seasonal cycle. Can J Res C 14:267–284

Levitt J, Scarth GW (1936b) Frost hardening studies with living cells II. Permeability in relation to frost resistance and the seasonal cycle. Can J Res C 14:285–305

Levitt J, Siminovitch D (1940) The relation between frost resistance and the physical state of the protoplasm. I. The protoplasm as a whole. Can J Res C 18:550–561

Lewis DA (1956) Protoplasmic streaming in plants sensitive and insensitive to chilling temperatures. Science 124:75–76

Lewis TL, Workman M (1964) The effect of low temperature on phosphate esterification and cell membrane permeability in tomato fruit and cabbage leaf tissue. Aust J Biol Sci 17:147–152

Li PH, Sakai A (1978) Plant cold hardiness and freezing stress – mechanisms and crop implications. Academic Press, New York

Linden CD, Wright KL, McConnell HM, Fox CF (1973) Lateral phase separations in membrane lipids and the mechanism of sugar transport in *Escherichia coli*. Proc Nat Acad Sci USA 70:2271–2275

Lineberger RD, Steponkus PL (1980) Cryoprotection by glucose, sucrose and raffinose to chloroplast thylakoids. Plant Physiol 65:298–304

Lovelock JE (1953) The haemolysis of human red blood cells by freezing and thawing. Biochim Biophys Acta 10:414–426

Lovelock JE (1957) The denaturation of lipid-protein complexes as a cause of damage by freezing. Proc R Soc London Ser A 147:427–33

Lyons JM (1973) Chilling injury in plants. Ann Rev Plant Physiol 24:445–466

Lyons JM, Asmundson CM (1965) Solidification of unsaturated/saturated fatty acid mixtures and its relationship to chilling sensitivity in plants. J Am Oil Chem Soc 42:1056–1058

Lyons JM, Breidenbach RW (1979) Strategies for altering chilling sensitivity as a limiting factor in crop production. In: Mussell H, Staples RC (eds) Stress physiology in crop plants. Wiley-Interscience, New York, pp 179–196

Lyons JM, Raison JK (1970) Oxidative activity of mitochondria isolated from plant tissue sensitive and resistant to chilling injury. Plant Physiol 45:386–389

Lyons JM, Wheaton TA, Pratt HK (1964) Relationship between the physical nature of mitochondrial membranes and chilling sensitivity in plants. Plant Physiol 39:262–269

Lyons JM, Raison JK, Graham D (1980a) Low temperature stress in crop plants: The role of the membrane. Academic Press, New York

Lyons JM, Raison JK, Steponkus PL (1980b) The plant membrane in response to low temperature: an overview. In: Lyons JM, Raison JK, Graham D (eds) Low temperature stress in crop plants: The role of the membrane. Academic Press, New York, pp 1–24

Margulies MM (1972) Effect of cold storage of bean leaves on photosynthetic reactions of isolated chloroplasts. Inability to donate electrons to photosystem II and relation to manganese content. Biochim Biophys Acta 267:96–103

Margulies MM, Jagendorf AT (1960) Effect of cold storage of bean leaves on photosynthetic reactions of isolated chloroplasts. Arch Biochem Biophys 90:176–183

Maximov NA (1912) Chemische Schutzmittel der Pflanzen gegen Erfrieren. Ber Dtsch Bot Ges 30:52–65; 293–305; 504–516

Maximov NA (1929) Internal factors of frost and drought resistance in plants. Protoplasma 7:259–291

Mazur P (1969) Freezing injury in plants. Ann Rev Plant Physiol 20:419–448

Mazur P (1970) Cryobiology: The freezing of biological systems. Science 168:939–949

Mazur P (1977a) The role of intracellular freezing in the death of cells cooled at supraoptimal rates. Cryobiology 14:251–272

Mazur P (1977b) Slow-freezing injury in mammalian cells. Proc Ciba Foundation Symposium on the freezing of mammalian embryos, London, pp 19–48

McCarty RE, Jagendorf AT (1965) Chloroplast damage due to enzymatic hydrolysis of endogenous lipids. Plant Physiol 40:725–735

McGlasson WB, Raison JK (1973) Occurrence of a temperature-induced phase transition in mitochondria isolated from apple fruit. Plant Physiol 52:390–392

McKersie BD, Thompson JE (1977) Lipid crystallization in senescent membranes from cotyledons. Plant Physiol 59:803–807

McKersie BD, Thompson JE (1979) Phase properties of senescing plant membranes. Biochim Biophys Acta 550:48–58

McMurchie EJ (1980) Temperature sensitivity of ion-stimulated ATPases associated with some plant membranes. In: Lyons JM, Raison JK, Graham D (eds) Low temperature stress in crop plants: The role of the membrane. Academic Press, New York, pp 163–176

McNaughton SJ (1965) Differential enzymatic activity in ecological races of *Typha latifolia* L. Science 150:1829–1830

McNaughton SJ (1966) Thermal inactivation properties of enzymes from *Typha latifolia* L. ecotypes. Plant Physiol 41:1736–1738

McWilliam JR, Naylor AW (1967) Temperature and plant adaptation. I. Interaction of temperature and light in the synthesis of chlorophyll in corn. Plant Physiol 42:1711–1715

McWilliam JR, Manokaran W, Kipnis T (1980) Adaptation to chilling stress in sorghum. In: Lyons JM, Raison JK, Graham D (eds) Low temperature stress in crop plants: The role of the membrane. Academic Press, New York, pp 491–505

Meryman HT (1968) Modified model for the mechanism of freezing injury in erythrocytes. Nature 218:333–336

Meryman HT (1971) Osmotic stress as a mechanism of freezing injury. Cryobiology 8:488–500

Meryman HT (1974) Freezing injury and its prevention in living cells. Ann Rev Biophys 3:341–363

Meryman HT, Williams RJ, Douglas MStJ (1977) Freezing injury from "solution effects": and its prevention by natural or artificial cryoprotection. Cryobiology 14:287–302

Miller RW, de la Roche IA, Pomeroy MK (1974) Structural and functional responses of wheat mitochondrial membranes to growth at low temperatures. Plant Physiol 53:426–433

Minamikawa T, Akazawa T, Uritani I (1961) Mechanism of cold injury in sweet potatoes. II. Biochemical mechanism of cold injury with species reference to mitochondrial activities. Plant Cell Physiol 2:301–309

Moline HE (1976) Ultrastructural changes associated with chilling of tomato fruit. Phytopathology 66:617–624

Molotkovsky YG, Zhestkova IM (1965) The influence of heating on the morphology and photochemical activity of isolated chloroplasts. Biochem Biophys Res Commun 20:411–415

Mukohata Y, Yagi T, Higashida M, Shinozaki K, Matsuno A (1973) Biophysical studies on subcellular particles. VI. Photosynthetic activities in isolated spinach chloroplasts after transient warming. Plant Cell Physiol 14:111–118

Murata N, Troughton JH, Fork DC (1975) Relationships between the transition of the physical phase of membrane lipids and photosynthetic parameters in *Anacystis nidulans* and lettuce and spinach chloroplasts. Plant Physiol 56:508–517

Niki T, Yoshida S, Sakai A (1978) Studies on chilling injury in plant cells. I. Ultrastructural changes associated with chilling injury in callus tissues of *Cornus stolonifera*. Plant Cell Physiol 19:139–148

Nobel PS (1974) Temperature dependence of the permeability of chloroplasts from chilling-sensitive and chilling-resistant plants. Planta 115:369–372

Nolan WG, Smillie RM (1976) Multitemperature effects of Hill reaction activity of barley chloroplasts. Biochem Biophys Acta 440:461–475

Nolan WG, Smillie RM (1977) Temperature-induced changes in Hill activity of chloroplasts isolated from chilling-sensitive and chilling-resistant plants. Plant Physiol 59:1141–1145

Olien CR (1967) Freezing stresses and survival. Ann Rev Plant Physiol 18:387–408

Olien CR (1974) Energies of freezing and frost desiccation. Plant Physiol 53:764–767

Olien CR (1977) Barley: patterns of response to freezing stress. USDA Tech Bull No 1558

Olienikova TV (1967) Effects of high temperature and light on the thermostability of cells of different crop varieties. In: Troshin AS (ed) The cell and environmental temperature. Pergamon Press, New York, pp 173–177

Palta JP, Li PH (1978) Cell membrane properties in relation to freezing injury. In: Li PH, Sakai A (eds) Plant cold hardiness and freezing stress – mechanisms and crop implications, Academic Press, New York, pp 93–115

Patterson BD, Graham D (1977) Effect of chilling temperatures on the protoplasmic streaming of plants from different climates. J Exp Bot 28:736–743

Patterson BD, Kenrick JR, Raison JK (1978) Lipids of chill-sensitive and -resistant *Passiflora* species: Fatty acid composition and temperature dependence of spin label motion. Phytochemistry 17:1089–1092

Patterson BD, Graham D, Paull R (1980) Adaptation to chilling: survival, germination, respiration and protoplasmic dynamics. In: Lyons JM, Raison JK, Graham D (eds) Low temperature stress in crop plants: The role of the membrane. Academic Press, New York, pp 25–35

Pearcy RW, Berry JA, Fork DC (1977) Effects of growth temperature on the thermal stability of the photosynthetic apparatus of *Atriplex lentiformis* (Torr.) Wats. Plant Physiol 59:873–878

Pike CS, Berry JA, Raison JK (1980) Fluorescence polarization studies of membrane phospholipid phase separations on warm and cool climate plants. In: Lyons JM, Raison JK, Graham D (eds) Low temperature stress in crop plants: The role of the membrane. Academic Press, New York, pp 305–318

Pomeroy MK, Andrews CJ (1975) Effect of temperature on respiration of mitochondria and shoot segments from cold-hardened and non-hardened wheat and rye seedlings. Plant Physiol 56:703–706

Quinn PJ, Williams WP (1978) Plant lipids and their role in membrane function. Prog Biophys Mol Biol 34:109–173

Raison JK (1973a) Temperature-induced phase changes in membrane lipids and their influence on metabolic regulation. Symp Soc Exp Biol 27:485–512

Raison JK (1973b) The influence of temperature-induced phase changes on the kinetics of respiratory and other membrane-associated enzyme systems. Bioenergetics 4:285–309

Raison JK (1974) A biochemical explanation of low-temperature stress in tropical and sub-tropical plants. In: Bieleski RL, Ferguson AR, Cresswell MM (eds) Mechanisms of regulation of plant growth. R Soc N Z B No 12, pp 487–497

Raison JK, Berry JA (1978) The physical properties of membrane lipids in relation to the adaptation of higher plants and algae to contrasting thermal regimes. Carnegie Inst Washington Yearb 77:276–282

Raison JK, Chapman EA (1976) Membrane phase changes in chilling-sensitive *Vigna radiata* and their significance to growth. Aust J Plant Physiol 3:291–299

Raison JK, McMurchie EJ (1974) Two temperature-induced changes in mitochondrial membranes detected by spin labelling and enzyme kinetics. Biochim Biophys Acta 363:135–140

Raison JK, Lyons JM, Thomson WW (1971a) The influence of membranes on the temperature-induced changes in the kinetics of some respiratory enzymes of mitochondria. Arch Biochem Biophys 142:83–90

Raison JK, Lyons JM, Mehlhorn RJ, Keith AD (1971b) Temperature-induced phase changes in mitochondrial membranes detected by spin labelling. J Biol Chem 246:4036–4040

Raison JK, Chapman EA, White PY (1977) Wheat mitochondria – oxidative activity and membrane lipid structure as a function of temperature. Plant Physiol 59:623–627

Raison JK, Berry JA, Armond PA, Pike CS (1980) Membrane properties in relation to the adaptation of plants to temperature stress. In: Turner NC, Kramer PJ (eds) Adaptation of plants to water and high temperature stress. Wiley-Interscience, New York, pp 261–273

Rajashekar C, Gusta LV, Burke MJ (1980) Membrane structural transitions: probable relation to frost damage in hardy herbaceous species. In: Lyons JM, Raison JK, Graham D (eds) Low temperature stress in crop plants. Academic Press, New York, pp 255–274

Rameshwar A (1971) High temperature injury in higher plants. Ph. D. Thesis, Cornell Univ, Ithaca, New York, 136 pp

Rameshwar A, Steponkus PL (1969) Reduction of high temperature injury in intact plants by cytokinins. Plant Physiol 44:S2

Santarius KA (1973) The protective effect of sugars on chloroplast membranes during temperature and water stress and its relationship to frost, desiccation and heat resistance. Planta 113:105–114

Santarius KA (1974) Seasonal changes in plant membrane stability as evidenced by the heat sensitivity of chloroplast membrane reactions. Z Pflanzenphysiol Bd 73S:448–451

Santarius KA (1975) Site of heat sensitivity in chloroplasts and differential inactivation of cyclic and noncyclic photophosphorylation by heating. J Thermal Biol 1:101–107

Santarius KA, Müller M (1979) Investigations on heat resistance of spinach leaves. Planta 146:529–538

Scarth GW (1941) Dehydration injury and resistance. Plant Physiol 16:171–179

Scarth GW, Levitt J (1937) The frost-hardening mechanism of plant cells. Plant Physiol 12:51–78

Scarth GW, Levitt J, Siminovitch D (1940) Plasma-membrane structure in the light of frost-hardening changes. Cold Spring Harbor Symp 8:102–109

Schäfers HA, Feierabend J (1976) Ultrastructural differentiation of plastids and other organelles in rye leaves with a high-temperature-induced deficiency of plastid ribosomes. Cytobiologie 4:75–90

Schreiber U, Armond PA (1978) Heat-induced changes of chlorophyll fluorescence in isolated chloroplasts and related heat-damage at the pigment level. Biochem Biophys Acta 502:138–151

Schreiber U, Berry JA (1977) Heat-induced changes of chlorphyll fluorescence in intact leaves correlated with damage of the photosynthetic apparatus. Planta 136:233–238

Schroeder CA (1963) Induced temperature tolerance of plant tissue *in vitro*. Nature 200:1301–1302

Schroeder CA (1967) Induction of temperature tolerance in excised plant tissue. In: Prosser CL (ed) Molecular mechanisms of temperature adaptation. Publ No 84, Amer Assoc Adv Sci, Washington, DC, pp 61–72

Semichatova OA, Bushuyeva TM, Nikulina GN (1967) The effect of temperature on respiration and oxidative phosphorylation of pea seedlings. In: Troshin AS (ed) The cell and environmental temperature. Pergamon Press, New York, pp 283–387

Shneyour A, Raison JK, Smillie RM (1973) The effect of temperature on the rate of photosynthetic electron transfer in chloroplasts of chilling-sensitive and chilling-resistant plants. Biochim Biophys Acta 292:152–161

Siminovitch D, Levitt J (1941) The relationship between frost resistance and the physical state of protoplasm. II. The protoplasmic surface. Can J Res 19:9–20

Siminovitch D, Scarth GW (1938) A study of the mechanism of frost injury to plants. Can J Res 16:467–481

Simon EW (1974) Phospholipids and plant membrane permeability. New Phytol 73:377–420

Skogqvist I, Fries N (1970) Induction of thermosensitivity and salt sensitivity in wheat roots and the effect of kinetin. Experimentia 26:1160–1162

Smillie RM (1976) Temperature control of chloroplast development. In: Bucher TH, Neupert W, Sebald W, Werner S (eds) Genetics and biogenesis of chloroplasts and mitochondria. Elsevier, North Holland, Amsterdam, New York, pp 103–110

Smillie RM (1979) Coloured components of chloroplast membranes as intrinsic membrane probes for monitoring the development of heat injury in intact tissues. Aust J Plant Physiol 6:121–133

Smillie RM, Nott R (1979a) Assay of chilling injury in wild and domestic tomatoes based on photosystem activity of chilled leaves. Plant Physiol 63:796–801

Smillie RM, Nott R (1979b) Heat injury in leaves of alpine, temperate and tropical plants. Aust J Plant Physiol 6:135–141

Steponkus PL (1971) Cold acclimation of *Hedera helix*: Evidence for a two phase process. Plant Physiol 47:175–180

Steponkus PL (1978) Cold hardiness and freezing injury of agronomic crops. Adv Agron 30:51–98

Steponkus PL (1979) Effects of freezing and cold acclimation on membrane structure and function. In: Mussell H, Staples RC (eds) Stress physiology of crop plants. Wiley-Interscience, New York, pp 143–158

Steponkus PL (1980) A unified concept of stress in plants? In: Rains DW, Valentine RC, Hollaender A (eds) Genetic engineering of osmoregulation. Plenum, New York, pp 235–255

Steponkus PL, Wiest SC (1978) Plasma mebrane alterations following cold acclimation and freezing. In: Li PH, Sakai A (eds) Plant cold hardiness and freezing stress – mechanisms and crop implications. Academic Press, New York, pp 75–91

Steponkus PL, Wiest SC (1980) Freeze-thaw induced lesions in the plasma mebrane. In: Lyons JM, Raison JK, Graham D (eds) Low temperature stress in crop plants: The role of the membrane. Academic Press, New York, pp 231–254

Steponkus PL, Rameshwar A, Cole FD (1970) Thermal inactivation of plant growth. Proc. XVIII Int Hortic Cong I:63

Steponkus PL, Garber MP, Myers SP, Lineberger RD (1977) Effects of cold acclimation and freezing on structure and function of chloroplast thylakoids. Cryobiology 14:303–321

Steponkus PL, Dowgert MF, Levin RL, Ferguson JF (1979a) Cryobiology of isolated plant protoplasts: IV cellular injury. Cryobiology 16:593–594

Steponkus PL, Dowgert MF, Roberts SR (1979b) Cryobiology of isolated plant protoplasts: VI. Influence of cold acclimation. Cryobiology 16:594

Stout DG, Steponkus PL, Cotts RM (1977) Quantitative study of the importance of water permeability in plant cold hardiness. Plant Physiol 60:374–378

Stout DG, Steponkus PL, Cotts RM (1978) Plasmalemma alteration during cold acclimation of *Hedera helix* bark. Can J Bot 56:196–205

Sullivan CY, Kinbacher EJ (1967) Thermal stability of Fraction I protein from heat-hardened *Phaseolus acutifolius* Gray, 'Tepary Buff'. Crop Sci 7:241–244

Taylor AO, Slack CR, McPherson HG (1974) Plants under climatic stress. VI. Chilling and light effects on photosynthetic enzymes of sorghum and maize. Plant Physiol 54:696–701

Torres-Pereira J, Melhorn R, Keith AD, Packer L (1974) Changes in membrane lipid structure of illuminated chloroplasts studies with spin-labeled and freeze-fractured membranes. Arch Biochem Biophys 160:90–99

Troshin AS (ed) (1967) The cell and environmental temperature. Pergamon Press, New York

van Hasselt PR (1972) Photo-oxidation of leaf pigments in *Cucumis* leaf discs during chilling. Acta Bot Neerl 21:539–548

van Hasselt PR (1974) Photo-oxidation of unsaturated lipids in *Cucumis* leaf discs during chilling. Acta Bot Neerl 23:159–169

Volger H, Heber U, Berzborn RJ (1978) Loss of function of biomembranes and solubilization of membrane proteins during freezing. Biochim Biophys Acta 511:455–469

Wade NL, Breidenbach RW, Lyons JM, Keith AD (1974) Temperature-induced phase changes in the membranes of glyoxysomes, mitochondria and proplastids from germinating castor bean endoperm. Plant Physiol 54:320–323

Went FW (1953) The effect of temperature on plant growth. Ann Rev Plant Physiol 4:347–362

Wiest SC, Steponkus PL (1977) Accumulation of sugars and plasmalemma alterations: factors related to the lack of cold acclimation in young roots. J Am Soc Hortic Sci 102:119–123

Wiest SC, Steponkus PL (1978a) Freeze-thaw injury to isolated spinach protoplasts and its simulation at above-freezing temperatures. Plant Physiol 62:599–605

Wiest SC, Steponkus PL (1978b) Freeze-thaw injury to wheat protoplasts. II. Membrane lesions. Agronomy Abstracts p 88

Wiest SC, Steponkus PL (1979) The osmometric behavior of human erythrocytes. Cryobiology 16:101–104

Wilson JM (1976) The mechanism of chill- and drought-hardening of *Phaseolus vulgaris* leaves. New Phytol 76:257–270

Wilson JM (1980) Drought resistance as related to low temperature stress. In: Lyons JM, Raison SK, Graham D (eds) Low temperature stress in crop plants: The role of the membrane. Academic Press, New York, pp 47–65

Wolfe J (1978) Chilling injury in plants – the role of membrane lipid fluidity. Plant Cell Environ 1:241–247

Wolfe J (1980) Some physical properties of membranes in the phase separation region and their relation to chilling damage in plants. In: Lyons JM, Raison JK, Graham D (eds) Low temperature stress in crop plants: The role of the membrane. Academic Press, New York, pp 327–335

Wolfe J, Bagnall D (1980a) Arrhenius plots – curves or straight lines. Ann Bot 45:485–488

Wolfe J, Bagnall D (1980b) Statistical tests to decide between straight line segments and curves as suitable fits to Arrhenius plots or other data. In: Lyons JM, Raison JK, Graham D (eds) Low temperature stress in crop plants. Academic Press, New York, pp 527–533

Yamaki S, Uritani I (1972) Mechanism of chilling injury in sweet potatoes. Part V. Biochemical mechanism of chilling injury with special reference to mitochondrial lipid components. Agr Biol Chem 36:47–55

Yarwood CE (1961) Acquired tolerance of leaves to heat. Science 134:941–942

Yarwood CE (1967) Adaptations of plants and plant pathogens to heat. In: Prosser CL (ed) Molecular mechanisms of temperature adaptation. Am Ass Adv Sci, Washington, DC, pp 75–89

Yoshida S, Niki T, Sakai A (1980) Possible involvement of the tonoplast lesion in chilling injury of cultured plant cells. In: Lyons JM, Raison JK, Graham D (eds) Low temperature stress in crop plants: The role of the membrane. Academic Press, New York, pp 275–290

Zavadskaya IG (1967) Changes in carbohydrate content of plants under heat hardening. In: Troshin AS (ed) The cell and environmental temperature. Pergamon Press, New York, pp 182–183

13 Ecological Significance of Resistance to Low Temperature

W. LARCHER and H. BAUER

CONTENTS

13.1 Introduction

Low temperatures reduce the biosynthetic activity of plants; they evoke disturbance in vital functions and productivity and they may inflict permanent injuries that finally bring about death. The survival capacity of a plant species or variety in a particular environment is determined by the specific limits to which its metabolic processes continue to function under low temperature stress and by its cold resistance, both of which are characteristics of its ecophysiological constitution (LARCHER 1968).

The topics of cold injury and cold resistance have been dealt with in an extensive body of literature which, according to the estimate of ALDEN and HERMANN (1971), already comprised 600 relevant publications in 1970. In the following decade the number of investigators concerned with cold resistance, and thus also the number of publications in the field, have rapidly increased. The situation in the first half of the century was reviewed by ULLRICH (1943), LEVITT (1956, 1958) and BIEBL (1962a). LEVITT (1980) provided a comprehensive account of cold resistance as a component of environmental resistance. Important results are reviewed and discussed by PARKER (1963), OLIEN (1967), MAZUR (1969), WEISER (1970), ALDEN and HERMANN (1971), HEBER and SANTARIUS (1973), LARCHER (1973, 1981a, 1981b), LYONS (1973), BURKE et al. (1976), STEPONKUS (1978), BURKE and STUSHNOFF (1979), CHRISTIANSEN (1979) and TUMANOV (1979). Proceedings of actual research on low temperature stress effects have been edited by LI and SAKAI (1978) and LYONS et al. (1979). Data lists on cold resistance are edited by ALTMAN and DITTMER (1973).

Few comprehensive accounts of impairment of life functions and productivity as a result of cold stress, in particular due to frost, are available. References can be found in LEVITT (1969, 1980), LARCHER (1973, 1981a) and BAUER et al. (1975).

The importance of the ecological aspects of low temperature effects was early recognized by ARTHUR PISEK *(1894–1975) and* RICHARD BIEBL *(1908–1974). They established ecological resistance research as we know it today, and it is to their memory that this chapter is dedicated.*

Terminology:
Low temperature stress: Any drop in temperature that can evoke reversible or irreversible functional disturbances or lethal injuries.
Chilling: The action of low positive temperatures.
Frost: Environmental temperatures below 0 °C.
Cold resistance: Ability to resist low temperature stress (chilling, frost) without injury. Contrary: *Cold susceptibility.*
Freezing resistance: Avoidance or tolerance of ice formation in plant tissues (LEVITT 1958).
Survival capacity: Ability of a plant to withstand the complex environmental stresses during adverse weather conditions (LARCHER 1973).

13.2 Cold as an Environmental Factor

Cold is a stress factor of widespread occurrence. The only places on the Earth where temperatures do not drop below +10 °C are the continually warm, humid lowland regions of the Amazon Basin, the Congo Basin, and parts of southeast Asia (see Fig. 13.7). For 64% of the Earth's land mass the mean minimum air temperature is below 0 °C, for 48% it is below −10 °C (HOFFMANN 1963). In Siberia, Alaska, and northwestern Canada, where trees still form stands, the temperature regularly sinks to −50 °C. An absolute temperature minimum of −68 °C has been recorded for Verkhoyansk, a Siberian lowland station, and one of −71 °C for Oimyakon, also in Eastern Siberia but situated in a high valley (HOFFMANN 1959; BORISOV 1965; LYDOLPH 1977). Absolute minima of −45 to −55 °C have been reported from coastal stations in the Antarctic and values of about −91 °C from the interior of the continent (ORVIG 1970).

In those polar regions and in peak regions of high mountains where the temperature never exceeds 0 °C (8.5% of the total land area of the earth) active life is only possible in habitats that offer thermic advantages. Information on the global occurrence of cold stress in the form of isotherm maps of extreme temperatures is given in HOFFMANN (1960).

Microclimate differs considerably from standard meteorological data. Although local climatic features, such as the accumulation of cold air on valley floors and in depressions, are reflected in the conventional measurements, temperature stratification of the air near the ground, such as occurs in connection with radiation frost, especially in places sheltered from the wind, are not. In such situations, the surface of the vegetation or of the ground itself cools down by several K more than the air at a height of 2 m, so that frost frequency is greater and the frost-free period shorter than in the open. Under conditions of negative radiation balance, freely radiating plant surfaces may be as much as 5 K cooler than the surrounding air (HÄCKEL 1978). Usually the minimum temperatures measured on plants during clear nights are 1–3 K below the minimum air temperature, depending upon the degree of exposure of the plants, the heat storage in organs, and the heat transfer from the surroundings.

The effect of chilling or frost on plants is not only governed by the magnitude of the drop in temperature but is to a large extent dependent upon the time at which the drop occurs and on how long the low temperature persist. *Episodic* outbursts of low temperatures follow in the wake of polar air masses and, if intensified by nocturnal heat loss, may result in spring and early autumn frosts in the temperate zone, in summer frosts in boreal and subarctic regions, in occasional outbreaks of cold in the subtropics, and in frost at any time of the year in temperate and tropical high mountains. The danger of this kind of weather-dependent stress, which can take effect within a few days or even hours, is that most plants have very little resistance to cold when in an active vegetative state and are unable to improve it adequately when faced with such sudden cold stress. *Periodic* cold stress is experienced seasonally so that plants have enough time to adapt gradually to this type of stress. Only in the form of severe or exceptionally long, uninterrupted periods of extremely low temperature does winter frost represent a danger to acclimatized plants.

13.3 Susceptibility and Adaptability of Plants to Injurious Low Temperature Stress

13.3.1 Chilling-Sensitive Plants

Chilling-sensitive plants suffer severe injury at low temperatures in the absence of freezing. The fact that most chilling injuries are chronic makes it difficult to determine numerical thresholds of cold susceptibility because both the critical temperature and the critical duration have to be taken into consideration (Fig.

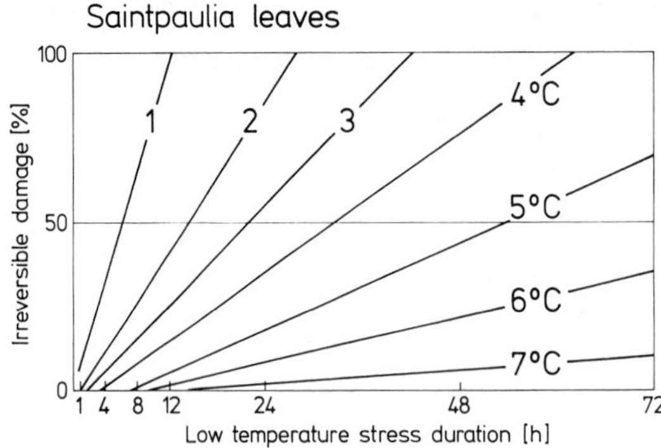

Fig. 13.1. Dose dependence of chilling injury in leaves of *Saintpaulia,* Rhapsody strain. (LARCHER and BODNER 1980)

13.1). In addition, comparative studies on chilling susceptibility have still to be made on a wide range of species: Most of the observations and measurements so far available concern cultivated and ornamental plants, vegetables and fruit in connection with cold storage, and microorganisms in connection with food preservation. Very little is known about the differences in chilling susceptibility between the various organs and tissues of one and the same plant.

13.3.1.1 Chilling-Sensitive Microorganisms and Thallophytes

Various species of bacteria such as *Escherichia coli, Aerobacter aerogenes, Pseudomonas pyocyanea, Salmonella typhimurium,* and *Bacillus subtilis* suffer severe or even lethal damage when exposed to shock-like cooling below +10 °C during their exponential phase of growth (MAZUR 1966; CHRISTOPHERSEN 1973). The position of the critical temperature is greatly influenced by adaptation to the environmental temperature prior to the onset of the cold shock and by nutritional conditions. Thermophilic fungi and actinomycetes are killed by several weeks of cooling at 10 to 12 °C, temperatures down to 5 °C can be tolerated only for shorter periods (NOACK 1912).

Algae sensitive to chilling are those inhabiting hot springs (such as *Anacystis nidulans, Synechococcus lividus* and *Cyanidium caldarium*) and seaweeds of tropical oceans. For many sublittoral algae of the Caribbean sea an exposure to temperatures below 5–10 °C for 12 h is lethal (BIEBL 1962b; see Table 13.2).

13.3.1.2 Chilling-Sensitive Cormophytes

Certain plants of tropical rain forests and mangroves, many cultivated and ornamental plants of tropical distribution, various tropical C_4 fodder grasses, vegetables of tropical origin and tropical seagrasses suffer visible injury after exposure to chilling temperatures (Table 13.1). Some of these species can adapt to chilling if the plant is conditioned for a few days to weeks at temperatures slightly above the critical stress temperature (CHRISTIANSEN 1979; LYONS and BREIDENBACH 1979).

Table 13.1. Susceptibility to low temperature stress of leaves of chilling-sensitive cormophytes (selected examples)

Species	Chilling injury at °C	Duration of chilling at the indicated temperature	Reference
1. Woody plants in their natural habitat (Puerto Rico)			
Guarea guara	4	24 h	BIEBL (1964)
Marcgravia sintenisii	4	24 h	BIEBL (1964)
Avicennia nitida	3	24 h	BIEBL (1964)
Marcgravia rectiflora	0.5–1	24 h	BIEBL (1964)
Cecropia peltata	0.5–1	24 h	BIEBL (1964)
Rhizophora mangle	0 –1	24 h	BIEBL (1964)
Psychotria berteriana	0 –1	24 h	BIEBL (1964)
2. Herbaceous plants in their natural habitat			
Pilea obtusata	<4	24 h	BIEBL (1964)
Ruellia coccinea	<4	24 h	BIEBL (1964)
Psychotria uliginosa	1	24 h	BIEBL (1964)
Peperomia hernandifolia	0–1	24 h	BIEBL (1964)
Passiflora edulis f. *flavicarpa*	0	5 d	PATTERSON et al. (1976)
Passiflora edulis	0	25 d	PATTERSON et al. (1976)
3. Cultivated plants of tropical origin			
Phalaenopsis cv. Pink Chiffon	7	<8 h	MCCONNELL and SHEEHAN (1978)
Episcia reptans	5	1 h	WILSON and CRAWFORD (1974); WILSON (1978)
Eranthemum tricolor	<4	2 d	MOLISCH (1897)
Impatiens sultani	<3	1.5 d	SPRANGER (1941)
Peperomia arifolia	<3	1.5 d	SPRANGER (1941)
Schismatoglottis pulchra	<3	2 d	SPRANGER (1941)
Scindapsus pictus	4.5	<4 d	MCWILLIAMS and SMITH (1978)
Maranta leuconeura	4.5	4 d	MCWILLIAMS and SMITH (1978)
Begonia stigmatosa	<4	5 d	MOLISCH (1897)
Piper decurrens	<3	5 d	SPRANGER (1941)
Peperomia argyrea	<4	5–7 d	MOLISCH (1897)
Lycopersicon esculentum	<5	>6 d	SEIBLE (1939)
Eranthemum tuberculatum	<4	11 d	MOLISCH (1897)
Zebrina pendula	<5	>14 d	SEIBLE (1939)
4. Tropical seagrasses			
Syringodium filiforme	2	1 h	MCMILLAN (1979)
Thalassia testudinum	2	4 h	MCMILLAN (1979)
Halodule wrightii	2	1 d	MCMILLAN (1979)

In some plant species only certain stages such as germination, the early phases of flower development, or senescent organs are susceptible to chilling. Germinating seeds and young seedlings of *Gossypium herbaceum, Glycine max,* and various species of *Phaseolus* are severely damaged if cooled below 12–15 °C (POLLOCK and TOOLE 1966; CHRISTIANSEN 1963, 1979; LAKHANOV and BALACH-

Table 13.2. Potential cold resistance of algae, lichens, and mosses in the hydrated state

Plant and habitat	Low temperature injury below °C	Reference
Algae		
Arctic seas		
Intertidal	− 8 to −28	BIEBL (1968, 1970)
Sublittoral	− 2 to − 4	BIEBL (1968, 1970)
Temperate seas		
Intertidal		
Chlorophyceae	− 8 to −25	TERUMOTO (1964); BIEBL (1958, 1972)
Phaeophyceae	−40 to −60	PARKER (1960)
Rhodophyceae	− 7 to − 8 (−70)	BIEBL (1939, 1958); TERUMOTO (1964); MIGITA (1966)
Sublittoral	− 2 to + 4	BIEBL (1958, 1970)
Tropical seas		
Intertidal	− 2 to +11	BIEBL (1962b)
Sublittoral	+ 3 to +14	BIEBL (1962b)
Arctic and Antarctic freshwaters		
Cyanophyceae	−70 to −196	HOLM-HANSEN (1963)
Phycophyta	−15 to − 30	HOLM-HANSEN (1963); BIEBL (1969)
Temperate lakes		
Cyanophyceae	(−25) −70 to −196	HOLM-HANSEN (1963)
Phycophyta	− 2 to − 20	TERUMOTO (1964); DUTHIE (1964); BIEBL (1967a); SCHÖLM (1968)
Tropical lakes	n.d.	
Hot springs[a]	[+15 to + 20]	SOEDER and STENGEL (1974); BROCK (1978)
Epiphytic and epipetric algae[a]	−70 to −196	KÄRCHER (1931); EDLICH (1936)
Soil algae[a]	−25 to −196	HOLM-HANSEN (1963)
Snow algae[a]	ca. −40	GÖPPERT (1878)
Lichens		
Arctic, Antarctic	−80 to −196	KAPPEN and LANGE (1972); RIEDMÜLLER-SCHÖLM (1974)
Desert, High mountains	−78	KAPPEN and LANGE (1970b, 1972)
Temperate zone	−50 and below	KAPPEN and LANGE (1972)
Mosses		
Arctic	−50 to −80	RIEDMÜLLER-SCHÖLM (1974)
Temperate zone		
Marchantiales	− 5 to −10	CLAUSEN (1964); DIRCKSEN (1964)
Jungermanniales	−10 to −15	CLAUSEN (1964); DIRCKSEN (1964)
Musci (Hydrophytes)	− 8 to −20	IRMSCHER (1912); DIRCKSEN (1964)
Musci (bogs)	− 7 to −15	DIRCKSEN (1964)

[a] Further experimental data needed
( ) Exceptional values
n.d. No experimental data available
[ ] Estimated from growth limitations

Table 13.2. (continued)

Plant and habitat	Low temperature injury below °C	Reference
Musci (forest floor)	(−10) −15 to −35 (−55)	IRMSCHER (1912); DIRCKSEN (1964); HUDSON and BRUSTKERN (1965); ANTROPOVA (1974)
Musci (epiphytic, epipetric)	−15 to −30[a]	IRMSCHER (1912)
Humid tropics	< 0 to − 7 (−16)	BIEBL (1964, 1967b)

KOVA 1978). Rice and sorghum are highly susceptible to cold shortly before flower initiation. Apparently, night temperatures of 13–16 °C at this stage lead to disturbance in the pollen mother cell development and hence to sterility (TORIYAMA 1976; BROOKING 1976). Extremely susceptible to chilling are ripening fruits of tropical and subtropical plants. Whether or not the observations with cold-stored fruit (cf. LUTZ and HARDENBURG 1968) are of ecological applicability, i.e., whether the same degree of cooling does the same amount of damage to fruit ripening on field-growing plants and what effect this might have on their progeny, still has to be investigated.

Chilling susceptibility does not only vary within a single genus (e.g., *Passiflora:* PATTERSON et al. 1976; *Psidium* and *Capsicum:* SMILLIE 1979), but even within one and the same species (e.g., varieties of rice: LIN and PETERSON 1975; ecotypes of seagrasses: MCMILLAN 1979). Genetic variability is the mechanism by means of which climatic stress can lead to the selection of the more resistant varieties. PATTERSON et al. (1978) found an altitudinal gradient in chilling susceptibility in the various subspecies and ecotypes of the wild tomato, *Lycopersicon hirsutum,* that thrive at different altitudes in the High Andes from sea level to 3,300 m. The sea level populations are as sensitive to chilling as the cultivated tomato, *L. lycopersicum,* whereas those at high altitudes have achieved the transition to chilling resistance.

13.3.2 Specific Differences in Freezing Resistance

13.3.2.1 Microorganisms, Thallophytes, and Poikilohydric Cormophytes

In an air-dry state most microorganisms, thallophytes, and mosses, as well as certain cormophytes, are able to survive immersion in liquid nitrogen. Unlimited resistance to low temperature has been demonstrated in air-dried bacteria and yeasts (MAZUR 1966; LOZINA-LOZINSKII 1974), aerial algae (EDLICH 1936), lichens (KAPPEN and LANGE 1970a), certain mosses (LIPMAN 1936) and ferns (KAPPEN 1966), and in the desiccation-tolerant phanerogams *Ramonda myconi* (KAPPEN 1966) and *Myrothamnus flabellifolia* (VIEWEG and ZIEGLER 1969). Dried seeds (STANWOOD and BASS 1978) and pollen (DAVIES and DICKINSON 1971; NATH and ANDERSON 1975) can also be absolutely resistant to freezing.

In the hydrated state, gram-negative bacteria are killed by exposure to −15 to −30 °C for periods of between several days and several weeks, whereas gram-positive bacteria, such as many soil bacteria, are freezing-tolerant (CHRISTOPHERSEN 1973; MAZUR 1966). Among the fungi, certain highly resistant species are capable of surviving temperatures of −40 to −192 °C in the vegetative state (KÄRCHER 1931; HWANG 1968). Wood-dwelling fungi and mycorrhizal fungi from high altitudes are especially resistant (LINDEGREN 1933; MOSER 1958). Some species of ectomycorrhizal fungi, on the other hand, are more sensitive to cold and their mycelia are damaged at −10 °C or survive but suffer appreciable disturbances in development (FRANCE et al. 1979). The resistance of microorganisms to freezing is closely dependent upon cell age and state of nutrition.

Among the algae, every grade of cold resistance is encountered (Table 13.2) sometimes showing good agreement with the thermal characteristics of the environment (e.g., marine algae: BIEBL 1962b, 1970). Lichens, with very few exceptions (*Roccella fucoides, Umbilicaria vellea*), are notable for an extremely high resistance to cold, regardless of their distribution, whereby the phycobiont is rather more sensitive than the mycobiont (KAPPEN and LANGE 1970b). Mosses and desiccation-tolerant cormophytes can, when in an active state and well provided with water, just like homeohydric plants, be either freezing-sensitive (e.g., certain tropical rain forest mosses and pteridophytes: BIEBL 1964), or freezing-tolerant in winter (mosses: HUDSON and BRUSTKERN 1965; ferns: KAPPEN 1964, 1965). It is remarkable that of the tropical mosses, Hymenophyllaceae, and ferns so far investigated, not a single chilling-sensitive species has been found (BIEBL 1964, 1967b).

13.3.2.2 Homeohydric Cormophytes

If plants are arranged in the order of their *potential frost resistance* (i.e., their maximal attainable resistance) they form a gradual series from extremely freezing-sensitive species up to plants with unlimited freezing-tolerance (see Tables 13.3 and 13.4). Species lists giving frost resistance of the vascular plants representative of biome and vegetation types are available for European forest trees and herbs (TILL 1956), for N. American and E. Asiatic forest trees (SAKAI and OKADA 1971; SAKAI and WEISER 1973; SAKAI 1978b, c), for the predominantly evergreen tree flora of the Pacific region (KUSUMOTO 1959; SAKAI 1971, 1972; SAKAI and WARDLE 1978), for mediterranean sclerophylls (LARCHER 1970), for tropical rain forest plants (BIEBL 1964), for subtropical grassland (ROWLEY et al. 1975), for ericaceous dwarf shrubs (Table 13.5), for high mountain plants (ULMER 1937; TYURINA 1957; SAKAI and OTSUKA 1970), and for the arctic tundra (BIEBL 1968; RIEDMÜLLER-SCHÖLM 1974).

Tropical and subtropical plants and the leaves of certain evergreen trees and shrubs of warm temperate coastal regions exhibit life-long susceptibility to freezing. Especially susceptible are the nonhardening species whose freezing point lies between −1 and −3 °C, such as certain C_4 grasses (ROWLEY et al. 1975; MILLER 1976; IVORY and WHITEMAN 1978) and of course all chilling-sensitive plants. Many plants showing sensitivity to freezing are nevertheless capable of reducing their frost sensitivity by 2 to 4 K. Cold adaptation has been demonstrated in the C_4-grasses *Paspalum dilatatum* and *Eragrostis curvula* (ROWLEY et al. 1975), in *Citrus*-species (YELENOSKY 1975), in *Persea americana*

Table 13.3. Potential frost resistance of herbaceous vascular plants

Plant group	Frost resistance (lowest temperature [°C] sustained without lethal injury)			Reference (examples)
	Leaves	Shoot apex	Subterraneous organs	
Pteridophytes				
Tropical ferns	0 to − 2			Biebl (1964)
Temperate ferns	−13 to −25		(−3) − 7 to −12	Kappen (1964)
Subarctic (*Lycopodium*)	−80			Riedmüller-Schölm (1974)
Graminoids				
Tropical and subtropical grasses	− 1 to − 4 (−6)	− 4 to − 6 (−10)		Rowley et al. (1975); Ivory and Whiteman (1978b)
Temperate grasses incl. cereals	(−6) −20 to −25 (−30)	− 5 to −30	− 7 to −20	Till (1956); Fowler et al. (1977); Fuller and Eagles (1978); Noshiro and Sakai (1979)
Steppe grasses	n.d.			
Alpine sedges	(−70)	(−70)	(−70)	Kainmüller (1975)
Arctic graminoids	n.d.			
Herbaceous dicotyledons				
Tropical herbs	(+2) − 1 to − 2			Biebl (1964)
Meadow plants and weeds	(−10) −15 to −25	−10 to −20	− 5 to −15	Schnetter (1965); Noshiro and Sakai (1979), and unpublished data
Temperate forest herbs	−10 to −20	−10 to −20	− 7 to −11	Till (1956)
Geophytes	− 5 to −12	ca. −10	− 7 to −14 (−18)	Till (1956); Goryshina (1972); Lundquist and Pellett (1976)
Halophytes	−10 to − 2	ca. −20	−10 to −20	Kappen (1969); Maier and Kappen (1979)
Winter-annual desert plants	− 6 to −10			Lona (1963)
Arctic and alpine rosette and cushion plants	(−15) −20 to −50 (−196)	−30 to −50 (−196)	−20 to −60 (−196)	Ulmer (1937); Sakai and Otsuka (1970); Kainmüller (1975)
Tropical high mountain plants	− 7 to −12			Larcher (1975)

n.d. No experimental data available
( ) Exceptional values

Table 13.4. Potential resistance of woody plants to low temperatures and frost (lowest

Plant group	Leaves	Buds	Flower buds
Tropical regions			
Forest and fruit trees	+ 4 to − 3	down to − 5	
Tropical palms	(−1) − 3 to − 5	[− 3 to − 5]	
Lianas	+ 4 to 0		
Mangroves	+ 4 to − 4		
Subtropical regions			
Evergreen trees and shrubs	(−2) − 4 to − 6 (−8)	− 6 to −12	
Drought deciduous trees		ca. −14	
Subtropical palms	− 5 to −12	down to −14	
Maritime-temperate regions			
Conifers	−10 to −20	−10 to −25	(−10)
Arcto-tertiary flora	−10 to −15	− 8 to −15	
Evergreen broadleaved trees and shrubs	− 7 to −15 (−20)	−10 to −18	down to −17
Mediterranean sclerophylls	− 5 to −12 (−15)	− 8 to −18	−10 to −16
Warm-temperate deciduous trees		−15 to −30	−15 to −30
Ericaceous heath shrubs	−15 to −30	−20 to −30	−15 to −25
Regions with severe winters			
Conifers	−40 to −70 (−196)	−30 to −70 (−196)	
Temperate deciduous trees		−25 to −35 (−60)	−25 to −40
Boreal deciduous trees		−30 to −80 (−196)	
Arctic and alpine dwarf shrubs	−30 to −50 (−80)	−20 to −40	

xy Stem resistance limited by deep supercooling of xylem
( ) Exceptional values
[ ] Estimated from observations of occasional frost injury

(SCORZA and WILTBANK 1976), in several species of *Eucalyptus* (ELDRIDGE 1968), and it has been observed in connection with the differentiation of ecotypes and related species at higher altitudes (*Setaria anceps:* IVORY and WHITEMAN 1978; tuber-bearing *Solanum:* PALTA and LI 1979).

The ability of plants living in seasonally cold climates to become freezing-tolerant is temporary. In woody plants with a seasonal pattern of activity and in freezing-tolerant C_3-grasses it is coupled with winter dormancy. But even plants that do not enter a phase of true dormancy, e.g., perennial herbaceous dicotyledons, are capable of developing freezing tolerance during winter.

temperatures [°C] sustained without lethal injury)

Stem	Roots	References for lists of species
down to −5		Biebl (1964); Sakai (1972, 1978b, c)
		Biebl (1964); Smith (1964); Larcher (1980b)
		Biebl (1964)
		Biebl (1964); McMillan (1975)
(−3) −6 to −15		Larcher (1971); Layton and Parsons (1972); Sakai (1972, 1978b); Yelenosky (1977)
ca. −15 to −20		Larcher (1971); Sakai (1978b)
		Larcher (1980a); Larcher and Winter (1982)
−15 to −30	−10 to −20	Parker (1960); Larcher (1954, 1970); Sakai and Okada (1971); Havis (1976); Sakai (1978b)
− 8 to −18		Sakai (1971)
−10 to −20	− 7 to − 9	Sakai (1972); Havis (1976); Sakai and Wardle (1978); Sakai and Hakoda (1979)
− 8 to −22 xy		Larcher (1954, 1970); Sakai (1978b)
−20 to −40 xy		Larcher (1970); Sakai (1971, 1972, 1978b, c); Sakai and Weiser (1973); Kaku and Iwaya (1978)
−15 to −35	−10 to −20	Till (1956); Havis (1976); Sakai and Miwa (1979)
−50 to −196	−20 to −35	Pisek and Schiessl (1947); Parker (1962); Sakai and Okada (1971); Havis (1976); Sakai (1978a, 1979b)
−30 to −50 xy	−15 to −25	Till (1956); Pisek (1958); Parker (1962); Sakai (1972, 1978c); Sakai and Weiser (1973); George et al. (1974)
down to −196		Sakai (1965); Sakai and Weiser (1973); Sakai (1978c)
−30 to −50 (−196)	−10 to −30	Ulmer (1937); Pisek and Schiessl (1947); Sakai and Otsuka (1970); Riedmüller-Schölm (1974); Larcher (1977)

Hardening, which results in the development of freezing tolerance, is a stepwise process (Tumanov 1962, 1979; Weiser 1970; Kacperska-Palacz 1978; Tyurina et al. 1978). At temperatures between +5 and 0 °C plants attain the first level of hardiness, making it possible for them to survive moderate frost. Progressive and persistent frost leads to complete hardiness and the plants now reach the limits of their potential freezing tolerance. Following stepwise freezing, boreal species of the genera *Abies, Picea, Pinus, Salix, Betula,* and *Ribes* become resistant to immersion in liquid nitrogen at −196 °C (Parker 1960; Krasavtsev 1960; Sakai 1960). Certain herbaceous plants of high mountains also exhibit unlimited freezing tolerance (SAKAI and OTSUKA 1970; KAINMÜLLER 1975).

If the frosts become less severe, resistance drops to a lower level but, with the recurrence of severe frost, resistance reverts to the higher level. This *responsive* adjustment of resistance takes place rapidly. During cold waves in late autumn and winter, tolerance increases noticeably in trees within only 1–2 days and the full effect is attained within 5–10 days (PISEK and SCHIESSL 1947; KRASAVTSEV 1960; MITTELSTÄDT 1965; SCHEUMANN 1968). In cereals the resistance rises within a week of exposure to continued frost, although several weeks are necessary for complete hardening (GUSTA and FOWLER 1979). Dehardening by warming to 15 to 20 °C takes place even faster and is usually complete within 1–2 days.

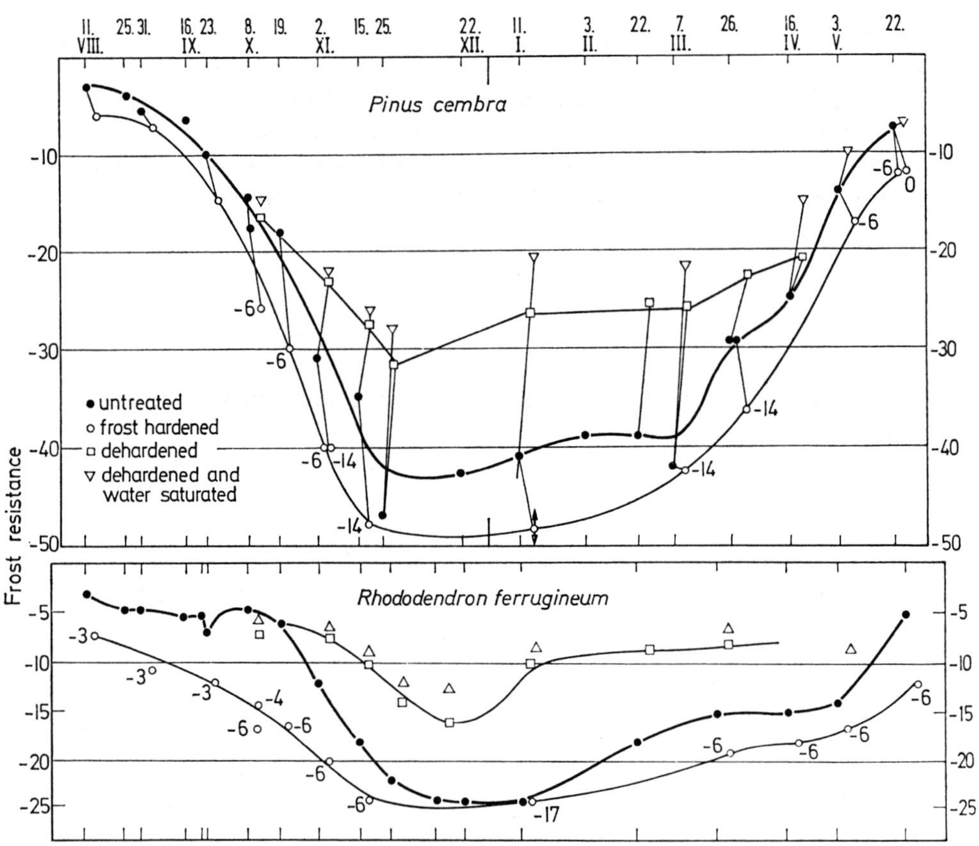

Fig. 13.2. Seasonal course of the *actual* frost resistance in the habitat, the *potential* resistance after cold hardening and the *minimal* resistance after artificial dehardening of leaves of *Pinus cembra* and *Rhododendron ferrugineum* from the alpine timber line. The difference between minimal and potential frost resistance is a measure of the degree to which the state of hardening can be influenced by weather conditions. (PISEK and SCHIESSL 1947)

Freezing tolerance is lost during the season of intensive growth, although there is some evidence that certain, probably inactive, regions of the shoot may retain their freezing tolerance even during the growth season, e.g., in plants of the high mountains such as *Silene acaulis*.

The specific ability to harden, which is the basis for the difference between individuals, ecotypes, varieties, and species with respect to potential resistance, is primarily genetically determined (for the inheritance of resistance characteristics see, e.g., SIMURA 1957; MURAWSKI 1961; WILNER 1965; LAW and JENKINS 1970). In addition, the hardening potential of a plant is influenced by its state of nutrition and health. Only a very low degree of hardening can be achieved by diseased or damaged plants (e.g., by hypoxia below snow and ice: RAKITINA 1970, 1977; ANDREWS and POMEROY 1977; by deicing salts: SUCOFF et al. 1976; by air pollution: KELLER 1978), by plants deficient in mineral nutrients (BIEBL 1962a; LEVITT 1980), or by plants that have been unable to accumulate sufficient carbohydrate reserves either on account of loss of foliage or as a result of a curtailed period of growth (KRAMER and WETMORE 1943; TUMANOV et al. 1972; STERGIOS and HOWELL 1977; NISSILA and FUCHIGAMI 1978). The readiness of response and the magnitude of short term adaptation in resistance are also specific features of species and varieties (PISEK and SCHIESSL 1947; SCHEUMANN and HOFFMANN 1967, Fig. 13.2). Measurement of the response amplitude of freezing tolerance provides a means of characterizing not only the capacity of a plant to harden but also its readiness to lower its resistance during intermittent periods of warm weather.

13.4 Ecological Aspects of Low Temperature Stress

Whether or not a plant species is able to establish itself in a particular locality depends upon a number of attributes:

1. its survival capacity, i.e., its ability to survive hazardous situations and adverse seasons;
2. its production capacity, which depends upon sufficient dry matter production in order to ensure growth and successful competition and
3. its reproductive capacity, i.e., its ability to propagate and multiply.

13.4.1 Survival Capacity

Survival is a highly complex phenomenon. Although winter survival capacity depends primarily on cold resistance, spatial and temporal mechanisms of stress exclusion, which help the plant to escape excessively adverse conditions, are also involved.

13.4.1.1 Temporal Stress Exclusion

The various phases in the life cycle and in the seasonal course of metabolic and developmental activities differ with respect to frost susceptibility and harden-

Table 13.5. Resistance spectrum for ericaceous dwarf shrubs (lowest temperature [°C] sus-

Species	Habitat	Minimal frost resistance during the growing season		
		Leaves	Stem	Flowers
Erica herbacea	Heath	−3 to −4		
Erica tetralix	Heath	−4		
Calluna vulgaris	Heath, Alpine	−4 to −5	−4 to −5	
Vaccinium myrtillus	Heath, Alpine	−4	−5 to −6	
Vaccinium uliginosum	Arctic, Alpine	−4 to −6	−5 to −7	
Vaccinium vitis idaea	Arctic, Alpine	−5	−8	
Ledum palustre	Arctic, Alpine	−5	−7	
Ledum groenlandicum	Arctic bog	−6		−3
Empetrum nigrum	Arctic, Alpine	−5 to −8	−5	
Arctostaphylos uva-ursi	Arctic, Alpine	−7		
Loiseleuria procumbens[a]	Arctic, Alpine	−5 to −6	−7 to −10	
Cassiope tetragona	Arctic	−8		
Cassiope lycopodioides	Alpine			
Chamaedaphne calyculata	Arctic			−9
Diapensia lapponica	Arctic	−9		

[a] *Loiseleuria* seedlings: −6 °C (ECCHER unpublished)
[ ] Central Alps, Europe
(N_2) Plant survived immersion in liquid nitrogen after prefreezing at −30 °C

ing ability (see Fig. 13.2). The exact timing of phenological events results in temporal frost exclusion. A striking example is provided by the autumn leaf-fall of the woody plants of winter-cold regions, by which both the danger of frost damage and of winter drought are reduced. Late resumption of vegetative growth and foliation reduce the hazard of damage from late spring frosts. This is of particular significance for mountain plants (KAINMÜLLER 1975; TYLER et al. 1978; LARCHER 1980b), bog ericads (READER 1979) and deciduous trees (TILL 1956). There is some evidence that the altitudinal and latitudinal distribution of woody plants are limited by inadequate synchronization of development with climatic seasonality (LANGLET 1937; EGUCHI et al. 1966; SMITHBERG and WEISER 1968; DIETRICHSON 1969).

tained without lethal injury)

Potential frost resistance during winter				Reference
Leaves	Buds	Stem	Below ground organs	
−18 to −23				ULMER (1937); SAKAI and MIWA (1979)
−18 to −20	−20	−15 to −20		TILL (1956); SAKAI and MIWA (1979)
−27 to −35	−30	−30	−20	LARCHER (1977); SAKAI and MIWA (1979)
		−15[−30]	[−30]	LARCHER (1977)
	−20	−50	−30	BIEBL (1968); SAKAI and OTSUKA (1970); LARCHER (1977)
−70 to −80	−30	−30	−20	SAKAI and OTSUKA (1970); RIEDMÜLLER-SCHÖLM (1974)
−70 to −80	−30	−50	−30	PARKER (1961); SAKAI and OTSUKA (1970); RIEDMÜLLER-SCHÖLM (1974)
				BIEBL (1968); READER (1979)
[−30] −70		−30	−30	ULMER (1937); BIEBL (1968); SAKAI and OTSUKA (1970); RIEDMÜLLER-SCHÖLM (1974)
[−30] −70				ULMER (1937); RIEDMÜLLER-SCHÖLM (1974); LARCHER and WAGNER (1976)
[−50] −70	−40	−40 to −60	−30	ULMER (1937); PISEK and SCHIESSL (1947); SAKAI and OTSUKA (1970); LARCHER and WAGNER (1976); LARCHER (1977)
			−30	BIEBL (1968)
−40		−30	−30	SAKAI and OTSUKA (1970)
−80				RIEDMÜLLER-SCHÖLM (1974); READER (1979)
−70 (N_2)		−70 (N_2)	−70	SAKAI and OTSUKA (1970)

13.4.1.2 Spatial Stress Exclusion

Spatial exclusion of winter stress consists of reduction of the vegetative parts to organs that are frost-resistant or have adequate winter protection as a result of life form adaptation (RAUNKIAER 1910). For judging the winter survival capacity, information on frost resistance of the overwintering parts of the plants is therefore necessary in addition to the usual data for leaves.

Of the vegetative organs the roots, rhizomes, and bulbs are the most sensitive; in winter they may even suffer damage at −5 °C. As a rule they are killed at −10 and −30 °C (CHANDLER 1954; BAUER et al. 1971; HAVIS 1976; LUNDQUIST and PELLETT 1976; see Tables 13.3, 13.4, 13.5; Figs. 13.3, 13.4). The

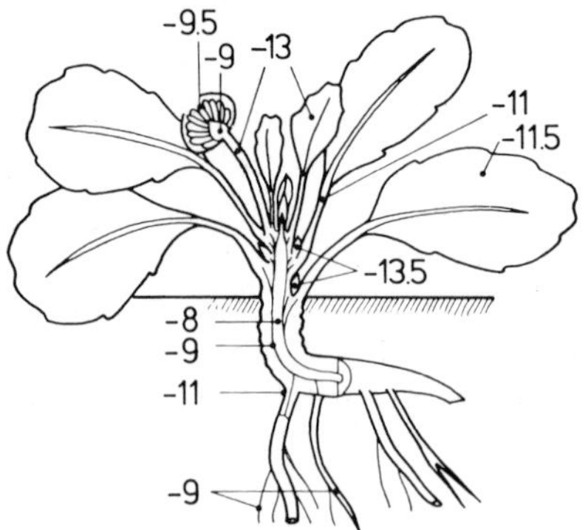

Fig. 13.3. The frost resistance of various organs and tissues of *Bellis perennis* in winter beneath 10 cm of snow. (After SOTTOPIETRA unpublished)

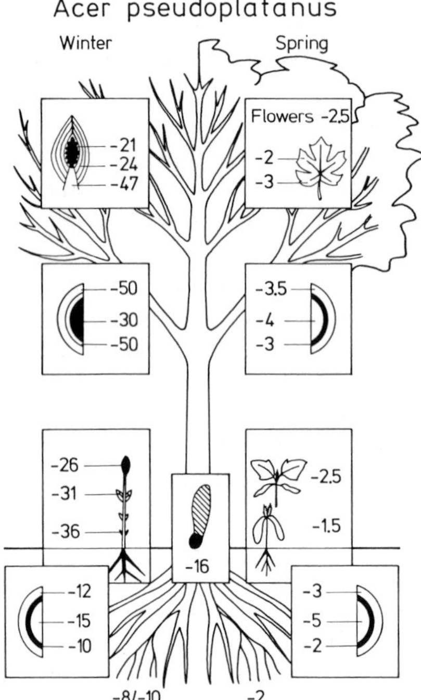

Fig. 13.4. Frost resistance of *Acer pseudoplatanus* in winter and during sprouting in spring. (After HARRASSER unpublished)

resistance of buds apparently differs according to their position on the twig: terminal buds are less resistant than lateral buds (MAIR 1968), and the basal reserve buds are particularly resistant. The most resistant tissue in the winter buds of dormant woody plants is the bud axis (LARCHER 1970; DEREUDDRE 1978; QUAMME 1978). In winter cereals and rosette plants the shoot apex is more resistant to freezing than the medullary parenchyma of the crown (OLIEN 1964; MARINI and BOYCE 1977). Woody stems are usually more resistant to cold than leaves and buds (cf. Table 13.4; Fig. 13.4). In fully frost-hardened twigs and trunks the cambium is the most resistant and the xylem the most sensitive tissue (LARCHER and EGGARTER 1960).

13.4.2 Low Temperature Effects on Production Processes

The responses of plants to cold stress, and above all to the periodically recurring complex of stress in winter, is not confined to the adaptive enhancement of cold resistance, but also includes low temperature effects on production processes. At moderate intensities cold stress causes a reversible disturbance in metabolic processes, but if its intensity (temperature or duration) exceeds a critical level, certain processes may come to a temporary or even permanent stop. Any impairment of carbon metabolism leads to a reduction in dry matter production and as a consequence retards growth.

13.4.2.1 Photosynthetic Activity During Exposure to Low Temperatures

At low temperatures the photosynthetic production is restricted in all plants by a reduced velocity of enzymatically catalyzed photosynthetic reactions. At temperatures below approximately 20 °C, C_4 photosynthesis is less efficient than C_3 photosynthesis due to a reduced quantum yield (EHLERINGER 1978). Therefore, the geographical abundance of C_4 species is closely correlated with the minimum temperature of the growing season (e.g., TEERI and STOWE 1976; BOUTTON et al. 1980). But there are also C_4 plants, such as the cool temperate saltmarsh grass *Spartina townsendii,* whose C_4 pathway is quite capable of functioning at 10 °C (THOMAS and LONG 1978).

In *chilling-sensitive plants,* such as most C_4 grasses of tropical origin, the activation energy for many chloroplast functions increases abruptly below 10 to 12 °C (STEPONKUS, Chap. 10.1, this Vol.), so that net photosynthesis gradually ceases between +5 and +10 °C (e.g., LUDLOW and WILSON 1971).

Chilling-tolerant plants, if appropriately adapted, sometimes achieve up to 60% of their optimal CO_2 uptake at temperatures between +5 and +10 °C (particularly evergreen species of temperate or subarctic regions; PISEK 1973). In these plants the CO_2 uptake is abruptly blocked when ice forms in the assimilatory organs, i.e., between −2 and −10 °C (PISEK et al. 1967). So far, no reports are available on experimental investigations into the cause of the abrupt cessation of CO_2 uptake when ice forms. Obviously the ice represents an enormous barrier to CO_2 diffusion: the diffusion rate of CO_2 in ice is more than four orders of magnitude lower than in a water layer of equal thickness. Some cryptogams are able to take up CO_2 even when frozen: mosses down to −10 °C, certain lichens even down to −22 °C. When lichens freeze the ice does not form a continuous crust, and thus presents no insuperable obstacle to CO_2 diffusion (LANGE 1965).

13.4.2.2 Photosynthetic Activity Following Low Temperature Stress

The degree to which dry matter production is impaired as a result of exposure to low temperatures depends, in addition to the immediate reduction in photosynthetic CO_2 uptake at low temperatures, upon the extent to which photosynthesis recovers after the cold stress and the speed with which recovery is achieved.

In *chilling-sensitive plants* such as *Cucumis sativus* the photosynthetic uptake of CO_2 in light after a cool night (about +2 to +5 °C) is resumed very slowly and does not attain its full intensity (TSCHÄPE 1970). Frequent cool nights prolong the inhibitory after-effect in maize, and it may take as long as 20 days from the last cool night before photosynthetic activity is fully restored (TEERI et al. 1977). Exposure of the leaves to strong irradiation during low temperature stress may lead to permanent impairment of the photosynthetic capacity (TAYLOR and ROWLEY 1971; LASLEY et al. 1979) or, if this is already manifest, may reinforce the inhibitory after-effects of the cold stress (e.g., KISLYUK and VAS'KOVSKY 1972). Irradiation during cold only has an injurious effect in the presence of oxygen (ROWLEY and TAYLOR 1972; VAN HASSELT 1972), which suggests that photo-oxidative damage to the photosynthetic apparatus is involved (VAN HASSELT and VAN BERLO 1980).

No single metabolic or physiological parameter is responsible for photosynthetic depression following chilling (TEERI et al. 1977). Besides the direct effects on chloroplast functions, chilling stress may also induce a water stress in the leaves which is especially pronounced if the roots are also cooled (e.g., TSCHÄPE 1970; CROOKSTON et al. 1974). This water stress leads to stomatal closure and is thought to be responsible for reduced CO_2 uptake following chilling in C_3 plants (CROOKSTON et al. 1974). In C_4 plants chilling injury to the photosynthetic apparatus may be independent of water stress.

Even in *chilling-tolerant plants,* low temperatures insufficient to cause freezing of the leaves may be followed by reduction of photosynthetic activity, although probably only if the plants were freezing-sensitive or in a freezing-sensitive state (examples for herbaceous plants: SOSINSKA et al. 1977; PEOPLES and KOCH 1978; for woody plants: LARCHER 1969a; SEELEY and KAMMERECK 1977; DREW and BAZZAZ 1979). As possible explanations PEOPLES and KOCH (1978) and DREW and BAZZAZ (1979) have discussed a feedback inhibition via carbohydrate accumulation and stomatal closure induced by water stress in the leaves, which results from reduced water uptake by cooled roots (e.g., KAUFMANN 1977).

If the leaves of woody plants in a freezing-tolerant state have been frozen, CO_2 uptake recovers very slowly after thawing, and remains inhibited for several hours or days (Fig. 13.5; BAUER et al. 1975). The less the degree of hardening of the plant and the greater the severity of the frost, the greater is the extent of the depression of photosynthetic capacity and the time required to return to normal (e.g., PISEK and KEMNITZER 1968; PHARIS et al. 1970). Several frosts in succession can bring about a complete cessation of CO_2 uptake (e.g., in timberline trees, TRANQUILLINI 1957; PISEK and WINKLER 1958). The post-freezing inhibition of CO_2 uptake is the result of the delayed and incomplete opening of the stomata (Fig. 13.5; ZELAWSKI and KUCHARSKA 1967; KOH et al. 1978), although the increase in stomatal resistance does not completely account for the reduced CO_2 uptake. The membrane-bound reactions in the chloroplasts are also impaired after frost (SENSER and BECK 1977; ÖQUIST et al. 1980).

In spinach and perhaps also in other herbaceous freezing-tolerant plants, it seems that nonlethal freezing has no after-effect on the post-thawing photosynthetic activity of intact leaves (KLOSSON and KRAUSE 1981). A rapid and complete restoration of photosynthetic capacity after frost is well-known from mosses and lichens (a survey of the literature can be found in KALLIO and KÄRENLAMPI 1975).

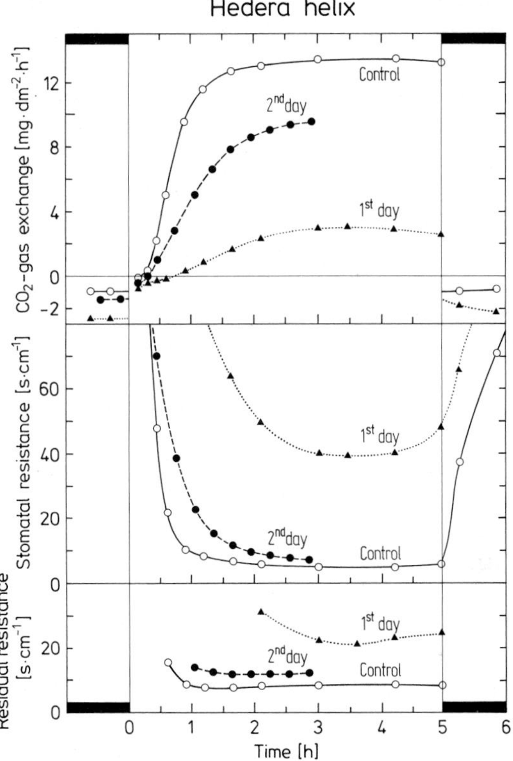

Fig. 13.5. CO_2-gas exchange, and stomatal and residual resistance to CO_2 transfer in darkness (*black bars*) and after exposure to light of *Hedera helix* following 4 h of frost stress at −9 °C. Figure shows the values of non-frost-treated controls and on the 1st and 2nd day after frost treatment. Conditions during measurement: 15 °C leaf temperature, 170 μmol m^{-2} s^{-1} PAR, 300 μl l^{-1} ambient CO_2 concentration and 0.4 to 0.5 s cm^{-1} resistance to the CO_2 transfer in the boundary layer. (BAUER unpublished)

13.4.2.3 Influence of Winter Dormancy and Frost Hardening on Photosynthetic Activity

Induction of winter dormancy and/or frost hardening by short days and/or low positive temperatures may lead to a decrease in photosynthetic capacity which is more pronounced in woody plants with deep dormancy (PISEK and WINKLER 1958; BAMBERG et al. 1967; ÖQUIST et al. 1980) than in herbaceous plants (BARTA and HODGES 1970; SOSINSKA et al. 1977). A rise in the stomatal resistance, observed in hardened conifers (NEILSON et al. 1972; ÖQUIST et al. 1980) and in ivy (BAUER unpublished), may be responsible for a larger part of the reduction in CO_2 uptake than the slight inhibition of photosynthetic electron transport observed in hardened conifers (SENSER and BECK 1977; ÖQUIST et al. 1980) but not in ivy and winter wheat (BARTA and HODGES 1970).

13.4.2.4 Photosynthetic Dry Matter Production During Winter

A question that repeatedly comes up for consideration is whether evergreen plants can photosynthesize in winter and produce dry matter (for a survey of literature see PISEK 1960; KOZLOWSKI and KELLER 1966; KRAMER and KOZLOWSKI 1979). Very little reliable experimental data is available, however. Most reports comprise estimations of the CO_2-balance from climate data and gas exchange parameters obtained in the laboratory or measurements of the dry matter increase of seedlings.

In regions with a mediterranean climate the photosynthetic capacity of evergreen species is usually only slightly lowered in winter (e.g., LARCHER 1961a; ECKARDT et al. 1975; DUNN 1975). At the northern limit of mediterranean sclerophylls in north Italy, however, the production in winter may be strongly curtailed by suboptimal temperatures and short days: LARCHER (1961b and 1981c) calculated that in December or January the net CO_2 input in *Olea europaea* and *Quercus ilex* amounts to only 20% to 40% of that of summer months. In southern France, too, the net CO_2 input of *Q. ilex* in winter attains only one fourth of the springtime value (ECKARDT et al. 1975).

In regions with a temperate or oceanic climate with relatively mild winters where frost periods are infrequent and sporadic, evergreen species also show an appreciable increase in dry weight during the winter. Dry weight of seedlings of *Pinus radiata* and *P. sylvestris* in west Wales (POLLARD and WAREING 1968) and of *Picea sitchensis* in southern Scotland (BRADBURY and MALCOLM 1978) almost doubles between October and April. Most of it occurs during October, late March, and April, but there is also a slight increase in weight in midwinter. Even the mild winter climate of the west coast of Norway permits a significant dry matter increase during the winter period in conifer seedlings (HAGEM 1962).

In regions with severe winters the photosynthetic activity is often completely extinguished during periods when the leaves freeze each night (e.g. PISEK and WINKLER 1958; PARKER 1961; NEGISI 1966; UNGERSON and SCHERDIN 1968). At the alpine timberline, the CO_2 balance of 6-year-old seedlings of *Pinus cembra* remains negative from mid-November to mid-May (TRANQUILLINI 1959). Occasional warm days during winter may even impair the CO_2 balance, since dark respiration is strongly enhanced after thawing of frozen tissues (e.g., BAUER et al. 1969; see also Fig. 13.5). Thus, evergreen woody plants may hardly be capable of utilizing their leaves for dry matter production during severe winters. Herbaceous winter annuals, however, are capable of exploiting mild weather for photosynthetic production during the whole winter (e.g., ZELLER 1951; KOH and KUMURA 1973; REGEHR and BAZZAZ 1976).

13.4.3 Low Temperature Effects on Propagation

The maintenance of a species in its natural habitat is essentially dependent on the fertility of adult individuals and the survival of seeds and seedlings. Thus, with respect to low temperature stress, the distribution range of a species is determined by the cold resistance of the reproductive organs, by the temperature conditions during embryogenesis and seed ripening, and by the freezing risk during germination and seedling establishment.

13.4.3.1 Susceptibility of Reproductive Organs

It is usually the reproductive organs of a plant that are most sensitive to chilling (e.g., male flower sterility in rice and sorghum; see Sect. 13.3.1.2) and frost.

Table 13.6 Frost resistance of flowers (selected examples; northern hemisphere)

Plant	Flowering time	First injury below °C	Reference
Viola wittrockiana	Winter	−15 to −18	SOTTOPIETRA (unpublished)
Senecio vulgaris	Winter	−15	SOTTOPIETRA (unpublished)
Anemone hepatica	Early spring	−11	TILL (1956)
Viburnum tinus	Autumn to spring	− 8	LARCHER (1970)
Rhamnus alaternus	Winter to spring	− 7	LARCHER (1970)
Rosmarinus officinalis	Winter to spring	− 7	LARCHER (1970)
Laurus nobilis	Early spring	− 5	LARCHER (1970)
Camellia vernalis	Winter to spring	− 5 to − 7	SAKAI and HAKODA (1979)
Corylus avellana ♂	February	−16	TILL (1956)
Corylus avellana ♀	April	− 4	TILL (1956)
Acer pseudoplatanus	April	− 2 to − 3	TILL (1956)
Quercus robur	April/May	− 2 to − 3	TILL (1956)
Fraxinus excelsior	April/May	− 2	TILL (1956)
Malus sylvestris	May	− 2 to − 3	PISEK (1958)
Pseudotsuga menziesii ♂	May	− 3	TIMMIS (1977)
Pseudotsuga menziesii ♀	May	− 2 to − 3	TIMMIS (1977)
Chamaedaphne calyculata	May	− 9	READER (1979)
Andromeda glaucophylla	May/June	− 5	READER (1979)
Kalmia polifolia	June	− 4	READER (1979)
Ledum groenlandicum	June/July	− 3	READER (1979)
Vaccinium macrocarpon	July	− 1	READER (1979)

The blossoms of most herbaceous and woody plants of the temperate zone freeze between −2 and −5 °C, those of plants flowering in winter to very early spring between −5 and −15 °C (Table 13.6). As a rule, the flower primordia in the buds of trees and shrubs are 2–5 K more sensitive than the apical meristem in leaf buds (cf. Table 13.4, Figs. 13.3, 13.4; LARCHER 1970; SAKAI 1979a; SAKAI and HAKODA 1979). Failure of propagation of *Tilia cordata* at the northern distribution limit is caused by insufficient pollen-tube growth due to low temperatures at flowering time and by incomplete seed development due to low summer temperatures (PIGOTT and HUNTLEY 1980).

13.4.3.2 Seedling Establishment

As a rule, germinating seeds and seedlings are sensitive in proportion to the intensity of growth (KEMMER and THIELE 1955; LEVITT 1956; CARY 1975). However, from the little information available, it seems that seedlings can acquire frost tolerance if their growth is interrupted by cold weather. Since the hardening capacity is not fully developed in seedlings and juvenile stages of certain plant species they do not achieve the same degree of resistance as adult individuals (LARCHER 1969b; FULLER and EAGLES 1978). As a consequence, if seedlings are likely to be exposed to frost, the survival capacity of the juvenile stages rather than the frost resistance of adult plants should be considered in ecological studies. Information on seedling resistance is particularly important in analyzing frost limitation of the distribution range of winter-germinating plants such as

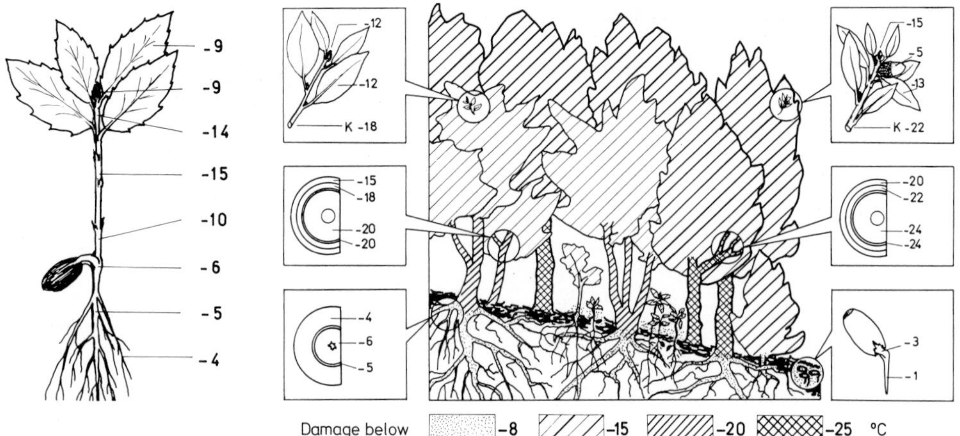

Fig. 13.6. Frost resistance of the various strata and age groups in a stand of *Quercus ilex*. (LARCHER and MAIR 1969)

most of the mediterranean species. This is impressively illustrated by a comparison of the various developmental stages in a population of the evergreen mediterranean oak *Quercus ilex* (Fig. 13.6). Even though the adult trees are able to thrive where temperatures of −12 °C occur each winter (e.g., in parks of the British Isles), natural propagation of the population is completely suppressed if winter temperatures regularly drop below −8 °C.

13.4.4 Cold Stress and the Distribution of Plants

The influence of cold stress on the limits of distribution becomes obvious when a comparison is made between the natural distribution of a species, or the areas in which it is successfully cultivated, and the area over which the plant could survive if the only limiting factor were its resistance to cold. At points where the *actual* limits of distribution coincide with those of the *potential* range with respect to low temperature resistance it is evident that cold stress can represent a limiting factor. The possibility of limits set by cold was already considered by DE CANDOLLE (1855). Further contributions on this topic have been reviewed by PARKER (1963), ALDEN and HERMANN (1971), and TUHKANEN (1980).

It is very probable that low winter temperatures are responsible for the northerly and altitudinal limits to the distribution of broadleaved evergreen woody plants of regions with mild winters (LAVAGNE and MUOTTE 1971; SAKAI 1971; SAKAI and WARDLE 1978; LARCHER 1981c) and of cacti (NOBEL 1980). The northern limits of the temperate deciduous forests of N. America and Eurasia, and of fruit cultivation, are set by the threshold for deep-supercooling of their xylem (−40 to −50 °C; GEORGE et al. 1974; QUAMME 1976; RAJASHEKAR and BURKE 1978; SAKAI 1978c). Episodic frosts may account in some places for the southern limits of the Miombo forests in Africa (ERNST 1971). In tropical

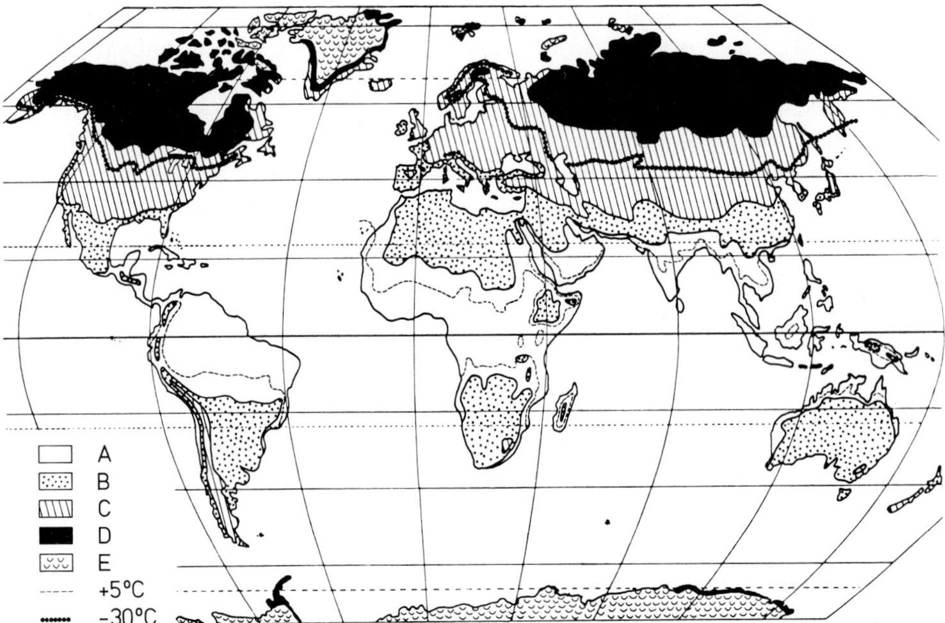

Fig. 13.7. Tentative map of low-temperature thresholds limiting plant distribution on the Earth. *A* frost-free zone; *B* zone with episodic frosts down to −10 °C; *C* zone with average annual minimum between −10 and −40 °C; *D* zone with average annual minimum below −40 °C; *E* polar ice; ---+5 °C lowest temperature isotherm; ...−30 °C average annual minimum isotherm. (Data of HOFFMANN 1960)

mountains episodic night frosts set the altitudinal limit for successful cultivation of coffee (SODERHOLM and GASKINS 1961) and potatoes (LI and PALTA 1978). Regularly occurring spring frosts have also been considered as the factor limiting the distribution of plants in the temperate and in the subarctic zone (FIRBAS 1949; KOTILAINEN 1950; TILL 1956).

In delimiting the potential range of distribution with respect to cold stress and cold resistance the absolute temperature minima measured in extremely cold winters a couple of times in each century cannot be employed. Although these cause considerable damage to the vegetation, the entire population of any species is seldom destroyed. It is the regularly recurring cold stress and long-term deterioration in climate, rather than catastrophic events, that set the limits to distribution. Thus, the means of the minima are better suited for comparisons with resistance data than the absolute extremes (EMBERGER 1932; LARCHER and MAIR 1969; BRISSE and GRANDJOUAN 1974; QUAMME 1976; SAKAI 1978d).

By employing indicator species with different degrees of cold resistance it is possible to subdivide larger regions into *winter survival zones* (U.S.A.: SKINNER 1962; Canada: OELLET and SHERK 1968; Japan: SAKAI and HAKODA 1979). Carrying the abstraction further, it might be attempted to design a global map of survival zones on the basis of characteristic stress temperatures (Fig. 13.7).

The following survival zones can be distinguished:

Zone A: The frost-free region on whose latitudinal and altitudinal limits the transition from chilling-sensitive to chilling-resistant plants must take place. The +5 °C isotherm limits the potential distribution of chilling-sensitive plants.

Zone B: Regions with episodic or periodic frost down to −10 °C. Plants that are sensitive to freezing throughout the year cannot exist beyond its limits.

Zone C: Regions with cold winters, which can only be survived by plants capable of becoming freezing-tolerant. The critical temperatures are taken to be average annual minima below −40 °C (near the threshold for deep-supercooling of xylem). Within Zone C the −30 °C isotherm marks a significant boundary since it apparently coincides with the limits of distribution of many herbaceous perennials that are unprotected in winter, as well as with the cold limits of propagation of many deciduous woody plants.

Zone D: The region with average winter minimum temperatures below − 40 °C can only be colonized by plants achieving freezing tolerance in all tissues, inclusive of bud meristem and xylem.

Zone E: Polar regions and the peaks of high mountains, where the temperature is so low throughout the entire year that the ground is permanently covered with snow and ice. Here, only those poikilohydric forms are found that can survive even the lowest temperatures of the Earth in a desiccated or frozen state.

13.5 Conclusions: Low Temperature Stress Ecology, a Challenging Task for Future Research

Research on cold resistance is concerned, on the one hand, with elucidating the factors responsible for death due to cold and for cold resistance, and on the other hand, with determining the specific resistance limits of the different plant species. As an *ecophysiological* discipline, cold resistance research uses both the results of basic studies and those of comparative studies, correlating this knowledge with data on the stress impact at particular habitats. An ecophysiological analysis must be comprehensive, covering each organ and each stage of development, as well as the seasonal and adaptive range of variation in resistance. So far, the ecological aspects of cold stress have received little attention as compared to that paid to basic research. Information on cold resistance or on the degree of endangerment of such ecologically characteristic plant types as desert or water plants, or parasites, is correspondingly sparse. Equally little is known of the range of variation in resistance of plants in different biomes, especially the tropics and subtropics. Of particular urgency is the need for field research, in order to verify laboratory experiences. Very little information is available concerning the effects of low, nonlethal temperatures on various physiological processes such as nutrition, production, and development. A wide and promising field of research awaits further investigation.

References

Alden J, Hermann RK (1971) Aspects of the cold-hardiness mechanism in plants. Bot Rev 37:37–142

Altman PL, Dittmer DS (1973) Biology data book, Vol. II, 2nd edn. Fed Am Soc Exp Biol, Bethesda

Andrews CJ, Pomeroy MK (1977) Changes in survival of winter cereals due to ice cover and other simulated winter conditions. Can J Plant Sci 57:1141–1149

Antropova TA (1974) Sezonnye izmeneniya kholodo- i teploustoichivosti kletok dvukh vidov mkhov. (The seasonal changes of cold and heat resistance of cells in two moss species). Bot Zh 59:117–122

Bamberg S, Schwarz W, Tranquillini W (1967) Influence of daylength on the photosynthetic capacity of stone pine (*Pinus cembra* L.). Ecology 48:264–269

Barta AL, Hodges HF (1970) Characterization of photosynthesis in cold hardening winter wheat. Crop Sci 10:535–538

Bauer H, Huter M, Larcher W (1969) Der Einfluß und die Nachwirkung von Hitze- und Kältestreß auf den CO_2-Gaswechsel von Tanne und Ahorn. Ber Dtsch Bot Ges 82:65–70

Bauer H, Harasser J, Bendetta G, Larcher W (1971) Jahresgang der Temperaturresistenz junger Holzpflanzen im Zusammenhang mit ihrer jahreszeitlichen Entwicklung. Ber Dtsch Bot Ges 84:561–570

Bauer H, Larcher W, Walker RB (1975) Influence of temperature stress on CO_2-gas exchange. In: Cooper JP (ed) Photosynthesis and productivity in different environments. Cambridge Univ Press, Cambridge, pp 557–586

Biebl R (1939) Über die Temperaturresistenz von Meeresalgen verschiedener Klimazonen und verschieden tiefer Standorte. Jahrb wiss Bot 88:389–420

Biebl R (1958) Temperatur- und osmotische Resistenz von Meeresalgen der bretonischen Küste. Protoplasma 50:218–242

Biebl R (1962a) Protoplasmatische Ökologie der Pflanzen. Wasser und Temperatur. Protoplasmatologia XII/1. Springer, Wien

Biebl R (1962b) Temperaturresistenz tropischer Meeresalgen. (Verglichen mit jener der Algen in temperierten Meeresgebieten). Bot Marina 4:241–254

Biebl R (1964) Temperaturresistenz tropischer Pflanzen auf Puerto Rico. Protoplasma 59:133–156

Biebl R (1967a) Temperaturresistenz einiger Grünalgen warmer Bäche auf Island. Botaniste 50:33–42

Biebl R (1967b) Temperaturresistenz tropischer Urwaldmoose. Flora 157:25–30

Biebl R (1968) Über Wärmehaushalt und Temperaturresistenz arktischer Pflanzen in Westgrönland. Flora Abt B 157:327–354

Biebl R (1969) Untersuchungen zur Temperaturresistenz arktischer Süßwasseralgen im Raum von Barrow, Alaska. Mikroskopie 25:3–6

Biebl R (1970) Vergleichende Untersuchungen zur Temperaturresistenz von Meeresalgen entlang der pazifischen Küste Nordamerikas. Protoplasma 69:61–63

Biebl R (1972) Studien zur Temperaturresistenz der Gezeitenalge *Ulva pertusa* Kjellmann. Bot Marina 15:139–143

Borisov AA (1965) Climates of the USSR (Engl. transl.) Oliver & Boyd, Edinburgh

Boutton TW, Harrison AT, Smith BN (1980) Distribution of biomass of species differing in photosynthetic pathway along an altitudinal transect in southeastern Wyoming grassland. Oecologia 45:287–298

Bradbury IK, Malcolm DC (1978) Dry matter accumulation by *Picea sitchensis* seedlings during winter. Can J For Res 8:207–213

Brisse H, Grandjouan G (1974) Classification climatique des plantes. Oecol Plant 9:51–80

Brock TD (1978) Thermophilic microorganisms and life at high temperatures. Springer, New York, Heidelberg, Berlin

Brooking IR (1976) Male sterility in *Sorghum bicolor* (L.) Moench induced by low night temperature. I. Timing of the stage of sensitivity. Aust J Plant Physiol 3:589–596

Burke MJ, Stushnoff C (1979) Frost hardiness: A discussion of possible molecular causes of injury with particular reference to deep supercooling of water. In: Mussel H, Staples R (eds) Stress physiology in crop plants. Wiley and Sons, New York, pp 197–225

Burke MJ, Gusta LV, Quamme HA, Weiser CJ, Li PH (1976) Freezing and injury in plants. Ann Rev Plant Physiol 27:507–528

Cary JW (1975) Factors affecting cold injury of sugarbeet seedlings. Agron J 67:258–262

Chandler WH (1954) Cold resistance in horticultural plants: a review. Proc Am Soc Hortic Sci 64:552–572

Christiansen MN (1963) Influence of chilling upon seedling development of cotton. Plant Physiol 38:520–522

Christiansen MN (1979) Physiological bases for resistance to chilling. Hort Sci 14:583–586

Christophersen J (1973) Basic aspects of temperature action on microorganisms. In: Precht H, Christophersen J, Hensel H, Larcher W (eds) Temperature and life. Springer, Berlin, Heidelberg, New York, pp 3–59

Clausen E (1964) The tolerance of hepatics to desiccation and temperature. Bryologist 67:411–417

Crookston RK, O'Toole J, Lee R, Ozbun JL, Wallace DH (1974) Photosynthetic depression in beans after exposure to cold for one night. Crop Sci 14:457–464

Davies MD, Dickinson DB (1971) Effects of freeze-drying on permeability and respiration of germinating lily pollen. Plant Physiol 24:5–9

De Candolle A (1855) Géographie botanique. Masson, Paris

Dereuddre J (1978) Effets de divers types de refroidissements sur la teneur en eau et sur la résistance au gel des bourgeons de rameaux d'Épicea en vie ralentie. Physiol Veg 16:469–489

Dietrichson J (1969) The geographic variation of springfrost resistance and growth cessation in Norway spruce (*Picea abies* (L.) Karst.). Norske Skofors. Vollebekk 94–106

Dircksen A (1964) Vergleichende Untersuchungen zur Frost-, Hitze- und Austrocknungsresistenz einheimischer Laub- und Lebermoose unter besonderer Berücksichtigung jahreszeitlicher Veränderungen. Dissertation Göttingen

Drew AP, Bazzaz FA (1979) Response of stomatal resistance and photosynthesis to night temperature in *Populus deltoides*. Oecologia 41:89–98

Dunn EL (1975) Environmental stresses and inherent limitations affecting CO_2 exchange in evergreen sclerophylls in mediterranean climates. In: Gates DM, Schmerl RB (eds) Perspectives of biophysical ecology. Ecological studies Vol. 12. Springer, Berlin, Heidelberg, New York, pp 159–181

Duthie HC (1964) The survival of desmids in ice. Brit Phyc Bull 2:376–377

Eckardt FE, Heim G, Methy M, Sauvezon R (1975) Interception de l'énergie rayonnante, échanges gazeux et croissance dans une fôret méditerranéenne à feuillage persistant (*Quercetum ilicis*). Photosynthetica 9:145–156

Edlich F (1936) Einwirkung von Temperatur und Wasser auf aerophile Algen. Arch Mikrobiol 7:62–109

Eguchi T, Sakai A, Usui G, Uehara T (1966) Studies of selection of frosthardy *Cryptomeria* I. Silvae Genetica 15:61–100

Ehleringer JR (1978) Implications of quantum yield differences on the distributions of C_3 and C_4 grasses. Oecologia 31:255–267

Eldrige KG (1968) Physiological studies of altitudinal variation in *Eucalyptus regnans*. Ecol Soc Aust Proc 3:70–76

Emberger L (1932) Sur une formule climatique et ses applications en botanique. La Météorologie 92:1–10

Ernst W (1971) Zur Ökologie der Miombo-Wälder. Flora 160:317–331

Firbas F (1949) Spät- und nacheiszeitliche Waldgeschichte Mitteleuropas nördlich der Alpen. Bd. 1: Allgemeine Waldgeschichte. Fischer, Jena

Fowler DB, Dvorak J, Gusta LV (1977) Comparative cold hardiness of several *Triticum* species and *Secale cereale* L. Crop Sci 17:941–943

France RC, Cline ML, Reid CPP (1979) Recovery of ectomycorrhizal fungi after exposure to subfreezing temperatures. Can J Bot 57:1845–1848

Fuller MP, Eagles CF (1978) A seedling test for cold hardiness in *Lolium perenne* L. J Agric Sci Camb 91:217–222

George MF, Burke MJ, Pellett HM, Johnson AG (1974) Low temperature exotherms and woody plant distribution. Hort Sci 9:519–522

Göppert HR (1978) as referred by Kol E (1968) Kryobiologie. I. Kryovegetation. Schweizerbart, Stuttgart

Goryshina TK (1972) Recherches écophysiologiques sur les plantes éphéméroides printaniéres dans les chênaies de la zone forêt-steppe de la russie centrale, Oecol Plant 7:241–258

Gusta LV, Fowler DB (1979) Cold resistance and injury in winter cereals. In: Mussell H, Staples R (eds) Stress physiology in crop plants. Wiley and Sons, New York, pp 159–178

Häckel H (1978) Modellrechnungen über die Temperaturen von Pflanzen in winterlichen Strahlungsnächten. Agric Meteorol 19:497–504

Hagem O (1962) Additional observations on the dry matter increase of coniferous seedlings in winter. Investigations in an oceanic climate. Medd Vestl Forstl Forsoeksstn 37:249–347

Havis JR (1976) Root hardiness of woody ornamentals. Hort Sci 11:385–386

Heber U, Santarius KA (1973) Cell death by cold and heat, and resistance to extreme temperatures. Mechanisms of hardening and dehardening. In: Precht H, Christophersen J, Hensel H, Larcher W (eds) Temperature and life. Springer, Berlin, Heidelberg, New York, pp 232–263

Hoffmann G (1959) Die mittleren jährlichen und absoluten Extremtemperaturen der Erde. III. Tabellen. Met Abh 8, Heft 4, Reimer, Berlin

Hoffmann G (1960) Die mittleren jährlichen und absoluten Extremtemperaturen der Erde. II. Ergebnisse. Met Abh 8, Heft 3. Reimer, Berlin

Hoffmann G (1963) Die höchsten und tiefsten Temperaturen auf der Erde. Umschau 63:16–18

Holm-Hansen O (1963) Viability of blue-green and green algae after freezing. Physiol Plant 16:530–540

Hudson MA, Brustkern P (1965) Resistance of young and mature leaves of *Mnium undulatum* (L.) to frost. Planta 66:135–155

Hwang S (1968) Investigation of ultra-low temperature for fungal cultures. I. An evaluation of liquid-nitrogen storage for preservation of selected fungal cultures. Mycologia 60:613–621

Irmscher E (1912) Über die Resistenz der Laubmoose gegen Austrocknung und Kälte. Jahrb wiss Bot 50:387–449

Ivory DA, Whiteman PC (1978) Effects of environmental and plant factors on foliar freezing resistance in tropical grasses. II. Comparison of frost resistance between cultivars of *Cenchrus ciliaris*, *Chloris gayana* and *Setaria anceps*. Aust J Agric Res 29:261–266

Kacperska-Palacz A (1978) Mechanism of cold acclimation in herbaceous plants. In: Li PH, Sakai A (eds) Plant cold hardiness and freezing stress. Academic Press, New York, pp 139–152

Kainmüller C (1975) Temperaturresistenz von Hochgebirgspflanzen. Anz math.-naturw Klasse Österr Akad Wiss:67–75

Kaku S, Iwaya M (1978) Low temperature exotherms in xylems of evergreen and deciduous broadleaved trees in Japan with references to freezing resistance and distribution range. In: Li PH, Sakai A (eds) Plant cold hardiness and freezing stress. Academic Press, New York, pp 227–239

Kallio P, Kärenlampi L (1975) Photosynthesis in mosses and lichens. In: Cooper JP (ed) Photosynthesis and productivity in different environments. Cambridge Univ Press, Cambridge, pp 393–423

Kappen L (1964) Untersuchungen über den Jahreslauf der Frost-, Hitze- und Austrocknungsresistenz von Sporophyten einheimischer Polypodiaceen. Flora 155:124–166

Kappen L (1965) Untersuchungen über die Widerstandsfähigkeit der Gametophyten einheimischer Polypodiaceen gegenüber Frost, Hitze und Trockenheit. Flora 156:101–116

Kappen L (1966) Der Einfluß des Wassergehaltes auf die Widerstandsfähigkeit von Pflanzen gegenüber hohen und tiefen Temperaturen, untersucht an Blättern einiger Farne und *Ramonda myconi*. Flora 156:427–445

Kappen L (1969) Frostresistenz einheimischer Halophyten in Beziehung zu ihrem Salz-, Zucker- und Wassergehalt im Sommer und Winter. Flora B 158:232–260

Kappen L, Lange OL (1970a) Kälteresistenz von Flechten aus verschiedenen Klimagebieten. Dtsch Bot Ges Neue Folge 4:61–65

Kappen L, Lange OL (1970b) The cold resistance of phycobionts from macrolichens of various habitats. Lichenologist 4:289–293

Kappen L, Lange OL (1972) Die Kälteresistenz einiger Makrolichenen. Flora 161:1–29

Kärcher H (1931) Über Kälteresistenz einiger Pilze und Algen. Planta 14:515–516

Kaufmann MR (1977) Soil temperature and drying cycle effects on water relations of *Pinus radiata*. Can J Bot 55:2413–2418

Keller Th (1978) Frostschäden als Folge einer „latenten" Immissionsschädigung. Staub Reinhalt Luft 1978:24–26

Kemmer E, Thiele I (1955) Frostresistenzprüfungen an keimenden Kernobstsamen. Züchter 25:57–60

Kislyuk IM, Vas'kovsky MD (1972) Vliyanie okhlazhdeniya list'ev ogurtsa na fotosintez i fotokhimicheskie reaktsii. (Effect of cooling of cucumber leaves on photosynthesis and photochemical reactions). Fiziol Rast 19:813–818

Klosson RJ, Krause GH (1981) Freezing injury in cold-acclimated and unhardened spinach leaves. Planta 151:339–346

Koh S, Kumura A (1973) Studies on matter production in wheat plant. I. Diurnal changes in carbon dioxide exchange of wheat plant under field conditions. Proc Crop Sci Soc Jpn 42:227–235

Koh S, Kumura A, Murata Y (1978) Studies on matter production in wheat plant. V. The mechanism involved in an after-effect of low night temperature. Jpn J Crop Sci 47:75–81

Kotilainen MJ (1950) Über die Frostschäden an wilden Pflanzen. Proc Finnish Acad Sci Letters 1948, Helsinki

Kozlowski TT, Keller Th (1966) Food relations of woody plants. Bot Rev 32:293–382

Kramer PJ, Kozlowski TT (1979) Physiology of woody plants. Academic Press, New York

Kramer JP, Wetmore TH (1943) Effects of defoliation on cold resistance and diameter growth of broad leaved evergreens. Am J Bot 30:428–431

Krasavtsev OA (1960) Zakalivanie drevesnykh rastenii k morozu. Trudy Konf. Fiziol ustoichivost rastenii. Nauka, Moskva, pp 229–234

Kusumoto T (1959) Physiological and ecological studies on the plant production in plant communities. 7. On the resistance of evergreen broad-leaved trees to cold temperature. Bull Educat Res Inst, Univ Kagoshima 11:48–55

Lakhanov AP, Balachkova NE (1978) Ustoivost zernobobovykh kultur k nizkim polozhitelnym temperaturam v protsesse ontogeneza rastenii. (Resistance of legumes to low positive temperatures in the process of plant ontogenesis). Fiziol Rast 25:592–597

Lange OL (1965) Der CO_2-Gaswechsel von Flechten bei tiefen Temperaturen. Planta 64:1–19

Langlet O (1937) Studier över tallens fysiologiska variabilitet och dess samband med klimatet. Medd Statens Skogsförs Anst 29:421–470

Larcher W (1954) Die Kälteresistenz mediterraner Immergrüner und ihre Beeinflußbarkeit. Planta 44:607–638

Larcher W (1961a) Jahresgang des Assimilations- und Respirationsvermögens von *Olea europaea* L. ssp. *sativa* Hoff. et Link., *Quercus ilex* L. and *Quercus pubescens* Willd. aus dem nördlichen Gardaseegebiet. Planta 56:575–606

Larcher W (1961b) Zur Assimilationsökologie der immergrünen *Olea europaea* und *Quercus ilex* und der sommergrünen *Quercus pubescens* im nördlichen Gardaseegebiet. Planta 56:607–617

Larcher W (1963) Winterfrostschäden in den Parks und Gärten von Arco und Riva am Gardasee. Veroeff Museum Ferdinandeum Innsbruck 43:153–199

Larcher W (1968) Die Temperaturresistenz als Konstitutionsmerkmal der Pflanzen. Dtsch Akad Landwirtschaftswiss, Tagungsbericht 100:7–21

Larcher W (1969a) The effect of environmental and physiological variables on the carbon dioxide gas exchange of trees. Photosynthetica 3:167–198

Larcher W (1969b) Zunahme des Frostabhärtungsvermögens von *Quercus ilex* im Laufe der Individualentwicklung. Planta 88:130–135

Larcher W (1970) Kälteresistenz und Überwinterungsvermögen mediterraner Holzpflanzen. Oecol Plant 5:267–286

Larcher W (1971) Die Kälteresistenz von Obstbäumen und Ziergehölzen subtropischer Herkunft. Oecol Plant 6:1–14 (1971)

Larcher W (1973) Gradual progress of damage due to temperature stress. Temperature resistance and survival. In: Precht H, Christophersen J, Hensel H, Larcher W (eds) Temperature and life. Springer, Berlin, Heidelberg, New York, pp 194–213

Larcher W (1975) Pflanzenökologische Beobachtungen in der Páramostufe der venezolanischen Anden. Anz Oesterr Akad Wiss Math-naturw Kl, 194–213

Larcher W (1977) Ergebnisse des IBP-Projekts „Zwergstrauchheide Patscherkofel". Sitz Ber Oesterr Akad Wiss, Math-naturw Kl, Abt I, 186:301–371

Larcher W (1980a) Untersuchungen zur Frostresistenz von Palmen. Anz Oesterr Akad Wiss Math-naturw Kl, Jg 1980, 3:1–12

Larcher W (1980b) Klimastreß im Gebirge – Adaptationstraining und Selektionsfilter für Pflanzen. Rheinisch-Westfäl Akad Wiss Vorträge, N 291:49–88. Westdeutscher Verlag, Opladen

Larcher W (1981a) Effects of low temperature stress and frost injury on plant productivity. In: Johnson CD (ed) Physiological processes limiting plant productivity. Butterworths, London, pp 253–269

Larcher W (1981b) Resistenzphysiologische Grundlagen der evolutiven Kälteakklimatisation von Sproßpflanzen. Plant Syst Evol 137:145–180

Larcher W (1981c) Low temperature effects on Mediterranean sclerophylls: An unconventional viewpoint: In: Margaris NS, Mooney HA (eds) Components of productivity of Mediterranean regions, Basic and applied aspects. Junk, The Hague, in press

Larcher W, Bodner M (1980) Dosisletalität-Nomogramm zur Charakteristik der Erkältungsempfindlichkeit tropischer Pflanzen. Angew Bot 54:273–278

Larcher W, Eggarter H (1960) Anwendung des Triphenyltetrazoliumchlorids zur Beurteilung von Frostschäden in verschiedenen Achsengeweben bei *Pirus*-Arten, und Jahresgang der Resistenz: Protoplasma 51:595–619

Larcher W, Mair B (1969) Die Temperaturresistenz als ökophysiologisches Konstitutionsmerkmal: 1. *Quercus ilex* und andere Eichenarten des Mittelmeergebietes. Oecol Plant 4:347–376

Larcher W, Wagner J (1976) Temperaturgrenzen der CO_2-Aufnahme und Temperaturresistenz der Blätter von Gebirgspflanzen im vegetationsaktiven Zustand. Oecol Plant 11:361–374

Larcher W, Winter A (1982) Frost susceptibility of palms: Experimental data and their interpretation. Principes 26, in press

Lasley SE, Garber MP, Hodges CF (1979) Aftereffects of light and chilling temperatures on photosynthesis in excised cucumber cotyledons. J Am Soc Hortic Sci 104:477–480

Lavagne A, Muotte P (1971) Premiéres observations chorologiques et phénologiques sur les ripisilves á *Nerium oleander* (Neriaies) en Provence. Ann Univ Provence Marseille 15:135–155

Law CN, Jenkins GA (1970) A genetic study of cold resistance in wheat. Gen Res Cambridge 15:197–308

Layton C, Parsons RF (1972) Frost resistance of seedlings of two ages of some southern Australian woody species. Bull Torrey Bot Club 99:118–122

Levitt J (1956) The hardiness of plants. Academic Press, New York

Levitt J (1958) Frost, drought, and heat resistance. Protoplasmatologia VIII/6. Springer, Wien

Levitt J (1969) Growth and survival of plant at extremes of temperature – A unified concept. In: Dormancy and survival 23rd. Symp Soc Exp Biol, pp 395–448

Levitt J (1980) Responses of plants to environmental stresses. Vol 1, Chilling, freezing, and high temperature stresses. 2nd edn, Academic Press, New York

Li PH, Palta JP (1978) Frost hardening and freezing stress in tuber-bearing *Solanum* species. In: Li PH, Sakai A (eds) Plant cold hardiness and freezing stresses. Academic Press, New York, pp 49–71

Li PH, Sakai A (eds) (1978) Plant cold hardiness and freezing stress. Mechanisms and implications. Academic Press, New York

Lin SS, Peterson ML (1975) Low temperature-induced floret sterility in rice. Crop Sci 15:657–660

Lindegren RM (1933) Decay of wood and growth of some Hymenomycetes as affected by temperature. Phytopathology 23:72–81

Lipman CB (1936) The tolerance of liquid air temperatures by dry moss protonema. Bull Torrey Bot Club 63:515–518

Lona F (1963) Caratteristiche termoperiodiche e di resistenza al freddo di alcune piante dei deserti circummediterranei. Giron Bot Ital 70:565–574

Lozina-Lozinskii LK (1974) Studies in cryobiology. Wiley and Sons, New York

Ludlow MM, Wilson GL (1971) Photosynthesis of tropical pasture plants. I. Illuminance, carbon dioxide concentration, leaf temperature, and leaf-air vapour pressure difference. Aust J Biol Sci 24:449–470

Lundquist V, Pellett H (1976) Preliminary survey of cold hardiness levels of several bulbous ornamental plant species. Hort Sci 11:161–162

Lutz JM, Hardenburg RE (1968) The commercial storage of fruits, vegetables, and florist and nursery stocks. Agric. Handbook No. 66, Washington, US Gov. Printing Office

Lydolph PE (1977) Climates of the Soviet Union. In: Landsberg HE (ed) World survey of climatology. Elsevier, Amsterdam, Vol. 7

Lyons JM (1973) Chilling injury in plants. Ann Rev Plant Physiol 24:445–466

Lyons JM, Breidenbach RW (1979) Strategies for altering chilling sensitivity as a limiting factor in crop production. In: Mussel H, Staples RG (eds) Stress physiology in crop plants. Wiley and Sons, New York, pp 179–196

Lyons JM, Graham D, Raison JK (eds) (1979) Low temperature stress in crop plants. The role of the membrane. Academic Press, New York

Maier M, Kappen L (1979) Cellular compartmentalization of salt ions and protective agents with respect to freezing tolerance of leaves. Oecologia 38:303–316

Mair B (1968) Frosthärtegradienten entlang der Knospenfolge auf Eschentrieben. Planta 82:164–169

Marini RP, Boyce BR (1977) Susceptibility of crown tissues of "Catskill" strawberry plants to low-temperature injury. J Am Soc Hortic Sci 102:515–516

Mazur P (1966) Physical and chemical basis of injury in single-celled microorganisms subjected to freezing and thawing. In: Meryman HT (ed) Cryobiology. Academic Press, London, pp 214–315

Mazur P (1969) Freezing injury in plants. Ann Rev Plant Physiol 20:419–448

McConnell DB, Sheehan TJ (1978) Anatomical aspects of chilling injury to leaves of *Phalaenopsis* Bl. Hort Sci 13:705–706

McMillan C (1975) Adaptive differentiation to chilling in mangrove populations. In: Walsh GE, Snedaker SC, Teas HJ (eds) Proc Int Symp Biol Managemt Mangroves, Univ Florida Press, Gainesville, pp 62–68

McMillan C (1979) Differentiation in response to chilling temperatures among populations of three marine spermatophytes, *Thalassia testudinum*, *Syringodium filiforme* and *Halodules wrightii*. Am J Bot 66:810–819

McWilliams EL, Smith CW (1978) Chilling injury in *Scindapsus pictus*, *Aphelandra squarrosa* and *Maranta leuconeura*. Hort Sci 13:179–180

Migita S (1966) Freeze-preservation of *Porphyra* thalli in viable state – II. Effect of cooling velocity and water content of thalli on the frost-resistance. Bull Fac Fisheries, Nagasaki Univ 21:131–138

Miller JD (1976) Cold tolerance in sugar cane relatives. Sugar y Azucar, March 1976

Mittelstädt H (1965) Beiträge zur Züchtungsforschung beim Apfel. VIII. Untersuchungen zur Frostresistenz an Sorten, Unterlagen und Zuchtmaterial. Züchter 35:311–327

Molisch H (1897) Untersuchungen über das Erfrieren der Pflanzen. Fischer, Jena

Moser M (1958) Der Einfluß tiefer Temperaturen auf das Wachstum und die Lebenstätigkeit höherer Pilze mit spezieller Berücksichtigung von Mykorrhizapilzen. Sydowia 12:386–399

Murawski H (1961) Beiträge zur Züchtungsforschung beim Apfel. VI. Untersuchungen über die Vererbung der Frostresistenz an Sämlingen der Sorten Glogierowka und Jonas Hannes. Züchter 31:52–57

Nath J, Anderson JO (1975) Effect of freezing and freeze-drying on the viability and storage of *Lilium longiflorum* L and *Zea mays* L. pollen. Cryobiology 12:81–88

Negisi K (1966) Photosynthesis, respiration and growth in 1-year-old seedlings of *Pinus densiflora, Cryptomeria japonica* and *Chamaecyparis obtusa*. Bull Tokyo Univ Forests 62:1–115

Neilson RE, Ludlow MM, Jarvis PG (1972) Photosynthesis in sitka spruce *Picea sitchensis* (Bong.) Carr. II. Response to temperature. J Appl Ecol 9:721–745

Nissila PC, Fuchigami LH (1978) The relationship between vegetative maturity and the first stage of cold acclimation. J Am Soc Hortic Sci 103:710–711

Noack K (1912) Beiträge zur Biologie der thermophilen Organismen. Jahrb Wiss Bot 51:593–648

Nobel PS (1980) Influences of minimum stem temperatures on ranges of cacti in southwestern United States and Central Chile. Oecologia 47:10–15

Noshiro M, Sakai A (1979) Freezing resistance of herbaceous plants. Low Temp Sci Ser B 37:11–18

Oellet CE, Sherk LC (1968) Zones de rusticité pour les plantes au Canada. Ministère de l'Agric Canada

Olien CR (1964) Freezings processes in the crown of "Hudson" barley, *Hordeum vulgare* (L. emend. Lam.) Hudson. Crop Sci 4:91–95

Olien CR (1967) Freezing stresses and survival. Ann Rev Plant Physiol 18:387–408

Öquist G, Brunes L, Hällgren J-E, Gezelius K, Hallén M, Malmberg G (1980) Effects of artificial frost hardening and winter stress on net photosynthesis, photosynthetic electron transport and RuBP carboxylase activity in seedlings of *Pinus silvestris*. Physiol Plant 48:526–531

Orvig S (1970) Climates of the polar regions. In: Landsberg HE (ed) World survey of climatology. Elsevier, Amsterdam, Vol 14

Palta JP, Li PH (1979) Frost hardiness in relation to leaf anatomy and natural distribution of several *Solanum* species. Crop Sci 19:665–671

Parker J (1960) Survival of woody plants at extremely low temperatures. Nature 187: 1133

Parker J (1961) Seasonal trends in carbon dioxide absorption, cold resistance, and transpiration of some evergreeens. Ecology 42:372–380

Parker J (1962) Seasonal changes in cold resistance and free sugars of some hardwood tree barks. For Sci 8:255–262

Parker J (1963) Cold resistance in woody plants. Bot Rev 29:123–201

Patterson BD, Murata T, Graham D (1976) Electrolyte leakage induced by chilling in *Passiflora* species tolerant to different climates. Aust J Plant Physiol 3:435–442

Patterson BD, Paull R, Smillie RM (1978) Chilling resistance in *Lycopersicon hirsutum* Humb, & Bonpl., a wild tomato with a wide altitudinal distribution. Aust J Plant Physiol 5:609–617

Peoples TR, Koch DW (1978) Physiological response of three alfalfa cultivars to one chilling night. Crop Sci 18:255–258

Pharis RP, Hellmers H, Schuurmans E (1970) Effects of subfreezing temperatures on photosynthesis of evergreen conifers under controlled environment conditions. Photosynthetica 4:273–279

Pigott CD, Huntley JP (1980) Factors controlling the distribution of *Tilia cordata* at the northern limits of its geographical range. III. Nature and causes of seed sterility. New Phytol, in press

Pisek A (1958) Versuche zur Frostresistenzprüfung von Rinde, Winterknospen und Blüte einiger Arten von Obsthölzern. Gartenbauwissenschaft 23:54–74

Pisek A (1960) Immergrüne Pflanzen. In: Ruhland W (ed) Handbuch der Pflanzenphysiologie. Springer, Berlin, Heidelberg, New York, Vol. V, Part II, pp 415–459

Pisek A (1973) Photosynthesis. In Precht H, Christophersen J, Hensel H, Larcher W (eds) Temperature and life. Springer, Berlin, Heidelberg, New York, pp 102–127

Pisek A, Kemnitzer R (1968) Der Einfluß von Frost auf die Photosynthese der Weißtanne (*Abies alba* Mill.). Flora B 157:314–326

Pisek A, Schießl R (1947) Die Temperaturbeeinflußbarkeit der Frosthärte von Nadelhölzern und Zwergsträuchern an der alpinen Waldgrenze. Ber Naturwiss- med Ver Innsbruck 47:33–52

Pisek A, Winkler E (1958) Assimilationsvermögen und Respiration der Fichte (*Picea excelsa* Link.) in verschiedener Höhenlage und der Zirbe (*Pinus cembra* L.) an der alpinen Waldgrenze. Planta 51:518–543

Pisek A, Larcher W, Unterholzner R (1967) Kardinale Temperaturbereiche der Photosynthese und Grenztemperaturen des Lebens der Blätter verschiedener Spermatophyten I. Temperaturminimum der Nettoassimilation, Gefrier- und Frostschadensbereiche der Blätter. Flora B 157:239–264

Pollard DFW, Wareing PF (1968) Rates of dry matter production in forest tree seedlings. Ann Bot 32:573–591

Pollock BM, Toole VK (1966) Imbibition period as the critical temperature sensitive stage in germination of lima bean seeds. Plant Physiol 41:221–229

Quamme HA (1976) Relationship of the low temperature exotherm to apple and pear production in North America. Can J Plant Sci 56:493–500

Quamme HA (1978) Mechanism of supercooling in overwintering peach flower buds. J Am Soc Hortic Sci 103:57–61

Rajashekar C, Burke MJ (1978) The occurrence of deep undercooling in the genera *Pyrus, Prunus,* and *Rosa*: A preliminary report. In: Li PH, Sakai A (eds) Plant cold hardiness and freezing stress. Academic Press, New York, pp 213–225

Rakitina ZG (1970) Vliyanie aeratsii na morozostojkost kornevykh sistem drevesnykh rasteniy. (Effect of aeration of the root frost-resistance in woody plants). Fiziol Rast 17:808–818

Rakitina ZG (1977) Vliyanie pritertoj ledyanoi korki na rasteniya ozimoi pshenitsy v zavistimosti ot usloviy ikh zatopleniya do zamorazhivaniya. (Effect of adjacent ice crust on winter wheat depending on conditions of its inundation prior to freezing). Fiziol Rast 24:403–411

Raunkiaer C (1910) Statistik der Lebensformen als Grundlage für die biologische Pflanzengeographie. Beih Bot Cbl 27/II:171–206d

Reader RJ (1979) Flower cold hardiness: a potential determinant of the flowering sequence exhibited by bog ericads. Can J Bot 57:997–999

Regehr DL, Bazzaz FA (1976) Low temperature photosynthesis in successional winter annuals. Ecology 57:1297–1303

Riedmüller-Schölm HE (1974) The temperature resistance of Alaskan plants from the continental boreal zone. Flora 163:230–250

Rowley JA, Taylor AO (1972) Plants under climatic stress. IV. Effects of CO_2 and O_2 on photosynthesis under high-light, low-temperature stress. New Phytol 71:477–481

Rowley JA, Tunnicliffe CG, Taylor AO (1975) Freezing sensitivity of leaf tissue of C_4 grasses. Aust J Plant Physiol 2:447–451

Sakai A (1960) Survival of the twig of woody plants at −196 °C. Nature 185:393–394

Sakai A (1965) Survival of plant tissue at super low temperatures. III. Relation between effective prefreezing temperatures and the degree of frost hardiness. Plant Physiol 40:882–887

Sakai A (1971) Freezing resistance of relicts from the arcto-tertiary flora. New Phytol 70:1199–1205

Sakai A (1972) Freezing resistance of evergreen and broad-leaf trees indigenous to Japan. J Jap For Soc 54:333–339

Sakai A (1978a) Low temperature exotherm of winter buds of hardy conifers. Plant Cell Physiol 19:1439–1446

Sakai A (1978b) Frost hardiness of flowering and ornamental trees. J Jap Soc Hortic Sci 47:248–260

Sakai A (1978c) Freezing tolerance of evergreen and deciduous broadleaved trees in Japan with reference to tree regions. Low Temp Sci Ser B 36:1–19

Sakai A (1978d) Freezing tolerance of primitive willows ranging to subtropics and tropics. Low Temp Sci Ser B 36:21–29

Sakai A (1979a) Deep supercooling of winter flower buds of *Cornus florida* L. Hortic Sci 14:69–70

Sakai A (1979b) Freezing avoidance mechanism of primordial shoots of conifer buds. Plant Cell Physiol 20:1381–1390

Sakai A, Hakoda N (1979) Cold hardiness of the genus *Camellia*. Am Soc Hortic Sci 104:53–57

Sakai A, Miwa S (1979) Frost hardiness of Ericoideae. Am Soc Hortic Sci 104:26–28

Sakai A, Okada S (1971) Freezing resistance of conifers. Silvae Genetica 20:53–100

Sakai A, Otsuka K (1970) Freezing resistance of alpine plants. Ecology 51:665–671

Sakai A, Wardle P (1978) Freezing resistance of New Zealand trees and shrubs. N Z J Ecol 1:51–61

Sakai A, Weiser CJ (1973) Freezing resistance of trees in North America with reference to tree regions. Ecology 54:118–126

Scheumann W (1968) Die Dynamik der Frostresistenz und ihre Bestimmung an Gehölzen im Massentest. Dtsch Akad Landwirt Wiss Berlin, Tagungsbericht 100:45–54

Scheumann W, Hoffmann K (1967) Die serienmäßige Prüfung der Frostresistenz einjähriger Fichtensämlinge. Arch Forstwesen 16:701–705

Schnetter ML (1965) Frostresistenzuntersuchungen an *Bellis perennis, Plantago media* und *Helleborus niger* im Jahresablauf. Biol Cbl 84:469–487

Schölm HE (1968) Untersuchungen zur Hitze- und Frostresistenz einheimischer Süßwasseralgen. Protoplasma 65:97–118

Scorza R, Wiltbank WJ (1976) Measurement of avocado cold hardiness. Hort Sci 11:267–268

Seeley EJ, Kammereck R (1977) Carbon fluxes in apple trees: Use of a closed system to study the effect of a mild cold stress on "Golden Delicious". J Am Soc Hortic Sci 102:282–286

Seible D (1939) Ein Beitrag zur Frage der Kälteschäden an Pflanzen bei Temperaturen über dem Gefrierpunkt. Beitr Biol Pflanz 26:289–330

Senser M, Beck E (1977) On the mechanisms of frost injury and frost hardening of spruce chloroplasts. Planta 137:195–201

Simura T (1957) Breeding of polyploid varieties of the tea plant with special reference to their cold resistance. Cytologia, Proc Int Genetics Symp 1956, pp 321–324

Skinner HT (1962) The geographic charting of plant climatic adaptability. 15th Int Cgr Hort, Nizza 1958, Proc 3:485–491

Smillie RM (1979) The useful chloroplast: A new approach for investigating chilling stress in plants. In: Lyons JM, Graham D, Raison JK (eds) Low temperature stress in crop plants. The role of the membrane. Academic Press, New York, pp 187–202

Smith D (1964) More about cold tolerance. Effects of a hard freezing upon cultivated palms during December 1962. Principes 8:26–39

Smithberg MH, Weiser CJ (1968) Patterns of variation among climatic races of red-osier dogwood. Ecology 49:495–505

Soderholm PK, Gaskins MH (1961) Evaluation of cold resistance in the genus *Coffea*. Coffee 3:40–45

Soeder C, Stengel E (1974) Physico-chemical factors affecting metabolism and growth rate. In: Stewart WDP (ed) Algal physiology and biochemistry. Biol Monogr, Blackwell, Oxford Vol 10, pp 714–740

Sosinska A, Maleszewski S, Kacperska-Palacz A (1977) Carbon photosynthetic metabolism in leaves of cold-treated rape plants. Z Pflanzenphysiol 83:285–291

Spranger E (1941) Das Erfrieren der Pflanzen über 0 °C mit besonderer Berücksichtigung der Warmhauspflanzen. Gartenbauwissenschaft 16:90–128

Stanwood PC, Bass LN (1978) Ultracold preservation of seed germ plasm. In: Li PH, Sakai A (eds) Plant cold hardiness and freezing stress. Academic Press, New York, pp 361–371

Steponkus P (1978) Cold hardiness and freezing injury of agronomic crops. Advances in Agronomy 30:51–98. Academic Press, New York

Stergios BG, Howell GS (1977) Effects of defoliation, trellis height, and cropping stress on the cold hardiness of Concord grapevines. Am J Enology and Viticulture 28:34–42

Sucoff E, Hong SG, Wood A (1976) NaCl and twig dieback along highways and cold hardiness of highway versus garden twigs. Can J Bot 54:2268–2274

Taylor AO, Rowley JA (1971) Plants under climatic stress. I. Low temperature, high light effects on photosynthesis. Plant Physiol 47:713–718

Teeri JA, Stowe LG (1976) Climatic patterns and the distribution of C_4 grasses in North America. Oecologia 23:1–12

Teeri JA, Patterson DT, Alberte RS, Castleberry RM (1977) Changes in the photosynthetic apparatus of maize in response to simulated natural temperature fluctuations. Plant Physiol 60:370–373

Terumoto I (1964) Frost resistance in some marine algae from the winter intertidal zone. Low Temp Sci B 22:19–28

Thomas SM, Long SP (1978) C_4 photosynthesis in *Spartina townsendii* at low and high temperatures. Planta 142:171–174

Till O (1956) Über die Frosthärte von Pflanzen sommergrüner Laubwälder. Flora 143:499–542

Timmis R (1977) Critical frost temperature for Douglas-fir cone buds. Can J For Res 7:19–22

Toriyama K (1976) Rice breeding for tolerance to climatic injury in Japan. In: Takahashi K, Yoshino MM (eds) Climatic change and food production. Univ Tokyo Press, pp 237–243

Tranquillini W (1957) Standortsklima, Wasserbilanz und CO_2-Gaswechsel junger Zirben (*Pinus cembra* L.) an der alpinen Waldgrenze. Planta 49:612–661

Tranquillini W (1959) Die Stoffproduktion der Zirbe (*Pinus cembra* L.) an der Waldgrenze während eines Jahres. II. Zuwachs und CO_2-Bilanz. Planta 54:130–151

Tschäpe M (1970) Untersuchungen über den Einfluß der Temperatur während der Dunkelperiode auf den Photosynthese-Gaswechsel in der nachfolgenden Lichtperiode. Flora 159:429–434

Tuhkanen S (1980) Climatic parameters and indices in plant geography. Acta Phytogeogr Suecica 67:5–110

Tumanov II (1962) Frost resistance of fruit trees. 16th Intern Hort Congr Bruxelles 1962, pp 737–743

Tumanov II (1979) Fiziologiya zakalivaniya i morozostoikosti rastenii. Izdat Nauka, Moskva

Tumanov II, Kuzina GV, Karnikova LD (1972) Vliyanie prodolzhitelnosti vegetatisii u drevesnykh rastenii na nakopleniye zapasnykh uglevodov i kharakter fotoperiodicheskoi reaktsii. (Effect of the duration of vegetation in trees on the accumulation of reserve carbohydrates and the character of photoperiodic reaction). Fiziol Rast 19:1122–1131

Tyler B, Borrill M, Chorlton K (1978) Studies in *Festuca*. X. Observations on germination and seedling cold tolerance in diploid *Festuca pratensis* and tetraploid *F. pratensis* var *apennina* in relation to their altitudinal distribution. J Appl Ecol 15:219–226

Tyurina MM (1957) Issledovanie morozostoikosti rastenii v usloviyakh vysokogorii Pamira. Akad Nauk Tadzhik SSR Stalinabad

Tyurina MM, Gogoleva GA, Jegurasdova AS, Bulatova TG (1978) Interaction between development of frost resistance and dormancy in plants. Acta Hort 81:51–60

Ullrich H (1943) Biologische Kältewirkungen und plasmatische Frostresistenz. Protoplasma 38:165–183

Ulmer W (1937) Über den Jahresgang der Frosthärte einiger immergrüner Arten der alpinen Stufe, sowie der Zirbe und der Fichte. Jahrb Wiss Bot 84:553–592

Ungerson J, Scherdin G (1968) Jahresgang von Photosynthese und Atmung unter natürlichen Bedingungen bei *Pinus silvestris* L. an ihrer Nordgrenze in der Subarktis. Flora 157:391–434

Van Hasselt PhR (1972) Photo-oxidation of leaf pigments in *Cucumis* leaf discs during chilling. Acta Bot Neerl 21:539–548

Van Hasselt PhR, Van Berlo HAC (1980) Photooxidative damage to the photosynthetic apparatus during chilling. Physiol Plant 50:52–56

Vieweg GH, Ziegler H (1969) Zur Physiologie von *Myrothamnus flabellifolia*. Ber Dtsch Bot Ges 82:29–36

Weiser CJ (1970) Cold resistance and injury in woody plants. Science 169:1269–1278

Wilner J (1965) The influence of maternal parent on frost-hardiness of apple progenies. Can J Plant Sci 45:67–71

Wilson JM (1978) Leaf respiration and ATP levels at chilling temperatures. New Phytol 80:325–334

Wilson JM, Crawford RMM (1974) The acclimatization of plants to chilling temperatures in relation to the fatty-acid composition of leaf polar lipids. New Phytol 73:805–820

Yelenosky G (1975) Cold hardening in *Citrus* stems. Plant Physiol 56:540–543

Yelenosky G (1977) The potential of *Citrus* to survive freezes. Proc Int Soc Citriculture 1:199–203

Zelawski W, Kucharska J (1967) Winter depression of photosynthetic activity in seedlings of Scots pine (*Pinus silvestris* L.) Photosynthetica 1:207–213

Zeller O (1951) Über Assimilation und Atmung der Pflanzen im Winter bei tiefen Temperaturen. Planta 39:500–526

14 Ecological Significance of Resistance to High Temperature

L. KAPPEN

CONTENTS

14.1 Introduction

Heat as a stress factor limiting the survival of plants has been recognized for some time, and reports about plants in hot environments were already critically discussed by SACHS (1864). At first plant response to heat was treated as a physiological problem (BĚLEHRÁDEK 1935, and cf. Chap. 12, this Vol.). The studies of HUBER (1935) and SAPPER (1935) were among the first in which ecological evidence for the heat resistance of plants received primary consideration. More recently this topic has been treated in several comprehensive books and reviews (PRECHT et al. 1955, 1973; LEVITT 1956, 1958, 1972, 1980; BIEBL 1962a; ALEXANDROV 1977).

From the ecological viewpoint, two major types of environments can be distinguished on the basis of thermal regime. One is characterized by permanently high or low temperatures, and the other by a range of temperature variations including extreme heat stress. The basis of plant adaptation in the first case depends solely on physiological properties of the cytoplasm (cf. Chaps. 10 and 11, this Vol.), whereas in the second case it depends on the extent to which plants normally living under moderate conditions can endure occasional heat stress. In this case, plant adaptation involves some degree of physiological tolerance, as well as mechanisms for avoiding heat stress.

14.2 Evidence of Heat Stress in Natural Environments

14.2.1 Thermal Stress Situations

Solar radiation, geothermic heat, and fire are the main natural sources of thermal stress on Earth. Metabolic heating plays a major role in compact organic masses such as compost, hay, or stored fruits. (Fire as an ecological factor and the impact of fire heat on plants is discussed in Chapter 16, this Volume). The thermal impact on the vegetation of hot volcanic soils has been only occasionally investigated, but much more is known about hot springs and their algal vegetation (BROCK 1978; cf. Chap. 11, this Vol.). In such permanently hot environments organisms must be adjusted to heat in all developmental phases. In the effluents of hot springs, the algal populations change from thermophilic to thermotolerant along a gradient of decreasing temperatures (e.g., FORSYTH 1977). Among the multicellular plants, particularly mosses (HESSELBO 1918) and some ferns can live almost directly adjacent to hot springs.

In addition to naturally occurring heat stress, industrial energy production has increasingly influenced natural ecosystems (CERNUSCA 1972). Particularly the thermal effluents of power plants are expected to disturb the vegetation of seas, rivers, and lakes (cf. Chap. 13, Vol. 12D).

Under conditions of open or sparse vegetation, solar radiation can strongly heat rock, soil, and peat surfaces. This in turn causes extreme thermal stress, particularly for plants living close to the soil surface (seedlings, procumbent plants, annuals, and small cryptogams). The strong heat absorption of soils high in humus content can create injurious conditions for seedlings even in high alpine regions (65–70 °C: LARCHER and WAGNER 1976) or in northern zones (DAHL 1963: 76 °C).

The same situation is found in bogs (SCHMEIDL 1965), in burnt soil (AHLGREN 1974), or in wood slash used as a substrate for fungi (LOMAN 1963).

In exposed locations of temperate regions, soil temperature extremes may reach between 65 and 75 °C (HUBER 1935; LANGE 1953), whereas forest soil in the spring does not warm beyond 40–50 °C (FIRBAS 1927). Arable soil in East Africa reached 60 °C and had a mean temperature of 39 °C during the growing season; the average in October was 44 °C (DAY et al. 1978). Desert soils are recorded to reach maximally 70–80 °C (BUXTON 1925; GATES et al. 1968).

14.2.2 Extreme Temperatures in Plant Organs

Thermal load in plants generally depends on the mass and conductivity of irradiated organs, their laminary boundary layer and transpiratory intensity

(HUBER 1935; GATES 1973; SMITH 1978) and the nature of the plant surface (cf. Chap. 1, this Vol.). Fruits and succulent plants may be subjected to the highest thermal stress.

Opuntia joints and thick leaves may reach about 20 K above ambient air temperatures (ANSARI and LOOMIS 1959; GATES et al. 1968; LARCHER 1980). With increasing succulence the "heating rate" decreases, but the amount of energy retained increases (MELLOR et al. 1964). Heat load is increased in succulents by Crassulacean Acid Metabolism, since stomata remain closed and transpiratory cooling is reduced during the day (COUTINHO 1969). Succulents other than CAM species remain somewhat closer to air temperature.

Leaves of nonsucculent plants in deserts reached maximally between 40.8 and 55 °C (HENRICI 1955; LANGE 1959). Leaf temperatures of *Tidestromia oblongifolia* in Death Valley (California) exceed 40 °C during almost the entire light period (PEARCY et al. 1971/72). Sclerophylls, e.g. plants of the mediterranean-type vegetation, heat up as much as desert plants (ROUSCHAL 1938; KREEB 1964). Such leaves reached 48 °C, an overtemperature of up to 20 K (LANGE and LANGE 1963; GATES 1965). Exposed tree trunks and stems experience the highest thermal load found in plants (HUBER 1935; DÖRR 1941).

The above thermal response type is termed an "overtemperature type" in comparison to the "undertemperature type" which comprises species which under similar conditions decrease their temperature below ambient air temperature via transpiratory cooling (cf. Sect. 14.3.1.2). An intermediate type maintains leaf temperatures close to air temperatures due to the small size of the assimilatory organs (LANGE 1959). Leaf temperature increases following wilting (MILLER and SAUNDERS 1923). It is also influenced by leaf color (DÖRR 1941). Another potential cause of overheating is a cover of dust or chemicals on leaves or fruits (ZSCHOKKE 1931; STEINHÜBEL 1966; ELLER 1977).

Algae in shallow water (FIRBAS 1931; BIEBL 1970) or terrestrial lower, poikilohydrous plants in open sites are particularly exposed to solar heating. The maximum temperatures of mosses and lichens reached 70 °C, and because of their low conductivity some of them heated much above the ground temperature (LANGE 1953, 1954; MACFARLANE and KERSHAW 1978).

14.2.3 Heat Injury Under Natural Conditions

It is difficult to discern the effects of heat in the field since a combination of factors may contribute to observed injury. Very often drought effects are involved. Thus, heat damage is only recognizable under particular conditions (WARTENBERG 1933). Heat injury in cryptogams has been observed only rarely but is very likely to occur, for example in or bordering on bog pools and hot steam vents (KAPPEN 1973; KAPPEN and SMITH 1980). There are several reports of heat damage in higher plants. Scorching of leaves and fruits, sunscald, sunburning, abscission of leaves, and growth retardation have been viewed as plant diseases which are due, at least predominantly, to thermal impact (WARTENBERG 1933; BARBER and SHARPE 1971). Heat-injured leaves of grape vine display red-brown spots, whereas drought-killed leaves are greenish

(ZSCHOKKE 1931). Heat injury of leaves in desert plants was observed in trees as well as in herbs (LANGE 1959; HELLMUTH 1971b; KARSHON and PINCHAS 1971; MACBRYDE et al. 1971). The small herb *Koenigia islandica* was widely killed in Scotland following two waves of hot weather. It was concluded that the distribution of this species is limited by a thermal border (DAHL 1963). Seedlings especially are in danger of suffering from high temperatures. The seedlings of various tree species are killed, due to injury of the narrow strip of bark around the stem base, when soil temperature exceeds 46 °C (BAKER 1929; FRANCO 1961).

Reduction in harvest due to periods of extreme summer heat was found in alfalfa in Arizona (PULGAR and LAUDE 1974). Cultivated plants frequently exhibit indirect effects of heat stress because their response to heat may differ from that of their microbial symbionts or parasites. For example, at soil temperatures above 30 °C young roots of *Pisum* were not receptive to inoculation by their N_2 fixing symbiont (FRINGS 1976; DAY et al. 1978).

Short heat shocks above 45 °C may inhibit the defensive reactions of plants against certain viral and fungal infections (YARWOOD 1965; AIST and ISRAEL 1977). Such heat-induced susceptibility to infection can be due to biochemical or physiological changes (WAGENBRETH 1968), as well as to changes in the physical nature of the leaf surface (YARWOOD 1977).

On the other hand, the negative effect of heat is often greater for the parasitic fungi and viruses than for the host plants. This is the basis for the "heat therapy" which is applied to agricultural plants (YARWOOD 1965).

14.3 Ecophysiological Responses of Plants to Heat Stress

The survival of a plant under stress conditions depends on the extent to which it is able to tolerate or to avoid the stress (cf. LEVITT 1972). Thus resistance in an ecophysiological sense includes both tolerance and avoidance of stress. Avoidance can be effected in various ways: through life form, morphology, quality of the leaf surface, or through physiological changes and responses. In general we can distinguish between constitutional (STOCKER 1956) and functional avoidance mechanisms. Tolerance, on the other hand, is a direct result of the degree to which a plant is able to withstand immediate heat stress on cellular and subcellular structures and functions.

14.3.1 Avoidance

14.3.1.1 Constitutional Avoidance of Heat Stress

Perhaps the relationship between incidence of heat and consequent avoidance is most clear in those plant life forms which survive periods of intense heat or fire in the form of (dormant) bulbs or tubers or other subterranean organs capable of vegetative reproduction, or which produce heat-resistant spores or seeds.

Plants persisting in a hot environment can avoid an excessive heat load by having small leaves with high convective heat exchange (GRIEVE and HELLMUTH 1970; PARKHURST and LOUCKS 1972), divided leaves, or leaves which can be torn (TAYLOR and SEXTON 1972). Certain desert plants (MOONEY et al. 1977a) display seasonal leaf dimorphism. Others have no leaves at all ("Rutensträucher"); their thin green axes remain generally close to the air temperature (LANGE 1959; STOCKER 1971).

Another means of avoiding overheating is through maintenance of a steep leaf angle against sun radiation in summer (DÖRR 1941; THOFELT 1975; MOONEY et al. 1977a).

The possibly protective effect of reflectance and especially the role of white hairs has been controversially discussed (HUBER 1935; HERZOG 1938; DÖRR 1941; GRIEVE and HELLMUTH 1970; ELLER 1979; cf. Chaps. 1 and 5, this Vol.). Whitewashing of tree trunks and spraying crops and fruits with artificial reflectants are methods used to prevent heat damage in hot climates (cf. BASNIZKI and EVENARI 1975). Plants with white or pale-colored fruit are selected for cultivation in warm countries.

Cacti have structural and morphological features which act to limit high heat load in several ways (HERZOG 1938; GIBBS and PATTEN 1970). Spines of *Opuntia bigelowii* reflect and absorb much energy, thus protecting the stem from overheating, and the small *Opuntia acanthocarpus* is very effectively cooled by convection due to its large surface to volume ratio. The angle of *Opuntia* pads proves optimal both for receiving radiative energy only during the cooler diurnal and annual periods and avoiding it under the steep sun at noon and in the summer. *Copiapoa haseltoniana* in Chile is an example of a combination of various ways of protection (MOONEY et al. 1977b).

14.3.1.2 Functional Avoidance

Functional avoidance mechanisms are those which reduce the heat load of the plant organ through energy consumption. The cooling effect of transpiration was early recognized (DUTROCHET 1839). Transpiratory cooling of leaves was observed under artificial conditions (SEYBOLD 1929), as well as under natural conditions (MILLER and SAUNDERS 1923; VOLODIN 1951 and others). According to early measurements, temperature reduction due to transpiratory cooling was small (1–3 K) and therefore it was regarded to be of no ecological relevance (HUBER 1935; CURTIS 1938). Plant leaves were considered to have normally higher-than-air temperatures under sunlight. However, according to calculations of RASCHKE (1956, 1960) and WOLPERT (1962) the energy loss due to transpiration amounts to one quarter of the incoming heat and could reach 2–8 times that resulting from convection (cf. GATES 1963).

The ecological importance of transpiratory cooling (cf. STOCKER 1954) in desert plants, keeping them below the killing temperature, was demonstrated by LANGE (1959; Table 14.1). The various dependencies of transpiratory capacity of plants on internal and external factors are discussed in Chapters 1 and

Table 14.1. Relations between air temperature, tissue temperature, and heat tolerance

Species	Air temperature (maximum measured) [°C]	Plant temperature (maximum measured when transpiring) [°C]	Potential maximum transpiratory cooling Δt[K]	Maximum heat tolerance (30 min exposure) measured [°C]	Reference
Phanerogams (leaves)					
Sahara					
Citrullus colocynthis	63.7	41.0	15.3	46	Lange (1959)
Chrozophora senegalensis	46.5	41.2	6.0	48	
Zygophyllum cf *fontanesii*	33.3	40.8	12.9[a]	50	
Phoenix dactylifera	45.0	52.3	11.4[a]	58	
Mediterranean					
Solanum melogena	27.1	32.0	13.3	44	Lange and Lange (1963)
Psoralea bituminosa	34.2	40.8	5.5	46	
Myrtus communis	35.6	44.6	2.0	52	
Quercus ilex	29.7 (27.8)	43.6 (36.7)	0.7 (−2.3)	54 (54)	
Alpine					
Loiseleuria procumbens	20–23	43		49	Larcher and Wagner (1976)
Arctostaphylos uva-ursi	22–25	44		51	
Lichens, mosses					
Cladonia rangiformis	26	54.9[b]		46.5	Lange (1953, 1954, 1955, 1969); Nörr (1974); Kappen (unpublished)
Cladonia furcata var. *palamaea*	27	65.5[b]		–	
Andreaea petrophila	19	47.0[b]		43–47	
Gymnomitrium obtusum	19	38.0[b]		43–47	
Ramalina maciformis	19.7	61.0[b]		38	

[a] above air temperature; [b] temperature in dry plant

7 of Volume 12B (cf. also Chap. 1, this Vol.). Under extreme temperature stress, sclerophyllous overtemperature species are also able to perform transpiratory cooling, at least under controlled conditions (LANGE 1962a). Thus the difference between under- and overtemperature species may be due to their different stomatal sensitivity to heat.

14.3.2 Tolerance

Response to heat is based ultimately on cell function and structure (cf. Chap. 12, this Vol.). In addition to this, one has to take into account reversible effects. Ability to recover rapidly is crucial, since functional disturbances such as depression of photosynthetic or protein metabolism reduces the productivity, growth, and competitive ability of a plant. Thus, not only lethal effects, but also the entire range of sublethal effects must be considered in estimating the physiological and ecological behavior of plants under heat stress. Investigations of such functional disturbances may also lead to the finding of significant indicators of plant vitality which could be used to predict the limits of heat tolerance.

14.3.2.1 The Upper Limit of Some Physiological Functions

Photosynthesis

The photosynthetic apparatus appears particularly sensitive to heat stress and has been frequently investigated (cf. BAUER et al. 1975). Considering the whole plant kingdom, the maximal upper limit (70–73 °C) of intact plant photosynthesis is found in photosynthetic bacteria and the Cyanophycea *Synechococcus lividus* (BROCK 1978; cf. Chap. 11, this Vol.). Among the eukaryotic plants, the maximum limit of net photosynthesis ranges between 35 and 61 °C (Table 14.2). Following the thermophilic algae, C_4-phanerogams display the highest tempera-

Table 14.2. Upper temperature limits of (net)-photosynthesis ($T_{p\,max}$) in plants (VIEWEG and ZIEGLER 1969; OECHEL et al. 1972; MEEKS and CASTENHOLZ 1972; TIESZEN 1973; LANGE et al. 1974; PEARCY et al. 1974; Bauer et al. 1975; LARCHER and WAGNER 1976; KALCKSTEIN 1976; RAKHIMOV 1976; BROCK 1978; MOONEY et al. 1978)

Species/Group of species	$T_{p\,max}$ [°C]
Synechococcus lividus	73–75
Thermophilic photosynthetic bacteria	70–73
Thermophilic algae	55–60
Tropical and subtropical C_4 species	50–61
Plants from arid and semi-arid hot regions	49–55
Evergreen trees from the mediterranean region and South Japan	43–48
Arctic and alpine dwarf shrubs	42–47
Deciduous trees from temperate regions	41–43
Arctic and alpine herbs	38–42 (45)
Boreal conifers	36–38
Tundra grasses	35–39

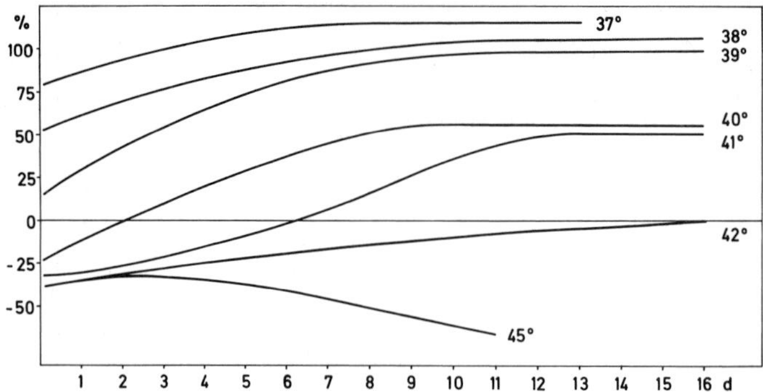

Fig. 14.1. Recovery of CO_2 exchange rate (in % of control) of the lichen *Cladonia rangiferina* in postculture following a 1 h heat exposure. (After LANGE 1965b)

ture limit of photosynthesis. They can remain active at temperatures which are injurious for most C_3 species. In thermophilic algae the difference between the limit of photosynthesis and that of heat tolerance is only from 1 to a few K, whereas in higher plants it ranges between 2 and 12 K (PISEK et al. 1968; LARCHER and WAGNER 1976).

In some cases a phytogeographical correlation between the difference in upper limit of photosynthesis and heat tolerance seems to exist. The upper limit of positive net photosynthesis may vary with the temperature regime as much as 10 K, particularly in hot climate and desert species (LANGE et al. 1974, 1978; PEARCY 1976; MOONEY et al. 1978), and from 1–3 K in temperate species (LUTOVA 1962; PISEK et al. 1968; LARCHER and WAGNER 1976). This adjustment of photosynthesis to higher temperatures seems possible only in species with a wide span between the upper limit of photosynthesis and that of heat tolerance.

If heat stress is sublethal (from about 1 to a few degrees K below lethal effects), the photosynthetic capacity is fully regained during a recovery period of from one to about 10 days, and even becomes adjusted to higher temperatures (see Fig. 14.1; LANGE 1965a, b; BAUER et al. 1975; LARCHER and WAGNER 1976). The recovery of photosynthesis is not influenced by temperatures between 10 and 30 °C (EGOROVA et al. 1978).

Respiration

Abrupt changes in the respiratory rate occur only in response to heat stress which is nearly lethal (lower plants: LANGE 1965b, GRINTAL 1976; higher plants: LARCHER 1973). In most cases a decrease or cessation of respiration was observed when the temperatures reached injurious levels (Fig. 14.2; cf. BAUER et al. 1975). If the stress was not lethal, the respiratory rate was either restored fully within several hours or it remained on a lower level, depending on the degree of cellular injury. Generally respiration is rather inert in its response to heat stress in contrast to its reaction to freezing and desiccation (BAUER 1972).

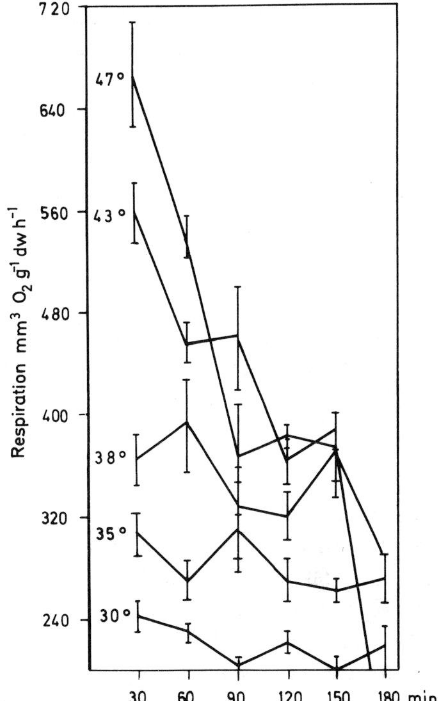

Fig. 14.2. Respiratory response to increasing time periods of heating in leaves of *Podophyllum peltatum*. Temperatures below 30 °C did not influence respiration rate. (After SEMIKHATOVA and DENKO 1960)

Consequences for the CO_2 Balance

If a heat-stressed plant is returned to normal conditions the photosynthetic capacity and other synthetic processes (cf., e.g., BERNSTAM 1974) rather than respiratory loss remain limiting for survival. The heat response in leaves and roots may be mediated by hormones, e.g., an increased ABA level resulting in reduced growth (ITAI et al. 1973). Injury was found to be translocated to other parts of the plant (YARWOOD 1961 b).

Growth inhibition and defects in the photosynthetic apparatus following heat stress were shown in algae (BRAWERMAN and CHARGAFF 1960) and in crop plants (ADELUSI and LAWANSON 1978). As a consequence of prolonged heat, photosynthesis of tree seedlings was diminished, thus decreasing their competitive ability (PISEK et al. 1968).

It is clear that long periods of unusual heat or transfer of nonadjusted plants to a hot environment seriously affect plant productivity, since the plants are obliged to exist close to the temperature of the upper compensation point of net photosynthesis (STRAIN and CHASE 1966). This is particularly of agronomic relevance.

14.3.2.2 The Limits of Vitality

Methods and Criteria for Measuring Heat Tolerance

Heat tolerance in plants is usually determined by exposing specimens to a series of different temperatures and noting any immediate or later effects. Since

Table 14.3. Heat-tolerance limits (due to exposure for indicated times to temperatures causing 10–50% damage) of the main plant groups. Comparison of hydrated and dry leaves or total plants

Plant species/groups	Tolerance limit				Reference
	Hydrated		Dry		
	Temperature [°C]	Time of exposure	Temperature [°C]	Time of exposure	
Thermophiles:					
Non photosynthetic bacteria	60–90	Life	No increase		CHRISTOPHERSEN (1955, 1973); TANSEY and BROCK (1978); BROCK (1978)
Photosynthetic bacteria	55–75	Life	No increase		TANSEY and BROCK (1978); BROCK (1978)
Cyanophyceae	55–75	Life	No increase		PEARY and CASTENHOLZ (1964); MEEKS and CASTENHOLZ (1972); BROCK (1978)
Algae	<57	Life	?		DOEMEL and BROCK (1971)
Fungi	<62	Life	?		TANSEY and BROCK (1978)
Psychrophiles:					
Bacteria, Fungi, Algae	10–40	10′–various	?		LEVITT (1972); ALEXANDROV (1977); MALCOLM (1969); SKINNER (1968)
Mesophiles:					
Bacteria, Actinomycetes	45–65	30′–60′	75–140	40′–50′	CHRISTOPHERSEN (1955); SZABO et al. (1964); TANSEY and BROCK (1978)
Cyanophyceae			<100	10′–6 h	MACENTEE et al. (1972)
Terrestrial algae	40–50	20′–30′	60	15′–3 h	EDLICH (1936)
Freshwater algae	34–50 (22–38)	30′ (12 h)	<105	10′–1 h	URL and FETZMANN (1959); BIEBL (1962a, 1967a); MCLEAN (1967)

Marine algae					
tidal	30–42 (24–40)	30′ (12 h)			SCHWENKE (1959); BIEBL (1962a, b, 1964, 1970); SCHÖLM (1966); SCHRAMM (1968)
littoral	32–40 (26–35) 35	30′ (12 h) 24 h	42	24 h	MONTFORT et al. (1957); FELDMANN and LUTOVA (1963)
sublittoral	24–38 (22–30)	30′ (12 h)			
Fungi	35–50 (70)	30′	70–100	30′	ZIMMERMANN and BUTIN (1973); SINGH (1970)
Lichens	33–46.5	30′–60′	70–100	30′	LANGE (1953, 1965b); KAPPEN (1973)
Bryophytes	(32–38)	(12 h)			SCHEIBMAIR (1938); LANGE (1955); BIEBL (1964, 1967c, 1968); CLAUSEN (1964); DIRCKSEN (1964); NÖRR (1974)
Hepaticae	39–45	30′	70–110	30′	
Mosses	41–51	30′	85–110	30′	
Cormophytes					
Poikilohydrous ferns and angiosperms	47–50	30′	50–60 (120?)	30′	SAPPER (1935); OPPENHEIMER and HALEVY (1962); KAPPEN (1964); VIEWEG and ZIEGLER (1969)
Homoiohydrous ferns	45–53	30′	48–55	30′	HAMMOUDA and LANGE (1962); KAPPEN (1964, 1965, 1966); LARCHER and WAGNER (1976)
Hot climate species and mediterranean species	44–59 (61)	30′			HUBER (1935); VOLODIN (1951); LANGE (1959); LARCHER (1961a); LANGE and LANGE (1963); BIEBL (1964); FELDMANN et al. (1970)
Temperate species	38–52	30′			SAPPER (1935); LANGE (1961); PISEK et al. (1968)
Arctic and alpine species	41–52 (60)	30′			SEMIKHATOVA et al. (1962); BIEBL (1968); KAINMÜLLER (1975); KJELVIK (1976); LARCHER and WAGNER (1976)
Water plants	(33) 38–44	30′			SAPPER (1935); BIEBL and MCROY (1971); STANLEY and MADEWELL (1976)

the heating effect is strongly time-dependent, the exposure time should be defined and standardized. The great differences among treatments reported in the literature complicate the comparison and interpretation of the results.

ALEXANDROV and his co-workers (cf. ALEXANDROV 1977), YARWOOD (1965) and WAGENBRETH (1965a) applied heat shocks of a few seconds or of 5 min, whereas others have exposed plants to heat for 30 or 60 min (cf. Table 14.3). MONTFORT et al. (1955) and BIEBL (1962b, 1970) exposed their algae for 1–12 h and longer.

The limit of tolerance is defined variously as the temperature at which initial disturbance of any function occurs, or at which the percentage of tissue damage is low (10–20%) or medium (LEVITT 1972), or at which an entire population of unicellular organisms is killed.

In heat-stress experiments small plants are usually exposed in their entirety (e.g., BIEBL 1939), but also entire larger plants may be exposed (e.g., SAPPER 1935; SHIRLEY 1936), or only attached parts of plants (e.g., ROUSCHAL 1939; BAUER 1972). If the plants are exposed in air within a heated chamber, a high humidity must be maintained to prevent lowering of leaf temperature by transpiratory cooling, and also loss of turgor which changes the tolerance limit. Thus many earlier results showing extremely high heat tolerance are of questionable validity (cf. the discussions of SAPPER 1935; LANGE 1959, 1965b). It is therefore more suitable to provide immediate contact between a liquid medium and the tested material (cf. LANGE 1961; KAPPEN 1964).

The criteria for evaluation of the heat-tolerance limit may be divided into metabolic depressions, cytological disturbances, and necroses (LEVITT 1972; ALEXANDROV 1977). The cytological criteria are well-suited to estimating reactions of small objects such as algae, fungi, or small tissues, but may not accurately reflect the situation of larger plant organs. Thus the amount of necrotic lesion in a leaf or a stem is an integrative measure of heat tolerance. For example, the black areas indicative of necrosis appear a few hours or 1–2 days after heat stress as a consequence of the oxidation of polyphenols (WAGENBRETH 1965b). The area affected depends on the intensity of heat stress and remains stable during a period of 2–3 weeks. Estimates of tolerance, then, may vary depending on whether the tolerance limit is assessed at 50% damage, which might involve a greater random variability, or whether it is fixed at a temperature immediately prior to first injury (e.g., BIEBL 1964), or at the temperature causing 10–20% injury (KAPPEN 1964: "vital tolerance limit").

Terminology

Referring to the reaction of protoplasmic streaming, ALEXANDROV (1977) distinguishes between a "primary" and a "secondary" heat resistance. The primary resistance reflects the genetically fixed response of the cytoplasm immediately subsequent to heat treatment, and is thus species-specific and indicative of its physiology. The secondary resistance involves repair effects resulting in the resumption of the cytoplasmic streaming. The degree of this recovery may vary with season, with hardiness, and even within the heating treatment. It is more indicative of plant vitality and shows the ecologically relevant adaptation of the plant. Consequently the comparison of plants according to the primary resistance can differ greatly from that respecting the secondary resistance (KAPPEN and LANGE 1968). In the first case responses correspond to what STOCKER (1956) has termed "plasmatic resistance" or LEVITT (1972) "heat tolerance". On the other hand, "constitutional resistance" (STOCKER 1956) or "heat resistance" (LEVITT 1972) include the protective reactions of plants which lead to avoidance of, or diminishing of the heat stress (cf. Sect. 14.3.1).

The Limits of Heat Tolerance in Active Plants

Prokaryonts. The heat tolerance limit of thermophilous organisms is frequently close to the temperature range of their highest activity (cf. Chap. 11, this Vol.). For example, *Oscillatoria* has its optimum growth at 64 °C but is killed when exposed to 68 °C for 10 min (BÜNNING and HERDTLE 1946).

Among the very thermophilous bacteria the heterotrophic species are most resistant (cf. Table 14.3). Chemolithotrophic bacteria (e.g., *Sulfolobus* above 90 °C) can become more tolerant than photosynthetic bacteria (limit 70–73 °C). The most resistant Cyanophyceae reach the same limit. One strain of *Synechococcus lividus* is still active at 73 °C and has its tolerance limit at 75 °C (PEARY and CASTENHOLZ 1964).

Mesophilic bacteria are killed within the range of 40 and 50 °C. Psychrophilic bacteria reach their heat tolerance limit at temperatures below 40 °C (SKINNER 1968). The lowest killing temperature (25 °C) was found in psychrophilic gram-negative rod on flounder eggs (HAGEN et al. 1964). *Micrococcus cryophilus* ceased growth at 25 °C (MALCOLM 1969). Heat killing of these organisms is obviously caused by denaturation or inactivation of the thermolabile proteins (LEVITT 1972).

Algae. Most of the eukaryotic plants are absent from environments which have temperatures permanently above 50 °C (cf. FOGG 1969). It is suggested that eukaryonts lack membranes which are both permanently thermostable and functional. Therefore the red alga *Cyanidium caldarium* is absolutely exceptional in having an upper temperature limit for photosynthesis at 57 °C (DOEMEL and BROCK 1971; TANSEY and BROCK 1978).

Those diatoms and green algae which have been frequently found in hot springs may not be active since their viability above 47 °C is not yet established (SOROKIN 1971; TANSEY and BROCK 1978). Freshwater algae in the eurythermic patches of raised bogs, which are preponderantly diatoms and Desmidiaceae, have an upper limit of 38 °C (12 h) or survive up to 40–45 °C when exposed for only 15–30 min (URL and FETZMANN 1959; SCHÖLM 1968). Some algae cannot survive even short exposure to temperatures warmer than 16–20 °C and the cryophilic *Koliella tatrae* is killed when cultivation temperature exceeds 10 °C (LEVITT 1972). The heat tolerance limit of soaked aerial algae is reached at 40–50 °C (EDLICH 1936). Within the same temperature range desert soil algae are inactivated (BERNER 1974). These algae survive hot periods only in the desiccated state.

The heat tolerance has been measured in about 200 marine algae species (cf. Table 14.3). In general, there is no difference between the main classes of red, brown and green algae. The heat tolerance of marine species is always highest in the tropics (35–40 °C, 12 h, BIEBL 1962b). The tolerance of red algae from sea levels between 2–25 m was most closely correlated with the prevailing water temperature (BIEBL 1967d). Among the algae of this zone were the most sensitive species (e.g., *Cryptopleura violacea*: 22 °C, 12 h; BIEBL 1970). Surprisingly, on the average, intertidal species are more tolerant to heat than ebb tide and tide pool species. They also display the greatest variation in tolerance limits among the species (16 K). *Dunaliella parva* from the Dead Sea tolerates 5-min heat treatments of up to 49 °C (GIMMLER et al. 1978).

Fungi. The maximum temperature limit for active life of thermophilic fungi reaches 61–62 °C (e.g., *Theoascus aurantiacus*, NOACK 1920). The heat tolerance in thermophilic compost fungi ranges between 59 and 68 °C*[,1] (TANSEY and BROCK 1978). Mesophilic fungi comprising phytopathogenic wood- and food-destroying species, are killed by temperatures between 39 and 60 °C (at 1–45 min exposure, cf., e.g., NIETHAMMER 1947; YARWOOD 1963a; DEVERALL 1965; NELSON and FAY 1974). Some species of soil fungi are much more tolerant (SINGH 1970). ZIMMERMANN and BUTIN (1973) demonstrated a close correlation between the heat tolerance (40–50 °C*) and habitat conditions of wood-inhabiting fungi. In the majority of desert fungi in the Negev (Israel) growth ceased at 40 °C (cf. FRIEDMANN and GALUN 1974).

Lichens. Lichens only rarely exist in hot, moist environments (soil temperatures up to 32 °C, cf. KAPPEN 1973). The heat tolerance of lichens from hot rain forests is not yet known. Lichens can grow abundantly in hot deserts if cool and moist conditions occur regularly (LANGE et al. 1970; KAPPEN et al. 1979). The tolerance limits of hot desert species as well as of temperate species range between 33 and 46 °C* (LANGE 1953; KAPPEN 1973). The heat tolerance of the European *Cladonia rangiformis* var. *pungens* is exceptionally high (46.5 °C*).

Bryophytes. Although bryophytes are mostly hygrophytic plants, they can be found in deserts, and around hot spots as well as in cool freshwater habitats. Bryophytes are frequently found on hot volcanic soils (up to 42 °C) or in close proximity to hot springs (HESSELBO 1918; LANGE 1973). These species must exist close to their heat-tolerance limit, because it does not exceed that of other mosses (KAPPEN and SMITH 1980). The heat-tolerance limits of tropical (BIEBL 1964), hot desert (VOLK 1959, unpublished), temperate (CLAUSEN 1964; DIRCKSEN 1964) and arctic species (BIEBL 1968) were not significantly different (39–51 °C*). Also no difference was found between hepatics and mosses. Some *Sphagnum* species are very heat-tolerant (41–44 °C*, DIRCKSEN 1964). Water mosses are more heat-sensitive than terrestrial species. The regenerative capacity of heat-damaged moss plants is very high (DIRCKSEN 1964).

Pteridophytes. Ferns are most frequently found in tropical and subtropical forests, and in hot climates. Only a few species grow in open habitats. Many epiphytic species and those colonizing exposed rocks are very tolerant to desiccation and, like mosses and lichens, survive high temperature stress in a dry or wilted state. The heat-tolerance limits of sporophytes of tropical and temperate species vary within the same range (44–50 °C*: LANGE 1959; BIEBL 1964;

1 Exposure time in the tolerance test of 30 min is marked with an asterisk (*)

KAPPEN 1964). Heat-tolerance limits often relate well to the typical habitats of the fern species.

Ferns cannot reproduce in habitats in which conditions favorable to the delicate gametophyte do not occur. The heat tolerance limits, between 39 and 45 °C* (10–20% damage, KAPPEN 1965), are comparable with those of bryophytes in such habitats. However, due to their high regenerative capacity their tolerance amplitude is still broader.

Spermatophytes. The maximum heat tolerance limits of terrestrial spermatophytes, between 42 and 60 °C*, correlate only roughly with the climate region of their habitats. However, within restricted plant groups, differences due to local climatic conditions are obvious (e.g., Amaryllidaceae, FELDMAN et al. 1970). This was also evident with *Acacia*, mediterranean sclerophylls, and mangrove species (LANGE and LANGE 1963; BIEBL 1964; GRIEVE and HELLMUTH 1970). In general, C_4 species are extremely heat-tolerant (cf. Table 10.2; ALEXANDROV 1977). The assumption that succulents have a high heat tolerance cannot be taken as a generalization (LANGE 1959). It obviously depends on the performance of CAM in these species (cf. Sect. 14.2.2). Plants capable of high transpiratory cooling, such as many malakophyllous species, have a low plasmatic heat tolerance and would suffer from heat stress if transpiration was prevented (cf. Table 14.1).

The heat tolerance of temperate, arctic, and alpine species can reflect their habitat conditions (SAPPER 1935; SEMIKHATOVA et al. 1962; KJELVIK 1976). On the other hand, the high heat tolerance of alpine plants or of the typical shade species *Taxus baccata*, appears unexpectedly (LANGE 1961; BIEBL 1968; LARCHER and WAGNER 1976). Submerged aquatic species are more sensitive to heat (38.5–42 °C*: SAPPER 1935; BIEBL and MCROY 1971) than the floating *Lemna minor* (44 °C*: STANLEY and MADEWELL 1976).

Several measurements demonstrate that heat tolerance differs significantly between different organs of one plant. Thus, in an ecological sense, survival depends on the resistance of those organs which maintain the existence of a plant or provide for its reproduction. In alpine dwarf shrubs, the strategy appears to be to keep the stems most resistant against heat (52–56 °C*) while the flowers are comparatively sensitive (42–46 °C*: LARCHER 1977). According to other findings the generative plant organs are the most heat-tolerant part of a plant (SAPPER 1935; HENKEL and MARGOLINA 1951; TANNER and GOLTZ 1972). Leaf injury results primarily from damage to the sensitive mesophyll, whereas the epidermis and especially the stomata are more tolerant (WEBER 1926).

14.3.2.3 Variability of the Heat Tolerance

Intraspecific Variations

Since heat tolerance in plants corresponds to the temperature regime of their habitat, it is expected that heat tolerance is capable of modification. This can be shown if samples of different stands are tested simultaneously. The heat-

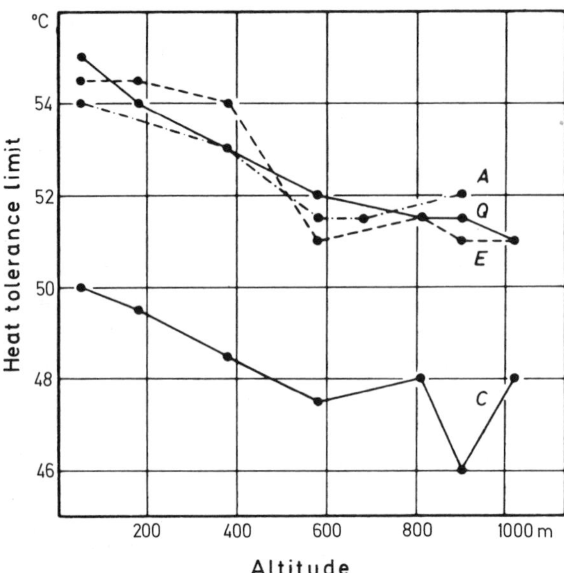

Fig. 14.3. Heat tolerance (temperatures causing 50% injury) in leaves of mediterranean plants in summer, depending on altitude of habitat (*abscissa*). *A Arbutus unedo; C Cistus salviaefolius; E Erica arborea; Q Quercus ilex.* (LANGE and LANGE 1962)

tolerance limits of sublittoral and littoral *Fucus vesiculosus* specimens differ by about 1 K (FELDMAN and LUTOVA 1963). The tolerance differences are extremely large in *Cladophora fracta* (10 K) and other species from an Austrian lake and a constant-temperature spring in Iceland (BIEBL 1967a; SCHÖLM 1968). Within the terrestrial plants, heat tolerance variation with habitat was apparent in some lichens and mosses in the desiccated state (LANGE 1953, 1954). In these cases it could not be decided whether the differences were due to plant modification or to the inheritable character of ecotypes. In other cases (Fig. 14.3; LANGE 1959; LANGE and LANGE 1962, 1963) differences in heat tolerance in thermally different environments are due to the actual thermoadaptation of the plants. Differences in heat tolerance between southern and northern populations of *Oxyria digyna* were maintained during equal cultivation and were therefore due to ecotypes (MOONEY and BILLINGS 1961). Similarly, *Zostera marina* was more tolerant in a tidepool than in a subtidal stand (BIEBL and MCROY 1971).

A comparison of closely related species with respect to their "primary resistance" shows correlations between heat tolerance and prevailing geographic distribution. Among 59 tested grass species (ALEXANDROV 1977), those of the Pooid subfamily, which generally occur in cool and temperate regions, are markedly less heat-tolerant (5 min heating to 43.3–45.7 °C) than members of the Panicoid subfamily (47.6–49.7 °C) which are common to subtropical and tropical regions. This difference exists as well when the plants are growing in places outside their original distribution areas. It may also demonstrate a requirement for a higher heat tolerance in the Panicoid C_4 species. On the other hand, a negative correlation between heat tolerance and the temperature conditions of the original habitat was also found (KARSHON and PINCHAS 1971).

Diurnal Variations

Diurnal changes in the heat tolerance of plants show very different patterns. LAUDE (1939) first observed that some crop plants were less injured by a 5 min exposure to 50 °C during noon than during the rest of the day; minimum tolerance was found in the early morning. Similar patterns were shown in grass species (BREGETOVA and POPOVA 1962a; ALEXANDROV 1977). YAZKULYEV (1964) interpreted the peak tolerance during midday (2–3 K increase) as an adjustive effect which is induced only when habitat temperatures surpass a certain level as occurs in summer (e.g., for *Arundo donax* 30 °C). Adjustment is also indicated when, during heat therapy against bean rust, the heat requirement is greater in the afternoon than in the early morning (YARWOOD 1963a).

On the other hand, short-day *Kalanchoë blossfeldiana* plants were less injured at 46 °C* if heat treatments occurred during the dark period rather than during the light period. Diurnal changes in heat tolerance were most pronounced in the leaf vertex, which is most sensitive to heat (SCHWEMMLE and LANGE 1959). This is not related to thermo-adjustment and may reflect the activity of the Crassulacean Acid Metabolism (cf. KLUGE and TING 1978, p. 151).

Another example of decreasing heat tolerance during midday (2 K; 14:00–16:00 h), shown in the algae *Spirogyra* and *Rhizoclonium*, was correlated with changes in the metabolic activity of the nucleus (SCHÖLM 1968). A different diurnal pattern of heat tolerance was exhibited by marine algae. Heat tolerance was higher during ebb tide than during high tide (LUTOVA et al. 1968; BIEBL 1969a).

Annual Variation

Seasonal variation in heat tolerance involves mostly changing developmental stages or aging of plants. Germinating and young plants or growing organs are generally (however, see LANGE 1967) more sensitive to heat than adult organs or plants (ILLERT 1924; LAUDE and CHAUGULE 1953; DIRCKSEN 1964; KAPPEN 1964; BULLOCK and COAKLEY 1978). For example, the difference between the tolerance limits of old and young leaves of *Ilex aquifolium* in spring was 8 K and in summer, 2 K (LANGE 1961).

Annual courses of heat tolerance are recorded in most plant groups: algae, bryophytes, pteridophytes and spermatophytes. Marine algae have apparently little (BIEBL 1962a; FELDMAN and LUTOVA 1963) or no annual variation in heat tolerance (MONTFORT et al. 1957). Algae from shallow freshwater habitats, however, undergo greater changes in heat tolerance (8–10 K) with maximum heat tolerance occurring in summer (SCHÖLM 1968).

Among the bryophytes and cormophytic plants, summer green species have varying heat tolerance during the growing season (BREGETOVA and POPOVA 1962b; cf. ALEXANDROV 1977). The heat tolerance increases either continuously to a maximum in the late season or has a second minimum in the fall. The temperature amplitude of the heat tolerance limits was less than 2 K in sensitive plants, and up to 4 K in more tolerant species. The different behavior of ferns clearly reflected their habitat selection (KAPPEN 1964).

The patterns of the annual courses of heat-tolerance limits in evergreen plants are very different and frequently do not correspond to the natural climate

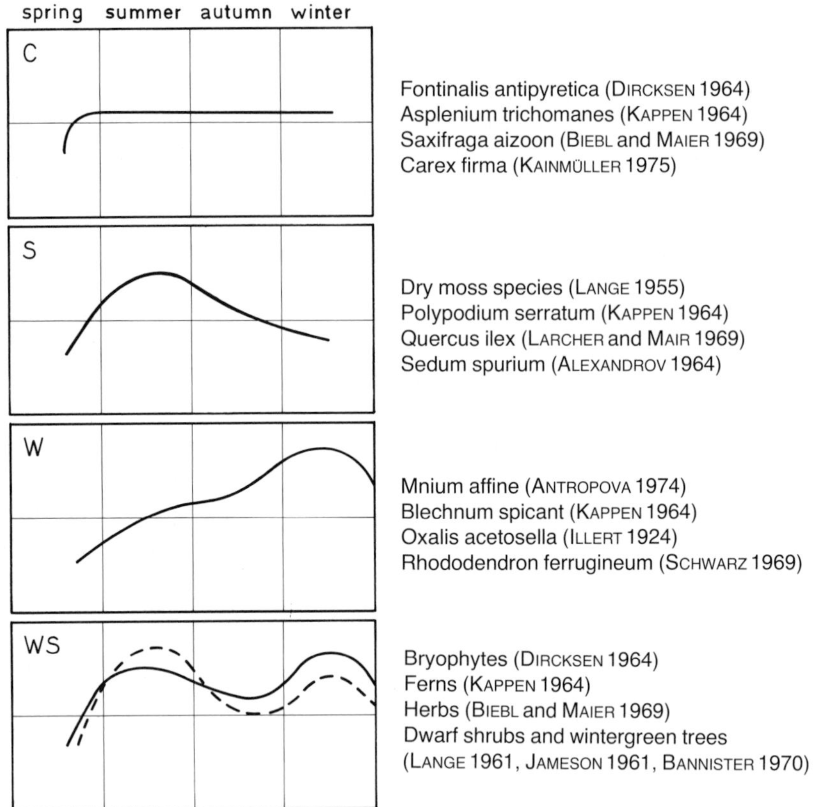

Fig. 14.4. General patterns of annual courses of heat tolerance in plant leaves of different response types (According to LARCHER et al. 1973); examples are indicated: *C* constant tolerance type; *S* summer peak type; *W* winter peak type; *WS* plants with increased heat tolerance in both winter and summer, having a higher peak either in summer (*broken line*) or in winter (*solid line*)

rhythm (cf. LARCHER et al. 1973). Common to all investigated species is the annual minimum heat tolerance in spring when the plants are rapidly growing. Four different response types can be distinguished (Fig. 14.4; cf. LANGE 1967; LARCHER et al. 1973). The first comprises a few species having, in addition to the spring drop in tolerance, an almost constant low or high heat-tolerance level ("C-type" according to LARCHER et al. 1973). The ecological significance of constant heat tolerance is obvious in *Fontinalis* and in other water plants growing in almost constant temperatures during the year, but not in terrestrial species. The ecological correspondence is clear, however, in most of the species displaying the "S-type" response (summer peak in heat tolerance), since they live in warm summer habitats, e.g., of mediterranean climatic regions, or areas within the southern Alps, the Caucasus, and Australia. A larger group of species develops high heat tolerance only in winter ("W-type"). It comprises typical shade plants, some alpine species, dwarf shrubs, and grasses. Most of these become very tolerant to freezing in winter. Their heat tolerance can be increased through low temperature cultivation or cold hardening (HIGHKIN 1959; KAPPEN 1964; MAGOMEDOV et al. 1972; ALEXANDROV 1977). In these cases heat and freezing tolerance appear to be linked to each other. This agrees with the fact that photosynthetic processes are more heat-tolerant in late fall (SANTARIUS 1974), or that diastase activity is more heat-tolerant in frost-hardened plants (JUNG and LARSON 1972).

A great number of plants exhibit a combination of an adaptive increase in heat tolerance in summer and another heat tolerance peak in winter ("WS-type": Fig. 14.4). The summer peak can be higher than the winter peak (e.g., *Dicranum scoparium, Asplenium septentrionale, Taxus baccata*), or lower (e.g., *Pellia epiphylla, Polypodium vulgare, Asarum europaeum*), or peaks are equal (*Mnium punctatum, Polystichum lobatum, Erica tetralix*). In most cases there is an apparent tendency for species with a more pronounced heat tolerance peak in summer to grow in warm open habitats.

Dependencies on the Time Periods of Heating: Hardening, Acclimation

The extensive study of COLLANDER (1924) illustrates the exponential nature of thermal death curves. From the ecological viewpoint the determination of long-term heat stress tolerance (several hours or days) best characterizes adaptation to life in a permanently hot or warm milieu. Heat stress for a restricted time (10–60 min) may characterize the situation of naturally occurring short periods of overheating which is found in terrestrial environments with large temperature oscillations. Heat shocks of a few minutes or even seconds are damaging mostly at very high temperatures which rarely occur under natural conditions. Such treatments give more information about the species-specific physiological nature of plant responses to heat (YARWOOD 1967; ALEXANDROV 1977; and others).

The upper limit of still noninjurious *long-term* heating (several hours or days) for sensitive plants such as most of the nonthermophilic algae, lichens, and water plants, may be between 22 and 40 °C (algae from the North Pacific were already killed when exposed to 24–26 °C for longer than 12 h; BIEBL 1970). Some strains of *Euglena gracilis* lost their ability to form green chloroplasts at growing temperatures above 34 °C (BRAWERMAN and CHARGAFF 1960). The air dry lichen *Peltigera canina* var. *praetextata* was damaged after 40 days exposure to 35 °C (MACFARLANE and KERSHAW 1978). The corresponding limits for terrestrial cormophytes may be between 38 and 45 °C. Seedlings of desert plants were killed if the temperatures exceeded 40 °C for longer time periods (FREEMAN et al. 1977).

On the other hand, longer exposure to raised but still tolerable temperatures increases the heat tolerance of plants through *adjustive processes*. Temperature acclimatization was early shown to occur in poikilothermic animals (cf. PRECHT et al. 1955; for microorganisms cf. Chap. 11, this Vol.) but its occurrence in plants was long doubted (LEVITT 1956; MONTFORT et al. 1957). SCHWENKE (1959) first succeeded in slightly increasing the heat tolerance of red algae by warm cultivation. Algae from the Barent Sea became 2–4 K more tolerant after exposure to 24–31 °C for a few hours (LUTOVA and FELDMAN 1960; freshwater

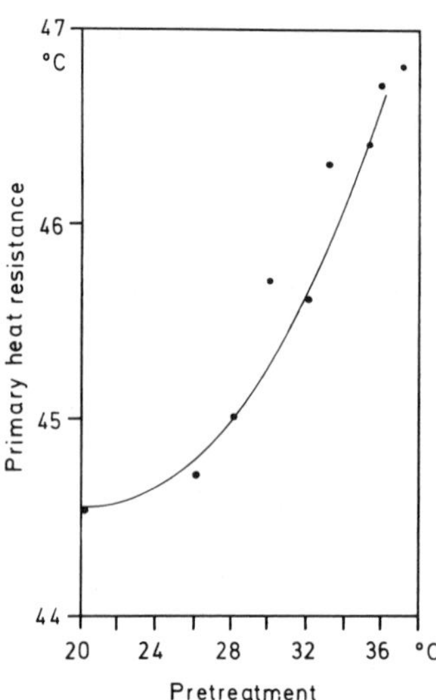

Fig. 14.5. Increases in heat tolerance of *Tradescantia fluminensis* following exposure to supraoptimal temperatures. Heat tolerance (*ordinate:* temperature which stops cytoplasmic motion after 5 min heating) resulting from 16–18 h pretreatment at different temperatures (see *abscissa*). (After ALEXANDROV 1964)

algae: SCHÖLM 1968). Within the bryophytes an adaptive increase of the heat tolerance by 2–3 K was shown (DIRCKSEN 1964; ANTROPOVA 1974).

Heat hardening (about 2 K) in higher plants following cultivation for 10 days at 28 °C was demonstrated by LANGE (1962a), LUTOVA and ZAVADSKAYA (1966) and BAKANOVA (1970). The concomitant adaptive increase in the thermostability of enzymes has been repeatedly demonstrated (cf. FELDMAN 1979). Within a certain range hardening increases with duration and level of the hardening temperature (Fig. 14.5; ALEXANDROV 1964, 1977; YARWOOD 1967; MAGOMEDOV et al. 1972). SHUKHTINA et al. (1978) showed experimentally that efficiency of hardening changes during the growing season in *Morus alba*.

The capacity for temperature adjustment in plants is obviously of survival advantage in a varying natural environment. This capacity is revealed in the appearance of a summer peak in heat tolerance and also in the irregular appearance of a heat-tolerance peak in extremely hot summers, after the threshold temperature for hardening has been reached (LANGE 1961; ALEXANDROV 1977). Consequently in the species of the "WS-type" the expression of the summer peak varies from year to year. This can be corroborated by the finding of a loss of heat tolerance (weakening) under suboptimal conditions (e.g., marine algae at 6° C: FELDMAN and LUTOVA 1963).

The transient state of increased heat tolerance which lasts generally for about 10–50 h following the hardening process, has been repeatedly shown (e.g., LUTOVA 1962; cf. SCHRÖDER 1963; YARWOOD 1967). A reduced hardening effect was observed still for 6 days (ALEXANDROV 1964). The adaptive increase in the upper limit of net photosynthesis was maintained for up to 20 days (BAUER 1978). Tolerance loss can be retarded by cool temperatures (cf. ALEXANDROV 1977).

Evidence of ecologically relevant hardening by *short heat shocks* has been shown (cf. YARWOOD 1963b; LUTOVA et al. 1977). For example, damage in bean leaves was three times less at 55 °C heating (5 min) if they were "predisposed" for 30 at 50 °C (YARWOOD 1961 a). While an increase in the "primary heat resistance" was detected already 4–5 s after the predisposition (LOMAGIN 1961; ZAVADSKAYA 1963), YARWOOD (1967) demonstrated that up to about 45 °C substantial hardening took place during the heat exposure itself if it lasted for more than a few minutes, whereas after shorter heat shocks hardening appeared only after a lag phase of 1–24 h. In the latter case the hardening process occurred under ambient conditions after and not during the hardening treatment.

Heat tolerance of plants was also raised by *repeated* heat shocks alternating with optimal temperatures (LAUDE and CHAUGULE 1953; cf. ALEXANDROV 1977). This treatment may even be similar to the naturally occurring changes of temperatures. However, the additional increase in tolerance is low compared with the effect after the first shock (cf. YARWOOD 1961 a). On the other hand, the heat tolerance continued to increase significantly if each subsequent heat shock temperature was somewhat increased (SHUKHTINA 1964).

Short shocks at supraoptimal temperatures can cause a *sensitization* to heat. Sensitization is also indicated when subsequent to a treatment at almost injurious temperatures plants suffer from high but normally noninjurious temperatures (SKOGQVIST 1974). This is also shown in fungi, algae, and bacteria (FRIES and SÖDERSTRÖM 1963). Sensitization as a transient effect is interpreted as "reversible injury" (ENGELBRECHT and MOTHES 1964) being a phase of repair during which an increase in heat tolerance is not possible (SKOGQVIST and FRIES 1970; ALEXANDROV 1977).

Isolated as well as attached leaves of *Populus deltoides* × *simonii* and *Gleditschia triacantha* responded to 90 s heating with a discontinuous curve having a first peak of injury between 42 and 45 °C, followed by a noninjurious temperature range, and finally injurious temperatures beyond 55 °C (WAGENBRETH 1965b). This type of response curve is interpreted as indicating first the appearance of a sensitizing time–temperature combination which is then followed by a combination causing thermoadaptation of the leaves. The *two-peaked* response curve did not appear in poplar and other species if they were tested in spring and early summer (KAPPEN and ZEIDLER 1977). It disappeared in winter again in the evergreen *Ligustrum*.

The two-phase response curve becomes more pronounced with decreasing time periods of heating, whereas it can no longer be detected if the heat treatment exceeds 10 min (Fig. 14.6). The ecological importance of injurious short-term heat stress has been shown only rarely. However, in support of its relevance one can cite the appearance of sudden beams of sun radiation especially under damp and calm conditions (sunscald), or the effect of sunflecks on forest ground vegetation.

Water Content

Seeds and *spores* are widely known to be highly tolerant of heat (70–150 °C, some bacterial spores even 195 °C) due to their low water content (JUST 1877; ZOBL 1944; CHRISTOPHERSEN 1955). Dry seeds can even be stimulated to a higher percentage of germination following one to several hours exposure to 70–110 °C (FIRBAS 1965; CHAWAN 1971).

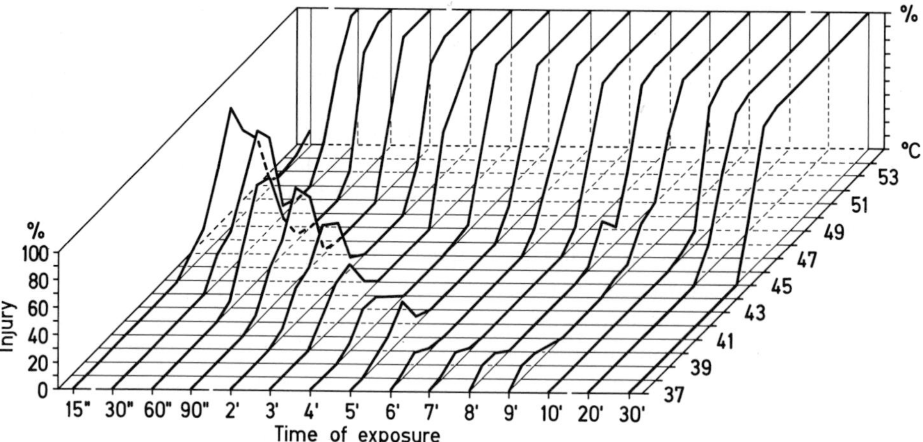

Fig. 14.6. Heat tolerance (in % of injury) in leaves of *Populus deltoides* × *simonii* in response to heat treatment at various temperature-time combinations from 37 to 53 °C and from 15 s to 30 min. The first phase of injurious effects disappears at exposure periods longer than 9 min. (KAPPEN and ZEIDLER 1977)

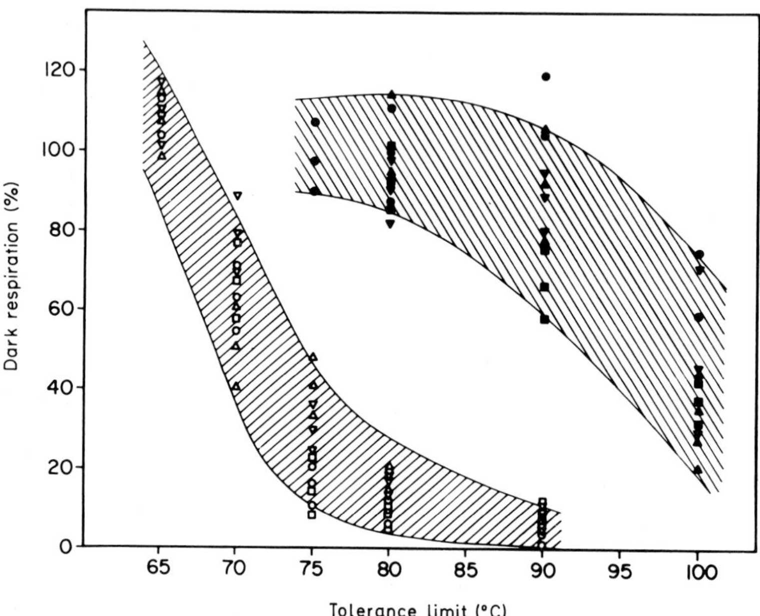

Fig. 14.7. Heat tolerance (dark respiration occurring after a 30-min heat exposure, in % of nontreated control) in dry thalli from two Central European lichen associations. Four species from a humid montane forest habitat, belonging to the Usneetum barbatae association (*open symbols*) are compared with four species from the Fulgensietum continentale (*black symbols*). (Modified, after LANGE 1953)

The ecological importance of wilting and drying for the heat tolerance of plants is particularly evident in terrestrial *cryptogams* which have a relatively low heat tolerance in the soaked state but exist close to strongly insolated surfaces. Their ecological adaptation results from their poikilohydric nature, which allows them to lose water rapidly under sun radiation, and their extreme tolerance to desiccation, since only in the desiccated state do they have high heat tolerance. Soil-inhabiting bacteria, actinomycetes, and blue-green algae, as well as fungi, are able to withstand 70–100 °C* maximally in the dry state (cf. Table 14.3; SZABO et al. 1964; SINGH 1970; MACENTEE et al. 1972; ZIMMERMANN and BUTIN 1973).

The difference in the heat-tolerance limits of dry and soaked terrestrial algae, lichens and bryophytes is 40–50 K (LANGE 1953, 1955; FRIEDMANN and GALUN 1974; NÖRR 1974). The annual amplitude in the heat tolerance limits of dry bryophytes is consequently much larger (8–23 K) than that of moist samples (2–4 K: DIRCKSEN 1964). The degree of heat tolerance in the dry state is frequently more closely related to the thermal conditions of the natural habitat than the heat tolerance of the soaked specimen (Fig. 14.7; LANGE 1953, 1955). It is therefore obvious that terrestrial cryptogams are adapted to existence in hot, but arid environments (cf. KAPPEN 1973; FRIEDMANN and GALUN 1974). Experimental evidence of survival over extended hot periods was shown with the desert liverwort *Riccia billardieri* (CHANDHOKE and SHARMA 1972), or with the xeric lichen *Peltigera canina* var. *rufescens* (MACFARLANE and KERSHAW 1978).

Even in a hydrophytic lichen (*Dermatocarpon aquaticum*) heat tolerance increased to 90 °C* when dry; however, wet cultivation decreases the heat resistance of mosses and lichens (LANGE 1953, 1955). Thermophilic algae–at least some species–are very sensitive to desiccation and their heat tolerance cannot be increased beyond its normally high level (FOGG 1969). This may be due to the fact that membrane stabilization resulting from thermo-adaptation is different from that due to adaptation to desiccation (cf. GRANIN et al. 1977).

According to some authors, species of the few highly desiccation-tolerant poikilohydric *cormophytes* may become as heat-tolerant as the lower cryptogams in the dry state (60–120 °C, CERNJAVSKY 1928; VIEWEG and ZIEGLER 1969). The heat-tolerance limit strongly depends on the water deficit of the leaves (Fig. 14.8) and is terminated in *Ramonda myconi* and ferns at the limit of the desiccation tolerance. Depending on the species the maximum increase in heat tolerance ranges between 4.5 and 8 K (KAPPEN 1966). Thus, such plants could not withstand more than 55–60 °C* (OPPENHEIMER and HALEVY 1962; SHARMA et al. 1977). The range of tolerance increase was nearly constant during the whole year (KAPPEN 1966).

In sclerophylls as well as in mesophytic plants the heat tolerance limit was increased by 1–3 K at a saturation deficit of the leaves between 10 and 30% (HAMMOUDA and LANGE 1962; BANNISTER 1970; LARCHER and WAGNER 1976: *Carex firma* 5 K). The reason why ZAVADSKAYA and DENKO (1966) or FALKOVA (1969) found a tolerance increase only in wilted xerophytes and not in mesophytic species, or in the case of barley (ZAVADSKAYA and SHUKHTINA 1974) only under dry cultivation, may be due to the method of vitality estimation (cf. KAPPEN

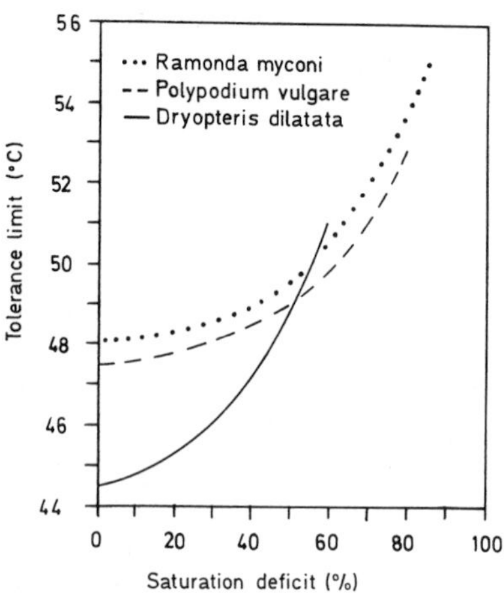

Fig. 14.8. Relationship between water-saturation deficit and heat tolerance (temperature causing 10% injury) in leaves of the poikilohydrous *Ramonda myconi* and two European fern species, in winter. (After KAPPEN 1966)

and LANGE 1968). It can be concluded that about 62 °C is the absolute upper heat-tolerance limit for cormophytic plants, which cannot be surpassed by removal of water of by any other treatment.

Osmotic agents in solutes can increase the heat tolerance in the same way as desiccation (algae: MCLEAN 1967; yeast: CORRY 1976; bryophytes: SCHEIBMAIR 1938; higher plant tissues: FELDMANN 1962; HELLMUTH 1971b). Plasmolysis without dehydration of the cytoplasm was not effective (DÖRING 1933; BOGEN 1948). The water loss causes shrinking of the membranes, which probably prevents the heat-induced injurious leaking of cellular solutions (GIBSON 1973).

Generally there is no close correlation between heat tolerance and actual desiccation tolerance and thus no coupling between both tolerances is apparent (DIRCKSEN 1964; KAPPEN 1964; BANNISTER 1970).

Visible Radiation and Photoperiod

Influences of light and darkness on the heat tolerance were already mentioned by ILLERT (1924), SAPPER (1935) and others. Short day periods induced an increase in heat tolerance in various plants already within 10 days (LANGE and SCHWEMMLE 1960; BIEBL 1967b). SCHWARZ (1969) demonstrated that the winter peak in heat tolerance of *Pinus cembra* was due to an endogenous component and short-photoperiod induction; but in *Rhododendron ferrugineum* it was due to the absence of long day influence. Herbaceous species either of the type having a pronounced winter peak (*Saxifraga aizoon*) or the type having a pronounced summer peak (*Sempervivum montanum*) in heat tolerance also respond to short photoperiods with increasing heat tolerance (BIEBL and MAIER 1969; MAIER 1971).

High light intensity causes an increase in heat tolerance either by compensating for a long-day-induced depression or by supporting the hardening effect of short photoperiods (MAIER 1971). It accelerates heat hardening (wheat and millet: OLEYNIKOVA 1965). However, in cotton its effect ceased after 15 min of subsequent darkening (VESELOVSKII et al. 1976). Light applied during heating was found to have a stabilizing effect on the heat tolerance of *Lemna minor*

(STANLEY and MADEWELL 1976). However, the light effects varied depending on the time of exposure and could also be detrimental (cf. LOMAGIN and ANTROPOVA 1966).

Ions, Nutrients, Salinity

The influence of ions on heat tolerance has been investigated primarily from the aspect of effects on the physiological state of the cytoplasm, however, it can also be of ecological importance as an edaphic effect. Calcium ions have been repeatedly shown to increase the heat tolerance of algae, bryophytes, tissues of terrestrial and water plants (SCHEIBMAIR 1938; BOGEN 1948; BAKANOVA 1970), or to stabilize thermophilic bacteria (CHRISTOPHERSEN 1973), whereas this is not due to its osmotic effect (cf. BARABALCHUK 1970). Apparently because of its lyophilic nature, Ca^{2+} causes a shrinking of the membranes (BOGEN 1948) which counteracts heat-induced leakage and thus prevents injury (TOPROVER and GLINKA 1976). Univalent alkali ions and nitrate cause swelling of the cytoplasm, and consequently decrease heat tolerance (SCHEIBMAIR 1938; BOGEN 1948; BARABALCHUK 1970). Thus, it is evident that high nutrient levels increase sensitivity to heat stress (SAPPER 1935; CARROLL 1943; STANLEY and MADEWELL 1976).

Algae as well as animals have been reported to be increasingly heat-tolerant as salinity of water increases (SCHWENKE 1959; BIEBL 1969b; ALEXANDROV 1977). The extremely halophilic green alga *Dunaliella parva* tolerates a 5-min heat treatment up to 49 °C in saline water, but in freshwater only up to 41 °C (GIMMLER et al. 1978). The triggering component is assumed to be either the content of Ca^{2+} in the seawater (cf. LEVITT 1956) or an osmotic effect (GIMMLER et al. 1978), which, however, was not evident in the experiments of BIEBL (1969b).

Among the terrestrial plants so far tested the halophytic species always display the highest heat tolerance (*Kochia planifolia* 50 °C*, under moisture stress 56 °C*; *Arthrocnemum bidens* 58 °C*: HELLMUTH 1971a, b; *Tamarix senegalensis* 50 °C*; *Nitraria schoberi* 51 °C*: LANGE 1959). Particularly in these cases an osmotic effect is very likely.

14.4 General Conclusions

Plant heat tolerance results from a complex of inheritable and environmentally adaptive responses, as well as from the actual physiological state. The ecological adaptation is most directly evident in thermophiles or in most psychrophiles in which physiological capacity is directly adjusted to the permanently extreme environment.

Water plants subjected to changing temperatures, salinity, and light conditions already display more complex responses to heat. Therefore a direct correlation between macroclimate and heat tolerance is only significant in species of the more constantly conditioned deeper regions of the ocean. On the other hand, it depends on the prevailing environmental factors influencing the heat-tolerance response (in shallow water heating and osmotic changes, in salt lakes salinity etc.). Algae increase their heat tolerance with increasing salt content. However, in general marine algae are not more tolerant than freshwater algae. If marine algal species are kept equally conditioned before heat-tolerance testing, their physiological response to heat corresponds closely to their natural habitat conditions (MONTFORT et al. 1955, 1957).

Amphibious and terrestrial plants experience still greater environmental fluctuations and extremes. In poikilohydrous, usually lower plants, changes in the water regime are most drastic. The advantage of this poikilohydric behavior is evident from the fact that such plants can live in environments where they become heated regularly to temperatures which would be lethal to them in the soaked, active state. Terrestrial cryptogams in the soaked state display a tolerance span between 30 and 50 °C in species living in the high arctic as well as in hot regions. It is striking that the heat tolerance of lichens, bryophytes, and fungi in the desiccated state relates well to climatic region or thermal character of the environment.

Resistance behavior is more complicated in homoiohydrous plants and direct correlations between plasmatic tolerance and climate are frequently not apparent. Extremely heat tolerant species exist in high arctic and alpine environments where they would rarely have need of this capacity. The appearance of a highly increased heat tolerance in winter seems to be of no ecological significance; nor is there an explanation for the nightly increase in heat tolerance in CAM plants. These examples demonstrate that heat tolerance is not simply a response to the environmental temperature regime, but also that it depends on various other exogenous and endogenous factors.

On the other hand, variation in heat tolerance related to differences in ecological conditions is obvious in species occurring in restricted areas and in uniform plant groupings or in ecotypes of one species.

The variation in heat tolerance between species and ecotypes can be modified by sensitization, hardening, or adjustment to different environmental conditions, or by developmental differences, or through effects of nutrition and ions in the different habitats. Halophytes generally display extremely high heat tolerance.

In addition to endogenous and exogenous factors governing plasmatic heat tolerance, functional and constitutional factors essentially influence plant response to heat. Therefore plants with low heat tolerance are not uncommon in extremely hot environments because they can avoid heat stress. The ecological importance of heat stress-avoiding mechanisms has recently become more and more evident and must be further evaluated. Delicate measuring methods are necessary to elucidate the role of reflection, leaf position, convection, transpiratory cooling and wilting effects, and also life form in general as a "strategy" against heat impact. Ephemerals and deciduous species display generally low, and perennials high, heat tolerance. This is especially obvious when closely related species are compared. A high heat tolerance appears essential for the CAM and C_4 plants because they have poor transpiratory cooling during the day, as well as for the sclerophylls which close stomata under hot, dry conditions. Moreover, a single plant does not react uniformly to heat stress. Its young leaves can be more susceptible than its older leaves. Plants keep their sensitive growing points sheltered and far enough from the ground which has a comparatively extreme temperature regime. Fruits, although highly heat-tolerant, most frequently suffer from heat stress. However, injury to fruits is of little importance for the survival of the plant as long as the seeds are not affected.

The determination of heat-tolerance limits by exposure to heat within a restricted time period is indicative of survival capacity. Such stress may occur in natural environments with strong heat oscillations. Short heat stress during diurnal maximum irradiation or from sunflecks may actually cause injury, sensitization, or even acclimation to heat. On the other hand, the measured values of sublethal or lethal heat tolerance limits may be indicative of the heat sensitivity of vital functions, which control productivity and growth. The threshold temperatures for disturbance of these processes are markedly lower and therefore are more frequent than lethal temperatures in natural environments. Although these reactions cannot easily be observed in the field, experimental data can help to explain the adaptive capacity and resistance of plants to heat.

References

Adelusi SA, Lawanson AO (1978) Heat-induced changes in dry weight, leaf size and number of chloroplasts per cell in maize and cowpea shoots. J Agr Sci 91:349–358

Ahlgren CE (1974) Effects of fires on temperate forests: North Central United States. In: Kozlowski TT, Ahlgren CE (eds) Fire and ecosystems. Academic Press, New York, pp 195–223

Aist JR, Israel HW (1977) Effects of heat-shock inhibition of papilla formation on compatible host penetration by two obligate parasites. Physiol Plant Pathol 10:13–20

Alexandrov VYa (1964) Cytophysiological and cytoecological investigations of resistance of plant cells toward the action of high and low temperature. Rev Biol 39:35–77

Alexandrov VYa (1977) Cells, molecules and temperature. Conformational flexibility of macromolecules and ecological adaptation. Ecological Studies Vol 21. Springer, Berlin, Heidelberg, New York

Ansari AQ, Loomis WE (1959) Leaf temperatures. Am J Bot 46:713–717

Antropova TA (1974) Temperature adaptation studies on the cells of some bryophyte species. Tsitologiya 16:38–42

Bakanova LV (1970) Relative heat resistance of leaves and spikelet glumes of certain cereal plants. Sov Plant Physiol 17:109–113

Baker FS (1929) Effect of excessively high temperatures on coniferous reproduction. J For 27:949–975

Bannister P (1970) The annual course of drought and heat resistance in heath plants from an oceanic environment. Flora 159:105–123

Barabalchuk KA (1970) Effects of calcium, manganese, magnesium and sodium ions on resistance of plant cells. Tsitologiya 12:609–621

Barber HN, Sharpe PJH (1971) Genetics and physiology of sunscald of fruits. Agric Meteorol 8:175–191

Basnizki J, Evenari M (1975) The influence of a reflectant on leaf temperature and development of the globe artichoke (*Cynara scolymus* L.). J Am Soc Hortic Sci 100:109–112

Bauer H (1972) CO_2-Gaswechsel nach Hitzestress bei *Abies alba* Mill. und *Acer pseudoplatanus* L. Photosynthetica 6:424–434

Bauer H (1978) Photosynthesis of ivy leaves (*Hedera helix*) after heat stress. I. CO_2-gas exchange and diffusion resistances. Physiol Plant 44:400–406

Bauer H, Larcher W, Walker RB (1975) Influence of temperature stress on CO_2-gas exchange. In: Cooper IBP (ed) Photosynthesis and productivity in different environments. Int Biol Prog, Vol 3, pp 557–586
Bělehrádek J (1935) Temperature and living matter. Protoplasma Monogr 8, Bornträger, Berlin
Berner T (1974) Studies on the ecophysiology of the hypolithic algae from the flint stones of the Meshash formation in the northern Negev, Israel. Thesis, Jerusalem
Bernstam VA (1974) Effects of supraoptimal temperatures on the myxomycete *Physarum polycephalum*. II. Effects on rate of protein and ribonucleic acid synthesis. Arch Mikrobiol 95:347–356
Biebl R (1939) Über Temperaturresistenz von Meeresalgen verschiedener Klimazonen und verschieden tiefer Standorte. Jahrb Wiss Bot 88:389–420
Biebl R (1962a) Protoplasmatische Ökologie der Pflanzen. Wasser und Temperatur. In: Protoplasmatologia, Handbuch der Protoplasmaforschung 12.1, Springer, Wien
Biebl R (1962b) Temperaturresistenz tropischer Meeresalgen. Bot Mar 4:241–254
Biebl R (1964) Temperaturresistenz tropischer Pflanzen auf Puerto Rico. Protoplasma 59:133–156
Biebl R (1967a) Temperaturresistenz einiger Grünalgen warmer Bäche auf Island. Le Botaniste Ser L:34–41
Biebl R (1967b) Kurztag-Einflüsse auf arktische Pflanzen während der arktischen Langtage. Planta 75:77–84
Biebl R (1967c) Temperaturresistenz tropischer Urwaldmoose. Flora 157:25–30
Biebl R (1976d) Protoplasmatische Ökologie. Naturwiss Rundsch 20:248–252
Biebl (1968) Über Wärmehaushalt und Temperaturresistenz arktischer Pflanzen in Westgrönland. Flora 157B:327–354
Biebl R (1969a) Untersuchungen zur Temperaturresistenz arktischer Süßwasseralgen im Raum von Barrow, Alaska. Mikroskopie 25:3–6
Biebl R (1969b) Studien zur Hitzeresistenz der Gezeitenalge *Chaetomorpha cannabina* (Aresch.) Kjellm. Protoplasma 67:451–472
Biebl R (1970) Vergleichende Untersuchungen zur Temperaturresistenz von Meeresalgen entlang der pazifischen Küste Nordamerikas. Protoplasma 69:61–83
Biebl R, Maier R (1969) Tageslänge und Temperaturresistenz. Oesterr Bot Z 117:176–194
Biebl R, McRoy CP (1971) Plasmatic resistance and rate of respiration and photosynthesis of *Zostera marina* at different salinities and temperatures. Mar Biol 8:48–56
Bogen HJ (1948) Untersuchungen über Hitzetod und Hitzeresistenz pflanzlicher Protoplaste. Planta 36:298–340
Brawerman G, Chargaff E (1960) A self-reproducing system concerned with the formation of chloroplasts in *Euglena gracilis*. Biochim Biophys Acta 37:221–229
Bregetova LG, Popova AI (1962a) Twentyfour hour rhythmics of changes of warm resistance in protoplasms of plants. Nekot Wopr Fotosynthesa i Bodonovo Regima Rast, Tematicheski Sornik 1:41–46
Bregetova LG, Popova AI (1962b) Heat resistance of the protoplasm of various types of the grass vegetation of Tadjikistan. Isdatel Akad Nauk Tadjik SSSR, Douchambe, Trudy II
Brock TD (1978) Thermophilic microorganisms and life at high temperatures. Springer, New York, Heidelberg, Berlin
Bünning E, Herdtle H (1946) Physiologische Untersuchungen an thermophilen Blaualgen. Z Naturforsch 1:93–99
Bullock JG, Coakley WT (1978) Investigation into the mechanisms and repair of heat damage to *Schizosaccharomyces pombe* growing in synchronous cultures. J Therm Biol 3:159–162
Buxton PA (1925) The temperature of the surface of deserts. J Ecol 12:127–134
Carroll JC (1943) Effects of drought, temperature and nitrogen on turf grasses. Plant Physiol 18:19–36
Cernusca A (1972) Energiebilanz natürlicher und künstlicher Ökosysteme. Umschau 72:628–630

Cernjavsky P (1928) Anabiosis of the *Ramondia nathaliae* Panc. Petr. J Soc Bot Russe 13:27–38

Chandhoke KR, Sharma K (1972) Adaptations of bryophytes against drought. 1: Effect of temperature on *Riccia billardieri*. Indian Sci Cong Ass Proc 59:313

Chawan DD (1971) Role of high temperature pre-treatments on seed-germination of desert species of *Sida* (Malvaceae). Oecologia 6:343–349

Christophersen J (1955) Microorganismen. In: Precht H, Christophersen J, Hensel H (eds) Temperatur und Leben. Springer, Berlin, Göttingen, Heidelberg, pp 178–328

Christophersen J (1973) Basic aspects of temperature action on microorganisms. In: Precht H, Christophersen J, Hensel H, Larcher W (eds) Temperature and life, Springer, Berlin, Heidelberg, New York, pp 3–86

Clausen E (1964) The tolerance of hepatics to desiccation and temperature. Bryologist 67:411–417

Collander R (1924) Beobachtungen über die quantitativen Beziehungen zwischen Tötungsgeschwindigkeit und Temperatur beim Wärmetod pflanzlicher Zellen. Commentat Biol 1:1–12

Corry JEL (1976) The effect of sugars and polyols on the heat resistance and morphology of osmophilic yeasts. J Appl Bacteriol 40:269–276

Coutinho LM (1969) Novas observações sobre a ocorrênia do "efeito de De Saussure" e suas relações com a suculência a temperatura folhear e os movimentos estomáticos. Bol Fac Fil Cien Letras Univ Sao Paulo Bot 24:77–102

Curtis OF (1938) Wallace and Clum, "leaf temperatures": A critical analysis with additional data. Am J Bot 25:761–771

Dahl E (1963) On the heat exchange of a wet vegetation surface and the ecology of *Koenigia islandica*. Oikos 14:190–211

Day JM, Roughley RJ, Eaglesham ARJ, Dye M, White SP (1978) Effect of high soil temperatures on nodulation of cowpea *Vigna unguiculata*. Ann Appl Biol 88:476–481

Deverall BJ (1965) Temperatures. In: Ainsworth GC, Sussman AS (eds) The Fungi. Vol I, Academic Press, New York, pp 534–550

Dircksen A (1964) Vergleichende Untersuchungen zur Frost-, Hitze- und Austrocknungsresistenz einheimischer Laub- und Lebermoose unter besonderer Berücksichtigung jahreszeitlicher Veränderungen. Diss Göttingen

Doemel WN, Brock TD (1971) The physiological ecology of *Cyanidium caldarium*. J Gen Microbiol 67:17–32

Döring H (1933) Beiträge zur Frage der Hitzeresistenz pflanzlicher Zellen. Planta 18:405–434

Dörr M (1941) Temperaturmessungen an Pflanzen des Frauensteins bei Mödling. Beih Bot Zentr 60:Abt A

Dutrochet M (1839) Récherches sur la température propre des végétaux. Ann Sci Nat 12:77–84

Edlich F (1936) Einwirkungen von Temperatur und Wasser auf aerophile Algen. Arch Mikrobiol 7:62–109

Egorova LI, Semikhatova OA, Yudina OS (1978) Influence of temperature on the reactivation of photosynthesis after heat injury. Bot Z 63:356–362

Eller BM (1977) Road dust induced increase of leaf temperature. Environ Pollut 13:99–107

Eller BM (1979) Die strahlungsökologische Bedeutung der Epidermisauflagen. Flora 168:146–192

Engelbrecht L, Mothes K (1964) Weitere Untersuchungen zur experimentellen Beeinflussung der Hitzewirkung bei Blättern von *Nicotiana rustica*. Flora 154:279–298

Falkova TV (1969) Influence of temperature of medium and water deficiency of leaves on protoplasm heat resistance in some species of Caprifoliaceae family. Byull Gos Nikit Bot Sada (4.II):83–88

Feldman NL (1962) The influence of sugars on the cell stability of some higher plants to heating and high hydrostatic pressure. Tsitologiya 4:633–643

Feldman NL (1979) Effect of heat hardening on thermostability of acid phosphatase from wheat leaves. J Therm Biol 4:41–45

Feldman NL, Lutova MI (1963) Variations de la thermostabilité cellulaire des algues en fonctions des changements de la température du millieu. Cah Biol Mar 4:435–458
Feldman NL, Artyushenko ZT, Shukhtina HG (1970) Cellular heat resistance in certain species of different genera of Amaryllidaceae. Bot Z 55:1678–1683
Firbas F (1927) Über die Bedeutung des thermischen Verhaltens der Laubstreu für die Frühjahrsvegetation des sommergrünen Laubwaldes. Beih Bot Zentr 44:179
Firbas F (1931) Untersuchungen über den Wasserhaushalt der Hochmoorpflanzen. Jahrb Wiss Bot 74:459–696
Firbas H (1965) Über die Resistenz von Samenarten gegen hohe Temperaturen und Beobachtungen an künstlich getrocknetem Saatgut. Saatgut Wirtsch 17:279–281
Fogg GE (1969) Survival of algae under adverse conditions. In: Woolhouse HW (ed) Symposia of the society for experimental biology, Nr. 23, Dormancy and survival, Cambridge Univ Press, pp 123–142
Forsyth DJ (1977) Limnology of lake Rotokawa and its outlet steam. N Z J Mar Freshwater Res 11:524–539
Franco CM (1961) Lesão do colo do cafeeiro, causada pelo calor. Bragantia 20:645–652
Freeman CE, Tiffany RS, Ried WH (1977) Germination responses of *Agave lecheguilla*, *A. parryi*, and *Fouquieria splendens*. Southwest Nat 22:195–204
Friedmann EI, Galun M (1974) Desert algae, lichens and fungi. In: Brown Jr GW (ed) Desert biology. Vol II, Academic Press, New York, pp 166–212
Fries N, Söderström I (1963) Induction of thermosensitivity in cells from various plants. Expt Cell Res 32:199–202
Frings JFJ (1976) The *Rhizobium* pea symbiosis as affected by high temperatures. Meded Landbouwhogesch Wageningen 76, 1–76
Gates DM (1963) Leaf temperature and energy exchange. Arch Meteorol Geophys Bioklimatol SerB 12:321–336
Gates DM (1965) Heat transfer in plants. Sci Am 1965:76–84
Gates DM (1973) Plant temperatures and energy budget. In: Precht H, Christopherson J, Hensel H, Larcher W (eds) Temperature and life. Springer, Berlin, Heidelberg, New York, pp 87–101
Gates DM, Alderfer R, Taylor E (1968) Leaf temperature of desert plants. Science 159:994–995
Gibbs JG, Patten DT (1970) Plant temperatures and heat flux in a Sonoran desert ecosystem. Oecologia 5:165–184
Gibson B (1973) The effect of high sugar concentrations on the heat resistance of vegetative micro-organisms. J Appl Bact 36:365–376
Gimmler H, Kühnl EM, Carl G (1978) Salinity dependent resistance of *Dunaliella parva* against extreme temperatures. I. Salinity and thermoresistance. Z Pflanzenphysiol 90:133–153
Granin AV, Pronina ND, Veselovskii VA (1977) Effect of dehydration and overheating on afterglow of the photosynthesis apparatus in poikilohydrous and homeohydrous plants. Fiziol Rast 24:1261–1268
Grieve BJ, Hellmuth EO (1970) Eco-physiology of Western Australian plants. Ecol Plant 5:33–68
Grintal AR (1976) Effect of temperature on rate of respiration in *Laminaria saccharina*. Bot Z 61:1608–1615
Hagen PO, Kushner DJ, Gibbons NE (1964) Temperature induced death analysis in a psychrophilic bacterium. Can J Microbiol 10:813–822
Hammouda M, Lange OL (1962) Zur Hitzeresistenz der Blätter höherer Pflanzen in Abhängigkeit von ihrem Wassergehalt. Naturwissenschaften 21:500
Hellmuth EO (1971a) Eco-physiological studies on plants in arid and semi-arid regions in Western Australia: III. Comparative studies on photosynthesis, respiration and water relations of 10 arid zone and 2 semi-arid zone plants under winter and late summer climatic conditions. J Ecol 59:225–259
Hellmuth EO (1971b) Eco-physiological studies on plants in arid and semi-arid regions in Western Australia: V. Heat resistance limits of photosynthetic organs of different seasons, their relation to water deficits and cell sap properties and the regeneration ability. J Ecol 59:365–374

Henkel PA, Margolina KP (1951) On the viscosity of cytoplasm and the (heat) drought resistance of vegetative and generative organs of plants. Dokl Acad Nauk SSSR 76:587–590

Henrici M (1955) Temperatures of Karroo plants. S Afr J Sci 51:245–248

Herzog F (1938) Formgestalt und Wärmehaushalt bei Sukkulenten. Jahrb Wiss Bot 87:211–243

Hesselbo A (1918) The Bryophyta of Iceland. In: The Botany of Iceland – pt II 4, Kopenhagen, London

Highkin HR (1959) Effect of vernalization on heat resistance in two varieties of peas. Plant Physiol 34:643–644

Huber B (1935) Der Wärmehaushalt der Pflanzen. In: Boas F (ed) Naturwissenschaft und Landwirtschaft 17, München

Illert H (1924) Botanische Untersuchungen über Hitzetod und Stoffwechselgifte. Bot Archiv 7:133–141

Itai C, Benzioni A, Ordin L (1973) Correlative changes in endogenous hormone levels and shoot growth induced by short heat treatments to the root. Physiol Plant 29:355–360

Jameson DA (1961) Heat and desiccation resistance of tissue of important trees and grasses of the pinyon-juniper type. Bot Gaz 122:174–179

Jung GA, Larson KL (1972) Cold, drought and heat tolerance. In: Hanson CH (ed) Agronomy 15, Alfalfa science and technology 24, Illus Maps Am Soc Agron, pp 185–209

Just L (1877) Über die Einwirkung höherer Temperaturen auf die Erhaltung der Keimkraft der Samen. Beitr Biol Pflanz 2:311–348

Kainmüller C (1975) Temperaturresistenz von Hochgebirgspflanzen. Oesterr Akad Wiss, Anz Math-Nat Kl 7:67–75

Kalckstein B (1976) Gaswechsel, Produktivität und Herbizidempfindlichkeit bei verschiedenen tropischen, subtropischen und europäischen Gramineen. Diss Univ Wien 133:203–243

Kappen L (1964) Untersuchungen über den Jahreslauf der Frost-, Hitze- und Austrocknungsresistenz von Sporophyten einheimischer Polypodiaceen (Filicinae). Flora 155:123–166

Kappen L (1965) Untersuchungen über die Widerstandsfähigkeit der Gametophyten einheimischer Polypodiaceen gegenüber Frost, Hitze und Trockenheit. Flora 156:101–115

Kappen L (1966) Der Einfluß des Wassergehaltes auf die Widerstandsfähigkeit von Pflanzen gegenüber hohen und tiefen Temperaturen, untersucht an Blättern einiger Farne und von *Ramonda myconi*. Flora 156B:427–445

Kappen L (1973) Response to extreme environments. In: Ahmadjian V, Hale ME (eds) The lichens III, 10, Academic Press, New York, pp 311–380

Kappen L, Lange OL (1968) Die Hitzeresistenz angetrockneter Blätter von *Commelina africana* – ein Vergleich zwischen zwei Untersuchungsmethoden. Protoplasma 65:119–132

Kappen L, Smith CW (1980) Heat tolerance of two *Cladonia* species and *Campylopus praemorsus* in a hot steam vent area of Hawaii. Oecologia 46:184–189

Kappen L, Zeidler A (1977) Seasonal changes between one- and two-phasic response of plant leaves to heat stress. Oecologia 31:45–53

Kappen L, Lange OL, Schulze E-D, Evenari M, Buschbom U (1979) Ecophysiological investigations on lichens of the Negev desert. VI. Annual course of the photosynthetic production of *Ramalina maciformis* (Del.) Bory. Flora 168:85–108

Karshon R, Pinchas L (1971) Variations in heat resistance of ecotypes of *Eucalyptus camaldulensis* Dehn. and their significance. Aust J Bot 19:261–272

Kjelvik S (1976) Varmeresistens og varmeveksling for noen planter, vesentlig fra Hardangervidda. Blyttia 34:211–226

Kluge M, Ting IP (1978) Crassulacean acid metabolism, analysis of an ecological adaptation. Ecological Studies 30. Springer, Berlin, Heidelberg, New York

Kreeb K (1964) Zur Methodik der NTC-Temperaturmessung und Tagesgänge der Blattemperatur bei immergrünen Macchienpflanzen Südfrankreichs. In: Beiträge zur Phytologie. Ulmer, Stuttgart

Lange B (1973) The *Sphagnum* flora of hot springs in Iceland Lindbergia 2:81–93

Lange OL (1953) Hitze- und Trockenresistenz der Flechten in Beziehung zu ihrer Verbreitung. Flora 140:39–97

Lange OL (1954) Einige Messungen zum Wärmehaushalt poikilohydrer Flechten und Moose. Arch Meteor, Geophysik Bioklimatol, Ser B 5:182–190

Lange OL (1955) Untersuchungen über die Hitzeresistenz der Moose in Beziehung zu ihrer Verbreitung. 1. Die Resistenz stark ausgetrockneter Moose. Flora 142:381–399

Lange OL (1959) Untersuchungen über Wärmehaushalt und Hitzeresistenz mauretanischer Wüsten- und Savannenpflanzen. Flora 147:595–651

Lange OL (1961) Die Hitzeresistenz einheimischer immer- und wintergrüner Pflanzen im Jahresverlauf. Planta 56:666–683

Lange OL (1962a) Über die Beziehungen zwischen Wasser und Wärmehaushalt von Wüstenpflanzen. Veroeff Geobot Inst Eidg Techn Hochsch Zuerich 37:155–168

Lange OL (1962b) Versuche zur Hitzeresistenz-Adaption bei höheren Pflanzen. Naturwissenschaften 49:20–21

Lange OL (1965a) The heat resistance of plants, its determination and variability. In: Methodology of plant eco-physiology: Proc Montpellier Symp

Lange OL (1965b) Der CO_2-Gaswechsel von Flechten nach Erwärmung im feuchten Zustand. Ber Dtsch Bot Ges 78:444–454

Lange OL (1967) Investigations on the variability of heat-resistance in plants. In: Troshin AS (ed) The cell and environmental temperature. Pergamon Press, Oxford, pp 131–141

Lange OL (1969) Experimentell-ökologische Untersuchungen an Flechten der Negev-Wüste I. CO_2-Gaswwechsel von *Ramalina maciformis* (Del.) Bory unter kontrollierten Bedingungen im Laboratorium. Flora 158:324–359

Lange OL, Lange R (1962) Die Hitzeresistenz einiger mediterraner Pflanzen in Abhängigkeit von der Höhenlage ihres Standortes. Flora 152:707–710

Lange OL, Lange R (1963) Untersuchungen über Blattemperaturen, Transpiration und Hitzeresistenz an Pflanzen mediterraner Standorte (Costa Brava, Spanien). Flora 153:387–425

Lange OL, Schwemmle B (1960) Untersuchungen zur Hitzeresistenz vegetativer und blühender Pflanzen von *Kalanchoë bloßfeldiana*. Planta 55:208–225

Lange OL, Schulze E-D, Koch W (1970) Experimentell-ökologische Untersuchungen an Flechten der Negev-Wüste. II. CO_2-Gaswechsel und Wasserhaushalt von *Ramalina maciformis* (Del) Bory am natürlichen Standort während der sommerlichen Trockenperiode. Flora 159:38–62

Lange OL, Schulze E-D, Evenari M, Kappen L, Buschbom U (1974) The temperature-related photosynthetic capacity of plants under desert conditions. I. Seasonal changes of the photosynthetic response to temperature. Oecologia 17:97–110

Lange OL, Schulze E-D, Evenari M, Kappen L, Buschbom U (1978) The temperature-related photosynthetic capacity of plants under desert conditions. III. Ecological significance of the seasonal changes of the photosynthetic response to temperature. Oecologia 34:89–100

Larcher W (1961) Jahresgang des Assimilations- und Respirationsvermögens von *Olea europea* L. ssp. *sativa* Hoff. et Link., *Quercus ilex* L. und *Quercus pubescens* Willd, aus dem nördlichen Gardaseegebiet. Planta 56:575–606

Larcher W (1973) Limiting temperatures for live functions. In: Precht H, Christophersen J, Hensel H, Larcher W (eds) Temperature and life. Springer, Berlin, Heidelberg, New York, pp 195–231

Larcher W (1977) Ergebnisse des IBP-Projekts „Zwergstrauchheide Patscherkofel". Sitz Ber Oesterr Akad Wiss, Math Nat Kl, Abt 1, Bd 186, Wien

Larcher W (1980) Ökologie der Pflanzen. Ulmer, Stuttgart

Larcher W, Mair B (1969) Die Temperaturresistenz als ökophysiologisches Konstitutionsmerkmal: 1. *Quercus ilex* und andere Eichenarten des Mittelmeergebietes. Oecol Plant 4:347–376

Larcher W, Wagner J (1976) Temperaturgrenzen der CO_2-Aufnahme und Temperaturresistenz der Blätter von Gebirgspflanzen im vegetationsaktiven Zustand. Oecol Plant 11:361–374

Larcher W, Heber U, Santarius KA (1973) Gradual progress of damage due to temperature stress. In: Precht H, Christophersen J, Hensel H, Larcher W (eds) Temperature and life. Springer, Berlin, Heidelberg, New York, pp 195–292

Laude HH (1939) Diurnal cycle of heat resistance in plants. Science 89:556–557

Laude HM, Chaugule BA (1953) Effect of state of seedling development upon heat tolerance in bromegrasses. J Range Manage 6:320–324

Levitt J (1956) The hardiness of plants. Agronomy 6, Academic Press, New York

Levitt J (1958) Frost, drought and heat resistance. Protoplasmatologia 6, Springer, Wien

Levitt J (1972) Responses of plants to environmental stresses. Academic Press, New York

Levitt J (1980) Responses of plants to environmental stresses. Vol. I. Chilling, freezing, and high temperatures. Academic Press, New York

Lomagin AG (1961) Changes in the resistance of plant cells after a short action of high temperature. Tsitologiya 3:426–436

Lomagin AG, Antropova TV (1966) Photodynamic injury to heated leaves. Planta 68:297–309

Loman AA (1963) The lethal effects of periodic high temperatures on certain lodgepole pine slash decaying Basidiomycetes. Can J Bot 43:334–338

Lutova MI (1962) The effect of heat hardening on photosynthesis and respiration on leaves. Bot Z 47:1761–1774

Lutova MI, Feldman NL (1960) A study of the ability of temperature adaptation in some marine algae. Tsitologiya 2:699–709

Lutova MI, Zavadskaya IG (1966) Effects of exposure of plants to different temperatures on the cell heat resistance. Tsitologiya 8:484–493

Lutova MI, Feldman NL, Drobyshev VP (1968) Changes in the thermoresistance of marine algae under the influence of environmental temperature. Tsitologiya 10:1538–1545

Lutova MI, Alexandrov VYa, Feldman NL (1977) Increase in cell thermostability of marine and fresh-water algae under the influence of superoptimal temperature. Tsitologiya 19:368–374

MacBryde B, Jefferies RL, Alderfer R, Gates DM (1971) Water and energy relations of plant leaves during period of heat stress. Oecol Plant 6:151–162

MacEntee FJ, Schreckenberg G, Bold HC (1972) Some observations on the distribution of edaphic algae. Soil Sci 114:171–179

MacFarlane JD, Kershaw KA (1978) Thermal sensivity in lichens. Science 201:739–741

Magomedov ZG, Tarusov BN, Doskosh YaE (1972) Optimisation of temperature regime for the hardening of plants. Vestn Mosk Univ Ser 6 Biol Pochvoved 27:113–116

Maier R (1971) Einfluß von Photoperiode und Einstrahlungsstärke auf die Temperaturresistenz einiger Samenpflanzen. Oesterr Bot Z 119:306–322

Malcolm NL (1969) Enzymatic bases of physiological changes in a mutant of the psychrophile *Micrococcus cryophilus*. Biochim Biophys Acta 190:337–346

McLean RJ (1967) Desiccation and heat resistance of the green alga *Spongiochloris typica*. Can J Bot 45:1933–1938

Meeks JC, Castenholz RW (1972) Photosynthesis at the upper temperature limit of the extreme thermophile *Synechococcus lividus*. J Phycol 8:18

Mellor RS, Salisbury FB, Raschke K (1964) Leaf temperatures in controlled environments. Planta 61:56–72

Miller EC, Saunders AR (1923) Some observations on the temperature of the leaves of crop plants. J Agr Res 26:15–43

Montfort C, Ried A, Ried I (1955) Die Wirkung kurzfristiger warmer Bäder auf Atmung und Photosynthese im Vergleich von eurythermen und kalt-stenothermen Meeresalgen. Beitr Biol Pflanz 31:349–375

Monfort C, Ried A, Ried I (1957) Abstufungen der funktionellen Wärmeresistenz bei Meeresalgen in ihren Beziehungen zu Umwelt und Erbgut. Biol Zentralbl 76:257–289

Mooney HA, Billings WD (1961) Comparative physiological ecology of arctic and alpine populations of *Oxyria digyna*. Ecol Monogr 31:1–29

Mooney HA, Ehleringer J, Björkman O (1977a) The energy balance of leaves of the evergreen desert shrub *Atriplex hymenelytra*. Oecologia 29:301–310

Mooney HA, Weisser PJ, Gulmon SL (1977b) Environmental adaptations of the Atacaman desert cactus *Copiapoa haseltoniana*. Flora 166:117–124

Mooney HA, Björkman O, Collatz CJ (1978) Photosynthetic acclimation to temperature

in the desert shrub, *Larrea divaricata*. I: CO_2-exchange characteristics of intact leaves. Plant Physiol 61:406–410
Nelson EE, Fay H (1974) Thermal tolerance of *Poria weirii*. Can J For Res 4:288–290
Niethammer A (1947) Technische Mycologie. Enke, Stuttgart
Noack K (1920) Der Betriebsstoffwechsel der thermophilen Pilze. Jahrb Wiss Bot 59:413–466
Nörr M (1974) Hitzeresistenz bei Moosen. Flora 163:388–397
Oechel WC, Strain BR, Odening WR (1972) Photosynthetic rates of a desert shrub, *Larrea divaricata* Cav. under field conditions. Photosynthetica 6:183–188
Oleynikova TV (1965) High temperature and light effects on the permeability of cells of spring cereal leaves. Sci Counc Cytol Problems Acad Nauk SSSR, pp 70–81
Oppenheimer HR, Halevy AH (1962) Anabiosis of *Ceterach officinarum* Lam. et D.C. Bull Res Counc Israel Sect D, 11:127–147
Parkhurst DF, Loucks OL (1972) Optimal leaf size in relation to environment. J Ecol 60:505–537
Pearcy RW (1976) Acclimation of photosynthesis and respiratory carbon dioxide exchange to growth temperature in *Atriplex lentiformis* (Torr.) Wats. Plant Physiol 59:795–799
Pearcy RW, Björkman O, Harrison AT, Mooney HA (1971/72) Photosynthetic performance of two desert species with C_4 photosynthesis in Death Valley, California. Carnegie Inst Washington Yearb 71:540–550
Pearcy RW, Harrison AT, Mooney HA, Björkman O (1974) Seasonal changes in net photosynthesis of *Atriplex hymenelytra* shrubs growing in Death Valley, California. Oecologia 17:111–121
Peary JA, Castenholz RW (1964) Temperature strains of a thermophilic blue-green alga. Nature 202:720–721
Pisek A, Larcher W, Pack I, Unterholzner R (1968) Kardinale Temperaturbereiche der Photosynthese und Grenztemperaturen des Lebens der Blätter verschiedener Spermatophyten. II. Temperaturmaximum der Nettophotosynthese und Hitzeresistenz der Blätter. Flora 158B:110–128
Precht H, Christophersen J, Hensel H (eds) (1955) Temperatur und Leben. Springer, Berlin, Göttingen, Heidelberg
Precht H, Christophersen J, Hensel H, Larcher W (eds) (1973) Temperature and life. Springer, Berlin, Heidelberg, New York
Pulgar CE, Laude HM (1974) Regrowth of alfalfa after heat stress. Crop Sci 14:28–30
Rakhimov GT (1976) After-effects of high temperatures on the Hill reaction as a function of age of some desert plants. UZB Biol Z 1:77–79
Raschke K (1956) Mikrometeorologisch gemessene Energieumsätze eines *Alocasia*-Blattes. Arch Meteorol Geophys Bioklimatol B 7:240–268
Raschke K (1960) Heat transfer between the plant and the environment. Ann Rev Plant Physiol 11:111–126
Rouschal E (1938) Zum Wärmehaushalt der Macchienpflanzen. Oesterr Bot Z 87:42–50
Rouschal E (1939) Die kühlende Wirkung des Transpirationsstromes in Bäumen. Ber Dtsch Bot Ges 57:53–66
Sachs J (1864) Über die obere Temperatur-Grenze der Vegetation. Flora 23:4–12, 24–29, 33–41, 65–75
Santarius KA (1974) Seasonal changes in plant membrane stability as evidenced by the heat sensivity of chloroplast membrane reactions. Z Pflanzenphysiol 73:448–451
Sapper I (1935) Versuche zur Hitzeresistenz der Pflanzen. Planta 23:518–556
Scheibmair G (1938) Hitzeresistenzstudien an Mooszellen. Protoplasma 29:394–424
Schmeidl H (1965) Oberflächentemperaturen in Hochmooren. Wetter Leben 17:87–97
Schölm HE (1966) Untersuchungen zur Wärmeresistenz von Tiefenalgen. Bot Mar 9:54–61
Schölm HE (1968) Untersuchungen zur Hitze- und Frostresistenz einheimischer Süßwasseralgen. Protoplasma 65:97–118
Schramm W (1968) Ökologisch-physiologische Untersuchungen zur Austrocknungs- und Temperaturresistenz an *Fucus vesiculosus* L. der westlichen Ostsee. Int Rev Ges Hydrobiol 53:469–510

Schröder CA (1963) Induced temperature tolerance of plant tissue in vitro. Nature 200:1301–1302

Schwarz W (1969) Der Einfluß der Photoperiode auf das Austreiben, die Frosthärte und die Hitzeresistenz von Zirben und Alpenrosen. Flora 159:258–285

Schwemmle B, Lange OL (1959) Endogen-tagesperiodische Schwankungen der Hitzeresistenz bei *Kalanchoë bloßfeldiana*. Planta 53:134–144

Schwenke H (1959) Untersuchungen zur Temperaturresistenz mariner Algen der westlichen Ostsee. I: Das Resistenzverhalten von Tiefenrotalgen bei ökologischen und nichtökologischen Temperaturen. Kiel Meeresforsch 15:34–50

Semikhatova OA, Denko EI (1960) Effect of temperature on the respiration of leaves. Exp Bot, Ser 4, 14:112–136

Semikhatova OA, Sakov WC, Gorbacheva GI (1962) Studies on the after-effect of temperature on the rate and dynamics of photosynthesis of *Polygonum sachalinense*. Exp Bot Ser 4, 15:25–42

Seybold A (1929) Die pflanzliche Transpiration. Ergeb Biol 5:29–165

Sharma P, Yadav AK, Bhardwaja TN (1977) Heat and drought resistance in *Actinopteris radiata* and *Adiantum caudatum*. Geobios (Jodhpur) 4:256–258

Shirley HL (1936) Lethal high temperatures for conifers, and the cooling effect of transpiration. J Agric Res 53:239–258

Shukhtina HG (1964) The influence of repeated heat-hardening on thermostability of plant cells. In: Collection of works (Sbornik) Cytological aspects of adaptation of plants to the environmental factors, Moscow, Leningrad, Izd Nauka, pp 26–30

Shukhtina GG, Yazkulyev A, Durdyev A (1978) Heat hardening of epidermal cells of *Morus alba* L. leaves under natural conditions. Bot Z 63:429–433

Singh SS (1970) Variations and survival of soil fungal flora on temperature treatments. Proc Nat Acad Sci India, Sect B 40:281–288

Skinner FA (1968) The limits of microbial existance. Proc R Soc London Ser B 171:77–89

Skogqvist I (1974) Induction of heat sensivity of wheat roots and its effects on mitochondria ATP, triglyceride and total lipid content. Exp Cell Res 86:285–294

Skogqvist I, Fries N (1970) Induction of thermosensivity and salt sensivity in wheat-roots (*Triticum aestivum*) and the effect of kinetin. Experientia 26:1160–1162

Smith WK (1978) Temperatures of desert plants: Another perspective of adaptability of leaf size. Science 211:614–616

Sorokin C (1971) Calefaction and phytoplankton. Bio Science 21:1153–1159

Stanley RA, Madewell CE (1976) Thermal tolerance of *Lemna minor* L. Circular Z 73, Tennessee Valley Authorithy, Musche Shoals, Alabama, USA, pp 1–16

Steinhübel G (1966) Wirkung einer Staubschicht auf die Überwärmung der Blattspreite bei direkter Insolation. Biologia Bratislava 21:277–294

Stocker O (1954) Der Wasser- und Assimilationshaushalt südalgerischer Wüstenpflanzen. Ber Dtsch Bot Ges 67:288–299

Stocker O (1956) Die Dürreresistenz. In: Ruhland W (ed) Handbuch der Pflanzenphysiologie. Springer, Berlin, Göttingen, Heidelberg, Vol. III, pp 696–741

Stocker O (1971) Der Wasser- und Photosynthese-Haushalt von Wüstenpflanzen der mauretanischen Sahara. II. Wechselgrüne, Rutenzweig- und stammsukkulente Bäume. Flora 160:445–494

Strain BR, Chase VC (1966) Effect of past and prevailing temperatures on the carbon dioxide exchange capacities of some woody desert perennials. Ecology 47:1043–1045

Szabo I, Marton M, Varga L (1964) Untersuchungen über die Hitzeresistenz, Temperatur und Feuchtigkeitsansprüche der Mikroorganismen eines mullartigen Waldrenzinabodens. Pedobiologica 4:43–64

Tanner CB, Goltz SM (1972) Excessively high temperatures of seed onion umbels. J Am Soc Hortic Sci 97:5–9

Tansey MR, Brock TD (1978) Microbial life at high temperatures: Ecological aspects. In: Kushner DJ (ed) Microbial life in extreme environments. Academic Press London. New York

Taylor SE, Sexton OJ (1972) Some implications of leaf tearing in Musaceae. Ecology 53:143–149

Thofelt L (1975) Studies on leaf temperature recorded by direct measurements and by thermography. Acta Univ Upsaliensis 12:1–143

Tieszen LL (1973) Photosynthesis and respiration in arctic tundra grasses: Field light intensity and temperature responses. Arct Alp Res 5:239–251

Toprover Y, Glinka Z (1976) Calcium ions protect beet root cell membranes against thermally induced changes. Physiol Plant 37:131–134

Url W, Fetzmann E (1959) Wärmeresistenz und chemische Resistenz der Grünalge *Gloeococcus bavaricus* Skuja. Protoplasma 50:471–482

Veselovskii VA, Leshchinskaya LV, Markarova EN, Veselova TV, Tarusov BN (1976) Effect of illumination of cotton leaves on heat resistance of the photosynthetic apparatus. Fiziol Rast 23:467–472

Vieweg GH, Ziegler H (1969) Zur Physiologie von *Myrothammus flabellifolia*. Ber Dtsch Bot Ges 82:29–36

Volodin AN (1951) The heat resistance of some xerophytes in their natural environment. Bjull Moscov Obsc Ispyt Prir N S Otdel Biol 56:72–80

Wagenbreth D (1965a) Durch Hitzeschocks induzierte Vitalitätsänderungen bei Laubholzblättern. Flora 156A:63–75

Wagenbreth D (1965b) Das Auftreten von zwei Letalstufen bei Hitzeeinwirkung auf Pappelblätter. Flora 156A:116–126

Wagenbreth D (1968) Die Wirkung von Hitzeschocks auf die Rostanfälligkeit von Pappelblättern. Phytopathol Z 61:87–97

Wartenberg H (1933) Kälte und Hitze als Todes- und Krankheitsursache der Pflanzen. In: Sorauer P (ed) Handbuch der Pflanzenkrankheiten, Bd I, 1 Die nichtparasitären und Virus-Krankheiten. 6 Aufl (Appel O ed) Parey, Berlin

Weber F (1926) Hitzeresistenz funktionierender Stomatanebenzellen. Planta 2:669–677

Wolpert A (1962) Heat transfer analysis of factors affecting plant leaf temperature. Significance of leaf hair. Plant Physiol 37:113–120

Yarwood CE (1961a) Acquired tolerance of leaves to heat. Science 134:941

Yarwood CE (1961b) Translocated heat injury. Plant Physiol 36:712–726

Yarwood CE (1963a) Heat therapy of bean rust. Phytopathology 53:1313–1316

Yarwood CE (1963b) Sensitization of leaves to heat. Adv Front Plant Sci (New Dehli) 7:195–204

Yarwood CE (1965) Temperature and plant disease. World Rev Pest Control 4:53–63

Yarwood CE (1967) Adaptation of plants and plant pathogens to heat. In: Prosser CL (ed) Molecular mechanisms of temperature adaptation. Am Ass Adv Sci, Washington, pp 75–89

Yarwood CE (1977) Heat-induced and cold-induced retention of inoculum by leaves. Phytopathology 67:1259–1261

Yazkulyev A (1964) The increase in cell thermostability of *Aristida karelini* (Trin. et Rupr.) Roshev. and *Arundo donax* L. under the influence of environmental temperature in natural conditions. In: Cytological aspects of adaptation of plants to the environmental factors. Acad. Sci. USSR Moscow Leningrad pp 3–25

Zavadskaya IG (1963) On the rate of increase of thermostability of plant cells after a short preliminary exposure to high temperature. Bot Z 48:755–758

Zavadskaya IG, Denko EJ (1966) The effect of dehydration on thermostability of plant cells. Bot Z 51:696–705

Zavadskaya IG, Shukhtina GG (1974) Effects of dehydration and elevated temperature on leaf cell thermo-resistance of a drought sensitive barley cultivar. Tsitologiya 16:950–955

Zimmermann G, Butin H (1973) Untersuchungen über die Hitze- und Trockenresistenz holzbewohnender Pilze. Flora 162:393–419

Zobl KH (1944) Untersuchungen über die Widerstandsfähigkeit von Pilzsporen gegen feuchte und trockene Hitze sowie deren chemische Zusammensetzung, verbunden mit Untersuchungen über die Anwendung von sulfonamidehaltigen Pilznährböden. Diss Würzburg

Zschokke A (1931) Sonnenbrand-, Hitzetod- und Austrocknungsschäden an Reben. Gartenbauwissenschaft 4:196–232

15 Wind as an Ecological Factor

P.S. Nobel

CONTENTS

15.1 Introduction

Wind affects plant growth, reproduction, distribution, death, and ultimately plant evolution. Some of the effects depend on the air boundary layers next to the aerial parts of a plant, across which gas and heat exchanges with the environment occur. Others relate to the mechanical deformation of the plant by the frictional drag of the moving air. Wind also disperses many types of particles (pollen, plant propagules, disease organisms) as well as moving gas molecules (CO_2, pollutants). Because of the many effects of wind, ranging from obvious crop or forest destruction during gales to subtle effects on a leaf boundary layer, the literature available is vast and covers many disciplines. Here a few of the fundamental principles underlying the interaction of plant parts with air currents will be developed and various consequences for plant physiological ecology will be discussed.

15.1.1 Concepts, Terminology

Wind describes the large-scale transport of air masses resulting from differences in air pressure. Such air movements reflect asymmetric interception of shortwave irradiation from the sun, leading to differential heating of the Earth's surface. In the lower kilometers of the atmosphere, land surface features such as mountains and canyons affect the wind-speed profile. Below about 10 m above the ground, wind speed is often most strongly influenced by plants, becoming completely arrested at their surfaces.

Wind speed tends to be higher in coastal regions, where it can average 7 m s^{-1} annually 10 m above the ground (GRACE 1977). Lake or sea breezes are usually twice as strong as land breezes, which blow in the opposite direction at night (OKE 1978). Inland, the highest winds occur near isolated mountains, and especially where air flow is funneled between mountains or through canyons. On average, wind speeds are lower at night.

Air flow can be laminar near surfaces, particularly at low wind speeds. However, turbulent flow where air movement is not parallel and orderly characterizes wind near natural vegetation. Air movement also exhibits temporal inhomogeneity. A useful parameter describing short-term variations in wind speed is the turbulence intensity (t):

$$t = \frac{\sqrt{(\bar{u} - u)^2}}{\bar{u}} \tag{15.1}$$

where $\bar{u}$ is the mean wind speed and u is its instantaneous value. Thus t represents the standard deviation divided by the mean wind speed. In vegetation under natural conditions t is often about 0.4 (CIONCO 1972; DENMEAD and BRADLEY 1973). For fairly uniform plant communities, t is fairly independent of height within the canopy (INOUE 1963; CIONCO 1972). However, if the leaves are concentrated in one layer, maximum turbulence intensity tends to occur there. As leaves are shed in a deciduous forest, t generally decreases (CIONCO 1972).

Growth chambers normally have low wind speed (O'LEARY and KNECHT 1974) as well as low turbulence intensity (DENMEAD and BRADLEY 1973). Extrapolation of results from growth chambers to the field is thus fraught with uncertainty as far as wind is concerned. This is particularly relevant to much current laboratory research on environmental effects on plants, where the influences of wind have often not been appreciated. For instance, low wind speeds in growth chambers lead to much taller kidney bean plants (*Phaseolus vulgaris* cv. Red Cherokee Bush) than when grown out-of-doors (JAFFE and BIRO 1979). Tall versus short plants respond differently to drought and other stresses, indicating that the morphological changes induced by wind can have other ecophysiological consequences.

For considerations within and above vegetation, turbulent air is described as moving in packets or eddies. Small eddies with diameters less than that of leaves may be important for gas exchange, and larger eddies may be responsible for mechanical loading on plants (GRACE 1977). Furthermore, eddies are crucial for movement of CO_2, H_2O, and other gases in and above plant canopies.

Air becomes vertically unstable if its temperature decreases rapidly with height. At the dry adiabatic lapse rate (9.8 °C decrease in temperature per km increase in altitude, OKE 1978), a rising parcel of dry air will cool by expansion due to the decrease in air pressure with no heat exchange with the surroundings, a case of neutral stability. Observed lapse rates are often -5 °C/km (GRACE 1977). Hot dry air results from air moving down in elevation, e.g., the Santa Ana condition in southern California, which can be desiccating to vegetation.

Forced convection usually dominates as a means of heat exchange from plant surfaces at most wind speeds encountered in the field (KREITH 1973; MONTEITH 1973). However, buoyancy effects also occur near vegetation leading to free convection of heat away from surfaces, which is particularly apparent in the desert (OKE 1978; FUCHS 1979). For instance, since the soil surface can be greatly heated in the desert, leading to a large temperature lapse rate in the air above, the lower part of the atmosphere can become convectively very unstable, leading to "dust devils" and the shimmering of objects.

Since air flow is retarded at any solid surface because of friction, wind speed profiles develop vertically away from the Earth's surface as well as away from plant parts. The region of reduced air movement immediately adjacent to a surface is referred to as the air boundary layer (SCHLICHTING 1968; MONTEITH 1973; GRACE 1977). Boundary layers next to leaves are often about 1 mm thick, and they significantly affect mass and heat exchange with the environment.

Momentum transfer is another consequence of air movement past plant surfaces. When the air flow is blocked or intercepted, form drag occurs. Even when the plant surface is parallel to the air flow, the air movement is slowed due to skin friction. The shearing stresses which develop cause the wind speed to decrease near vegetation. Thus vegetation also affects the wind.

15.1.2 Wind Speed Profiles

A useful relationship describing the variation in wind speed [u(z)] with height (z) above a large, horizontal, uniform canopy under stable (neutral) atmospheric conditions is:

$$u(z) = \frac{u_*}{k} \ln\left(\frac{z-d}{z_0}\right) \tag{15.2}$$

where u_* is the shearing or friction velocity, k is the von Karman constant (about 0.41), z_0 is the roughness length, and d is the zero plane displacement (GATES 1962; COWAN 1968; BRADLEY and FINNIGAN 1973; ROSENBERG 1974; GRACE 1977; FUCHS 1979; cf. also Chap. 1, this Vol.). For uniform vegetation, z_0, which is essentially an expression of the drag exerted by the wind passing over the vegetation, is often about 10% of the length of surface protuberances, but it is considerably less for sparse desert vegetation. Plant communities with a high z_0 exert more drag on the moving air. Generally d is 60 to 70% of the canopy height, since most of the friction-producing part of the plants occurs near that level (COWAN 1968; THOM 1971; MONTEITH 1973; GRACE 1977; OKE 1978). Equation (15.2) requires averaging over a time interval much longer than the characteristic periods of fluctuation in local wind speed. The most important point for the present discussion is the logarithmic relationship between height and wind speed.

Air flow within plant canopies is complicated by the three-dimensional architecture of the plants. Consequently, the one-dimensional flux analyses often applied to crop plants are rarely appropriate for natural ecosystems. Form drag by the leaves is responsible for most of the shearing stresses, and hence the distribution of leaves in space determines the local wind patterns. Wind speed does not necessarily decrease toward the ground, as air in an open forest can tunnel under the branches and hence the wind speed can be greater there than further up in the canopy. When all the variations are included, the difficulties in specifying the wind speed at particular plant surfaces become apparent (BRADLEY and FINNIGAN 1973).

15.2 Boundary Layers and Environmental Exchanges

One of the most widely studied effects of wind on plants is the transfer of heat and mass across the air boundary layers surrounding leaves. Since wind speed affects the thickness of the boundary layer and boundary layers affect the temperature of the plant part, any plant process depending on temperature can be affected by wind speed.

15.2.1 Description, Fluxes

The boundary layer is a region dominated by the shearing stresses originating at some surface. Adjacent to a leaf is a laminar sub-layer of air, whose movement is parallel to the leaf surface. Diffusion away from the leaf surface is by molecular motion in this region. Further from the surface the boundary layer becomes turbulent and the transfer process is eddy-assisted. An eddy will move more or less as a unit and therefore will carry all the molecules in it at the same rate, i.e., differences in diffusion properties between molecular species will become obliterated.

Instead of trying to describe separately transfer processes across the boundary layer in the laminar and turbulent portions, both of which change in thickness across a leaf's surface, an effective or equivalent boundary layer thickness (δ) averaged over the whole leaf surface is used (often referred to as the displacement air boundary layer). Since there is no sharp discontinuity of wind speed from the air adjacent to the leaf and the free airstream, the definition of boundary layer thickness is somewhat arbitrary and hence it is generally defined operationally. For instance, the flux of heat (J_H) is

$$J_H = k\,(T_s - T_a)/\delta \tag{15.3}$$

where k is the thermal conductivity of air, T_s is the leaf surface temperature, and T_a is the air temperature in the free stream outside the boundary layer. From Fick's first law of diffusion, the flux of substance j (J_j) can be represented as follows:

$$J_j = D_j\,\Delta c_j/\delta \tag{15.4}$$

where D_j is the diffusion coefficient of substance j in air and Δc_j is its concentration drop across the boundary layer (NOBEL 1974a).

As indicated above, movement across the boundary layer is partly by molecular diffusion and partly by eddy diffusion. Thus D_j averaged over δ is intermediate between the value for molecular diffusion and that for eddy diffusion, where all molecules are carried along together. For instance, D_{H_2O}/D_{CO_2} for Eq. (15.4) is not 1.56 (the ratio of molecular diffusion coefficients) or 1.00 (the ratio of eddy diffusion coefficients), but is found empirically to be about 1.35 (MONTEITH 1973). More generally, D_j in Eq. (15.4) for species j is about $(D_{H_2O}/D'_j)^{2/3}$, where D_{H_2O} is the molecular diffusion coefficient of water (2.5×10^{-5} m^2 s^{-1} at 20 °C) and D'_j is the molecular diffusion coefficient of species j. Similarly, D_{H_2O}/k is empirically about 1.10 for boundary layers next to leaves (THOM 1968; GRACE 1977). Such definitions of D_j and k allow us to use the same δ for fluxes of heat [Eq. (15.3)] or mass [Eq. (15.4)], which facilitates comparison of boundary layer effects on exchanges at plant surfaces.

For many studies of heat and mass transfer, relationships between parameters are presented in terms of dimensionless numbers (cf. Chap. 1, this Vol.). The size of objects is represented by a characteristic dimension (d), which for a flat rectangular plate is often the length in the direction of the wind and for a cylinder or sphere is the diameter. This leads to the Nusselt number, d/δ. The ratio of inertial forces to viscous forces is the Reynolds number, ud/ν, where ν is the kinematic viscosity for the fluid (1.5×10^{-5} m^2 s^{-1} for dry air at 20 °C, MONTEITH 1973). At low Reynolds numbers viscous forces dominate and the flow tends to be laminar, while at high Reynolds numbers the flow becomes turbulent. Turbulence sets in at a lower Reynolds number for real leaves compared to flat plates (PERRIER et al. 1973). The effects of wind on transfer processes are represented by relationships between these dimensionless numbers, which is the basis for determining the values of δ presented below. At very low wind speeds and large $T_s - T_a$, free convection dominates forced convection and the Reynolds number should be replaced by the Grashof number, which takes into account buoyancy forces (GATES 1962; MONTEITH 1973; NARAIN and UBEROI 1973).

15.2.2 Magnitudes

By combining the data of various workers, a useful but approximate empirical expression for δ of flat leaves under field conditions is

$$\delta = 4\sqrt{d/u} \tag{15.5}$$

where d is the characteristic dimension in m, u is the wind speed in m s^{-1}, and δ is the displacement air boundary layer thickness in mm (COWAN 1968; PARLANGE et al. 1971; PEARMAN et al. 1972; MONTEITH 1973; NOBEL 1974a). Although d is often defined as the length or the mean length across a flat surface in the direction of the wind, a more precise definition is the downwind length of a flat rectangular plate that has the same convective heat transfer as the surface in question (PARKHURST et al. 1968; GATES and PAPAIN 1971; TAYLOR 1975).

For a variety of reasons, boundary layer thicknesses predicted from theory are often about twofold greater than those determined experimentally [Eq. (15.5)] on leaves (RASCHKE 1956; PARLANGE et al. 1971; PEARMAN et al. 1972; GRACE and WILSON 1976). Instead of determining δ using Eq. (15.5), the boundary layer conductance for water vapor (D_{H_2O}/δ) is often estimated by placing a wet piece of blotting paper in the shape of the leaf in question in the air stream and measuring the water loss, J_{H_2O}; δ is then calculated using Eq. (15.4).

Although most studies of boundary layers for plants have dealt with relatively flat surfaces, primarily leaves, many plant parts such as stems, branches, inflorescences, fruits, and even certain leaves (e.g., onion) represent sizable three-dimensional objects. Air flow is blocked by such "bluff bodies", and its streamlines are diverted around the objects. For plant surfaces in the shape of cylinders or spheres of diameter d, estimates of boundary layer thickness under turbulence intensities appropriate for field conditions are as follows (NOBEL 1974b, 1975):

$$\delta \cong 5.8\sqrt{d/u} \quad \text{cylinder} \tag{15.6}$$

$$\delta \cong 2.8\sqrt{d/u} + \frac{0.25}{u} \quad \text{sphere} \tag{15.7}$$

Units in Eqs. (15.6) and (15.7) are the same as for Eq. (15.5).

As turbulence intensity increases [Eq. (15.1)], boundary layer thickness tends to decrease. In going from extremely low turbulence intensities typical of wind tunnels (e.g., 0.02 or less) to those encountered in the field (0.2 to 1.0), δ may decrease about 10 to 50% for bluff bodies, with the smaller decreases occurring at lower Reynolds numbers (KESTIN 1966; SCHLICHTING 1968; NOBEL 1974b; NOBEL 1975; KOWALSKI and MITCHELL 1975). Turbulence intensity apparently has less effect on the effective boundary layer thickness for flat plates (KESTIN 1966; GRACE and WILSON 1976).

To discuss the influence of wind speed on transfer processes, the relative conductance (or its reciprocal, the resistance) of the air boundary layer must be related to the other conductances (or resistances) in the overall pathway. From Eq. (15.3), the heat conductance is k/δ and from Eq. (15.4) the mass conductance is D_j/δ. Since the boundary layer is the only conductance occurring in the heat flux pathway, δ has more influence on heat than on mass fluxes, where stomatal, cuticular, and internal leaf conductances also must be considered. Specifically, the conductance of the boundary layer of leaves generally ranges from 10 to 100 mm s^{-1} for a wind speed of 1 m s^{-1}, while the total water vapor conductance for open stomata is 1 to 10 mm s^{-1} and the total CO_2 conductance rarely exceeds 3 mm s^{-1} (MONTEITH 1973; NOBEL 1974a).

15.2.3 Leaf Heat Exchange

Wind affects the temperature of a leaf by the convection of sensible heat [Eq. (15.3)] and the loss of latent heat. The latter equals J_{H_2O} [Eq. (15.4)] times the latent heat of vaporization (GATES 1962; NOBEL 1974a; GRACE 1977). An increase in wind speed leads to a thinner air boundary layer [Eqs. (15.5–15.7)] and thus the leaf moves closer to air temperature. For example, for a leaf 50 mm across with open stomata (water vapor conductance = 10 mm s^{-1}) in an exposed location (shortwave irradiation of 1,000 W m^{-2}) and an air temperature of 20 °C with 50% relative humidity, the leaf temperature goes from 31 to 24 °C as the wind speed is raised from 0.2 to 2 m s^{-1} (GATES and PAPAIN 1971). Effects of wind on leaf temperature are less at higher wind speeds, for lower absorbed shortwave irradiation, and for smaller leaves.

The thinner the boundary layer, the more transfer processes across it are encouraged. Lobing of leaves tends to reduce the effective characteristic dimension of leaves and thus to reduce δ [Eq. (15.5)]. Indeed, the more finely divided sun leaf of *Quercus alba* dissipated heat more readily than did a shade leaf with its smoother outline (VOGEL 1968, 1970). Also, boundary layers tended to be thinner for the deeply lobed "okra" cotton (*Gossypium hirsutum*) than for normal varieties (BAKER and MYHRE 1969). The attribute of smaller characteristic dimensions is also exemplified by microphyllous plants in arid environments. Leaves of such plants rely on convective heat exchange rather than transpiration to prevent their temperatures from rising above air temperature (GATES et al. 1968; TAYLOR 1975). The leaves can thus remain at sublethal and/or photosynthetically more optimal temperatures without the expenditure of much water.

As the angle between the wind direction and leaf surface increases, the effective boundary layer thickness for pinnate leaves decreases, thus facilitating sensible heat exchanges by 10 to 100% (PARKHURST et al. 1968; VOGEL 1970; BALDING and CUNNINGHAM 1976). Since a set of pinnate leaves will have smaller characteristic dimensions than a simple entire leaf of the same total area, pinnate leaves will tend to have a thinner boundary layer for this reason also. Leaf flutter, which can be induced by wind, has been found to have relatively little effect on δ for heat transfer (RASCHKE 1956; PARKHURST et al. 1968; PARLANGE et al. 1971).

The boundary layer is actually not uniform over a leaf surface, but rather is thinner at the leading edge. Such a region will be closer to air temperature [Eq. (15.3)]. Thus, nocturnal frost formation tends to occur first in the center of leaves, which cool considerably below air temperature due to net loss of infrared radiation, especially on clear nights.

Leaf pubescence can trap air and hence lead to a thicker layer of unstirred air, i.e., the displacement boundary layer begins near the top of the hairs instead of at the leaf surface. For *Verbascum thapsus,* the dense layer of hairs decreased convective heat exchange with the environment and reduced the water vapor conductance (WUENSCHER 1970). If the hairs were widely spaced, they might increase turbulence near the leaf surface and thereby increase convective losses. Pubescence also affects radiation absorption by the leaf, which in turn affects leaf temperature and transpiration (EHLERINGER and MOONEY 1978).

15.2.4 Leaf Gas Exchange

The effects of wind on transpiration are complicated, since both the boundary layer thickness [Eqs. (15.5–15.7)] and leaf temperature [which directly affects Δc_{H_2O}, Eq. (15.4)] are influenced. Changes in leaf temperature affect stomatal opening (HALL et al. 1976), which in turn affects transpiration. Increasing the wind speed can reduce the water vapor content of the air adjacent to the leaf, which may induce stomatal closure, since stomata can respond to the local humidity conditions (GRACE et al. 1975; HALL et al. 1976). Thus the initial effects of wind on transpiration can be extremely complicated (cf. Vol. 12 B, Chap. 7, Chap. 1, this Vol.). For instance, below an air temperature of 35 °C increasing wind speed increased transpiration of *Xanthium strumarium* under high irradiance, while above 35 °C increasing wind speed decreased transpiration (DRAKE et al. 1970). In general, for leaves above air temperature, increases in wind speed tend to decrease transpiration when latent heat loss is larger than sensible heat loss (e.g., under high shortwave irradiation and substantial stomatal conductance) and to increase it when sensible heat loss dominates (GRACE 1977; MONTEITH 1973). Similarly, wind affects photosynthesis through changes in both boundary layer thickness and leaf temperature. Since leaf temperature changes can affect not only the stomatal conductance, but also the mesophyll CO_2 conductance, wind effects on photosynthesis involve even more considerations than for transpiration.

Transient responses of transpiration to wind are common. As wind speed increases above about 1 m s^{-1}, transpiration is often initially increased, but then it decreases, reflecting partial stomatal closure in response to tissue water deficits (MARTIN and CLEMENTS 1935; SATOO 1962; TRANQUILLINI 1969; CALDWELL 1970; DAVIES et al. 1974). Such a response may be a mechanism for avoiding the damaging effects of desiccation. Stomata of *Cytisus scoparius* ssp. *maritimus*, which grows on exposed sea cliffs, closed within 1 h in response to an abrupt change from calm conditions to a windspeed of 0.4 m s^{-1}, while stomata of ssp. *scoparius*, which naturally occurs in protected habitats, were hardly affected over 8 h (DAVIES et al. 1978). Thus ssp. *maritimus* can conserve water under the adverse conditions it is periodically exposed to, giving it a competitive advantage in exposed locations. However, when wind speed is increased partial stomatal closure and greatly reduced transpiration occur for leaves of *Rhododendron ferrugineum*, which grows in protected ravines, but there is very little effect on well-watered *Pinus cembra*, which grows on windswept ridges (CALDWELL 1970). For Sitka spruce (*Picea sitchensis*), the increase in boundary layer conductance with wind speed, which is most significant at the lower wind speeds, is accompanied by some stomatal closure (apparently as a response to a lowering of the humidity next to the leaf) and so the overall transpiration rate is not markedly affected (GRACE et al. 1975).

Boundary layer effects may explain the increase in photosynthesis as wind speed increases for apple (AVERY 1977). However, differences in boundary layer thickness for deeply lobed versus normal cotton leaves had no detectable effect on photosynthesis, possibly because the boundary layer conductance was much larger than the stomatal conductance (BAKER and MYHRE 1969).

Wind can also exert other effects on the gas exchange of leaves. For instance, oscillations as well as wind interception can induce bulk flow of air through leaves (WOOLEY 1961; SHIVE and BROWN 1978). However, this probably has only relatively small effects on transpiration or photosynthesis, except possibly at very high wind speeds (DAY and PARKINSON 1979). Respiration in the dark of nine different species increased 20 to 40% in a matter of minutes when the wind speed was raised from 1.8 to 7.2 m s^{-1}, which led to fairly violent shaking of the plants (TODD et al. 1972). Although the cause is not clear, increases in respiration may be related to mechanical effects of the wind, since mechanical stimulation of plants can increase respiration (JAFFE 1980).

15.2.5 Bluff Bodies

As indicated above, many plant parts cannot be treated as flat plates as far as boundary layer considerations are concerned. These bluff bodies can be extremely important for plant gas exchange, e.g., considerable photosynthesis occurs for the branches of *Cercidium floridum* (ADAMS and STRAIN 1969), *Populus tremuloides* (STRAIN and JOHNSON 1963) and the stems of *Cytisus scoparius* (DAVIES et al. 1978). On the windward side of such bluff bodies a laminar sublayer of the boundary layer occurs, while on the leeward side the boundary layer tends to separate from the surface, adverse velocity gradients develop, vortices are shed, and the air movement becomes quite complicated (SCHLICHTING 1968; KREITH 1973; OKE 1978). Even so, an average thickness of the displacement boundary layer can be determined [Eqs. (15.6) and (15.7)]. Indeed, Eq. (15.7) has been successfully used to describe the boundary layer thickness for heat flux from grape berries at various wind speeds (SMART and SINCLAIR 1976).

The water vapor conductance of the boundary layer for bluff bodies is usually greater than the total water vapor conductance of the stomata and cuticle. For example, using Eq. (15.6) for cylinders exposed to a wind speed of 0.5 m s^{-1}, the water vapor conductance of the boundary layer was eightfold greater than the total water vapor conductance for the base of the cylindrical inflorescence of *Xanthorrhoea australis* ssp. *australis* (4.1 cm diameter), 6-fold greater for leaf of *Allium cepa* (2.1 cm diameter), 11-fold greater for the inflorescence of *Scirpus validus* (0.9 cm diameter), and over 80-fold greater for the needles of *Pinus radiata* (0.1 cm diameter) (Nobel 1974b).

Although the boundary layer conductance was considerably greater than the water vapor conductance for approximately spherical fruits on over 30 species and varieties, this was not the case for the fruiting bodies of a series of Basidiomycetes (Nobel 1975). At a wind speed of 0.2 m s^{-1}, which is actually rather high near the ground where the approximately spherical fruiting bodies occur, the water vapor conductance of the boundary layer averaged 60% less than the total water vapor conductance of *Scleroderma australe* (4.0 cm diameter), *Lycoperdon polymorphum* (2.4 cm diameter), and *Lycoperdon perlatum* (1.4 cm diameter). The basidiocarps, which have no stomata, have a high tissue water vapor conductance, and thus the boundary layer conductance can become the main controller of transpiration under natural conditions.

15.2.6 Whole Plant Considerations

When mass transfer over the longer path lengths involved in the gas phase of an entire plant canopy is considered, the coefficient in flux equations like Eq. (15.4) becomes the much larger eddy diffusion coefficient, K (Nobel 1974a; Grace 1977). At the top of the canopy K is often 0.05 to 0.2 $m^2 s^{-1}$ and it decreases (sometimes more or less linearly) to the value of molecular diffusion coefficients (about 0.00002 $m^2 s^{-1}$) at the ground, reflecting the wind speed profile [cf. Eq. (15.2)]. Specifically, K can be approximately proportional to wind speed, although it also depends on the wind speed gradient and other factors (Thom 1971; Monteith 1973; Nobel 1974a; Grace 1977). The conductance (K/distance) of air in the upper half of vegetation is often comparable in magnitude to the conductance of air boundary layers of leaves in that region. Lower down K decreases, and considerable CO_2 and water vapor gradients can occur near the ground, cf. the large nocturnal buildup of respiratory CO_2 in the lower part of the canopy for the usually reduced wind speeds at night (Rosenberg 1974; Nobel 1974a). Calculations using K can predict CO_2 and water vapor fluxes within plant communities (Uchijima 1970; Thom 1971). However, the influence of plant structure on the eddy diffusion coefficient and the dependence of K on wind speed have so far hardly been studied in natural ecosystems (e.g., Landsberg and Jarvis 1973), in large measure because the conditions for applying the one-dimensional analysis rarely exist in nature.

The low wind speed at the soil surface can have special consequences for plant development. Meristem temperatures are more important than leaf, soil, or air temperatures in controlling the rate of growth of *Zea mays* cv. Earliking and *Lolium perenne* cv. S24 (Watts 1972; Peacock 1975). Since the basal meristem of such plants occurs near the soil surface where the wind speed is low, boundary layer effects may cause the temperature to be quite different than that measured in standard meteorological weather boxes (for discussion of heat damages of plants near the ground cf. Chap. 14, this Vol.).

An optimum wind speed for plant growth is often observed, with boundary layer effects becoming limiting at lower wind speeds and plant water deficits at higher wind speeds (WADSWORTH 1960). Optimal wind speeds vary from 0.3 to 1 m s^{-1}, depending on plant species and soil water potential. For instance, wind tunnel studies have indicated that growth increased about 10% up to wind speeds of approximately 0.5 m s^{-1} (WADSWORTH 1959; YABUKI and MIYAGAWA 1970). At wind speeds of 3 to 4 m s^{-1}, plant growth is often reduced about 30% (GRACE 1977). For rape (*Brassica napus*), WADSWORTH (1959) indicated that the optimal speed decreased as the plants became taller. Effects at higher wind speeds can actually be very species-dependent. For *Festuca arundinacea* raising the wind from 1.0 to 3.5 m s^{-1} lowered the photosynthetic rate after a few days, which reflected a decreased mesophyll conductance (GRACE and THOMPSON 1973). TRANQUILLINI (1969) showed that gas exchange by conifers decreased only slightly up to wind speeds of 15 m s^{-1}, but hardwoods decreased markedly beginning at about 7 m s^{-1}, which correlated with the presence of the coniferous species at windier sites. The decrease in both transpiration and photosynthesis at higher wind speeds apparently reflected a local desiccation of the leaf surfaces.

Increased air movement at the top of the canopy increases the eddy diffusion near the photosynthetic surfaces. For a dwarf shrub community 3-cm in height growing in a windswept alpine habitat, increasing the wind speed at canopy height from 0.5 to 2.5 m s^{-1} (the latter corresponds to a wind speed of 25 m s^{-1} at 2 m, which occurs frequently there) increased the CO_2 conductance of the entire canopy over threefold (GRABHERR and CERNUSCA 1977).

15.3 Frictional Interactions

Frictional interactions with the wind occur for all aerial plant parts. Certain adaptations to such stresses become particularly apparent for the coconut palm (*Cocos nucifera*), which often is exposed to gale-force winds. Specifically, the stem and petioles are flexible with many longitudinal fibers of great tensile strength. Engineering analysis has shown that the rather semi-circular cross-section of the petiole leads to high bending resistance and also high resistance to torque (WAINWRIGHT et al. 1976). Also, the leaflets can orient with the wind, thereby greatly reducing the force the plant must withstand.

15.3.1 Drag Coefficients

Form drag occurs when air is decelerated by a bluff object. If air flow were entirely stopped, the force transmitted per unit area of surface normal to the wind direction would be $\frac{1}{2}\rho u^2$, where ρ is the air density. However, much of the air flow is deflected around the object, leading to a lesser force that is dependent on the object's shape. Such considerations are quantified by the drag coefficient (c_d):

$$c_d = \frac{\text{actual force}}{\frac{1}{2}\rho u^2 A} \tag{15.8}$$

where A is the area of the object projected in the direction of the wind. Equations (15.2) and (15.8) are interrelated, since c_d for the stand as a whole equals $2 u_*^2/u^2$ (COWAN 1968; FUCHS 1979). Also, the sum of the drag by all the leaves and other plant parts is the total drag of the vegetation.

For fairly flat leaves, c_d is often 0.1 to 0.4 (COWAN 1968; GRACE 1977). The force per area on a rigid thin flat plate at zero angle of attack can be about 0.03 times $\frac{1}{2}\rho u^2$, which represents the limit for skin friction without any contribution from form drag (THOM 1968). Leaf pubescence can increase the frictional interaction with the moving air and thereby increase the drag coefficient. Also, c_d is higher when the concave side of a leaf is facing the wind. The arrangement of leaves in bunches reduces the average drag coefficient due to a mutual sheltering effect (THOM 1971; FUCHS 1979). For cylinders at right angles to the wind direction and for spheres, c_d can be 0.4 to 1.2 (LANDSBERG and THOM 1971; MONTEITH 1973). The higher value for these bluff bodies than for most flat plates reflects the fact that the air pressure in the downstream wake of a bluff body is lower than upstream and this contributes to the form drag.

Although it might be expected that the force of wind on trees would increase as the square of the wind speed [Eq. (15.8)], it often increases in a rather linear fashion. This occurs when the area of plant surfaces projected in the wind direction decreases as the tree becomes more streamlined at higher wind speeds (RAYMER 1962; WAINWRIGHT et al. 1976; GRACE 1977). The flexibility of petioles, twigs, and the leaves themselves is crucial for reducing A and hence the force that must be sustained by the stem in such cases. In general, leaves with a large attack angle will have a large A in Eq. (15.8), and thus they will absorb most of the momentum transferred from the moving air to the stand of plants.

15.3.2 Air-borne Particles

The dispersal and deposition of air-borne particulate matter has widespread importance for plant disease epidemiology, pollination, and propagule distribution. Deposition on plant surfaces is favored by high winds, large particles, and abrupt deflection of the air flow, such as by small objects (BELOT and GAUTHIER 1975; GRACE 1977). For instance, nearly 50% of small spores directed at the narrow glumes and feathery stigmas of grasses will impact (GREGORY 1973).

15.3.2.1 Release and Dispersal

In general, thicker boundary layers protect spores and pollen from being blown off a surface (GREGORY 1951). Conidia of *Helminthosporium maydis,* race T, the pathogen of southern corn leaf blight, were not blown off of *Zea mays* (Northrup King PX 446) until the wind speed exceeded 5 m s^{-1}; about 10% were removed in 15 s at 8 m s^{-1} and 50% at 12 m s^{-1} (AYLOR 1975). The thinner boundary layer at the leading edge of a corn leaf caused 84% of conidia there to be released in 15 s at a wind speed of 6 m s^{-1}, but very few were released about 10 mm away where the boundary layer had thickened (WAGGONER 1973).

After assisting in their release, wind also serves to disperse particles. For a large series of composite achenes, SHELDON and BURROWS (1973) found that the maximum dispersal distance was proportional to wind speed. About 40% of pollen from timothy (*Phleum pratense*) or ragweed (*Ambrosia*) released approximately 1 m above the ground was still air-borne 60 m away (RAYNOR et al. 1972). Pollen reaches higher altitudes and hence moves further horizontally under convectively unstable conditions.

Wind also affects spatial patterns of dispersal of litter, e.g., it can accumulate in wind-protected regions such as cliff ridges or at the base of desert shrubs like *Larrea tridentata*. For various reasons, such locations become favorable sites for seedling establishment. Finally, wind can lead to the dispersal of whole plants, such as tumbleweeds.

Wind pollination occurs in certain relatively primitive groups of plants as well as in advanced ones such as the Compositae. It is favored in dry or cold climates, where pollinators may be scarce (STEBBINS 1974). Also, wind pollination is more common in open windy regions (LEVIN and KERSTER 1974), although it also occurs for conifers in closed stands. Wind pollination is rather imprecise and presumes that a high incidence of pollen grains will occur on stigmatic surfaces near the source of the pollen grains. It dominates in Gramineae, Cyperaceae, and Juncaceae (FÆGRI and VAN DER PIJL 1979). Wind-pollinated flowers tend to be small and are frequently clustered in dense inflorescences. Nectar tends to be less, and the sexes are often on different inflorescences or even different plants (STEBBINS 1974). In some cases, pollen is mechanically ejected into the moving air, overcoming the low wind speeds next to the surfaces of anthers.

Wind-dispersed seeds tend to be light, often with structures enhancing form drag (LEVIN and KERSTER 1974; BURROWS 1975). Dispersal is favored by release far above the ground, the maximum distance traveled often being rather proportional to the height of release (SHELDON and BURROWS 1973; SALISBURY 1976). Plants using wind dispersal with its random broadcasting tend to produce numerous seeds.

15.3.2.2 Deposition

Air-borne particles moving toward a bluff body will be carried toward the surface for some distance due to their own momentum, even though the air flow is deflected around the object. Because of momentum considerations, heavier particles in higher winds will cross the boundary layer and impact the bluff body more readily (GREGORY 1973). For *Pinus sylvestris* and *Quercus sessiliflora*, deposition increased more than 20-fold as the aerosol diameter increased from 2 to 9 μm (BELOT and GAUTHIER 1975). Deposition efficiency can be approximately proportional to wind speed. Sticky or wet surfaces can increase deposition up to about tenfold (CHAMBERLAIN and CHADWICK 1972), which may be particularly relevant to pollen adhering to stigmatic surfaces. For sticky cylinders, efficiency of trapping the spores of the club moss *Lycopodium* increased with wind speed and decreased with diameter, e.g., it went from 33 to 66% as wind speed was raised from 1.1 to 5.8 m s^{-1} for a d of 3.2 mm and decreased from 66 to 18% as d was then raised to 20 mm (GREGORY 1951). Deposition was greatest along the stagnation line (midline of cylinder facing the wind) and was much less on the leeward side.

Deposition varies with position within a canopy and also with the type of vegetation. Deposition of fog increases with surface roughness [z_0, Eq. (15.2)], and so forests will intercept more fog than grassland. Larger particles would be deposited by the wind near the top of the canopy. Leaves on the windward side of trees near oceans contained more Cl^- than on the leeward side, presumably as a result of the preferential interception of sea spray on the windward side (BOYCE 1954). Although only a few percent of spores or pollen directed toward a particular plant surface will generally impact, a tree with its large leaf area will often intercept most of the air-borne particles directed at it. Deposition of pollutants on plants is also highly influenced by the wind, varying from the large-scale vegetation disturbance downwind from a smelter to more localized roadside effects.

15.3.3 Mechanical Damage

At certain wind speeds mechanical damage to plants can occur. This can be due to skin friction and form drag, but also may reflect the abrasive action of leaves rubbing together or wind-blown particles (WAISTER 1972; GRACE 1977). For instance, abrasive action by windblown soil reduces the yield for many agricultural crops (ARMBRUST et al. 1974). Sporadic high winds in gales (defined as wind speeds of 17 to 21 m s^{-1}, 62 to 74 km h^{-1}, or 39 to 46 miles h^{-1} on the Beaufort scale) can break twigs off trees and have other mechanical effects of far greater consequence than their relative infrequency would indicate. Similarly, wind gusts are often more important than the mean wind speed in mechanical damage, e.g., in uprooting trees. Eddies and other aspects of wind structure are also involved in "lodging" (permanent displacement of stems from their upright position) of certain crops.

15.3.3.1 Leaf and Plant Abrasion

Mechanical damage to leaf surfaces caused by wind can alter the cuticular wax and thereby increase cuticular transpiration on *Festuca arundinacea* var. S170 (THOMPSON 1974; GRACE 1974). As the wind speed was increased from

1.0 to 3.5 m s^{-1}, the grass blades frequently rubbed against each other. Upon the higher wind speed continuing on subsequent days, the transpiration rate increased and leaf water deficits developed as a result of the surface wear (GRACE 1974).

Substances more readily enter leaf surfaces which have been abraded. Abrasion can lead to entry of viruses (SILL et al. 1954) and may be important for salt damage to plants in windswept coastal areas (BOYCE 1954). For instance, wind led to scratches on the leaf surfaces and other evidence of minor abrasion, which was correlated with a greater entry of Cl^- if the plants were located in a region having sea sprays. Also, leaves of the dominant plants from coastal regions showed less damage to the leaf epidermis under given wind conditions, perhaps due to their short stiff petioles which prevented rubbing of one plant part against another (BOYCE 1954). For winter wheat, the viable leaf area progressively decreased as the amount of windblown sand increased, necrosis occurring across the entire leaves (ARMBRUST et al. 1974).

Seedlings can be particularly vulnerable to abrasive damage, because windblown soil is generally more common when ground cover is less, a time favorable for seedling establishment. Such damage limits the establishment of various rangeland grasses on sandy soils (FRYREAR et al. 1973); wind alone had little effect, and the seedlings became more tolerant to sand damage after about 6 weeks. Damage to seedlings of green bean, which are particularly sensitive to the abrasive action of sand, increased fairly linearly with windspeed and duration of exposure (SKIDMORE 1966). Yield of winter wheat (WOODRUFF 1956) and various Great Plains grasses (LYLES and WOODRUFF 1960) depended on the total amount of soil striking the plant. For many crops the prevention of damage by windblown sand and soil using windbreaks or control of soil erosion is an economic necessity.

If the mechanical damage to the cuticle occurs when the leaves are still expanding, recovery from surface abrasions can occur, as has been noted in *Chrysanthemum segetum* and *Eucalyptus* species (MARTIN and JUNIPER 1970). Recovery of whole plants from abrasive damage apparently is better under mild water stress, probably because the stress leads to morphological or anatomical changes in the leaves which causes them to resist the wind damage better (WHITEHEAD 1963; FRYREAR 1971). For example, plants of *Helianthus annuus* var. Pole Star exposed to wind speeds of 18 m s^{-1} survived five times longer under dry compared to wet conditions (WHITEHEAD 1963).

15.3.3.2 Leaf Tearing, Removal

Wind-caused leaf tearing is common for the large leaves of certain plants. For instance, wind often tears leaves of Musaceae, resulting in no apparent damage other than a temporarily increased rate of water loss. The torn leaves have a smaller characteristic dimension and hence a thinner boundary layer. This can prevent their rising to such high temperatures in direct sunlight that photosynthesis is depressed and heat stress occurs, as is the case for untorn leaves of *Heliconia latispatha* (TAYLOR and SEXTON 1972).

Sometimes wind induces leaf folding, such as the whiplash motion of grass lamina in the wind (THOMPSON 1974). The cuticle can be broken in such folded regions, leading to greater water loss, wilting, chlorosis, and even necrosis of the distal tissue. The necrotic region can then be removed by the wind. Because of form drag, wind can also remove healthy leaves on petioles, especially after some of the abscission layer has developed. Removal of green leaves will not always reduce the total photosynthesis of a plant, because the photosynthetic rates of the remaining leaves often increase, which can compensate (ARMBRUST et al. 1974; GRACE 1977).

15.3.3.3 Lodging

Lodging of grasses usually reflects excessive bending at ground level. Stem lodging reflects bending or breaking at the lower culm internodes and can be reversible if cell elongation is not complete. Root lodging describes intact culms leaning at the crown level. Root lodging occurs in moist soils, while stem lodging typically occurs for plants held rigidly by a dry hard upper soil layer. Stem lodging requires wind speeds of 15 to 30 m s^{-1}, considerably greater than for root lodging (PINTHUS 1973).

Lodging is discouraged when form drag is low (short plants, small heads of grain), culms are thick and stiff, and roots spread adequately (GRACE 1977). The torque on a stem caused by wind increases toward the basal portion, and hence the properties of the basal region are critical for lodging (PINTHUS 1973). For instance, fungal infestations of the lower internodes promote lodging even under mild wind. Considerable success has been achieved in breeding for lodging resistance in wheat, barley, and oats, often by selecting for reduced culm length (PINTHUS 1973).

15.3.3.4 Windthrow

Windthrow of trees becomes likely in "storms" (wind speed on the Beaufort scale at 10 m above the ground of 25 m s^{-1}). Uprooting generally considerably exceeds stem breaking, particularly where rooting is shallow. Trees taller than 5 m can be particularly vulnerable to windthrow (GRACE 1977), reflecting the wind speed profile and greater surface area for drag forces [Eq. (15.8)]. A surface-rooting habit has developed for the Sitka spruce (*Picea sitchensis*) introduced into upland Britain, thus making it quite susceptible to windthrow on soils subject to waterlogging (SANDERSON and ARMSTRONG 1978). For virgin forests of northern Maine, windthrow was most common on the shallow-rooted spruce (*Picea* spp.) and fir (*Abies balsamea*); moderate windthrow occurred every few hundred years and large-scale windthrow about every 1,200 years (LORIMER 1977). Windthrow can be greater upon fertilizer application to crops, which leads to increased crown growth, or where root penetration deep into the soil is hindered (WAISTER 1972). Root disease can also make trees vulnerable to windthrow, e.g., 86% of the blowdown in a stand of quaking aspen (*Populus tremuloides*) in Colorado were infected with a root rot (*Fomes applanatus*), while only 5% of those remaining standing were so infected (LANDIS and EVANS 1974).

Windthrow can markedly change the species composition of forests. The blowdown of the dominant trees leads to a succession favoring certain colonizers. Species present on the soil mound of uprooted trees can be quite different from those in the adjacent hollow created (SKVORTSOVA and ULANOVA 1977). In fact, wind is one of the major factors in the ecology of forests, creating gaps in the canopy by blowing down trees, creating special microhabitats, and then further influencing subsequent succession by distributing pollen, spores, and seeds.

15.3.4 Windbreaks

Windbreaks have been used to decrease wind speed near crops and thereby to reduce mechanical damage and ameliorate microclimatic conditions. Intense eddying generally occurs immediately downwind of the windbreak, which is often a row of closely spaced trees (also referred to as a shelter belt). A solid wall tends to cause violent eddies, and hence most artificial windbreaks are porous.

The mean wind speed downwind from the windbreak is reduced about 30% for a distance of about 15 to 20 times the windbreak height (ROSENBERG 1974; GRACE 1977). Besides mechanical protection, the lowering of evapotranspiration seems to be the most important effect of windbreaks. Immediately downwind of the windbreak, air and soil temperature, as well as water vapor concentrations, tend to be higher. This can increase crop yield 10 to 30%, but results are extremely variable (GRACE 1977).

15.4 Wind Effects on Growth and Development

Although not entirely separable from boundary layer and frictional effects, wind has other influences on plant growth. Windier sites tend to have shorter vegetation. Specifically, wind effects are regarded as responsible for the short plants occurring in the alpine tundra as well as the procumbent form of plants on coastal dunes (JAFFE 1980). In addition to being a common ecological observation in exposed wind-swept habitats, this dwarfing effect of wind has implications for growth chamber versus field plants, since most growth chambers are designed to have a maximum wind speed near 0.5 m s^{-1} (O'LEARY and KNECHT 1974).

15.4.1 Morphological and Anatomical Observations

Higher wind speeds tend to lead to shorter plants, as already mentioned, and these plants often have more xeromorphic leaves (MARTIN and CLEMENTS 1935; WHITEHEAD 1957; SATOO 1962; LARSON 1965). For reasons discussed above, increased wind can lead to increased transpiration and hence water stress, which may help account for certain xeromorphic features. Specific results will next be examined for the morphological effects on leaves, branches, and stems.

15.4.1.1 Leaves

The morphology and anatomy of leaves are affected by the wind speed during development. Leaves of *Zea mays* developing under a wind speed of 15 m s^{-1} were thicker and apparently had slightly lower stomatal conductances than did leaves developing in still air (WHITEHEAD and LUTI 1962). Leaf area of *Helianthus annuus* var. Pole Star grown at a wind speed of 15 m s^{-1} was threefold lower than for 0.4 m s^{-1} (WHITEHEAD 1962, 1963). Experiments on *Festuca arundinacea* cv. S170 showed no effect of wind (1.0 m s^{-1}) on leaf thickness, but leaf length was decreased about 10%; such nonfully expanded leaves had somewhat higher stomatal frequencies and the stomatal conductance was increased over plants grown under calm conditions (GRACE and RUSSELL 1977). When the wind was increased to 7.4 m s^{-1}, the rate of increase of leaf extension was reduced about 25% for *Festuca arundinacea* and *Lolium perenne* cv. S23 and the leaves became thicker. Although some abrasive damage seemed to be sustained by the leaves, their water status was apparently not affected and the photosynthetic rate per unit leaf area actually increased, probably as a reflection of the increase in leaf thickness (RUSSELL and GRACE 1978).

Leaves of *Festuca arundinacea* grown at wind speeds of 1 m s^{-1} were stiffer and had a Young's modulus (the ratio of stress to strain) about 60% higher than for development under calm conditions (GRACE and RUSSELL 1977). Also, leaves on the wind-treated plants exhibited slightly more elastic behavior, returning to within 7° of the original position after a 45° deflection, while leaves on plants grown under calm conditions returned only within 12°.

15.4.1.2 Branches

"Flag trees" are commonly observed, where branches are permanently swept to the leeward. The deformation indicates the direction and, with less fidelity, the speed of the wind (GRACE 1977; NOGUCHI 1979). This provides information relevant to windthrow, as well as pollen and seed dispersal.

Flagging occurred in response to night winds for valleys that had uphill air movement during the day and downhill cold air drainage at night (HOLROYD 1970). Flagging in coastal areas exposed to sea spray apparently results as the seaward side is preferentially killed, while buds on the leeward side continue to grow, resulting after several growing periods in a leeward curvature of the stem (BOYCE 1954). In contrast to the amalgam of internodes produced by different meristems leading to such coastal flagging, secondary cell walls can be formed while a twig is wind-entrained in a particular direction, leading to stem curvature resulting from the activity of a single meristem.

15.4.1.3 Stems, Trunks

Plants often respond to wind by developing greater mechanical strength. JACOBS (1954) noted that free-swaying trees of *Pinus radiata* grew 50% more in diameter near their base over a 2-yr period compared to trees where wind sway was prevented by guy wires. Such stayed trees tended to blow over or break when re-exposed to natural conditions. Woody trees can develop a trunk eccentric on the leeward side of the prevailing wind direction due to thicker walled tracheids there, as shown by LARSON (1965) for *Larix laricina*. Freely swaying trees had nearly 50% more radial growth near their base and were 20% shorter than stayed trees. The retardation of stem elongation and increase in girth caused by wind can be mainly accounted for by the development of shorter cells with thicker cell walls (JAFFE 1981). In the laboratory, at least 24 h of deflection in a particular direction was necessary to induce the formation of reaction wood in *Aesculus hippocastanum* (CASPERSON 1960). Such a condition could occur within weeks in nature by a prevailing wind blowing for a few hours each day from a particular direction.

Buttresses at the base of tree trunks are more frequent on the windward side (HENWOOD 1973). They thus must act not in compression but rather in tension, contrary to the original meaning of the word "buttress." Roots are often more common on the windward side, and so they also must act in tension as far as preventing mechanical deformation of trees is concerned (NAVEZ 1930). Since the compressional strength of wood is less than its tensile strength, roots and buttresses on the windward side of a tree are more effective in resisting wind-caused upsetting moments than if they were on the leeward side. Buttresses are more common on large dominant trees that grow on shallow soils or that have no major taproot. They occur where wind loading is frequent, especially if the trees are stressed by winds when young (HENWOOD 1973).

Spacings between the nodes along stems are also affected by wind. Mechanical rubbing of internodes, which could be caused by wind, reduced elongation of young plants of six species; elongation could stop within minutes of rubbing and could require days before normal growth resumed (JAFFE 1973). Experiments on kidney beans in the field demonstrated that the decrease in internodal elongation and increase in radial enlargement (representing more secondary xylem) was fairly linear with wind speed up to 4.5 m s^{-1} (HUNT and JAFFE 1980). For *Helianthus annuus* var. Pole Star, WHITEHEAD (1962) demonstrated that the internodal distance was reduced threefold as the wind during a growing period of 30 days was increased from 0.4 to 15 m s^{-1}. He suggested that the ecological significance of such results was that a plant exposed throughout its development would phenotypically adjust to minimize damage and thus become more competitive.

15.4.2 Shaking Experiments

The shaking of plants caused by wind gusts appears to have important effects on plant morphology. Ten daily 4.7 m s^{-1} gusts lasting 10 s each reduced stem growth of kidney beans by 40% (JAFFE 1976). In experiments mimicking wind blasts produced by vehicular traffic, leaves of aspen (*Populus tremula*) were reduced in size; leaf area began to be affected at 4 m s^{-1}, was reduced about 50% at 6 m s^{-1}, and over 70% at 10 m s^{-1} (FLÜCKIGER et al. 1978). Mechanical shaking of stems for only 30 s per day for 27 days reduced growth in height by 70 to 80% for glasshouse-grown *Liquidambar styraciflua* (NEEL and HARRIS 1971). Xylem vessels were shorter and thinner in the shaken saplings. Wind-induced movement out-of-doors caused a similar reduction of growth in height. Shaking for 30 s per day reduced the increase in height of *Zea mays* by 50%, with recovery of full growth rate occurring in 3 days (NEEL and HARRIS 1972). Similar shaking reduced shoot extension 37% in *Cucurbita melopepo* (TURGEON and WEBB 1971). After 35 days with 30-s shaking, stem length of tomato (*Lycopersicon esculentum* cv. Bonny Best) was reduced 42%, node number was reduced 13%, and development of lateral branches was increased (MITCHELL et al. 1975). Also, 10 s daily shaking reduced the height of tomatoes about 5% per week compared to unshaken controls (WHEELER and SALISBURY 1979). Shaking caused an increase in radial growth for the petioles of *Cucurbita melopepo* (TURGEON and WEBB 1971). In summary, shorter plants with thicker stems result from mechanical shaking, which can be caused by wind.

Agronomic implications of mechanical shaking have been recognized. For instance, increasing the spacing between nursery stock led to sturdier trees of larger trunk area, presumably because of increased wind sway (HARRIS et al. 1972). Even an automated shaking device for greenhouse plants has been described (BEYL and MITCHELL 1977).

15.4.3 Hormonal Response to Mechanical Stress

Since growth can stop within minutes and a lag phase of a few days can occur between cessation of mechanical agitation and resumption of normal growth, a plant growth substance was suspected to be involved in the response (TURGEON and WEBB 1971; NEEL and HARRIS 1972). For instance, rubbing of plant parts against each other, as can occur in the wind, was proposed by JAFFE (1973) as leading to ethylene production and a resulting decrease in internodal growth. JAFFE and BIRO (1979) showed that growth retardation and stem enlargement are apparent in one day for kidney bean, while the electrical resistance of the tissue changes within minutes after the mechanical stimulation. Comparable effects were obtained by applying "ethrel", an ethylene precursor, to the apical meristem. It was proposed that ethylene reduces stem elongation by inhibiting basipetal auxin transport. This apparently leads to reduction in cell elongation without affecting the cell number along the stem axis. The increased radial growth upon mechanical stimulation was due to a slight increase in the diameter of certain cells together with lateral cell division (JAFFE and BIRO 1979; JAFFE 1980).

In certain locations, a prevailing wind may bend a tree and thus provide the mechanical stimulation necessary for the formation of reaction wood (JAFFE 1980). Such reaction wood can form near the base of trees at high auxin levels, which would result in fatter, shorter vegetation (LARSON 1965). Of the two types of reaction wood, compression wood forms under high auxin on the underside of leaning stems of gymnosperms and tension wood is produced on the upper side under low auxin for angiosperms (WILSON and ARCHER 1977).

Mechanical stress reduced auxin-dependent growth in etiolated pea (*Pisum sativum* cv. Alaska) seedlings, an action that may also have been mediated by ethylene (MITCHELL 1977). For a detailed discussion of hormonal responses to mechanical stresses see the Chapter by JAFFE in Volume 11 of this series.

15.5 Synthesis and Future Research

In the last decades, more attention has been focused on particular effects of wind on plants. Such plant interactions with the wind encompass studies in many disciplines and many research approaches. By affecting both the boundary layer thickness and temperature, wind has multiple influences on transpiration and photosynthesis of leaves. To be useful for understanding natural ecosystems, modeling efforts of gas exchange will need more data on such parameters as the eddy diffusion coefficient, as well as three-dimensional formulations.

The dwarfing of vegetation by wind is an extremely interesting interaction between microclimate and plant hormones. Many of the cellular and organ-level details of the plant response remain to be elucidated. Extrapolations from growth chamber to the field must take into account the effect of wind speed and turbulence intensity on plant morphology.

Future environmental changes will also influence the ecophysiology of plants. Wind is crucial for the movement of air pollutants and for establishing the thickness of the boundary layers that they must cross to enter plants. The year-by-year increase in atmospheric CO_2 levels, currently nearly 1% annually, will affect plant growth (see Chap. 15, Vol. 12D). It remains to be seen whether such changes in growth will make the plants more susceptible to wind damage.

The interaction of many factors must be considered in prediciting the effect of wind on plants under natural conditions. All exchanges with the atmospheric environment and the temperatures of all plant parts are influenced by wind speed. Mechanical interactions affect plant structure, longevity, fertilization, and seed dispersal. Hopefully this chapter serves to bring into focus some of the processes involved.

Acknowledgment. Financial support from Department of Energy Contract DE-AM03-76-SF00012 is gratefully acknowledged.

References

Adams MS, Strain BR (1969) Seasonal photosynthetic rates in stems of *Cercidium floridum* Benth. Photosynthetica 3:55–62

Armbrust DV, Paulsen GM, Ellis R Jr (1974) Physiological responses to wind- and sand-blast-damaged winter wheat plants. Agron J 66:421–423

Avery DJ (1977) Maximum photosynthetic rate – A case study in apple. New Phytol 78:55–63

Aylor DE (1975) Force required to detach conidia of *Helminthosporium maydis*. Plant Physiol 55:99–101

Baker DN, Myhre DL (1969) Effects of leaf shape and boundary layer thickness on photosynthesis in cotton (*Gossypium hirsutum*). Physiol Plant 22:1043–1049

Balding FR, Cunningham GL (1976) A comparison of heat transfer characteristics of simple and pinnate leaf models. Botan Gaz 137:65–74

Belot Y, Gauthier D (1975) Transport of micronic particles from atmosphere to foliar surfaces. In: de Vries DA, Afgan NH (eds) Heat and mass transfer in the biosphere, Part 1, Transfer processes in the plant environment. Wiley and Sons, New York, pp 583–591

Beyl CA, Mitchell CA (1977) Automated mechanical stress application for height control of greenhouse chrysanthemum. HortScience 12:575–577

Boyce SG (1954) The salt spray community. Ecol Monographs 24:29–67

Bradley EF, Finnigan JJ (1973) Heat and mass transfer in the plant-air continuum. In: Proceedings of the first Australasian conference on heat and mass transfer, Monash University, Melbourne, Australia, pp 55–77

Burrows FM (1975) Wind-borne seed and fruit movement. New Phytol 75:405–418

Caldwell MM (1970) The effect of wind on stomatal aperture, photosynthesis, and transpiration of *Rhododendron ferrugineum* L. and *Pinus cembra* L. Cbl Ges Forstwesen 87:193–201

Casperson G (1960) Über die Bildung von Zellwänden bei Laubhölzern 1. Feststellung der Kambiumaktivität durch Erzeugen von Reaktionsholz. Ber Dtsch Bot Ges 73:349–357

Chamberlain AC, Chadwick RC (1972) Deposition of spores and other particles on vegetation and soil. Ann Appl Biol 71:141–158

Cionco RM (1972) Intensity of turbulence within canopies with simple and complex roughness elements. Boundary-Layer Meteorol 2:453–465

Cowan IR (1968) Mass, heat and momentum exchange between stands of plants and their atmospheric environment. Q J R Meteorol Soc 94:523–544

Davies WJ, Kozlowski TT, Pereira J (1974) Effect of wind on transpiration and stomatal aperture of woody plants. Bull R Soc New Zealand 12:433–438

Davies WJ, Gill K, Halliday G (1978) The influence of wind on the behaviour of stomata of photosynthetic stems of *Cytisus scoparius* (L.) Link. Ann Bot 42:1149–1154

Day W, Parkinson KJ (1979) Importance to gas exchange of mass flow of air through leaves. Plant Physiol 64:345–346

Denmead OT, Bradley EF (1973) Heat, mass and momentum transfer in a wheat crop. In: Proceedings of the first Australasian conference on heat and mass transfer, Monash University, Melbourne, Australia, pp 25–30

Drake BG, Raschke K, Salisbury FB (1970) Temperatures and transpiration resistances of *Xanthium* leaves as affected by air temperature, humidity, and wind speed. Plant Physiol 46:324–330

Ehleringer JR, Mooney HA (1978) Leaf hairs: effects on physiological activity and adaptive value to a desert shrub. Oecologia 37:183–200

Fægri K, van der Pijl L (1979) The principles of pollination ecology, 3rd rev edn. Pergamon Press, Oxford

Flückiger W, Oertli JJ, Flückiger-Keller H (1978) The effect of wind gusts on leaf growth and foliar water relations of aspen. Oecologia 34:101–106

Fryrear DW (1971) Survival and growth of cotton plants damaged by windblown sand. Agron J 63:638–642

Fryrear DW, Stubbendieck J, McCully WG (1973) Grass seedling response to wind and windblown sand. Crop Sci 13:622–625

Fuchs M (1979) Atmospheric transport processes above arid-land vegetation. In: Goodall DW, Perry RA (eds) Arid-land ecosystems: Structure, functioning and management, Vol. 1. Cambridge Univ Press, Cambridge, pp 393–408

Gates DM (1962) Energy exchange in the biosphere. Harper and Row, New York

Gates DM, Papain EL (1971) Atlas of energy budgets of plant leaves. Academic Press, New York

Gates DM, Alderfer R, Taylor E (1968) Leaf temperatures of desert plants. Science 159:994–995

Grabheer G, Cernusca A (1977) Influence of radiation, wind, and temperature on the CO_2 gas exchange of the alpine dwarf shrub community *Loiseleurietum cetrariosum*. Photosynthetica 11:22–28

Grace J (1974) The effect of wind on grasses. I. Cuticular and stomatal transpiration. J Exp Bot 25:542–551

Grace J (1977) Plant responses to wind. Academic Press, London

Grace J, Russell G (1977) The effect of wind on grasses III. Influence of continuous drought or wind on anatomy and water relations in *Festuca arundinacea* Schreb. J Exp Bot 28:268–278

Grace J, Thompson JR (1973) The after-effect of wind on the photosynthesis and transpiration of *Festuca arundinacea*. Physiol Plant 28:541–547

Grace J, Wilson J (1976) The boundary layer over a *Populus* leaf. J Exp Bot 27:231–241

Grace J, Malcolm DC, Bradbury IK (1975) The effect of wind and humidity on leaf diffusive resistance in Sitka spruce seedlings. J Appl Ecol 12:931–940

Gregory PH (1951) Deposition of air-borne *Lycopodium* spores on cylinders. Ann Appl Bot 38:357–376

Gregory PH (1973) The microbiology of the atmosphere, 2nd edn. Leonard Hill, Aylesbury

Hall AE, Schulze E-D, Lange OL (1976) Current perspectives of steady-state stomatal responses to environment. In: Lange OL, Kappen L, Schulze E-D (eds) Water and plant life, Ecological Studies Vol 19, Springer, Berlin-Heidelberg-New York, pp 169–188

Harris RW, Leiser AT, Neel PL, Long D, Stice NW, Maire RG (1972) Spacing of container-grown trees in the nursery. J Am Soc Hortic Sci 97:503–506

Henwood K (1973) A structural model of forces in buttressed tropical rain forest trees. Biotropica 5:83–93

Holroyd EW III (1970) Prevailing winds on Whiteface mountain as indicated by flag trees. For Sci 16:222–229

Hunt ER Jr, Jaffe MJ (1980) Thigmomorphogenesis: the interaction of wind and temperature in the field on the growth of *Phaseolus vulgaris* L. Ann Bot 45:665–672

Inoue E (1963) On the turbulent structure of airflow within crop canopies. J Meteorol Soc Jpn, Ser II, 41:317–326

Jacobs MR (1954) The effect of wind sway on the form and development of *Pinus radiata* D Don. Aust J Bot 2:35–51

Jaffe MJ (1973) Thigmomorphogenesis: the response of plant growth and development to mechanical stimulation. Planta 114:143–157

Jaffe MJ (1976) Thigmomorphogenesis: a detailed characterization of the response of beans (*Phaseolus vulgaris* L.) to mechanical stimulation. Z Pflanzenphysiol 77:437–453

Jaffe MJ (1980) Morphogenetic responses of plants to mechanical stimuli or stress. BioScience 30:239–243

Jaffe MJ (1981) Wind and other mechanical effects in the development and behavior of plants, with special emphasis on the role of hormones. In: Encyclopedia of Plant Physiology (new series), Springer, Berlin, Heidelberg, New York, in press

Jaffe MJ, Biro R (1979) Thigmomorphogenesis: the effect of mechanical perturbation on the growth of plants, with special reference to anatomical changes, the role of ethylene, and interactions with other environmental stresses. In: Mussell H, Staples R (eds) Stress physiology in crop plants, Wiley and Sons, New York, pp 25–59

Kestin J (1966) The effect of free-stream turbulence on heat transfer rates. In: Irvine TF Jr, Hartnett JP (eds) Advances in heat transfer, Vol 3. Academic Press, New York, pp 1–32

Kowalski G, Mitchell JW (1975) Heat transfer from spheres in the naturally turbulent, outdoor environment. Am Soc Mech Eng Winter Ann Meeting, 1975. Paper no. 75-WA/HT-57. Am Soc Mech Eng, New York, pp 1–7

Kreith F (1973) Principles of heat transfer, 3rd edn. Intext Educational Publishers, New York

Landis TD, Evans AK (1974) A relationship between *Fomes applanatus* and aspen windthrow. Plant Dis Rep 58:110–113

Landsberg JJ, Jarvis PG (1973) A numerical investigation of the momentum balance of a spruce forest. J Appl Ecol 10:645–655

Landsberg JJ, Thom AS (1971) Aerodynamic properties of a plant of complex structure. Q J Roy Meteorol Soc 97:565–570

Larson PR (1965) Stem formation of young *Larix* as influenced by wind and pruning. For Sci 11:412–424

Levin DA, Kerster HW (1974) Gene flow in seed plants. In: Dobzhansky T, Hecht MK, Steere WC (eds) Evolutionary biology, Vol 7, Plenum Press, New York, pp 139–220

Lorimer CG (1977) The presettlement forest and natural disturbance cycle of northeastern Maine. Ecology 58:139–148

Lyles L, Woodruff NP (1960) Abrasive action of windblown soil on plant seedlings. Agron J 52:533–536

Martin EV, Clements FE (1935) Studies of the effect of artificial wind on growth and transpiration in *Helianthus annuus*. Plant Physiol 10:613–636

Martin JT, Juniper BE (1970) The cuticles of plants. Edward Arnold, London

Mitchell CA (1977) Influence of mechanical stress on auxin-stimulated growth of excised pea stem sections. Physiol Plant 41:129–134

Mitchell CA, Severson CJ, Woot JA, Hammer PA (1975) Seismomorphogenic regulation of plant growth. J Am Soc Hortic Sci 100:161–165

Monteith JL (1973) Principles of environmental physics. Elsevier, New York

Narain JP, Uberoi MS (1973) Combined forced and free-convection over thin needles. Int J Heat Mass Transfer 16:1505–1512

Navez A (1930) On the distribution of tabular roots in *Ceiba* (Bombacaceae). Proc Nat Acad Sci US 16:339–344

Neel PL, Harris RW (1971) Motion-induced inhibition of elongation and induction of dormancy in Liquidambar. Science 173:58–59

Neel PL, Harris RW (1972) Tree seedling growth: Effects of shaking. Science 175:918–919

Nobel PS (1974a) Introduction to biophysical plant physiology. W.H. Freeman, San Francisco

Nobel PS (1974b) Boundary layers of air adjacent to cylinders: Estimation of effective thickness and measurements on plant material. Plant Physiol 54:177–181

Nobel PS (1975) Effective thickness and resistance of the air boundary layer adjacent to spherical plant parts. J Exp Bot 26:120–130

Noguchi Y (1979) Deformation of trees in Hawaii and its relation to wind. J Ecol 67:611–628

Oke TR (1978) Boundary layer climates. Wiley and Sons, New York

O'Leary JW, Knecht GN (1974) Raising the maximum permissible air velocity in controlled environment plant growth chambers. Physiol Plant 32:143–146

Parkhurst DF, Duncan PR, Gates DM, Kreith F (1968) Wind-tunnel modelling of convection of heat between air and broad leaves of plants. Agric Meteorol 5:33–47

Parlange J-Y, Waggoner PE, Heichel GH (1971) Boundary layer resistance and temperature distribution on still and flapping leaves. I. Theory and laboratory experiments. Plant Physiol 48:437–442

Peacock JM (1975) Temperature and leaf growth in *Lolium perenne* II. The site of temperature perception. J Appl Ecol 12:115–123

Pearman GI, Weaver HL, Tanner CB (1972) Boundary layer heat transfer coefficients under field conditions. Agric Meteorol 10:83–92

Perrier ER, Aston A, Arkin GF (1973) Wind flow characteristics on a soybean leaf compared with a leaf model. Physiol Plant 28:106–112

Pinthus MJ (1973) Lodging in wheat, barley, and oats: the phenomenon, its causes, and preventive measures. Adv Agron 25:209–263

Raschke K (1956) Über die physikalischen Beziehungen zwischen Wärmeübergangszahl, Strahlungsaustausch, Temperatur und Transpiration eines Blattes. Planta 48:200–238

Raymer WG (1962) Wind resistance in conifers. In: Aerodynamics Division Report 1008, Nat Phys Lab, Teddington, England, pp 1–5

Raynor GS, Ogden EC, Hayes JV (1972) Dispersion and deposition of timothy pollen from experimental sources. Agric Meteorol 9:347–366

Rosenberg NJ (1974) Microclimate: the biological environment. Wiley and Sons, New York

Russell G, Grace J (1978) The effect of wind on grasses V. Leaf extension, diffusive conductance, and photosynthesis in the wind tunnel. J Exp Bot 29:1249–1258

Salisbury E (1976) Seed output and the efficacy of dispersal by wind. Proc Roy Soc London B 192:323–329

Sanderson PL, Armstrong W (1978) Soil waterlogging, root rot and conifer windthrow: oxygen deficiency or phytotoxicity? Plant Soil 49:185–190

Satoo T (1962) Wind, transpiration, and tree growth. In: Kozlowski TT (ed) Tree growth. Ronald Press, New York, pp 299–310

Schlichting H (1968) Boundary layer theory, 6th edn. McGraw-Hill, New York

Sheldon JC, Burrows FM (1973) The dispersal effectiveness of the achene-pappus units of selected Compositae in steady winds with convection. New Phytol 72:665–675

Shive JB Jr, Brown KW (1978) Quaking and gas exchange in leaves of cottonwood (*Populus deltoides*, Marsh.). Plant Physiol 61:331–333

Sill WH Jr, Lowe AE, Bellingham RC, Fellows H (1954) Transmission of wheat streak-mosaic virus by abrasive leaf contacts during strong winds. Plant Dis Rep 38:445–447

Skidmore EL (1966) Wind and sandblast injury to seedling green beans. Agron J 58:311–315

Skvortsova EB, Ulanova NG (1977) Certain aspects of the influence of blowdowns on forest biogeocenosis. Soil Sci Bull, Moscow Univ 32:1–5

Smart RE, Sinclair TR (1976) Solar heating of grape berries and other spherical fruit. Agric Meteorol 17:241–259

Stebbins GL (1974) Flowering plants: evolution above the species level. Harvard Univ Press, Cambridge.

Strain BR, Johnson PL (1963) Corticular photosynthesis and growth in *Populus tremuloides*. Ecology 44:581–584

Taylor SE (1975) Optimal leaf form. In: Gates DM, Schmerl RB (eds) Perspectives of biophysical ecology, Ecological Studies Vol 12, Springer, Berlin, Heidelberg, New York, pp 73–86

Taylor SE, Sexton OJ (1972) Some implications of leaf tearing in Musaceae. Ecology 53:143–149

Thom AS (1968) The exchange of momentum, mass, and heat between an artificial leaf and the airflow in a wind-tunnel. Q J R Meteorol Soc 94:44–55

Thom AS (1971) Momentum absorption by vegetation. Q J R Meteorol Soc 97:414–428

Thompson JR (1974) The effect of wind on grasses II. Mechanical damage in *Festuca arundinacea* Schreb. J Exp Bot 25:965–972

Todd GW, Chadwick DL, Tasi S-D (1972) Effect of wind on plant respiration. Physiol Plant 27:342–346

Tranquillini W (1969) Photosynthese und Transpiration einiger Holzarten bei verschieden starkem Wind. Zentralbl Gesamte Forstwes 86:35–48

Turgeon R, Webb JA (1971) Growth inhibition by mechanical stress. Science 174:961–962

Uchijima Z (1970) Carbon dioxide environment and flux within a corn crop canopy. In: Prediction and measurement of photosynthetic productivity. PUDOC, Wageningen, pp 179–196

Vogel S (1968) 'Sun leaves' and 'shade leaves': Differences in convective heat dissipation. Ecology 49:1203–1204

Vogel S (1970) Convective cooling at low airspeeds and the shapes of broad leaves. J Exp Bot 21:91–101

Wadsworth RM (1959) An optimum wind speed for plant growth. Ann Bot 23:195–199

Wadsworth RM (1960) The effect of artificial wind on the growth-rate of plants in water culture. Ann Bot 24:200–211

Waggoner PE (1973) The removal of Helminthosporium maydis spores by wind. Phytopathology 63:1252–1255

Wainwright SA, Biggs WD, Currey JD, Gosline JM (1976) Mechanical design in organisms. Wiley and Sons, New York

Waister PD (1972) Wind damage in horticultural crops. Hortic Abstr 42:609–615

Watts WR (1972) Leaf extension in *Zea mays* II. Leaf extension in response to independent variation of the temperature of the apical meristem, of the air around the leaves, and of the root-zone. J Exp Bot 23:713–721

Wheeler RM, Salisbury FB (1979) Water spray as a convenient means of imparting mechanical stimulation to plants. HortScience 14:270–271
Whitehead FH (1957) Wind as a factor in plant growth. In: Hudson JP (ed) Control of the plant environment. Butterworth Scientific, London, pp 84–95
Whitehead FH (1962) Experimental studies of the effect of wind on plant growth and anatomy II. *Helianthus annuus*. New Phytol 61:59–62
Whitehead FH (1963) Experimental studies of the effect of wind on plant growth and anatomy IV. Growth substances and adaptive anatomical and morpholigical changes. New Phytol 62:86–90
Whitehead FH, Luti R (1962) Experimental studies of the effect of wind on plant growth and anatomy I. *Zea mays*. New Phytol 61:56–58
Wilson BF, Archer RR (1977) Reaction wood: induction and mechanical action. Ann Rev Plant Physiol 28:23–43
Woodruff NP (1956) Wind-blown soil abrasive injuries to winter wheat plants. Agron J 48:499–504
Wooley JT (1961) Mechanisms by which wind influences transpiration. Plant Physiol 36:112–114
Wuenscher JE (1970) The effect of leaf hairs of *Verbascum thapsus* on leaf energy exchange. New Phytol 69:65–73
Yabuki K, Miyagawa H (1970) Studies on the effect of wind speed upon the photosynthesis. (2) The relation between wind speed and photosynthesis. J Agric Meteorol 26:137–141 (in Japanese with English summary)

16 Fire as an Ecological Factor

P.W. RUNDEL

CONTENTS

16.1 Introduction

The subject of fire as an ecological factor is an exceedingly broad and complex one. The literature on fire in nature currently numbers hundreds of papers annually and seems to be growing at an exponential rate. It is certainly impossible to compress even a small amount of the available literature on aspects of fire as an ecological factor into a review of this size. For this reason this chapter will emphasize the effects of fire on ecosystem properties that are important for plant growth and development and on the influence of fire on growth and reproductive characteristics of plants. The final portion of the chapter will very briefly discuss the role of fire as an ecological factor in four major types of ecosystems – coniferous forests, grasslands, mediterranean-climate shrub and related shrublands and tropical forests.

16.2 Fire Effects on Ecosystem Processes

16.2.1 Fire Temperatures

Many of the fire-induced changes in the chemical, physical, and biological nature of soils are the direct result of the degree and duration of soil heating. These heating properties are highly variable, and it has proved to be extremely difficult to determine accurately thermal conductivity and heat transfer in soils during fires in natural fuels. Among the variables which must be considered are fuel type, the nature of the litter layer (thickness, density, and moisture content), fire intensity (rate of heat release per unit of ground area), and soil characteristics (texture, organic matter content, moisture content).

In general, soil temperatures reached during fires are directly related to the biomass of fuel which undergoes combustion. The highest soil surface temperatures recorded have occurred with fires in heavy slash of coniferous forests. Temperatures in excess of 1,000 °C have been measured just above the duff layer in stands of *Pinus banksiana* (SMITH 1970) and in *Pseudotsuga menziesii* (ISAAC and HOPKINS 1937). Temperatures reached 320 °C at 2.5 cm below the soil surface in the *Pseudotsuga* slash. Very high temperatures have also been reported for burning windrows of *Eucalyptus* and log piles, with peak temperatures reaching over 660 °C just below the soil surface and levels above 100 °C extending to depths below 22 cm (ROBERT 1965; HUMPHREYS and LAMBERT 1965; CROMER and VINES 1966).

The majority of forest fires have much lower temperatures than the extremes associated with slash and log piles. The maximum temperature recorded at 2.5 cm below the soil surface in an extremely hot *Eucalyptus* forest fire was 275 °C, while more typical *Eucalyptus* fires resulted in temperatures of 175 °C at this depth (BEADLE 1940). Soil surface temperatures reached only 93 °C in a prescribed burn in a mixed conifer forest in California although duff temperatures reached 260 °C (AGEE 1973). Ground fires in *Pinus palustris* savannas in the southeastern United States reach only 135 °C just below the soil surface (HEYWARD 1938). STARK (1977) set up a qualitative scale for Douglas fir/larch forests in Montana relating estimates of fire intensity and relative litter consumption to surface soil temperatures: intense fires reaching 300 °C at the soil surface consume all of the litter present; medium intensity fires at 180–300 °C consume approximately one-half of available litter, while light fires which just scorch litter reach less than 180 °C. Since the degree of litter consumption is strongly dependent on many complex variables, this scale may not be applicable for other forest types.

Peaks and durations of soil temperature associated with chaparral fires have been studied in considerable detail (DEBANO et al. 1977). Typical profiles of maximum temperature with soil depth are shown in Fig. 16.1 for intense, moderate, and light chaparral burns. Peak soil surface temperatures in rapidly burning chaparral wildfires may reach over 700 °C, a hotter temperature than that reached in most forest fires (Fig. 16.2), but this heat penetrates very little into the soil because of the rapid speed of movement associated with these chaparral fires. Such detailed measurements of fire temperatures associated with

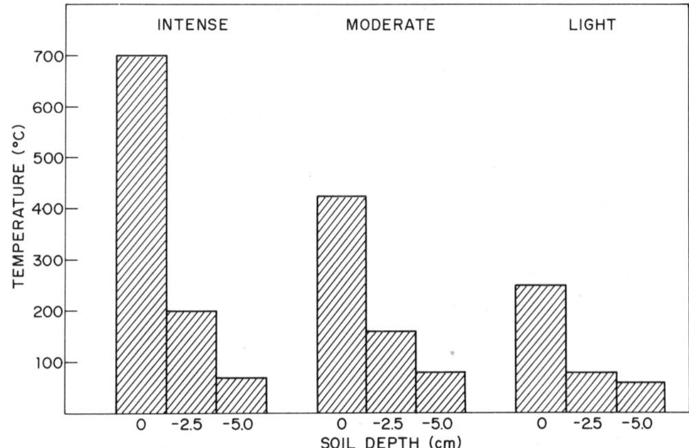

Fig. 16.1. Profiles of maximum temperature with soil depth for three intensities of chaparral fires. (DEBANO et al. 1977)

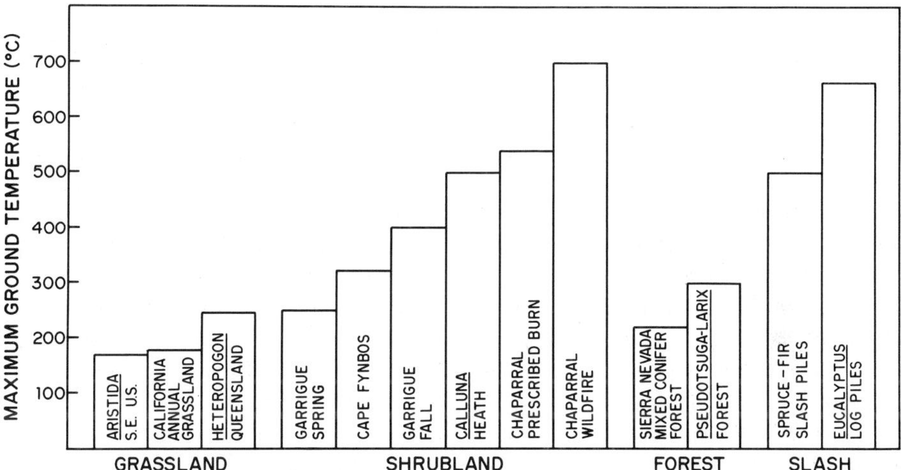

Fig. 16.2. Comparative maximum ground surface temperatures for grassland, shrubland, forest, and slash fires

other types of shrubland areas have not been made, but the similarities of above-ground biomass and canopy structure in shrub communities from other Mediterranean-climate regions of the world and from shrublands in the South-eastern United States suggest that comparable temperatures probably occur.

In contrast to the high temperatures of woody plant communities, grassland fires are relatively cool. Fires in annual grasslands have been found to produce maximum soil surface temperatures ranging from 90–180 °C (BENTLEY and FENNER 1958). Surface temperatures under burning *Heteropogon contortus* in Australia may reach 245 °C (SCOTTER 1970).

The direct effects of high fire temperatures on soil conditions are felt in many ways. Perhaps the most significant of these effects is the combustion of organic matter and subsequent deposition of ash on the soil surface. The nutritional consequences of this combustion will be discussed later in this chapter. Destruction of protective layers of duff and litter by intense fires have commonly been associated with increased erosion, particularly in areas of high topographic diversity. In special circumstances fires may actually promote physical weathering of rock surfaces through heating effects. WRIGHT and HEINSELMAN (1973) suggest that fires in northern Minnesota may be the most significant of all rock-weathering processes.

Changes in surface albedo and thermal conductivity following fires often produce very pronounced effects in post-fire surface temperatures. Temperatures at shallow soil depths may be up to 10 °C higher on open burned sites than comparable unburned sites (HENSEL 1923; BEATON 1959a, 1959b; SCOTTER 1963; NEAL et al. 1965; DAUBENMIRE 1968; SHANNON and WRIGHT 1977). In grassland soils this heating effect is usually short-lived, but in forest soils where ash is incorporated slowly such temperature effects may be highly significant in influencing chemical and biological processes.

16.2.2 Hydrology, Erosion, and Other Geomorphic Processes

The significance of fire in influencing geomorphic processes is a complex topic in which both time-scale of fire effect and individual ecosystem responses must be considered. On the short-term time scale of a single fire, hydrological and geomorphologic influences on ecosystem dynamics are largely determined through direct alteration of vegetation and soil properties. The long-term significance of such fire-induced effects are a function of the relative importance of fire as a mechanism of disturbance in a given ecosystem. On the long-term time scale of landscape development, landforms may influence fire pattern and behavior. Where precipitation, frost-heaving, and other disturbance processes are predominate geomorphic factors, fire is less important.

Since direct fire impact on geomorphic processes is a function of alteration of plant and/or soil cover, fire intensity, and frequency are important factors. Equally important is the sensitivity of an ecosystem to disturbance. Thus steep, youthful landscapes where plant cover provides important stabilizing influences on rates of geomorphic processes are highly subject to fire disturbance. Older, heavily eroded landscapes with little topography have geomorphic processes much less subject to perturbation.

Combustion of plant cover by fire has several direct effects on hydrologic and geomorphic processes. One such effect is the loss of organic litter layers which help serve to protect soil from the full range of erosional processes. The loss of this layer not only increases the susceptibility of surface soil layers to erosion, but also influences soil temperature, since organic matter is an effective thermal insulator. This process may be particularly significant in permafrost areas with seasonal freeze-thaw cycles in surface layers. Studies in the Alaskan taiga have shown that loss of insulating organic matter layers may increase the depth and duration of seasonal thawing for a period of at least 15 years following fire (VIERECK 1973a). Increased depth of thawing may result in subsidence of soil areas, alteration of local drainage patterns, and accelerated soil fluction activity in taiga and tundra soils (VIERECK 1973a; SWANSTON in SWANSON 1981).

Permafrost melting may be strongly affected by boreal forest fires. Removal of insulating mats of mosses and lichens by fire may cause gradual melting and reduction of permafrost layers, but this effect is a temporary one (VIERECK 1973b).

Complex hydrologic responses to fire may also occur in nonpermafrost soils. DYRNESS et al. (1957) showed that hot ground fires could reduce water storage capacity of surface organic matter. Loss of litter layers, development of water-repellant soil layers (as described below), soil compaction, and heat fusing of soil surface layers are all fire effects which may influence hydrologic cycles (ROWE and COLMAN 1951; DYRNESS et al. 1957; HELVEY et al. 1976; RICE 1973; ANDERSON et al. 1976; CAMPBELL et al. 1977). Decreased infiltration following fire has been generally reported in studies from a variety of ecosystems (see review in WELLS et al. 1979). Studies in Missouri forests found that burning reduced infiltration rates by 38% while litter removal by raking lowered them by only 18%. Combustion or fire-induced mortality of leaves and fine branches of vegetation alters hydrologic cycles by limiting interception and associated stem flow of precipitation and reducing evapotranspiration (ANDERSON et al. 1976; CAMPBELL et al. 1977). The net hydrologic response to all of these many factors is generally increased soil water storage and runoff following fire.

Surface erosion following fire may be caused by dry sliding or surface creep (ANDERSON et al. 1959; KRAMMES 1960, 1965; FRANKLIN and ROTHACHER 1962), by rill and sheet erosion (SARTZ 1953; RICE 1973; GRIFFIN 1978), by wind (BLAISDELL 1953) and by cycles of needle ice formation and melt (SWANSON 1981). Interactions of vegetation, climate, topography, and soil conditions determine the relative importance of the processes. Coarse-textured soils are most subject to dry debris movement on steep slopes while fine-textured soils are susceptible to erosion by wind and sheet flow (SWANSON 1981).

Fire-induced effects on hydrologic processes of ground water storage and runoff have direct influence on streamflow characteristics. Both increased peak and total streamflow rates have been documented following fires in a variety of temperate and semi-arid ecosystems (RICH 1962; STOREY et al. 1964; ANDERSON et al. 1976; HELVEY et al. 1976; CAMPBELL et al. 1977). BERNDT (1971) observed the reduction in streamflow by one-half on the day of an intense forest fire in central Washington. Over the subsequent two rainless weeks the streamflow increased to levels above normal and diurnal oscillations were minimal, reflecting reduced upstream transpiration following the fire.

Since sediment movement and storage in stream channels is regulated by the amount and position of large organic debris, fire may have variable effects on sediment flow in these channels. Fire may have the immediate effect of increasing debris loads and therefore increase channel storage capacity. Fire may also decrease debris loading through combustion and increase the potential for aggrading the channel through major flushing events. Studies of coniferous forest ecosystems in the Pacific Northwest of the United States indicate that debris loading of streams may decrease for about a century following large fires as residual organic matter from the original forest stand decomposes in situ. Nearly a century is required for post-fire stand development to reach a point where down logs and other debris begin to build up in channels (SWANSON and LIENKAEMPER 1978).

SWANSON (1981) has discussed the possible effects of fire on geomorphic processes of sediment movement within individual watersheds. Displaced sediments may find temporary storage areas following fire in hillslope depressions and dammed behind large organic debris on hillslopes and in channels. Since most studies of fire impact on erosion have dealt with total sediment yield from a watershed, the significance of such temporary local transport and accumulation of sediments is poorly known.

The formation of water-repellent soil layers has been shown to be an important effect of fires, particularly in chaparral areas of California and Arizona and in some coniferous forest areas (DEBANO et al. 1967, 1977; SCHOLL 1975; CAMPBELL et al. 1977; DEBYLE 1973; SALIH et al. 1973; DYRNESS 1976). The hydrophobic compounds which lead to the formation of water-repellent soil layers occur in accumulated litter on the soil surface before a fire. When hot fires consume this litter and produce steep thermal gradients in the upper soil layers, an efficient distillation process occurs which leads to the volatilization and subsequent condensation of these hydrophobic compounds on individual soil particles at several centimeters depth in the soil profile (Fig. 16.3). The depth and thickness of the water-repellent soil is a function of the intensity of the fire and the physical characteristics of the soil (SAVAGE 1974; SCHOLL 1975; DEBANO et al. 1976). Much of the increased rill, sheet, and mass movement forms of water-induced erosion following fire are thought to be influenced by the formation of water-repellent soil layers. However, very few data are available to judge the relative importance in erosion of the formation of these soil layers in relation to loss of litter layers and loss of vegetation stabilization on slopes.

High elevation forest fires on steep slopes may influence snow avalanche activity by eliminating the functions of vegetation in snow stabilization and alterations of patterns of snow accumulation (MUNGER 1911). The formation of areas of avalanche initiation on such steep slopes may directly impact unburned forests downslope. Repeated avalanching may suppress revegetation of extensive avalanche track areas.

One of the most notable areas for major impact of fire on hydrologic and geomorphic processes is the chaparral region of Southern California. Here, geologically active landscapes have produced very steep mountain slopes which are affected by intense fires and by the formation of hydrophobic soil conditions. Sediment cycle–fire relationships have been investigated in detail in small watersheds in the mountains of Southern California (STOREY et al. 1964; SCOTT and WILLIAMS 1978). Based on these studies, SWANSON (1981) has estimated that sediment yield in the first year following a fire is 30 times the rate for stable, vegetated slopes. After this first year these rates of sediment loss drop steadily but do not return to a steady-state level for about 8 years. On this

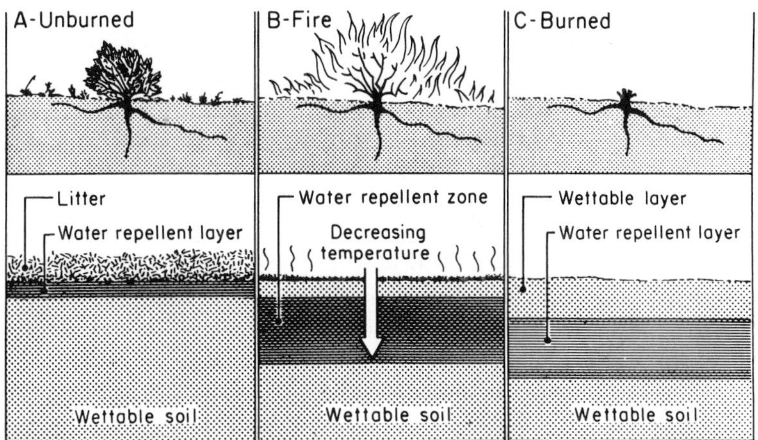

Fig. 16.3. Patterns of development of water-repellent soil layers following fire in Californian chaparral. (DEBANO et al. 1977)

basis, given a mean fire cycle of 25 years, SWANSON (1981) calculates that over 70% of the long-term sediment yield from steep chaparral slopes is the result of fire impact. RICE (1973) estimates an even greater percentage of fire-induced sediment yield. Estimates by SCOTT and WILLIAMS (1978) for a small steep watershed in Los Angeles County suggest a soil erosion rate of 23 cm per century from all causes.

Numerous studies have documented short-term erosional effects following fire in chaparral areas (ROWE 1941; SINCLAIR 1954; KRAMMES 1960, 1965; HIBBERT et al. 1974). The initial phase of erosion operates through gravity-activated landslides and debris movement. Debris transport in this manner may amount to 200–4,300 kg ha^{-1} during and immediately following a fire (ANDERSON et al. 1959), and may remain high for three months (KRAMMES 1960). With the onset of winter rains, a second phase of erosion is commonly initiated. One form of sediment and debris movement occurs as overland flow as the soil surface is exposed to splash from raindrops. Water-repellent soil layers increase the importance of erosional loss of sediments as precipitation intensity increases. Accumulations of debris and landslide sediments in channels at the base of steep slopes provide readily transported sediments as high discharge rates occur.

Studies with experimental burns provide data indicating the importance of slope in determining sediment yield from chaparral areas. DEBANO and CONRAD (1976) found that surface erosion was 2.6 times greater on a 50% slope than on a 20% slope. Erosional levels following fire on the steeper slope were 7,300 kg ha^{-1} in comparison to 210 kg ha^{-1} on an unburned control slope.

Chaparral fire studies relating intensity of fire to sediment yield are generally lacking, but a positive relationship is expected. PASE and LINDEMUTH (1971) found little erosion on lightly burned chaparral slopes in Arizona, but moderate erosion in areas where greater fire intensities had consumed more than 40% of the litter coverage.

Many coniferous forest areas of the western United States have major fire-induced impacts on hydrologic and geomorphic processes. Rates of post-fire erosion are highly variable, however, due to topography, fire intensity, and precipitation intensity. Forest litter layers are critically important in controlling runoff and erosion (LOWDERMILK 1930; WELLS et al. 1979). Studies by MEGAHAN and MOLITOR (1975) in Idaho demonstrated that erosion rate increased directly with fire intensity. Erosion rates of granitic soils also increased in proportion to this intensity.

CAMPBELL et al. (1977) estimated that ponderosa pine watersheds in Arizona lost 23.9 kg ha^{-1} of soil in six months due to unusually heavy rains which followed an intense wildfire. With reestablishment of groundcover the following year, erosion was insignificant. Rapid revegetation of burned slopes is a key factor in minimizing erosional loss of sediments in any ecosystem. For this reason, seasonality of fire is an important factor in predicting its geomorphic effects.

Pseudotsuga forests in the western Cascade Range of Oregon experience infrequent fires and much less dramatic erosional effects following fire. Assuming a 200-year fire frequency, SWANSON (1980) estimates that accelerated erosion due to fire accounts for about 20% of long-term sediment yield. Human perturbations such as road construction and clearcutting may alter this pattern, however, and produce mass erosion (SWANSTON and SWANSON 1976).

Available data on the magnitude of fire-induced soil erosion in the southeastern United States have been summarized by RALSTON and HATCHELL (1971). With the exception of a single study of hardwoods burned semi-annually in the Piedmont of North Carolina, erosional rates in all burned stands were less than 0.3 cm per century.

16.2.3 Fire Effects on Nutrient Cycling

One of the most significant ecosystem consequences of fire is its effect on nutrient flow through existing biogeochemical cycles. The large body of literature describing fire influences on patterns of nutrient cycling has been discussed in several recent reviews (VIRO 1974; WELLS et al. 1979; WOODMANSEE and WALLACH 1981; CHAPIN and VAN CLEVE 1981). WOODMANSEE and WALLACH (1981) categorize three types of fire responses. Primary responses caused by the direct effects of fire are volatilization of nutrients and ash deposition. Secondary responses, considered to be those brought on by abiotic factors include wind erosion, dissolution of ions, surface runoff, nutrient retention on soil colloids, leaching of ions out of the root zone, volatilization of NH_3, and precipitation inputs. The tertiary effects are those mediated by biological processes subsequent to fire. These include microbial uptake, plant uptake and nitrogen transformations (nitrification, denitrification, and nitrogen fixation).

In the following discussion the effects of fire on important nutrients including nitrogen, phosphorus, potassium, calcium, and magnesium are considered individually. The major emphasis here is on the primary responses of volatilization of nutrients and ash deposition.

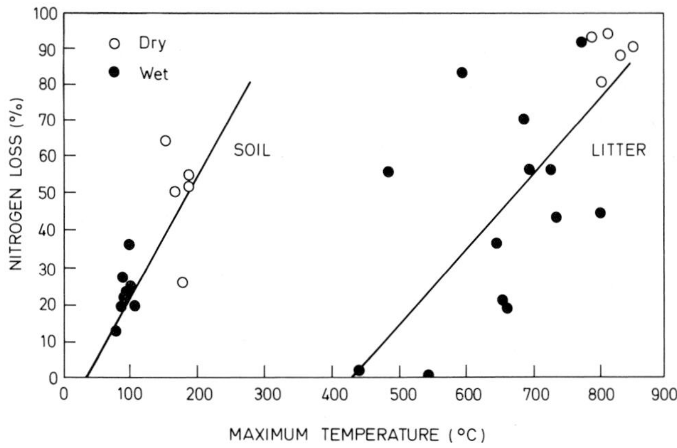

Fig. 16.4. Volatilization of nitrogen from soil and litter compartments of chaparral systems in relation to maximum fire temperature. (DEBANO et al. 1979)

Nitrogen cycling can be very strongly affected by fire in a variety of manners influencing plant, litter, and soil compartments. One effect is that nitrogen is readily volatilized by heating during a fire. DEBELL and RALSTON (1970) observed volatile losses of nitrogen of 58–85% in a study of a variety of forest fuel types. WHITE et al. (1973) found that 100% of the nitrogen in plant and litter material could be lost at fire temperatures above 500 °C. At temperatures of 400–500 °C they found potential losses of 75–100%, at 300–400 °C losses of 50–75%, and at 200–300 °C losses up to 50%. Below 200 °C they observed no measurable nitrogen volatilization. Experimental studies using soil slabs from chaparral plots have provided careful measurements of the effect of temperature (measured at 1 cm soil depth) on the volatilization of nitrogen from soil and litter compartments (DEBANO et al. 1979; Fig. 16.4).

Data on nitrogen volatilization during fires have produced quite variable results and differing manners of expressing data compound the problem. DE BANO and CONRAD (1978) reported that 10% of the total nitrogen from plant, litter, and upper soil layers was lost in a prescribed chaparral burn. In a later study with chaparral soils and litter layers alone they found that about 67% of the total nitrogen was lost during intense burns over dry soils but less than 25% was lost when soil and litter were moist (DEBANO et al. 1979). Laboratory studies simulating fires in *Calluna*-dominated heath communities in England have reported that approximately 70% of the nitrogen in plant biomass is volatilized under temperature conditions of 500–800 °C (ALLEN 1964). Since about 30–50% of the available nitrogen in these heaths is in the standing crop, the expected loss of up to 45 kg ha^{-1} would be about 25% of the available nitrogen. CHRISTENSEN (1977) found a 70% loss of nitrogen during burning from the biomass of a *Pinus palustris* savanna in North Carolina. This would be a nitrogen loss of 33 kg ha^{-1}. WELCH and KLEMMEDSON (1975) reported a 20% loss of total ecosystem nitrogen in a wildfire in *Pinus ponderosa*. Many other estimates of quantitative losses of nitrogen by fire are available in the literature: 140 kg N ha^{-1} lost in a prescribed burn of *Pinus ponderosa* (KLEMMED-

SON et al. 1962); 669 and 750 kg N ha^{-1} from *Pseudotsuga menziesii* slash burning (ZAVITKOWSKI and NEWTON 1968; YOUNGBERG and WOLLUM 1976); 112 kg N ha^{-1} from a prescribed burn in *Pinus taeda* (WELLS et al. 1979); and 907 kg N ha^{-1} from a severe wildfire in a coniferous forest in Washington.

Despite major volatilization of nitrogen in fires, the available forms of nitrogen are commonly higher on burned than unburned sites. This condition results from the rapid mineralization of litter and associated enrichment of the soil (e.g. WELLS 1942; GARREN 1943; CHRISTENSEN and MULLER 1975; ST. JOHN and RUNDEL 1976; STARK and STEEL 1977; DEBANO and CONRAD 1978). The nature of this enrichment is highly variable, however, due to complex interactions of volatilization and change in chemical form (LEWIS 1974). RUNDEL (1981b) reported rapid increases in ammonium-nitrogen following fire in *Adenostoma* chaparral, but no change in nitrate-nitrogen levels. CHRISTENSEN (1977) found only 1% of the nitrogen in ash added to pine savanna soils following fire in the coastal plain of North Carolina was in extractable forms. He suggested that ash may act as a reservoir for gradual mineralization and release of available forms, but this reservoir is subject to leaching effects. ORME and TEEGE (1976) have reported variable increases in available forms of nitrogen depending on the seasonality and severity of fires in northern Idaho. Cool spring burns increased soil concentrations of nitrate and ammonium by 3 and 1.5 times respectively, while more severe fall burns increased nitrate by 20 times and ammonium by 3 times.

Even when total soil nitrogen is reduced greatly following fires, post-burn soil conditions favor increased soil microbial activity and promote nitrogen fixation by free-living microorganisms (AHLGREN and AHLGREN 1960; JORGENSEN and WELLS 1971; CHRISTENSEN and MULLER 1975; DUNN et al. 1979). The effect is hypothetically due to combined influences of pH shifts toward neutral from acidic conditions (VIRO 1974; CHRISTENSEN and MULLER 1975; ST. JOHN and RUNDEL 1976) and breakdown of allelopathic compounds (CHRISTENSEN and MULLER 1975). Fertilization of burned chaparral soils with nitrogen does not produce increased growth (DEBANO and CONRAD 1974; VLAMIS and GOWANS 1961).

In addition to free-living nitrogen fixers, many shrub species with symbiotic fixation are greatly increased in numbers following fires. This effect may be due to native legumes in pine forests of the southeastern United States (CHEN et al. 1975; CUSHWA and REED 1966; CUSHWA and MARTIN 1969) and for both legumes and *Ceanothus* in western coniferous forests (ORME and TEEGE 1976; YOUNGBERG and WOLLUM 1976; HARTESVELDT et al. 1975). Estimates of post-fire nitrogen fixation due to *Ceanothus* are 72 kg N ha^{-1} yr^{-1} for pine stands for the first 10 years following a wildfire and 108 kg N ha^{-1} yr^{-1} for a fir stand over the same period following slash burning (YOUNGBERG and WOLLUM 1976). If 750 kg N ha^{-1} are lost by volatilization during a severe fire, *Ceanothus* fixation alone could replace this in about 7 years. Much lower levels of post-fire nitrogen fixation by *Ceanothus* are reported for the Oregon Cascades, with 35 years required to restore nitrogen to pre-fire levels given no other inputs (ZAVITKOWSKI and NEWTON 1968). KUMMEROW et al. (1978) suggested that chaparral stands in Southern California with about 30% *Ceanoth-*

us cover have only about 0.1 kg N ha^{-1} yr^{-1} of symbiotic fixation. This amount is sufficient to replace nitrogen lost in runoff but is only a small factor in the overall nitrogen balance. No data are currently available on rates of nitrogen fixation in leguminous plants which increase in abundance following fires in the southeastern United States. Their small biomass in relation to *Ceanothus* suggests that their contribution of nitrogen is relatively small.

Phosphorus cycles are also influenced by fire, but this effect is less well studied. Significant quantitites of phosphorus in litter and in plant canopies are often lost during fires, apparently physically transported away as fine ash particles. CHRISTENSEN (1977) estimated a 46% loss of total phosphorus from standing biomass following fire in a southern pine savanna. LLOYD (1971) also showed high losses due to fire in English heath communities, but ALLEN (1964) found no effect. DEBANO and CONRAD (1978) found only a 2% loss of total phosphorus from the ecosystem pool (plant, litter, upper soil horizon) following a chaparral fire. They suggested that virtually all of the total phosphorus in the standing crop biomass was added to the soil as ash, but this amount was only 10% of the ecosystem phosphorus pool. They did not speculate on the causes of the loss of 11.7 kg ha^{-1} of phosphorus from the upper mineral soil. ORME and TEEGE (1976) found a soil phosphorus increase of 200–500% following fires in northern Idaho.

Available forms of phosphorus in soil generally increase following fires but a lack of standardization in methods of measuring "available" phosphorus makes it difficult to compare literature studies. VIRO (1974) reported small increases in soluble phosphorus for the first 3 years following fire. WAGLE and KITCHEN (1972) found a 32-fold increase in available soil phosphorus after a prescribed burn in *Pinus ponderosa,* but CAMPBELL et al. (1977) reported much smaller increases for this forest type. ST. JOHN and RUNDEL (1976) found small increases following fire in a mixed conifer forest, but the relationship of burned to unburned plots showed a definite seasonal cycle. CHRISTENSEN and MULLER (1975) found large increases in available phosphorus in burned chaparral soils, but their seasonal data does not show clear patterns. Despite this general pattern of increased availability of phosphorus, exceptions have been described (ISAAC and HOPKINS 1934). Some of these apparently contradictory results may be explained by differential binding of phosphate ions in soils and by variables in fire intensity. Very hot fires such as those studied by ISAAC and HOPKINS (1934) increase losses of phosphorus, although the mechanisms are poorly studied. DONAGHEY (1969) reported increases in available phosphorus of coniferous forest soil from the Sierra Nevada at temperatures up to 500 °C, but losses above this level. WHITE et al. (1973) suggest that greatest phosphorus availability occurs in soils heated to less than 200 °C, while HOFFMAN (1966) found greatest availability after heating to 200–500 °C.

Soil levels of cations almost invariably increase following fire. CHRISTENSEN (1977) found that 60% of the standing crop of potassium in vegetation is returned to the soil as ash. Nearly 60% of this 6.3 kg ha^{-1} of potassium was readily soluble, resulting in an initial doubling of pre-fire levels. This increased potassium concentration was rapidly reduced to levels comparable to unburned soils within a few months, however, apparently due to leaching. Studies of chaparral soils have also reported large initial increases in potassium following fire. CHRISTENSEN and MULLER (1975) recorded 43 kg ha^{-1} of potassium added as ash, with the effect increasing extractable soil potassium by over 50%. Investigations by DEBANO and CONRAD (1978) calculated an addition of 43 kg ha^{-1} of potassium in ash following chaparral fires, but noted a loss (export) of 12% of the total ecosystem pool of potassium (48 kg ha^{-1}) possibly due to volatilization. An additional 27 kg ha^{-1} were lost by erosion following fire, 30% of this dissolved in water. Numerous studies in other ecosystems have also documented increases in exchangeable potassium in post-fire soils (ISAAC and HOPKINS 1934; AUSTIN and BAISINGER 1955; HATCH 1960; HOFFMAN 1966; LEWIS 1974; ST. JOHN and RUNDEL 1976; STARK 1977). However, a few studies have recorded small decreases in total potassium following fire (REYNOLDS and BOHNING 1956; ORME and TEEGE 1976; CAMPBELL et al. 1977). Studies by WHITE et al. (1973) suggest that as for phosphorus, high fire temperatures above 500 °C may cause volatilization of potassium.

Calcium and magnesium commonly behave quite similarly under fire conditions. CHRISTENSEN (1977) reports that 80% (16.4 kg ha^{-1}) of the standing crop potassium are returned to the soil as ash following pine savanna fires. As with potassium the initially high post-fire soil concentrations of these cations return to pre-fire levels within a few months. He found that secondary divergences between burned and unburned soils occurred during the first spring and summer following the initial spring burn, suggesting the possibility that ash and charcoal may act as reservoirs for these cations and allow slow release, possibly subject to microbial activity. DEBANO and CONRAD (1978) found that 45 kg ha^{-1} of calcium and 5.3 kg ha^{-1} of magnesium were transported to soils as ash following a chaparral fire, but even larger amounts were lost through post-fire erosion in debris and runoff (67 kg ha^{-1} and 32 kg ha^{-1}, respectively). These findings are difficult to interpret, but if they are correct they indicate that cations were lost from soil and litter as well as from ash.

Numerous other studies have also reported increases in calcium and magnesium following fire (AHLGREN and AHLGREN 1960; HATCH 1960; SCOTTER 1963; AUSTIN and BAISINGER 1955; LEWIS 1974; ST. JOHN and RUNDEL 1976; STARK 1977). A few studies have found no change or actual decrease in soil concentrations of these cations following fires (CHRISTENSEN and MULLER 1975; VIRO 1974). These latter results are probably related to changes in the cation exchange capacity which is commonly decreased in fires as organic matter content of soils is reduced. Cases where cation exchange capacity is reduced and excess cations deposited in ash provide conditions in which these ions may be readily leached out of the rooting zone in ground water (FINN 1943; STARK 1977).

Increase of soil pH following fire is a consistent effect which occurs as a result of basic cations released by the combustion of organic matter. The degree to which pH is shifted and the length of time it remains higher than under pre-fire conditions are a function of the original soil pH and organic matter content, the amount of ash produced and its chemical composition, and the amount of regional precipitation (LUTZ 1956; METZ et al. 1961; WELLS 1971; GRIER 1975). Hot slash fires in coniferous forests in the western United States may commonly raise the pH 1–2 units from about 5 or 6 to 7 or more

(Isaac and Hopkins 1937; Tarrant 1956; St. John and Rundel 1976). Much smaller increases occur in eastern pine forests where fire intensity is lower. Soils under *Pinus taeda* stands in South Carolina subject to annual burning for 20 years changed pH from 4.2 to 4.6 in the upper mineral soil, while burning at 4–5 year intervals did not effect pH (Wells 1971). Fourteen years of annual burning in stands of *Pinus resinosa* and *Pinus strobus* in Connecticut increased mineral soil pH from 4.3 to 5.0 (Lunt 1950). Varying frequencies of fires in *Pinus resinosa* stands in Minnesota increased pH of mineral soil from 5.3 to 5.5 (Alban 1977). Christensen (1977) found no pH change following fire in *Pinus palustris* savannas. Rangeland fires where small amounts of organic matter undergo combustion produce no significant changes in soil pH (Owensby and Wyrill 1973).

In addition to the direct effects of fire on nutrient availability discussed above, there are a number of indirect mechanisms which also act to influence the relative post-fire availability of nutrients. One such mechanism is the accelerated fall and decomposition of litter which has been observed following fire (Stark 1977; Grigal and McColl 1977). Increased post-fire rates of ammonification and/or nitrification have also been described following fires as microbial activity increases (Christensen 1973; St. John and Rundel 1976; Rundel 1981b). Alterations of soil pH following fire may either increase or decrease the availability of phosphate and certain cations (Daubenmire 1968; Viro 1974). Finally, fire-induced destruction of soil organic matter may decrease soil cation exchange capacity (Chapin and Van Cleve 1981). These indirect effects, often ignored in many studies of fire–nutrient relationships, may be significant for several years following a fire and be more important in determining increased plant growth than the direct effects of nutrient release in ash (Chapin and Van Cleve 1981).

Post-fire nutrient uptake by plants is often very rapid. In some cases this uptake may be sufficient to prevent nutrient loss in runoff to adjacent watersheds (McColl and Grigal 1977). Available studies of post-fire nutrient losses to adjacent watersheds (Helvey 1972; Helvey et al. 1976; Clayton 1976; Schindler et al. 1978) generally show a pattern of a small increased rate of loss for months or a few years at most, but this increase may be within ranges of normal annual variability (Wright 1981).

Luxury consumption of nutrients (absorption in excess of those immediately required for metabolism) appears to occur in post-fire growth of many species (Lay 1957; Steward and Ornes 1975; Rundel and Parsons 1980). While the ecological advantage of luxury consumption of nutrients in many low-nutrient environments is obvious, the mechanism and form of nutrient storage has not been investigated. The storage of polyphosphates in the roots of Australian heath plants may be such a mechanism.

16.2.4 Soil Microbiology

Because soil microorganisms are strongly influenced by physical and chemical changes in soil characteristics, it is not surprising that fire may have a strong influence on these organisms. Changes in soil pH and direct heating effects are particularly important in this regard. AHLGREN (1974) has discussed the impact of fire on soil microbiology in considerable detail and thus only a brief review of this topic will be attempted here.

Available data on the effects of fire on soil microbiology suggest that microhabitat characteristics, fire intensity, and sampling methodology are all important variables. Very intense fires clearly have a strong effect on microorganisms. Light annual burns have little effect on soil microorganisms, however (JORGENSEN and HODGES 1970). A major impact of hot fires may be a temporary sterilization of surface soil layers (AHLGREN and AHLGREN 1965; RENBUSS et al. 1973). It has been suggested that the accelerated post-fire growth associated with ash piles where intense heat occurred results from reduced competition between vascular plant roots and microorganisms for available nutrients (RENBUSS et al. 1973). The rate of regrowth of bacterial populations may be quite slow (DUNN et al. 1979) or occur rapidly with the first post-fire rains (AHLGREN 1974).

Fires typically increase pH significantly, as previously described, and this effect favors bacterial population growth over fungal populations growth. It has been frequently hypothesized that high soil nitrification rates which are commonly observed following fire result from increased activities of populations of *Nitrosomonas* and *Nitrobacter* (AHLGREN and AHLGREN 1960; CHRISTENSEN and MULLER 1975; ST. JOHN and RUNDEL 1976). Experimental studies with chaparral soils by DUNN et al. (1979), however, found that populations of *Nitrosomonas* and *Nitrobacter* remained low for 12 months following fire. They hypothesized that heterotrophic nitrifiers were of primary importance in these soils in the first year following fire.

Increased bacterial activity following fire may be due in part to a variety of environmental effects in addition to pH shift (MARGARIS 1977). One such effect is the addition of readily available forms of nutrient and organic substances in the ash layer following fire. Ammonium increases are notably important in this regard. Rates of organic matter decomposition may also be increased following fire as soil particle size breaks down, and allow greater direct contact between enzymes and organic matter (GREENWOOD 1968).

Quantitative studies of the seasonal variations of microbial populations following fires are few in number and often suffer from the problem of noncomparable methodology. CHRISTENSEN and MULLER (1975) and MARGARIS (1977) have documented numbers of soil bacteria and fungi in burn and control plots for Californian chaparral and Greek phrygana communities. DUNN et al. (1979) report population dynamics of four groups of microorganisms for a period of 14 months following a series of experimental burns using chaparral soil slabs.

The differential ability of soil microorganisms to endure soil heating results from a complex interaction of such factors as maximum temperature, duration of heating and soil moisture content, as well as biological variation in the thermostability of proteins and membrane lipids (WELKER 1976). For bacteria, which are generally more resistant to heating than fungi, the water content of bacterial spores is important, with moderately dry spores more heat-resistant than either very wet or very dry spores (MURRELL and SCOTT 1966). Lethal temperatures for heterotrophic bacteria in chaparral soils are reached at 210 °C in dry soil, but most are killed by temperatures above 150 °C (DUNN and DEBANO 1977). In wet soil, rapid death begins at 50 °C and no bacteria survive beyond 110°. AHLGREN and AHLGREN (1965) reported corresponding lethal temperatures for bacteria in forest soils.

Actinomycetes behave very similarly to bacteria in chaparral soils with death at 125 °C in dry soil and 110 °C in wet soil. Studies of actinomycetes in steam sterilization studies, however, have reported that these microorganisms are generally more resistant to heating than bacteria (BOLLEN 1969; MALOWANY and NEWTON 1947).

Nitrifying bacteria appear to be somewhat more sensitive to soil heating than typical heterotrophic bacteria. DUNN and DEBANO (1977) reported that *Nitrosomonas* and *Nitrobacter* group bacteria were killed in dry soil at temperatures of 140 °C, but only 75 and 50 °C respectively in wet soil. This sensitivity of nitrifying bacteria to heating may be significant in recovery of low nitrogen ecosystems following fire. DUNN et al. (1979) found that intense burning over wet soils inhibited *Nitrosomonas* and *Nitrobacter* populations for over 12 months. They hypothesized that increased concentrations of available forms of nitrogen which have commonly been observed following fires result from action by other heterotrophic microorganisms.

Considerable research has been done on the direct effects of fire on macrofungi, but no clear pattern of response is apparent (AHLGREN and AHLGREN 1965; PETERSEN 1970; WICKLOW 1975; WIDDEN and PARKINSON 1975). Work on chaparral soils indicates that typical fungal heat-tolerance limits are 155 °C in dry soil and 100 °C in wet soil (DUNN and DEBANO 1977). Differential sensitivity is clearly present, however. Normal saprophytic fungi prevail up to temperatures of 120 and 60 °C in dry and wet soils respectively, but above this level of heat a group of "heat shock" fungi become dominant. Reduced competition and more alkaline soil pH following fires favors growth and reproduction of these fire-successional fungi (EL-ABYAD and WEBSTER 1968a, b; WICKLOW 1975).

An important area for future research will be the impact of fire on basidiomycetes responsible for mycorrhizal associations with vascular plant roots. TARRANT (1956) found reduced levels of mycorrhizae in seedlings of *Pseudotsuga menziesii* following slash fires. Few other observations have been made, however.

16.3 Fire Effects on Plant Growth and Development

16.3.1 Fire-Stimulated Flowering and Growth

Fire-stimulated flowering and increased growth in both herbaceous and woody plants is a common phenomenon in many areas of the world. This process has been most commonly described for monocotyledons, in particular grasses. OLD (1969) described detailed aspects of fire effects on grassland species in Illinois, where flowering rates increase ten times after burning. A similar response has been described in a variety of other studies of North American grassland species (e.g., CURTIS and PARTCH 1948; EHRENREICH 1959; HADLEY and KIECKHEFER 1963; RATCLIFFE 1964; KLETT et al. 1971; WEIN and BLISS 1973; PEET et al. 1975). DAUBENMIRE (1968) has summarized the literature on the effects of fire on flowering in grasses.

The most carefully documented studies of fire-stimulated growth and flowering have been carried out with the snow tussock (*Chionochloa rigida*) in New Zealand. Spring burning stimulates growth for two seasons, after which time growth rates decline to those equivalent of unburned plants (O'CONNOR and POWELL 1963; MARK 1965). The first autumn following spring burning is characterized by initiation of large numbers of floral buds which flower during the following spring (ROWLEY 1970). Following this profuse flowering there is a significant depression in subsequent flowering for up to 14 years (MARK 1965).

The stimulation of growth and flowering in grasses following fire has commonly been attributed to greater soil temperatures on burned areas which allow an earlier initiation of spring growth than on unburned sites (EHRENREICH 1959; EHRENREICH and AIKMAN 1963; OLD 1969; PEET et al. 1975). Although plants on burned soils initiate growth earlier, develop faster, and mature earlier, significant differences in rates of productivity may not necessarily occur (EHRENREICH and AIKMAN 1963).

Increased flowering rates following fire in prairie grasses in Illinois may result from increased photosynthetic efficiency of grass canopies resulting from reduction in biomass (OLD 1969). Other authors have considered that increased carbohydrate levels brought about by early initiation of growth may be the primary factor in increased floral production (EHRENREICH and AIKMAN 1957, 1963). MARK (1965, 1969) suggested that the decline of flowering following the first post-fire year in *Chionochloa* may be associated with depletion of carbohydrate reserves, but more recent data have shown that rapid growth after fire does not result in a buildup of carbohydrate levels in the first season and that subsequent reductions in growth-promoted carbohydrate accumulations (PAYTON and MARK 1979). Thus carbohydrate levels in *Chionochloa* reflect rather than determine post-fire growth and show little response to flowering.

More favorable nutrient conditions following fire may be an important consideration in increased rates of growth and flowering. Increased foliage concentrations of nitrogen, phosphorus, potassium, calcium, and/or magnesium have been found in post–fire growth of grassland species in Illinois (OLD 1969), the Florida Everglades (STEWART and ORNES 1975), New Zealand (WILLIAMS and MEURK 1977), and tundra communities (WEIN and BLISS 1973). Increased foliar concentrations of nitrogen and phosphorus may reflect that increased availability of nutrients following fire more than increased soil concentrations. The basis for increased availability has been discussed earlier in this chapter.

In comparison to woody plant communities where significant quantities of limiting nutrients are released to the soil by ash following fire (MILLER et al. 1955; CHRISTENSEN and MULLER 1975; ST. JOHN and RUNDEL 1976; DE BANO and CONRAD 1978; CHRISTENSEN 1977) nutrient return is relatively low following fire in most grasslands due to the relatively small biomass. OLD (1969) found no increase in productivity of Illinois prairie grasses following fire which could be attributed to ash nutrients. WEIN and BLISS (1973) reported that amounts of nutrients released from fires in *Eriophorum* tundra communities were very low in comparison to levels of fertilization required to produce increased productivity. Ash returned to the soil following fires in snow tussock grasslands in New Zealand similarly does not affect leaf growth or stem growth in *Chionochloa*. Contrary to this general pattern of results, CURTIS and PARTCH (1950) found that ash deposition following fires was responsible for increased flower production in *Andropogon gerardi* in the Great Plains, but had no effect on plant height. EHRENREICH and AIKMAN (1963) reported increases in levels of available soil phosphorus of up to 12 times following fire.

PAYTON and MARKS (1979) concluded that the factors which stimulate post-fire growth and flowering of *Chionochloa* and subsequent long-term depression of these processes are complex and still largely obscure. Nutrient availability, presence of mature tillers, and related vegetative growth phases may all be important.

Increased flowering and growth following fire is well-documented for many groups of monocotyledons in addition to grasses. Many South African geophytes in the Amaryllidaceae, Iridaceae, and Orchidaceae concentrate the majority of their flowering in the first year following fynbos fires (HALL 1959; KRUGER 1977). Fall burning induces flowering in 50% of the stalks of *Watsonia pyramidata* (Iridaceae), while spring burns and unburned plots have flowering in only about 5% of the stalks (KRUGER 1977). It has been suggested that the fire-stimulated flowering of these species may be due to some combination of increased soil temperature, reduced competition, and greater diurnal fluctuations in temperature (MARTIN 1966). Fire-induced flowering in *Asparagus* in South Africa is earlier than flowering without fire (KRUGER 1977) indicating a possible importance of soil temperature conditions.

Fires also stimulate flowering in species of *Xanthorrhoea*, the tree grass of Australia (GILL and INGWERSEN 1976). Clipping of crowns, as well as injecting of ethylene-releasing compounds, also stimulated flowering, suggesting the possibility that fires induce flowering through chemical effects. GILL (1981) has hypothesized that the prolific flowering at irregular intervals which is induced by fires may be an important factor in the success of seed set of *Xanthorrhoea* due to the effect of predator satiation.

16.3.2 Fire-Stimulated Resprouting

Resprouting individuals exhibit a variety of mechanisms of bud protection to ensure vegetative regrowth following fire. These mechanisms include bud protection by bark, by dense leaf bases, and by soil (GILL 1977). The role of bark in bud protection results from the thickness and insulating qualities of the bark. Many tree species are able to survive crown fires which consume much of their canopies because buds survive beneath thick bark on trunks and large branches. Loss of foliage initiates growth in these meristems and rapid regrowth of foliage occurs. This pattern of resprouting is termed epicormic growth. The ability of bark to insulate underlying cambium and meristems from heat damage has been thought by many authors to be largely a function of the thickness of the bark (MARTIN 1963; KAYLL 1966; REIFSNYDER et al. 1967; MCARTHUR 1968). A number of studies, however, have suggested that other bark variables such as density, structure, composition, and moisture content may also be important factors in determining thermal diffusivity through bark (STICKEL 1941; GILL and ASHTON 1968; VINES 1968; HARE 1975). Despite the well-known fact that bark thickness is highly variable within single species and between species, there has been little attempt to correlate bark characteristics with fire regimes. ST. JOHN (1976) found that thick bark was generally associated with trees whose above-ground tissues survived fire, while woody plants which resprouted from ground level or which were killed by fire had relatively thin bark.

Because stringy outer bark of many species of *Eucalyptus* is highly flammable, fire frequency is an important consideration in the relative protection which bark may offer. Repeated fires before the bark thickness has been restored to adequate levels for protection may cause mortality of trees (GILL 1977). A similar effect may occur from heat-stimulated abscission of bark layers which is present in some tree species. Since many trees are unable to regenerate fire-damaged bark layers, subsequent fires very frequently kill cambial layers associated with poorly protected sections of the outer radii of trees. The resulting fire-scars are the basis for much of our knowledge of past frequencies of fire (ARNO and SNECK 1977). In large trees, fire scars may penetrate well into the heartwood of living trees. Such scars in *Sequoiadendron* have been shown to restrict available cambium for water transport and thus decrease water potentials for damaged trees under conditions of water stress (RUNDEL 1973).

Rates of crown recovery in epicormically sprouting species have not been well studied, but available data suggest that regrowth may be very rapid. GILL (1978) showed that low trees of *Eucalyptus dives* reestablished pre-fire leaf areas within a single year and had restored normal branching patterns in three years. Relatively rapid crown recovery takes place in *Sequoia sempervirens* and many evergreen species of *Quercus* in California.

Many species of plants are known to survive fire through bud protection offered by dense leaf bases. This form of protection, most notable in monocotyledons, may occur at ground level as in many tussock grasses or on aerial stems. In perennial bunch grasses dense bases of old stems provide a strong protection for meristems. Because of the lack of a cambial layer, regrowth of both roots and shoots occurs from abundant meristems on the short internodes of the stems at ground level. Unlike dicotyledons which must develop entire new leaves, monocotyledons can renew growth of old leaves from the protected leaf bases. Because of this characteristic, perennial bunch grasses are extremely efficient at rapidly resprouting following fires.

Morphologically similar tufted growths of leaves are found on aerial stems of many monocotyledons as well. Here again moist, densely packed leaf bases protect meristems from damage. The dispersed nature of the vascular system of stems in these monocotyledons also increases fire resistance (GILL 1977). Examples of these adaptations to fire can be seen in the genus *Xanthorrhoea* in Australia (GILL and INGWERSEN 1976), in many tropical Palmae (GILL 1977), and tropical *Pandanus* (VOGL 1969).

Although rare outside the monocotyledons, other vascular plant examples of bud protection by tufted leaf tissues have been described. The most notable example of such adaptations is the seedling stage of *Pinus palustris* in the coastal plain of the southeastern United States, where densely packed needles protect the apical meristem from light ground fires which kill competing hardwood seedlings. Other examples of similar meristem protection occur in tree ferns and cycads.

Subterranean protection of dormant buds, a widespread adaptation for fire survival in plants, has been summarized in some detail by GILL (1977). Soil is an effective insulator of heat (AL NAKSHABANDI and KOHNKE 1965) and thus only a few centimeters of soil are usually sufficient to protect underground buds from heat damage during fires (FLINN and WEIN 1977). Underground buds occur on rhizomes in ferns and many flowering plants. Many herbaceous geophytes have bulbs or tubers which serve the same function following fire. In woody dicotyledons it is characteristic for buds at the base of major stems to resprout following fire. AXELROD (1975) hypothesizes that resprouting is an ancestral trait among woody angiosperms. Such resprouting behaviour may not necessarily be an adaptation to fire, however, since frost damage, drought, heavy grazing, or mechanical damage produce the same response.

Many woody plants in ecosystems subject to frequent fire have evolved expanded woody root tissues with masses of adventitious buds at their stem base. These structures are referred to as lignotubers, root crowns, or basal burls. Lignotubers are best known from mallee and heath communities of Australia where they are very widespread (KERR 1925). The majority of *Eucalyptus* species have lignotubers with the trait notably absent only in tall, rapidly growing species on rich substrates. GILL (1975) suggests that the lignotuber in Australian *Eucalyptus* is an adaptive trait for recovery and persistence under stresses of many types including fire. Australian families notable for the occurrence of lignotubers are the Myrtaceae, Casuarinaceae, Proteaceae, Leguminosae, Tremandraceae, Sterculiaceae, and Dilleniaceae (GARDNER 1957). The greatest frequency and prominence of plants with lignotubers occurs in the most stressful habitats, particularly those with Mediterranean climates.

Lignotubers in California chaparral are much more restricted in occurrence taxonomically. Basal rootcrowns are produced by all resprouting species of *Arctostaphylos* and most species of *Ceanothus*, as well as in *Adenostoma* and shrubby *Aesculus*. Although virtually all woody shrubs in the mediterranean-climate regions of Chile and Europe resprout following fire, lignotubers are virtually unknown. Lignotubers are relatively common in some plant groups within the South African fynbos (KRUGER 1977) but are surprisingly rare in the Proteaceae in comparison to Australia.

The physiological significance of lignotuber resprouting following fire has not been studied in sufficient detail to generalize about its adaptive role. It is known that lignotubers contain large stores of inorganic nutrients and starch (JONES and LAUDE 1960; BAMBER and MULLETTE 1978; MULLETTE and BAMBER 1978), and probably water. The significance of lignotubers may therefore be to provide storage of resources to enable rapid regrowth of shoots or to maintain root systems through extended periods of drought or environmentally induced dormancy.

Since the great majority of woody dicotyledons are able to resprout following fires without lignotubers, the high frequency of these structures in many mediterranean-climate regions is not easy to explain. KEELEY (1981) suggests that the combined stresses of summer dry mediterranean climates and fire seasonality may be important. It is interesting to note that the major part of carbon allocation to roots in *Heteromeles arbutifolia* in the California chaparral takes place in the fall after the peak of fire occurrence (MOONEY and CHU 1974). If this pattern of carbon allocation is typical of other mediterranean-climate shrubs, lignotubers may be important in buffering root systems against loss of carbohydrates in the event of summer fires. Loss of growing roots could seriously restrict the ability of shrubs to take up available soil nutrients and thereby might limit rates of post-fire resprouting. Investigations of recovery of rhizome carbohydrate storage in *Ilex glabra* in pine savannas of the southeastern United States have shown that carbohydrates returned to prefire levels within one year of the time of burn, regardless of the season of the fire or whether fires had burned in one or two successive years (HUGHES and KNOX 1964).

Root structures very similar to lignotubers have been reported from tropical and subtropical savannas in Africa. WHITE (1977) concluded that these did not evolve primarily in response to fire but rather as an adaptation to poor soil conditions. Analogs of lignotubers occurring on woody species in the arid cerrado vegetation of Brazil, however, are thought to be associated with fire survival (EITEN 1972).

The ability of nonlignotuber woody plants to resprout following fire or other disturbance is a function of the age of the stand. Studies of *Calluna vulgaris* have established that regeneration is best in stands 6–10 years in age and declines steadily in older stands to a minimum level of less than 10% above 25 years (KAYLL and GIMINGHAM 1965; MILLER and MILES 1970). The causes of this declining capacity for regeneration have been hypothesized to be a deterioration of the capacity of stem bases to produce shoots due to increased woody growth around potential meristems (MOHAMED and GIMINGHAM 1970). Forestry experience with oaks similarly suggests that frequency of stump sprouting decreases as stem size increases.

A final group of variables to consider in resprouting are fire characteristics. Seasonality of fire may be an important factor in determining resprouting ability (JONES and LAUDE 1960; HUGHES and KNOX 1964; CREMER 1973). Fire intensity may also be significant (NAVEH 1974) but there are few data available on this aspect. Lastly, fire frequency has obvious importance (GRANO 1970). Regular annual burning is a common management technique to eliminate undesirable woody plants from grazing lands.

16.3.3 Fire-Stimulated Germination

Many woody plants utilize specialized systems of seed storage and fire-stimulated germination to maintain species survival. In many but not all of such species resprouting does not occur. Instead, dense stands of seedlings germinate in many recently burned plant communities, the result of stores of seeds maintained in the soil or in specialized woody cones or fruits on woody plant stems. Hardcoated seeds of a number of chaparral shrubs, including species of *Arctostaphylos, Ceanothus,* and *Adenostoma,* have been studied in California. In the first two genera, species are characteristically either reseeders or resprouters following fire, but not both. The scarification requirement of such seeds for germination is met by the intense heat of a fire (WRIGHT 1931; STONE and JUHREN 1953; QUICK 1935; HADLEY 1961; BERG 1975). The soil storage levels of seeds are clearly the result of dynamic fluctuations of inputs of seeds from the canopy and outputs due to granivory or erosion. Seed burial by ants may be very important in minimizing granivory (BERG 1975). Few data are available on the magnitudes of these soil storage compartments for seeds, but KEELEY (1977) has compared seed dynamics for paired resprouting and reseeding species of both *Arctostaphylos* and *Ceanothus.* Seed stores of *Acacia cyclops* in South Africa reach 25,000 m^{-2} (TAYLOR 1977).

Flushes of herbaceous growth following fires in Californian chaparral also result from germination of soil stores of seeds. MULLER et al. (1968) have suggested that release from allelopathy is the major factor in promoting this lush growth rather than direct fire-stimulated germination. This interpretation is clearly true in some instances, but its broad-scale application needs more supporting data.

Since conditions for successful seedling establishment are particularly favorable following fire, many woody plants have become adapted to on-plant storage of seeds with fire-stimulated dispersal mechanisms. Some of the best examples of such mechanisms are the serotinous or close-coned pines which retain seeds in cone scales held shut by resins. The heat of a fire melts the resins and releases the seeds. In most instances such release only occurs with fires hot enough to kill the parent plant. Unlike other pines, where cones are situated on small branches and regularly shed, closed-cone pines have the cones on major branches or even the trunk, and these cones remain permanently attached. Good examples of closed-cone pines are *Pinus attenuata, P. muricata, P. remorata* and *P. contorta* ssp. *contorta* in California (VOGL et al. 1977; VOGL 1973), *P. banksiana* in the Great Lakes region (TEICH 1970), *P. pungens* in the eastern United States (ZOBEL 1970; BARDEN 1979), and *P. halepensis* and *P. brutia* in the Mediterranean region of Europe (NAVEH 1974).

Perhaps the most interesting example of a closed-cone pine, however, is *Pinus contorta* from the Rocky Mountains. This species is notable because variable populations of both serotinous and open cones are found over its range, and the mechanism of formation of the necessary resin bond to promote serotiny is under relatively simple genetic control. Commonly areas with frequent fires in recent history support a majority of serotinous-coned individuals. Long fire-free periods, however, select for open-coned individuals (TEICH 1970; LOTAN 1976).

Many other examples of seed storage in perennial reproductive structures on woody plants have been described in some detail. A number of species of Proteaceae, for example, have this characteristic. *Banksia ornata* in Australia sheds massive numbers of seeds from woody fruits following fire (SPECHT et al. 1958). Only about 1% of the fruits on intact mature plants open naturally and fire is required to release most of the seeds on even dead branches (GILL 1977). The closely related *B. marginata,* which may occur sympatrically, has woody fruits which open without heat and which do not store seeds (GILL 1977). In *Hakea* in Australia seeds are retained in woody fruits through the life of the plant, but these fruits open upon the death of their parent branch. Under natural conditions, however, fire is usually the cause of such death. In the large shrub genus *Leucadendron* in South Africa over half of the species retain ripe seed within woody fruiting heads (WILLIAMS 1972). Most of this group of species have flat-winged fruits that disperse in the wind. The most remarkable among these is *L. platyspermum* which germinates in the woody fruit following fire. Dispersal occurs as the emerging radicle forces the seed from the fruit, and winds carry the broadly winged fruit up to 50 m. The cotyledons are green and the radicle is actively growing by the time the seedling touches the ground. WILLIAMS (1972) notes that while the mature seeds are palatable to birds, the seedlings appear to be chemically protected against herbivory. Seeds of the other group of *Leucadendron,* which shed their seeds annually, are all hard, nut-fruited species.

The genus *Eucalyptus* frequently utilizes on-plant seed storage. Although the majority of species shed some seed annually, partial storage does occur and fires release the equivalent of two or more years of seeds (GILL 1977). The importance of stored seed in the crown of *Eucalyptus regnans* is described in considerable detail by GILL (1981), who integrated data on seed production, seed predation, and fire ecology. He suggested the possibility that fire-stimulated release of seeds not only enhances seedling survival by allowing seedling development to occur on ash beds or bare mineral soil, but by producing predator satiation as well. In addition to *Eucalyptus,* many other genera of Myrtaceae, including *Melaleuca, Callistemon,* and *Angophora,* store seeds.

Closed-cone morphologies for seed retention are present in other conifers in addition to the pines. Members of the Cupressaceae in many areas of the world including Australia (GARDNER 1957) and California (VOGL et al. 1977) store seed in closed cones and release it following fire. The fire ecology of *Cupressus forbesii* in Southern California and Baja California has been studied in detail by ZEDLER (1977). *Sequoiadendron giganteum* (Taxodiaceae) maintains viable seeds in cones for up to 20 years (HARTESVELDT et al. 1975). While fires rarely kill the long-lived mature sequoias, the convective heat transfer from ground fires dries out cones sufficiently to cause seed release. Seed storage capacity is immense in these trees with estimates of approximately 7,000 closed cones with an average of 200 seeds per cone in a mature tree (HARTESVELDT et al. 1975).

Because the majority of species of woody plants which store seed in the soil or in perennial fruits are killed by fires, reproduction at a relatively early age has been selected for. ST. JOHN (1976) showed that mean age of initiation of seed production is significantly lower in seed storage species than in woody species which resprouted following fire or which survived most fires. *Pinus banksiana, P. clausa, P. murrayana, P. strobus,* and *P. virginiana,* all of which are commonly killed by fires, begin reproduction at about 5 years of age (FOWELLS 1965). Most pines which survive typical habitat fires (e.g., *Pinus echinata, P. jeffreyi, P. palustris, P. ponderosa, P. resinosa,* and *P. taeda*) do not produce seeds until at least 20 years of age. Only 3–4 years are required for seeds of *Leucospermum* (Proteaceae) in South Africa to reach reproductive maturity.

16.3.4 Plant Disease

Fire has also been shown to have an effect on many pathogenic organisms in a forest and shrubland ecosystem. Although only limited data are available to assess the physiological and ecological implications of fire-pathogen interactions, recent surveys suggest that these effects may be very significant (HARDISON 1976; HARVEY et al. 1976; PARMETER 1977). The direct effects of fire on pathogens result from two major factors – destruction of inoculum on decaying duff and litter, and direct inhibition or stimulation of pathogen populations (PARMETER 1977). Numerous pathogenic fungi are associated with duff and litter. Infection of *Fomes annosus,* an important stump rot pathogen that persists for years in decaying stumps, is reduced in pine stands which are burned before thinning (FROELICH and DELL 1967). Fire-damaged trees, however, may be subject to increased *Fomes* infection (BASHAM 1957). Damping-off fungi are also killed by fire (COOPER 1965). PARMETER (1977) suggests that the frequently observed success of seedlings on ash beds may be due in large part to the removal of seed decay, damping-off, and root-rot fungi. In addition to these effects on pathogens in litter and duff, fire may also reduce stand infection by other pathogens in living tissues. A good example of this effect is the reduction of inoculum of *Scirrhia acicola* (brown needle spot) on seedlings of *Pinus palustris* in the southeastern United States (CROCKER and BOYER 1975; MAPLE 1976). Fire may also kill galls of rusts (SIGGERS 1949).

Diseases of individual trees have clearly been shown to increase following fire damage. Fire scars commonly create infection sites for many pathogens, particularly heart-rot fungi (HARVEY et al. 1976). Reduced vigor of fire-damaged plants leads to increased susceptibility to canker diseases (DEARNESS and TANSBROUGH 1934) and to insect attack (HARE 1961).

While data are lacking, there are good suggestions that fire may affect the rate of spread of many pathogenic diseases by destroying populations of insect vectors (CANTLON and BUELL 1952). Opposite effects on "vectors" may also occur since populations of many alternate hosts of rusts are stimulated by fire. This has been shown for fusiform rusts with *Quercus* (CZABATOR 1971) and white pine blister rust with *Ribes* (DAVIES and KLEHM 1939; QUICK 1962).

Many chemical substances produced by combustion processes during fires have been shown to either stimulate or inhibit pathogen populations. Many variables are involved with these effects including the chemical nature of the fuel, the intensity of the fire, and the environmental conditions for the deposition and retention of chemical compounds produced. Smoke from fires has an inhibiting effect on spore germination and infection of a number of plant pathogens (MELCHING et al. 1974; PARMETER and UHRENHOLDT 1975, 1976). The critical chemical constituents which cause such effects have not been investigated, but pyrolysis of wood is known to produce a variety of phenolic compounds. However, inhibitory chemicals associated with heated soils may be rapidly detoxified by soil microorganisms (ROVIRA and BOWEN 1966; ZAK 1971). The spore production, growth, and fruiting of many higher fungi, particularly ascomycetes, may be stimulated by fire through either heat or chemical effects (SUSSMAN and HALVORSON 1966; AHLGREN 1974). This group includes several pathogenic fungi which may attack new seedlings in burned areas (HARDISON 1976; HARVEY et al. 1976).

Fires also have very strong indirect effects on pathogens through their impact on plant community structure. Overmature stands of forest trees promote disease activity, but regular occurrence of natural fires may provide an important cleansing influence on these stands. Exclusion of natural fire may result in extensive damage of old even-age stands of trees by *Fomes annosus* (FELIX et al. 1974). In uneven-age stands, fires tend to selectively burn dead and damaged trees. Fire often destroys dwarf mistletoe-infected trees, for example, as woody witches' brooms of infected tissue on branches provide ladders for ground fires to ascend into the crowns (PARMETER 1977). The fire ecology of dwarf-mistletoes has been discussed in detail (ALEXANDER and HAWKSWORTH 1975; WICKES and LEAPHART 1976). ROTH (1966) has suggested that the tendency for fire to destroy trees laden with dwarf mistletoe may have interfered with the evolution of resistance to infection. Such an argument seems, however, to ignore the evolutionary implications of fire adaptation.

Other effects of fires on community structure have been hypothesized to stimulate development of pathogens. Since pure stands of forest species are well known to have increased susceptibility to disease, fire-induced perpetuation of such pure stands may promote disease activity. Succession in the absence of natural fires would tend to promote increased diversity in such stands. PARMETER (1977) suggests that periodic intense fires result in the perpetuation of mosaics of even-age stands where "islands" of pathogens provide inoculum for infection of new stands. While this may be true, there is increasing evidence that restoration of natural fire frequencies and intensities in coniferous forests of the western United States will reduce levels of pathogens. Recent policies in fire exclusion have reduced fire frequency but caused large increases in fire intensities as litter and other fuels have accumulated. It is these uncommon catastrophic fires which have produced extensive even-age stands with little or no fire-induced thinning.

16.4 Fire Regimes in Natural Plant Communities

16.4.1 Coniferous Forest Ecosystems

Fire plays a major role in the community functions of virtually all coniferous forest ecosystems. In addition to influences on hydrologic and biochemical cycles in these ecosystems, fire has very direct effects in determining community structure and diversity. These community effects are strongly related to the frequency and intensity of fires within a specific community type (HEINSELMAN 1981; KILGORE 1981).

Frequent, light ground fires with less than a 25-year cycle are characteristic of coniferous forests with rapid buildup of flammable fuels on the ground surface and regular inputs of lightning for ignition. Such conditions prevail in the pine savannas of the coastal plain of the southeastern United States where the highly flammable ground cover of *Aristida* commonly burns at frequencies of 2–8 years (WELLS 1928; CHRISTENSEN 1981). These fires kill hardwood seedlings and help maintain pine dominance. Low intensity ground fires are also common at frequent intervals in ponderosa pine forests and mixed conifer forests in many areas of the western United States (COOPER 1960; KILGORE and TAYLOR 1978; MCNEIL and ZOBEL 1979). Natural fires in these forest types burn relatively small areas, producing mosaics of uneven-age stands as small as a few hectares or less.

Montane lodgepole pine and Douglas fir forests in the Rocky Mountains have evolved under variable fire frequencies. Regimes of both frequent, light ground fires and infrequent, intense crown fires have promoted the development of such fire-adapted species as *Pinus contorta, Pseudotsuga menziesii, Larix occidentalis, Pinus monticola, P. ponderosa* and *Populus tremuloides* over *Picea, Abies, Tsuga,* and *Thuja* (HABECK and MUTCH 1973; WEAVER 1974).

The vast boreal forest region of Canada and Alaska is dominated over much of its area by a fire regime of infrequent crown fires (over 100-year period) which characteristically burn very large areas with high intensity (ROWE and SCOTTER 1973; VIERECK 1973b; HEINSELMAN 1980). Many of these fires burn areas of thousands or even hundreds of thousands of hectares.

16.4.2 Grassland Ecosystems

The majority of grassland ecosystems of the world are environments highly conducive to fire (DAUBENMIRE 1968; VOGL 1974; KUCERA 1981). Even brief drought periods are usually sufficient to dry grasses to a level where they are readily combustible. These flammable fuels, together with a low contiguous canopy, provide ideal conditions for ignition. Low humidities and moderate winds which commonly are present during the dry season increase the potential for fire spread. While the dry above-ground stems of perennial grasses are readily consumed in a fire, the major portion of living biomass is at or below the ground surface. Since grassland fires are relatively cool in comparison to fires in woody fuels, these basal and belowground tissues are protected from damage.

Lightning is the basic source of fires in temperate grasslands. Regular drought periods lead to annual buildups of combustible fuels. Dry thunderstorms with frequent lightning strikes but little precipitation provide the ignition potential for these fuels, most commonly in spring and in late fall and winter. Studies of natural grassland fires suggest that a fire cycle of every 2–6 years is typical. Natural lightning ignitions of fire in tropical grasslands are much rarer, however.

Vegetation boundaries between grasslands and forest communities are almost always strongly influenced by patterns of fire frequency and intensity. The occurrence of fire kills woody plant seedlings and young saplings, thereby opening up canopies and increasing survival of fire-resistant perennial grasses. Increased grass coverage provides fuels to provide the potential for more frequent fires, and thus a cycle of regular fires at intervals of a few years is initiated. Under these conditions grasslands expand at the expense of woody plants. In the absence of fire, however, woody plants invade the margins of grasslands.

16.4.3 Mediterranean-Climate Shrub and Related Shrublands

Evergreen sclerophyll shrublands of the five mediterranean-climate regions of the world are all subject to regular natural fires. Structurally similar heathlands and shrublands outside this climatic region also share this characteristic. Fire is clearly a dominant environmental factor in influencing all chaparral communities (BISWELL 1974). While it is difficult to precisely determine natural frequencies of chaparral fires, most speculation leads to estimates of frequencies of 10–40 years.

Very little is known of the natural fire frequency of matorral communities in central Chile. Indirect evidence, however, suggests that natural fires are much less common than those in other mediterranean-climate regions (RUNDEL 1981a). Convective thunderstorms and associated lightning are rare in central Chile, providing relatively few opportunities to initiate combustion under natural circumstances.

Both shrublands (maquis and garrigue) and pine forests of the Mediterranean region are strongly influenced by fire as an environmental factor (NAVEH 1974). The interpretation of the role of fire in vegetation patterning is much more difficult in this region than in other mediterranean-climate areas, however, because of the long history of major human disturbance. Only a small fraction of present fires are caused by lightning.

Although human activities have been an important source of South African fynbos fires for at least 100,000 years, natural fire ignitions caused by lightning are still the most important sources of wildfires in remote areas (KRUGER 1979). The western Cape Region with pure mediterranean-climate conditions has the majority of natural fires in summer, but fires usually occur from September to April, with winter burns rare. Where biseasonal rainfall patterns occur further east, fires are biseasonal in distribution, with most occurring from December to March and again from June to September. Natural fire frequencies are thought to be between 6 and 30–40 years.

Estimates of natural fire frequency for Australian heathland communities are largely conjectural. At least two years of regrowth following fire are necessary for heathlands to carry even low intensity fires. At the other extreme, old-growth heathland communities appear to be senescent at 40–50 years of age. While fires may burn through coastal heathlands in southern Queensland at frequencies as low as five years, natural fire frequencies due to lightning ignitions are somewhat longer. SPECHT (1972) suggests natural frequencies of 30–50 years in South Australia.

Treeless dwarf shrub heaths are a characteristic fire-type community through much of northern Europe and Canada. In northern Europe where the fire ecology of heathlands is best studied, *Calluna vulgaris* is the dominant species and may account for over 90% of stand biomass (GIMINGHAM 1972).

16.4.4 Tropical Forest Ecosystems

Many tropical vegetation zones have had a long history of fire, particularly grasslands and savannas. A variety of evidence, however, suggests that fire frequencies before the advent of man were relatively low and that the present pattern of community distribution has been strongly influenced by human use of fire. Open dry forests and grasslands with seasonal drought periods are readily subject to burning, but there are abundant indications that such areas have dramatically expanded their ranges in the tropics during the past few thousand years (BATCHELDER and HIRT 1966).

While thunderstorms with associated lightning are relatively abundant in most tropical regions, this lightning ignites relatively few fires. One important factor in this low natural fire frequency is that heavy precipitation regularly accompanies lightning storms and prevents ignition from occurring. BATCHELDER (1966) also suggests that anthropogenic burning used during the dry season to prepare grasslands for cultivation and grazing effectively reduces available dry fuels before the onset of the lightning season when natural fires could be ignited.

The relative distribution of tropical seasonal forests and adjacent savannas has been strongly influenced by fire. Although there is commonly a sharp transition line between these vegetation types, this is a dynamic transition which is rapidly expanding the areas of savannas at the expense of the forest vegetation. The process of "savannization" has been described in detail by many authors. Fires originating in the grass layer of savannas during the dry season penetrate the edges of forests, killing seedlings and small saplings. As a result of greater penetration of light, grasses begin to become established in openings and at the forest edge. If fires again occur the following year, more young trees are killed and grass invasion continues. As this cycle repeats there is a steady increase in fire intensity, inflicting increasing mortality on mature forest trees. Dense grass cover and steady soil changes, as well as frequent fires, prevent reproduction of the forest tree species. Slash and burn agriculture, of course, accelerates this process. Since grasses and shrubs are the early colonizers when cultivated land is abandoned, regular fires will prevent the reestablishment of forest vegetation by maintaining dominance of these flammable species.

16.5 Conclusions

Despite a large body of literature describing aspects of the role of fire as an ecological factor it is clear that there are many gaps in our understanding of such influences. We know that such vegetation characteristics as moisture content, chemical flammability, and canopy form and structure all interact to determine fire frequency, as well as does fuel buildup. Such fuel buildup may be due to natural causes or to a variety of disturbances such as drought, insect damage, or storm damage. Most available data on those subjects are only qualitative, however. We know that fire influences geomorphic processes such as surface erosion, soil mass flow, channel scouring, and avalanche occurrence, but there is a poor understanding of the cyclic interactions of fire with these processes.

Recent experimental studies of fire–nutrient relationships have done much to increase our understanding on the role of fire in influencing nutrient cycles. The literature still contains much apparently conflicting data, however. Some of this inconsistency is clearly due to poor experimental design and to a lack of comparable methods of analysis. The significance of post-fire nutrient immobilization by soil microbes is still poorly studied.

The large number of papers on the influence of fire on growth and development of herbaceous species, particularly grasses, has provided many quantitative examples of fire stimulation of these factors. We still lack a detailed understanding of the interaction of processes which stimulate post-fire growth and flowering. Recent analyses of the evolutionary significance of fire-stimulated reproductive strategies in woody plants have done much to establish the adaptive nature of reseeding and resprouting strategies.

The largest body of fire literature is that which describes the fire regimes of natural plant communities. While a detailed discussion of this literature is beyond the scope of this review, it is clear that much of the available data remain anecdotal in nature. An understanding of the frequency and intensity of natural fires is lacking for many fire-influenced areas of the world.

References

Agee JK (1973) Prescribed fire effects on physical and hydrologic properties of mixed-conifer forest floor and soil. Water Resource Center, Univ California, Contrib Rep 143, p 57

Ahlgren CE (1974) The effect of fire on soil organisms. In: Kozlowski TT, Ahlgren CE (eds) Fire and Ecosystems. Academic Press, New York pp 47–72

Ahlgren IF, Ahlgren CE (1960) Ecological effects of forest fires. Bot Rev 26:483–533

Ahlgren IF, Ahlgren CE (1965) Effects of prescribed burning on soil microorganisms in a Minnesota jack pine forest. Ecology 46:304–310

Alban DH (1977) Influence on soil properties of prescribed burning under mature red pine. USDA Forest Serv Res Pap NC-139, p 8

Alexander ME, Hawksworth FG (1975) Wildland fires and dwarf mistletoes: a literature review of ecology and prescribed burning. USDA For Serv Gen Tech Rep RM-14, p 12

Allen SE (1964) Chemical aspects of heather burning. J Appl Ecol 1:347–367
Al Nakshabandi G, Kohnke H (1965) Thermal conductivity and diffusivity of soils as related to moisture tension and other physical properties. Agr Meteor 2:221–229
Anderson HW, Coleman GB, Zinke PJ (1959) Summer slides and winter scour... dry-wet erosion in southern California. USDA For Serv Tech Pap PSW-36, p 12
Anderson HW, Hoover MD, Reinhart KG (1976) Forests and water: effects of forest management on floods, sedimentation, and water supply. USDA For Serv Gen Tech Rep. PSW-18, p 115
Arno SF, Sneck KM (1977) A method for determining fire history in coniferous forests of the mountain west. USDA For Serv Gen Tech Rep INT-42
Austin RC, Baisinger DH (1955) Some effects of burning on forest soils of Western Oregon and Washington. J For 53:275–280
Axelrod DI (1975) Evolution and biogeography of Madrean-Tethyan sclerophyll vegetation. Ann Missouri Bot Garden 62:280–334
Bamber RK, Mullette KJ (1978) Studies of the lignotubers of *Eucalyptus gummifera* (Gaertn. and Hochr.) II Anatomy. Aust J Bot 26:15–22
Barden LS (1979) Serotiny and seed viability of *Pinus pungens* in the southern Appalachians. Castanea 44:44–47
Basham JT (1957) The deterioration by fungi of jack, red, and white pine killed by fire in Ontario. Can J Bot 35:155–172
Batchelder RB, Hirt HF (1966) Fire in tropical forests and grasslands. US Army Natick Lab Tech Rep 67–41-ES, p 380
Beadle NCW (1940) Soil temperatures during forest fires and their effect on the survial of vegetation. J Ecol 28:180–192
Beaton JD (1959a) The influence of burning on the soil in the timber range area of Lac Le Jeune, British Columbia. I. Physical properties. Can J Soil Sci 39:1–5
Beaton JB (1959b) The influence of burning on the soil in the timber range area of Lac Le Jeune, British Columbia. II. Chemical properties. Can J Soil Sci 39:6–11
Bentley JR, Fenner RL (1958) Soil temperatures during burning related to postfire seedbeds on woodland range. J For 56:737–740
Berg AR (1975) *Arctostaphylos* Adans. In: Seeds of Woody Plants in the United States. USDA Agric Handbook No 450:228–231
Berndt HW (1971) Early effects of forest fires on streamflow characteristics. USDA For Serv Res Note PNW-148, p 9
Biswell HH (1956) Ecology of California Grasslands. J Range Manag 9:19–24
Biswell HH (1974) Effects of fire on chaparral. In: Kozlowski TT, Ahlgren CE (eds) Fire and Ecosystems. Academic Press, New York, pp 321–364
Blaisdell JP (1953) Ecological effects of planned burning of sagebrush-grass range on the upper Snake River plains. USDA Tech Bull 1975, p 39
Bollen GJ (1969) The selective effect of heat treatment on the microflora of a greenhouse soil. Neth J Plant Pathol 75:15–163
Campbell RE, Baker MB, Folliott PF, Larson FR, Avery CC (1977) Wildfire effects on a ponderosa pine ecosystem an Arizona case study. USDA For Serv Res Pap RM-191, p 12
Cantlon JE, Buell MF (1952) Controlled burning – its broader ecological aspects. Bartonia 26:48–52
Chapin FC, Van Cleve K (1981) Plant nutrient absorption under differing fire regimes. In: Mooney HA, Bonnicksen TM, Christensen NL, Lotan JE, Reiners WA (eds) Fire Regimes and Ecosystem Properties. USDA For Serv Gen Tech Rep WO-26, pp 301–321
Chen M, Hodgkins EJ, Watson WJ (1975) Precribed burning for improving pine production and wildlife habitat in the hilly coastal plain of Alabama. Alabama Agric Exp Sta Bull 473, p 19
Christensen NL (1973) Fire and the nitrogen cycle in California chaparral. Science 181:66–68
Christensen NL (1977) Fire and soil-plant nutrient relations in a pine-wiregrass savanna on the Coastal Plain of North Carolina. Oecologia 31:27–44
Christensen NL (1981) Fire regimes in southeastern forests. In: Mooney HA, Bonnicksen

TM, Christensen NL, Lotan JE, Reiners WA (eds) Fire Regimes and Ecosystem Properties. USDA For Serv Gen Tech Rep WO-26, pp 112–136

Christensen NL, Muller CH (1975) Effects of fire on factors controlling plant growth in *Adenostoma* chaparral. Ecol Monogr 45:29–55

Clayton JL (1976) Nutrient gains to adjacent ecosystems during a forest fire: an evaluation. For Sci 22:162–166

Cooper CF (1960) Changes in vegetation, structure, and growth of southwestern pine forests since white settlement. Ecol Monogr 30:129–164

Cooper DW (1965) The coast redwood and its ecology. Univ Calif Agr Ext Serv Berkeley, p 20

Cremer KW (1973) Ability of *Eucalyptus regnans* and associate evergreen hardwoods to recover from cutting or complete defoliation in different seasons. Aust For Res 6:9–22

Crocker TC, Boyer WD (1975) Regenerating longleaf pine naturally. USDA for Serv Res Pap 50–105, p 21

Cromer RN, Vines RG (1966) Soil temperatures under a burning windrow. Aust For Res 2:29–34

Curtis T, Partch ML (1948) Effect of fire on the competition between blue grass and certain prairie plants. Am Midland Natur 39:437–443

Curtis JT, Partch MC (1950) Some factors affecting flower production in *Andropogon gerardi*. Ecology 31:488–489

Cushwa CT, Martin RE (1969) The status of prescribed burning for wildlife mamagement in the Southeast. North Am Wildlife Nat Res Conf 34:419–428

Cushwa CT, Reed JB (1966) One prescribed burn and its effect on habitat of the Powhatan game management area. USDA For Serv Res Note SE-61, p 2

Czabator FJ (1971) Fusiform rust of southern pines – a critical review. USDA For Serv Res Pap SO-65, p 39

Daubenmire R (1968) Ecology of fire in grasslands. Adv Ecol Res 5:209–266

Davis KP, Klehm KA (1939) Controlled burning in the western white pine types. J For 37:399–407

Dearness J, Jansbrough JR (1934) *Cytospora* infection following fire injury in Western British Columbia. Can J Res 10:125–128

DeBano LF, Conrad CE (1974) Effect of a wetting agent and nitrogen fertilizer on establishment of ryegrass and mustard on a burn watershed. J Range Manag 27:57–60

DeBano LF, Conrad CE (1976) Nutrients lost in debris and runoff water from a burned chaparrel watershed. Proc Third Int Agency Sedimentation Conf 3:13–27

DeBano LF, Conrad CE (1978) The effect of fire on nutrients in a chaparral ecosystem. Ecology 59:489–497

DeBano LF,Osburn JF, Krammes JS, Letey J (1967) Soil wettability and wetting agents – our current knowledge of the problem. USDA For Serv Res Pop PSW-43, p 13

DeBano LF, Savage SM, Hamilton DM (1976) The transfer of heat and hydrophobic substances during burning. Soil Sci Am J Soc 4:779–782

DeBano LF, Dunn PH, Conrad CE (1977) Fire's effect on physical and chemical properties of chaparral soils. In: Mooney HA, Conrad CE (tech coord) Proc Symp on the Environmental Consequences of Fire and Fuel Management in Mediterranean Ecosystems, USDA For Serv Gen Tech Rep WO-3. pp 65–74

DeBano LF, Eberlein GE, Dunn PH (1979) Effects of burning on chaparral soils. I. Soil nitrogen. Soil Sci Soc Am J 43:504–509

DeBell DS, Ralston CW (1970) Release of nitrogen by burning light forest fuels. Soil Sci Soc Am Proc 34:936–938

DeByle NV (1973) Broadcast burning of logging residues and water repellancy of soil. Northwest Sci 47:77–87

Donaghey JL (1969) The properties of heated soils and their relationships to giant sequoia (*Sequoiadendron giganteum*) germination and seedling growth. MS Thesis Calif State Univ San Jose

Dunn PH, DeBano LF (1977) Fire's effect on the biological properties of chaparral soils. In: Mooney HA, Conrad CE (tech coord). Proc Symp on the Environmental Consequences of Fire and Fuel Management in Mediterranean Ecosytems. USDA For Serv Gen Tech Rep WO-3, pp 75–84

Dunn PH, DeBano LF, Eberlein GE (1979) Effects of burning on chaparral soils. II. Soil microbes and nitrogen mineralization. Soil Sci Soc Am J 43:509–514

Dyrness CT (1976) Effects of wildfire on soil wettability in the High Cascades of Oregon. USDA For Serv Res Paper PNW-202, p 18

Dyrness CT, Youngberg CT, Ruth RH (1957) Some effects of logging and slash burning on physical soil properties in the Corvallis watershed. USDA For Serv Pac NW For Range Exp Sta Res Pap 19, p 15

Ehrenreich JH (1959) Effect of burning and clipping on growth of native prairie in Iowa. J Range Manag 12:133–137

Ehrenreich JH, Aikman JM (1957) Effects of burning on seedstalk production of native prairie grasses. Proc Iowa Acad Sci 64:205–212

Eiten G (1972) The cerrado vegetation of Brazil. Bot Rev 90:201–341

El-Abyad MSH, Webster J (1968a) Studies on pyrophilous discomycetes. I. Comparative physiological studies. Trans Brit Mycol Soc 41:353–367

El-Abyad MSH, Webster J (1968b) Studies on pyrophilous discomycetes. II. Competition. Trans Brit Mycol Soc 41:369–375

Felix LS, Parmeter JR, Uhrenholdt B (1974) *Fomes annosus* as a factor in the management of recreational forests. Proc Fourth Int Conf *Fomes annosus*. USDA For Serv Wash DC pp 2–7

Finn RF (1943) The leaching of some plant nutrients following the burning of forest litter. Black Rock Forest Pap 1, 128–134

Flinn MA, Wein RW (1977) Depth of underground plant organs and theoretical survival during fire. Can J Bot 55:2550–2554

Fowells HA (1965) Silvics of Forest Trees of the United States. USDA For Serv Agr Handb 271

Franklin JF, Rothacher JS (1962) Are your seedlings being buried? Tree Planters Notes 51:7–9

Froehlich RC, Dell TR (1967) Prescribed fire as a possible control for *Fomes annosus*. Phytopathology 57:811

Gardner CA (1957) The fire factor in relation to the vegetation of Western Australia. West Aust Natur 5:166–173

Garren KH (1943) Effects of fire on vegetation of the southeastern United States. Bot Rev 9:617–654

Gill AM (1975) Fire and the Australian flora: a review. Aust For 38:4–25

Gill AM (1977) Plants' traits adaptive to fires in the Mediterranean land ecosystems. In: Mooney HA, Conrad CE (tech coord) Symp Environmental Consequences of Fire and Fuel management in Mediterranean Ecosystems. USDA For Serv Gen Rep WO-3, pp 17–26

Gill AM (1978) Crown recovery of *Eucalyptus dives* following wildfire. Aust For 41:207–244

Gill AM (1981) Fire-adaptive traits of vascular plants. In: Mooney HA, Boonicksen TM, Christensen NL, Lotan JE, Reiners WA (eds) Fire Regimes and Ecosystem Properties. USDA For Serv Gen Tech Rep WO-26, pp 208–230

Gill AM, Ashton DH (1968) The role of bark type in relative tolerance to fire of three central Victorian eucalypts. Aust J Bot 16:491–498

Gill AM, Ingwersen F (1976) Growth of *Xanthorrhoea australis* R. Br. in relation to fire. J Appl Ecol 13:195–203

Gimingham GH (1972) Ecology of Heathlands. Chapman and Hall, London

Grano CX (1970) Eradicating understory hardwoods by repeated prescribed burning. USDA For Serv Res Pap, pp 50–56

Greenwood DJ (1968) Measurement of microbial metabolism in soil. In: Gray TRG, Parkinson D (eds) The Ecology of Soil Bacteria. Liverpool Univ Press, Liverpool, pp 338–157

Grier CC (1975) Wildfire effects on nutrient distribution and leaching in a coniferous ecosystem. Can J For Res 5:599–607

Griffin JR (1978) The marble-cone fire ten months later. Fremontia 6:8–14
Grigal DF, McColl JG (1977) Litter decomposition following forest fire in northeastern Minnesota. J Appl Ecol 14:531–538
Habeck JR, Mutch RW (1973) Fire-dependent forests in the northern Rocky Mountains. Q Res 3:408–424
Hadley EB (1961) Influence of temperature and other factors on *Ceanothus megacarpus* seed germination. Madroño 16:132–138
Hadley EB, Kieckhefer BJ (1963) Productivity of two prairie grasses in relation to fire frequency. Ecology 44:389–395
Hall AV (1959) Observations on the distribution and ecology of Orchidaceae in the Muizenberg Mountains, Cape Peninsula. J S Afr Bot 25:265–268
Hardison JR (1976) Fire and flame for plant disease control. Ann Rev Phytopathology 14:355–379
Hare RC (1961) Heat effects on living plants. USDA For Serv SE For Exp Sta Occ Pap 183, p 32
Hare RC (1975) Contribution of bark to fire resistance of southern trees. J For 63:248–251
Hartesveldt RJ, Harvey HT, Shellhammer HS, Stecker RE (1975) The giant sequoia of the Sierra Nevada. National Park Service, Washington DC
Harvey AR, Jurgensen MF, Larsen MJ (1976) Intensive fiber utilization and prescribed fire: effects on the microbial ecology of forests. USDA For Serv Gen Tech Rep INT-28, p 46
Hatch AB (1960) Ash bed effects in westen Australia forest soils. Bull For Dept W Aust No. 64
Heinselman M (1981) Fire and the distribution and structure of northern ecosystems. In: Mooney HA, Bonnicksen TM, Christensen NL, Lotan JE, Reiners WA (eds) Fire Regimes and Ecosystem Properties. USDA For Serv Gen Tech Rep WO-26, pp 7–57
Helvey JD (1972) First-year effects of wildfire on water yield and stream temperature in northcentral Washington. Watersheds in Transition, Proc Am Water Res Ass, pp 308–317
Helvey JD, Tiedemann AR, Fowler WB (1976) Some climatic and hydrologic effects of wildfire in Washington state. Proc Tall Timbers Fire Ecol Conf 15:201–222
Hensel RL (1923) Effect of burning on vegetation in Kansas pastures. J Agr Res 32:631–644
Heyward F (1938) Soil temperatures during forest fires in the longleaf pine region. J For 36:478–491
Hibbert AR, Davis EA, Scholl DG (1974) Chaparral conversion potential of Arizona: Part I. Water yield response and effects on other resources. USDA For Serv Res Pap RM-126, p 36
Hoffman GR (1966) Ecological studies of *Funaria hygrometrica* Hedw in eastern Washington and northern Idaho. Ecol Monogr 36:157–180
Hughes RH, Knox FE (1964) Response of gallberry to seasonal burning. USDA For Serv Res Note SE-21, p 3
Humphreys FR, Lambert MJ (1965) An examination of a forest site which has exhibited the ash-bed effect. Aust J Soil Res 3:81–94
Isaac LA, Hopkins HG (1937) The forest soil of the Douglas fir region and the changes wrought upon it by logging and slash burning. Ecology 18:264–279
Jones MB, Laude HM (1960) Relationship between sprouting in chamise and the physiological condition of the plant. J Range Manag 13:210–214
Jorgensen JR, Hodges CS (1970) Microbial characteristics of a forest soil after twenty years of prescribed burning. Mycologia 62:721–726
Jorgensen JR, Wells CG (1971) Apparent nitrogen fixation in soil influenced by prescribed burning. Soil Sci Soc Am Proc 35:806–810
Kayll AJ (1966) A technique for studying the fire tolerance of living tree trunks. Can Dept For Publ 1012
Kayll AJ, Gimingham CH (1965) Vegetative regeneration of *Calluna vulgaris* after fire. J Ecol 53:729–734
Keeley JE (1977) Seed production, seed populations in soil and seedling production after fire for two congeneric pairs of sprouting and nonsprouting chaparral shrubs. Ecology 59:820–829

Keeley JE (1981) Fire regimes and reproductive cycles. In: Mooney HA, Bonnicksen TM, Christensen NL, Lotan JE, Reiners WA (eds) Fire Regimes and Ecosystem Properties. USDA For Serv Gen Tech Rep WO-26, pp 231–277

Keeley JE, Zedler PH (1978) Reproduction of chaparral shrubs after fire: a comparison of sprouting and seeding strategies. Am Midl Nat 99:142–161

Kerr LR (1925) The lignotubers of eucalypt seedlings. Proc R Soc Victoria 37:79–97

Kilgore BM (1971) The role of fire in managing red fir forests. Trans N Am Wildlife Nat Res Conf 36:405–416

Kilgore BM (1981) Fire in ecosystem distribution and structure: western forests and shrublands. In: Mooney HA, Bonnicksen TM, Christensen NL, Lotan JE, Reiners WA (eds) Fire Regimes and Ecosystem Properties. USDA For Serv Gen Tech Rep WO-26, pp 58–89

Kilgore BM, Taylor D (1978) Fire history of a sequoia-mixed conifer forest. Ecology 59:29–42

Klemmedson JO, Shultz AM, Jenny H, Biswell HH (1962) Effect of prescribed burning of forest litter on total soil nitrogen. Soil Sci Soc Am Proc 26:200–202

Klett WE, Hollingsworth D, Schuster JL (1971) Increasing utilization of weeping lovegrass by burning. J Range Manag 24:22–24

Krammes JS (1960) Erosion from mountain side slopes after fire in Southern California. USDA For Serv Pac SW For Range Exp Sta Res Note 171, p 7

Krammes JS (1965) Seasonal debris movement from steep mountainside slopes in southern California. Proc Third Fed Interagency Sed Conf USDA Misc Publ 970, pp 85–89

Kruger FJ (1977) Ecology of Cape fynbos in relation to fire. In: Mooney HA, Conrad CE (tech coord) Proceedings of the Symposium on the Environmental consequences of Fire and Fuel management in mediterranean Ecosystems. USDA For Serv Tech Rep WO-3, pp 230–244

Kruger FJ (1979) South African heathlands. In: Specht RL (ed) Heathlands and related shrublands of the world, A descriptive studies. Amsterdam: Elsevier, Amsterdam, pp 19–80

Kucera CL (1980) Grasslands and fire. In: Mooney HA, Bonnicksen TM, Christensen NL, Lotan JE, Reiners WA (eds) Fire Regimes and Ecosystem Properties. USDA For Serv Gen Tech Rep WO-26, pp 90–111

Kummerow J, Alexander JV, Neel JW, Fishbeck K (1978) Symbiotic nitrogen fixation in *Ceanothus* roots. Am J Bot 65:63–69

Lay DW (1957) Browse quality and the effects of prescribed burning in southern pine forests. J For 55:342–349

Lewis WM (1974) Effects of fire on nutrient movement in a South Carolina pine forest. Ecology 55:1120–1127

Lloyd PS (1971) Effects of fire on the chemical status of herbaceous communities of the Derbyshire Dales. J Ecol 59:261–273

Lotan JE (1976) Cone serotiny-fire relationships in lodgepole pine. Proc Tall Timbers Fire Ecology Conf 14:267–278

Lowdermilk WC (1930) Influence of forest litter, run-off, percolation and erosion. J For 28:474–491

Lunt HA (1950) Liming and twenty years of litter raking and burning under red and white pine. Soil Sci Soc Am Proc 15:381–390

Lutz HJ (1956) The ecological effects of forest fires in the interior of Alaska. USDA Tech Bull 1133, p 121

McArthur AG (1968) The fire resistance of eucalypts. Ecol Soc Aust Proc 3:83–90

McColl JG, Grigal DF (1977) Nutrient changes following a forest wildfire in Minnesota: effects in watersheds with different soils. Oikos 28:105–112

McNeil RC, Zobel DB (1979) Vegetation and fire history of a ponderosa pine-white fir forest in Crater Lake National Park. Northwest Sci

Malowany SN, Newton JD (1947) Studies on steam sterilization of soils. I. Some effects on physical, chemical and biological properties. Can J Res C25:189–208

Maple WR (1976) How to estimate long-leaf seedling mortality before control burns. J For 74:517–518

Margaris NS (1977) Decomposers and the fire cycle in Mediterranean-type ecosystems.

In: Mooney HA, Conrad CE (tech coord) Proc Symposium on the Environmental Consequences of Fire and Fuel Management in Mediterranean Ecosystems. USDA For Serv Gen Tech Rep WD-3, pp 37–45

Mark AE (1965) Flowering, seeding and seedling establishment of narrowleaved snow tussock, *Chionochloa rigida*. N Z J Bot 3:180–193

Mark AE (1969) Ecology of snow tussocks in the mountain grasslands of New Zealand. Vegetatio 18:289–306

Martin ARH (1966) The plant ecology of the Grahamstown Nature Reserve. II. Some effects of burning. J S Afr Bot 32:1–39

Martin RE (1963) A basic approach to fire injury of tree stems. Tall Timbers Fire Ecol Conf Proc 2:151–162

Megahan WF, Molitor DC (1975) Erosional effects of wildfire and logging Idaho. In: Watershed manag Symp Irrig and Drainage Div Amer Soc Cir Eng Logan Utah pp 423–444

Melching JS, Stanton JR, Koogle DL (1974) Deleterious effects of tobacco smoke on germination and infectivity of spores of *Puccinia graminis* tritici and on germination of spores of *Puccinia striiformis, Pyricularia oryzae,* and an *Alternaria* species. Phytopathology 64:1143–1147

Metz LJ, Lotti T, Klawitter RA (1961) Some effects of prescribed burning on coastal plain forest soil. USDA For Serv Sta Pap SE-133:p 10

Miller GR, Miles J (1970) Regeneration of heather [*Calluna vulgaris* (L. Hull)] at different ages and seasons in northeast Scotland. J Appl Ecol 7:51–60

Miller RB, Stout JD, Lee KE (1955) Biological and chemical changes following scrub burning on a New Zealand hill soil. NZ Sci Tech 37:290–313

Mohamed TF, Gimingham CH (1970) The morphology of vegetative regeneration in *Calluna vulgaris*. New Phytol 69:743–750

Mooney HA, Chu C (1974) Seasonal carbon allocation in *Heteromeles arbutifolia*, a California evergreen shrub. Oecologia 14:295–306

Muller CH, Hanawalt RB, McPherson JK (1968) Allelopathic control of herb growth in the fire cycle of California chaparral. Bull Torrey Bot Club 95:225–231

Mullette NJ, Bamber RK (1978) Studies of the lignotubers of *Eucalyptus gummifera* (Gaertn. & Hochr.). III. Inheritance and chemical composition. Aust J Bot 26:23–28

Munger TT (1911) Avalanches and forest cover in the northern Cascades. USDA For Serv Circ 1973, p 12

Murrell WG, Scott WJ (1966) The heat resistence of bacterial spores at various water activities. J Gen Microbiol 43:411–425

Naveh Z (1974) The ecology of fire in Israel. Proc Tall Timbers Fire Ecol Conf 13:131–170

Neal JL, Wright E, Bollen WB (1965) Burning Douglas fir slash: physical, chemical and microbial effects in the soil. Oregon State Univ For Res Lab Res Pap 1:1–32

O'Connor KF, Powell AJ (1963) Studies on the management of snow-tussock grassland I. The effects of burning, cutting and fertiliser on narrow-leaved snow-tussock (*Chionochloa rigida* (Raoul) Zotov) at a mid-altitude site in Canterbury, New Zealand. NZ J Bot 6:354–367

Old S (1969) Microclimate, fire, and plant production in an Illinois prairie. Ecol Monogr 39:355–384

Orme ML, Teege TA (1976) Emergence and survival of redstem (*Ceanothus sanguineus*) following prescribed burning. Proc Tall Timbers Fire Ecol Conf 14:391–420

Owensby CE, Wyrill JB (1973) Effects of range burning on Kansas Flint Hills soils. J Range Manag 26:185–188

Parmeter JR (1977) Effects of fire on pathogens. In: Mooney HA, Conrad CE (tech coord) Proc Symposium on the Fire and Fuel Management in Mediterranean Ecosystems. USDA For Serv Gen Tech Rep WO-3, pp 58–64

Parmeter JR, Uhrenholdt B (1975) Some effects of pine-needle or grass smoke on fungi. Phytopathology 65:28–31

Parmeter JR, Uhrenholdt B (1976) Effects of smoke on pathogens and other fungi. Proc Montana Tall Timbers Fire Ecol Conf and Fire and Land Management Symp 14:299–304

Pase CP, Lindemuth AW (1971) Effects of prescribed fire on vegetation and sediment in oak-mountain mahogany chaparral. J For 69:800–805

Payton IJ, Mark AF (1979) Long-term effects of burning on growth, flowering, and carbohydrate reserves in narrow-leaved snow tussock (*Chionochloa rigida*). NZ J Bot 17:43–54

Peet M, Anderson R, Adams M (1975) Effect of fire on big bluestem production. Am Midl Nat 94:15–26

Petersen PM (1970) Danish fireplace fungi. An ecological investigation on fungi in burns. Dan Bot Ark 27:1–97

Quick CR (1935) Notes on the germination of *Ceanothus* seeds. Madroño 3:135–140

Quick CR (1962) Resurgence of a gooseberry population after fire in mature timber. J For 60:100–103

Ralston CW, Hatchell GE (1971) Effects of prescribed burning on physical properties of soil. In: Prescribed Burning Symp Proc Asheville NC. USDA For Serv Southeast For Exp Sta pp 68–85

Ratcliffe DA (1964) Mires and bogs. In: Burnitt JH (ed) The Vegetation of Scotland. Oliver and Boyd, Edinburgh, pp 426–478

Reifsnyder WE, Herrington LP, Spalt KW (1967) Thermophysical properties of bark of shortleaf, longleaf, and red pine. Yale Univ. Sch For Bull. 70

Renbuss M, Chilvers GA, Pryer LD (1973) Microbiology of an ashbed. Proc Linn Soc NS W 97:302–310

Reynolds HG, Bohning JW (1956) Effects of burning of a desert grasshrub range in southern Arizona. Ecology 37:769–777

Rice RM (1973) The hydrology of chaparral watersheds. In: Proc Symp Living with the chaparral. Riverside Calif Sierra Club, pp 27–33

Rich LR (1962) Erosion and sediment movement following a wildfire in a ponderosa pine forest of central Arizona. USDA For Serv Rocky Mtn For Range Exp Sta Res Note 76, p 12

Robert WB (1965) Soil temperatures under a pile of burning logs. Aust For Res 1:21–25

Roth LF (1966) Foliar habit of ponderosa pine as a heritiable basis for resistance to dwarf mistletoe. In: Gerold HD, McDermott RE, Schreiner EJ, Winleski JA (eds) Breeding Pest-Resistant Trees. Pergamon Press, New York pp 221–228

Rovira AD, Bowen GD (1966) The effect of microorganisms upon plant growth. II. Detoxication of heat sterilized soils by fungi and bacteria. Plant Soil 25:129–142

Rowe JS, Scotter GW (1973) Fire in the boreal forest. Q Res 3:444–464

Rowe PB (1941) Some factors of the hydrology of the Sierra Nevada foothills. Nat Res Counc Am Geophys Union Trans 1941, pp 90–101

Rowe PB, Colman RA (1951) Disposition of rainfall in two mountain areas of California. USDA for Serv Tech Bull 1018, p 84

Rowley J (1970) Effects of burning and clipping on temperature, growth, and flowering of narrow-leaved snow tussock. N Z J Bot 8:264–2B2

Rundel PW (1973) The relationship between basal fire scars and crown damage in giant sequoia. Ecology 54:210–213

Rundel PW (1981a) The matorral zone of central Chile. In: di Castri F, Goodall D, Specht RL (eds) Mediterranean Sclerophyll Ecosystems. Elsevier, Amsterdam, pp 175–201

Rundel PW (1981b) The impact of fire on nutrient cycles. In: Kruger FJ, Mitchell DT, Jarvis JUM (eds) Symposium on Nutrients as Determinants of the Structure and Functioning of Mediterranean-Type Ecosystems. The Role of Nutrients. Springer Verlag, Heidelberg (in press)

Rundel PW, Parsons DJ (1980) Nutrient changes in two chaparral shrubs along a fire-induced age gradient. Am J Bot 67:51–58

St John TV (1976) The dependence of certain conifers on fire as a mineralizing agent. PhD Diss Univ Calif Irvine

St John TV, Rundel PW (1976) The role of fire as a mineralizing agent in a Sierran coniferous forest. Oecologia 25:35–45

Salih MSA, Taha FK, Payne GF (1973) Water repellency of soils under burned sagebrush. J Range Manag 26:330–331

Sartz RS (1953) Soil erosion on a fire-denuded forest area in the Douglas-fir region. J Soil Water Conserv 8:279–281
Savage SM (1974) Mechanisms of fire-induced water repellency in soil. Soil Sci Soc Am Proc 38:682–657
Schindler DW, Newbury RW, Beaty KL, Prokopowich J, Ruszczynski T, Dalton JA (1978) Effects of a windstorm and forest fire on chemical losses from forested watersheds and on the quality of receiving streams. J Fish Res Board Can 35
Scholl DG (1975) Soil wettability and fire in Arizona chaparral. Soil Sci Soc Am Proc 39:356–361
Scott KM, Williams RP (1978) Erosion and sediment yields in the Transverse Ranges, Southern California. US Geol Servey Prov Paper 1030, p 38
Scotter DR (1970) Soil temperatures under grass fires. Aust J Soil Res 8:273–279
Scotter GW (1963) Effects of forest fires on soil properties in northern Saskatchewan. For Chron 39:412–421
Shannon SH, Wright HA (1977) Effects of fire, ash, and litter on soil nitrate, temperature, moisture and tobosa grass production in the rolling plains. J Range Manag 30:266–270
Siggers PV (1949) Fire and southern fusiform rust. USDA For Path Spec Rel 33, p 4
Sinclair JD (1954) Erosion in the San Gabriel mountains of California. Am Geophys Union Trans 35:264–268
Smith DW (1970) Concentration of soil nutrients before and after fire. Can J Soil Sci 50:17–29
Specht RL (1972) The Vegetation of South Australia, 2nd edn Gov Printer Adelaide, p 328
Specht RL, Rayson P, Jackman ME (1958) Dark Island heath (Ninetymile Plain, South Australia). VI. Pyric succession: changes in composition, coverage, dry weight, and mineral status. Aust J Bot 6:59–88
Stark NM (1977) Fire and nutrient cycling in a Douglas-fir/larch forest. Ecology 58:16–30
Stark NM, Steele R (1977) Nutrient content of forest shrubs following fire. Am J Bot 64:1218–1224
Steward KK, Ornes WH (1975) The autecology of sawgrass in the Florida everglades. Ecology 56:162–171
Stickel PW (1941) On the relation between bark character and the resistance to fire. USDA For Serv NE For Exp Sta Tech Note 39
Stone EC, Juhren G (1953) Fire-stimulated germination. Calif Agric 7:13–14
Storey HC, Hobba RL, Rosa JM (1964) Hydrology of forest lands and range lands. In: Chow VT (ed) Handbook of Applied Hydrology. McGraw-Hill, New York
Sussman AS, Halvorson HD (1966) Spores, their Dormancy and Germination. Harper and Row, New York
Swanson FJ (1981) Fire and geomorphological processes. In: Mooney HA, Bonnicksen TM, Christensen NL, Lotan JE, Reiners WA (eds). Fire Regimes and Ecosystem Properties. USDA For Serv Gen Tech Rep WO-26, pp 401–420
Swanson FJ, Lienkaemper GW (1978) Physical consequences of large organic debris in Pacific Northwest streams. USDA For Serv Gen Tech Rep PNW-69, p 12
Swanston DN, Swanson FJ (1976) Timber harvesting, mass erosion, and steepland forest geomorphology in the Pacific Northwest. In: Coates DR (ed) Geomorphology and Engineering. Dowden, Hutchinson and Ross, Stroudsburg, Penn, pp 199–221
Tande GF (1977) Forest fire history around Jasper Townsite, Jasper National Park, Alberta. MS Thesis Univ Alberta Edmonton
Tarrant RF (1956) Effects of slash burning on some soils of the Douglas fir region. Soil Sci Soc Am Proc 20:408–411
Taylor HC (1977) Aspects of the ecology of the Cape of Good Hope Nature Preserve in relation to fire and conservation. In: Mooney HA, Conrad CE (tech coord) Proc Sym on the Environmental Consequences of Fire and Fuel Manag in Mediterranean Ecosystems. USDA For Serv Gen Tech Rep WO-3, pp 483–487
Teich AH (1970) Cone serotiny and inbreeding in natural populations of *Pinus banksiana* and *P. contorta*. Can J Bot 48:1805–1809
Viereck LA (1973a) Ecological effects of river flooding and forest fires on permafrost

in the taiga of Alaska. In: Permafrost: North Am Contr to Second Inter Con N A Sci Washington DC pp 60–67
Viereck LA (1973b) Wildfire in the taiga of Alaska. Q Res 3:465–495
Vines RG (1968) Heat transfer through bark, and the resistance of trees to fire. Aust J Bot 16:499–514
Viro PJ (1974) Effects of forest fire on soil. In: Kozlowski TT, Ahlgren CE (eds) Fire and Ecosystems. Academic Press, New York pp 7–45
Vlamis J, Gowans KD (1961) Availability of nitrogen, phosphorus and sulfur after brush burning. J Range Manag 14:38–40
Vogl RJ (1969) The role of fire in the evolution of the Hawaiian flora and vegetation. Proc Tall Timbers Ecol Conf 9:5–60
Vogl RJ (1973) Fire in the southeastern grasslands. Proc Tall Timbers Fire Ecol Conf 12:175–198
Vogl RJ (1974) Effects of fire on grasslands. In: Kozlowski TT, Ahlgren CE (eds) Fire and Ecosystems. Academic Press, New York
Vogl TJ, Armstrong WP, White KL, Cole KL (1977) The closedcone pines and cypresses. In: Barbour MG, Major J (eds) Terrestrial Vegetation of California. Wiley-Interscience, New York pp 295–358
Wagle RF, Kitchen JH (1972) Influence of fire on soil nutrients in a ponderosa pine type. Ecology 53:119–125
Weaver H (1974) Effects of fire on temperate forests. Western United States. In: Kozlowski TT, Ahlgren CE (eds) Fire and Ecosystems. Academic Press, New York pp 279–319
Wein RW, Bliss LC (1973) Changes in arctic *Eriophorum* tussock communities following fire. Ecology 54:845–852
Welch TG, Klemmedson JO (1975) Influence of the biotic factor and parent material on distribution of nitrogen and carbon in ponderosa pine ecosystems. In: Bernier B, Winget CW (eds) Forest Soils and Forest Land Manag. Les Presses de L'Univ, Laval, Quebec pp 159–178
Welker NE (1976) Microbial endurance and resistance to heat stress. In: Grey TRG, Postgate JR (eds) The survival of vegetative microbes. Cambridge Univ Press, Cambridge, pp 241–277
Wells BW (1928) Plant communities of the coastal plain of North Carolina and their successional relations. Ecology 9:230–242
Wells BW (1942) Ecological problems of the southeastern United States coastal plain. Bot Rev 8:533–561
Wells CG (1971) Effects of prescribed burning on soil chemical properties and nutrient availability. In: Prescribed Burning Symp Proc USDA For Serv SE For Exp Sta Asheville NC, pp 86–89
Wells CG (1979) Effects of fire on soil. USDA For Serv Gen Tech Rep WO-7, p 34
White EM, Thompson WW, Gartner FR (1973) Heat effects on nutrient release from soils under ponderosa pine. J Range Manag 26:22–24
White F (1977) The underground forests of Africa: a preliminary review. Gardens Bull, Singapore 29:57–71
Wickes EF, Leaphard CD (1976) Fire and dwarf mistletoe (*Arceuthobium* spp.) relationships in the northern Rocky Mountains. Proc Montana Tall Timbers Fire Ecol Conf and Fire and Land Manag Symp 14:279–298
Wicklow DT (1975) Fire as an environmental cue initiating ascomycete development in a tallgrass prairie. Mycologia 67:853–862
Widden P, Parkinson D (1975) The effects of a forest fire on soil microfungi. Soil Biol Biochem 7:125–338
Williams IJM (1942) A revision of the genus *Leucadendron* (Proteaceae) Contr Bolus Herb 3:1–425
Williams PA, Meurk CD (1977) The nutrient value of burnt tall-tussock. J Tussock Grasslands Mountain Lands Inst 34:63–66
Woodmansee RG, Wallach LS (1981) Effects of fire regimes on biogeochemical cycles. In: Mooney HA, Bonnicksen TM, Christensen NL, Lotan JE, Reiners WS (eds)Fire Regimes and Ecosystem Properties. USDA For Serv Gen Tech Rep WO-26, pp 379–400

Wright E (1931) The effect of high temperatures on seed germination. J For 29:679–687
Wright HA (1974) Effect of fire on southern mixed prairie grasses. J Range Manag 27:417–419
Wright HE (1981) The role of fire in land-water interactions. In: Mooney HA, Bonnicksen TM, Christensen NL, Lotan JE, Reiners WA (eds) Fire Regimes and Ecosystem Properties. USDA For Serv Gen Tech Rep WO-26, pp 421–444
Wright HE, Heinselman ML (1973) The ecological role of fire in natural conifer forests of western and northern North America. Introduction. Q Res 3:319–328
Youngberg CT, Wollum AG (1976) Nitrogen accretion in developing *Ceanothus velutinus* stands. Soil Sci Soc Am Proc 40:109–112
Zak B (1971) Detoxication of autoclaved soil by a mycorrhizal fungus. USDA For Serv Res Note PNW-159, p 4
Zavitkowski J, Newton M (1968) Ecological importance of snowbrush *Ceanothus velutinus* in the Oregon Cascades. Ecology 49:1134–1145
Zedler PH (1977) Life history attributes of plants and the fire cycle: a case study in chaparral dominated by *Cupressus forbesii*. In: Mooney HA, Conrad CE (tech coord) Symposium on the Environmental Consequences of Fire and Fuel Management in Mediterranean-climate Ecosystems, pp 451–458
Zobel DB (1969) Factors affecting the distribution of *Pinus pungens*, an Appalachian endemic. Ecol Monogr 39:303–333

17 The Soil Environment

P. BENECKE and R.R. VAN DER PLOEG

CONTENTS

17.1 Introduction: Soil – a Three-Phase System

Soil is such a common word, that probably most people will not stop to wonder what it precisely means. If one looks up pertinent literature, one is likely to become quite bewildered by the variety of definitions offered. In spite of this there are suggestions for categorizing, depending on one's reference point. The capability of soils to support plant life is an essential part of one definition, while development processes by which the top layer of the Earth is incessantly transformed can be more important to another. Soils as porous, structured bodies allowing specific physical, chemical, and biological processes may be the predominant feature to still others, as in this chapter. To be sure, definitions based on these preferences do not contradict but complement each other, underlining the complex nature and make-up of soils.

The intimate interaction of the litho-, atmo-, hydro-, and biosphere results in two main processes: the weathering of the top layer of the lithosphere and the formation of a new substrate called the pedosphere, or soil. Among the changes that the "parent material" of the lithosphere undergoes is the decrease in weight per unit volume. Consolidated rock has an average density close

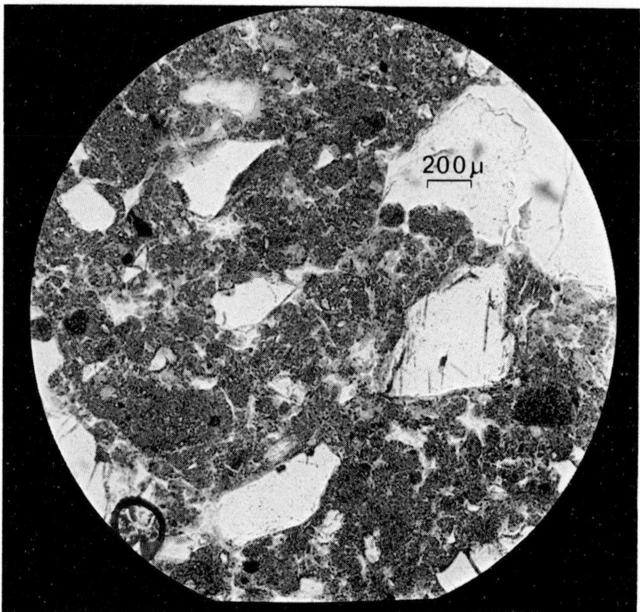

Fig. 17.1. Thin section of a clay soil (clay 53%, silt 11%, sand 36%, $\rho_b = 1.34$ g/cm^3) showing the microstructure. Sand particles (*light color* and *clear boundary*) are embedded in a disperse system of fine soil particles and micro-aggregates intermixed with fissures and pores (*light color* with a somewhat *diffuse boundary*, due probably to migrating substances). The diameter of the photograph corresponds to 2.7 mm. (Courtesy H. FÖLSTER 1964)

to 2.65 g/cm^3, but the bulk density ρ_b of unconsolidated soil material (oven-dry to 105 °C) is about 1.45 g/cm^3. This means that, since the solid phase density of the weathered and possibly altered individual mineral grains remains the same as that of the unweathered bedrock, the volume fraction occupied by the solid matter of a soil has values around (1.45)/(2.65) or 0.55. Thus a considerable pore space has been created as a result of soil formation processes. This pore space is intimately intermixed with the solid "phase" (Fig. 17.1), thus explaining the notion of the soil being a *disperse* three-phase system. The porous phase is normally occupied partly by a gaseous and partly by a liquid phase, the latter always being in the finer pores. Disperse refers to the extremely small, partly submicroscopical dimensions within which the phases change in space. In terms of bulk volumetric relationships the gaseous and the liquid phase are changing continuously, whereas the volume fraction of the solid phase remains relatively stable, though it may change to some degree due to swelling and shrinking, as well as to biological activity and soil cultivation.

17.2 The Solid Phase

17.2.1 Mechanical Composition – Grain Size Distribution

Many of the physical aspects of soils are closely related to the mechanical composition, in particular the texture being a mixture of particles ranging in diameter from <0.002 mm to 2.0 mm. Particles >2.0 mm are not attributed to the "fine earth", but are specified as a soil's content of stones and gravel. The boundary at 0.002 mm separates the clay from the silt (Fig. 17.3), which are two fractions with quite different properties.

Clay is capable of binding particles together. In this way aggregates are formed that can be classified according to their varying but characteristic shape (see Sect. 17.2.2). They constitute an essential feature of soil structure. Soil consistence is another property that largely depends on the clay content. It means (*Soil Taxonomy* 1975) the degree and kind of cohesion and adhesion or the resistance to deformation and rupture. It depends strongly on the water content of the soil. A further important soil property, its cation exchange capacity[1] (C.E.C.), is predominantly a function of the clay content.

Silt is too coarse to flocculate and thus form lasting aggregates. Soils with high silt content are nevertheless very fertile, provided sufficient clay is present (>15%). Such soils exhibit excellent water dynamics, being able to store large amounts of water available to plants without allowing the soil air to become too low (see Sect. 17.2.2). The main disadvantage of silty soils is their erodibility by wind and water. Furthermore, they show a tendency to puddle at the surface under the influence of heavy rainfall with subsequent formation of crusts that may seriously hinder gas exchange and lead to reduction processes in the main rooting zone. Those disadvantages, however, are of less importance under permanent vegetation cover.

Sand, the coarsest fraction within the soil separates (particles <2.0 mm), is characterized by low water-holding and cation exchange capacity, thus being normally the main constituent of poor soils. Corresponding to the low water-holding capacity there is always ample air present in the soil (high air capacity), if free drainage downward is provided. Like silt, sand grains do not flocculate and remain in "single-grain structure" as long as cementing or sticking substances are absent. But they are far less subject to wind or water erosion and show little tendency to form crusts at the top.

The solid phase of a soil normally consists of an intimate mixture of clay, silt, and sand. Textural classes have been defined (*Soil Taxonomy* 1975) each of which can be found in Fig. 17.2. A new term "loam" is used there, defined according to *Soil Taxonomy* (1975) as soil material that contains 7 to 27% clay, 28 to 50% silt and less than 52% sand. For further reference the texture of "typical" soils is displayed in Fig. 17.3 in the form of cumulative grain size distribution curves. They are chosen so as to exhibit the texture-dependent physical soil properties in the most distinctive way.

---

1 Amount of cations in milliequivalents that are adsorbed in exchangeable form per 100 g of solid soil matter

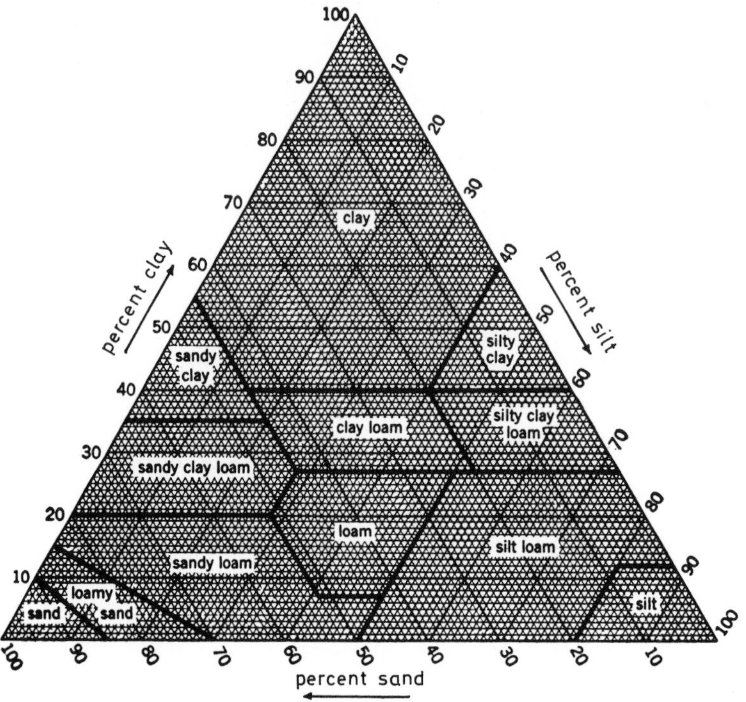

Fig. 17.2. Texture triangle showing the percentages of clay, silt, and sand in the soil classes. (*Soil Taxonomy* 1975)

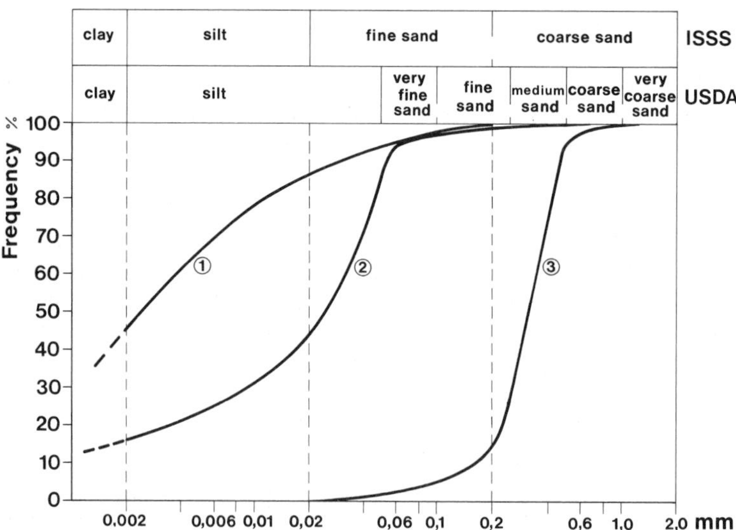

Fig. 17.3. Cumulative grain size distribution of three soils. Size limits of soil separates according to the *International Soil Science Society* (1975) and *Soil Taxonomy* (1975). Following the Soil Taxonomy scheme (USDA) soil *1* is a silty clay, soil *2* is a silt loam, and soil *3* is a sand

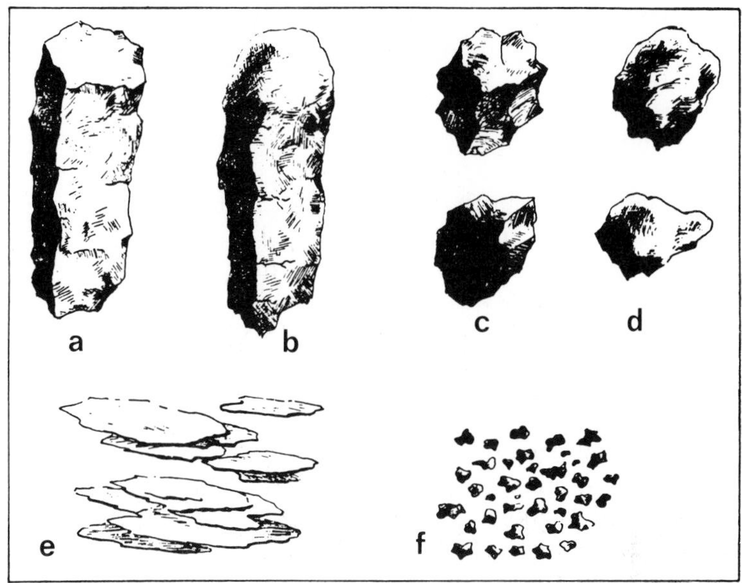

Fig. 17.4a–f. Drawings illustrating some of the types of soil structure: **a** prismatic; **b** columnar; **c** angular blocky; **d** subangular blocky; **e** platy; **f** granular. (*Soil Taxonomy* 1975)

17.2.2 Structure

"Soil structure" refers to the aggregation of primary soil particles into compound particles, or clusters of primary particles, which are separated from adjoining aggregates by surfaces of weakness. The exteriors of some aggregates have thin, often dark-colored, surface films which perhaps help to keep them apart. Other aggregates have surface and interiors of like color, and the forces holding the aggregates together appear to be wholly internal (*Soil Taxonomy* 1975). More simply, soil structure is often defined as the spatial arrangement of the solid phase of an undisturbed soil (BENECKE 1966). This includes "structureless" soils, consisting of single-grained or simply coherent material. Examination of most soils, however, reveals that they are structured in a characteristic way. Types of soil structure often met are shown in Fig. 17.4. A full description should include not only the structure type or shape of aggregates, but also their size, the distinctness and grade of aggregation, the way they are packed together (giving information about the form, width, and continuity of the interspace). Furthermore the degree of adhesion between and cohesion within the aggregates should be observed. Doing this often reveals that the aggregates consist of smaller peds[2] and that primary (smallest), secondary, and tertiary peds can be distinguished (BENECKE 1966, BUOL et al. 1973). A full description also includes the visible voids, cracks, fissures, worm holes and/or pores. In fact, in terms of water and air dynamics, the description of the structure aids primarily in understanding and assessing the nature of

2 A "ped" is according to *Soil Taxonomy* (1975) an individual natural soil aggregate

the pathways for these two phases. The importance of soil structure in soil classification and soil productivity can scarcely be overemphasized.

17.2.3 Soil Profile

To complete the representation of a soil as a physical environment, the distribution of structural features throughout the soil profile must be considered. A soil profile is a vertical section of the soil through all its horizons and extending into the parent material. A soil horizon, in turn, may be defined as a layer of soil, approximately parallel to the land surface and differing from adjacent genetically related layers in physical, chemical, and biological properties or characteristics such as color, structure, texture, and consistency (*Soil Sci Soc America* 1975).

Nature and intensity of soil-forming processes change generally with depth. Top layers are commonly most strongly influenced by biological activities and have the highest rooting density, highest humus content, and often spheroidal aggregates of high porosity (crumbs). The structure type of the lower layers depends largely on their history, particularly with respect to the moisture regime and the content and nature of the clay fraction. Peds generally become coarser with depth, are less subdividable into smaller peds, and are packed closer, thus reducing pathways for low tension water (see Sect. 17.2.4.1) and air (see Fig. 17.6). The physical conditions of single-grained or coherent horizons are closely correlated with bulk density ρ_b which may differ between 1 and 2 g/cm^3, and the degree of cohesion.

Of course, intergradations of the structural types are encountered and often it requires some experience to recognize the varying structure as a continuous system of the whole soil. But there is no doubt that this is the most efficient way to assess the physical conditions of a soil in the field and, for that matter, find typical spots to extract undisturbed soil samples for further physical examination in the laboratory as discussed in the subsequent sections.

17.3 Water and Air in the Pore System

The pore system consists of voids with different diameters ranging over several orders of magnitude from <0.0001 mm to >0.1 mm and even more for worm holes and interspaces between some aggregates. The pore system can be considered to be continuous in all directions but does have dead-end pores and cavities, too. Incorporation of dead-end pores into mathematical models on solute movement in soils resulted in better agreement with experimental data (BOAST 1973; VAN GENUCHTEN et al. 1976, 1977a, 1977b). Continuity is often more pronounced in one direction (higher transmissivity for water and air), frequently the vertical (anisotropic pore system). The voids are of irregular shape with permanently varying dimensions.

Forces at the particle surface attract ions as well as water molecules (due to their dipole character). These forces decline rapidly with distance from the particle surface, vanishing after about 10 nm. Thus small amounts of water in the pore system will cover the particle surface as a film. Additional water will cause the films to grow thicker and eventually flow together to form menisci. Capillary forces now come into play, allowing further storage of water against gravitational forces. Accordingly, water in the pore system of soils is subject to at least two space-dependent forces, adsorptive or capillary forces and gravitational forces. This implies that it requires work to move water, meaning that potential energy is associated with soil water. Its magnitude changes as a function of the distance from particle surfaces and the vertical distance from an arbitrary reference level. The potential energy per unit quantity (volume, mass, or weight) of soil water due to surface forces of the soil "matrix" constitutes the matric potential Ψ_m (or tension). Similarly, the gravitational potential, Ψ_g, is given by the relative height of the quantity of water in the gravitational field. If solutes are present in the soil water another potential arises, called the osmotic potential Ψ_s. Like the matric potential it opposes water withdrawal from the soil, as by roots.

All soil water potentials can be expressed in units of pressure. Matric potential appears only in the "unsaturated" soil zone (the pore space is only partly filled with water), and has always negative pressure values. A water table would mark the transition into the "saturated" zone and indicate the horizontal plane where the pressure is zero or atmospheric. Beneath the water table the pressure in the soil water is positive (hydrostatic pressure) and is designated by the pressure potential Ψ_p. (For a more detailed discussion of the potential concept of soil water see any textbook on soil physics, e.g., ROSE 1966; CHILDS 1969; HILLEL 1971; MARSHALL and HOLMES 1979; HANKS and ASHCROFT 1980. Water in the soil-plant-atmosphere continuum will be treated in Chap. 1, Vol. 12B.)

17.3.1 Soil Water Characteristic

If water is withdrawn from an originally water-saturated volume of soil, the matric potential decreases continuously from zero pressure to -22 MPa (-220 bar) for a dry soil (in equilibrium with a relative vapor pressure of 0.85) and to less than -100 MPa (-10^3 bar) when the soil is dried to constant weight at 105 °C. Thus there is a continuous relationship between water content and matric potential, called the soil moisture characteristic. It can be obtained by using a pressure membrane apparatus (see RICHARDS 1941, or any textbook on soil physics). Figure 17.5 shows the moisture characteristic curves of three "typical" soils whose texture is depicted in Fig. 17.3.

The soil water characteristic curve can be interpreted in different ways. Since the matric potential is related to the pore diameter, each fraction of extracted soil water represents an equal volume of pores with a corresponding diameter range (see Fig. 17.3).

The ecological significance is given by subdividing the total pore space into three main fractions: (1) coarse pores that allow water to drain off within

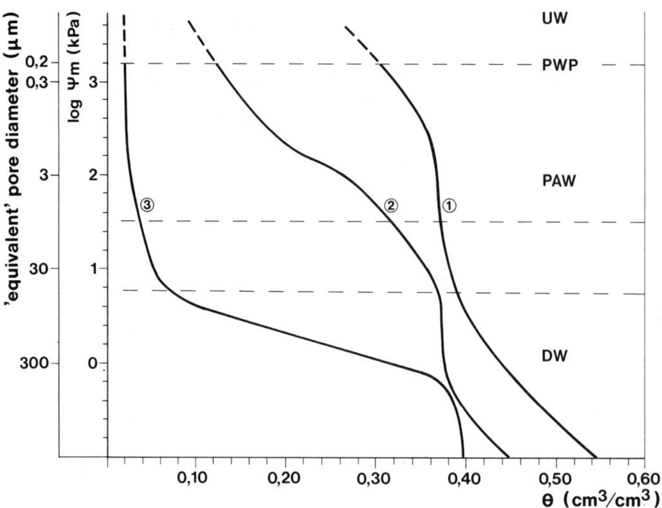

Fig. 17.5. Soil water characteristic curves for the same soils as in Fig. 17.3 *1* silty clay; *2* silt loam; *3* medium sand. Ψ_m is the matric potential (negative pressure), Θ is the volumetric water content. Equivalent pore diameters are given according to the law of capillary rise of water

a few days, free drainage downward provided, (2) medium pores that keep water against gravity in a form available to plants, and (3) fine pores, filled with water that is neither draining nor available to plants. These three fractions are designated by DW, PAW, and UW, respectively in Fig. 17.3. The DW-fraction is also called air capacity of the soil. Its lower boundary is somewhat dependent on the local moisture regime and the drainage conditions and may differ between matric potential values of −6 and −30 kPa (frequently −10 kPa is used). The volume fraction of the air capacity should be >0.12 of the soil bulk volume in the main rooting zone and >0.06 in the subsoil (*Forstliche Standortsaufnahme* 1978; MÜLLER et al. 1970; FEDDES et al. 1978). HILLEL (1971) states that soil aeration is likely to become limiting to plant growth when the air-filled porosity falls below about 0.10 (see also Fig. 17.9).

Within the PAW-fraction the matric potential covers the range from −10 kPa to −1.5 MPa (−15 bar) which means that water withdrawal is increasingly opposed as the deficit builds up. This is reflected in reduced transpiration (see Fig. 17.10) and assimilation rates. Water held in the soil at matric potential below −1.5 MPa, the "permanent wilting point" (PWP), is not able to keep most mesophytic plants alive (*Soil Taxonomy* 1975; KRAMER 1969; MARSHALL 1959).

In terms of a soil's ability to support plant life, the total volume of the PAW-fraction must be known. To find it, the PAW-fractions of each horizon of the soil profile down to a depth of at least 1 m are summed up. Figure 17.6 depicts that a medium-textured soil is capable of holding the rainfall equivalent of more than 200 mm in a form available to plants in the upper meter. According to experience, as well as to results of model calculations, water

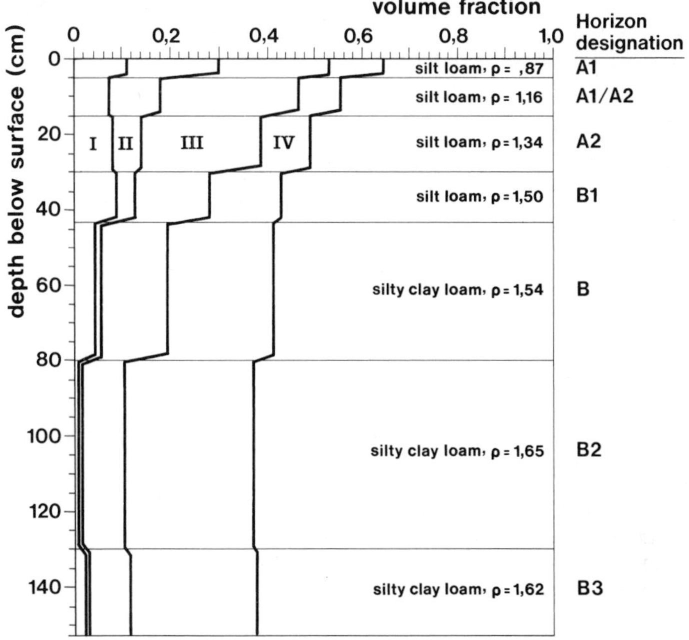

Fig. 17.6. Volume fractions of pore space (porosity) and solid matter as a function of depth of a gray-brown podzolic soil with slight tendency to become waterlogged. Parent material is loess. Vegetation cover is a timber-sized spruce stand. The pore space is subdivided (in accordance with Fig. 17.5) such that *I* stands for $\Psi_m > -6$ kPa, *II* for $-6\text{ kPa} > \Psi_m > -25$ kPa, *III* for $-25\text{ kPa} > \Psi_m > -1.5$ MPa, and *IV* for $\Psi_m < -1.5$ MPa. *I* and probably a small part of *II* correspond to the air capacity, the remaining part of *II* and *III* correspond to the amount of water held against gravity in a form available to plants. *IV* contains the unavailable water fraction. ρ_b designates the bulk density (g/cm^3) of the dry soil

from still deeper horizons is often available, partly because of the roots growing deep enough, partly because of capillary rise. So a total of about 300 mm of water is available, which meets the annual transpiration rate of a crop under moderate climatic conditions. As can be seen from Fig. 17.5 soils differ over a wide range in their capacity to hold water in a form available to plants. RENGER (1971) examined a large set of soils and found that with respect to a 10-cm-thick soil layer this range runs from 5 to 26 mm.

17.4 Water Movement

Points on the soil moisture characteristic curve reflect a static equilibrium situation, where all forces acting on the soil water balance each other and no water moves. In a natural soil such a situation is a rare exception because changes of matric potential occur almost incessantly. They are caused mainly by input and output of water (precipitation, evaporation, transpiration, drainage) leading locally to an increase or decrease of the water content and thus to a change in matric potential (cf. Fig. 17.5). In this way energy gradients arise that cause the water to move as will be described subsequently.

17.4.1 Darcy's Law and Hydraulic Conductivity

Liquid water movement in the unsaturated zone of soils is governed by spatial differences of the hydraulic potential, Ψ_h, defined as the sum of the matric and the gravitational potential

$$\Psi_h = \Psi_m + \Psi_g \text{ (Pa or cm).} \tag{17.1}$$

While the gravitational potential is an invariable linear function of height stretching only over a small interval within a soil the matric potential may vary between wide limits, particularly near the soil surface.

The flux rate, J, is proportional to the hydraulic gradient $d\Psi_h/ds$ by a coefficient K, called hydraulic conductivity. The water moves in the direction of decreasing hydraulic potential:

$$J = -K\frac{d\Psi_h}{ds} \quad \left(\frac{m^3}{m^2 s} = \frac{m}{s}\right) \tag{17.2}$$

where s is a distance in arbitrary direction (m). For convenience (to avoid higher negative exponents) units for K are often given in cm/day (1 cm/day = 1.1574×10^{-7} m/s). DARCY (1856) found this relationship experimentally for water-saturated conditions, with K a constant for a given porous medium. Adopting this equation for unsaturated soils, as is generally done (MARSHALL 1959; ROSE 1966; KIRKHAM and POWERS 1972; HILLEL 1971), means an important difference, namely K is now a function of the degree of water saturation, expressed either as water content or matric potential (cf. Fig. 17.5)

$$K = (K(\Psi_m(\Theta)) \tag{17.3}$$

where Θ is the volume fraction of the bulk soil containing water.

The nature of K can be visualized by putting the hydraulic gradient $d\Psi_h/ds$ in Eq. (17.2) equal to 1. Since this gradient can be made dimensionless by expressing potentials in m (1 m = 9.8 kPa) (ROSE 1966, or most textbooks on soil physics), K becomes equal to J, thus showing that it can be interpreted as the amount of water passing through a unit cross-section of the soil (not only the pores) per unit of time at a hydraulic gradient equal to 1 m m^{-1} (corresponding to 9.8 kPam^{-1}). $K(\Psi_m)$-curves of the soils whose texture and soil moisture characteristic, $\Psi_m(\Theta)$, have been depicted in Figs. 17.3 and 17.5 are shown in Fig. 17.7. It may be noticed that the range of matric potential is limited to values > -15 kPa but includes the most important part for bulk water movement (as may be seen from the corresponding K values).

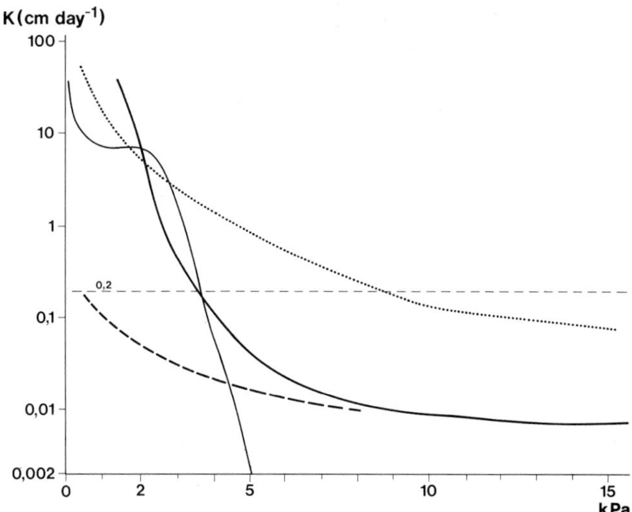

Fig. 17.7. Conductivity as a function of matric potential for the same soils as in Figs. 17.3 and 17.5, with the exception that two silty clay soils are shown, of which one (*thick line*) has a better structure (cf. text) than the other (*dashed line*). The *dotted line* represents the silt loam and the *thin solid line* the sand

The rates of decrease of hydraulic conductivity are generally extremely large in the presence of high matric potentials (low tension), thus reflecting the dominating effect of the macroscopic soil structure on soil water mobility. An instructive example is given by the two clay soils in Fig. 17.7. Both have angular blocky aggregates (cf. Fig. 17.4), but the aggregates are better developed and are separated by bigger and more continuous interspaces in the soil with higher conductivity. This conductivity-increasing effect of structure seems to be limited to a range of matric potentials being > -8 kPa, where both curves merge. Further decrease of matric potential has little effect on hydraulic conductivity. This can be explained by the entirely different nature of the pore system within the clay aggregates as compared to interspaces between the aggregates. As can be seen from Fig. 17.5, pores equivalent to diameters from 30 to 3 μm or a matric potential range from -10 to -100 kPa are almost absent. After the interspaces have emptied of water the water content and thus the hydraulic conductivity remain practically constant until the matric potential drops below about -100 kPa.

While clay keeps most of its water when in the low-tension range, sand allows a rapid drainage at matric potentials smaller than about -3 kPa accompanied by a correspondingly fast decrease of the hydraulic conductivity (cf. Figs. 17.5 and 17.7).

In ecological terms the conductivity function of the medium-textured soil is the most favorable one. It allows for high drainage rates as long as air supply, e.g., for root respiration, is insufficient. The conductivity then gradually decreases, thus keeping as much water as possible in the soil that is easily available to plants. (This is further illustrated in the next section.)

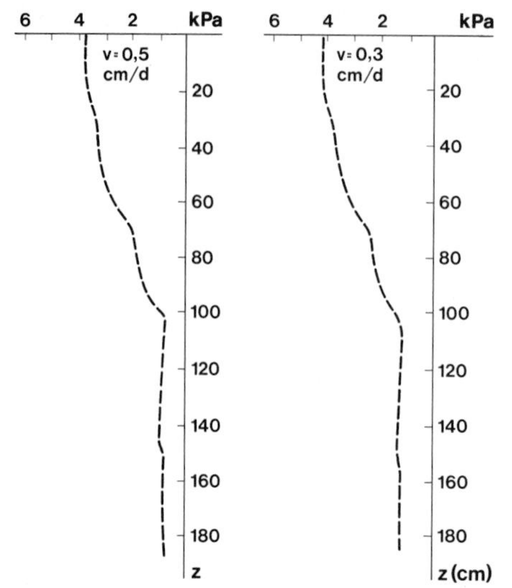

Fig. 17.8. Matric potential profiles in equilibrium with approximately constant rainfall during a low evaporation period in autumn.
Average rainfall rates are 0.5 and 0.3 cm/day, causing an approximately constant flux J of 0.5 and 0.3 cm/day throughout the soil. Since the hydraulic conductivity as a function of matric potential decreases with soil depth, higher matric potentials are prevailing in the deeper horizons to allow the same seepage rates as in the upper horizons. (Benecke and Van der Ploeg 1979)

17.4.2 Soil Water Dynamics

17.4.2.1 Liquid Water Movement

Darcy's law [Eq. (17.2)] describes a constant flux, J, of water through soil. Under field conditions approximately constant fluxes throughout a soil may occasionally occur, but the rule is nonstationary water movement where the flux rates change with time and depth and are accompanied by changes in water storage.

Figure 17.8 reflects a steady-state situation and illustrates some physical as well as ecological principles for soil. The soil consists of five layers with boundaries at 35, 75, 100, and 150 cm depth. The hydraulic conductivities are such that, for a given matric potential, they decrease stepwise from layer to layer. Thus, to allow for the same flux rate at every depth, the matric potential must increase. As Fig. 17.8 shows, this increase takes place in the lower part of each layer, whereas no gradient in Ψ_m exists in the upper part. So in the upper parts the steady-state downward water movement takes place under a unit hydraulic gradient, due solely to the action of gravity, whereas the hydraulic gradient drops increasingly below unity in the lower parts to compensate for increased conductivity. The reduction of the hydraulic gradient is facilitated by a matric potential gradient building up in the opposite (upward) direction.

Ecologically it can be inferred that the air content decreases with depth and that the total air volume decreases with increasing flux rates. This may eventually even lead to a build-up of a free water table requiring a flux rate of > 1.0 cm/day as, in the present case, occasionally occurs during the snow melt period. It may be pointed out that the soil of Fig. 17.8 that has been intensively investigated (Benecke 1979) provides an optimal physical environment for plant growth.

As has been mentioned above, soil water movement in the unsaturated soil zone occurs mostly under nonstationary conditions. This more general case is fundamentally described by the law of conservation of matter (CHILDS 1969; KIRKHAM and POWERS 1972, and most textbooks on soil physics)

$$\frac{\partial J}{\partial s} = -\frac{\partial \Theta}{\partial t} \; (s^{-1} \text{ or day}^{-1}) \tag{17.4}$$

where t is time in seconds or days. Substitution of Eq. (17.2) gives

$$\frac{\partial \Theta}{\partial t} = \frac{\partial}{\partial s}\left(K \frac{\partial \Psi_h}{\partial s}\right) \tag{17.5}$$

with K a function of Ψ_m or Θ [cf. Eq. (17.3)].

Equations (17.4) and (17.5) state that the change of water storage in a soil volume within a time interval is equal to the difference of the amount of water leaving this volume minus the amount entering it during the same time interval. In this form sinks or sources that may withdraw from or add to the water in the soil volume under consideration are disregarded. Sinks, in particular, may be present as roots.

Equation (17.5) has to be solved for specified initial and boundary conditions, which, due to their complexity, seldom allow analytical solutions. The boundary conditions are given by the input and output variables and/or flow conditions at the surfaces of the soil volume. The most important input variable is the amount of precipitations that enters (infiltrates) the soil. Output variables are evaporation from the soil surface, transpiration, and lateral and vertical drainage, of which the transpiration is generally added as a sink term to Eq. (17.5). All these fluxes crossing the boundary planes have to be measured or determined by appropriate algorithms. Examples for determining evaporation flux are given by BLACK et al. (1969), GARDNER and GARDNER (1969), STAPLE (1974), BEESE et al. (1978).

Numerous papers deal with the development, testing, and application of mathematical models for water uptake by roots, e.g., NIMAH and HANKS (1973a and b), MOLZ and REMSON (1974), HILLEL et al. (1975), HILLEL et al. (1976), HILLEL (1977), FEDDES et al. (1976), HERKELRATH et al. (1977), BEESE et al. (1978), DE JONG and CAMERON (1979). These models generally include, besides physical soil properties, the amount (or length) and distribution of fine roots. Their growing rate, effectiveness, and probably other relevant features may also be included. Basic parts are commonly the rate of potential evaporation (PENMAN 1948, 1963) and the tension of the soil water as parameters controlling transpiration. The sink term used by VAN DER PLOEG (1978) (Fig. 17.9) shows that the maximum possible transpiration occurs if, as a rough average, $-10\,\text{kPa} > \Psi_m > -50\,\text{kPa}$. For Ψ_m increasing beyond -10 kPa water uptake by roots quickly approaches zero rates. This matric potential, called "anaerobis point" by FEDDES et al. (1978), may not be the same for different soils, since it depends on the rate of gas exchange through the water-free pores, i.e., on the gas diffusion coefficient (see Sect. 17.4.2.2). If Ψ_m drops below about -50 kPa,

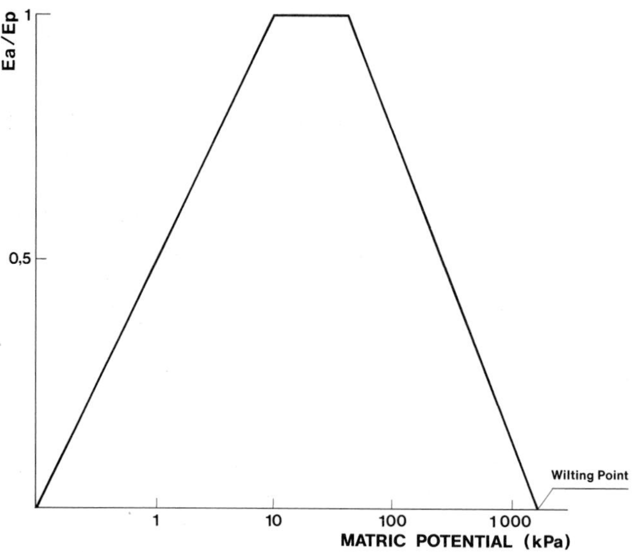

Fig. 17.9. Ratio of the actual (*Ea*) and potential (*Ep*) transpiration for crop surfaces, as a function of the matric potential in the root zone. The graph exhibits in a general way the shape of the sink term. (VAN DER PLOEG 1978)

water withdrawal is increasingly opposed by matric surface forces, decreasing hydraulic conductivity, and possibly spatial effects.

Thus transpiration is reduced over a wide Ψ_m range, becoming zero again at −1.5 MPa. The point at −50 kPa is, according to FEDDES et al. (1978), a maximum value that may decrease to −100 kPa with decreasing potential evaporation (often called the evaporative demand of the atmosphere). DENMEAD and SHAW (1962) found even wider limits between −20 and −300 kPa for the lowest matric potential that allows unrestricted transpiration.

These models do not take into account plant-dependent resistance to evaporation as do CAMACHO-B et al. (1974), ELFVING et al. (1972), and ROSE et al. (1976) by using the effect of leaf water potential on stomatal resistance to transpiration. More detailed models consider the soil-plant-atmosphere system as a physical continuum to simulate the reduction of transpiration, as for instance SHAWCROFT et al. (1974), REED and WARING (1974) and EHLERINGER and MILLER (1975).

17.4.2.2 Water Vapor Movement

Vapor pressure gradients as driving forces for diffusion of water vapor in soils can be caused by gradients in osmotic potential, matric potential, and temperature. Generally, only temperature gradients are considered to have practical significance.

Fick's law is applied in a modified form (to allow for a reduced cross-sectional area and tortuous pathways) for gaseous diffusion in the partly water filled pore system of soils:

$$J_v = -D_0\, a\, \varepsilon_g \frac{d\rho_v}{ds} \tag{17.6}$$

where J_v is the vapor flux (kg m^{-2} s^{-1}), D_0 is the diffusion coefficient through air (m^2 s^{-1}), a=0.66 according to PENMAN (1940), ε_g is the gas or vapor containing fraction of the soil bulk volume and ρ_v is the vapor density (kg m^{-3}).

Actually a is not likely to be constant, but to vary with water content and particle shape, particularly due to poorer connections between adjacent pores with decreasing ε_g. Details can be found in MARSHALL and HOLMES (1979), who give a brief literature review on the impedance of gaseous diffusion in porous mediums.

Temperature increases the vapor pressure and, at the same time, the matric potential. Thus the overall response of soil water under a temperature gradient is to move toward a colder region. But this, in turn, increases the water content and thus the matric potential in the colder region. GURR et al. (1952) used enclosed soil columns (loam) kept under a constant temperature difference at both ends for 5 days. By means of dissolved chloride they could distinguish liquid and vapor water movement. They found that condensation of vapor at the cold end caused a matric potential gradient in the direction of the warm end that was greater than the matric potential gradient due to the temperature gradient. Liquid water therefore moved against the temperature gradient toward the warm end, thus starting a circulation. Similar experiments were conducted by ROLLINS et al. (1954). In both cases vapor movement was found to be four to five times greater than expected according to Fick's law, most likely due to a partially liquid transfer, as has been described by PHILIP and DE VRIES (1957).

According to PHILIP and DE VRIES water movement in soils under a temperature gradient occurs as vapor movement in gas-filled pores, subsequent condensation at a liquid water meniscus and re-evaporation from the opposite side of the water-filled pore exposed to the next gas-filled pore. The temperature gradient across gas-filled pores should be larger than the mean temperature gradient due to different thermal conductivities in air and water, thus allowing a faster transfer than by vapor diffusion alone.

The main effects of temperature gradients under natural conditions are diurnal and seasonal changes of water content in the upper soil horizons. ROSE (1968) found after watering diurnal changes of up to 0.06 cm^3 water per cm^3 bulk soil in the top 1.3 cm layer of a soil due to the strong daily temperature wave in arid Australia. ABRAMOVA (1963, quoted in MARSHALL and HOLMES 1979) during a seasonal cycle has measured movements of 13 mm of water downward in summer and upward in winter. MARSHALL and HOLMES (1979) conclude that, in general, vapor movement in response to a temperature gradient is overridden by the greater fluxes in the liquid phase during infiltration, drainage, evaporation, and water uptake by roots.

17.5 Soil Temperature

Temperature affects chemical as well as biological processes in the soil. FITZPATRICK (1971) reports that for every 10 °C rise in soil temperature, reactions occur two to three times faster (up to 20 °C). KRAMER (1969) attributes reduced root growth and metabolic activity of root cells, as well as increased resistance to movement of water into roots, to low soil temperatures. Water uptake may thus be at 10 °C only about 20% of the rate at 25 °C, but there are species differences in tolerance of low soil temperatures.

Temperature fluctuates in the soil in diurnal and annual cycles. In the daytime part of the solar radiation is absorbed by the soil resulting in an increase of temperature. During nighttime long-wave energy is radiated off the land surface, causing the temperature in the soil to sink. Principles and pattern of this heat exchange are discussed in great detail by GEIGER (1961).

One way, probably the most important, in which heat flow takes place, is by thermal conduction due to molecular exchange of energy in the direction of decreasing temperature:

$$\frac{dQ}{dt} = -kA\frac{\partial T}{\partial z} \quad \text{(Fourier's heat flow law)} \tag{17.7}$$

Q is the quantity of heat that passes through a soil from z to $z+\delta z$ (J), k is the thermal conductivity ($J\ cm^{-1}\ s^{-1}\ °C^{-1}$), A is the cross-sectional area of the soil (cm^2), and T is the temperature (°C).

The thermal conductivity, k, can vary considerably in soils, predominantly with water content and soil structure. SMITH and BEYERS (1938, quoted in ROSE 1966), found that k for dry soils increased linearly with the volume fraction of the solid phase. This indicates that heat conduction takes place mainly through the points of contact between the particles. Increasing water content particularly affects initially dry soils because it provides conductivity-increasing bridges between particles.

Equation 17.7 describes a constant heat flux without change in heat storage. To meet the changing conditions in soils, energy conservation principles are applied in a theoretical approach:

$$\frac{\partial T}{\partial t} = \kappa\frac{\partial^2 T}{\partial z} \quad (°C\ s^{-1}) \tag{17.8}$$

where $\kappa = k/c_v$ is the thermal diffusivity ($cm^2\ s^{-1}$), and c_v the volumetric heat capacity of the moist soil ($J\ cm^{-3}\ °C^{-1}$). According to Eq. (17.8) changes in soil temperature will occur the more rapidly the greater the value of the thermal diffusivity κ. It should be noticed, however, that the thermal diffusivity depends in a somewhat involved manner on the water content since both the thermal conductivity, k, and the heat capacity, c_v, increase with water content. Figure 17.10 gives a general idea of these relationships (after RIDER 1957). DE VRIES (1952) proposed an instructive formula for calculating the heat capacity

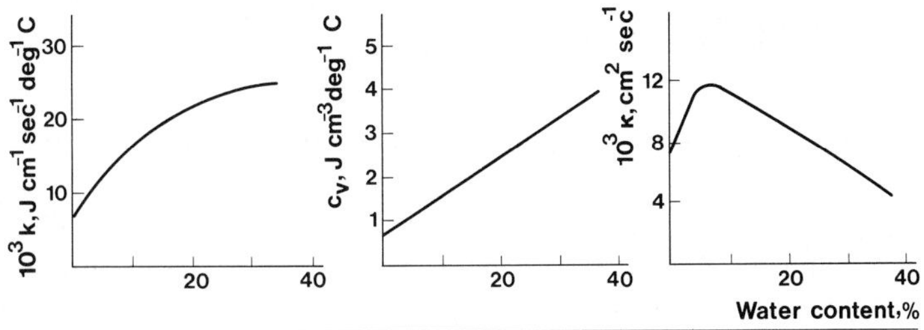

Fig. 17.10. Generalized curves illustrating the variation of thermal conductivity k, heat capacity c_v and thermal diffusity κ with water content in soil. (After RIDER 1957)

$$c_v = 0.46\ \Theta_m + 0.6\ \Theta_o + \Theta_w \tag{17.9}$$

where Θ_m, Θ_o, and Θ_w are volume fractions of mineral material, organic matter, and water, respectively.

KIRKHAM and POWERS (1972) solved Eq. (17.8) assuming sinusoidal temperature variation at the soil surface (with cycle length of a day or a year) and constant thermal diffusivity κ. The solution predicts both increased damping of amplitude and increased phase delay with depth. According to MARSHALL and HOLMES (1979) these predictions are well satisfied by actual observations. This shows that the mechanism as depicted in Eq. (17.8), despite simplifying assumptions, is in accordance with the general features of propagation of temperature waves into soil due to periodic daily and annual fluctuations in insolation (ROSE 1966).

17.6 Conclusions

Soil is a disperse system of three phases, solid, liquid, and gaseous, the latter two occupying the pore space. The solid phase is characterized by its grain size distribution and structure. Soil development results in a soil profile. It consists of soil horizons which have more or less equal properties but differ from each other in one or more features, particularly the structure. Water is held in the pore space by adsorptive and capillary forces giving rise to the matric potential, Ψ_m, of the soil water. Gravity is another force acting on the soil water. It causes the gravitational potential Ψ_g. In the presence of solutes an osmotic potential, Ψ_s, exists.

The matric potential decreases with the water content in a way that is mainly dependent on soil texture and structure. This relationship, called the soil moisture characteristic, may be used to subdivide the total pore space in coarse pores that cannot hold water against gravity, medium pores in which water can be stored in a form available to plants, and fine pores, the water of which is not available to plants.

The principles of water movement can be seen from Darcy's law. Complications arise in the "unsaturated" zone, where water and air are both present in the soil pores, because the hydraulic conductivity changes with water content or, in a different way, with matric potential.

To describe liquid water movement accompanied by water storage or depletion in the soil pores, Darcy's law is combined with the equation of continuity. This general flow equation is basic for most dynamic soil water models. To account for water withdrawal by roots a sink term is added.

Water movement in vapor form occurs mainly under a temperature gradient. Complications arise because the matric potential is also affected by temperature, and may be further affected if the vapor recondenses. Vapor transfer may be increased by interaction with liquid soil water.

The main effect of temperature is its influence on chemical and biological processes that may occur several times faster with rising temperatures. Diurnal and seasonal fluctuations, damping out with depth, are characteristic for soils.

17.7 Symbols and Abbreviations

Symbol	Description	Unit
ρ_b	bulk density of soil (g of solid matter per cm^3 of undisturbed soil)	
ρ	density of fresh soil	($g\ cm^{-3}$)
Θ	volume fraction of the bulk soil, generally that of water, (or indicated otherwise)	($cm\ cm^{-3}$)
J	volume of water crossing unit area per unit time, or flux rate,	($m\ s^{-1}$ or $cm\ day^{-1}$)
	or Joule	(1 J = 1 Nm)
K	hydraulic conductivity	($m\ s^{-1}$ or $cm\ day^{-1}$)
t	time	(s or day)
s	distance in arbitrary direction	(cm)
z	distance in vertical direction	(cm)
A	area	(cm^2)
J_v	water vapor flux density	($g\ cm^{-2}\ s^{-1}$)
ρ_v	density of vapor	($g\ cm^{-3}$)
D_0	molecular diffusion coefficient of water vapor through air	($cm^2\ s^{-1}$)
ε_g	gas-filled volume fraction of the bulk soil	($cm^3\ cm^{-3}$)
T	temperature	(°C)
Q	quantity of heat	(J)
k	thermal conductivity	($J\ cm^{-1}\ s^{-1}\ °C^{-1}$)
κ	thermal diffusivity	($cm^2\ s^{-1}$)
c_v	volumetric heat capacity	($J\ cm^{-3}\ °C^{-1}$)
DW	coarse soil pores from which water can drain within a few days, as volume fraction of the bulk soil	($cm^3\ cm^{-3}$)
PAW	medium soil pores that hold water in a form available to plants, as volume fraction of the bulk soil	($cm^3\ cm^{-3}$)
UW	fine soil pores that hold water in a form not available to plants, as volume fraction of the bulk soil	($cm^3\ cm^{-3}$)
PWP	permanent wilting point	(1.5 MPa)

References

Abramova MM (1963) Movement of water vapour in soil. Pochvovedenie 10:49–63 and Sov Soil Sci 952–963

Beese F, van der Ploeg RR, Richter W (1978) Der Wasserhaushalt einer Löß-Parabraunerde unter Winterweizen und Brache – Computermodelle und ihre experimentelle Verifizierung. Z Acker Pflanzenbau 146:1–19

Benecke P (1966) Die Geländeansprache des Bodengefüges in Verbindung mit der Entnahme von Stechzylinderproben für Durchlässigkeitsmessungen. Z Kulturtechn Flurbereinig 7:91–104

Benecke P (1979) Der Wasserumsatz eines Buchen- und eines Fichtenwald-Ökosystems im Hochsolling. Habilitationsschrift, Forstliche Fakultät der Universität Göttingen

Benecke P, van der Ploeg RR (1979) Das hydrologische Verhalten ungesättigter Bodenschichten am Beispiel forstlicher Ökosysteme. Z Pflanzenernaehr Bodenkd 142:131–136

Black TA, Gardener WR, Thurtell GW (1969) The prediction of evaporation, drainage and soil water storage for a bare soil. Soil Sci Soc Am Proc 33:655–660

Boast CW (1973) Modeling the movement of chemicals in soils by water. Soil Sci 115,3:224–230

Buol SW, Hole FD, McCracken RJ (1973) Soil Genesis and Classification. The Iowa State Univ Press, Ames, USA

Camacho-B SE, Kaufmann MR, Hall AE (1974) Leaf water potential response to transpiration by citrus. Physiol Plant 31:101–105

Childs EC (1969) An introduction to the physical basis of soil water phenomena. Wiley and Sons London

Darcy H (1856) Les fontaines publiques de la ville de Dijon. Paris, Dalmont

Denmead OP, Shaw HR (1962) Availability of soil water to plants as affected by soil moisture and meteorological conditions. Agron J 54(5), 385–390

Ehleringer JR, Miller PC (1975) A simulation model of plant water relations and production in the Alpine Tundra, Colorado. Oecologia 19:177–193

Elfving DC, Kaufmann MR, Hall AE (1972) Interpreting leaf water potential measurements with a model of the soil-plant-atmosphere continuum. Physiol Plant 27:161–168

Feddes RA, Kowalik P, Kolinska-Malinka K, Zaradny H (1976) Simulation of field water uptake by plants using a soil water dependent root extraction function. J Hydrol 31:13–26

Feddes RA, Kowalik PJ, Zaradny H (1978) Simulation of field water use and crop yield. PUDOC, Centre for Agricultural Publ and Document, Wageningen, Netherlands

FitzPatrick EA (1971) Pedology. Oliver & Boyd, Edinburgh, Great Britain

Fölster H (1964) Die Pedi-Sedimente der südsudanesischen Pediplane, Herkunft und Bodenbildung. Pedologie XIV, 1:64–84

Forstliche Standortsaufnahme (1978) Herausgeber: Arbeitskreis Standortskartierung in der Arbeitsgemeinschaft Forsteinrichtung. Landwirtschaftsverlag GmbH, Münster-Hiltrup, FRG

Gardner HR, Gardner WR (1969) Relation of water application to evaporation and storage of soil water. Soil Sci Soc Am Proc 33:192–196

Geiger R (1961) Das Klima der bodennahen Luftschicht. Vieweg & Sohn, Braunschweig

Genuchten M Th van, Wierenga P (1976) Mass transfer studies in sorbing porous media. I. Analytical solutions. Soil Sci Soc Am Proc 40:473–480

Genuchten M Th van, Wierenga P (1977a) Mass transfer studies in sorbing porous media. II. Experimental evaluation with tritium (^3H_2O). Soil Sci Soc Am Proc 41:272–278

Genuchten M Th van, Wierenga P, O'Connor GA (1977b) Mass transfer studies in sorbing porous media. III. Experimental evaluation with 2,4,5,-T. Soil Sci Soc Am Proc 41:278–285

Gurr CG, Marshall TJ, Hutton JT (1952) Movement of water in soil due to a temperature gradient. Soil Sci 74:335–345

Hanks RJ, Ashcroft GL (1980) Applied Soil Physics. Advanced Series in Agricultural Sciences 8. Springer, Berlin, Heidelberg, New York

Herkelrath WN, Miller EE, Gardner WR (1977) Water uptake by plants: II. The root contact model. Soil Sci Soc Am Proc 41:1039–1043

Hillel D (1971) Soil and water. Academic Press, New York
Hillel D (1977) Moisture extraction by root systems and the concurrent movement of water and salt in the soil profile. In: Hillel D (ed) Computer simulation of soil water dynamics. Int Develop Res Centre Box 8500, Ottawa, Canada. K1G 3H9 Head Office: 60 Queen Street
Hillel D, van Beek CGEM, Talpaz H (1975) A microscopic-scale model of soil water uptake and salt movement to plant roots. Soil Sci 120:385–399
Hillel D, Talpaz H, van Keulen H (1976) A microscopic-scale model of water uptake by a nonuniform root system and of water and salt movement in the soil profile. Soil Sci 121:242–255
Jong R de, Cameron DR (1979) Computer simulation model for predicting soil water content profiles. Soil Sci 128:41–48
Kirkham D, Powers WL (1972) Advanced Soil Physics. Wiley & Sons, New York
Kramer PJ (1969) Plant and Soil Water Relationsships. McGraw-Hill, New York
Marshall TJ (1959) Relations between water and soil. Technical Communication No. 50. Commonwealth Bureau of Soils, Harpenden
Marshall TJ, Holmes JW (1979) Soil Physics. Cambridge Univ Press
Molz FJ, Remson I (1974) Extraction term models of soil moisture use by transpiring plants. Water Resour Res 6:1346–1356
Müller W, Renger M, Benecke P (1970) Bodenphysikalische Kennwerte wichtiger Böden, Erfassungsmethodik, Klasseneinteilung und kartographische Darstellung. Geol Jahrb Reihe F 99,2:13–70
Nimah MN, Hanks RJ (1973a) Model for estimating soil water, plant and atmospheric interrelations: I. Description and sensivity. Soil Sci Soc Am Proc 37:522–527
Nimah MN, Hanks RJ (1937b) Model for estimating soil water plant, and atmospheric interrelations: II. Field test of model. Soil Sci Soc Am Proc 37:528–532
Penman JL (1940) Gas and vapour movements in the soil. 1. Diffusion of vapours through porous solids. J Agric Sci 30:437–462
Penman HL (1948) Natural evaporation from open water, bare soil and grass. Proc Soc London Ser A 193:120–145
Penman HL (1963) Vegetation and Hydrology. Technical Communication No. 53, Commonwealth Bureau of Soils Harpenden. Commonwealth Agricultural Bureaux, Farnham Royal, Bucks, England
Philip JR, de Vries DA (1957) Moisture movement in porous materials under temperature gradients. Trans Am Geophys Union 38:222
Ploeg RR van der (1978) Entwicklung zweidimensionaler Modelle für den Wasserumsatz im Boden hängiger Fichtenstandorte des Harzes. Habilitationsschrift, Forstl Fakultät, Universität Göttingen
Ploeg RR van der, Beese F, Strebel O, Renger M (1978) The Water balance of a sugar beet crop: A model and some experimental evidence. Z Pflanzenernaehr Bodenkd 141:313–328
Reed KL, Waring RH (1974) Coupling of environment to plant response: a simulation model of transpiration. Ecology 55:62–72
Renger M (1971) Die Ermittlung der Porengrößenverteilung aus der Körnung, dem Gehalt an organischer Substanz und der Lagerungsdichte. Z Pflanzenernaehr Bodenkd 130:53–67
Richards LA (1941) A pressure membrane extraction apparatus for soil solution. Soil Sci 51:377–386
Rider NE (1957) A note on the physics of soil temperature. Weather 12:241
Rollins RL, Spangler MG, Kirkhan D (1954) Movement of soil moisture under a thermal gradient. Highw Res Board Proc Washington 33:492–508
Rose CW (1966) Agricultural Physics, Pergamon Press, London
Rose CW (1968) Water transport in soil with a daily temperature wave. 1. Theory and experiment. Aust J Soil Res 6:31–44
Rose CW, Byrne GF, Hansen GK (1976) Water transport from soil through plant to atmosphere: a lumped-parameter model. Agric Meteorol 16:171–184

Shawcroft RW, Lemon ER, Allen Jr LH, Stewart DW, Jensen SE (1974) The soil-plant-atmosphere model and some of its predictions. Agric Meteorol 14:287–307

Soil Science Society of America (1975) Glossary of soil science terms. Soil Sci Soc Am Madison, Wisconsin

Soil Taxonomy (1975) Agric Handb No 436, Soil Cons Ser USDA

Smith WO, Byers HG (1938) The thermal conductivity of dry soils of certain of the great soil types. Soil Sci Soc Am Proc 3:13

Staple WJ (1974) Modified Penman equation to provide the upper boundery condition in computing evaporation from soil. Soil Sci Soc Am Proc 38:837–839

Vries DA de (1952) A nonstationary method for determining thermal conductivity of soil in situ. Soil Sci 73:83–89

Author Index

Page numbers in *italics* refer to the references

Taxonomic Index

Page numbers in **bold face** refer to pages on which tables or figures appear

Subject Index

Page numbers in **bold face** refer to tables or figures

Encyclopedia of Plant Physiology

New Series · Editors: A. Pirson, M. H. Zimmermann

Physiological Plant Ecology

Editors of Volume 12A–D: O. L. Lange, P. S. Nobel, C. B. Osmond, H. Ziegler

In Preparation
Volume 12 B:

Water Relations and Carbon Assimilation

ISBN 3-540-10906-4

Contents:
Introduction. – J. B. Passioura: *Fundamentals of Plant Water Relations: Water in the Soil-Plant-Atmosphere Continuum.* – M. T. Tyree, P. Jarvis: *Fundamentals of Plant Water Relations: Water in Tissues and Cells.* – P. E. Weatherley: Water Uptake by Roots. – P. Rundel: *Water Uptake by Organs Other Than Roots.* – M. H. Zimmermann, J. A. Milburn: *Transport and Storage of Water.* – J. Schönherr: *Resistances of Plant Surfaces to Water Loss: Transport Properties of Cutin, Suberin and Associated Lipids.* – E.-D. Schulze, A. E. Hall: *Stomatal Responses, Water Loss and CO_2-Assimilation Rates of Plants in Contrasting Environments.* – A. E. Hall: *Mathematical Models of Plant Water Loss and Plant Water Relations.* – T. C. Hsiao, K. T. Bradford: *Physiological Responses to Moderate Water Stress.* – J. D. Bewley, J. Krochko: *Desiccation Resistance.* – W. Tranquillini: *Frost Drought and its Ecological Significance.* – D. Koller, A. Hadas: *Environmental Aspects of the Germination of Seeds.* – A. S. Sussman, H. A. Douthit: *Environmental Aspects of the Germination of Spores.* – R. Crawford: *Physiological Responces to Flooding.* – C. B. Osmond, H. Ziegler, K. Winter: *Functional Significance of Different Path Ways of CO_2 Fixation in Photosynthesis.* – G. D. Farquhar, S. von Caemmerer: *Modelling of Photosynthetic Response to Environmental Conditions.* – I. R. Cowan: *Water Use and Optimization of Carbon Assimilation.* – E.-D. Schulze: *Plant Life Forms and their Carbon, Water and Nutrient Relations.*

Volume 12 C:

Responses to the Chemical and Biological Environment

ISBN 3-540-10907-2

Contents:
Introduction. – M. G. Pitman, U. Lüttge: *The Ionic Environment and Plant Ionic Relations.* – G. Wyn-Jones: *Osmoregulation.* – H. Greenway, R. Munns, G. Kirst: *Halotolerant Eukaryotes.* – A. D. Brown: *Halotolerant Prokaryotes.* – M. Runge: *Physiology and Ecology of Nitrogen Nutrition.* – I. H. Rorison: *Influence of Limestone, Silicates and Soil pH on Plants.* – H. W. Woolhouse: *Toxicity and Tolerance in the Responses of Plants to Metals.* – A. Gibson, D. C. Jordan: *Ecophysiology of N_2-fixing Systems.* – K. Moser, K. Hanselwandter: *Ecophysiology of Mycorrhizal Symbioses.* – B. Feige: *Ecophysiology of Lichen Symbioses.* – W. Höll: *Marine Plant-Animal Symbioses.* – U. Lüttge: *Ecophysiology of Carnivorous Plants.* – P. R. Atsatt: *Host-Paratite Interactions in Higher Plants.* – A. J. Gibbs: *Virus-Ecology – 'Struggle' of the Genes.* – St. Vogel: *Ecophysiology of Zoophilic Pollination.* – D. H. Janzen: *Ecophysiology of Fruits and their Seeds.* – S. J. McNaughton: *Physiological and Ecological Implications of Herbivory.* – E. Newman: *Interactions between Plants.*

Volume 12 D:

Ecosystem Processes: Mineral Cycling, Productivity and Pollution

ISBN 3-540-10908-0

Contents:
Introduction. – B. Richards, J. L. Charley: *Nutrient Allocation in Plants: Cycling in Terrestrial Ecosystems.* –A. Melzer, Ch. Steinberg: *Nutrient Cycling in Freshwater Ecosystems.* – T. R. Parsons, P. J. Harrison: *Nutrient Cycling in Marine Ecosystems.* – W. T. Penning de Vries: *Modelling of Growth and Production.* – W. Loomis: *Productivity of Agricultural Systems.* – L. L. Tieszen, J. K. Detling: *Productivity of Tundra and Grassland.* – J. Dhleringer, H. Mooney: *Productivity of Desert and Mediterranean Climate Plants.* – P. Jarvis: *Productivity of Temperate Deciduous and Evergreen Forests.* – E. Medina, E. Klinge: *Productivity of Tropical Forest and Tropical Woodland.* – W. R. Boynton, C. H. Hall, P. G. Falkowski, C. W. Keefe, W. A. Kemp: *Phytoplankton Productivity in Aquatic Ecosystems.* – W. Urbach, K. Pfister: *Effects of Biocides and Artificially Introduced Growth Regulators: Physiological Basis.* – R. Schubert: *Effects of Biocides and Artificially Introduced Growth Regulators: Ecological Implications.* – A. Kohler, A. Labus: *Ecophysiological Effects of Aquatic Pollutants, Including Eutrophication and Waste Heat.* – I. Ziegler, M. H. Unsworth: *Ecophysiological Effects of Atmospheric Pollutants.* – D. Gates, B. R. Strain: *Ecophysiological Effects of Changing Atmospheric CO_2 Concentration.* – W. D. Billings: *Man's Influence on Ecosystem Structure, Operation, and Ecophysiological Processes.*

Springer-Verlag
Berlin Heidelberg New York